2024 RELEASE

Introducing Communication Theory

Analysis and Application

Richard West
Emerson College, Boston, MA

Lynn H. Turner
Marquette University, Emerita, Milwaukee, WI

INTRODUCING COMMUNICATION THEORY: ANALYSIS AND APPLICATION, 2024 RELEASE

Published by McGraw Hill LLC, 1325 Avenue of the Americas, New York, NY 10019. Copyright ©2024 by McGraw Hill LLC. All rights reserved. Printed in the United States of America. Previous editions © 2021, 2018, and 2014. No part of this publication may be reproduced or distributed in any form or by any means, or stored in a database or retrieval system, without the prior written consent of McGraw Hill LLC, including, but not limited to, in any network or other electronic storage or transmission, or broadcast for distance learning.

Some ancillaries, including electronic and print components, may not be available to customers outside the United States.

This book is printed on acid-free paper.

1 2 3 4 5 6 7 8 9 LKV 29 28 27 26 25 24

ISBN 978-1-266-54871-0 (bound)
MHID 1-266-54871-8 (bound)
ISBN 978-1-266-90385-4 (loose-leaf)
MHID 1-266-90385-2 (loose-leaf)

Associate Portfolio Manager: *Katie Reuter*
Marketing Manager: *Kim Schroeder-Freund*
Lead/Senior Content Project Manager: *Melissa Leick/Laura Payne*
Manufacturing Project Manager: *Sandy Ludovissy*
Content Licensing Specialist: *Prineeth Priyan*
Compositor: *Aptara®, Inc.*

All credits appearing on page or at the end of the book are considered to be an extension of the copyright page.

Library of Congress Cataloging-in-Publication Data

Names: West, Richard L., author. | Turner, Lynn H., author.
Title: Introducing communication theory : analysis and application /
   Richard L. West, Emerson College, Lynn H. Turner, Marquette University.
Description: 2024 Release | New York, NY : McGraw Hill LLC, [2024] |
   Includes bibliographical references and index.
Identifiers: LCCN 2023032389 (print) | LCCN 2023032390 (ebook) | ISBN
   9781266548710 (hardcover) | ISBN 9781266856983 (hardcover) | ISBN
   9781266903854 (spiral bound) | ISBN 9781266903731 (ebook) | ISBN
   9781266898105 (ebook other)
Subjects: LCSH: Information theory. | Communication.
Classification: LCC Q360 .W47 2024 (print) | LCC Q360 (ebook) | DDC
   003/.54–dc23/eng/20231025
LC record available at https://lccn.loc.gov/2023032389
LC ebook record available at https://lccn.loc.gov/2023032390

mheducation.com/highered

Brief Contents

Part One   Foundations

COMMUNICATION, THEORY, AND RESEARCH

Part Two   Empirical/Post-Positivist Theories

Organization of "Introducing Communication Theory: Analysis and Application" 69

INTRAPERSONAL: THE SELF AND MESSAGES

INTERPERSONAL: RELATIONSHIP DEVELOPMENT

GROUPS, TEAMS, AND ORGANIZATIONS

THE MEDIA

CULTURE AND DIVERSITY

Part Three   Interpretive Theories

INTRAPERSONAL: THE SELF AND MESSAGES

INTERPERSONAL: RELATIONSHIP DEVELOPMENT

GROUPS, TEAMS, AND ORGANIZATIONS

Contents

The Media

Chapter 11   Agenda Setting Theory 197

Preface

We're pleased and proud that *Introducing Communication Theory: Analysis and Application* remains among the leading college texts in communication theory. This new release is testimony to the commitment that students and teachers have made to the book and to the enterprise of communication theory. The previous seven editions demonstrate that communication theory courses are vibrant, that teachers of communication understand the importance of theoretical thinking, and that both instructors and students appreciate the consistent and organized template we employ throughout. This text explores the practical, engaging, and relevant ways in which theory operates in our lives. *It is written primarily for students who have little or no background in communication theory.*

We originally wrote the book because we thought that students need to know how theorizing helps us understand ourselves, as well as our experiences, relationships, media, environment, and culture. We also wrote this book because we believe that students should have a text that relates theory directly to their lives. We felt that some books insulted the student and trivialized theory while other books were written at a level that was far too advanced for undergraduates.

In this book, we take great care to achieve the following additional objectives:

- Familiarize students with the principles and central ideas of important theories they are likely to encounter in the communication discipline.
- Demystify the notion of theory by discussing it in concrete and unequivocal ways.
- Provide students with an understanding of the interplay among theory, communication, and application.
- Introduce students to the research process and the role of theory within this process.
- Assist students in becoming more systematic and thoughtful critical thinkers.

The new release of this book maintains its original focus of introducing communication theory to students in an accessible, appealing, and consistent way. We believe that students understand material best when it is explained in a clear, direct way through a number of realistic and applicable examples. Our hope is that students will take away a basic knowledge of, and appreciation for, communication theory from reading our text.

The theories in communication studies have roots in both communication and in other fields of study. This interdisciplinary orientation is reflected in the selection of the various theories presented in the text. We not only include the unique contributions of communication theorists, but also theories with origins in other fields of study, including psychology, sociology, biology, education, business, and philosophy. Communication theorists have embraced the integration of ideas and principles forged by their colleagues across many disciplines. Yet, the application, influence, and inherent value of communication are all sustained by the theorists in this text. In other words, although theories cut across various academic disciplines, their relevance to communication remains paramount and we articulate this relevancy in each theory chapter. We do not presume to speak for the theorists; we have distilled their scholarship in a way that we believe represents and honors their hard work. Our overall goal is to frame their words and illustrate their theories with practical examples so their explication of communication behaviors becomes accessible for students. Consider us their theoretical ambassadors, working to make their words come to life.

Together, we have nearly 70 years of experience in teaching communication theory. During this time, we have learned a great deal. *Introducing Communication Theory: Analysis and Application* utilizes and applies all that we as teachers have learned from our students. We continue to be indebted to both students and colleagues whose suggestions and comments have greatly influenced this newest release. In fact, many of these observations are found throughout the book!

The Challenges of Teaching and Learning Communication Theory

The instructor in a communication theory course may face several challenges that they don't encounter in other courses. First, because many students think of theory as distant, abstract, and obscure, teachers must overcome these potentially negative connotations. Negative feelings toward the subject can be magnified in classrooms where students represent a variety of ages and socioeconomic, cultural, and linguistic backgrounds. *Introducing Communication Theory* addresses this challenge by offering a readable and pragmatic guide that integrates content with examples, capturing the essence and elegance of theory in a straightforward manner. In addition, the book takes an incremental approach to learning about theory, resulting in a thoughtful and appropriate learning pace. In our decades of teaching this course, we have found that students cannot be overwhelmed with too much information at once. Therefore, we unpack theory in reasonable and digestible ways.

A second challenge associated with teaching and learning communication theory relates to preconceived notions of research: Students often view scholarship as difficult or remote. This book demonstrates to students that they already possess many of the characteristics of researchers, such as curiosity and ambition. Regardless of whether students are plugged into their social media, we believe they will be pleasantly surprised to know that they operate according to many personal theories every day. Once students begin to revise their misconceptions about research and theory, they are in a position to understand the principles, concepts, and theories contained in this book.

A third challenge of teaching and learning communication theory is capturing the complexity of a theory. In essence, theories must be presented in an approachable way without oversimplifying the theoretical process. To address this problem, instructors often present a skeletal version of a theory and then fill in the missing pieces with personal materials. By providing a variety of engaging examples and applications reflecting a wide range of classroom demographics, *Introducing Communication Theory* facilitates such an approach.

A final challenge relates to a theory's genesis and today's students. In this technological age, students look for and usually crave a "tech angle" to communication theory. Although many theories were conceptualized decades ago, in each chapter, we have provided the most recent research that represents a theory–technology nexus. Further we have added dozens of student comments throughout the book speaking to how the theories can be applied to technology, such as social media, texting, and so forth. In addition, we've added a discussion question at the end of each chapter that specifically relates to this technological framework.

Major Changes in Content in the New Release

As we do in every new release, we have edited and modified *each and every chapter* to reflect our continued emphasis on making theory more approachable.

Most importantly, *the entire book continues to reflect the template most useful to students*.

In the past, we relied on an approach that was context-specific. Yet, after considering how students learn, looking at the foundational information, and reviewing comments by colleagues across the country, we were struck by the narrowness of this approach. We found ourselves "forcing" a complex theory into a particular context, sometimes neglecting the fact that many of the theories fall across several contexts. Regrettably, many other books discuss theory by employing this approach, often resulting in a confusing or misguided presentation for students.

Therefore, the reorganization of the book adheres to a commonly-accepted division found in the field: Approaches to Knowing, or better known as Empirical, Interpretive, and Critical-Cultural approaches. Each "Approach to Knowing" is elaborated in **Chapter 3**, allowing students to see the relationship of a foundational chapter with the theory chapters.

In this new release, we strengthened and streamlined each chapter, and in several cases, reorganized the chapter to make the material more accessible. Further, we've updated many of our opening vignettes to make them aligned with the communication challenges that students face in their lives. And, of course, we've rigorously updated each theory in keeping with the current research and changes in the theorists' thinking.

Representative Chapter Changes

The decision to revise a successful book is made with a great deal of thought. We wish to retain the areas of a book that have received favorable feedback. Yet, because we wish to demonstrate that communication theory is "alive" each day, we simultaneously wish to include contemporary applications. Each theory chapter has undergone revision to make the content more recent, examples more compelling, material more organized, and critiques more balanced. Here's a sample of specific changes in content, examples, and organization undertaken in various chapters:

Chapter 1 (Thinking About Communication: Definitions, Models, and Ethics) adds information on human resources, employment opportunities, and possessing effective communication skills

Chapter 2 (Thinking About the Field: Traditions and Contexts) includes the application of information to the possibility of peace related to the Ukraine-Russia war

Chapter 3 (Thinking About Theory and Research) expands on qualitative research methods

Chapter 4 (Expectancy Violations Theory) includes the importance of personal space and COVID-19

Chapter 5 (Uncertainty Reduction Theory) adds new information applying URT to international contexts

Chapter 7 (Social Penetration Theory) adds content related to the role of personal stress and COVID-19 as well as new information regarding dating apps (e.g., Bumble)

Chapter 8 (Social Information Processing Theory) contains new information on online health information and perceived trustworthiness of that information

Chapter 9 (Structuration Theory) continued to make it more practical in tone and added information on ChatGPT and social media to advance the contemporariness of the theory

Chapter 10 (Organizational Information Theory) presents new clarification on the relationship between sensemaking and storytelling

Chapter 11 (Agenda Setting Theory) has been significantly reorganized and also highlights the history of the theory and the 3-part process of agenda setting

Chapter 12 (Spiral of Silence Theory) provides information on genetically-modified food, food safety, and "speaking out"

Chapter 13 (Uses and Gratifications Theory) is now comprised of an expanded section on the history of the theory with additional attention paid to media effects

Chapter 16 (Coordinated Management of Meaning) includes an insert pertaining to the confluence of Facetime, Zoom, face-to-face communication, and meaning coordination

Chapter 18 (Groupthink) presents new information on groupthink practices and TikTok

Chapter 19 (Organizational Culture Theory) includes a contemporary interpretation of "culture" from a human capital lens

Chapter 23 (Media Ecology Theory) enhances the relevancy of the theory by including information on "mass-mindedness" and the January 6, 2020 insurrection at the U.S. Capitol

Chapter 26 (Cultivation Theory) continued discussion on the relevance of CT in a changing media landscape

Chapter 27 (Cultural Studies) contains new information on the continued dominance of television as a source of information for older citizens

Features of the Book

To accomplish our goals and address the challenges of teaching communication theory, we have incorporated a structure that includes number of special features and learning aids into the new release:

- *Part One, Foundations.* The first three chapters of the book continue to provide students a solid foundation for studying the theories that follow. This groundwork is essential in order to understand how theorists conceptualize and test their theories. **Chapters 1** and **2** define and interpret communication and provide a framework for examining the theories. We present several traditions and contexts in which theory is customarily categorized and considered. **Chapter 3** provides an overview of the intersection of theory and research. This discussion is essential in a theory course and also serves as a springboard for students as they enroll in other courses. In addition, we present students with a template of various evaluative components that we apply in each of the subsequent theory chapters.

- *Theories and theoretical thinking.* Updated coverage of **all** theories. Separate chapters on each of the theories provide accessible, thorough coverage for students, and offer flexibility to instructors. This updating results in a more thoughtful, current, and applicable presentation of each theory. As noted earlier, in many cases, we have provided the most recent information of the influences of culture and/ or technology upon a particular theory, resulting in some very compelling discussions and examples.

- *Chapter-opening vignettes.* Each chapter begins with an extended vignette, which is then integrated throughout the chapter, providing examples to illustrate the theoretical concepts and claims. We have been pleased that instructors and students point to these vignettes as important applications of sometimes complex material. These stories/case studies help students understand how communication theory plays out in the everyday lives of ordinary people. These opening narratives also help drive home the important points of the theory. In addition, the real-life tone of each vignette entices students to understand the practicality of a particular theory.

- *A structured approach to each theory.* Every theory chapter is self-contained and includes a consistent format that begins with a story, followed by an introduction, a summary of theoretical assumptions, a description of core concepts, and a critique (using the criteria established in Part One). This consistency provides continuity for students, ensures a balanced presentation of the theories, and helps ease the retrieval of information for future learning experiences. Instructors and students have found this template to be quite valuable because it focuses their attention on the key elements of each theory.

- *Students Talking boxes.* These boxes, featured in every chapter, present both new and returning student comments on a particular concept or theoretical issue. The comments, extracted from journals in classes we have taught, illustrate the practicality of the topic under discussion and also show how theoretical issues relate to students' lives. This feature illustrates how practical theories are and how much their tenets apply to our everyday lived experiences. It also allows readers to see how other students taking this course have thought about the material in each chapter. Some of the boxes are students' general observations about the theories as they compare the assertions made to their own lived experiences. Other boxes, called Students Talking Tech, specifically focus on students' reflections on how the theory explained (or failed to explain) the impact of technology on communication.

- *Visual template for theory evaluation.* At the conclusion of each theory chapter, a set of criteria for theory evaluation (presented in **Chapter 3**) is employed. In addition, the theory's context, scholarly tradition (based on Robert Craig's typology), and approach to knowing are presented on visual charts.

- *Theory at a Glance boxes.* In order for students to have an immediate and concise understanding of a particular theory, we incorporate this feature near the beginning of each theory chapter. Students will have these brief explanations and short summaries before reading the rest of the chapter, thereby allowing them to have a general sense of what they are about to encounter.

- *Theory into Practice boxes.* We include this feature to provide further application of the information contained in the chapter. We identify a conclusion or two from the theory and then provide a real-world application of the particular claim. This feature sustains our commitment to enhancing the pragmatic value of a theory.

- *Afterword: ConnectingQuests.* This section of the book provides students with an integration of the various theories in order to see the interrelationships between and among theories. We believe that theories cut across multiple contexts. To this end, students are asked questions that address the intersection of theories. For instance, to understand "decision making" from two theoretical approaches, students are asked to compare the concept and its usage in both Groupthink and Structuration Theory.

- *Visuals.* To increase conceptual organization and enhance the visual presentation of content, we have provided several tables and figures throughout the text. Further, we have provided cartoons to provide another engaging reading option. Many chapters have visual illustrations for students to consider, helping them to understand the material. These visuals provide a clearer sense of the conceptual organization of the theories, and they support those students who best retain information visually.

- *Bolded key terms.* Throughout each chapter, key terms provide students immediate access to unfamiliar concepts and their meanings. Each key term is defined in a box in the margin near its presentation so students have the definitions immediately accessible to them.

- *Appendix.* At the end of the book, all of the theories are listed with a short paragraph summarizing their main points.

In addition to the aforementioned features, several new additions and updates exist in the 2024 release of *Introducing Communication Theory*:

- **NEW *DISCUSSION QUESTIONS.*** Our opening stories are woven throughout each chapter and to "bookend" this thread, we include a discussion question about the application of material to the vignette. We also include a question specifically addressing the tech application of the chapter content ("Do-It-Your-Selfie"), making the theory more relevant to today's technological environment.

- **NEW *TIMELY EXAMPLES.*** To ensure that communication theory remains relevant to all generations of students, the book includes relevant and contemporary topics, including many hashtag activism movements such as #MeToo and #BlackLivesMatter, among others. In addition, examples related to immigration, Title IX, food insecurity, climate change, health care, student debt, hate speech, school shootings, and many more are integrated for students to consider as they unpack the complexity of each theory. As we've discussed various theoretical concepts, we've also invoked celebrities including Taylor Swift, Pink, Bruno Mars, Jack Nicholson, and others. Finally, when appropriate, we've included mediated applications of the material from podcasts, television shows, literature, and movies.

- **NEW *INTEGRATION OF OVER 150 NEW REFERENCES.*** The explosion in communication research is reflected in the incorporation of dozens of new studies, essays, and books that help students understand the theory or theoretical issue. In this release, we have included the references immediately following an individual chapter. This allows for an efficient reflection of a particular source. Further, we also provide students with easy access to a citation by integrating an APA format (the accepted writing style in the communication field) so that they can see the relevancy and currency of a theory. When appropriate, we also have provided URLs for useful websites.

- **UPDATED *THEORY-INTO-ACTION.*** Students are introduced to further applications of the various theories and theoretical concepts by applications taken from current events and media stories.

- **UPDATED** *STUDENTS TALKING*. These boxes, featured in every chapter, present both new and returning student comments on a particular concept or theoretical issue. The comments, extracted from journals in classes we have taught, illustrate the practicality of the topic under discussion and also show how theoretical issues relate to students' lives. This feature illustrates how practical theories are and how much their tenets apply to our everyday lived experiences. It also allows readers to see how other students taking this course have thought about the material in each chapter.
- **UPDATED** *STUDENTS TALKING TECH.* The feature has been added to reflect students' comments about social media and technology pertaining to various theoretical issues. Student comments related to Snapchat, Facebook, LinkedIn, TikTok, X, YouTube, Weibo, Instagram, and Slack, among others, are spliced throughout the book to demonstrate students' understanding and application of the theories to contemporary communication contexts.

Organization of the Content

Part One, Foundations, provides a conceptual base for the discrete theory chapters in Parts Two, Three, and Four. **Chapter 1** begins by introducing the discipline and describing the process of communication. **Chapter 2** provides the prevailing traditions and contexts that frame the communication field. In this chapter, we focus on one approach to the ways in which communication theory can be considered. The chapter then turns to primary contexts of communication, which frame the study of communication in most academic settings across the country. **Chapter 3** explores the intersection of theory and research. In this chapter, we provide students an understanding of the nature of theory and the characteristics of theory. The research process is also discussed, as are perspectives that guide communication research. Our goal in this chapter is to show that research and theory are interrelated and that the two should be considered in tandem as students read the individual chapters. **Chapter 3** also provides a list of evaluative criteria for judging theories as well as for guiding students toward assessment of each subsequent theory chapter.

With Part One establishing a foundation, the remaining chapters introduce students to 27 different theories, each in a discrete, concise chapter. Each is also discussed within a particular Approach to Knowing (Post-Positive, Interpretive, and Critical) identified in **Chapter 3** and emphasized in an insert between the Foundational and Theory chapters. Each of the three theoretical parts places the theories into one of the following contexts: a) intrapersonal, b) interpersonal, c) groups, teams, and organizations, d) the public, e) the media, and f) culture and diversity. In this way, students see how theorists combine a preferred approach to knowing with thinking about communication within a context.

Proctorio

Remote Proctoring & Browser-Locking Capabilities

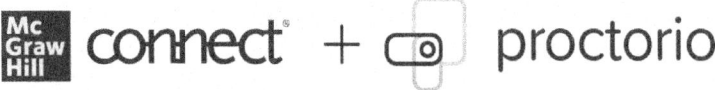

Remote proctoring and browser-locking capabilities, hosted by Proctorio within Connect, provide control of the assessment environment by enabling security options and verifying the identity of the student.

Seamlessly integrated within Connect, these services allow instructors to control the assessment experience by verifying identification, restricting browser activity, and monitoring student actions.

Instant and detailed reporting gives instructors an at-a-glance view of potential academic integrity concerns, thereby avoiding personal bias and supporting evidence-based claims.

 ReadAnywhere® App

Read or study when it's convenient with McGraw Hill's free ReadAnywhere® app. Available for iOS and Android smartphones or tablets, give users access to McGraw Hill tools including the eBook and SmartBook® or Adaptive Learning Assignments in McGraw Hill Connect®. Students can take notes, highlight, and complete assignments offline—all their work will sync when connected to Wi-Fi. Students log in with their Connect username and password to start learning—anytime, anywhere!

OLC-Aligned Courses

Implementing High-Quality Instruction and Assessment through Preconfigured Courseware

In consultation with the Online Learning Consortium (OLC) and our certified Faculty Consultants, McGraw Hill has created preconfigured courseware using OLC's quality scorecard to align with best practices in online course delivery. This turnkey courseware contains a combination of formative assessments, summative assessments, homework, and application activities, and can easily be customized to meet an individual instructor's needs and desired course outcomes. For more information, visit **https://www.mheducation.com/highered/olc**.

Test Builder in Connect

Available within McGraw Hill Connect®, Test Builder is a cloud-based tool that enables instructors to format tests that can be printed, administered within a Learning Management System, or exported as a Word document. Test Builder offers a modern, streamlined interface for easy content configuration that matches course needs, without requiring a download.

Test Builder allows you to:

- access all test bank content from a particular title.
- easily pinpoint the most relevant content through robust filtering options.
- manipulate the order of questions or scramble questions and/or answers.
- pin questions to a specific location within a test.
- determine your preferred treatment of algorithmic questions.
- choose the layout and spacing.
- add instructions and configure default settings.

Test Builder provides a secure interface for better protection of content and allows for just-in-time updates to flow directly into assessments.

Writing Assignment

Available within McGraw Hill Connect®, the Writing Assignment tool delivers a learning experience to help students improve written communication skills and conceptual understanding. Assign, monitor, grade, and provide feedback on writing more efficiently and effectively.

Evergreen

Content and technology are ever-changing, and it is important that you can keep your course up to date with the latest information and assessments. That's why we want to deliver the most current and relevant content for your course, hassle-free.

Introducing Communication Theory: Analysis and Application is moving to an Evergreen delivery model, which means it has content, tools, and technology that is updated and relevant, with updates delivered directly to your existing McGraw Hill Connect® course. Engage students and freshen up assignments with up-to-date coverage of select topics and assessments, all without having to switch editions or build a new course.

Create

Your Book, Your Way

McGraw Hill's Content Collections Powered by Create® is a self-service website that enables instructors to create custom course materials—print and eBooks—by drawing upon McGraw Hill's comprehensive, cross-disciplinary content. Choose what you want from our high-quality textbooks, digital products, articles, cases, and more. Combine it with your own content quickly and easily, and tap into other rights-secured, third-party content such as cases, articles, readings, cartoons, and labs. Content can be arranged in a way that makes the most sense for your course, and you can select your own cover and include the course name and school information as well. Choose the best format for your course: color print, black-and-white print, or eBook. The eBook can be included in your Connect course and is available on the free ReadAnywhere® app for smartphone or tablet access as well. When you are finished customizing, you will receive a free digital copy to review in just minutes! Visit McGraw Hill Create®— **www.mcgrawhillcreate.com** — today and begin building!

Reflecting the Diverse World Around Us

McGraw Hill believes in unlocking the potential of every learner at every stage of life. To accomplish that, we are dedicated to creating products that reflect, and are accessible to, all the diverse, global customers we serve. Within McGraw Hill, we foster a culture of belonging, and we work with partners who share our commitment to equity, inclusion, and diversity in all forms. In McGraw Hill Higher Education, this includes, but is not limited to, the following:

- Refreshing and implementing inclusive content guidelines around topics including generalizations and stereotypes, gender, abilities/disabilities, race/ethnicity, sexual orientation, diversity of names, and age.
- Enhancing best practices in assessment creation to eliminate cultural, cognitive, and affective bias.
- Maintaining and continually updating a robust photo library of diverse images that reflect our student populations.
- Including more diverse voices in the development and review of our content.
- Strengthening art guidelines to improve accessibility by ensuring meaningful text and images are distinguishable and perceivable by users with limited color vision and moderately low vision.

A complete course platform

Connect enables you to build deeper connections with your students through cohesive digital content and tools, creating engaging learning experiences. We are committed to providing you with the right resources and tools to support all your students along their personal learning journeys.

65%
Less Time Grading

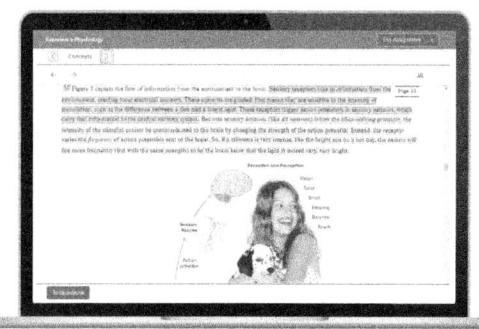

Laptop: Getty Images; Woman/dog: George Doyle/Getty Images

Every learner is unique

In Connect, instructors can assign an adaptive reading experience with SmartBook® 2.0. Rooted in advanced learning science principles, SmartBook 2.0 delivers each student a personalized experience, focusing students on their learning gaps, ensuring that the time they spend studying is time well spent. **mheducation.com/highered/connect/smartbook**

Study anytime, anywhere

Encourage your students to download the free ReadAnywhere® app so they can access their online eBook, SmartBook® 2.0, or Adaptive Learning Assignments when it's convenient, even when they're offline. And since the app automatically syncs with their Connect account, all of their work is available every time they open it. Find out more at **mheducation.com/readanywhere**

"I really liked this app— it made it easy to study when you don't have your textbook in front of you."

Jordan Cunningham, a student at *Eastern Washington University*

Effective tools for efficient studying

Connect is designed to help students be more productive with simple, flexible, intuitive tools that maximize study time and meet students' individual learning needs. Get learning that works for everyone with Connect.

Education for all

McGraw Hill works directly with Accessibility Services departments and faculty to meet the learning needs of all students. Please contact your Accessibility Services Office, and ask them to email **accessibility@mheducation.com**, or visit **mheducation.com/about/accessibility** for more information.

Affordable solutions, added value

Make technology work for you with LMS integration for single sign-on access, mobile access to the digital textbook, and reports to quickly show you how each of your students is doing. And with our Inclusive Access program, you can provide all these tools at the lowest available market price to your students. Ask your McGraw Hill representative for more information.

Solutions for your challenges

A product isn't a solution. Real solutions are affordable, reliable, and come with training and ongoing support when you need it and how you want it. Visit **supportateverystep.com** for videos and resources both you and your students can use throughout the term.

Updated and relevant content

Our new Evergreen delivery model provides the most current and relevant content for your course, hassle-free. Content, tools, and technology updates are delivered directly to your existing McGraw Hill Connect® course. Engage students and freshen up assignments with up-to-date coverage of select topics and assessments, all without having to switch editions or build a new course.

Acknowledgments

A successful book of this nature owes its existence to efforts made by others in addition to the listed authors. And, some people who have helped with this book may not even realize the debt we acknowledge here. We would like to thank all those who have helped us as we worked our way through this large project. First, many professors and students have written to us, providing important clarification and examples. We've included their ideas, suggestions, and impressions throughout the new release.

In addition, our work rests on the shoulders of the theorists whose creations we profile in this book. We are grateful for their creative thinking, which allows us to understand and begin to predict the complexities of the communication process. We worked hard to try to capture their insights and conclusions and convert these thoughts for introductory students in theory.

Further, our insights represent the discussions that we have had with our communication theory students and colleagues over the years. Several parts of this book are based on student input at both of our institutions. Students have contributed to this book in both direct and indirect ways.

Textbook writers understand that no book is possible without the talents and commitment of both an editorial and production team. We extend our deep appreciation and admiration to those who have made our words come to life in various ways:

Sarah Remington, Executive Portfolio Manager
Katie Reuter, Associate Portfolio Manager
Melissa Leick, Lead Content Project Manager
Prineeth Priyan, Content Licensing Specialist

Finally, the project development team was facilitated by the keen eyes of Rashmi Tickyani, Adina Lonn, and with the support of Meghan McDonaugh.

Our production team remains among the best group of experts with whom we've worked in our years of writing! We thank them for their patience, precision, and guidance.

As is customary in each book he writes, Rich would like to acknowledge his mother Bev. Sadly, she passed away before this book came to fruition. But, her inspiration related to tenacity, patience, and other-centeredness will live on in Rich's life. He is forever grateful for her positive influence. Rich would also like to thank his husband, Chris, who knows precisely when to make things less intense and more relaxing. No two individuals have influenced his life more profoundly than Bev and Chris!

Lynn would like to thank her entire family for invaluable lessons in communication theory and practice. And always, Lynn is grateful for the memory of her loving parents whose steadfast support and encouragement of her scholarship, and all of her interests, sustain her in every project she undertakes. Friends and colleagues provided great support and have taught her many valuable lessons about scholarship and communication theory.

Finally, we thank the manuscript reviewers who gave their time and expertise to keep us on track in our interpretation of the ideas of others. We are grateful for their careful reading and insightful suggestions, which expanded and clarified our thinking in many ways. Our text is a much more useful product because of the comments and suggestions of the following reviewers who have shaped this book over the past many editions:

Amy Hubbard, *University of Hawaii at Manoa*

Anestine Theophile-Lafond, *Montgomery College*

Anna Laura Jansma, *University of California, Santa Barbara*

Anne M. Nicotera, *George Mason University*

Bryan Horikami, *Salisbury University*

Cam Brammer, *Marshall University*

Christine Armstrong, *Northampton Community College - Monroe*

Christine North, *Ohio Northern University*

Chrys Egan, *Salisbury University*

Deborah Smith-Howell, *University of Nebraska*

Debra Mazloff, *University of St. Thomas*

Denise Solomon, *University of Wisconsin*

Edward T. Funkhouser, *North Carolina State University*

Elizabeth M. Perse, *University of Delaware*

Emily Cripe, *Kutztown University*

Greg G. Armfield, *New Mexico State University*

Holly H. Bognar, *Cleveland State University*

J. Dean Farmer, *Campbell University*

Jack Baseheart, *University of Kentucky*

James D. Robinson, *University of Dayton*

James Gilchrist, *Western Michigan University*

Jamie Byrne, *Millersville University*

Janet Skupien, *University of Pittsburgh*

Javette Grace Hayes, *California State University, Fullerton*

Jeffrey D. Brand, *North Dakota State University*

Jimmie Manning, *University of Nevada, Reno*

Joann Keyton, *University of Kansas*

Jodi Hallsten Lyczak, *Illinois State University*

John R. Baldwin, *Illinois State University*

Jon Conlogue, *Westfield State University*

Jon Smith, *Southern Utah University*

Juan Liu, *Wayne State University*

Kate Joeckel, *Bellevue University*

Katy Wiss, *Western Connecticut State University*

Kelly Jones, *Pitt Community College*

Kent Drummond, *University of Wyoming*

Kevin Wright, *University of Memphis*

Kyle Tusing, *University of Arizona*

Laura Jansma, *University of California-Santa Barbara*

Libby McGlone, *Columbus State Community College*

Linda M. Pledger, *University of Arkansas*

Lisa Hanasono, *Bowling Green State University*

Lisa Hebert, *Louisiana State University*

Madeline M. Keaveney, *California State University-Chico*

Mark Zeigler, *Florida State University*

Martha J. Haun, *University of Houston*

Mary Ann Renz, *Central Michigan University*

Matthew McAllister, *Virginia Tech*

Melanie Laliker, *Bridgewater College of Virginia*

Michael Barberich, *University at Albany, SUNY*

Nora Madison, *Chestnut Hill College*

Patricia Rockwell, *University of Southwestern Louisiana*

Ralph Thompson, *Cornell University*

Randall S. Chase, *Dixie State University*

Randy K. Dillon, *Southwest Missouri State University*

Raphael Mazzone, *University of Maryland*

Rebecca Dumlao, *East Carolina University*

Rebecca W. Tardy, *University of Louisville*

Reed Markham, *Salt Lake Community College*

Rita L. Rahoi-Gilchrest, *Winona State University*

Robert Wawee, *University of Houston-Downtown*

Robert William Wawee, *University of Houston Downtown*

Scott Guest, *Bowling Green State University*

Shaun Cashman, *Pfeiffer University*

Sheryl Bowen, *Villanova University*

Sue Barnes, *Fordham University*

Sunny Lie Owens, *California State Polytechnic University Pomona*

Susan Jarboe, *San Diego State University*

Tami Spry, *St. Cloud State University*

Thomas Feeley, *State University of New York, Geneseo*

Yashu Chen, *San Diego State University*

Yifeng Hu, *The College of New Jersey*

About the Authors

Rich West is a Professor in the Department of Communication Studies at Emerson College in Boston. Rich received his BA and MA from Illinois State University and his PhD from Ohio University. Rich has been teaching since 1984, and his teaching and research interests range from family diversity to teacher–student communication. He began teaching communication theory as a graduate student and has taught the class in lecture format to more than 200 students. Rich is a past recipient of the Outstanding Alumni Award in Communication at Illinois State University and Ohio University. He is also a two-time recipient of the Presidential Citation for Excellence in Service from the National Communication Association (NCA). He is also a recipient of the National Communication Association's Samuel L. Becker Award for Distinguished Service. Rich is a member of several editorial boards in communication journals and co-editor of *the Routledge Handbook of Communication and*

Rich West

Bullying. NCA has also recognized him with the Samuel L. Becker Award for Distinguished Service to the discipline. Rich is also the recipient of the Distinguished Service of the Eastern Communication Association, where he also received recognition as a Teaching Fellow and Research Fellow. In 2008, Rich assumed the ECA Presidency.

Lynn H. Turner is Professor Emerita in Communication Studies at Marquette University in Milwaukee, Wisconsin. Lynn received her BA from the University of Illinois and her MA from the University of Iowa, and she received her PhD from Northwestern University. She has taught communication theory and research methods to undergraduates and graduates in the Diederich College of Communication at Marquette since 1985. Prior to coming to Marquette, Lynn taught at Iowa State University and in two high schools in Iowa. Her research interests include interpersonal communication, family communication, and gendered communication. She is the recipient of several awards, including Marquette's College of Communication Research Excellence Award, and the Book of the Year award from the Organization for the Study of Communication, Language, and Gender for her book with Patricia Sullivan, *From the Margins to the Center: Contemporary Women and Political Commu-nication*. Lynn is a past president of the Central States Communication Association

Lynn H. Turner

and was recognized for her contributions in service and research by CSCA as a member of their Hall of Fame. In 2022, Lynn was honored as a Distinguished Scholar by the National Communication Association.

Rich and Lynn, together, are coauthors of dozens of essays and articles in the communication field. In addition, the two have served as guest coeditors of the *Journal of Family Communication* a few times, focusing on diversity and the family. They have coauthored several books, including *Gender and Communication, Perspectives on Family Communication, IPC, Understanding Interpersonal Communication,* and *Introduction to Communication*. The two have coedited the *Family Communication Sourcebook* (Sage, 2006; Winner of the Outstanding Book Award by the National Communication Association), and *The Sage Handbook of Family Communication*. Further, they are co-recipients of the Bernard J. Brommel Award for Outstanding Scholarship and Service in Family Communication. Finally, both recognize the uniqueness and the honor to have served as consecutive presidents of the National Communication Association (Lynn in 2011; Rich in 2012), "the oldest and largest organization in the world promoting communication scholarship and education" (www.natcom.org).

CHAPTER 1
Thinking About Communication: Definitions, Models, and Ethics

"I suppose all of us get accustomed to look at what we are doing in a certain way and after a while have a kind of 'trained incapacity' for looking at things in any other way."

—Marie Hochmuth Nichols

The Hernandez Family

José and Angie Hernandez have been married for almost 30 years, and they are the parents of three children who, until recently, had been out of the house for years. But, a layoff at the company where their son Eddy worked forced the 24-year-old to return home until he could get another job. The job market after the recession was still not moving along fast enough, particularly for a man with limited skills and no college education.

At first, Eddy's parents were glad that he was home. His father was proud of the fact that his son wasn't embarrassed about returning home, and his mom was happy to have him help her with some of the mundane tasks at home. In fact, Eddy showed both José and Angie how to Zoom with their friends and also put together a family website. His parents were especially happy about having a family member who was "tech-savvy" hanging around the house.

But the good times surrounding Eddy's return soon ended. Eddy brought his cell phone to the table each morning, marring the Hernandezes' once-serene breakfasts. The clicking sound of texting and his incessant looking down undermined an otherwise calm beginning to the day. In addition, José and Angie's walks each morning were complicated because their son often wanted to join them. At night, when they went to bed, the parents could hear Eddy FaceTiming with his friends, sometimes until 1:00 A.M. When Eddy's parents thought about communicating their frustration and disappointment, they quickly recalled how tough it must be for their son in his situation. He was looking for a job, and they didn't want to upset him too much. The Hernandezes tried to figure out a way to communicate to their son that although they loved having him around, they wished that he would get a job soon and leave the house. They simply wanted some peace, privacy, and freedom, and their son was getting in the way. It wasn't a feeling either one of them liked, but it was their reality.

They considered a number of different approaches. In order to get the conversation going, a friend of José and Angie gave them a few rental apartment website links to forward on to Eddy. Recently, the couple's frustration with the situation took a turn for the worse. Returning from one of their long walks, they discovered Eddy on the couch, hung over from a party held the night before at his friend's house. When José and Angie confronted him about his demeanor, Eddy shouted, "Don't start lecturing me now. Is it any wonder that none of your other kids call you? It's because you don't know when to stop! Look, I got a headache and I really don't need to hear it right now!" José snapped, "*Get out of my house. Now!*" Eddy left the home, slamming the front door behind him. Angie stared out of the window, wondering when or if they would ever hear from their son again.

The value of communication has been lauded by philosophers (*"Be silent or say something better than silence"*–Pythagoras), writers (*"The difference between the right word and the almost right word is the difference between lightning and a lightning bug"*–Mark Twain), performing artists (*"Any problem, big or small, in a family usually starts with bad communication"*–Emma Thompson), business leaders (*"Writing is great for keeping records and putting down details, but talk generates ideas"*–T. Boone Pickens), motivational speakers (*"The quality of your communication is the quality of your life"*–Tony Robbins), talk show hosts (*"Great communication begins with connection"*–Oprah), and even reality TV superstars (*"Why not share my story?"*–Kim Kardashian). Perhaps one of the most lasting of all words came from a 1967 film (*Cool Hand Luke*): *"What we got here is a failure to communicate"*–a quotation that has subsequently been stated in such diverse settings as in the movie *Madagascar*, the song "Civil War" by Guns N' Roses, and television shows *NCIS, Modern Family, Law and Order: SVU,* and *Abbott Elementary*. It's clear that nearly all cross sections of a Western society view communication as instrumental in human relationships. And to be sure, regardless of where we live around the globe, we can't go through a day without communication.

In the most fundamental way, communication depends on our ability to understand one another. Although our communication can be ambiguous (*"I never thought I'd get this gift from you"*), as we suggested above, one primary and essential goal in communicating is understanding. Our daily activities are wrapped in conversations with others. Yet, as we see with the Hernandez family, even those in close relationships can have difficulty expressing their thoughts.

Being able to communicate effectively is highly valued in the United States. Corporations have recognized the importance of communication. HSI, for instance, a human resources and operations leader in jobs, notes that professions related to risk management and environmental health and safety require outstanding communication skills (HSI, n.d.). In 2023 Forbes predicted that because 1 billion jobs will be transformed by technology in four powerhouse economies alone (the U.K., Australia, Canada, the U.S.), communication remains the most sought-after skill that employers seek (Meister, 2023). Interestingly, as early as the late 1960s, doctor–patient communication has been a topic of concern in research (Korsch, Gozzi, & Francis, 1968). More recent literature shows that effective doctor–patient communication is essential for the recovery of patients (Kittleson, 2023). Finally, in the classroom, researchers have concluded that affirming feedback/ student confirmation positively affects student learning (Titsworth, Mazer, Goodboy, Bolkan, & Myers, 2015), and in athletics, this confirming communication influences athlete motivation and competitiveness (Cranmer, Gagnon, & Mazer, 2020). And, with respect to cross-platform messaging sites such as WhatsApp, individuals in intergenerational families report its use helps to make communicating to various family members both realistic and practical (Taipale, 2019). Make no mistake about it: abundant evidence underscores the fact that communication is an essential, pervasive, and consequential behavior in our society. Human communication is the essence of what it means to be alive.

As a student of communication, you are uniquely positioned to determine your potential for effective communication. To do so, however, you must have a basic understanding of the communication process and of how communication theory, in particular, functions in your life. We need to be able to talk effectively, for instance, to a number of very different types of people during an average day: roommates, teachers, ministers, salespeople, family members, friends, automobile mechanics, and health care providers, among many others. And, we must be able to adapt and accommodate to the technology that is ever-present in our ongoing relationships with others.

Communication opportunities fill our lives each day. However, we need to understand the whys and hows of our conversations with others. For instance, why do two people in a relationship feel a simultaneous need for togetherness and independence? Why do some women feel ignored or devalued in conversations with men? Why does language often influence the thoughts of others? How do media influence people's behavior?

To what extent can social media affect the communication between and among people? These and many other questions are at the root of why communication theory is so important in our society and so critical to understand.

Defining Communication

Our first task is to create a common understanding for the term *communication*. Defining communication can be challenging because it's a term that has been used by a wide assortment of people—from politicians to evangelical preachers to our parents. It is also an all-encompassing term and invoked with different motivations in mind. A friend might think everything is communication, while you might think that it occurs only with mutual understanding. A long-standing interpretation related to communication has been advanced by scholars. Sarah Trenholm (1991), for instance, notes that although the study of communication has been around for centuries, it does not mean communication is well understood. In fact, Trenholm interestingly illustrates the dilemma when defining the term. She states, "Communication has become a sort of 'portmanteau' term. Like a piece of luggage, it is overstuffed with all manner of odd ideas and meanings. The fact that some of these do fit, resulting in a conceptual suitcase much too heavy for anyone to carry, is often overlooked" (p. 4).

We should note that there are many ways to interpret and define communication—a result of the complexity and richness of the communication discipline. Imagine, for instance, taking this course from two different professors. Each would have their own way of facilitating the material, and each classroom of students would likely approach communication theory in a unique manner. Ideally, the result would be two exciting and distinctive approaches to studying the same topic.

Students Talking Tech: Maddy

 My own way of defining communication would have to include how I met my current boyfriend. I would never be with him if it wasn't for social media and Bumble. The site let me—as a woman—make the first move. When I heard about this app, I thought, "Finally!" I was sick of guys who were looking for "now" rather than "now and later!" My boyfriend and I talked online and then over the phone, and then we met. The whole process was something I controlled, which made it easier and more comfortable for me. I can't imagine that I would've had any chance to even meet this guy, let alone communicate with him, if Bumble didn't help me start that process.

This uniqueness holds true with defining communication. Scholars tend to see human phenomena from their own perspectives, something we delve into further in **Chapter 2.** In some ways, researchers establish boundaries when they try to explain phenomena to others. Communication scholars may approach the interpretation of communication differently because of differences in scholarly

> **communication** a social process in which individuals employ symbols to establish and interpret meaning in their environment

values. With these caveats in mind, we offer the following definition of *communication* to get us pointed in the same direction. **Communication** is a social process in which individuals employ symbols to establish, interpret, and co-create meaning in their environment(s). We necessarily draw in elements of mediated communication as well in our discussion, given the importance that communication technology plays in contemporary society. With that in mind, let's define five key terms in our perspective: *social, process, symbols, meaning,* and *environment* (**Figure 1.1**).

First, we believe that communication is a social process. When interpreting communication as **social,** we mean to suggest that it involves people and interactions, whether face-to-face or online. This necessarily includes two people, who act as senders and receivers. Both play an integral role in the communication process. When communication is social, it involves people who come to an interaction with various intentions, motivations, and abilities. To suggest that communication is a **process** means that it is ongoing and unending. Communication is also dynamic, complex, and continually changing. With this view of communication, we emphasize the dynamics of making meaning. Therefore, communication has no definable beginning and ending. For example, although José and Angie Hernandez may tell their son that he must leave the house, their discussions with him and about him will definitely continue well after he leaves (e.g., "*What do we do now?*"). In fact, the conversation they have with Eddy today will most likely affect their communication with him tomorrow. Similarly, our past communications with people have been stored in our minds and have affected our conversations with them.

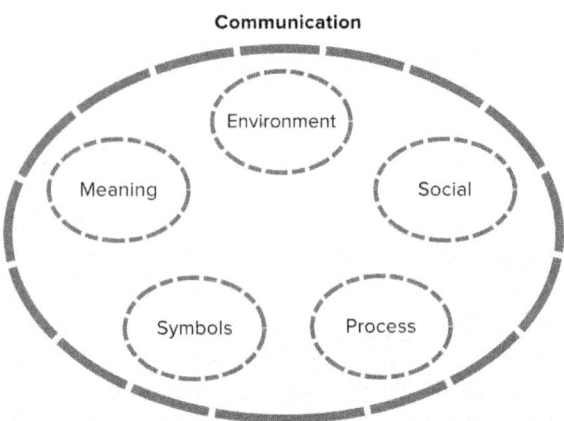

Communication

Environment · Meaning · Social · Symbols · Process

Figure 1.1   Key Terms in Defining Communication

social the notion that people and interactions are part of the communication process

process ongoing, dynamic, and unending occurrence

The process nature of communication also means that much can happen from the beginning of a conversation to the end. People may end up at a very different place once a discussion begins. This is exemplified by the frequent conflicts that roommates, spouses, and siblings experience. Although a conversation may begin with absolute and inflexible language, the conflict may be resolved with compromise. All of this can occur in a matter of minutes. While online, one text can prompt a great deal of conflict between two people, only exacerbated by the fact that texting often leads to confusion and a lack of clarity.

Individual and cultural changes affect communication. Conversations between siblings, for example, have shifted from the 1950s to today. Years ago, siblings rarely discussed the impending death of a parent or the need to take care of an aging parent. Today, it's not uncommon to listen to even young people talking about senior care, home health care, and even cremation arrangements. Perceptions and feelings can change and may remain in flux for quite some time.

Some of you may be thinking that because the communication process is dynamic and unique it may be nearly impossible to study. However, C. Arthur VanLear (1996) argues that because the communication process is so dynamic, researchers and theorists can look for patterns over time. He concludes that "if we recognize a pattern across a large number of cases, it permits us to 'generalize' to other unobserved cases" (p. 36). Or, as communication pioneers Paul Watzlawick, Janet Beavin, and Don Jackson (1967) suggest, the interconnectedness of communication events is critical and pervasive. Thus, it is possible to study the dynamic communication process.

To help you visualize this process, imagine a continuum where the points are unrepeatable and irreversible. The communication field employed the historical spiral or helix to explain this process (**Figure 1.2**). In doing so, two conclusions emerged: (1) communication experiences are cumulative and are influenced by the past, and (2) because present experiences inevitably influence a person's future, communication is nonlinear. Communication, therefore, can be considered a process that changes over time and among interactants.

A third term associated with our definition of communication is *symbols*. A **symbol** is an arbitrary label or representation of phenomena. Words are symbols for concepts and things—for example, the word *love* represents the idea of love; the word *chair* represents a thing we sit on. Labels may be ambiguous, may be both verbal and nonverbal, and may occur in face-to-face and mediated communication. Symbols are usually agreed on within a group but may not be understood outside of the group. In this way, their use is often arbitrary. For instance, most college students understand the phrase "preregistration is closed"; those outside of college may not understand its meaning. Further, there are both **concrete symbols** (the symbol represents an object) and **abstract symbols** (the symbol stands for a thought or idea).

Figure 1.2   **Communication Process known as a Helix**

Even the innocuous X symbol—the hashtag—resonates in a number of fields, particularly in politics. Think, for instance, of the thousands of X posts that President Trump sent before and during his presidency, even though most of his posts represented the "politics of debasement" (Ott, 2017, p. 58). Further, in-depth political reporting and discussion are fast becoming rare in politics, and "the more candidates used Twitter to broadcast their thoughts, the more people retweeted them, spreading their messages and journalists mentioned tweets in their election coverage" (Buccoliero, Bellio, Crestini, & Arkoudas, 2018, p. 88). The "politics of COVID-19,"

symbol arbitrary label given to a phenomenon

concrete symbol symbol representing an object

abstract symbol symbol representing an idea or thought

meaning what people extract from a message

too, found a home on X. In fact, in one study, elected leaders used X to communicate a great deal of misinformation about the value of vaccines (Pérez-Curiel, Rúas-Araújo, & Rivas-de-Roca, 2022). This, in turn, fostered mistrust in any public health campaigns related to vaccinations.

In addition to process and symbols, meaning is central to our definition of communication. **Meaning** is what people extract from a message. In communication episodes, messages can have more than one meaning and even multiple layers of meaning. Without sharing some meanings, we would all have a difficult time speaking the same language or interpreting the same event. Judith Martin and Tom Nakayama (2018) point out that meaning has cultural consequences:

> [W]hen President George W. Bush was about to go to war in Iraq, he referred to this war as a "crusade." The use of this term evoked strong negative reactions in the Islamic world, due to the history of the Crusades nearly 1,000 years ago.... While President Bush may not have knowingly wanted to frame the Iraq invasion as a religious war against Muslims, the history of the Crusades may make others feel that it is. (p. 70)

Clearly, not all meaning is shared, and people do not always know what others mean. In these situations, we must be able to explain, repeat, and clarify. For example, if the Hernandezes want to tell Eddy to move out, they will probably need to go beyond telling him that they just need their "space." Eddy may perceive "needing space" as simply wanting him to stay out of the house two nights a week. Furthermore, his parents will have to figure out what communication "approach" is best. They might believe that being direct may be best to get their son out of the house. Or they might fear that such clear communication is not the most effective strategy to change Eddy's behavior. Regardless of how José and Angie Hernandez communicate their wishes, without sharing the same meaning, the family will have a challenging time getting their messages across to one another.

The final key term in our definition of communication refers to the multiple environments related to communication. An **environment** is the situation or context in which communication occurs. The environment includes a number of elements, including time, place, historical period, relationship, and a speaker's and listener's cultural backgrounds. You can understand the influence of environments by thinking about your beliefs and values pertaining to socially significant topics such as marriage equality, physician-assisted suicide, and immigration into the United States. If you have had personal experience with any of these topics, it's likely your views are affected by your perceptions.

> **environment** situation or context in which communication occurs
>
> **models** simplified representations of the communication process

The environment can also be mediated. By that, we mean that communication takes place with technological assistance. All of us have communicated in a mediated environment, namely, through email, chat rooms, or social networking sites, such as WhatsApp or Snap. These mediated environments influence the communication between two people in that people in electronic relationships are (usually) challenged in observing the impact of many nonverbal communication behaviors, such as eye contact, vocal characteristics, and even body movements. We may "think" a message has meaning, but technology may be influencing how we interpret that message.

Models of Understanding: Communication as Action, Interaction, and Transaction

Communication theorists create **models,** or simplified representations of complex interrelationships among elements in the communication process, which allow us to visually understand a sometimes complex process. Models help us weave together the basic elements of the communication process. Although there are many communication models, we discuss the three most prominent ones here (linear, interactional, and transactional). In discussing these models and their underlying approaches, we wish to demonstrate the manner in which communication has been conceptualized over the years. We conclude our discussion by proposing a fourth model that infuses technology and other elements into our discussion. We term this the holistic model.

Communication as Action: The Linear Model

In 1949, Claude Shannon, a Bell Laboratories scientist and professor at the Massachusetts Institute of Technology, and Warren Weaver, a consultant on projects at the Sloan Foundation, described communication as a linear process. They were concerned with radio and telephone technology and wanted to develop a model that could explain how information passed through various channels. The result

> **linear model of communication** one-way view of communication that assumes a message is sent by a source to a receiver through a channel

was the conceptualization of the **linear model of communication.** The linear model has been depicted by writers in this way:

Who?

says What?

in what Channel?

to Whom?

The linear approach to human communication comprises several key elements, as **Figure 1.3** demonstrates. A **source,** or transmitter of a message, sends a **message** to a **receiver,** the recipient of the message. The receiver is the person who makes sense out of the message. All of this communication takes place in a **channel,** which is the pathway to communication. Channels frequently correspond to the visual, tactile, olfactory, and auditory senses. Thus, you use the visual channel when you see your roommate, and you use the tactile channel when you hug your parent.

Communication also involves **noise,** which is anything not intended by the informational source. There are four types of noise. First, **semantic noise** pertains to the slang, jargon, or specialized language used by individuals or groups. For instance, when Jennifer received a medical report from her ophthalmologist, the physician's words included phrases such as "ocular neuritis," "dilated funduscopic examination," and "papillary conjunctival changes." This is an example of semantic noise because outside of the medical community, these words have limited (or no) meaning. **Psychological noise** refers to a communicator's prejudices, biases, and predispositions toward another or the message. **Physical, or external, noise** exists outside of the receiver. To exemplify these two types, imagine listening to participants at a political rally. You may experience psychological noise listening to the views of a politician whom you do not support, and you may also experience physical noise from the people nearby who may be protesting the politician's presence. Finally, **physiological noise** refers to the biological influences on the communication process. Physiological noise exists if you or a speaker is ill, fatigued, or hungry.

Although this view of the communication process was highly respected many years ago, the approach is very limited for several reasons. First, the model presumes that there is only one message in the communication process. Yet we all can point to a number of circumstances in which we send several messages at once. Think about, for example, the numbers of times you find yourself texting someone and talking on the phone with another person! Second, as we have previously noted, communication does not have a definable beginning and ending. Shannon and Weaver's model adopts this mechanistic orientation. Furthermore, to suggest that communication is simply one person speaking to another oversimplifies

Dan Reynolds/CartoonStock Ltd

source originator of a message

message words, sounds, actions, or gestures in an interaction

receiver recipient of a message

channel pathway to communication

noise distortion in channel not intended by the source

semantic noise linguistic influences on reception of message

psychological noise cognitive influences on reception of message

physical (external) noise bodily influences on reception of message

physiological noise biological influences on reception of message

the complex communication process. Listeners are not so passive, as we can all confirm when we are in heated arguments with others. Clearly, communication is more than a one-way effort and has no definable middle or end.

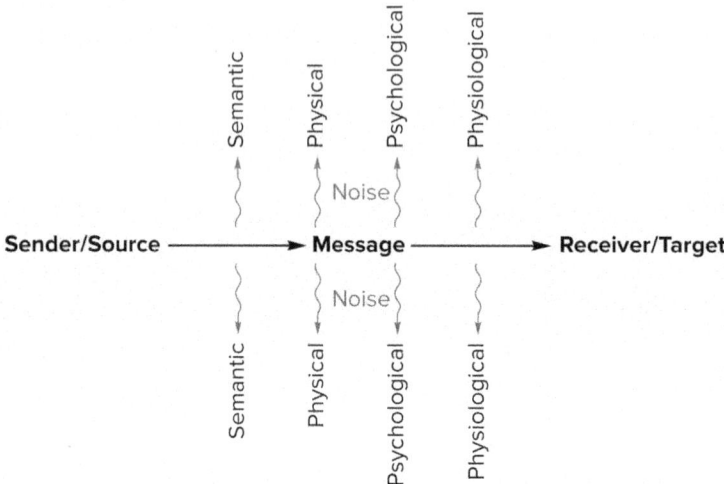

Figure 1.3   Linear Model of Communication

Communication as Interaction: The Interactional Model

The linear model suggests that a person is only a sender or a receiver. That is a particularly narrow view of the participants in the communication process. Wilbur Schramm (1954), therefore, proposed that we also examine the relationship between a sender and a receiver. He conceptualized the **interactional model of communication,** which emphasizes the two-way communication process between communicators (**Figure 1.4**). In other words, communication goes in two directions: from sender to receiver and from receiver to sender. This circular process suggests that communication is ongoing. The interactional view illustrates that a person can perform the role of either sender or receiver during an interaction, but not both roles simultaneously.

One element essential to the interactional model of communication is **feedback,** or the response to a message. Feedback may be verbal or nonverbal, intentional or unintentional. Feedback helps communicators to know whether or not their message is being received and the extent to which meaning is achieved. In the interactional model, feedback takes place after a message is received, not during the message itself. By now, we all realize that with such technologies as FaceTime, feedback takes place both during and after an interaction.

interactional model of communication view of communication as the sharing of meaning with feedback that links source and receiver

feedback a subprocess of calibration; information allowing for change in the system

To illustrate the critical nature of feedback and the interactional model of communication, consider our opening example of the Hernandez family. When Eddy's parents find him on the couch drunk, they proceed to tell Eddy how they feel about his behavior. Their outcry prompts Eddy to argue with his parents, who, in turn, tell him to leave their house immediately. This interactional sequence shows that there is an alternating nature in the communication between Eddy and his parents. They see his behavior and provide their feedback on it, Eddy listens to their message and responds, then his father sends the final message telling his son to leave. We can take this even further by noting the door slam as one additional feedback behavior in the interaction.

A final feature of the interactional model is a person's **field of experience,** or how a person's culture and experiences influence their ability to communicate with another. Each person brings a unique field of experience to each communication episode, and these experiences frequently influence the communication between people. For instance, when two people come together and begin dating, the two inevitably bring their fields of experience into the relationship. One person in this couple may have been raised in a large family with several siblings, while the other may be an only child. These experiences (and others) will necessarily influence how the two come together and will most likely affect how they maintain their relationship.

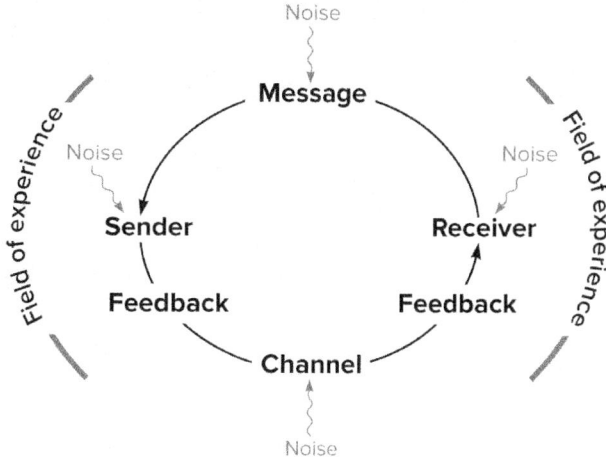

Figure 1.4 **Interactional Model of Communication**

Like the linear view, the interactional model has been criticized. The interactional model suggests that one person acts as sender while the other acts as receiver in a communication encounter. As you have experienced, however, people communicate as both senders and receivers in a single encounter. But the prevailing criticism of the interactional model pertains to the issue of feedback. The inter-

field-of-experience a factor in how NDGMs choose communication strategies when interacting with DGMs, it is the totality of an individual's life experiences

actional view assumes two people speaking and listening, but not at the same time. But what occurs when a person sends a nonverbal message during an interaction? Smiling, frowning, or simply moving away from the conversation during an interaction between two people happens all the time. For example, in an interaction between a mother and her daughter, the mother may be reprimanding her child while simultaneously "reading" the child's nonverbal behavior. Is the girl laughing? Is she upset? Is she even listening to her mother? Each of these behaviors will inevitably prompt the mother to modify her message. And as we pointed out earlier, feedback during FaceTime can occur at all times. These criticisms and contradictions inspired development of a third model of communication.

Communication as Transaction: The Transactional Model

The **transactional model of communication** (Barnlund, 1970, 2017) underscores the simultaneous sending and receiving of messages in a communication episode, as **Figure 1.5** shows. To say that communication is transactional means that the process is cooperative; the sender and the receiver are mutually responsible for the effect

transactional model of communication view of communication as the simultaneous sending and receiving of messages.

and the effectiveness of communication. In the linear model of communication, meaning is sent from one person to another. In the interactional model, meaning is achieved through the feedback of a sender and a receiver. In the transactional model, people build shared meaning. Furthermore, what people say during a transaction is greatly influenced by their past experience. So, for instance, at a college fair, it's likely that a college student will have a great deal to say to a high school senior because of the college student's experiences in class and around campus. A college senior will, no doubt, have a different view of college than, say, a college sophomore, due in large part to their past college experiences.

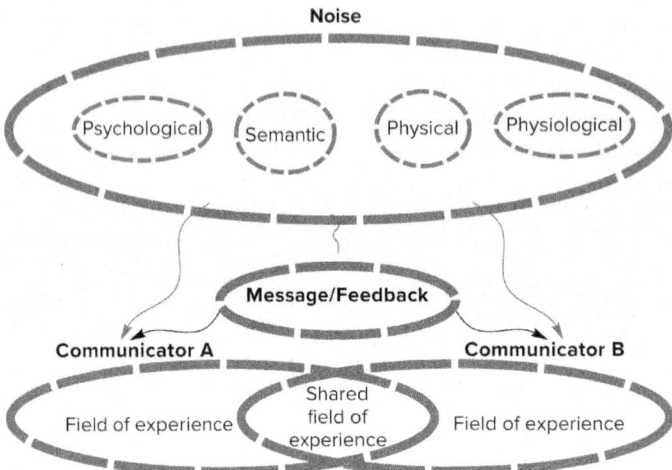

Figure 1.5 Transactional Model of Communication

Transactional communication requires us to recognize the influence of one message on another. One message builds on the previous message; therefore, there is an interdependency between and among the components of communication. Think of this as "shared meaning." A change in one causes a change in others. Furthermore, the transactional model presumes that as we simultaneously send and receive messages, we attend to both verbal and nonverbal elements of a message. In a sense, communicators negotiate meaning. For instance, if a friend asks you about your family background, you may use some private language that your friend doesn't understand. Your friend may make a face while you are presenting your message, indicating some sort of confusion with what you've said. As a result, you will most likely back up and define your terms and then continue with the conversation. This example highlights the degree to which two people are actively involved in a communication encounter. The nonverbal communication is just as important as the verbal message in such a transactional process.

Earlier we noted that the field of experience functions in the interactional model. In the transactional model, the fields of experience exist, but overlap occurs. That is, rather than person A and person B having separate fields of experience, eventually the two fields merge (see **Figure 1.5**). This was an important addition to the understanding of the communication process because it demonstrates an active process of understanding. That is, for communication to take place, individuals must build shared meaning. For instance, in our earlier example of two people with different childhoods, the interactional model suggests that they would come together with an understanding of their backgrounds. The transactional model, however, requires each of them to understand and incorporate the other's field of experience into their life. For example, it's not enough for Julianna to know that Paul has a prior prison record; the transactional view holds that she must figure out a way to put his past into perspective. Will it affect their current relationship? How? If not, will Julianna discuss it with Paul? The transactional model takes the meaning-making process one step further than the interactional model. It assumes reciprocity, or shared meaning.

Communication Models of the Future

As we move further into the 21st century, we have to ask the question: Are these models sufficient as we examine human communication? We already know that communication models are usually incomplete and unsuitable for all purposes (Perse & Lambe, 2016). The answer is fairly complex. First, the proliferation of new social networking sites (SNS), for example, and their influence upon communication, demand that communication models integrate technological discussions. Second, traffic to SNS has grown exponentially over the past few years with about 72 percent of the public using social networking (Pew Research Center, 2021) up from 7 percent in 2005. The diversity of these sites—from Reddit to LinkedIn to Instagram to Snapchat—suggests that no simple model will be possible.

To this end, we suggest that the holistic model of communication might be on the horizon (Turner & West, 2019). This approach emphasizes some of the foundational elements found in our definition and the other models. The **holistic model of communication** underscores communication as a coherent combination of environment, shared technology experience, and communication effect. We address these elements below (**Figure 1.6**).

> **holistic model of communication** view of communication suggesting that communication occurs in a context, with overlapping fields of experiences, and having an effect

> **context** the general environment in which communication takes place

First, we believe that all communication occurs in a **context,** or an environment in which a message is sent. Context is complex and includes more than the tangible; it can be cultural, historical, and/or situational. We briefly address each type of context below.

> **cultural context** the environment in which unique cultural patterns of communication take place

The **cultural context** pertains to the various patterns of communication that are unique to a particular culture. Whether we're addressing their rules, roles, or norms, cultures both in the United States and across the globe are idiosyncratic, and we cannot ignore this distinctness when talking about the communication process. Imagine, for instance, talking to a colleague. Culture always influences the communication that takes place between and among people. If you've emailed a potential date before the date, that communication may influence how the date will evolve. We return to a more comprehensive discussion of the impact of culture on communication in **Chapter 2.** For now, it's simply important to note that the cultural context influences people's communication.

In the **historical context,** messages are understood in relationship to the historical period in which they are exchanged, underscoring the process-centered nature of communication, which we identified earlier. For each of you, for instance, sending a text or an email is commonplace. But, think about the sci-fi nature of such a message if you lived in the 1940s! The notion of what it meant to be "unemployed" during the Great Depression is vastly different from the interpretation of "unemployment" today. In fact, the word "underemployed" is often used more frequently to avoid the negative meaning related to not having sufficient income to live. The word was never fully understood in the early 1900s!

> **historical context** the environment in which communication can be understood via its historical period

> **situational context** the tangible environment in which communication occurs

The **situational context** is the tangible environment in which communication occurs—the train on the way to your job, the breakfast bar, and the inside of a mosque are examples of situational contexts. Environmental conditions such as overhead lighting, room temperature, and room size are components of this context. Further, the social and emotional climates are also associated with this context. For example, to what degree are the communicators friendly/unfriendly or supportive/unsupportive? Think also about the consequences of talking about marriage equality to an audience of LGBTQ families and to a group of Orthodox Jews.

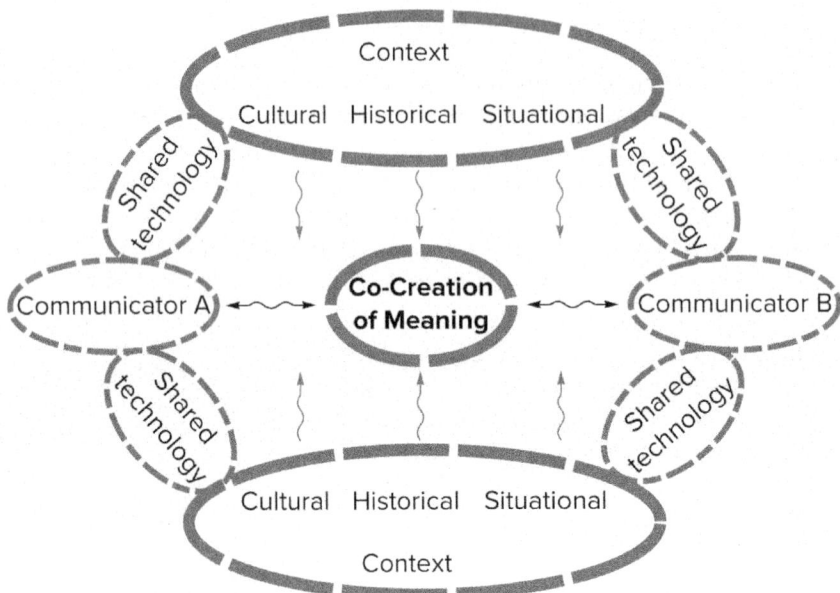

Figure 1.6   Holistic Model of Communication

When you examine **Figure 1.6,** you will note that the holistic model shows each communicator carries with them a technological field of experience, expanding upon a concept identified in the interactional model. The *technological* field of experience refers to a person's use of technology as it influences or is influenced by their culture, past experiences, personal history, and/or heredity. The infusion of technology in this model distinguishes it from the other three. First, communicators employ technology with a variety of goals in mind, namely, to stay in touch, to stay up to date, to network, to meet new people, to share opinions, and so forth. Contemporary models of communication must include technology, as we noted earlier, to understand the nuances of SNS, in particular, and their value to the communication process.

Moreover, the holistic model shows a common technological field of experience between communicator A and communicator B. This overlap between fields of experience is where messages are exchanged. Thus, the model suggests more than sending a Snap post; for communication to be achieved, someone must comment. That communication can either be direct (*"I love your Snap!"*) or indirect (*"Will someone tell this person that they're nuts?!"*). So, to co-create meaning, a comment–response dynamic must take place.

Finally, the holistic model shows that all communication generates some type of **effect,** or a result, coming from the communication encounter. Effect suggests that something evolved from the conver-

effect a condition that inevitably follows a causative condition

sation. For example, if you hear a commercial for a Walmart product and go out and buy that product, you have experienced an effect. If you are a member of a task group at work and you have a really productive meeting, you might feel more confident that you can get the job finished on time than you did before the meeting. That is an effect. Effects range in their magnitude; some are large (you and a friend stop speaking to each other), while others are more minor (you feel better about the choice of car you recently purchased), but they always exist in any communication encounter.

Thinking about the holistic model and its relationship to platforms such as Snapchat will be an ongoing process for years to come. Clearly, Shannon, Weaver, Schramm, and Barnlund could never have envisioned such technology. We're sure that in the not-so-distant future, we will have an abundance of research on the influences of these technological influences on the communication process.

You now have a basic understanding of how we define communication, and we have outlined the basic elements and a few communication models. Recall this interpretation as you read the book and examine the various theories. It is probable that you will interpret communication differently from one theory to another. Remember that theorists set boundaries in their discussions about human behavior, and, consequently, they often define *communication* according to their own view. One of our goals in this book is to enable you to articulate the role that communication plays in a number of different theories.

Thus far, we have examined the communication process and unpacked the complexity associated with it. We have identified the primary models of communication, trying to demonstrate the evolution and maturation of the communication field. We now explore a component that is a necessary and vital part of every communication episode: ethics.

Ethics and Communication

In the movie *The Insider,* which is based on a true story, the lead character's name is Jeffrey Wigand. The real-life Wigand was a former tobacco scientist who violated a contractual agreement and exposed a cigarette maker's efforts to include addictive ingredients in all cigarettes. The movie shows Wigand as a man of good conscience with the intention of telling the public about the company and its immoral undertakings. Wigand clearly believed that saving lives was the right and only thing to do, and he made his actions fit his beliefs: he acted on his ethics.

In this section, we examine **ethics,** or the perceived rightness or wrongness of action or behavior. Ethics is a type of moral decision making (Carter, 2020), and determining what is right or wrong

ethics perceived rightness or wrongness of an action or behavior

is influenced by society's rules and laws. For example, although some may believe Wigand's efforts were laudable, others may note that Wigand apparently knew what was going on when he signed a contract prohibiting him from disclosing company secrets. Furthermore, the murkiness of ethics is evidenced when one considers that Wigand made a lot of money before disclosing what was occurring.

The United States and other democracies are built on standards of moral conduct, and these standards are central to a number of institutions and relationships. Because ethical standards tend to shift according to historical period, the environment, the conversation, and the people involved, ethics can be difficult to understand. Let's briefly discuss ethical issues as they pertain to cultural institutions; a more comprehensive explanation of ethics and moral decision making can be found elsewhere (see Rachels and Rachels, 2023).

To begin, George Cheney, Debashish Munshi, Steve May, and Erin Ortiz (2010) posit the following: "Communication, as both a discipline and an 'interdiscipline' or field, is poised to play a unique role in advancing discussions of ethics because the field offers an array of concepts and principles attuned to the examination of ethics" (p. 1). Their words resonate throughout this discussion.

Let's start here by asking why we should understand ethics, next explain ethics as it relates to society, and finally, explain the intersection of ethics and communication theory. As you think about this information, keep in mind that ethical decision making is culturally based. That is, what we consider to be ethical and appropriate in one society is not necessarily a shared value in another society. For instance, though many in the United States can identify with the plight of the Hernandez family, you should know that in many cultures, having a son return to his family of origin is revered and would not pose the problems that the Hernandezes are experiencing.

Why study ethics? The response to this question could easily be another question: why not study it? Ethics permeates all walks of life and cuts across gender, race, class, sexual identity, and spiritual/religious affiliation, among others. In other words, we cannot (and should not) escape ethical principles that guide our lives. Ethics is part of virtually every decision we make, regardless of our cultural heritage. Moral development is part of human development, and as we grow older, our moral code undergoes changes well into adulthood. Ethics is also what prompts a society toward higher levels of integrity and truth. Elaine Englehardt (2001) observes that "we don't get to 'invent' our own system of ethics" (p. 2), which means that we generally follow a given cultural code of morality. And, Ken Andersen (2005) argues that without an understanding and an expression of ethical values, society will be disadvantaged: "Violating the norms of ethical communication is, I believe, a major factor in the malaise that has led many people to withdraw from the civic culture whether of their profession, their associations, their political arena" (p. 14).

From a communication perspective, ethical issues surface whenever messages potentially influence others. Consider, for instance, the ethics associated with telling your professor that you couldn't turn in a paper on time because a member of your family is ill, when such an illness doesn't exist. Think about the ethics involved if you take an idea of a coworker and present it to your boss as if it were your own. Consider the ethical consequences of going out on several dates with someone and choosing not to disclose a past felony for assault, or of posing as someone other than yourself on Tinder, or of X events that are deceptive. Television, too, carries ethical implications. For example, can television promote racial tolerance and harmony and simultaneously present portrayals of cultural groups in stereotypic and offensive ways? We continue this discussion of ethics by identifying some of the institutions whose ethical standards have been the subject of much conversation. Business and industry, religion, entertainment, education, medicine, politics, and technology are just a few of the many fields that have been prone to ethical lapses and have been challenged in communicating messages of integrity (**Table 1.1**).

Table 1.1 Examples of Ethical Decision Making in the United States

INSTITUTION	EXAMPLES OF ETHICAL ISSUES
Business and industry	Should CEOs be given pay raises in companies that are not profitable?
Religion	Should Catholic churches allow priests to counsel couples who are about to be married?
Entertainment	Does viewing violence in movies prompt violence in society?
Higher education	Should student fees go to activist groups on campus?
Medicine	Can pharmaceutical companies be held responsible for the opioid crisis?
Politics	Should political candidates make promises to citizens?
Technology	Should X be prohibited from sharing any of your personal data with advertisers?

Students Talking: Caitlyn

I could go on and on about how my high school dealt with unethical situations. We had one kid smoking in the bathroom, but nothing happened to him because he was the son of a school board member. We had a girl who had a cheat sheet for her math midterm, but because she admitted to it, the teacher did nothing about it. Even our principal was caught with another married parent. The school board just asked him to leave. It's like there are no ethics anymore.

Business and Industry

Perhaps no cultural institution has been under more ethical suspicions of late than "corporate America." Unethical behavior in corporations has reached proportions never before seen. In fact, many of these scandals prompted the Occupy Wall Street protest movements in 2012, the rise of (then) two little-known U.S. senators from New England: Elizabeth Warren and Bernie Sanders, and even a 2023 bill in the U.S. Congress called the "Ending Corporate Greed Act."

Because a corporation is usually obsessed about its reputation, companies have tried to hide costs, use creative accounting practices, commit accounting fraud, and a plethora of other ethical breaches. In fact, in the movie *The Big Short*, an ex-physician invests more than $1 billion of investors' money into credit defaults in the home mortgage industry. While some argue this is legal, the decision to take advantage of an impending mortgage crisis is clearly an unethical business practice. Other examples are not Hollywood-based but found around the globe: the former head of the World Bank engineers a job promotion and salary increase for his longtime companion; WorldCom declares bankruptcy after the discovery of an $11 billion accounting "error"; hackers steal data from nearly 150 million consumers because Equifax chose not to update its security systems; Trump University in New York, a defunct for-profit education company, is sued by former students who claim that they were duped by the organization because it did nothing to educate them about real estate; Volkswagen, the world's biggest automaker, admits to rigging diesel emissions tests in the United States and Europe; Enron inflates earnings reports and hides billions in debt, while increasing salaries of its executives; the founder of Adelphia Communications and his two sons commit bank and securities fraud, leading to the company's demise; and Boeing ignores safety upgrades to its 737 to cut costs. Finally, the Bernie Madoff investment securities scandal includes Madoff bilking nearly $64 billion from over 4,500 clients. The list of business scandals has been especially prominent. But with the advent of Corporate Ethics Statements, congressional legislation requiring public accountability, improved transparent accounting practices, and increased accountability to stockholders, most businesses have begun to improve their ethical standing. Of course, much, much more needs to be done to eliminate lingering levels of distrust.

Religion and Faith

Both Eastern and Western civilizations have stressed ethics in their moral traditions. For instance, according to Taoism, no one exists in isolation, and, therefore, empathy and insight will lead to truth. For the Buddhist, being moral requires that one use words that elicit peace and avoid gossip, self-promotion, anger, argument, and lying. From a Western perspective, many ethical issues derive from early Greek civilization. One Greek philosopher, Aristotle, first articulated the principle of the Golden Mean. He believed that a person's moral virtue stands between two vices, with the middle, or the mean, being the foundation for a rational society. For instance, when the Hernandezes are deciding what to say to their son, Eddy, about overstaying his welcome, their Golden Mean might look like this:

EXTREME	GOLDEN MEAN	EXTREME
lying	truthful communication	reveal everything

The Judeo-Christian religions are centered as well on questions of ethics. In fact, there is an online publication devoted solely to Religion and Ethics (**https://www.pbs.org/wnet/religionandethics/**). Christianity is founded on the principle of example—that is, live according to God's laws and set an example for others. However, some believe that such moral standards are not uniquely religious. For those not affiliated with organized religion, the secular values of fairness and justice and working toward better relationships are important as well. Affiliating with a religion may also pose some ethical difficulty if a person does not subscribe to a number of its philosophies or orientations. For instance, people who believe that Catholic priests should be able to marry will have a difficult time reconciling that value with Catholic law that prohibits such marriages.

Despite efforts to retain ethicality, religious institutions have had a number of ethical challenges over the years. Ministers frequenting prostitutes, pedophilic priests, drug abuse by church leaders, and sexual immorality among parishioners and pastors are just a few of the dozens of religious scandals that have caused outrage. Fortunately, many religious bodies are now developing clear ethical statements on appropriate behavior and clarifying the consequences of ethical violations.

Entertainment

The entertainment industry has also been intimately involved in dialogues about ethics and communication. Often, a circular argument surfaces with respect to Hollywood: does Hollywood reflect society or does Hollywood shape society? Many viewpoints are raised in these arguments, but three seem to dominate. One belief is that Hollywood has a responsibility to show the moral side of an immoral society; movies should help people escape a difficult reality, not relive it. A second opinion is that Hollywood should create more nonviolent and nonsexual films so that all family members can watch a movie. Unfortunately, critics note, films like *Polar, Tokyo Gore Police, Django Unchained, The Night Comes for Us, My Heart Can't Beat Unless You Tell it To,* and *Slaxx* tend to exacerbate violent attitudes in young people. A third school of thought is that Hollywood is in show *business,* and therefore making money is what moviemaking is all about. Regardless of whether you agree with any of these orientations, the entertainment industry will continue to reflect *and* influence changing moral climates in the United States. Some might say that Hollywood leading the charge in any conversation on ethics is, in itself, a question of ethics.

Higher Education

A third cultural institution that has been charged with questionable ethics is higher education. First, as you know, colleges and universities across the United States teach introductory courses in ethics, and these are required courses in many schools.

Yet, despite this interest, many schools have lost their own moral compass. For instance, colleges and universities face an ethical choice about reporting crime statistics on their campus. Despite the fact that they are required to report campus crime via the Jeanne Clery Act (named after a Lehigh University student who was murdered in 1986), some campuses fear the bad publicity. As a result, crime is "contextualized" (*"Our campus is in a large city; there's a lot of crime everywhere"* or *"Our campus is in a rural setting; we may have a few problems, but we're still relatively safe"*). On a logistic level, another ethical decision arises when schools are required to identify the enrollment patterns for legislative and financial support. Frequently, schools report statistical increases in enrollment over a number of years when in reality the school system has simply adopted a new way of counting heads. And, perhaps one of the most compelling examples of how higher education clashed with ethics came in 2019. The college admissions cheating scandal ensnared dozens of wealthy parents and several elite schools, including Georgetown, Stanford, and the University of Southern California. In particular, parents paid a college placement firm to help their kids cheat on college entrance exams and to falsify athletic records. Parents were instructed to donate to a bogus charity as part of the scheme, and the "donations" ranged from $200,000 to $6.5 million. Clearly, although they are called institutions of "higher" learning, some would argue that many campuses have reached new "lows" in ethical decision making.

Medicine

A fourth institution concerned with ethics and communication is the medical community. Specifically, with advances in science changing the cultural landscape, bioethical issues are topics of conversation around the dinner table. Physician-assisted suicide is one example of medicine at the center of ethical controversy. To some, the decision to prolong life should be a private one, made by the patient and their doctor. To others, society should have a say in such a decision.

Medical decisions can become publicly debated far from the hospital bedside. Late-term abortions, stem cell therapy, drug-enhanced athletes, medicinal marijuana, executed prisoners' organ use, and physician-assisted suicide are all topics that demonstrate the interrelationship among ethics, politics, and medicine. This topic has resonated sufficiently in our society that there is an abundance of professional journals, websites, and publications dedicated to the interface of ethics and medicine. The American Society of Law, Medicine, and Ethics is one such organization dedicated to ethical decision making.

Theory-Into-Action

 Ethics, as we noted above, permeates a number of different Eastern and Western cultures. And, perhaps the most critical professional arenas where being ethical is important are in medicine. While medical journals have talked about ethical discussions being crucial in medical communities, the practices are often concerning. Whether it's the overuse of tests, harmful treatments, overpriced prescription drugs, disclosing a patient's confidential information, romantic relationships between health care professionals and their patients, over-prescribing opioids, or a number of other areas, medical personnel are often forgetting to prioritize their patients' well-being. The "Hippocratic Oath," taken by most doctors and still taught in most medical schools, needs further recalibration to require that the "proper conduct" related to patients' health is always primary.

Politics

It is difficult to disentangle the topic of ethics from politics. The two are often viewed as incompatible. We are living in cynical times, and opinion polls consistently show that the public's view of political leaders rates lower than the public's view of paying taxes. The scandals associated with politics relate to lobbyists, campaign financing, infidelity, deception, conflicts of interest, cover-ups, bribery, conspiracy, tax evasion, and denying the reality of election results. Political scandals have been in the news since, well, the beginning of politics. It may seem, however, that they have grown in number over the past few years. Whether it's the resignation of the CIA director because of his extramarital affair, sexual harassment allegations levied at a U.S. president and the discovery of classified presidential documents in private residences, or the hateful social media postings by political candidates, the list goes on and on.

The ethical problems associated with politics may never go away, despite efforts to establish ethics commissions and oversight advisory boards in state and national governments. Though many of us wish to be optimistic in thinking that we are cultivating a new generation of political leaders who are ethical beings, there are those who are not as hopeful. Still, with nonprofit organizations such as Public Interest Research Group (PIRG) and the Government Accountability Project—groups that expose ethical shortcomings in our government and its leaders—and with aptly named government offices such as the U.S. Office of Government Ethics, there may be cause for increased confidence in the future.

Technology

Technology is at the center of many ethical debates today. Armed with a copy of the First Amendment, proponents of free speech say the internet, for example, should not be censored. Free speech advocates stress that what is considered inappropriate can vary tremendously from one person to another, and consequently, censorship is arbitrary. Consider, for instance, the U.S. Supreme Court decision protecting "virtual" child pornography on the internet. Noting that the Child Pornography Protection Act was overly broad, the justices felt that banning computer-generated images of young people was unjustified. In addition, social networking sites such as Facebook, with more than 3 billion global users every day, are prone to ethical problems. How much information is too much? Teenagers letting others know where they live, blogs that divulge too much information about a family's financial situation, and would-be employers looking over the shoulders of users are just a few of the many ethical challenges characterizing the online world.

As the United States is clearly more reliant on technology than ever before, ethical issues will continue to arise. Lying about one's identity online, downloading copyrighted material, inviting young people into violent and hate-filled websites or video games, and watching executions on the internet are all examples of potential technological ethics dilemmas.

Some Final Thoughts

The relationship between communication and ethics is intricate and complex. Yet, the public is able to assess the ethical compass of various professions in their midst (see **Table 1.2**). Public discourse requires responsibility. We presume that political leaders will tell the truth and that spiritual leaders will guide us by their example. Yet we know that not all elected officials are honest and that not all religious leaders set a spiritual standard. Organizations are especially prone to ethical dilemmas. For instance, whistle-blowing, or any strategic decision to reveal ethically suspicious behavior, can have lasting implications. In fact, it was a whistleblower who first reported the activities of President Trump in Ukraine, resulting in his impeachment. Unethical practices do little to garner trust in people.

Table 1.2 Honesty and Ethics of Professions Ranking

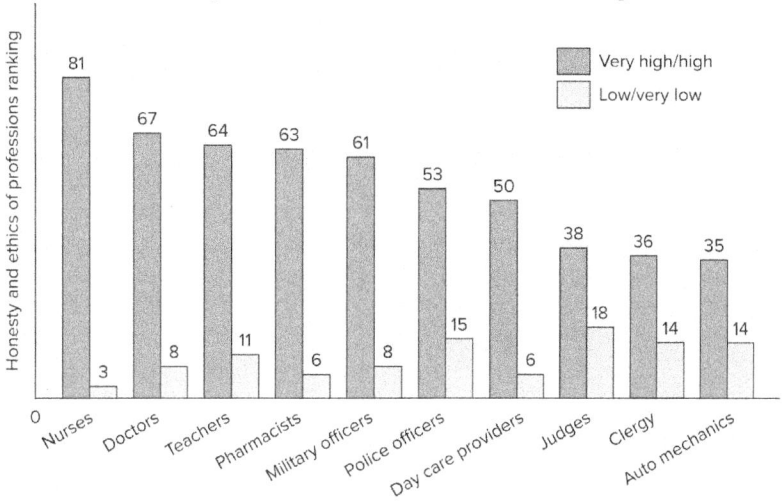

% No opinion not shown; sorted by Very high/High

Adapted from (**https://news.gallup.com/poll/388649/military-brass-judges-among-professions-new-image-lows.aspx**, 2021).

Although our focus has been rather general, you should keep in mind the ethical dimensions of communication theory as well. As readers of theory, we have a responsibility and obligation to being open to the complexity, relevance, and timeliness of communication theory, and we are ethically compelled to give a fair hearing to the ideas of others. Rob Anderson and Veronica Ross (2002) point out six important ethical strategies to consider when reading communication theory:

1. Remain open to being persuaded by the statements of others.
2. Remain willing to try out new ideas that may be seen by others as mistakes, and invite others to experiment also.
3. Accept that multiple perspectives on reality are held as valid by different people, especially in different cultural contexts.
4. Attempt to test any tentatively held knowledge.
5. Live with ambiguity, but become less tolerant of contradiction.
6. Evaluate knowledge claims against personal experience and the everyday concrete pragmatics of what works. (p. 15)

To provide you even further evidence of the interplay between communication and ethics, we urge you to review the National Communication Association's "Credo for Ethical Communication" (Appendix A). This document, developed and endorsed by communication teachers and researchers, includes nine statements that suggest, among other things, that we should be advocates for honesty and that we should condemn communication that degrades others. Deciding whether these principles are possible in an increasingly diverse and technologically connected world remains critical.

Finally, in addition to these suggestions, we add one more directly related to the book you're about to read: if a theory is a bit difficult to understand at first, don't gloss over it. Delve into the explanation once more to gain a clearer picture of the theorists' intentions. You may feel inexperienced or unprepared to challenge these theories. Yet we offer the theories for review, application, and comment and want you to ask questions about them. Although we would like to think that all theorists are open and receptive to multiple ways of knowing, the reality remains that theory construction is bound by culture, personality, time, circumstance, and the availability of resources. As students of communication theory, we all must be willing to ask some difficult questions and probe some confusing areas.

The Value of Understanding Communication Theory

Although we've alluded to this throughout the chapter, we'd like to emphasize the importance of communication theory to all of our lives. We realize that many of you may not be immediately aware of the value of this topic. Therefore, we want to give you a glimpse into the significance of communication theory. Remember: you can understand all sorts of issues in life (e.g., divorce, a job interview, your self-concept) through communication theory. As you read and understand each chapter, you will likely develop your own personal understanding of the importance of communication theory. That is, you'll likely develop your own theoretical lens of understanding human behavior. We encourage you to be open as you explore this exciting area.

Understanding Communication Theory Cultivates Critical Thinking Skills

One important value you glean from studying communication theory relates to your critical thinking skills. Without doubt, as you read and reflect on the theories in this book, you will be required to think critically about several issues. Learning how to apply the theory to your own life, recognizing the research potential of the theory, and understanding how a particular theory evolved will be among your responsibilities in this course. In addition, understanding communication can aid in your skill set. If only every Hollywood celebrity going through a divorce had read **Chapter 25** (Relational Dialectics Theory), perhaps there would be a better understanding of relational pushes and pulls. The various decision-making groups of global governments might be more effective in-group if they had read about Groupthink (**Chapter 18**). Even motivational speakers would find value in what Aristotle had to say (The Rhetoric; **Chapter 20**). These skills notwithstanding, as Daniel Muzio (2022) notes, many theories and communication traditions have a Western cultural bias, so we need to be cautious in a universal application of the theory. These activities require that you cultivate your critical thinking skills—skills that will help you on the job, in your relationships, and as you try to understand the world around you.

Understanding Communication Theory Helps You to Recognize the Breadth and Depth of Research

In addition to fostering critical thinking skills, being a student of communication theory will help you appreciate the richness of research across various fields of study. Regardless of what your current academic major is, the theories contained in this book are based on the thinking, writing, and research of intellectually curious people who have drawn on the scholarship of numerous disciplines. For instance, as you read about the Relational Dialectics Theory (**Chapter 25**), you will note that many of its principles originate in philosophy. Groupthink (**Chapter 18**) originated in foreign policy decision making. In Communication Accommodation Theory (**Chapter 24**), you will find that many of its principles have their foundation in cultural anthropology. And, Face-Negotiation Theory (**Chapter 14**) was influenced by research in sociology. As you pursue a particular degree, keep in mind that much of what you are learning is a result of theoretical thinking and this thinking is often interdisciplinary in nature.

Understanding Communication Theory Helps You to Make Sense of Personal Life Experiences

Understanding communication theory also helps you to make sense of your life experiences. It's impossible to find a theory in this book that does not in some way relate to your life or to the lives of people around you. Communication theory aids you in understanding people, media, and events and helps you answer important questions. Have you ever been confused about why some men speak differently from some women? Chances are that reading Muted Group Theory (**Chapter 28**) will help you understand why that may be the case. Do the media promote a violent society? Cultivation Theory (**Chapter 26**) will likely help you answer that question. What role does technology play in society? Social Information Processing Theory responds to that question. And what happens when someone stands too close while talking to you? Expectancy Violations Theory (**Chapter 4**) explores and explains this type of behavior. Some of you may have entered this course thinking that communication theory has limited value in your life. You will see that much of your life and your experiences in life will be better understood because of communication theory.

Communication Theory Fosters Self-Awareness

Thus far, we have observed that learning about communication theory helps your critical thinking skills, informs you about the value of research across different fields of study, and aids you in understanding the world around you. One additional reason to study theories of communication pertains to an area that is likely to be most important in your life—you. Learning about who you are, how you function in society, the influence you are able to have on others, the extent to which you are influenced by (social) media, how you behave in various circumstances, and what motivates your decisions are just a handful of the possible areas that are either explicitly or implicitly discussed in the theories you will be introduced to in this book. We are not suggesting that you should be central to whether a theory has relevance. Rather, we are stating that theories usually relate directly to you—your thoughts, values, behaviors, and background. For instance, Social Penetration Theory (**Chapter 7**) will help you consider the value of self-disclosing in your relationships. **Chapter 15** (Symbolic Interaction Theory) will assist you in thinking about the meaning of the various symbols surrounding you. In addition, in **Chapter 30,** Co-Cultural Theory will help you understand your cultural connections with others who function in dominant society. Yet these are not "self-help" communication theories; they do not provide easy answers to difficult questions. What you will encounter are theories that will help you as you try to understand yourself and your surroundings.

You are about to embark on an educational journey that is likely new to you. We hope you persevere in unraveling and unpacking the various issues related to communication theory. The journey may be different, challenging, and tiring at times, but it will always be applicable to your life. Stick with it!

Conclusion

In this chapter, we introduced you to the communication process. We presented our definition of communication and provided you various elements embedded in that process. In addition, we identified three prevailing models of communication: the linear model, the interactional model, and the transactional model. Moreover, to address the infusion of technology, we also proposed the holistic model of communication. The chapter also addressed ethics and its relationship to communication theory. Finally, we provided several reasons why it is important for you to study communication theory.

You now have an understanding of the communication process and some sense of how complex it can be. As you read the many theories in this book, you will be able to view communication from a variety of perspectives. You will also gain valuable information that will help you understand human behavior and give you a new way to think about our society. We continue this examination in **Chapter 2** when we present the traditions in communication study and important contexts in which communication takes place.

Discussion Starters

Case-In-Point: Identify the conditions that led to the blowup between Eddy Hernandez and his parents. Do you believe that Eddy and his parents were trying to handle his situation in an ethical way? Why or why not?

Try-It-Your-Selfie: Provide some examples of unethical social media posts/profiles. What is your impression of those who post pictures or words that lack an ethical compass?

1. Do you believe that all slips of the tongue, conversational *faux pas*, and unintentional nonverbal behaviors should be considered communication? Why or why not? What examples can you provide to justify your thoughts?

2. Explain why the linear model of communication was so appealing years ago. Describe the appeal of both the transactional and holistic models using current societal events.

3. Discuss the value of looking at communication theory from a variety of different disciplinary angles, including psychology, medicine, and politics.

4. Discuss how different family members might have different fields of experience.

References

Andersen, K. (2005). *Recovering the civic culture: The imperative of ethical communication.* Pearson.

Anderson, R., & Ross, V. (2002). *Questions of communication: A practical introduction to theory.* Bedford/St. Martins.

Barnlund, D. C. (1970). A transactional model of communication. In K. K. Sereno & C. D. Mortensen (Eds.), *Foundations of communication theory* (pp. 83–102). Harper.

Barnlund, D. C. (2017). A transactional model of communication. In C. D. Mortensen (Ed.), *Communication theory* (pp. 47–57). Routledge.

Buccoliero, L., Bellio, E., Crestini, G., & Arkoudas, A. (August 2018). Twitter and politics: Evidence from the US presidential elections 2016. *Journal of Marketing Communications, 26*(1), 88–114.

Carter, S. (2020). *Everything to lose: Doing the right thing when the stakes are high.* David C. Cook.

Cheney, G., Munshi, D., May, S., & Ortiz, E. (2010). *Handbook of communication ethics.* Routledge.

Cranmer, G. A., Gagnon, R. J., & Mazer, J. P. (2020). Division-I Student-Athletes' affective and cognitive responses to receiving confirmation from their head coach. *Communication & Sport, 8*(2), 262–285.

Englehardt, E. E. (2001). *Ethical issues in interpersonal communication.* Harcourt.

HSI. (n.d.). Top soft skills for safety leaders and EHS professionals. Retrieved on May 25, 2023 from https://hsi.com/blog/top-soft-skills-for-safety-leaders-and-ehs-professionals.

Kittleson, M. (2023). *Mastering the art of patient care.* Springer Nature.

Korsch, B. M., Gozzi, E. K., & Francis, V. (1968). Gaps in doctor-patient communication. *Pediatrics, 42,* 855–871.

Martin, J., & Nakayama, T. (2018). *Intercultural communication in contexts* (7th ed.). McGraw Hill.

Meister, J. (2023, January 10). Top ten HR trends for the 2023 workplace. *Forbes.* https://www.forbes.com/sites/jeannemeister/2023/01/10/top-ten-hr-trends-for-the-2023-workplace/?sh=26a17f365933.

Muzio, D. (2022). Re-conceptualizing management theory: How do we move away from Western-centered knowledge? *Journal of Management Studies, 59*(4), 1032–1035.

Ott, B. L. (2017). The age of Twitter: Donald J. Trump and the politics of debasement. *Critical Studies in Media Communication, 34*(1), 59–68.

Pérez-Curiel, C., Rúas-Araújo, J., & Rivas-de-Roca, R. (2022). When politicians meet experts: Disinformation on Twitter about COVID-19 vaccination. *Media and Communication, 10*(2), 157–168.

Perse, E. M., & Lambe, J. (2016). *Media effects and society.* Routledge.

Pew Research Center. (2021, April 7). *Social media fact sheet.* https://www.pewresearch.org/internet/fact-sheet/social-media/.

Rachels, J., & Rachels, S. (2023). *The elements of moral philosophy.* McGraw Hill.

Schramm, W. L. (1954). *The process and effects of mass communication.* University of Illinois Press.

Taipale, S. (2019). The big meaning of small messages. In S. Taipale (Ed.), *Intergenerational connections in digital families* (pp. 87–101). Springer.

Titsworth, S., Mazer, J. P., Goodboy, A. K., Bolkan, S., & Myers, S. A. (2015). Two meta-analyses exploring the relationship between teacher clarity and student learning. *Communication Education, 64*, 187–207.

Trenholm, S. (1991). *Human communication theory.* Prentice Hall.

Turner, L. H., & West, R. (2019). *An introduction to communication.* Cambridge University Press.

VanLear, C. A. (1996). Communication process approaches and models: Patterns, cycles. In J. H. Watt & C. A. VanLear (Eds.), *Dynamic patterns in communication processes* (pp. 15–70). Sage.

Watzlawick, P., Beavin, J. B., & Jackson, D. D. (1967). *Pragmatics of human communication: A study of interactional patterns, pathologies, and paradoxes.* W.W. Norton.

CHAPTER 2
Thinking About the Field: Traditions and Contexts

"Theory, in my understanding of theory, is a form of discourse."

—Robert T. Craig

Jenny and Lee Yamato

As the 18-year-old daughter of a single parent, Lee Yamato knows that life can be difficult. She is the only child of Jenny Yamato, a Japanese American woman whose husband died from a heart attack several years ago. Jenny raised her daughter in Lacon, a small rural town in the South. It was stressful being a single mom, and Jenny was sometimes the target of overt racist jokes. As a waitress, Jenny knew that college would be the way to a better life for her daughter. She saved every extra penny and worked at the children's library for several months to bring in extra income. Jenny knew that Lee would get financial aid in college, but despite her small salary, she also wanted to be able to help her only child with college finances.

As Lee finished her senior year in high school, she knew that before too long, she would be leaving to attend a public university. Unfortunately, the closest college was over 200 miles from Lacon. Lee had mixed feelings about her move. She was very excited to get away from the small-town gossip, but she also knew that leaving Lacon meant that her mom would have to live by herself. Being alone could be devastating to her mom, Lee thought. Still, Lee recognized that her education was her first priority and that in order to get into veterinary school, she would have to stay focused. Thinking about her mom would only make the transition more difficult and could sour her first year as a college student.

Jenny, too, felt ambivalent about Lee leaving. When Jenny's husband died, she didn't think that she could raise a 13-year-old by herself. In fact, given that single moms were not looked at with a lot of respect, Jenny felt even more vulnerable. However, her own tenacity and determination had paid off, and she was extremely proud that her child was going away to college. Like her daughter, though, Jenny felt sad about Lee's departure. She felt as if her best friend was leaving her, and she couldn't imagine her life without her daughter. She knew that they would text or FaceTime or talk on the phone, but none of those could replace the hugs, the laughter, and the memories.

On the day that Lee was to leave, Jenny gave her a box with chocolate chip cookies, some peanut butter cups (Lee's favorite), a photo album of Lee and Jenny during their camping trip to Arizona, and a shoe box. Inside the shoe box were old letters that Jenny and her husband had written to each other during their courtship. Jenny wanted Lee to have the letters to remind her that her dad's spirit lives on in her. When Lee looked at the first letter, she put it down, hugged her mom, and cried. She then got into her car and slowly drove away, leaving her house and her only true friend behind.

The communication discipline is vast, and its depth is reflected in the lives of people across the United States, people like Lee and Jenny Yamato. Their relationship is obviously a close one, marked now by a common and often emotional point in a family's development: college. As the two begin to adapt to a new type of relationship characterized by distance, their communication will also take on new levels of importance. The Yamato family will likely communicate with an appreciation for the full impact that communication can have on their lives.

Let's begin our discussion of the communication field by looking at seven traditions in communication. We will then examine various settings in which communication occurs. These two approaches guide this chapter. The first approach (the seven traditions) is theoretical in nature; the second framework (the seven contexts) is more practical in its approach. We describe each in the following pages.

Chapter 1 provided a foundation for conceptualizing what communication is and understanding the complexity of the communication process. In this chapter, we further unravel the meaning of communication by articulating two primary models for the field. We first begin with a brief history.

A Historical Briefing

It's impossible to capture the essence and evolution of the communication field in a book of this nature. We encourage you, however, to look at alternative sources for detailed information on how it came to be (e.g., Eadie, 2020; Gehrke & Keith, 2015). For now, however, in addition to providing you various traditions and contexts, we also describe the following historical markers for you to get a glimpse into the field's breadth and depth.

Since the beginning of recorded history, communication has been on the minds of philosophers, teachers, scholars, practitioners, poets, and people of all backgrounds. According to Sherry Morreale and Matt Vogl (1998), "systematic comment on communication goes back at least as far as *The Precepts of Kagemni and Ptah-Hopte* (3200-2800 B.C.)" (p. 4). Clearly, the long and storied history of the communication discipline has evolved over the centuries, and this narrative has affected billions of people along the way. Again, however, we reiterate that this information is simply a snapshot to give you an overview of our legacy.

Classical Origins (466 B.C.–A.D. 400)

This period is best characterized by communication in ancient Greece, a society in which oral communication skills were revered and speaking well was viewed as a practical and necessary skill. This reverence for oral communication coincided with the spread of democracy throughout the Greek world. The ideals of democracy placed a premium on learning to communicate effectively as well as understanding the nature of persuasion. Indeed, citizens used communication daily, most notably in pleading their own judicial cases related to land disputes. Citizens also needed to be articulate as they argued their political values while running for office or as they served as jurors or adjudicators of city boundaries (Golden, Berquist, Coleman, & Sproule, 2011). Practical elements of communication were pivotal at this time (Trenholm, 2020).

All of these sorts of public speaking activities required some knowledge in persuasion, which, as we will learn in **Chapter 20**, was the central cornerstone of democracy in early Greek times.

The Post-Classical Period (A.D. 400–1600)

What we are identifying as the post-Classical period comprises two major epochs of Western history: the Middle Ages and the Renaissance. During this period, which lasted from approximately 400 to 1400, Christianity became a critical component of living. With the rise in power of religion came the decline of practices seen as irreligious or pagan. Thus, oral communication was only needed to reveal the will of God by preaching well.

During the Renaissance (1400-1600), people turned again to an interest in the Classical roots of rhetoric. The Renaissance provided a focus on the individual (in place of the group or institution focus fostered in the Middle Ages) and on art. Communicating persuasively was seen as an art. Despite this more hospitable climate for rhetorical thought, the Renaissance was a relatively quiet time in the development of rhetoric.

The Modern Period (1600–1942)

In the early part of the modern period, the world became more secular, and religiosity did not play as substantial a role as it did in earlier times (McHendry et al., 2020). This led to embracing more scientific examination and less religious influence in those viewpoints. Consequently, this interest in science led to the rise in the empirical method, which paved the way for seeing communication as a social science. But not all those who studied communication during this time were interested in the same things.

First, during that time, some modern thinkers continued to foster the thinking of Classical scholars such as Aristotle, applying its principles to contemporary situations of the day. Others, however, were applying scientific approaches to the study of communication, creating the field as a social science. Still others in this period were primarily interested in style and presentation. The early 19th century saw the rise of what is called the *elocution movement*. This movement also harkened back to the Classical period, but its emphasis was strictly on elevating the notion of speech delivery. Those in the elocution movement were most interested in the nonverbal aspects of oral communication. They were quite interested in their prescriptions for effective gestures and vocal behaviors (such as pitch, volume, and speaking rate). Emerson College in Boston and Northwestern University in Illinois were two primary examples of where the elocution movement took hold and both established schools of oratory (speech communication). Interestingly, your authors have affiliations with both, in that Rich is currently a professor at Emerson, and Lynn received her doctorate at Northwestern. We both are clearly immersed in communication's legacy!

Students Talking: Jackson

I had NO idea that communication can be traced all the way back to ancient times! I always wanted to be a COMM major, but I didn't know that my major pretty much started with Aristotle. I've read the short history and looked up some more stuff because I eventually want to get a master's. I'm amazed at how much the field has changed over the years.

The Contemporary Period (1942–Present)

Today, the field of communication honors both its Aristotelian legacy and contemporary ways of thinking about communication. In addition, communication now embraces more multidisciplinary ways of looking at phenomena, including research imported from such fields as psychology, education, sociology, anthropology, and business, among others. Further, the diversity of the discipline can best be exemplified by looking at two of the largest organizations dedicated to communication: the National Communication Association (NCA) and the Association for Education in Journalism and Mass Communication (AEJMC). The National Communication Association (**www.natcom.org**) is the nation's largest organization dedicated to the teaching and study of communication. This organization boasts over 60 distinct subdivisions, from Communication and Sport to Family Communication to Communication and Social Justice/Activism. While some lament that the organization has grown quite expansive in how it views communication, others believe that the disparate areas of teaching and research reflect the changes that have occurred in the United States and around the globe. In other words, the NCA (and the field itself) has worked to understand and unpack the complexities of our world, from the communication implications related to terrorism to the impact that communication has upon children of alcoholic parents. This pragmatism of "disseminating communication practices" (Craig, 2018) characterizes a great deal of what will likely continue to be an ethical imperative in communication studies.

Another organization immersed in the study and practice of communication—specifically journalism and mass communication—is AEJMC (**www.aejmc.org**). AEJMC is not as expansive as NCA, as it comprises over 20 divisions and interest groups. Yet, the mission of the organization pertains to enhancing the integrity of curriculum, promoting a cultural understanding, and advancing research. From advertising to law and policy, AEJMC works to become an influential association for media practitioners and media policymakers.

Seven Traditions in the Communication Field

When you encounter the various theories in this book, you need to first understand that each theorist operates within a particular framework. In other words, someone doesn't just wake up one morning and say: "Today, I'm going to be a theorist." The process of theory building is complex (as we note in **Chapter 3**), and theorists typically adhere to a specific way of thinking about life.

Robert Craig (Craig, 1999, 2018; Craig & Muller, 2007) outlines how theorists develop communication theory by articulating one of the more intellectually valuable ways. Craig believed that communication theory is an often unwieldy area of study and, to this end, conceptualized various categories to aid our understanding of it and to provide further evidence that theory construction is not a vacuous enterprise. Craig and Muller note that trying to make sense of communication theory is often complicated because of different intellectual styles in the field. A classification system for understanding communication theory, then, helps us to break down the challenges related to understanding theory. After all, that is a primary goal of this book and this course!

Craig interprets the following approach as "traditions" to highlight the belief that theoretical development doesn't just occur naturally. Indeed, theorizing in communication is a deliberative, engaging, and innovative experience that happens over time. As Craig and Muller (2007) point out, "Theorists invent new ideas to solve problems they perceive in existing ideas in a particular tradition" (p. xiii). And, although traditions suggest adhering to a historical preference, traditions change frequently and, like communication, are dynamic. Think, for instance, about how social media have already become part of a "tradition" in many countries across the globe!

So let's examine the seven traditions of communication theory advanced by Craig (1999). To honor the integrity of each tradition yet avoid irrelevant detail for this section of the chapter, we will provide you an overview of each tradition. If you'd like additional details, you are encouraged to consult Craig's research. See **Figure 2.1** for more information on the traditions.

The Rhetorical Tradition: Communication and the Art of Public Speaking

At the heart of the rhetorical tradition is what Craig notes as the "practical art" (p. 73) of talk. Rhetoric, as you will learn later in this chapter, refers to the available means of persuasion. Therefore, this tradition suggests that we are interested in public address and public speaking and their purposes in a society. Regardless of speaking venue, effective public speakers understand that they need advice on their speech/speaking. This tradition includes the ability to reflect on different viewpoints before arriving at a personal view. It is the usefulness of the rhetorical tradition that remains attractive to researchers, theorists, and practitioners.

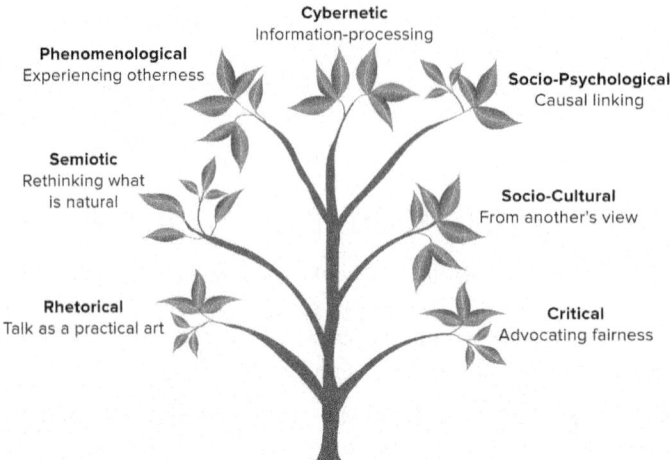

Figure 2.1 Traditions in Communication Theory

The rhetorical tradition necessarily involves elements pertaining to language and the audience. Nikki DiGregorio (2022), for example, discusses the linguistic challenges for many gay and lesbian couples. First, consider the differences in these phrases: "same-sex marriages" and "marriage equality." We're sure most of you use the latter phrase. Why? Also, to make this point even more compelling: how does identifying as cisgender differ from identifying as non-binary? In part, the non-binary nature of sexual identity has ushered in some complexities in word choice. Some couples wish to avoid the heteronormative nature of husband and wife and instead, elect to have more unique titles, including "hersbands," "wusbands," and even "support staff." The tradition also includes a discussion pertaining to audience appeals; for example, how do audience members respond to emotions? To what extent does the power of language move people to emotional and decisive action? How are we influenced or swayed by the appeals by social media? What role does personal example play in having others accept our point of view? What effect does speaking to a large group of people have on the perceptions or actions of that group? Or, to what extent does the rhetorical tradition challenge the common belief that "telling the plain truth is something other than the strategic adaptation of a message to an audience"? (Craig, 2007, p. 73). Answering such questions is not easy and such questions are often the domain of philosophers. Yet, they are important if we are to consider the value and historical significance of looking at the communication process through a rhetorical lens.

The Semiotic Tradition: Communication and the Co-Creation of Signs

Simply put, **semiotics** is the study of signs. Signs are part of a social life; they stand for something else. Children laughing and running

semiotics the study of signs

around is a *sign* of play. A ring on the ring finger of the left hand is typically a *sign* of a married individual. An adult crying in a funeral home can be a *sign* of sadness. Most common among these signs are words, or what we generally consider language usage. According to the semiotic tradition, meaning is achieved when we share a common language. As noted in **Chapter 1**, people arrive at a communication exchange with various fields of experience and values placed on these experiences. But words are arbitrary and have no intrinsic meaning. Consequently, achieving commonality in meaning is more difficult than first imagined, particularly if one is using language that is not recognized or valued by another.

Semiotics suggests that what we think of as "natural" or "obvious" in public discourse needs to be considered in context. That is, our values and belief structures are often a result of what has been passed down from one generation to another (a tradition). What was considered to be a "given" years ago may simply not be that way today. Semiotics challenges the notion that words have appropriate meanings. Actually, words change as the people using those words change. Consider, for example, the use of the words *war* and *peace* in the 1940s, the 1960s/1970s, the 1990s, and today. There are also likely to be multiple meanings of these two words if, for example, someone lost a family member in one of these wars and if another person protests war on a regular basis. Your views of the Ukraine–Russia war or the Gaza-Israeli war perceptions of world peace will also affect your interpretations. Now think about the phrase *single parent*. In the 1950s, we rarely found the word either in our conversations or in print. However, as time evolved, the divorce rate soared and marriage was not a "default" choice; being a single mom (like Jenny Yamato) or single dad is now commonplace.

The Phenomenological Tradition: Communication and the Analysis of the Everyday

Let's explore the term *phenomenology,* a concept derived from the field of philosophy. In the simplest of terms, **phenomenology** is a personal interpretation of everyday life and activities.

phenomenology a personal interpretation of everyday life and activities

Craig (2007) believes that the phenomenological tradition is marked by communication that he contends is an "experience of otherness" (p. 79). What this means is that a person tries to attain authenticity by eliminating biases in a conversation. Many phenomenologists believe that an individual's system of beliefs should not influence the dialogue taking place. As you're probably figuring out, this is quite challenging, or, as Craig points out, is a "practical impossibility" (p. 80). Consider, for example, the challenge of people who communicate with people from different cultural backgrounds. Craig notes that many phenomenological ideas are especially applicable to issues pertaining to gender, identity, class, sexual identity, and religion. More recently, scholars (e.g., Miike & Yin, 2022) have implored communication theory researchers to infuse their work with global viewpoints and approaches. Such an undertaking, they contend, will result in a more accurate understanding of human behavior around the world.

The Cybernetic Tradition: Communication and the Processing of Information via Feedback

Today, we would state that this tradition is most closely aligned with artificial intelligence (AI), one of the most recent tech evolutions around the world. Communication as information science was first introduced by Shannon and Weaver, two scholars associated with the linear model we discussed in **Chapter 1**. Recall that this model's fundamental shortcoming pertained to the fact that human communication is not as simplistic as linearity suggests. Nonetheless, what Shannon and Weaver believed was that communication involves noise. Cybernetics in particular looks at problems such as noise in the communication process. But it goes further. Cybernetics tries to unravel the complexities of message meaning by underscoring the unpredictability of the feedback we receive. In this tradition, communication is seen as a bond that connects several parts of a system. Think about the family as a system and how communication serves to connect the different family members, regardless of their locations. In most cases, connection occurs via cell phones.

By advocating a cybernetic approach, communication theorists are embracing an expansive view of communication. As Craig (2007) states: "It is important for us as communicators to transcend our individual perspectives, to look at the communication process from a broader, systemic viewpoint, and not to hold individuals responsible for systemic outcomes that no individual can control" (p. 82). In other words, the cybernetic tradition asks us to understand that communication is not only information processing, but also that individuals enter into communication settings with different abilities in that information processing. Fields of study which have been influenced by cybernetics include game theory, psychology, architecture, and artificial intelligence (Ashby, 2015).

The Socio-Psychological Tradition: Communication and the Science of Human Behavior

In this book, you will encounter many communication theorists who represent the scientific approach we detail in **Chapter 3**. This tradition suggests that research can discover "truth" by systematic investigation. In this capacity, think of "truth" as being aligned with how things and people exist. Those who adhere to the socio-psychological tradition support a cause–effect model. That is, communication theory is examined from a view that holds that someone's behavior is influenced by something else—something social psychologists call a "variable." Craig (2007) believes that underlying this tradition is the assumption that our own communication patterns and the patterns of others vary from one person to another. It is up to the social psychologist to unravel the relationship among these patterns. And these scholars use statistical reasoning in their quest to understand human behavior.

An early advocate of the socio-psychological tradition was Carl Hovland. Hovland, a Yale psychologist, examined attitude change and investigated the extent to which long- and short-term recall influences an individual's attitudes and beliefs. In the 1950s—long before personal computers came into existence—Hovland also was the first to experiment with computer simulations and the learning process (Briñol, Petty, & Guyer, 2019). His work and the work of other social psychologists underscored the importance of experimental research and trying to understand causal links. It is this scientific evidence for human behavior that continues to pervade much communication theorizing from this tradition.

The Socio-Cultural Tradition: Communication and Socially Constructed Reality

The essence of the socio-cultural tradition can be summed up this way: "Our everyday interactions with others depend heavily on preexisting, shared cultural patterns and social structures" (Craig, 2007, p. 84). The core of the socio-cultural tradition suggests that individuals are parts of larger groups who have unique rules and patterns of interaction. To theorize from this tradition means to acknowledge and become sensitive to the many kinds of people who occupy this planet. Theorists should not instinctively nor strategically "group" people without concern for individual identity.

Socio-cultural theorists may advocate that we abandon the binary "you/me" or "us/them" approach to understanding people. Instead, appealing to the *co-creation* of social order/reality is a worthier goal for consideration. As people communicate, they produce, maintain, repair, and transform reality (Carey, 1989). Dialogue and interaction must be characterized by an understanding of what Craig (2007) terms "voice" (p. 84), an individual point of view that inevitably finds its way into everyday conversation. Further, our view of the world can be shaped by the words we use, and the words we use can be shaped by the world we see. Consider the power of language in communicating your reactions to a group of people. For instance, a difference exists between the words "undocumented immigrant" and "illegal alien." Your use of either phrase, too, likely communicates your perceptions of and feelings about this marginalized population.

The Critical Tradition: Communication and Questioning the Status Quo

Individuals who are concerned with injustice, oppression, power, and linguistic dominance are those who would likely identify themselves as critical theorists. Critiquing the social order and imposing structures or individuals on that order are at the heart of critical theory. Among the theorists most known for protesting social order is philosopher and political economist/revolutionary Karl Marx. Marx believed that power in society has been hijacked by institutions that have no real concern for the working class. In his book *The Communist Manifesto*, Marx and his colleague Friedrich Engels (1848) contend that the history of a society is best understood by looking at the class struggles in that society, an issue that continues to resonate across the globe.

Critical theorists look at language and how it creates imbalances and power differentials in society. Critical theorists also find that openly questioning the assumptions that guide a society is legitimate. In doing so, communicators expose the beliefs and values that guide their decision making and actions. As is suggested in our opening story of Jenny and Lee Yamato, it appears that Jenny felt that as a single mom, she could never achieve the level of respect afforded to other family types. Critical theorists would attempt to unravel how a society defines freedom, equality, and reason, three qualities identified by Craig (2007), in order to understand Jenny's experiences. Who or what are the principal forces on social order? How does one achieve the freedom to express one's will? And, what does "freedom" even mean? These and a host of other questions are at the core of the critical tradition.

Putting It All Together

This discussion provides you one way of looking at the texture of the communication field. Communication theory, as you will discover, is not vacuous. Scholars generally enter into the theory-building process with particular positions, some of which influence the direction of the theories they construct and refine. The preceding traditions are rooted in communication and we will consider each as we introduce you to the various theories. As we alluded to earlier, theory is not always so "clean" and therefore, you will witness the hybrid of a few traditions along the way.

In addition, Craig (1999) notes that "communication theory is logically open to new traditions" (p. 150). He suggests that four are opportunistic to consider for future theory consideration: communication employed to connect with others and providing voice, particularly to and with women (*the Feminist Tradition*); communication that embodies performance in that it creates rituals, relationships, and meanings (*the Aesthetic Tradition*); communication that suggests every message has an exchange value equivalent to its meaning (*the Economic Tradition*); and communication that reveals a community's dependency on faith (*the Spiritual Tradition*). Which ones do you think are currently gaining more attention as you think about what's taking place in the United States and around the world?

With this backdrop, we now wish to explore a more practical framework from which to view communication theory. We turn our attention to the various contexts, or environments, of communication from which research and theory develop.

Seven Contexts in the Communication Field

We have already underscored the challenge in explaining communication. So, in order to make the communication field and the communication process more understandable and manageable, we now look at the various contexts of communication. Recall from **Chapter 1** that **contexts**

contexts environments in which communication takes place are environments in which communication takes place. They also provide a backdrop against which researchers and theorists can analyze phenomena. Our discussion of context focuses on **situational contexts**, a concept we examined in the holistic model in **Chapter 1**.

Earlier we noted that the communication field is very diverse and offers various research opportunities. This can be a bit cumbersome, and at times even communication scholars lament the wide array of options. Still, there seems to be some universal agreement on the fundamental contexts of communication. In fact, most communication departments are built around some or all of the following seven communication contexts: intrapersonal, interpersonal, small group/

> **situational contexts** environments that are limited by such factors as the number of people present, the feedback, the space between communicators, among others

team, organizational, public/rhetorical, mass/media, and cultural (**Figure 2.2**). Keep in mind, however, that communication departments in colleges and universities across the United States divide themselves uniquely. Some, for instance, include mass communication in a Department of Communication whereas others may have a separate Department of Mass Communication. Some schools have a Department of Communication and include every context therein. This variety underscores that the discipline is permeable and that boundary lines among the contexts are not absolute.

Intrapersonal Communication

As you review theories in this book, keep in mind that a theory may focus on how individuals perceive their own behavior. At times, too, a theory may delve into how that behavior affects or influences another. At the root of this thinking is intrapersonal communication.

> **intrapersonal communication** communication with oneself

Intrapersonal communication theorists frequently study the role that cognition plays in human behavior. **Intrapersonal communication** is communication with oneself. It is an internal dialogue and may take place even in the presence of another individual. Intrapersonal communication is what goes on inside your head even when you are with someone. Intrapersonal communication is usually more repetitive than other communication; we engage in it many times each day.

This context is also unique from other contexts in that it includes those times when you imagine, perceive, daydream, and solve problems in your head. Intrapersonal communication is much more than talking to oneself. It also includes the many attributions or inferences you may make about another person's behavior. For instance, an employer may want to know why an employee arrives late to work and looks disheveled each day. The supervisor may believe that the worker's tardiness and demeanor are a result of some domestic problems. In reality, the employee may have another job in order to pay for their child's college tuition. Or, think about what went through your mind as you were deciding where to get your college degree. Most of you probably weighed the pros (reputation of school, quality of degree program, etc.) and the cons (tuition, anticipated student debt, etc.). Although you ultimately may have discussed this decision with others, the initial conversations are typically internal dialogues, and these internalized voices can vary tremendously from one person to another.

As noted earlier, intrapersonal communication is distinguished from other contexts in that it allows communicators to make attributions about themselves. People have the ability to assess themselves. From body image to work competencies, people are always making self-attributions. You may have thought seriously about your own strengths and shortcomings in a number of situations. For example, do you find that you are an excellent roommate but not so excellent as a statistics student? Are there times when you feel that you are a trusted friend, but not so trusted in your own family?

CONTEXT

INTRAPERSONAL
Communication
with oneself

INTERPERSONAL
Face-to-face
communication

SMALL GROUP/TEAM
Communication with
a group of people

ORGANIZATIONAL
Communication within
and among large and
extended environments

PUBLIC/RHETORICAL
Communication to a large
group of listeners
(audience)

MASS/MEDIA
Communication to an
audience through
mediated forms

CULTURAL
Communication between
and among members
of different cultures

**SOME RESEARCH AND THEORETICAL
CONCERNS**

Impression formation and decision
making; symbols and meaning;
observations and attributions; ego
involvement and persuasion

Relationship maintenance strategies;
relational intimacy; relationship control;
interpersonal attraction

Gender and group leadership; group
vulnerability; groups and stories; group
decision making; task difficulty

Organizational hierarchy and power;
culture and organizational life; employee
morale; opinions and worker satisfaction

Communication apprehension; delivery
effectiveness; speech and text criticism;
ethical speechmaking; popular culture
analysis

Use of media; affiliation and television
programming; social media deception;
media and need fulfillment; effects of
dating sites on relationship initiation

Culture and rule-setting; power differentials
in marginalized communities; hegemony;
ethnocentrism

Figure 2.2 Contexts of Communication

Although some people may believe that talking to oneself is a bit peculiar, Virginia Satir (1988) believes that these internal dialogues may help individuals bolster their **self-esteem**—the degree of positive orientation people have about themselves. Often, intrapersonal communication is difficult; it requires individuals to accept their

self-esteem the degree of positive
orientation people have about
themselves

accomplishments and frequently to confront their fears and anxieties. Looking in a mirror can be both enlightening and frightening. Of course, mirrors can also be distorting. Jenny Yamato, for example, may think that her world is over once her daughter leaves for college. The reality for the vast majority of parents, however, is that they survive the "loss." As a single parent, Jenny may think that she is incapable of moving on without Lee. Once Lee leaves, however, Jenny may find that she is more empowered living alone.

Overall, then, the research in intrapersonal communication focuses a great deal on the cognitions, symbols, and intentions that individuals have. Without recognizing oneself, it is difficult to recognize another, leading us to the next context, interpersonal communication.

Interpersonal Communication

From its beginnings, **interpersonal communication** referred to face-to-face communication between people. Sometimes referred to as *dyads*, couples, or pairs, interpersonal relationships involve two people. Still, today's views of interpersonal communication necessarily incorporate a technological lens (e.g., dating websites, gaming, etc.). Several theories that you will read about in this book have their origins in the interpersonal context. This context is rich with research and theory and is often seen as the most expansive of all of the contexts. Investigating how relationships begin, are maintained, and dissolve characterizes much of the interpersonal context.

> **interpersonal communication**
> face-to-face communication between people

One reason researchers and theorists study relationships is that relationships are not only ubiquitous, but also complex and diverse. For instance, you may find yourself in several types of relationships (family, friend, roommate, romantic partner, office colleague) right now. Interacting within each of these relationships affords communicators a chance to maximize the number of channels (visual, auditory, tactile, olfactory) used during an interaction. In this context, these channels function simultaneously for both interactants: Think about how a parent holds a crying child or how a nursing assistant hugs a resident of a long-term care facility to calm their anxieties.

We noted above that the interpersonal context itself comprises many related subcontexts, including friendships, family, long-term marriages, and physician–patient relationships, among many others. In addition, researchers are interested in a host of issues and themes within those relationships, including risk, power, teasing, gossip, liking, attraction, and emotions.

Finally, relationships that have not been studied enough, including trans relationships, cohabiting relationships, and long-distance relationships, are being investigated at a rapid pace in the communication field, and we will see how students like you view the influence of technology as you read and review each chapter. As you can see, researchers have framed some very diverse and exciting work within the interpersonal communication context, and studying relationships and what takes place within them has broad appeal. Relationships can be larger than two people, however, underscoring our third communication context: small group/team communication.

Small Group/Team Communication

A third context of communication is the small group or team. **Small group/team communication** refers to communication between and among at least three people who work together to achieve some common purpose. Small group/team research focuses on task groups as opposed to the friendship and family groups found in the interpersonal

> **small group/team communication**
> communication between and among at least three people who work together to achieve some common purpose

context. Communication theory centering on small groups frequently concerns the dynamic nature of small groups, including group roles, boundaries, and trust.

Researchers disagree about how many people make up a small group/team. Some scholars argue that the optimal number for a small group/team is five to seven members, whereas others put no limit on the maximum number of members. Of course, the size of the overall population will always influence the number.

The number in a group is not as important as the implications of that number. The more people, the greater the opportunity for more personal relationships to develop. This may influence whether small groups stay focused on their goals and whether group members are satisfied with their experiences (Fujishin, 2023). A classic study (Kephart, 1950) revealed that as the size of the group increases, the number of relationships increases substantially. With a three-person group, then, the number of potential relationships is 6; with a seven-person group, there are 966 possible relationships! When there are too many group members, there is a tendency for cliques to form. However, large numbers of group members may result in additional resources not present in smaller groups.

People are influenced by the presence of others. For example, some small groups and teams are very cohesive, which means having a high degree of togetherness and a common bond. This **cohesiveness** may influence whether the group functions effectively and efficiently. In addition, the small group/team context affords individuals a chance to gain multiple perspectives on an issue. That is, in the intrapersonal context, an individual views events from their own perspective; in the interpersonal context there are more perspectives. In the small group/team context, many more people have the potential to contribute to the group's goals. In problem-solving groups, or task groups in particular, many perspectives may be advantageous. This exchange of multiple perspectives results in **synergy,** and explains why small groups may be more effective than an individual at achieving goals.

cohesiveness the degree of togetherness between and among communicators

synergy the intersection of multiple perspectives in a small group

networks communication patterns through which information flows

roles positions of group members and their relationship to the group

Networking and role behavior are two important components of small group/team behavior. **Networks** are communication patterns through which information flows, and networks in small groups answer the following question: who speaks to whom and in what order? The patterns of interaction in small groups may vary significantly. For instance, in some groups the leader may be included in all deliberations, whereas in other groups, members may speak to one another without the leader. The small group/team context is made up of individuals who take on various **roles,** or the positions of group members and their relationship to the group. These roles may be very diverse, including task leader, passive observer, active listener, recorder, and so forth.

Before we close our discussion of small groups, we should point out that as with the interpersonal communication context, research on small groups spans a variety of areas that include meeting management, emotional intelligence, and public school classroom gossip.

Working in small groups seems to be a fact of life in society. All across the world, citizens in numerous locations, including classrooms, corporate offices, and government buildings, rely on teams and small groups, even as we increase our reliance on technology. In fact, this technology has found its way into group experiences, namely in the form of telecommuting and web conferencing. Still, (perhaps it's because we are students of communication or perhaps it's because we understand relational development in myriad ways), we feel that person-to-person contact will never go out of style. Computers may crash, email may freeze up, and cell phones may run out of charge, but people will continue to function and communicate in small groups. And, these groups are enmeshed in organizational life, a context we now examine.

Organizational Communication

It is important to distinguish between small group/team communication and organizational communication. **Organizational communication** pertains to communication within and among large, extended environments. This communication is extremely diverse in

organizational communication communication within and among large, extended environments

that organizational communication necessarily entails interpersonal encounters (supervisor–subordinate conversations), public speaking opportunities (presentations by company executives), small group/team situations (a task group preparing a report), and mediated experiences (internal memos, email, and instant messaging platforms such as Slack). Organizations, then, are groups of groups. Theories of organizational communication are generally concerned with the functionality of the organization, including its climate, rules, and personnel.

What distinguishes this context from others is that a clearly defined hierarchy exists in most organizations. **Hierarchy** is an organizing principle whereby things or persons are ranked one above the other. For an example of the hierarchy in many colleges and universities,

hierarchy an organizing principle whereby things or people are ranked one above the other

see **Figure 2.3**. Does your school follow the same hierarchy? Dennis Mumby and Tim Kuhn (2019) acknowledge that organizations are traditionally hierarchical in nature and that there are almost no uniform ideas about structure, values, and division of labor. Organizations are unique in that much of the communication taking place is highly structured, and role playing is often specialized and predictable. Employees and employers alike are clear in their chain of command. Unlike in the interpersonal context, several modes of communication can substitute for face-to-face interaction, including email and teleconferencing.

The uniqueness of organizational communication is also represented by the research and theory conceptualized in this context. Many of the present-day organizational communication theories had their origins in a series of studies conducted in the mid-1920s to early 1930s. These studies, known as the **Hawthorne experiments,** were significant

Hawthorne experiments a set of investigations that ushered in a human relations approach to organizations

influences on modern theory in that they inaugurated the human relations approach to organizations (Mannevuo, 2018). Researchers at the Western Electric Hawthorne plant in suburban Chicago were interested in determining the effect of lighting levels on employee productivity. Interestingly, results of this research indicated that not only did the environmental conditions influence employee output, but so did the interpersonal relationships with other employees and supervisors. One conclusion arising from these studies was that organizations should be viewed as social entities; to speed up production, employers must consider workers' attitudes and feelings. These studies were among the first to put a human face on the impersonal corporate world (Roethlisberger & Dickson, 1939).

Although the human relations approach has enjoyed a great deal of theoretical and research attention, today there are a number of additional organizational orientations, including cultural systems and scientific management. Further, organizational (communication) theory and research today address various eclectic issues, including uncertainty on the job, whistle-blowing, workplace rumors, sexual harassment, and workplace bullying.

Figure 2.3   **Example of Hierarchy in Higher Education**

What is important to glean from this discussion is that, like other contexts, the organizational context has a rich tradition. The Hawthorne studies of human behavior on the job have led today's researchers and theorists to expand their perspectives of organizations and organizational life. At times, these organizational issues turn into opportunities to stake a public claim to a particular topic. At the heart of this is the fifth context we wish to explore: public/rhetorical communication.

Public/Rhetorical Communication

The fifth context is known as the **public communication** context, or the dissemination of information from one person to a large group. This is not a new context; speech presentations have existed since the beginning of time and continue today. Tony Robbins, Magic Johnson, Suze Orman, Dorie Clark, and Judy Shepard are just a few of the contemporary public figures who are in high demand as public speakers.

public communication the dissemination of information from one person to many others (audience)

In public speaking, speakers usually have three primary goals in mind: to inform, to entertain, or to persuade. This latter goal—persuasion—is at the core of rhetorical communication. Many of the principles of persuasion—including audience analysis, speaker credibility, and the verbal and nonverbal delivery of a message—are necessarily part of the persuasive process. As you reflect on your own public speaking experiences, remember that in actuality, you have been following rhetorical strategies rooted in early Greek and Roman days. How people have constructed their persuasive speeches has been the focus of study for nearly 2,500 years.

Effective public speakers owe their success to early rhetorical principles, a topic that we discussed earlier. For our purposes here, and as we noted earlier in the chapter we define **rhetoric** as a speaker's available means of persuading their audience. This definition was advanced many centuries ago by Aristotle. Rhetoric has been described as an art that brings together speakers and audience (Herrick, 2018). The study of rhetoric is expansive and can include the study of presidential inaugural addresses, rhetorical analyses of speeches on YouTube, Facebook posts of politicians, and speeches related to slavery, among many others.

> **rhetoric** a speaker's available means of persuasion

One area in the public/rhetorical context that has received significant scholarly attention is **communication apprehension** (CA), or the general sense of fear of speaking before an audience. Pioneering and valuable research undertaken by James McCroskey and Virginia Richmond, over the years, has been quite informative as the communication field tries to understand the challenges of speech anxiety. You will recall that boundaries between and among the contexts are often blurred, and CA research is one example of that blurring. Although CA is a public speaking concern, it focuses on intrapersonal issues. Furthermore, researchers have advanced ways to reduce communication apprehension. The public communication/rhetorical context addresses the confluence of theory, research, and skills. There are times when our public speaking is expanded to mediated venues, comprising the essence of our sixth context: mass/media.

> **communication apprehension** a generalized fear or anxiety regarding communicating in front of others

Students Talking Tech: Connor

We talked about how we have technology everywhere, and my generation grew up with things like Facebook and FaceTime. But we hardly ever talked about how some people actually fear using technology. There are things like privacy, bullying, stalking, and other things that make even some of my friends afraid to post anything that is personal. I get it. But I think my generation is accused of sharing everything, and I can tell you that's not always the case.

Mass/Media Communication

The sixth context is the mass communication or mediated context, which targets large audiences. First, we need to define a few terms. **Mass media** refers to the channels, or delivery modes, for mass messages. Mass media include newspapers, tablets, TV, radio, computers, and so forth. **Mass communication** refers to communication to a large audience via one of these channels of communication. Although mass communication frequently refers to "traditional" venues (e.g., newspapers), we expand our discussion to include **new media,** which encompasses computer-related technology. This communication technology includes the internet, including emailing, blogging, and texting; the influence of social networking sites

> **mass media** channels or delivery modes for mass messages
>
> **mass communication** communication to a large audience via various channels (e.g., radio, internet, television, etc.)
>
> **new media** computer-related technology

(Instagram and TikTok) on communication; cell phone usage; and high-definition (HD) television. For our purposes, we look at mass communication pertaining to communication to a large audience via multiple channels. This context, therefore, includes both the channel and the audience.

Like each of the preceding contexts, the mass communication context is distinctive. It allows both senders and receivers to exercise control. Sources such as a News Feed editor or a television broadcaster make decisions about what information should be sent, and receivers have control over what they decide to read, listen to, watch, or review. Suppose, for instance, that you are an advertiser who has slotted an expensive television commercial featuring Venus and Serena Williams. You've paid the Williams sisters well, and yet, to determine whether their endorsement has made a difference in sales, you have to wait for the numbers to come in. You have control over the choice of the endorsers, but the audience also simultaneously has control over what they watch and what they ultimately buy.

Some, like theorist Stuart Hall (see **Chapter 27**), suggest that mass media inherently serve the interests of the elite, especially big business and multinational corporations, who, Hall suggests, fund much of the research in mass communication. Many studies, however, are not underwritten by corporate sponsorship. They reflect the growing diversity of mass communication researchers and theorists. A myriad of topics using a mass media framework have been studied, including the portrayal of sexual themes on prime-time television, visual learning during COVID-19, the pervasiveness of coverage of terrorist organizations on television, political incivility on talk shows, and an analysis of *The Colbert Report*. As you can see, a wide array of research studies characterize this context.

As we write this, some of our comments may already be out of date. In fact, the rapidity of change may even render some of our technological references as obsolete! Mass communication quickly changes, and what was promised as a marvelous advance today is often considered outdated tomorrow. Because of the pervasiveness and availability of mass media in our society, media theorists will have to deal with the impact of media on the communication process itself. Some researchers (e.g., Turkle, 2015) suggest that computers help (re)define the way we conceive of ourselves. This redefinition may have an inevitable impact on the communication process. Furthermore, although a large number of homes and businesses subscribe to new technologies, a gap will always exist between those who have the resources and accessibility and those who do not. While such media remain a part of nearly any culture around the globe, we now discuss the cultural communication context—our seventh—to underscore the comprehensiveness of the communication environments.

Cultural Communication

The final communication context we wish to examine is cultural communication. To begin, we should define what we mean by *culture*. There are many definitions of culture and we don't wish to elaborate too much on the wide variation in interpreting this term. Therefore, for our purposes, **culture** can be viewed as a "community of meaning and a shared body of local knowledge" (Gonzalez & Chen, 2015, p. 5). **Cultural communication,** in particular, refers to communication between and among individuals whose cultural

culture a community of meaning with, among other things, a shared body of knowledge

cultural communication communication between and among individuals whose cultural backgrounds vary

backgrounds vary. These individuals do not necessarily have to be from different countries. In a diverse country such as the United States, one can experience cultural communication variation within one state, one community, and even one block. It's not uncommon in many parts of this society, for instance, to see two people from different cultural backgrounds speaking to each other. Urban centers, in particular, can be exciting cultural arenas where communication takes place between members of different co-cultures. **Co-cultures** are groups of individuals who are part of the same larger culture, but who—through unity and individual identification around such attributes as race, ethnicity, sexual identity, religion, and so forth—create opportunities of their own. The word *co-culture* is now widely accepted in the academic community as a replacement for *subculture*, a term suggesting that one culture has dominance over another culture. And in **Chapter 30**, we delve deep into this area by examining a cultural theory.

> **co-cultures** groups of individuals who are part of the same larger culture, but who can be classified around various identities (i.e., race, sex, age, etc.)

Cultural communication is a historically important academic context, with its beginnings traced back only to the 1950s (Leeds-Hurwitz, 1990). Nonetheless, much exciting work has been done since then. The growth of this area of study can be attributed to the growth across organizational cultures, with more U.S. companies doing business abroad. In addition, technological availability, population shifts, and genuine efforts to understand other cultures contribute to the growing interest and frequent conversations pertaining to this context. Some of these dialogues are still difficult, nearly 50 years after the signing of the Civil Rights Act.

The cultural context differentiates itself from other contexts in a few ways. First, as you may have determined, this context is the only context that specifically addresses culture. Although some contexts, such as the organizational context, comprise research on racial and ethnic cultures, this work is often ancillary, with culture being examined for its effects on the organization, for example. In the cultural context, however, researchers and theorists purposely explore the interactions and events between and among people of different cultures. Second, study in this context means that researchers inherently accept the fact that human behavior is culturally based. In other words, culture structures how we perceive and how we act.

Gonzalez and Chen (2015) state that when studying culture and communication, it's important to "invite *experience*" (p. 3) into the research arena. And, you will quickly see that although culture is a context in and of itself, every theory we discuss in this book both is affected by and affects culture.

Collating the Contexts

In discussing these seven contexts, we have provided you with a basic category system for dividing the broad field of communication. These seven categories help us discuss the communication process more clearly and specifically. Yet the template is not perfect, and as you have probably noted in our discussion, there is often overlap among the categories. For instance, when people belong to an online cancer support group, their communication has elements of at least four contexts: intrapersonal, interpersonal, small group/team, and mass communication. Thus, we caution you against viewing these categories as completely exclusive and distinctive from one another.

Finally, as alluded to earlier, technology similarly pervades each of these contexts and consequently, communication is affected. That is, one or more contexts are influenced by technology. For example, online dating sites are not always researched with a mediated lens; it is a topic that is also both intrapersonal (e.g., the personal decision to go online) as well as interpersonal (the communication between two people).

In fact, as you review these theories, you should be asking yourself whether any important questions can be asked about the impact of technology. And, despite the pervasiveness of technology in all of our lives, sometimes its influence is muted. As Nancy Baym (2012) and her colleagues suggest, legitimate reasons need to exist to justify the inclusion of technology and new media in research. Baym states that it's important for researchers and theorists to rationalize the need to study communication technology; explaining phenomena is much more important than exploring it.

Theory-Into-Action

 We have noted in our discussion that the mass/media context comprises outlets from newspapers to social media platforms. And, among the most addictive mediated platforms is Facebook. Facebook has served a valuable function in (re)connecting old friends, relatives, and colleagues; it has also been accessed for news, to share pictures, and to offer opinions and views. And, with all of this engagement, Facebook's use has also resulted in some people becoming obsessed with its usage. Some telltale signs include showing anxiety when restricted from using the site, establishing fake friends, reducing activities outside of Facebook browsing, and substituting virtual dates for real dates. Facebook Addiction Disorder (FAD) has been recognized as a legitimate and cautionary tale of how generations can be too immersed in an environment not designed for long-term obsession and compulsiveness.

Conclusion

This chapter provided you two frameworks to use as you try to understand the communication field. We began by exploring the seven established traditions of communication theory that are inclusive of the theories you're about to read in this book. These seven traditions include Rhetorical, Semiotic, Phenomenological, Cybernetic, Socio-Psychological, Socio-Cultural, and Critical. We next examined the primary contexts of communication: Intrapersonal, Interpersonal, Small Group/Team, Organizational, Public/Rhetorical, Mass/ Media, and Cultural. As various theories are introduced to you, you will see that many of them fall neatly in these divisions. Other theories will cut across traditions and contexts.

You should now be able to discern the uniqueness of the communication discipline. As you read the next chapter, on theory building, you will begin to link the communication process with the theoretical process. These preliminary chapters offer important foundations to draw on as you encounter the theories presented later in the book. Before reading each theory, we hope you return to these chapters to refresh your thinking.

Discussion Starters

 Case-In-Point: Lee and Jenny Yamato's experiences fall across several of the contexts we identified in this chapter. Which of the contexts are in play in their story?

Try-It-Your-Selfie: Determine how at least two social networking sites (e.g., LinkedIn, Facebook, TikTok) can be applied to all seven communication traditions. Provide an example of each.

1. Which context of communication is most relevant to you as you consider your career choice? Why?

2. How would you illustrate—with a picture—the overlap between and among the various communication traditions?

3. Explain how politics can influence each of the communication contexts.

4. Suppose you were asked to differentiate between a tradition and a context. What would guide your thinking in your response?

References

Ashby, W. R. (2015). *An introduction to cybernetics*. Martino Books.

Baym, N. K. (2012). Fans or friends?: Seeing social media audiences as musicians do. *Participations, 9*(2), 286–316.

Baym, N., Campbell, S. W., Horst, H., Kalyanaraman, S., Oliver, M. B., Rothenbuhler, E., & Miller, K. (2012). Communication theory and research in the age of new media: A conversation from the CM Café. *Communication Monographs, 79*(2), 256–267.

Briñol, P., Petty, R. E., & Guyer, J. J. (2019). A historical view on attitudes and persuasion. In *Oxford Research Encyclopedia of Psychology*. Oxford University Press.

Carey, J. W. (1989). *Communication as culture: Essays on media and society*. Unwin Hyman.

Craig, R. T. (1999). Communication theory as a field. *Communication Theory, 9*, 119–161.

Craig, R. T. (2007). Communication theory as a field. In R. T. Craig & H. L. Muller (Eds.), *Theorizing communication: Readings across traditions* (pp. 63–98). Sage.

Craig, R. T. (2018). For a practical discipline. *Journal of Communication, 68*(2), 289–297.

Craig, R. T., & Muller, H. L. (Eds.). (2007). *Theorizing communication: Readings across traditions*. Sage.

DiGregorio, N. (2022). Language use and the social affordances of marriage: An exploration of the experiences of cisgender gay men. In A. Hoy (Ed.), *The social science of same-sex marriage* (pp. 255–270). Routledge.

Eadie, W. (2020). *Becoming an academic discipline of communication, 1964–1982*. Lexington Books.

Fujishin, R. (2023). *The art of communication: Improving your fundamental communication skills*. Rowman & Littlefield.

Gehrke, P. J., & Keith, W. M. (Eds.). (2015). *A century of communication studies: The unfinished conversation*. Routledge.

Golden, J. L., Berquist, G., Coleman, W. E., & Sproule, J. M. (2011). *The rhetoric of Western thought* (10th ed.). Kendall-Hunt.

Gonzalez, M., & Chen, V. (2015). *Our voices: Essays in culture, ethnicity, and communication*. Oxford University Press.

Herrick, J. A. (2018). *The history and theory of rhetoric: An introduction*. Routledge.

Kephart, W. M. (1950). A quantitative analysis of intragroup relations. *American Journal of Sociology, 60*, 544–549.

Leeds-Hurwitz, W. (1990). Notes on the history of intercultural communication: The foreign service institute and the mandate for intercultural training. *Quarterly Journal of Speech, 76*, 262–281.

Mannevuo, M. (2018). The riddle of adaptation: Revisiting the Hawthorne studies. *The Sociological Review, 66*, 1242–1257.

Marx, K., & Engels, F. (1848). *The communist manifesto*. Penguin Books.

McHendry, G., Thorpe, M. E., Kurr, J. A., Golden, J. L., Berquist, G., Coleman, W., & Sproule, J. M. (2020). *The rhetoric of Western thought: From the Mediterranean world to the global setting*. Kendall-Hunt.

Miike, Y., & Yin, J. (Eds.). (2022). *The handbook of global interventions in communication theory*. Routledge.

Morreale, S. P., & Vogl, M. W. (1998). *Pathways to careers in communication.* National Communication Association.

Mumby, D., & Kuhn, T. A. (2019). *Organizational communication: A critical introduction.* Sage.

Roethlisberger, F. L., & Dickson, W. (1939). *Management and the worker.* Wiley.

Satir, V. (1988). *The new peoplemaking.* Science and Behavior Books.

Trenholm, S. (2020). *Thinking through communication: An introduction to the study of human communication.* Routledge.

Turkle, S. (2015). *Reclaiming conversation: The power of talk in the digital age.* Penguin.

CHAPTER 3
Thinking About Theory and Research

"There is nothing so practical as a good theory."

—Kurt Lewin

Rolanda Nash

Rolanda Nash hurried to class after work. She always seemed to be running late. Since she decided to divorce Anton and move from Sheridan, Wyoming, to Chicago, things were confusing. After her relationship with Anton, she felt she would never trust another man again. Meanwhile, she had to complete six credits to graduate and keep the new job she had landed in Chicago. In addition to doing her schoolwork, Rolanda was working 30 hours a week for one of her professors, Dr. Stevens. Dr. Stevens was testing a theory called Communication Accommodation Theory that focused on how and when people made their own communication sound like their conversational partner's (a process of accommodation). According to the theory, when someone wants to get another's approval, there's a higher chance they'll mirror the other's talk. Dr. Stevens wanted to observe communication accommodation in an organizational setting. The professor sent Rolanda into two different organizations to tape naturally occurring conversations between subordinates and managers.

It was very challenging for Rolanda to capture natural conversations. Although Stevens had obtained permission for her to record conversations in the organizations, some people recognized her and seemed to clam up when she appeared. In addition, neither of the two organizations employed many African Americans and Rolanda felt she stuck out as she walked through the hallways filled with white people. But she was used to that. In most of her classes, she was the only African American woman. She had been hoping Chicago would offer more diversity.

If she could get enough conversations to satisfy Dr. Stevens, she could go home to tackle her English assignment. Stevens hadn't really told her how many conversations she needed, but Rolanda was hoping 10 would be enough because that's all she had gotten in five days of recording. Dr. Stevens had mentioned last week that when Rolanda was finished taping, she would probably be sending her back to the organizations to do some follow-up interviews with the people she had taped. Rolanda wondered how that would work out.

Rolanda is involved in theorizing and researching about complex communication interactions both in her work and in her personal life. Although not everyone does research for a living, often people wonder to themselves, or ask one another: how can we stop arguing so much? Why are we successful in communicating sometimes and not at other times? How can we be better communicators? Will social media unite us or divide us? We can provide answers to these kinds of questions with theory, because as Heidi Muller and Robert Craig (2007) observe, "Theorizing is a formalized extension of everyday sense-making and problem solving" (p. ix). Theories allow us to see how organizations and individuals handle crisis and communication. Furthermore, when we make observations and compare them to theory, we're doing research. Theory and research are inextricably linked. Paul Reynolds (2016) points out that some researchers begin with theory (theory-then-research) whereas others begin with research (research-then-theory), but all researchers need to think about both theory and research.

In this book we are discussing theory and research as professionals use them in their work. Yet, all of us in daily life think like researchers, using implicit theories to help us understand those questions we mentioned previously, as well as many others. Fritz Heider (1958) referred to everyday interactors engaging in theoretical thinking as "naïve psychologists" or what we might call *implicit theorists.* Whenever we pose an answer to one of our questions (i.e., if we suggest that maybe we are really fighting over power and control and not what color to paint the living room), we're engaging in theoretical thinking.

Sometimes, an implicit theorist and a professional work in the same ways. First, as we have just mentioned, they are similar because both puzzle over questions encountered through observations and both seek answers for these questions. Both also set up certain criteria that define what an acceptable answer might be. For instance, when Ely wonders why his college roommate talks so much, he might decide on the following criteria for an answer: the answer has to apply to all communication contexts (online, face-to-face, and so forth); and the answer has to make sense (Ely wouldn't accept an answer stating that his roommate comes from another planet where talk is more highly valued than here on earth, for instance). When Ely (and social scientists) find answers that satisfy their criteria, they generalize from them and may apply them to other situations that are similar. If Ely concludes that his roommate is insecure and talks to cover up his insecurity, he may determine that others he meets who talk more than he does are also insecure. We'll return to the notion of an implicit theorist a bit later in the chapter.

In all those processes, everyday communicators follow the basic outline advanced by social science. However, there are differences between them as well. Social scientists systematically test theories whereas nonscientists test selectively. Ely merely makes observations of people he knows, whereas a social scientist will try to observe some kind of systematic sample of the population. In addition, Ely will probably easily accept evidence that agrees with his theory about the relationship between insecurity and talking, and tend to ignore evidence that contradicts it. Researchers are more rigorous in their testing and are more willing to amend theories, incorporating information arising from inconsistencies to create a revised formulation of the theory.

In this chapter, we build on what you already know as an implicit theorist and prepare you for reading about the theories in the text by providing the following: (1) a definition of *theory* that maps the term onto intellectual traditions and explains how assumptions affect the process of theorizing, (2) a list of evaluative criteria for assessing a theory, and (3) a brief description of the research process.

Defining Theory: What's in a Name?

Generally speaking, a **theory** is an abstract system of concepts with indications of the relationships among these concepts that help us to understand a phenomenon. Stephen Littlejohn, Karen Foss, and John Oetzel (2021) suggest this abstract system is derived through systematic observation. Donald Stacks and Michael Salwen (2014)

theory an abstract system of concepts and their relationships that help us to understand a phenomenon

suggest that theory "is like a map for exploring unexplored territories" (p. 4). Jonathan H. Turner (1986) defines *theory* as "a process of developing ideas that can allow us to explain how and why events occur" (p. 5). Turner's definition focuses on the nature of theoretical thinking without specifying exactly what the outcome of this thinking might be. William Doherty and his colleagues (1993) elaborate on Turner's definition by stating that theories are both process and product: "Theorizing is the process of systematically formulating and organizing ideas to understand a particular phenomenon. A theory is the set of interconnected ideas that emerge from this process" (p. 20). In this definition, the authors do not use Turner's word *explain* because the goals of theory can be more numerous than simply explanation, a point we explore later in this chapter.

In this brief discussion, you have probably noticed that different theorists approach the definition of *theory* somewhat differently. The search for a universally accepted definition of theory is a difficult, if not impossible, task. In part, the difficulty in defining theory is due to the many ways in which a theory can be classified or categorized. We'll refine our definition here by examining the components and goals of theories.

As you work toward understanding theory and the various theoretical perspectives in this book, understand that a "Eurocentric cultural bias" problem exists (Craig, 2013, p. 42), a notion we introduced earlier in the book. We need to frame our understanding of the current discussion with the belief that "communication research around the world has relied on Western theories and methods, and the global discourse on communication theory has been excessively one-sided" (Craig, p. 43). With this preliminary information in mind, we now turn our attention to interpreting theory further via its components and goals. We close this section by providing you a more aesthetic approach—metaphor—for understanding theory.

Components of Theories

To understand theory as a whole, we need to understand the component parts that make up theories. Theories are composed of several key parts, the two most important of which are called concepts and relationships. **Concepts** are words or terms that label the most important elements in a theory. Concepts in some of the theories we discuss in this text include *cohesiveness* (Groupthink), *self* (Symbolic Interaction Theory), and *scene* (Dramatism).

concepts labels for the most important elements in a theory

"I'm between theories right now. I'm very vulnerable."

Christopher Burke/CartoonStock Ltd

A concept often has a specific definition that is unique to its use in a theory, which differs from how we would define the word in everyday conversation. For example, the concept "cultivation" used in Cultivation Theory (see **Chapter 26**) refers specifically to the way media, especially television, create a picture of social reality in the minds of media consumers. This use of the term differs from using it to mean hoeing your garden or developing an interest, skill, or friendship. It's always the task of the theorist to provide a clear definition of the concepts used in the theory.

Concepts may be nominal or real. **Nominal concepts** are those that are not observable, such as *democracy* or *love*. **Real concepts** are observable, such as *text messages* or *spatial distance*. As we'll discuss later in the chapter, when researchers use theory in their studies, they must turn all the concepts into something concrete so that they can be observed. It's much easier to do this for real concepts than for nominal ones.

nominal concepts concepts that are not directly observable

real concepts concepts that are directly observable

Relationships specify the ways in which the concepts in the theory are combined. For example, in **Chapter 1** we presented three established models of the process of communication. In each model, the concepts appear to be similar. What is different is the relationship specified among them. In the first model, the relationship is a linear one where one concept relates to the second, which then relates to the next, and so forth. In the second model, the posited relationship is interactive, or two way. The third model illustrates mutual influence (transaction), where all the concepts are seen as affecting one another simultaneously.

relationships the ways in which the concepts of a theory relate to one another

Goals of Theories

We can also clarify the definition of theory by understanding its purposes. In a broad sense, the goals of theory include explanation, understanding, prediction, and social change; we are able to *explain* something (why Rolanda and Anton's marriage ended, for example) because of the concepts and their relationships specified in a theory. We are able to *understand* something (Rolanda's distrust of men) because of theoretical thinking. In addition, we are able to *predict* something (how Rolanda will respond to other men she meets) based on the patterns suggested by a theory. Finally, we are able to effect *social change* or empowerment (altering the institution of marriage so that it more completely empowers both partners, for example) through theoretical inquiry.

Although some theories try to reach all these goals, most feature one goal over the others. Rhetorical theories, some media theories, and many interpersonal theories seek primarily to provide explanation or understanding. Others (e.g., traditional persuasion and organizational theories) focus on prediction. Still others (e.g., some feminist and other critical theories) have as their central goal to change the structures of society. This means actually bringing about a change in the social norms and structures, not simply improving individual lives. To understand the difference, think about a theory of conflict management that may help people understand how to engage in conflict more productively, thus enriching their lives. Yet, the theory does nothing to change the underlying structures that promoted the conflict in the first place.

In the previous discussion, we suggested that theories act as lenses to help researchers interpret concrete experiences and observations. Janet Yerby (1995) refers to theories as "the stories we have developed to explain our view of reality" (p. 362). This line of thinking prompts us to ask what motivates a scholar to choose one theory (or lens) over another in their work. The answer to this question comes from an examination of the "approaches to knowing" that scholars bring to their work before they begin to do research.

Approaches to Knowing: How Do You See (and Talk About) the World?

Scholars (e.g., Treadwell, 2016; White, Martin, & Adamsons, 2019; Zhou & Sloan, 2015) have discussed how researchers think and talk about the world. Most of these scholars have identified three general approaches: positivistic or empirical, interpretive, and critical. We will discuss each of these below, and remember, because we wish for you to have a comprehensive understanding of each, we're presenting each approach in detail. Most researchers, however, do not subscribe to all the details we present.

The Positivistic, or Empirical, Approach

The **positivistic, or empirical, approach** assumes that *objective* truths can be uncovered and that the process of inquiry that discovers these truths can be, at least in part, value-neutral. The notion of neutrality means that researchers are observers and do not affect what is happening; they merely record their observations during a study. This tradition advocates the methods of the natural sciences, with the goal of constructing general laws governing human interactions. An empirical researcher strives to be objective and works for **control**, or direction over the important concepts in the theory. In other words, when researchers make observations, they carefully structure the situation so that only one element varies. This enables researchers to make relatively definitive statements about that element. As Leslie Baxter and Dawn Braithwaite (2008) observe, the researcher's task in the empirical approach is "to deduce testable hypotheses from a theory" (p. 7). In other words, the positivistic approach moves along the theory-then-research model to which Paul Reynolds (2016) referred. Graham Bodie and Susanne Jones (2012) illustrate this process because they conducted a study to test which one of three different theoretical models of supportive communication best explained their results. Most researchers today would not call themselves positivistic, however, because the requirements of control and the researcher's neutrality in this tradition are too strict. Contemporary researchers call their approach **post-positivism**, a term meaning an empirical approach that acknowledges that the researcher cannot be completely value-free in any study and cannot control all the variables involved. Post-positivism allows for more subjectivity and rejects total objectivity as being unattainable (Caughlin & Wilson, 2022). Researchers adopting post-positivism acknowledge that they can influence the study in many (unintended) ways.

positivistic/empirical approach an approach assuming the existence of objective reality and value-neutral research

control direction over the important concepts in a theory

post-positivism an approach following most of the assumptions of positivism but relaxing the requirements of objectivity and neutrality somewhat

The Interpretive Approach

The **interpretive approach** views truth as *subjective* and co-created by the participants, with the researcher clearly being one of the participants. There is less emphasis on objectivity in this approach than in the empirical approach because complete objectivity is seen as impossible. However, this does not mean that research in this approach has to rely totally on what participants say with no outside judgment by the researcher. The interpretive researcher believes that values are relevant in the study of communication and that researchers need to be aware of their own values and to state them clearly for readers, because values will naturally permeate the research. These researchers are not concerned with control and the ability to generalize across many people as much as they are interested in rich descriptions about the people they study. This emphasis

interpretive approach an approach viewing truth as subjective and stressing the participation of the researcher in the research process

on rich description leads interpretive researchers to put a lot of focus on the voices of their participants and quote their comments extensively (Darder, 2019). For interpretive researchers, theory is best induced from the observations and experiences the researcher shares with and/or hears from the respondents.

The Critical Approach

In the **critical approach**, an understanding of knowledge relates to *power*. Critical researchers believe that those in power shape knowledge in ways that perpetuate the status quo. Thus, powerful people work at keeping themselves in power, while silencing minority voices questioning the distribution of power and the power holders' version

> **critical approach** an approach stressing the researcher's responsibility to change the inequities in the status quo

of truth. Feminists and Marxists, among others, work from this intellectual vantage point. For critical researchers, it's important to change the status quo to resolve power imbalances and give voice to those silenced by the power structure. Kent Ono (2011) observes that critical theory "often emerges out of the everyday life experiences of women, people of color, and members of LGBTQ communities" (p. 9). Claudia Anguiano and her colleagues (Anguiano, Milstein, De Larkin, Chen, & Sandoval, 2012) provide a concrete example of Ono's statement. They examined environmental inequity using Latin Critical Race Theory. They wished to promote Hispanic activism around environmental justice issues, and sought with their research "to translate theoretical knowledge into practical strategies for better policymaking" (p. 137).

Some critical theorists, notably Stuart Hall (1981), whose work we feature in **Chapter 27**, have commented that power imbalances may not always be the result of intentional strategies on the part of the powerful. Rather, ideology, or "those images, concepts, and premises which provide the frameworks through which we represent, interpret, understand and 'make sense' of some aspect of social existence" (Hall, p. 31), is often "produced and reproduced" accidentally. For example, this may come about when certain images of masculinity are effective in selling a product. When advertisers observe this success, they continue creating ads with these images. In this fashion, the images of masculinity become entrenched in society. Thus, although the powerful are interested and invested in staying in power, they may not be fully aware of what they do to silence minority voices.

For a visual understand of the three approaches to knowing, look at **Table 3.1**.

Table 3.1 Three Approaches to Knowing

	EMPIRICAL	INTERPRETIVE	CRITICAL
Goal	Explanation of world	Probe the relativism of world	Change the world
Engagement of researcher	Separate	Involved	Involved
Application of theory	To generalize about many like cases	To illuminate the individual case	To critique a specific set of cases

Approaches to Knowing: What Questions Do You Ask About the World?

Each of the three approaches to knowing provides different answers to questions about the nature of reality (researchers call this **ontology**), questions about how we know things (known as **epistemology**), and questions about what is worth knowing (or what researchers call **axiology**). We'll briefly address each of these terms (ontology, epistemology, and axiology), and suggest how the three approaches to knowing treat them differently.

> **ontology** the study of the nature of reality which shapes the background understanding for theorizing about human communication
>
> **epistemology** questions about how we go about knowing things
>
> **axiology** questions about what is worth knowing

Ontology is the study of being and nonbeing, or in other words, the study of reality. The word *ontology* comes from the Greek language and means the science of being or the general principles of being. Pat Arneson (2009b) states that ontology is "the study of what it means to be human, which shapes the background understanding for theorizing about human communication" (p. 697). This definition focuses on the idea that ontology gives us a certain vision of the world and on what constitutes its important features. Ontology is called the first philosophy because it is not possible to philosophize until the nature of reality is determined. Often questions of ontology cluster around how much free will people have. Researchers who see the world from an empirical approach believe that general laws govern human interactions. Thus, they also believe that people don't have a lot of free choice in what they do—people are predictable because they follow the laws of human behavior which, to a large extent, determine their actions. A researcher's job is to *uncover* what is already out there in reality. This differs from researchers with an interpretive bent, who would allow that people do have free choice, and see a researcher's job as co-creating reality with research participants. Finally, critical researchers see both choice and constraint in the power structures they wish to change.

The questions surrounding epistemology focus on how we go about knowing; what counts as knowledge is intimately related to ontology. Epistemology looks at "nature, scope, and limits of human knowledge" (Arneson, 2009a, p. 350). How researchers see the world, truth, and human nature necessarily influences how they believe they should try to learn about these things. The approach (positivistic, interpretive, or critical) Dr. Stevens took in our opening example would affect her way of collecting information, an epistemologic choice. For instance, if Dr. Stevens researched as a positivist, she would institute many more controls than we described in our opening case study. Furthermore, the number of observations would not be left to chance or to Rolanda's schedule. Dr. Stevens would have calculated the number of conversations she needed to support the statistics she'd use to test relationships among status and communication accommodation. If Dr. Stevens operated in the interpretive tradition, she would not be content with her own analysis of the conversations. She might invite the participants to read the transcripts of their conversations so that they could tell her whether they were trying to accommodate with what they were saying on the tape. Stevens would probably be interested in the participants' explanations for why they changed (or did not change) their speech patterns as they conversed with superiors or subordinates in the workplace. Using a critical approach, Dr. Stevens might bring some of the following questions to her research: how is the relationship between workers of differing statuses communicatively constructed? Does convergence happen unequally based on status? Are there status differences other than occupation that impact communication accommodation? How can we change the prevailing power structures to improve the inequities we observe in the workplace?

Axiology focuses on the place of values in theory and research. The empirical position on axiology is that science must be value free. However, most researchers do not take this extreme position and accept that some subjectivity, in the form of values, informs the research process: the position we described as post-positive (Merrigan & Huston, 2019). Research and theory-building are, indeed, axiological efforts in that the topic selection and the approach to studying that topic are usually value-laden undertakings. The question is not *whether* values should permeate theory and research but *how* they should.

As you've probably noticed, axiology has a relationship to the three ways of knowing we discussed previously: avoiding values as much as possible in research (empirical), recognizing how values influence the entire research process (interpretive), and advocating that values should be closely intertwined with scholarly work (critical). The first stance argues that the research process consists of many stages and that values should inform some of these stages, but not others. For example, the part of the research enterprise that focuses on theory choice must be informed by the values of the researcher. Scholars choose to view a research problem through the lens that they believe most accurately describes the world. Thus, some researchers choose theoretical frameworks that are consistent with an ontology of free choice, whereas others choose frameworks that are more "lawlike" and deterministic. Yet, when they test these theories (the verification stage), they must eliminate "extra-scientific values from scientific activity" (Popper, 1976, p. 97). As you can see, this stance proposes a very limited role for values.

The second position argues that it's not possible to eliminate values from any part of theorizing and research. In fact, some values are so embedded in researchers' culture that researchers are unconscious that they even hold them. Sandra Bem (1993), for instance, observes that much of the research on differences between women and men was influenced by biases existing at the time. Many feminist scholars argue that social science itself suffers from a male bias (Harding, 1991). Some scholars examining African American issues make the same observations about the European American biases that permeate much of social scientific research (Kazeroony & du Plessis, 2019). Thomas Nakayama and Robert Krizek (1995) point out that communication researchers often take white for granted as the default race. Thus, the values and assumptions held by those with a European American perspective are never highlighted, questioned, or acknowledged; they simply inform a scholar's process (Gunaratne, 2010). Yoshitaka Miike (2019) takes this one step further to advocate using Asiacentric (meaning rooted in Asian thinking) theories rather than Eurocentric ones.

The final position argues that not only are values unavoidable, but they are desirable in research. Earlier in this chapter we mentioned social change as one goal of theory. Those who embrace this goal are called critical theorists. Critical theorists advocate seeing theory and research as political acts that call on scholars to change the status quo. Thus, scholars must contribute to changing conditions rather than simply reporting conditions (**Table 3.2**).

Table 3.2   Relationship Between How You See the World and Questions You Ask About It

	EMPIRICAL	INTERPRETIVE	CRITICAL
Ontology	No free choice	Free choice	Choice restrained by power
Epistemology	Theory first. Control study	Research first. Co-create study	Critique power. Seek change
Axiology	Reduce role of values	Acknowledge values	Celebrate values

Approaches to Knowing: How Do We Go About Theory Building?

When researchers seek to create theory, they are guided by all of the issues we've just discussed: their general approach to knowing things (empirical, interpretive, or critical) and the answers to questions about truth or reality, gathering information, and values (ontology, epistemology, and axiology). In addition, social science provides some guidelines about how to create theory, sometimes called *meta-theories*. We will review three traditional guidelines: covering law, rules, and systems. Covering law and rules represent two extremes, whereas systems provide an intermediate position between the extremes. Remember that few scholars take the extreme positions sketched out here. Rather, these positions form benchmarks from which researchers anchor their own stances on questions of communication.

Covering law meta-theory seeks to explain an event in the real world by referring to a general law. Researchers applying covering law believe that communication behavior is governed by forces that are predictable and generalizable. The **rules meta-theory** is at the other end of the ontological continuum, and holds that communication behavior is rule governed, not lawlike. Rules differs from covering law in that researchers using rules meta-theory to construct their specific theory admit the possibility that people are free to change their minds, to behave irrationally, to have idiosyncratic meanings for behaviors, and to change the rules. Ultimately, their differences focus on the concept of choice. Covering law explains human choices by seeking a prior condition (usually a **cause**) that determines the choice that is made (usually an **effect**). From the rules model, rule following results from a choice made by the follower, but does not necessarily involve antecedent conditions or any aspect of the cause–effect logic of covering law.

Covering law meta-theory a guideline, or **meta-theory,** for creating theory requiring that theories provide a general law that is universal and invariant

rules meta-theory a guideline for creating theory that builds human choice into explanations

cause an antecedent condition that determines an effect

effect a condition that inevitably follows a causative condition

A third view, the **systems meta-theory,** subscribes somewhat to the beliefs of the rules approach while also suggesting that people's free will may be constrained by the system in which they operate. Further, this meta-theory acknowledges the impossibility of achieving what covering law requires: laws about human communication that are invariant and general. Systems meta-theory proposes assumptions that are more easily met than those of covering law (Monge, 1973). We now examine each of the three meta-theories in more detail and provide an overview in **Table 3.3**.

systems meta-theory a guideline for creating theory that acknowledges both human choice and the constraints of the systems involved

Covering Law Meta-Theory

The term *covering law* was first introduced by William Dray (1957), a historian who defined it by saying "explanation is achieved, and only achieved, by *subsuming what is to be explained under a general law*" (p. 1 [emphasis in original]). Some covering law explanations refer to universal laws that state all *x* is *y*. These laws are not restricted by time or space. However, as new information comes to light, even laws have to be modified. Covering law explanations do not always have to be cause and effect. They may also specify relationships of coexistence. We have a causal relationship when we say, for example, that self-disclosures by one person cause self-disclosures from a relational partner. A claim of coexistence merely asserts that two things go together—that is, when one person self-discloses, the other does, too—but it does not claim that the first self-disclosure causes the second. It's possible that social norms of reciprocity cause the second self-disclosure or that both disclosures are caused by the environment (an intimate, dimly lit bar or consuming more alcohol than usual).

Table 3.3 Guidelines (Meta-theories) for Communication Theory Construction

META-THEORY	DESCRIPTION/EXAMPLE
Covering Law	Covering law theorists hold that there are fixed relationships between two or more events or objects. Example: whenever Linda speaks, Bob interrupts her; this is a lawlike statement that expresses a relationship between Linda and Bob. These statements are commonly referred to as if–then statements.
Rules	Rules theorists contend that much of human behavior is a result of free choice. People pick the social rules that govern their interactions. Example: in an interaction between coworkers, much of their conversation will be guided by rules of politeness, turn taking, and so on.
Systems	Systems theorists hold that human behavior is part of a system of interrelated parts. Example: think of a family as a system of family relationships rather than individual members. This illuminates the complexity of communication patterns within the family and the family's relationship to society at large.

Critical attributes of covering law explanations are that they provide an explicit statement of a boundary condition and that they allow **hypotheses**, testable predictions of relationships, of varying levels of specificity, to be generated within this boundary condition. Furthermore, because the system is deductive, complete confirmation of theories is never possible. There will always be unexamined instances of the theory.

hypotheses testable predictions of relationships between concepts that follow the general predictions made by a theory

The type of covering law that we have just described is considered outdated by most social scientists (Bostrom, 2004). Most researchers today recognize that this type of universal law is unrealistic. Instead, researchers might strive for "probabilistic laws," or statements we can predict with a certain degree of probability. For example, as Berger (1977) asserts, "We can predict with a certain probability that if males and females with certain eye colors have large numbers of children, a certain proportion of those children will have a certain eye color. However, we are *not* in a position to predict what the eye color of a *particular* child will be" (p. 10).

Overall, covering law instructs researchers to search for lawlike generalizations and regularities in human communication. These lawlike generalizations may be culturally bound or may have some other complex relationship with culture. Covering law offers a theory-generating option that aims for complete explanation of a phenomenon. The law, in effect, governs the relationships among phenomena.

Students Talking Tech: Amelia

So, I've been thinking about how I use Snapchat. I wanted to call how I use it a "theory," but after reading the chapter, I'm thinking it's more of an idea that could become a theory. It goes like this: I only Snap when my friends ask me to. I never, ever Snap myself. I got to thinking that if I wanted to develop a theory about this, I'd have to start thinking about how many times I do this in, say, a week. Then, I have to figure out who asks me the most to Snap them. Then, I need to come up with some information on facial recognition. Then, I figure I'd have to get a hypothesis or two. I think I'm working out something from the rules meta-theory, but it could be covering law. Frankly, all of this is making me tired! How do theorists do all of this thinking and still stay sane?!

Rules Meta-Theory

This approach assumes that people are typically engaged in intentional, goal-directed behavior and are capable of acting rather than simply being acted upon. We can be restricted by previous choices we have made, by the choices of others, and by cultural and social conditions, but we are conscious and active choice makers. Further, human behavior can be classified into two categories: activities that are stimulus–response behaviors (termed **movements**) and activities that are intentional choice responses (termed **actions**) (Cushman & Pearce, 1977). Rules theorists contend that studying actions is most relevant to theorists.

movements activities based on stimulus–response

actions activities based on intentional choice responses

Rules theorists look inside communities or cultures to get a sense of how people regulate their interaction with others (Shimanoff, 1980). Rules do not require people to act in a certain way; rather, rules refer to the standards or criteria that people use when acting in a particular setting. For example, when two people meet, they normally don't begin talking at an intimate level of exchange. Rather, there is an agreed-upon starting point, and they will delve further into intimacy if the two see the relationship as having a future. The process of meeting another is guided by rules, although these rules are rarely verbally identified by either person. Don Cushman and Barnett Pearce (1977) believed that if the relationship evolves, the rules guiding interactions change. Rules, then, are important benchmarks for the direction of an interaction. **Table 3.4** illustrates how rules guide initial encounters of peers in the United States.

Several researchers (Lull, 1982; Van den Bulck, Custers, & Nelissen, 2016; Wolf, Meyer, & White, 1982) have used a rules-based theoretical framework to study media and family television-viewing behaviors. James Lull (1982) identified three types of rules that govern family television watching. First are **habitual rules**, which are nonnegotiable and are usually instituted by the authority figures in the family. When Lucy and Marie tell their children that there can be no television until all homework is checked over by one of them and declared finished for the night, they are establishing a habitual rule.

habitual rules nonnegotiable rules that are usually created by an authority figure

Table 3.4 Rules Governing Initial Peer Encounters

In the first fifteen minutes of an encounter:	In the second fifteen minutes:
Politeness should be observed.	Politeness should be observed.
Demographics should be exchanged.	Likes and dislikes can be discussed.
Partners should speak in rough equivalence to each other.	One partner can speak more than another, but avoid dominance.
Interruptions and talk-overs should be minimal.	More interruptions can be tolerated, but avoid dominance.

Parametric rules are also established by family authority figures, but they are more negotiable than habitual rules. For example, the Marsh family may have a rule that members can engage in extended talk only during commercial breaks when they are viewing television. Yet, if something exciting has happened to one member, they may negotiate to talk about it during the program itself.

parametric rules rules that are set by an authority figure but are subject to some negotiation

Finally, Lull identified **tactical rules**, or rules that are understood as a means for achieving a personal or interpersonal goal, but are unstated. For example, if Rob and Jeremy are watching television together and Rob likes Jeremy, he may tune in to Jeremy's favorite show even though he himself wouldn't have chosen that program. He follows the tactical rule of maintaining relational harmony with his partner.

tactical rules unstated rules used to achieve a personal or interpersonal goal

Overall, a rules meta-theory instructs researchers to discover the rules that govern particular communication contexts and construct theoretical statements around these rules. The rules perspective offers a theory-generating option that aims for a satisfying explanation of a specific communication situation. The theorist would normally begin with a typology of the rules that govern the situation and move from those to statements connecting the rules and specifying the conditions affecting the rules.

Systems Meta-Theory

Systems thinking in communication is derived from General Systems Theory (GST), which is both a theory of systems in general—"from thermostats to missile guidance computers, from amoebas to families" (Whitchurch & Constantine, 1993, p. 325)—and a program of theory construction. Systems thinking captured the attention of communication researchers because it changed the focus from the individual to an entire family, a small group/team, or an organization. This shift reconceptualized communication for scholars and helped them to think innovatively about experience and interaction in groups. Further, systems thinking replaced the stringent assumptions of covering law with more realistic ones. Systems thinking requires systemic, nonuniversal generalizations, does not depend on inductive reasoning, separates the logical from the empirical, allows alternative explanations for the same phenomenon, and permits partial explanations (Monge, 1973).

Systems thinking rests on several properties, including wholeness, interdependence, hierarchy, boundaries, calibration/feedback, and equifinality. We will explain each of these properties briefly.

Wholeness   The most fundamental concept of the systems meta-theory is **wholeness**. It states that a system can't be fully comprehended by a study of its individual parts in isolation from one another. To understand the system, it must be seen as a whole. Wholeness suggests that we learn more about a couple, for example, by analyzing their interactions together than we do by analyzing one partner's motivations or statements alone.

Interdependence   Because the elements of a system are interrelated, they exhibit **interdependence**. This means that the behaviors of system members co-construct the system, and all members are affected by shifts and changes in the system. Virginia Satir (1988) compares the family to a mobile to illustrate how this principle applies to families. We might expect that when elderly parents decide to sell the family home and move to a small condo, their decisions will affect all of their children.

Hierarchy   All systems have levels, or **subsystems**, and all systems are embedded in other systems, or **suprasystems** (Hofkirchner, 2019).

wholeness a fundamental property of systems theory stating that systems are more than the sum of their individual parts

interdependence a property of systems theory stating that the elements of a system affect one another

subsystems smaller systems that are embedded in larger ones

suprasystems larger systems that hold smaller ones within them

hierarchy a property of systems theory stating that systems consist of multiple levels

Thus, systems are a **hierarchy**, a complex organization. Each of the subsystems can function independently of the whole system, but each is an integral part of the whole. Subsystems generally shift and change over time, but they may potentially become extremely close and turn into alliances or coalitions that exclude others. For example, if one parent confides a great deal in a son whereas the other talks to one of their daughters, two coalitions may form in the family, making interactions more strained and troubled. Hierarchy, in the context of an organization, is mapped in **Figure 3.1**.

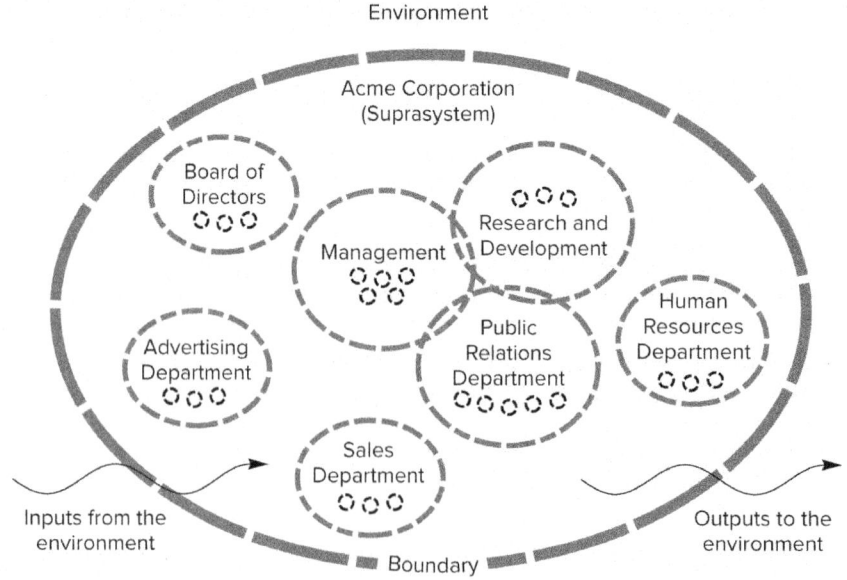

Figure 3.1 **Suprasystems, Systems, and Subsystems in the Acme Corporation**

Boundaries   Implicit in the preceding discussion about hierarchy and complexity is the notion that systems develop **boundaries** around themselves and the subsystems they contain. Because human systems are open systems (it is not possible to completely control everything that comes into or goes out from them), these boundaries are relatively permeable: they have **openness**. Thus, although the managers of a General Motors plant in Ohio may wish that their employees did not know about the strike at a General Motors plant in Michigan, they will be unable to prevent information and communication from passing through the boundary around their organizational system. And, it's fair to say that technology usually influences the boundaries in any human system.

boundaries a property of systems theory stating that systems construct structures specifying their outer limits

openness the acknowledgment that within all human systems the boundaries are permeable

calibration a property of systems theory stating that systems periodically check the scale of allowable behaviors and reset the system

Calibration/Feedback   All systems need stability and constancy within a defined range (Watzlawick, Beavin, & Jackson, 1967). **Calibration**, or checking the scale, and subsequent **feedback** to change or stabilize the system, allow for control of the range. The thermostat provides a common example illustrating this process. Home heating is usually set at a certain temperature, say, 65 degrees. The thermostat will allow a temperature range around 65 before changing anything. Therefore, if the thermostat is set for 65 and the temperature is 65 plus or minus 3 degrees, nothing happens. If the temperature drops below 62 degrees, the heat goes on; if it rises above 68 degrees,

feedback a subprocess of calibration; information allowing for change in the system

morphogenic a process that occurs when a system recalibrates (or changes)

homeostatic a term for a stable system that isn't changing

the furnace shuts off. In this way, the heating system remains stable. However, if conditions change in the house (e.g., the family insulates the attic), the thermostat may need to be recalibrated or set at a slightly lower temperature to accommodate the change. After insulating, the house may feel comfortable if the temperature is set at 63 degrees.

Changing the standard (moving the temperature from 65 to 63 degrees) is accomplished through feedback. Feedback, in systems thinking, is positive when it produces change (the thermostat is set differently) and negative when it maintains the status quo (the thermostat remains at 65). When systems change they are called **morphogenic**, and when they stay the same they are called **homeostatic**.

Equifinality   Open systems are characterized by the ability to achieve the same goals through different means, or **equifinality** (von Bertalanffy, 1968). This principle applies to human groups in two ways. First, a single group can achieve a goal through many different routes. For example, if a manager wants to increase productivity, they can

equifinality a property of systems theory stating that systems can achieve the same goals through different means

raise wages, threaten the workers with firing, hire a consultant, or do some combination of these. There are several ways the manager can reach the goal. Additionally, equifinality implies that different groups can achieve the same goal through multiple pathways. For instance, PK Computer Systems may achieve profitability by adopting a casual organizational culture, whereas Western Communication Systems may achieve profitability by demanding a more formal workplace.

Overall, a systems meta-theory instructs researchers to search for holistic explanations for communication behavior. Systems thinking offers a theory-generating option that aims to model the phenomenon as a whole, admitting the possibility for change from a variety of outside influences.

Evaluating Theory: Determining Theory Effectiveness

As you read the communication theories in the text, you'll need some standards for judging their value, worth, and effectiveness. Communication theory should not be viewed as "good" versus "bad." All theories have unique strengths and shortcomings; therefore, we should not dismiss or embrace any theory without understanding all of its various parts. The following criteria are generally accepted as useful measures for evaluating communication theory: scope, logical consistency, parsimony, utility, testability, heurism, and the test of time. We will discuss each of them briefly.

Scope   **Scope** refers to the breadth of communication behaviors covered by a theory. It is somewhat similar to the level of generality notion we discussed earlier. Boundaries are the limits of a theory's scope. Although theories should explain enough of communication to be meaningful, they should also have clear boundaries specifying their

> **scope** a criterion for evaluating theories; refers to the breadth of communication behaviors covered in the theory

limits. Uncertainty Reduction Theory (URT), for example, which we discuss in **Chapter 5**, originally was bounded by initial encounters between strangers. In some ways this suggests a rather limited scope for the theory. However, although the duration of initial encounters is short, it's true that people spend a great deal of time throughout their lives meeting and conversing with new people. Thus, the scope of the theory may seem a bit broader upon reflection.

Logical Consistency   Simply put, theories should make sense and have an internal logical consistency so they are clear and not contradictory. The claims made by the theory should be consistent with the assumptions of the theory. **Logical consistency** means that the theory "hangs together" and doesn't contradict itself, either by

> **logical consistency** a criterion for evaluating theories; refers to the internal logic in the theoretical statements

advancing two propositions that are in conflict with each other or by failing to operate within the parameters of its assumptions. When reviewing a theory with several assumptions, reflect on the theory's logical consistency to ensure that the assumptions do not contradict each other.

Parsimony   **Parsimony** refers to the simplicity of the explanation provided by the theory. Theories should contain only the number of concepts necessary to explain the phenomenon under consideration. If a theory can explain a person's communication behavior satisfactorily by using one concept (such as expectancy violations),

> **parsimony** a criterion for evaluating theories; refers to the simplicity of the explanation provided by the theory

that is more useful than having to use many concepts. However, because theories of communication and social behavior are dealing with complex phenomena, they may have to be complex themselves. Parsimony requires simplicity without sacrificing completeness. In **Chapter 18**, for instance, we outline Groupthink. Although the theory's name is precisely the concept under study, a number of issues and themes exist related to the theory.

Utility   This criterion refers to the theory's usefulness, or practical value. A good theory has **utility** when it tells us a great deal about communication and human behavior. It allows us to understand some element of communication that was previously unclear. It

> **utility** a criterion for evaluating theories; refers to the theory's usefulness or practical value

weaves together pieces of information in such a way that we are able to see a pattern that was previously unseen. To a large extent, the utility criterion asks the following question: "Can real-world applications be found?"

Testability   **Testability** refers to our ability to investigate a theory's accuracy. One of the biggest issues involved in testability concerns the clarity of the theory's central concepts. For example, as we discuss in **Chapter 6**, Social Exchange Theory is predicated on the concepts of costs and rewards. The theory predicts that people will engage in behaviors that they find rewarding and avoid behaviors that are costly. However, the theory defines costs and rewards in a circular fashion: Behaviors that people engage in repeatedly are rewarding, and those that they avoid are costly. You can see how difficult it is to test the central prediction of Social Exchange Theory given this circular definition.

> **testability** a criterion for evaluating theories; refers to our ability to test the accuracy of a theory's claims

Heurism   **Heurism** refers to the amount of research and new thinking that is stimulated by the theory. Theories are judged to be good to the extent that they generate insights and new research. Although not all theories produce a great deal of research, an effective theory prompts research activity. For example, the theory we discuss in **Chapter 27**, Cultural Studies, came from many diverse disciplines and has stimulated research programs in English, anthropology, social psychology, and communication.

> **heurism** a criterion for evaluating theories; refers to the amount of research and new thinking stimulated by the theory
>
> **test of time** a criterion for evaluating theories; refers to the theory's durability over time

Test of Time   The final criterion, the **test of time**, can be used only after some time has passed since the theory's creation. Is the theory still generating research or has it been discarded as outmoded? Deciding whether a theory has withstood the test of time is often arbitrary. For instance, if a theory was conceptualized and tested in the 1970s, but has remained dormant in the literature decades, but is now being reintegrated into research, has this theory satisfied the test of time? Judging a theory based on this criterion is often a subjective process. Furthermore, it is not a criterion that can be used to assess a new theory (see **Table 3.5** for a review of each of the seven criteria just discussed).

Table 3.5   Criteria for Evaluating Communication Theories

CRITERIA	QUESTIONS TO CONSIDER
Scope	What are the boundaries of the theory's explanation?
Logical consistency	Do the claims of the theory match its assumptions? Do the principles of the theory contradict each other?
Parsimony	Is the theory as simple as it can be to explain the phenomenon under consideration?
Utility	Is the theory useful or practical?
Testability	Can the theory be shown to be false?
Heurism	Has the theory been used in research extensively to stimulate new ways of thinking about communication?
Test of time	How long has the theory been used in communication research?

These criteria have been general standards for evaluating theories for some time, but our changing communication environment may require us to add to or revise this list.

For instance, technology is likely one area that will enter into your evaluation of a theory. You might be inclined to simply dismiss a theory because it doesn't "relate" to your experiences online or with social media. Yet, the consequences of doing that are profound. What happens when a theory fails to accommodate the technological influence that happens today in human communication? Do we reject the theory? Accept it with caution? Modify it to suit today's world? How do we go about retaining the authenticity of the original theory if we want to make it more relevant to, say, TikTok or Instagram? Would you be comfortable altering a theory by Aristotle? Does every theory have to integrate technology, or should there be a specific subset of theories that deal only with the interface of technology and communication? These questions and others will continue to alter how we evaluate communication theories going forward.

Theory and Metaphor

One of Shakespeare's most famous lines is "All the world's a stage and all the men and women merely players." While we're not interested in exploring *As You Like It*, what this line does is provide a metaphor comparing life to theater. This quote, like all good metaphors, helps us understand life a little better by seeing how it relates to things we know about theater. We wish to briefly address four metaphors that we believe will help you further understand communication theory: theories as gardens, theories as family, theories as marathons, and theories as spider webs. Like all metaphors, each one of these comparisons draws our attention to some aspects of theory while glossing over other aspects. The metaphor you like best will probably be the one that features the elements of theoretical thinking you believe are most important.

Theories as Gardens: This image suggests that communication theorists see their theoretical developments as needing constant attention, primarily through research and refinement. Like a garden, theories need to be pruned to ensure that they are parsimonious, and after a theory has been refined, its "growth" can be immeasurable. Think, for instance, about a theory conceptualized over 2,500 years ago: Aristotle's Rhetoric (**Chapter 20**). Very few communication scholars have "thrown shade" on the theory, they have instead cultivated its value by demonstrating its application over the centuries. Like a perennial, most theories start off small, but over the years grow exponentially.

Theories as Family: Communication theories can be understood by employing a family metaphor. That is, like family members, theories are accepted at some times and rejected at other times. There is also periodic conflict between and among theorists, as there is in family life—disagreement might result over a theory's assumptions, for instance. This notion imagines that a "theoretical family" in communication is dynamic and ever-changing, much like families in general. However, as you review the theories in this book, keep in mind that even those who study a particular theory are part of that theory's family. So, it's not uncommon for a particular theory to be utilized and expanded by scholars who were part of the original theorist's "family." Expectancy Violations Theory (**Chapter 4**), for example, is authored by Judee Burgoon, and a great deal of the research undertaken using that theory has been authored by her former students and close colleagues.

Theories as Marathons: For those of you who have ever run a marathon, you know that it takes a great deal of preparation, practice, discipline, and tenacity. Theory-building is the same, and the theorists found in this book have spent years in "training." As we noted in an earlier chapter, no communication theorist wakes up and says, "Today, I'll write a theory!" Rather, some theories may start from the personal experience of the theorist. Stuart Hall, the theorist associated with Cultural Theory (**Chapter 27**), was born in Jamaica, and during the evolution of his theory, he noted that this cultural backdrop was influential in how he created his theory. Overall, scholars see their theories as able to withstand the "heat" of criticism or the "cold" shoulder. Theorists work hard, and their view of "winning" is to be accepted as legitimate contributors to theory construction.

Theories as Spider Webs: Communication theory, like a spider web, is resilient. Theory is architecturally complex, sometimes so complex that you will have to read a chapter a few times to fully comprehend the crux of the theory. And like most spider webs, theories are built to "catch"—whether it's the eye of another scholar, a colleague, or a student just like you. Further, spider webs have been used in a number of different ways, from clotting blood to forming parts of a telescope. Theories such as Uses and Gratifications Theory (**Chapter 13**) have also been employed in a multitude of ways, from explaining elements of television to social media.

The Research Process

Any introduction to communication theory must necessarily include a discussion of research. You already know that the two processes—theory-building and research generation—are unique although interrelated, as our opening chapter story illustrated. Thus, we need to provide you some sense of the research process for you to have a foundation from which to draw as we discuss each theory.

Our discussion will necessarily be brief here. We know many of you will take an entire class devoted to the study of research methods, so here we simply give you an idea of how important theory and research are to one another. Although we could delve into a variety of areas in communication research, we maintain our focus on the objective (quantitative) and subjective (qualitative) efforts undertaken by researchers.

Students Talking: Miles

 My uncle recently got back from serving in Iraq, and I know that he has PTSD. I'd like to do a study on military people who are on "the front lines" and what happens to their relationships when they return home. My aunt and uncle always argue now, and before he left, they never yelled. I know it'd be stupid to study this using statistics. I'd rather "survey his voice" to see what his experiences were. I can see how personal history can affect how researchers develop theory.

Objective (Quantitative) Communication Research

Communication theory after Aristotle has mainly grown from quantitative thinking. That is, many of the theories you read about have their roots in experiments and the empirical approach. We will briefly address several themes related to this research orientation. Should you wish to find out more about quantitative communication research, you should look at additional information (e.g., Wrench, Thomas-Maddox, Richmond, & McCroskey, 2015). At the beginning of this chapter, Dr. Stevens's study illustrated theory-then-research,

deductive logic moving from the general (the theory) to the specific (the observations)

inductive logic moving from the specific (the observations) to the general (the theory)

an approach we identified earlier in the chapter. Rolanda's transcripts would be used to test what Communication Accommodation Theory predicts about communication behaviors in the workplace. Dr. Stevens wanted to see if the speculations she made based on the theory's logic held true in the conversations that Rolanda taped. This traditional process, follows **deductive logic** in that Stevens moved from the general (the theory) to specific instances (the actual conversations gathered in two workplaces). If Stevens had used **inductive logic** (moving from specific to general), she would have asked Rolanda to record many more conversations. Stevens would have refrained from hypothesizing, or guessing, about what she might find in advance of the data collection. Then she and Rolanda would have listened to their tapes, trying to find some type of pattern that best explained what they heard. Finally, Stevens would have generalized based on her observations.

After Stevens had hypothesized about what she would find in the workplace regarding accommodations between workers and managers based on the theory, she then had to **operationalize** all the

operationalize making an abstract idea measurable and observable

concepts. This means she needed to specify how she would measure the concepts that were important to her study. In this process, Dr. Stevens turned the abstract concepts of the theory into concrete variables that could be observed and measured. For example, status difference is a critical notion in the theoretical framework, so Stevens specified to Rolanda how she should measure this. In this case, measurement was based on job title. Rolanda had to discover the job title for each of the people she observed and then compare those titles to a chart Stevens had given her classifying job titles into two categories: "supervisor" and "subordinate." This seems like a fairly straightforward means to operationalize the notion of status, but there may be instances where it's not a perfect operationalization. For instance, a lower-level employee who has worked for the company for many years might hold more status than a middle manager who has only recently arrived and is just learning the corporate culture. Additionally, women managers often report some problems with achieving the status expected from their job title. You can see how nominal concepts that are more complex and abstract, such as love and intimacy, would be even more difficult to operationalize than occupational status.

A next step in quantitative investigation and the traditional scientific model sent Rolanda into the two organizations to make **observations** and collect **data** (in this case, the conversations and the job titles). When Rolanda returned with the tapes, Dr. Stevens would have to **code** the conversations, again using operationalizations for various concepts related to the theory. Some types of data do not need extensive coding to analyze. For example, if Dr. Stevens operationalized status based on income and then provided respondents with a survey asking them to indicate the category for their salary, these data would not need the same type of coding required in the taped conversations. The income categories could simply be numbered consecutively. In contrast, the conversations have to be listened to repeatedly to determine whether a given comment converges with or diverges from the comment preceding it.

observations focused examination within a context of interest; may be guided by hypotheses or research questions

data the raw materials collected by the researcher to answer the questions posed in the research or to test a hypothesis

code convert raw data to a category system

Although some researchers approach their work strictly as hypothesis testers and some approach it more as theory generators, in practice most weave back and forth between the two. Furthermore, Wallace (1983) has argued that two types of research exist: pure and applied (**Figure 3.2**). In **pure research**, researchers are guided by knowledge-generating goals. They are interested in testing or generating theory

pure research research to generate knowledge

applied research research to solve a problem or create a policy

for its own sake and for the sake of advancing our knowledge in an area. In **applied research**, researchers wish to solve specific problems with the knowledge they or other researchers have generated. **Figure 3.2** illustrates the relationship between these two types of research goals and processes.

In our example of Dr. Stevens's research, we see her performing pure research. If a specific organization hired Dr. Stevens to consult with them to improve employee morale, however, her research would become applied. Theory and practice are intertwined, and pure and applied research are not unrelated processes. In

Figure 3.2, the arrows running between the two types of research show this interrelationship. Without the other, each type of research would be conducted in a vacuum (Pettey, Bracken, & Pask, 2017).

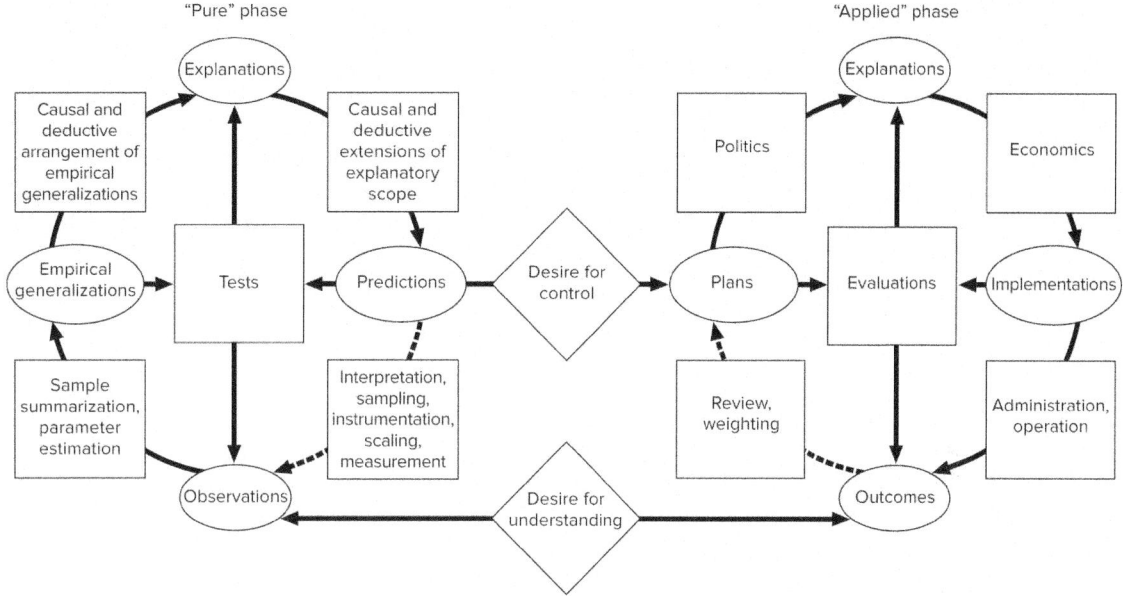

Figure 3.2 The Procedures of Scientific Analysis

Subjective (Qualitative) Communication Research

Similar to quantitative design, we could write a great deal related to qualitative communication research. In fact, an entire journal in the communication field is dedicated to this topic (*Qualitative Research Reports in Communication*). Therefore, we necessarily truncate our discussion and focus on a number of different methods that can be undertaken using this approach, knowing that you can find further information elsewhere (e.g., Lindlof & Taylor, 2019). Instead of delineating all of the numerous methods related to qualitative inquiry, we focus our attention on the primary characteristics of this type of research. As a result, you will be able to see how a particular communication theory is influenced by qualitative thinking. We address four different characteristics: context-situated, inductive, emergent, and privileged-centered.

When we say that qualitative research is context-situated, similar to what we noted in **Chapter 2**, we are saying that individuals must be studied in their naturalistic setting and accommodate the time, space, and location. Qualitative scholars do not believe that objectivity can be achieved, and in most cases, objectivity is not the goal (unlike quantitative researchers, who believe that objectivity is a worthy, if unachievable, goal). If, for instance, Marla is interested in studying roommate rivalry, she should talk to roommates inquiring about their interactions without presuming any measurement or template. Marla will need to understand conflict, taking into account when it happened, why, where, and with whom. In other words, she will have to examine conflict really listening to what her participants tell her.

Researchers using a qualitative lens often operate inductively by collecting and drawing general claims about human behavior. However, in this process, every effort is made to honor the voice of the participant, even while trying to develop some conclusions. So, let's look at Marla's research on roommates again. If she interviewed 30 people with roommates and found out that most of the sources of their conflict were related to perceptions that one person was paying more money for bills related to their shared housing than the other, she could draw conclusions to generate ideas or claims about the participants, which would lead her eventually to a theory of roommate conflict involving bill paying and perceived inequities.

A third characteristic of qualitative research is that it is emergent. When we note that research is emergent we mean that as questions are posed, they must be responsive to the setting. Again, this is an organic process in that researchers must be aware of respondent reactions and manage appropriately. So, when Marla finds that some of the interviewees want to talk about the high cost of housing in their towns rather than roommate conflict, Marla will have to encourage them to focus on the question about conflict. However, she also must simultaneously recall this response diversion as she analyzes her results, because her goal was to be open to the participants' voices and concerns.

Embedded in many qualitative research studies is the notion that participants possess privilege. This privilege-centeredness means that answers can ramble and that any "power" differences between the researcher and the participant are diminished. So, as Marla listens to roommates talk about their conflicts, she shouldn't look at her watch for time limits, she should avoid telling her participants to "stay focused on the question," and she must take into consideration all of the dimensions of their responses. In doing so, she will be encouraging the authenticity so often sought out in qualitative research.

The preceding four characteristics are among many that are considered by qualitative researchers. We hope that as you review the theories in this text, you'll see these characteristics underlying the way some of the theories are constructed.

Let's discuss a few more points relative to quantitative and qualitative research. In conducting quantitative research, one key difference between professional researchers and implicit ones rests on the definition of two terms: reliability and validity. Researchers say that something has **reliability** when you can get the same results over time. For example, if Rolanda visited the organizations in two years and her observations there yielded the same results that they do now, her observations would be called reliable. You can imagine many reasons reliability is difficult to obtain. For instance, if there had been a big turnover in personnel at one of the organizations, achieving reliability would be difficult. Professional researchers conduct statistical tests to judge reliability. Implicit scientists usually operate as though their observations are reliable without ever testing for it.

> **reliability** the stability and predictability of an observation

Reliability is important, but validity is even more critical to the empirical research process. This is the case because observations can be reliable even if they are not valid, but the opposite is not true; validity assumes reliability. To draw useful conclusions from research, observations must be both reliable and valid. **Validity** refers to the fact that the observation method actually depicts what it is supposed to. For instance, Dr. Stevens was interested in communication accommodation, so she was having Rolanda listen to workplace conversations in hallways to find it. If people engaged in extensive conversations where Rolanda could tape them, then probably the observations were able to measure the concept of interest. But, what if Rolanda taped a lot of casual greetings that didn't show much of anything important to the notion of communication accommodation? Would Dr. Stevens be correct in concluding that people do not accommodate their communication according to status? Maybe not, if a lot of accommodating was going on behind closed doors where Rolanda didn't go. In that case, the measurement (taped hallway conversations) was not valid because it didn't reflect what the researcher was interested in. Again, professional researchers are concerned about the validity of their observations and work diligently to demonstrate validity. Implicit researchers don't think too much about validity unless they somehow discover that they have been basing their generalizations on a mistaken notion.

> **validity** the truth value of an observation

In qualitative research a different standard is applied to assess the value and rigor of a study. In particular, four areas are proposed: credibility, transferability, dependability, and confirmability. The credibility criterion requires the research participant to assess the legitimacy of the findings. For instance, if members of the "Black Lives Matter" movement were interviewed for a study on discrimination, only the members participating in the study could provide credence to the research. Transferability is the extent to which a study's results can be generalized to other settings or contexts. Qualitative researchers tend to avoid using "reliability" in their findings because they often don't find it possible to measure the same thing twice with the same results. Therefore, they often contend that dependability is more appropriate; researchers need to account for the dynamism of human behavior and that it continually changes. Finally, when research can be corroborated by others, it is confirmable. Confirmable research stands up to challenges or modifications to the original study. Although there is some disagreement in the research community whether or not a different set of expectations should exist in qualitative research, there is value, nevertheless, in knowing that not all scholars think and act alike in their research.

Overall, the research process is similar for implicit and professional researchers, but professional researchers are more rigorous at every step of the process. Both draw conclusions based on their findings, and ultimately we are convinced by the arguments each advances about the strength of their process. When we believe that the results are based on good (reliable and valid) observations and careful logic, we accept the findings.

We have briefly outlined a discussion of the primary ways to understand human behavior. Regardless whether a quantitative or qualitative approach is embraced, there is room for a great deal of creativity in the research process. You will encounter this as you read and understand the theories in this book.

Conclusion

This chapter introduced the concepts of theory and research and discussed their usefulness for examining communication behaviors. We have provided an initial definition of *theory*, noting that the key components of any theory consist of concepts and the relationships among those concepts; then we explored some of the goals of theory. We discussed the frameworks for theories, or three approaches to knowing: empirical, interpretive, and critical. Each of these approaches answers questions about truth (ontology), gathering information (epistemology), and values (axiology) somewhat differently. Further, we discussed how theories can be created using three different guidelines or meta-theories: covering law, systems, and rules. We offered a list of criteria to help with evaluating different theories and offered metaphors to consider when trying to unpack the complexities of theory. We explained the research process briefly noting that some researchers follow a process that's objective while others are more subjective. Objective researchers tend to use deduction, moving from theory to observation. Subjective researchers are generally inductive, first gathering observations and then moving to theorize about those observations. Both types of researchers are interested in the rigor and truthfulness of their work. This emphasis on rigor sets professional researchers apart from everyday, or implicit, theorists. As we seek to understand communication, we turn to theory both to help us organize the information that research provides and to stimulate our thinking so we can do informed research.

With these three preliminary chapters completed, you are well on your way to understanding the foundation that is necessary to become a critical consumer of communication theory. You will find yourself reflecting back to **Chapters 1–3** continually as you review each theory and as you begin to talk about theory. We have mapped the theory territory, as previously noted, and now it's up to you to navigate this exciting journey upon which you're about to embark.

Discussion Starters

 **Case-In-Point:** Rolanda Nash has a number of different concerns as she works with Dr. Stevens. From having enough time to gather her interviews to trying to capture natural conversations, she clearly has a lot on her "to-do" list. If you were advising her, what sort of advice would you offer her as a research assistant?

Try-It-Your-Selfie: Create a question about communication and social media. What overarching theory do you think will help you answer this question? Do you have to come up with a completely new theory, or are there some existing theories that might be of help in answering your question?

1. Provide some examples of ways you think like a theorist in your daily life.

2. What is the difference between inductive and deductive logic? Provide some examples of your everyday use of both induction and deduction. When would you be likely to use induction and when deduction in answering a communication question through research?

3. How do a researcher's beliefs about the world actually affect the research process? Use at least two examples to defend your point of view.

4. Suppose you were interested in surveying your classmates about their impressions of diversity on campus. What sort of process would you undertake?

References

Anguiano, C., Milstein, T., De Larkin, I., Chen, Y., & Sandoval, J. (2012). Connecting community voices: Using a Latino/a critical race theory lens on environmental justice advocacy. *Journal of International and Intercultural Communication, 5*(2), 124–143.

Arneson, P. (2009a). Epistemology. In S. Littlejohn & K. Foss (Eds.), *Encyclopedia of communication theory* (pp. 349–352). Sage.

Arneson, P. (2009b). Ontology. In S. Littlejohn & K. Foss (Eds.), *Encyclopedia of communication theory* (pp. 695–698). Sage.

Baxter, L. A., & Braithwaite, D. O. (2008). Relational dialectics theory. In L. A. Baxter & D. O. Braithwaite (Eds.), *Engaging theories in interpersonal communication: Multiple perspectives* (pp. 349–361). Sage.

Bem, S. (1993). *The lenses of gender: Transforming the debate on sexual inequality.* Yale University Press.

Berger, C. R. (1977). The covering law perspective as a theoretical basis for the study of human communication. *Communication Quarterly, 25*(1), 7–18.

Bodie, G. D., & Jones, S. M. (2012). The nature of supportive listening II: The role of verbal person centeredness and nonverbal immediacy. *Western Journal of Communication, 76*(3), 250–269.

Bostrom, R. N. (2004). Empiricism, paradigms, and data. *Communication Monographs, 71*(3), 343–351.

Caughlin, J. P., & Wilson, S. R. (2022). Multiple goals theories: From message production to evaluation. In D. O. Braithwaite & P. Schrodt (Eds.), *Engaging theories in interpersonal communication: Multiple perspectives* (3rd ed., digital ed.). Taylor & Francis.

Craig, R. T. (2013). Constructing theories in communication research. In P. Cobley & P. J. Shulz (Eds.), *Theories and models of communication* (pp. 39–58). Walter De Gruyter.

Cushman, D. P., & Pearce, W. B. (1977). Generality and necessity in three types of human communication theory: Special attention to rules theory. In B. Ruben (Ed.), *Communication yearbook 1* (pp. 173–182). Transaction Books.

Darder, A. (2019). Decolonizing interpretive research. In A. Darder (Ed.), *Decolonizing interpretive research: A subaltern methodology for social change* (pp. 3-36). Routledge.

Doherty, W. J., Boss, P. G., LaRossa, R., Schumm, W. R., & Steinmetz, S. K. (1993). Family theory and methods. In P. G. Boss, W. J. Doherty, R. LaRossa, W. R. Schumm, & S. K. Steinmetz (Eds.), *Sourcebook of family theories and methods: A contextual approach* (pp. 325-352). Plenum.

Dray, W. (1957). *Laws and explanation in history*. Oxford University Press.

Gunaratne, S. A. (2010). De-westernizing communication/social science research: Opportunities and limitations. *Media, Culture & Society, 32*(3), 300-473.

Hall, S. (1981). The whites of their eyes: Racist ideologies and the media. In G. Bridges & R. Brunt (Eds.), *Silver linings: Some strategies for the eighties* (pp. 28-52). Lawrence and Wishart.

Harding, S. (1991). *Whose science, whose knowledge? Thinking from women's lives*. Cornell University Press.

Heider, F. (1958). *The psychology of interpersonal relations*. Wiley.

Hofkirchner, W. (2019). Social relations: Building on Ludwig von Bertalanffy. *Systems Research and Behavioral Science, 36*(3), 263-273.

Kazeroony, H. H., & du Plessis, Y. (2019). *Diversity and inclusion: A research proposal framework*. Routledge.

Lindlof, T. R., & Taylor, B. C. (2019). *Qualitative communication research methods*. Sage.

Littlejohn, S. W., Foss, K. A., & Oetzel, J. G. (2021). *Theories of human communication* (12th ed.). Waveland Press.

Lull, J. (1982). How families select television programs: A mass-observational study. *Journal of Broadcasting, 26*(4), 801-811.

Merrigan, G., & Huston, C. L. (2019). *Communication research methods* (4th ed.). Oxford University.

Miike, Y. (2019). Intercultural communication ethics: An Asiacentric perspective. *The Journal of International Communication, 25*(2), 159-192.

Monge, P. R. (1973). Theory construction in the study of communication: The system paradigm. *Journal of Communication, 23*, 5-16.

Muller, H. L., & Craig, R. T. (2007). Introduction. In R. T. Craig & H. L. Muller (Eds.), *Theorizing communication: Readings across traditions* (pp. ix-xviii). Sage.

Nakayama, T. K., & Krizek, R. L. (1995). Whiteness: A strategic rhetoric. *Quarterly Journal of Speech, 81*(3), 291-309.

Ono, K. A. (2011). Critical: A finer edge. *Communication & Critical/Cultural Studies, 8*(1), 93-96.

Pettey, G. R., Bracken, C. C., & Pask, E. B. (2017). *Communication research methodology: A strategic approach to applied research*. Routledge.

Popper, K. (1976). The myth of the framework. *Boston Studies in the Philosophy of Science, 245*, 35-62.

Reynolds, P. D. (2016). *A primer in theory construction*. Routledge.

Satir, V. (1988). *The new peoplemaking*. Science and Behavior Books.

Shimanoff, S. (1980). *Communication rules: Theory and research*. Sage.

Stacks, D. W., & Salwen, M. B. (2014). Integrating theory and research: Starting with questions. In D. W. Stacks & M. B. Salwen (Eds.), *An integrated approach to communication theory* (pp. 1-12). Sage.

Treadwell, D. T. (2016). *Introducing communication research: Paths of inquiry*. Sage.

Turner, J. H. (1986). *The structure of sociological theory,* 4th ed. Dorsey Press.

Van den Bulck, J., Custers, K., & Nelisson, S. (2016). The child-effect in the new media environment: Challenges and opportunities for communication research. *Journal of Children and Media, 44*(1), 30-38.

Von Bertalanffy, L. (1968). *General system theory: Foundations, development*. George Braziller.

Wallace, W. L. (1983). *Principles of scientific sociology*. Aldine.

Watzlawick, P., Beavin, J. B., & Jackson, D. D. (1967). *Pragmatics of human communication: A study of interactional patterns, pathologies, and paradoxes*. W. W. Norton.

Whitchurch, G. G., & Constantine, L. L. (1993). Systems theory. In P. G. Boss, W. J. Doherty, R. LaRossa, W. R. Schumm, & S. K. Steinmetz (Eds.), *Sourcebook of family theories and methods: A contextual approach* (pp. 325–352). Plenum.

White, J. M., Martin, T., Adamsons, K. (2019). *Family theories*. Sage.

Wolf, M. A., Meyer, T. P., & White, C. (1982). A rules-based study of television's role in the construction of social reality. *Journal of Broadcasting, 26*(4), 813–829.

Wrench, J., Thomas-Maddox, C., Richmond, V., & McCroskey, J. (2015). *Quantitative research methods for communication: A hands-on approach*. Oxford University.

Yerby, J. (1995). Family systems theory reconsidered: Integrating social construction theory and dialectical process. *Communication Theory, 5*(4), 339–365.

Zhou, S., & Sloan, W. D. (2015). *Research methods in communication*. Vision Press.

Organization of "Introducing Communication Theory: Analysis and Application"

Based on feedback from the way we organized the previous edition, we continue this unique approach to provide you a clear application of the templates we introduced in **Chapters 2** and **3.** We therefore organize the communication theories according to Tradition, Context, and Approach to Knowing. At the end of each chapter, we work to further your applied understanding by providing a visual snapshot of each category. The following categorization depicts the entire book and how each theory falls under a particular category. We think this approach makes for a more relevant and compelling book architecture based on the foundations we've given you in the first three chapters.

Empirical/Post-Positivist Theories

Intrapersonal: The Self and Messages
Expectancy Violations Theory

Interpersonal: Relationship Development
Uncertainty Reduction Theory
Social Exchange Theory
Social Penetration Theory
Social Information Processing Theory

Groups, Teams, and Organizations
Structuration Theory
Organizational Information Theory

The Media
Agenda Setting Theory
Spiral of Silence Theory
Uses and Gratifications Theory

Culture and Diversity
Face-Negotiation Theory

Interpretive Theories

Intrapersonal: The Self and Messages
Symbolic Interaction Theory
Coordinated Management of Meaning

Interpersonal: Relationship Development
Communication Privacy Management Theory

Groups, Teams, and Organizations
Groupthink
Organizational Culture Theory

The Public

The *Rhetoric*

Dramatism

The Narrative Paradigm

The Media

Media Ecology Theory

Culture and Diversity

Communication Accommodation Theory

Critical Theories

Interpersonal: Relationship Development

Relational Dialectics Theory

The Media

Cultivation Theory

Cultural Studies

Culture and Diversity

Muted Group Theory

Feminist Standpoint Theory

Co-Cultural Theory

CHAPTER 4
Expectancy Violations Theory

*Based on the research of **Judee Burgoon***

"One of the things that always intrigued me in communication was looking at things that are counter-intuitive; things that challenge the basic truisms that everybody holds and that we should all buy into."

—Judee Burgoon

Margie Russo

As she prepared for her interview with the prestigious Kane Polling Agency, Margie Russo felt confident that she would be able to handle any questions posed to her. As a 44-year-old mother of three young children, she felt that her life experiences alone would help her respond to difficult questions. She was a Girl Scout leader, served as treasurer of the Parent Teacher Organization at the middle school, and worked part time as an executive assistant. She knew these experiences would be invaluable as she answered questions in her interview.

Despite her confidence, Margie suddenly felt anxious about her interview with Alyssa Mueller, the polling company's human resources representative. When told by the office assistant that Ms. Mueller was ready to see her, Margie approached Alyssa's office, knocked on her door, and went into the room. When she was still more than 10 feet away from the big desk, Alyssa looked up and asked, "You're Ms. Russo?" Margie responded, "I am." Alyssa replied, "Well, please ... c'mon over here and sit down and let's chat a bit."

As Margie approached her interviewer, an uneasy feeling fell over her, a nervousness that she had never experienced before and had certainly not expected. Alyssa could sense Margie's anxiety and asked if she could get her some coffee or tea. "No, thank you," said Margie. "Well, why don't you sit down?" asked Alyssa.

Margie really wanted this job. She had been preparing for the interview with her husband, who asked her a number of different kinds of questions the night before. She didn't want to lose her chance at getting this job.

As the two sat and discussed the job and its responsibilities, Margie's mind began to wander. Why was she so nervous? She had been around people, and she knew that she had expertise for the job. Yet Margie was very nervous, and she had butterflies in her stomach.

Alyssa focused on the duties for which Margie would be responsible and to whom she would report. As she spoke, Alyssa walked around her office a bit, at times leaning on the side of her desk in front of Margie's chair. Alyssa had a number of different questions remaining but wanted Margie to speak. She asked her whether she had seen a good movie recently. "Oh, sorry, Alyssa," *Margie* replied. "I just don't have time for movies."

"I guess I should have figured that out," said Alyssa. "You really are a busy person. I'm very impressed by how you seem to manage so many things at one time. Your children are very lucky. I'm sure finding free time is pretty much impossible as I know myself!"

"Oh, I'm fine, thanks. I do get some free time, but I try to spend as much time as I can with my children." Margie was feeling more relaxed as she began to talk about how busy she was helping her two daughters sell Girl Scout cookies. She then talked about her ability to juggle several things at once.

Alyssa responded, sitting next to the job candidate: "That's great! Let's talk some more about how you handle deadlines."

It was apparent that as the two talked, Margie became more comfortable speaking to Alyssa.

Eventually, she dismissed her nervousness. She observed Alyssa's eye contact and her forward body positioning as she and her interviewer continued talking. And, while she wasn't prepared to say that she would be employed by Kane, Margie felt that her connection with Alyssa was one of the better conversations she had had in her job journey.

An important part of any discussion of communication is the role of nonverbal communication. While we tend to focus on the importance of what a person actually says, we can simultaneously be affected by what is not said. In fact, what we do in a conversation (or how we say something) can be more important than what we actually say (Burgoon, Manusov, & Guerrero, 2022; Palczewski, DeFrancisco, & McGeough, 2022). To understand nonverbal communication and its effects on messages in a conversation, Judee Burgoon conceptualized Expectancy Violations Theory (Burgoon, 1978; 2016). Since that time, Burgoon and a number of her associates have studied various messages and the influence of nonverbal communication on message production. Burgoon (1994) discusses the intersection of nonverbal communication and message production when she states that "nonverbal cues are an inherent and essential part of message creation (production) and interpretation (processing)" (p. 239). The theory was originally called the Nonverbal Expectancy Violations Theory, but Burgoon later dropped the word *nonverbal* because the theory now examines issues beyond the domain of nonverbal communication, something that we will explore a bit later in the chapter. Nonetheless, from its early beginnings in the late 1970s, EVT has been a leading theory in identifying the influence of nonverbal communication on behavior and is a theory that addresses how people respond to unexpected communication (Smith, 2021).

Our chapter-opening story of Margie Russo and Alyssa Mueller represents the essence of the theory. Margie entered the conversation with her interviewer with a sense of trepidation, and once their brief interaction was under way, she began to feel uneasy about the manner in which the space between them changed. Alyssa moved closer to Margie during the interview, prompting an awkwardness in Margie. However, once the conversation centered on Margie's children, she did not view Alyssa nor her closeness as a threat to her confidence. She considered Alyssa's forward body positioning and her ongoing eye contact, too, evidencing a connection between the two.

EVT suggests that people hold expectations about the nonverbal behavior of others. Burgoon contends that unexpected changes in conversational distance between communicators are arousing and frequently ambiguous. Interpreting the meaning behind an expectancy violation depends on how favorably the "violator" is perceived. Returning to our opening scenario, in many interviews, the interviewer is not typically expected to lean on a desk in front of the job candidate or walk from behind a desk to sit next to a job candidate. When this occurred, Margie initially became uncomfortable. It was only after Alyssa began to talk about the movies that Margie began to feel more at ease. In other words, she started to view Alyssa in a more favorable light.

In our discussion thus far, we've been using examples from nonverbal communication, primarily distance. Burgoon's (1978) early writing on EVT integrated specific instances of nonverbal communication, namely, personal space and people's expectations of conversational distance. When the theory was first conceptualized, space was a core concept of the theory. Although the theory has expanded beyond personal space, it is important to provide you an understanding of Burgoon's original thinking before you can understand the expansion of the theory. Because spatial violations constitute a primary feature of the theory, let's discuss the importance of various spatial distances before we delve further.

Space Relations

The study of a person's use of space is called **proxemics.** Proxemics includes the way people use space in their conversations as well as perceptions of another's use of space. Many people take spatial relations between communicators for granted, yet, as we have concluded elsewhere (West & Turner, 2023), people's use of space can seriously affect their ability to achieve desired goals. Spatial use can influence meaning and message, and people's spaces have intrigued researchers for some time.

proxemics study of a person's use of space

Theory Into Practice • Expectancy Violations Theory

Theoretical Claim: Understanding various proxemic differences helps us to manage our expectations of another's behavior.

Practical Implication: Jessie and Carla are roommates, but each was born and raised in a different country (Jessie was born in Venezuela and Carla was born in Canada). Living in a small dorm room, the two found out early on that they would have a hard time finding private "space." Yet, when each goes out, the two have different experiences and expectations. Jessie generally doesn't mind when people stand close to her to talk; Carla does. Even when talking to their family, Carla wants to keep a distance that is maintained with friends or classmates. Jessie has no expectation of distance, except for those whom she first meets and especially guys she may be interested in dating.

Burgoon (1978) starts from the premise that humans have two competing needs: affiliation and personal space. **Personal space,** according to Burgoon, can be defined as "an invisible, variable volume of space surrounding an individual which defines that individual's preferred distance from others" (p. 130). Burgoon and other Expectancy Violations writers believe that people simultaneously desire to stay in close proximity to others, but also desire some distance. This is a perplexing but realistic dilemma for most of us. Few people can exist in isolation, and yet people prefer their privacy at times.

personal space individual's variable use of space and distance

Proxemic Zones

Burgoon's EVT has been informed by the pioneering and classic work of anthropologist Edward Hall (1966, 1992). After studying North Americans (in the Northeast), Hall claimed that four proxemic zones exist—intimate, personal, social, and public—and each zone is used for different reasons. Hall includes ranges of spatial distance and the behaviors that are appropriate for each zone. Further, as you think about each, keep in mind that this research is based on Western ways of maintaining space. As our text underscores throughout, culture influences nearly every communication encounter, a point we return to in this chapter. We highlight the proxemic zones in **Figure 4.1.**

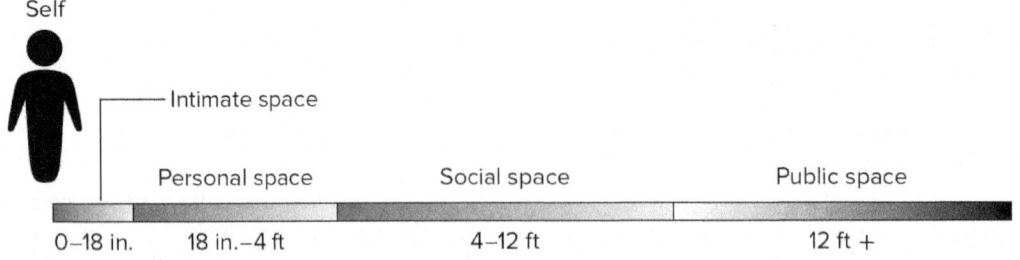

Figure 4.1 Proxemic Zones

Intimate Distance   This zone includes behaviors that exist in a range encompassing 0–18 inches. Hall (1966) notes that this includes behaviors that range from touch (for instance, making love) to being able to observe a person's facial characteristics. Whispers, for instance, carried out in this **intimate distance** range have the ability

> **intimate distance** very close spatial zone spanning 0–18 inches, usually reserved for those with whom we share personal feelings

to become influential in the communication process. Hall believes it's interesting that when U.S. citizens find themselves in intimate surroundings but are not with intimate partners, they often attempt to create a nonintimate experience. Consider what happens in an elevator. People usually fix their eyes on the ceiling, the buttons, or the door as the elevator passes floor after floor. People generally keep their hands at their side or grasp some object, usually their cell phone. Hall finds it both perplexing and amusing that many people expend so much energy extracting themselves from intimate distances. Margie Russo in our opening story seems to be uncomfortable with the intimate distance created by Alyssa. If Margie weren't in an interview, would she remove herself from the situation? In this circumstance, does the context of an interview impact her perceptions of personal space expectation? Further, it's important to point out that some invasions of personal space may be construed as sexual harassment, regardless of the intent. For this reason, we need to remain sensitive to the various perceptions of intimate distance.

Personal Distance   This zone includes those behaviors that exist in an area ranging from 18 inches to 4 feet. According to Hall (1966), **personal distance** encompasses being as close as holding another's hand to keeping someone at arm's length. You may find that most, if

> **personal distance** spatial zone of 18 inches to 4 feet, reserved for family and friends

not all, of the intimate relationships you have are within the closest point of the personal distance zone. Personal distance is likely to be used for your family and friends. The farthest point—4 feet—is usually reserved for less personal relationships, such as retail clerks. Hall indicates that in the personal distance zone, the voice is usually moderate, body heat is detectable, and breath and body odor may be perceptible.

Social Distance   With a proxemic range spanning 4–12 feet, the **social distance** category characterizes many conversations in U.S. culture, for instance, between and among coworkers. Hall (1966) contends that the closer social distance is usually reserved for those in a casual social setting, for example, a cocktail party or a dorm

> **social distance** spatial zone of 4–12 feet, reserved for more formal relationships such as those with coworkers

floor meeting. Although the distance seems a bit far, Hall reminds us that we are able to perceive skin and hair texture in the close phase of this category. The far phase is associated with individuals who have to speak louder than those in the close phase. In addition, the far phase can be considered to be more formal than the close phase. The far phase of social distance allows people to carry on simultaneous tasks. For instance, receptionists are able to carry on with their work while they converse with approaching strangers. It is possible, therefore, to monitor another person while completing a task.

Public Distance   The range encompassing 12 feet and beyond is considered to be **public distance.** The close phase of public distance is reserved for fairly formal discussions, for instance, in-class discussions between teachers and students. Public figures usually are at the far phase (around 25 feet or more). As you may have determined, it is difficult to read facial reactions at this point, unless media enhancements (for instance, large-screen projection) are used in the presentation. Whereas the close phase characterizes teachers in a classroom, the far phase includes teachers in a lecture hall. Also, actors use public distance in their performances. Consequently, their actions and words are exaggerated. Teachers and actors, however, are just two of the many types of people who use public distance in their lives.

> **public distance** spatial zone of 12 feet and beyond, reserved for very formal discussions such as between professor and students in class

Territoriality

Before we close our discussion of personal space, we explore an additional feature: **territoriality,** or a person's ownership of an area or object. Frequently, we lay claim to various spatial areas that we want to protect or defend. People decide that they want to erect fences, put on nameplates, or designate spaces as their own (e.g., Nev's room, Mom's car, etc.). Three types of territories exist: primary, secondary, and public (Altman, 1975; Lyman, 1990). **Primary territories** signal an individual's exclusive domain. For instance, one's own workshop or computers are primary territories. In fact, many people put their names on their primary territories to further signify ownership. **Secondary territories** signal some sort of personal connection to an area or object. Secondary territories are not exclusive to an individual, but the individual feels some sort of association to the territory. For instance, many graduate students feel that a campus library is their secondary territory; they don't own the building, but they frequently occupy a space in the building. **Public territories** involve no personal affiliations and include those areas that are open to all people—for example, beaches, parks, movie theaters, and public transportation areas.

> **territoriality** person's ownership of an area or object
>
> **primary territories** signal a person's exclusive domain over an area or object
>
> **secondary territories** locations that signal a person's affiliation with an area or object
>
> **public territories** locations that signal open spaces for everyone, including beaches and parks

Territoriality is frequently accompanied by prevention and reaction (Burgoon et al., 2022). That is, people may either try to prevent you from entering their territory or respond once the territory is invaded. Some gangs use territorial markers such as urban graffiti in a neighborhood to prevent other gangs from invading their turf. Knapp et al. (2014) note that if prevention does not work in defending one's territory, a person may react in some way, including getting both physically and cognitively aroused. In sum, humans typically stake out their territory in four primary ways: markers (marking our spot), labels (identification symbols), offensive displays (demonstrating aggressive looks and behaviors), and tenure (being there first and staying the longest) (Knapp et al., 2014).

Our elaborated discussion of space has relevance to EVT not only because the theory has roots in proxemics, but because it has direct application to the distances previously discussed. EVT assumes that people will react to space violations. To this end, our expectations for behavior will vary from one distance to another. That is, people have a sense of where they want others to place themselves in a conversation. For instance, consider Margie and Alyssa from our opening story. Just as Margie has expectations for Alyssa's behavior in an interview, Alyssa, too, expects Margie to behave in a predictable way. Alyssa expects Margie to maintain a comfortable distance as well. She does not expect Margie to come into the office, put her briefcase on the desk, and pull a chair up next to Alyssa. According to EVT, if Margie's behavior is unexpected and Alyssa evaluates her behavior negatively, Alyssa may become more concerned with the expectancy violation than with Margie's credentials. The proxemic zones proposed by Hall, then, are important frameworks to consider when interpreting another's behavior.

Thus far, we have introduced you to how personal space is associated with EVT. The theory has evolved over the years, with Burgoon and other EVT proponents clarifying their original findings and concepts. Although we provide her pioneering theory in detail in this chapter, we note Burgoon's most recent revisions and updates as well. To further explore the theory, we will first provide the basic assumptions of the theory and then examine a number of issues associated with the theory.

Theory at a Glance • Expectancy Violations Theory

Expectancy Violations Theory is concerned primarily with the structure of nonverbal messages. It asserts that when communicative norms are violated (per a society's rules), the violation may be perceived either favorably or unfavorably, depending on the perception the receiver has of the violator. Violating another's expectations is a communication strategy that may be used rather than conforming to another's expectations.

Assumptions of Expectancy Violations Theory

Expectancy Violations Theory (EVT) concerns how messages are presented to others and the kinds of behaviors others undertake during a conversation. According to Burgoon (2016), the theory "arose out of an effort to resolve conflicting views of proxemics in human interactions" (p. 1). In addition, three assumptions guide the theory:

- Expectancies drive human interaction.
- Expectancies for human behavior are learned.
- People make predictions about nonverbal behavior.

The first assumption states that people carry expectancies in their interactions with others. In other words, expectancies drive human interaction. **Expectancies** can be defined as the cognitions and behaviors anticipated and prescribed in a conversation with another person. Further, expectancies arise from cultural norms, a point we elaborate upon a bit later. Expectancies, therefore, necessarily include individuals' nonverbal and verbal behavior. In her early writings of EVT, Burgoon (1978) notes that people do not view others' behaviors as random; rather, they have various expectations of how others should think and behave. Reviewing the research by Burgoon and her associates, Tim Levine and his colleagues (2000) suggest that expectancies are a result of social norms, stereotypes, hearsay, and the idiosyncrasies of communicators.

expectancies thoughts and behaviors anticipated in conversations

Consider, for instance, our opening story of Margie Russo and Alyssa Mueller. If you were the interviewer, what sort of expectations would you have for the nonverbal and verbal behavior of the interviewee? Of course, we cannot ignore that not all people have the capacity to see, speak, or touch, and in these situations, the interview will take on a less traditional approach. Still, we expect an interviewer to own a specific level of confidence, a give-and-take conversational flow, and active listening skills. Interviewees would also be expected to keep a reasonable distance from the interviewer during the interview process. Many people in the United States do not want people whom they do not know to stand either too close or too far away from them. Whether it's in an interview situation or even a discussion between two people who have a prior relationship, Burgoon and other EVT scholars (e.g., White, 2021) argue that people enter interactions with a number of expectations about how a message should be delivered and how the messenger should deliver it. See **Figure 4.2** for factors that influence a person's expectations.

Individual communicator factors (*gender identity, personality, age, appearance, reputation*)

Relational factors (*prior relational history, status differences, levels of attraction and liking*)

Context factors (*formality/informality, social/task functions, environment restrictions, cultural norms*)

Expectancies

Figure 4.2   **Influences on Expectancies**

Students Talking: MJ

I can't begin to tell you enough how hard it is these days to be non-binary. I'm okay in college with it, but in my family they don't get it. I think they expect me to act a certain way because I was born a male. But, I don't like categories and I can't stand that people will talk to me in a way that assumes I identify as a guy. I think some people change their messages with me once I tell them that I'm non-binary.

Judee Burgoon and Jerold Hale (1988) contend that two types of expectations exist: pre-interactional and interactional. **Pre-interactional expectations** include the types of interactional knowledge and skills the communicator possesses *before* they enter a conversation. People do not always understand what it takes to enter and maintain a conversation. Some conversationalists may be very argumentative, for example, and others may be extremely passive. Most people do not expect such extreme behavior in their dialogues with others. **Interactional expectations** pertain to an individual's ability to carry out the interaction itself. Most people expect others to maintain appropriate conversational distance. In addition, in communicating with others, listening behaviors such as prolonged eye contact are frequently expected. These and a host of other behaviors are important to consider when examining the role of expectations before and during an interaction.

pre-interactional expectations the knowledge or skills a communicator brings to an interaction

interactional expectations an individual's ability to carry out the interaction

Of course, as we continue to note, depending on the cultural background of communicators, these behaviors can vary tremendously from one person to another. In addition, as alluded to earlier, whether our expectations are met will usually be influenced by the culture in which we live and by whether we have internalized cultural patterns for conversation expectations.

This leads us to our second assumption of EVT—that people learn their expectations from both the culture at large and the individuals in that culture. For instance, the U.S. culture teaches that a physician–patient relationship should be underscored by professional respect for each other. Although not explicitly stated in most doctor's offices, physicians typically have higher social status than their patients. For instance, we expect our doctors to be knowledgeable about health-related issues, to be able to communicate matters clearly, and to be available if patients are concerned or confused about a topic. The physician–patient relationship is just one example of a co-culture teaching its citizens that expectations exist in a particular relationship. A number of societal institutions (the family, the media, business and industry, and so forth) are central in prescribing what cultural patterns to follow. These at-large cultural prescriptions ultimately may be followed by individuals in conversation with each other.

Individuals within a culture are also influential in communicating expectations. Burgoon and Hale (1988) remark that differences based on our prior knowledge of others, our relational history with them, and our observations are important to consider. For instance, Alyssa Mueller's past experiences with prospective employees influence how she perceives an interaction and her expectations of job candidates in an interview (relational history). In addition, expectations result from our observations. In one family, for instance, standing very close to one another is a family norm, and yet this norm is not shared in other families. Interesting scenarios occur in conversations between individuals with different norms; (cultural) expectations for conversational distance vary and may influence perceptions of the interaction or have other consequences.

The third assumption pertains to the predictions people make about nonverbal communication. Later in this chapter we note that EVT theorists have applied the notion of expectancies to verbal behavior. Nonetheless, the original statement of EVT related specifically to nonverbal behavior. To this end, it's important to point out a belief inherent in the theory: people make predictions about another's nonverbal behavior.

In later writings of EVT, Judee Burgoon and Joseph Walther (1990) expanded the original understanding of EVT via personal space to other areas of nonverbal communication, including touch and posture. They suggest that the attractiveness of another influences the evaluation of expectancies. In conversations, people do not simply attend to what another is saying. As you will learn in this chapter, nonverbal behavior affects the conversation, and this behavior prompts others to make predictions.

Let's use an example to explain this third assumption a bit further. Suppose someone whom you feel is attractive starts to make direct eye contact with you at a café. At first, you may feel a bit odd with the prolonged stare. But because you are attracted to the person, that initial awkwardness may fade into comfort. Then you may begin to surmise that the person is interested in you because you see a decrease in physical distance between the two of you. This example illustrates the fact that you were making predictions (e.g., the person is attracted to you) based on their nonverbal behavior (e.g., eye contact and personal space). Before you begin to believe your own projection of attractiveness, however, keep in mind that your reaction may be either misguided or simply wrong. Despite your level of confidence, nonverbal communication is frequently ambiguous and is open to multiple interpretations (Palczewski et al., 2022).

With these assumptions in mind, let's now turn our attention to additional concepts and features of EVT: arousal, threat threshold, violation valence, and communicator reward valence. These four areas are critical to understand and to fully capture the essence of the theory. Further, the four, independently, have garnered considerable scholarly attention.

Theory-Into-Action

Nearly all of us can recall the challenges related to personal space during the COVID-19 pandemic. We experienced some people bumping into us, some strategically moving away from us, and still others believing that personal space had nothing to do with the virus transmission. Each of these spatial encounters underscores various impressions of the personal space that we and others expect—even in public venues. During the pandemic, people, at times, flinched when others touched them (i.e., zero personal space distance) and at other times, some were confused by why others were standing so far away. Nearly each of these unforeseen and unexpected encounters suggests one thing: Our personal space remains a valuable commodity and we have unwritten rules for distance related to our comfort, safety, and protection.

Arousal

Burgoon originally felt that deviations from expectations have consequences. These deviations, or violations, have what is called "arousal value" (Burgoon, 1978, p. 133). By this she means that when a person's expectations are violated, the person's interest or attention

arousal increased interest or attention when deviations from expectations occur

is aroused, and they use a particular mechanism to cope with the violation. When **arousal** occurs, one's interest in or attention to the deviation increases and one pays less attention to the message and more attention to the source of the arousal (Bachman & Guerrero, 2006; Le Poire & Burgoon, 1996). Burgoon and Hale (1988) later termed this "mental alertness" or an "orienting response," in which attention is diverted toward the source of the deviation. Norman Wong (2018) suggests that people focus on the relational meaning of the messages communicated by another.

A person may be both cognitively and physically aroused. **Cognitive arousal** is an alertness or an orientation to a violation. When we are cognitively aroused, our intuitive senses become heightened. **Physical arousal** includes those behaviors that a communicator employs during an interaction—such as moving out of uncomfortable speaking distances, adjusting one's stance during an interaction, and so forth. Most EVT studies have investigated cognitive arousal (via self-report

cognitive arousal mental awareness of deviations from expectations

physical arousal bodily changes as a result of deviations from expectations

surveys), yet little research has examined physiological arousal. One interesting study that examined physical arousal in conversation was undertaken by Beth Le Poire and Judee Burgoon (1996). Specifically, they asked college students to engage in a practice medical interview. During the interaction, the researchers studied heart rate, skin temperature, and pulse volume changes every five seconds while they assessed expectancy violations. Only heart rate and pulse volume demonstrated any statistical significance. Results indicated that after subjects registered cognitive arousal to a violation, they first experienced heart rate decreases and pulse volume increases. This was followed by pulse volume decreases. In sum, people notice when others are not adhering to interaction expectations. Arousal remains a complicated but important part of EVT. As you can see, arousal is more than simply recognizing when someone commits a violation.

Threat Threshold

Once arousal exists, threats may occur. A second key concept associated with EVT is **threat threshold,** which Burgoon (1978) defines as the "distance at which an interactant experiences physical and

threat threshold tolerance for distance violations

physiological discomfort by the presence of another" (p. 130). In a sense, the threat threshold is a tolerance for distance violation. Burgoon maintains that "when distance is equated with threat, closer distances are perceived as more threatening and farther distances as less threatening" (p. 134). Distance is interpreted as a (statement) of threat from a communicator.

People may either reward or punish a threat. Burgoon arrives at this conclusion by consulting the research on liking and attraction. This research suggests that closer distances are reserved for people we like or to whom we are attracted. Some people don't mind when others stand close to them; their threat threshold, therefore, is high. Others become very uncomfortable around those who stand too close; for them, the threat threshold is low. So, for instance, if you are attracted to a person you see each morning at Dunkin' Donuts, your threat threshold will likely be high as they talk to you and come closer to you as your conversation progresses.

During this same interaction, however, you may discover that this is not the sort of person you want to hang out with, and you may find your threat threshold getting smaller. Burgoon notes that the size of the threshold is based on how we view the initiator of the threat. Once a violation occurs, however, we again interpret the violation, although Burgoon later decided that the threat threshold is not necessarily associated with the other communicator.

Violation Valence

Throughout this chapter, we have emphasized that when people speak to others, they have expectations. Many of these expectations are based on social norms of the other person. When expectations are violated, however, many people evaluate the violation on a valence. **Violation valence** refers to the positive or negative assessment of an unexpected behavior. Violation valence focuses on the deviation of an expectation.

Violation valence requires making sense of a violation through interpretation and evaluation (Burgoon & Hale, 1988). Quite simply, communicators try to interpret the meaning of a violation and decide whether they like it. If, for instance, a professor is speaking very close

> **violation valence** perceived negative or positive assessment of an unexpected behavior

to you, you may interpret the behavior as an expression of superiority or intimidation. Consequently, the violation valence would be negative. Or you may view the violation as something positive; you might think the professor is demonstrating a sense of connection. Your violation valence, then, would be positive. Most of the research in the area of violations suggests that violations are likely to have a negative impact on close relationships (e.g., Cohen, 2010).

To better understand the violation valence, consider two situations between coworkers Noland and Rick. Standing in the break room, Noland begins to talk about his phone call to his wife this morning. As he discusses his conversation about where they decided to go for vacation, Noland begins to close the distance between him and Rick. Rick feels very uncomfortable with the distance in that his expectations for spatial distance between coworkers is violated. In other words, Rick is negatively aroused by Noland's distance behavior. A different situation, however, might prompt a different reaction. Imagine that Noland corners Rick to tell him that he heard that the company was laying off 20 percent of its workers within a few months. Because Rick was recently hired by the company, he might be positively aroused and allow Noland to violate his personal space. Most likely, he will positively evaluate Noland and allow the violation to take place.

It may be odd for you to think that violations can be viewed positively. Yet there are many such examples. For instance, let's think about the context of our chapter-opening story: the job interview. It's clear that the job candidate who is able to convince the interviewer that they are the most qualified is usually the person who gets the job. Most job interviews are very structured and have an agreed-on informal process.

Most job candidates follow the interview script and do not violate anyone's expectations. At times, though, candidates do not follow the script; they violate expectations. Although some interviewers may think these candidates are too independent, others (say, in startups and tech companies) may see them as creative, bold, and original. These may be the qualities that ultimately get the person the job. Thus, the violation was not expected in an interview, yet resulted in a favorable impression.

Communicator Reward Valence

What happens when our expectations are not met in a conversation with another? Burgoon believes that when people depart, or *deviate*, from expectations, how that deviation is received depends on the reward potential of others. Let's explain this a bit further. Burgoon, along with Deborah Coker and Ray Coker (1986), notes that not all violations of expected behavior necessarily yield neg-

communicator reward valence the sum of the positive and negative characteristics of a person and the potential for them to carry out rewards or punishments

ative perceptions. Specifically, the researchers offer the following: "In cases where behaviors are ambiguous or have multiple interpretations, acts committed by a high-reward communicator may be assigned positive meanings, and the same acts committed by a low-reward communicator may be assigned negative meanings" (p. 498). Communicators can offer each other a number of rewards, including smiles, head nods, physical attractiveness, attitude similarity, socioeconomic status, credibility, and competence. In our opening story, Alyssa Mueller's demeanor in asking about Margie's children was apparently viewed as reward behavior because Margie's nervousness eventually subsided. Burgoon thinks people have the potential to either reward or punish in conversations and maintains that people bring both positive and negative characteristics to an interaction. She terms this **communicator reward valence**.

Burgoon holds that the concept of reward includes a number of characteristics that allow a person to be viewed favorably or unfavorably. She states that "highly regarded communicators, such as those with high status, reputed expertise, purchasing power, physical attractiveness, similarity to partner, or who give positive feedback, have more favorable meanings ascribed to their nonverbal behavior than those with lower reward valence regardless of actions" (2015, p. 8). According to EVT, interpretations of violations depend on the communicator and their value. So, for instance, Margie Russo may not view Alyssa's close proximity as a positive deviation from expected behavior in an interview. Yet Alyssa's behavior was more positively received because of other characteristics, namely, her courteous manner and interest in Margie's children.

Let's apply this idea to how eye contact functions in a number of different contexts. In the United States and other cultures, a prolonged stare from a person on public transportation is probably not going to be received favorably, but it may be received favorably from one's romantic partner. If a keynote speaker at a dinner banquet looked above the listeners' heads, many people would be bewildered by this lack of eye contact. But when strangers pass on the street, lack of eye contact seems to be expected. Or, think about your response to receiving a constant stare from your supervisor or a coworker. As we infer here, cultural differences can influence things. A wife who avoids eye contact while telling her husband that she loves him may elicit a different evaluation than if she had direct eye contact, but this interpretation varies across cultural groups. Some (e.g., Irish Americans) would expect another to look directly at them when saying something very personal, such as "I love you." Others in several Eastern and Caribbean cultures, however, may find direct eye contact inappropriate or uncomfortable. In each of these contexts, violations of expected eye behavior may be interpreted differently according to how we receive the communicator.

Students Talking Tech: Brooke

There are times when I find myself getting into arguments on social media with people I would never view favorably. I mean, most of these people are way too ignorant of what is going on in the world. But, every now and then, I think that the "rewards" that someone has would make me comfortable for that person to violate my expectations. I'm thinking about someone who has the same political stripes as me or someone who is not obsessed with looking perfect or someone who loves their family.... These things are really important to me, and if I was in front of someone with these qualities, I'd be fine.

Integration and Critique

Judee Burgoon's Expectancy Violations Theory is one of the few theories specifically focused on what people expect—and their reactions to others—in conversations. Unequivocally, the research has adhered to a quantitative approach since its inception in the 1970s. The assumptions and core concepts clearly demonstrate the importance of nonverbal messages and information processing. EVT also enhances our understanding of how expectations influence conversational distance. The theory uncovers what takes place in the minds of communicators and how communicators monitor nonverbal (and verbal) behavior during their conversations. Among the criteria for evaluating a theory, four seem especially relevant for discussion: scope, utility, testability, and heurism.

Integration

Communication Tradition	Rhetorical | Semiotic | Phenomenological | Cybernetic | **Socio-Psychological** | Socio-Cultural | Critical
Communication Context	**Intrapersonal** | **Interpersonal** | Small Group/Team | Organizational | Public/Rhetorical | Mass/Media | Cultural
Approach to Knowing	**Positivistic/Empirical** | Interpretive/Hermeneutic | Critical

Critique

Evaluation Criteria	Scope | Logical Consistency | Parsimony | **Utility** | **Testability** | **Heurism** | Test of Time

Scope

At first glance, the scope of this theory may appear to be too broad; nonverbal communication is an expansive area. Yet Burgoon's theory has parameters in that she originally conceptualized one category of nonverbal communication as she articulated her theory: personal space. She has investigated and expanded her research to include other nonverbal behaviors such as eye gaze, yet her original work was clear in scope.

Utility

The practicality of EVT is apparent. Burgoon's theory presents advice on how to elicit favorable impressions and discusses the implications of space violations, a topic that affects countless conversations. In particular, researchers who employ EVT in their work have investigated topics that make a difference in the lives of people, whether related to relationship challenges (Wright & Roloff, 2015) or cell phone usage (Miller-Ott & Kelly, 2015). This sort of research suggests that it has fulfilled the criterion of utility.

Testability

Some scholars (e.g., Sparks & Greene, 1992) have criticized the clarity of concepts in Burgoon's theory, suggesting that testability may be problematic. Sparks and Greene comment that self-perceptions of arousal are not valid measures. They specifically note that Burgoon and her associates failed to establish valid indices of observers' ratings and believe "we should not accept the claim about the validity of any nonverbal index until that validity has been demonstrated" (p. 468). This intellectual debate may appear trivial to you, yet recall that arousal is a key component of EVT. Burgoon and Le Poire (1993) responded to this criticism by first claiming that Sparks and Greene did not fairly reflect the objectives of Burgoon's research. In fact, the two authored a provocative essay in which they opined about the scholarly credibility of Sparks and Greene and their criticism (i.e., *A Reply from the Heart: Who Are Sparks and Greene and Why Are They Saying All These Horrible Things?*). Additionally, Burgoon and Le Poire (1993) contend that because arousal is such a complicated and layered concept, their approach to defining arousal remains valid.

Still, several years later, even Burgoon (2015) suggests that the testability of the theory merits attention. In particular, she states that EVT lacks testing across large demographic groups and in non-Western cultures. Although she believes there are consistencies and applications across various groups, much more investigation is needed. Generally speaking, EVT is a testable theory. In **Chapter 3**, we noted that testability requires that theorists be specific in their concepts. In fact, Burgoon (1978) is one of a few theorists who clearly defines her terms; as she refined her theory, she also clarified past ambiguities. In doing so, she presents a foundation from which future researchers might continue to draw and replicate her claims.

Heurism

The scholarship related to Expectancy Violation Theory has proliferated over the decades since EVT's inception. The theory has been incorporated into a myriad of studies that span a number of diverse topic areas. For instance, EVT has been incorporated into an array of interesting and important areas, including political humor (Walther-Martin, 2015), obesity (Schyns, Roefs, Mulkens, & Jansen, 2016), men's hereditary cancer (Dean, Rauscher, Gomez, & Fischer, 2019), and birth-family contacts by adoptees (Anzur, 2023). The theory has also been introduced into studies related to social media "likes" on various X and Facebook posts (Tang, 2022), artificial intelligence chatbots (Lew & Walter, 2023), and the interplay between racial expectations and African American employees (Wayne, Sun, Kluemper, Cheung, & Ubaka, 2022).

Closing

EVT is an important theory because it offers a way to link behavior and cognitions. It is one of the few communication theories that offers us a better understanding of our need for both other people and personal space. For that, Burgoon's work continues to be critical and groundbreaking in the communication discipline.

Discussion Starters

 Case-In-Point: In addition to distance behaviors, what other nonverbal behaviors are present in interview situations like the one between Margie and Alyssa?

Try-It-Your-Selfie: Thinking about how EVT relies on nonverbal communication, discuss how expectancies on X, for instance, differ from your expectancies of your best friend when you're face-to-face.

1. Explain how EVT might inform research and thinking on touch behavior. For instance, does the theory help us to understand the difference between appropriate and inappropriate touch? Explain with examples.

2. Provide some nonverbal expectations that you have learned as a result of your cultural identity. Discuss what similarities and differences exist.

3. How do you suppose arousal manifests itself in conversations between supervisors and employees? Identify a few arousal mechanisms.

4. Suppose you want to study expectancy violation in school. How might you begin to investigate violations? Be as specific as possible, and identify some methods for studying expectations.

References

Altman, I. (1975). *The environment and social behavior: Privacy, personal space, territory, and crowding.* Brooks/Cole.

Anzur, C. K. (2023). "Do you really want to be disappointed?": Adoptees' expectations and violations during birth family contact. *Communication Quarterly, 71*(2), 175–194.

Bachman, G. F., & Guerrero, L. K. (2006). Forgiveness, apology, and communicative responses to hurtful events. *Communication Reports, 19,* 45–56.

Burgoon, J. K. (Spring 1978). A communication model of personal space violations: Explication and an initial test. *Journal of Human Communication Research, 4*(2), 129–142.

Burgoon, J. K. (1994). Nonverbal signals. In M. L. Knapp & G. R. Miller (Eds.), *Handbook of interpersonal communication* (pp. 229–285). Sage.

Burgoon, J. K. (2015). Expectancy violations theory. In C. R. Berger & M. E. Roloff (Eds.), *The international encyclopedia of interpersonal communication* (pp. 1–9). John Wiley & Sons.

Burgoon, J. K. (2016). Expectancy violations theory. In C. R. Berger & M. E. Roloff (Eds.), *The international encyclopedia of interpersonal communication* (pp. 1–9). John Wiley & Sons.

Burgoon, J. K., & Hale, J. L. (1988). Nonverbal expectancy violations: Model elaboration and application to immediacy behaviors. *Communication Monographs, 55,* 58–79.

Burgoon, J. K., & Le Poire, B. A. (1993). Effects of communication expectancies, actual communication, and expectancy disconfirmation on evaluations of communicators and their communication behavior. *Human Communication Research, 20*(1), 67–96.

Burgoon, J. K., & Walther, J. B. (1990). Nonverbal expectancies and the evaluative consequences of violations. *Human Communication Research, 17,* 232–265.

Burgoon, J. K., Coker, D. A., & Coker, R. A. (June 1986). Communicative effects of gaze behavior: A test of two contrasting explanations. *Journal of Human Communication Research, 12*(4), 495-524.

Burgoon, J. K., Manusov, V., & Guerrero, L. (2022). *Nonverbal communication.* Routledge.

Cohen, E. L. (2010). Expectancy violations in relationships with friends and media figures. *Communication Research Reports, 27*(2), 97-111.

Dean, M., Rauscher, E., Gomez, E., & Fischer, C. (2019). Expectations versus reality: The impact of men's expectancy violations in conversations with healthcare providers about BRCA-related cancer risks. *Patient Education and Counseling, 102*, 1650-1655.

Hall, E. T. (1966). *The hidden dimension.* Anchor/Doubleday.

Hall, E. T. (1992). *An anthropology of everyday life.* Doubleday/Anchor Books.

Knapp, M. L., Hall, J. A., & Horgan, T. G. (2014). *Nonverbal communication and human interaction.* Cengage/Wadsworth.

Le Poire, B. A., & Burgoon, J. K. (1996). Usefulness of differentiating arousal responses within communication theories: Orienting response or defensive arousal within nonverbal theories of expectancy violation. *Communication Monographs, 63*(3), 208-230.

Levine, T. R., Anders, L. N., Banas, J., Baum, K. L., Endo, K., Hu, A. D., & Wong, N. H. (2000). Norms, expectations, and deception: A norm violation model of veracity judgments. *Communications Monographs, 67*(2), 123-137.

Lew, Z., & Walther, J. B. (2023). Social scripts and expectancy violations: Evaluating communication with human or AI Chatbot interactants. *Media Psychology, 26*(1), 1-16.

Lyman, S. M. (1990). *Civilization: Contents, discontents, malcontents, and other essays in social theory.* University of Arkansas Press.

Miller-Ott, A., & Kelly, L. (2015). The presence of cell phones in romantic partner face-to-face interactions: An expectancy violation theory approach. *Southern Communication Journal, 80*(4), 253-270.

Palczewski, C. H., DeFrancisco, V. P., & McGeough, D. (2022). *Gender and communication: A critical introduction.* Sage.

Schyns, G., Roefs, A., Mulkens, S., & Jansen, A. (2016). Expectancy violation, reduction of food cue reactivity and less eating in the absence of hunger after one food cue exposure session for overweight and obese women. *Behaviour Research and Therapy, 76*, 57-64.

Smith, S. A. (2021). *Expectancy violations theory.* Salem Press.

Sparks, G. G., & Greene, J. O. (1992). On the validity of nonverbal indicators as measures of physiological arousal: A response to Burgoon, Kelley, Newton, and Keeley-Dyreson. *Human Communication Research, 18*, 445-471.

Tang, J. L. (2022). Are you getting likes as anticipated? Untangling the relationship between received likes, social support from friends, and mental health via expectancy violation theory. *Journal of Broadcasting & Electronic Media, 66*(2), 340-360.

Walther-Martin, W. (2015). Media-generated expectancy violations: A study of political humor, race, and source perceptions. *Western Journal of Communication, 79*, 492-507.

Wayne, S. J., Sun, J., Kluemper, D. H., Cheung, G. W., & Ubaka, A. (2022). The cost of managing impressions for Black employees: An expectancy violation theory perspective. *Journal of Applied Psychology, 108*, 208-224.

West, R., & Turner, L. H. (2023). *Interpersonal communication.* Sage.

White, C. H. (2021). Expectancy violations theory and interaction adaptation theory: From expectations to interactions. In D. Braithwaite & P. Schrodt (Eds.), *Engaging theories in interpersonal communication* (pp. 158–170). Routledge.

Wong, N. (2018). Well that was unexpected: Effect of intimacy and commitment on responses to an interpersonal expectancy violation. *Studies in Media and Communication, 6*(2), 45–56.

Wright, C. N., & Roloff, M. E. (2015). You should just know why I'm upset: Expectancy violation theory and the influence of mind reading expectations (MRE) on responses to relational problems. *Communication Research Reports, 32,* 10–19.

CHAPTER 5
Uncertainty Reduction Theory

*Based on the research of **Charles Berger** and **Richard Calabrese***

"Human relationships are fraught with uncertainty. From initial encounters between complete strangers to enduring close relationships, individuals experience uncertainties about their relationship partners as individuals and uncertainties about their relationships."

—Charles Berger

Gia Banks and Lucas Rogers

Gia Banks and Lucas Rogers take the same philosophy class at Urban University but until today, they really had not spoken together, although they saw each other in class every Monday, Wednesday, and Friday. Lucas had noticed Gia and found her attractive, and he wondered why she never spoke up in class. Gia thought Lucas was cute and had made some good comments in class discussions. Today, as Gia was leaving the classroom, she noticed Lucas staring at her from the corner of the room where he sat with his friends. Although Gia had thought about getting to know Lucas, she felt a little uncomfortable about having him stare so hard at her, and she hurried to get out of the classroom and away from his gaze.

Unfortunately, her friend Maggie stopped her at the doorway with a question about the assignment for next week, and so Gia and Lucas reached the hallway at the same time. There was an awkward pause as they smiled uncertainly at each other. Lucas cleared his throat and said, "Hi. That was a pretty interesting lecture in class today, wasn't it?" Gia shrugged, smiled back, and replied, "I'm not sure I get what's going on in there. I'm majoring in engineering, and this is just an elective for me. Sometimes I think I should have taken bowling instead." Lucas smiled and said, "I'm a communication studies major myself, and this class relates to a lot of what we've talked about in some of my other classes, so it's OK for me. But I guess I'd have the same reaction you're having if I got stuck in an engineering class! I probably couldn't engineer my way out of a paper bag." The two laughed for a minute. There was an awkward silence and finally Gia said, "Gotta run. Catch you later," and hurried off down the hall.

Lucas walked to his next class wondering if they would talk again, if Gia was putting him down, if she thought he had been rude about her major, if she liked him, if he liked her, or if he cared. Gia, for her part, was kicking herself for sounding like an idiot the first time she'd talked to Lucas. "Why in the world would I say I should have taken bowling?" she asked herself. "Lucas probably thinks I'm totally stupid!" But then she wondered why he'd been staring so intently at her before. It was a little creepy—maybe it didn't matter whether he thought she was dumb, she might not want to get to know him after all. Gia sighed to herself. It was confusing.

Sometimes called Initial Interaction Theory, Uncertainty Reduction Theory (URT) was originated by Charles Berger and Richard Calabrese in 1975. It continues to be an important theory, because, as Leanne Knobloch (2008) states, "Everyday life is infused with uncertainty" (p. 133). Berger (2016) makes a similar point by stating: "Attempting to reduce uncertainty is a pervasive and vital activity across a wide range of human endeavors" (p. 1). Berger and Calabrese's goal in constructing this theory was to explain how communication is used to reduce uncertainties between strangers engaging in their first conversation together. The theory explains that we all accomplish uncertainty reduction by activating two primary processes: proactive and retroactive. When using **proactive processes,** a person can predict the most likely actions another might take out of many alternatives. So, for example, when Lucas uses proactive processes, he thinks of all the alternative things that Gia might do after their encounter and predicts which one she's likely to actually do. **Retroactive processes** help Lucas explain Gia's behavior after she does something (for instance, if she ignores him when they see each other next, Lucas would need retroactive processes to figure out why she did that) (Evuleocha & Ugbah, 2018).

Sometimes proactive processes are called *prediction*, and retroactive processes are called *explanation*. Berger (2011) comments on URT, saying:

> The main supposition underlying the theory is that when strangers meet, they are faced with myriad uncertainties about each other's attitudes, beliefs, values and potential actions. In the service of predicting and, in some cases explaining, each other's beliefs and actions so that communicative choices can be made, individuals seek to reduce their uncertainties by acquiring information about each other (p. 215).

You can see how our opening example of Lucas and Gia illustrates Berger and Calabrese's basic contention that when people first meet, their uncertainty levels are high. Lucas doesn't know Gia, so he doesn't have a background that'll help him interpret her comments to him. Nor is he certain about what will happen the next time they see each other. Further, there could be many possible explanations for Gia's behavior. All of these factors mean that Lucas will be uncertain. Berger and Calabrese use the ideas of theorists Claude E. Shannon and Warren Weaver (1949) as a foundation for URT. Shannon and Weaver note in their information theory that uncertainty exists whenever the number of possible alternatives in a given situation is high and the likelihood of their occurrence is relatively equal. Conversely, they say, uncertainty is decreased when the alternatives are limited in number and/or there is an alternative that is usually chosen.

prediction or proactive processes the ability to forecast one's own and others' behavioral choices

explanation or retroactive processes the ability to interpret the meaning of behavioral choices

For example, when Teresa walks into her Spanish I classroom on the first day of class and the person sitting nearest the door smiles at her, Teresa has a few alternative explanations for this behavior. The person could be friendly, trying to get to know her, squinting in the sunlight, or mistaken in thinking she knows Teresa. Because a college classroom is often governed by a norm of friendliness and because the alternative explanations are few in number, Teresa will probably decide the smile was one of friendly welcoming, reducing her uncertainty fairly easily. But if Teresa walked into a job interview and found another candidate in the waiting room with her who glanced her way and smiled, the alternative explanations would be more numerous. They would contain all of the above possibilities and others, including that the person is sizing her up as competition, the person thinks she's weak competition, the person is trying to get her to let her guard down, and so forth. These increased alternatives will increase uncertainty, causing Teresa to attempt to reduce it. Berger and Calabrese theorize that communication is the vehicle by which people reduce their uncertainty about one another. In turn, reduced uncertainty creates the conditions for the development of interpersonal relationships.

After Berger and Calabrese (1975) originated their theory, it was slightly elaborated (Berger, 1979; Berger & Bradac, 1982). The current version of the theory suggests that there are two types of uncertainty in initial encounters: cognitive and behavioral. Our cognitions refer to the beliefs and attitudes that we and others hold. **Cognitive uncertainty,** therefore, refers to the degree of uncertainty associated with those beliefs and attitudes. When Lucas wonders whether Gia was ridiculing his major and whether he really cares,

cognitive uncertainty degree of uncertainty related to cognitions

behavioral uncertainty degree of uncertainty related to behaviors

self-disclosure personal messages about the self disclosed to another

he experiences cognitive uncertainty. **Behavioral uncertainty**, on the other hand, pertains to "the extent to which behavior is predictable in a given situation" (Berger & Bradac, 1982, p. 7). Because we have cultural rituals for small talk, Gia and Lucas probably have an idea of how to behave during their short conversation. If one of them had violated the ritual by either engaging in inappropriate **self-disclosure** (revealing private information about oneself to another) or totally ignoring the other, their behavioral uncertainty would have increased. People may be cognitively uncertain, behaviorally uncertain, or both before, during, or following an interaction.

In addition, Berger and Calabrese (1975) and Berger (2016) theorized that uncertainty is related to several concepts rooted in communication and relational development: verbal output, nonverbal warmth (such as pleasant vocal tone and leaning forward), information seeking (asking questions), self-disclosure, reciprocity of disclosures, similarity, and liking. URT has been described as an example of original theorizing in the field of communication (Miller, 1981) because it employs concepts (such as information seeking and self-disclosure) that are specifically relevant to studying communication behavior. URT attempts to place communication as the cornerstone of human behavior, and to this end a number of assumptions about human behavior and communication underlie the theory.

Theory at a Glance • Uncertainty Reduction Theory

When strangers meet, their primary focus is on reducing their level of uncertainty in the situation because uncertainty is uncomfortable. People use the processes of proactive uncertainty reduction (prediction) and retroactive uncertainty reduction (explanation) to reduce their discomfort. There are two main types of uncertainty that people may experience. They may be unsure of how to behave (or how the other person will behave), and they may also be unsure of what they think of the other person and what the other person thinks of them. High levels of uncertainty are related to a variety of verbal and nonverbal behaviors.

Assumptions of Uncertainty Reduction Theory

As we have mentioned in previous chapters, theories are frequently grounded in assumptions that reflect the worldview of the theorists. URT is no exception. The following assumptions frame this theory:

- People experience uncertainty in interpersonal settings, and it generates cognitive stress.
- When strangers meet, their primary concern is to reduce their uncertainty and increase predictability.
- Interpersonal communication is a developmental process that occurs through stages, and it's the primary means of uncertainty reduction.
- The quantity and nature of information that people share change through time.
- It's possible to predict people's behavior in a lawlike fashion.

We will briefly address each assumption. First, in a number of interpersonal settings, people feel uncertainty. Because differing expectations exist for different interpersonal occasions, it's reasonable to conclude that people are uncertain or even nervous about meeting others. Online dating site designers confirm that one of their main concerns is to help people reduce their uncertainties about online dating in general and their site in particular (Jung, Roh, Yang, & Biocca, 2017). In addition, let's consider the case of Lucas and Gia. Although there are a great many cues in the environment that can help Lucas and Gia make sense out of their interaction, there are complicating factors as well. For example, Lucas may have noticed Gia hurrying to leave the room. There may be several alternative explanations for this behavior, including another class that is a distance away, a general predisposition toward hurrying, having to go to the bathroom, feeling faint and wanting fresh air, wanting to avoid meeting Lucas at the door, and so forth. Given all these alternatives, it's likely that Lucas (or anyone in his situation) feels uncertain about how to interpret Gia's behavior. Further, this assumption asserts that uncertainty is an aversive state because it results in cognitive stress. As Berger and Calabrese (1975) state, "When persons are unable to make sense out of their environment, they usually become anxious" (p. 106). The theory assumes that it takes a great deal of emotional and psychological energy to remain uncertain, and people would prefer not to experience that.

The next assumption underlying URT advances the proposition that when strangers meet, two concerns are important: reducing uncertainty and increasing predictability (Solomon, 2016). URT suggests that information seeking is a primary method to reduce uncertainty and attain some sort of predictability. Information seeking usually takes the form of asking questions. Think about the last time you had an initial encounter with someone in an interpersonal setting. More than likely a great deal of time in this interaction was occupied with questions and answers (e.g., Where are you from? What is your major? Do you live on campus?, etc.) This process can be quite engaging, and many people do this unconsciously. Jessica Deyo and her colleagues (2011) found that in speed-dating contexts, participants' behaviors adhered to this assumption albeit in a speeded-up fashion. The first minute of the speed date was filled with questions and answers focused on demographic information. One study (Mongeau, Jacobsen, & Donnerstein, 2007) found that reducing uncertainty was cited as a primary goal in dating.

The third assumption of URT states both that interpersonal communication is a process involving developmental stages, and that interpersonal communication is the primary means people have for reducing uncertainty. According to Berger and Calabrese (1975), generally speaking, most people begin interactions in an **entry phase,** defined as the beginning stage of a communication encounter between strangers. The entry phase is guided by implicit and explicit rules and norms, such as responding in kind when someone says, "Hi! How are you doing?" (e.g., *"Fine, how are you?"*). Individuals then enter the second stage, called the **personal phase**, or the stage where the interactants start to communicate

entry phase the beginning stage of an interaction between strangers

personal phase the stage in a relationship when people begin to communicate more spontaneously and personally

exit phase the stage in a relationship when people decide whether to continue or leave

more spontaneously and to reveal more idiosyncratic information. The personal phase can occur during an initial encounter, but it is more likely to begin after repeated interactions. The third stage, the **exit phase**, refers to the stage during which individuals make decisions about whether they wish to continue interacting with this partner in the future. Although all people do not enter a phase in the same manner or stay in a phase for a similar amount of time, Berger and Calabrese believe that this universal framework explains how interpersonal communication is used to reduce uncertainties and to shape and reflect the development of interpersonal relationships.

The fourth assumption underscores the nature of time. It also focuses on the fact that interpersonal communication is developmental, as the third assumption states. Uncertainty reduction theorists believe that initial interactions are a key element in the developmental process. To illustrate this assumption, consider the experiences of Miranda, who spent a few anxious minutes by herself before entering the YWCA to attend her first meeting of a grief support group for recently widowed partners. She immediately felt more comfortable when Danielle came over to introduce herself and welcome her to the group. As the two exchanged information with each other, Miranda felt more confident. As they talked, Miranda reduced her uncertainties about what the other members of the support group would be like. Meeting Danielle and feeling somewhat confident about her helped Miranda feel better about the whole process.

The final assumption underlying URT indicates that people's behavior can be predicted in a lawlike fashion. Recall from **Chapter 3** that theorists have some guidelines or meta-theories to help them in the job of theory construction. One of the guidelines we reviewed was covering law, which assumes that human behavior is regulated by generalizable principles that function in a lawlike manner. Although there may be some exceptions, in general people behave in accordance with these laws. The goal of a covering law theory is to lay out the laws that will explain how we communicate. As you might imagine, covering law theorists have a difficult task. Although some aspects of the natural world may operate under laws, the social world is much more variable. That is why covering laws in the social sciences are called "lawlike." A pattern is outlined, but the deterministic notion implied with natural laws is relaxed a bit. Still, even to approach the goal of lawlike statements is daunting. Thus, theories like URT begin with what may seem like commonsense observations in order to establish regularities that govern people's behaviors. Covering law theories are constructed to move from statements that are presumed to be true (or axioms) to statements that are derived from these truisms (or theorems).

Key Concepts of Uncertainty Reduction Theory: The Axioms and Theorems

URT is a covering law theory, so it is based on **axioms**, which are statements that are presumed to be true, and *theorems* that are derived from the axioms. Each one of the theorems is identified as an area for research exploration (Berger, 2016; Berger & Calabrese, 1975). We explore the axioms and then the theorems below.

> **axioms** truisms drawn from past research and common sense

Axioms of Uncertainty Reduction Theory

Berger and Calabrese based their axiomatic approach to constructing URT on earlier researchers such as Herbert Blalock (1969), who argued that causal relationships should be stated in the form of an axiomatic truism. Each axiom presents a relationship between uncertainty (the central theoretical concept) and one other concept. URT originally posited seven axioms. To understand each, we refer back to our chapter-opening example of Gia and Lucas.

> **Axiom 1:** As the amount of verbal communication between strangers increases, the level of uncertainty for each interactant in the relationship decreases. As uncertainty is further reduced, the amount of verbal communication increases. (The verbal communication axiom.)

Regarding Lucas and Gia's situation with reference to the verbal communication axiom, the theory maintains that if they talk more to each other, they will become more certain about each other. Furthermore, this is a reciprocal relationship, so as they get to know each other better, they will talk more with each other.

Axiom 2: As nonverbal affiliative expressiveness increases, uncertainty levels decrease in an initial interaction. In addition, decreases in uncertainty level will cause increases in nonverbal affiliative expressiveness. (The nonverbal expressiveness axiom.)

If Gia and Lucas express themselves to each other in a warm nonverbal way and smile at one another, they will grow more certain of each other, and as they do this, they will increase their nonverbal affiliation with each other: They may become more facially animated, or they may engage in more prolonged eye contact. The two might even touch each other in a friendly fashion as they begin to feel more comfortable with each other.

Axiom 3: High levels of uncertainty cause increases in information-seeking behavior. As uncertainty levels decline, information-seeking behavior decreases. (The information-seeking axiom.)

This axiom, which we will discuss more later, is one of the more provocative propositions associated with URT. It suggests that Gia will ask questions and otherwise engage in information seeking as long as she feels uncertain about Lucas. The more certain she feels, the less information seeking she will do. The same would apply to Lucas.

Axiom 4: High levels of uncertainty in a relationship cause decreases in the intimacy level of communication content. Low levels of uncertainty produce high levels of intimacy. (The intimacy axiom.)

Because uncertainty is relatively high between Gia and Lucas in our chapter-opening story, they engage in small talk with no real self-disclosures. The intimacy of their communication content is low, and their uncertainty level remains high. The fourth axiom asserts that if they continue to reduce the uncertainty in their relationship, then their communication will consist of higher levels of intimacy.

Axiom 5: High levels of uncertainty produce high rates of reciprocity. Low levels of uncertainty produce low levels of reciprocity. (The reciprocity axiom.)

According to URT, as long as Gia and Lucas remain uncertain about each other, they will tend to mirror each other's behavior. For example, after Gia shares that she is lost in the class and that she is an engineering major, Lucas reveals his major to her and admits that he would probably have troubles in engineering classes. Immediate reciprocation of that sort (I tell you where I'm from and you tell me where you're from) is a hallmark of initial encounters. When people talk more with each other and develop their relationship more, they trust that reciprocity will be made at some point (if I don't tell you something that mirrors your communication today, I'll probably do so the next time we talk or the time after that). With this in mind, strict reciprocity is replaced by an overall sense of reciprocity in our relationship.

Axiom 6: Similarities between people reduce uncertainty, whereas dissimilarities increase uncertainty. (The similarity axiom.)

Because Gia and Lucas are both European American college students of roughly the same age at Urban University, they may have similarities that reduce some of their uncertainties about each other immediately. Yet they are different sexes and have different majors—dissimilarities that may contribute to their uncertainty level.

Axiom 7: Increases in uncertainty level produce decreases in liking; decreases in uncertainty produce increases in liking. (The liking axiom.)

As Gia and Lucas reduce their uncertainties, they typically will increase their liking for each other. If they continue to feel highly uncertain about each other, they probably will not like each other very much. Although axioms are presumed true and, thus, not tested directly in studies, this axiom has received some indirect empirical support. In a study examining the relationship between communication satisfaction and uncertainty reduction, James Neuliep and Erica Grohskopf (2000) found that participants playing interviewers in an organizational role play were more likely to feel positively toward the participants playing the job seekers (and more likely to hire them) when their uncertainty was low. The seven axioms and their relationships are summarized in **Table 5.1**.

Table 5.1   Axioms of Uncertainty Reduction Theory

AXIOM	MAIN CONCEPT	RELATIONSHIP	RELATED CONCEPT
1.	↑ Uncertainty	Negative	↓ Verbal Communication
2.	↑ Uncertainty	Negative	↓ Nonverbal Affiliative Expressiveness
3.	↑ Uncertainty	Positive	↑ Information Seeking
4.	↑ Uncertainty	Negative	↓ Intimacy Level of Communication
5.	↑ Uncertainty	Positive	↑ Reciprocity
6.	↓ Uncertainty	Negative	↑ Similarity
7.	↑ Uncertainty	Negative	↓ Liking

Theorems of Uncertainty Reduction Theory

Berger and Calabrese combined all seven axioms in every possible pairwise combination to derive 21 theorems (see **Table 5.2**). **Theorems** are theoretical statements that are derived from axioms. Unlike axioms, theorems are not assumed to be true; they are the-

theorems theoretical statements derived from axioms, positing a relationship between two concepts

oretical relationships. They need to be tested to see if the assumed relationship actually holds up under scrutiny. Theorems also suggest a relationship between two concepts. For instance, because the amount of verbal communication is negatively related to uncertainty (Axiom 1) and uncertainty is negatively related to intimacy levels of communication (Axiom 4), then verbal communication and intimacy levels are positively related (Theorem 3; see **Table 5.2**). You can generate the other 20 theorems by combining the axioms using the deductive formula above. You need to use the rule of multiplication for multiplying positives and negatives. For example, if two variables have a positive relationship with a third, they are expected to have a positive relationship with each other. If one variable has a positive relationship with a third, whereas the other has a negative relationship with the third, they should have a negative relationship with each other. Finally, if two variables each have a negative relationship with a third, they should have a positive relationship with each other.

Table 5.2 Theorems of Uncertainty Reduction Theory Deduced from Axioms

AXIOM	MAIN CONCEPT	RELATIONSHIP	RELATED CONCEPT
1.	↑ Verbal Communication	Positive	↑ Nonverbal Affiliative Expressiveness
2.	↑ Verbal Communication	Negative	↓ Information Seeking
3.	↑ Verbal Communication	Positive	↑ Intimacy Level of Communication
4.	↑ Verbal Communication	Negative	↓ Reciprocity
5.	↑ Verbal Communication	Positive	↑ Similarity
6.	↑ Verbal Communication	Positive	↑ Liking
7.	↑ Nonverbal Affiliative Expressiveness	Negative	↓ Information Seeking
8.	↑ Nonverbal Affiliative Expressiveness	Positive	↑ Intimacy Level of Communication
9.	↑ Nonverbal Affiliative Expressiveness	Negative	↓ Reciprocity
10.	↑ Nonverbal Affiliative Expressiveness	Positive	↑ Similarity
11.	↑ Nonverbal Affiliative Expressiveness	Positive	↑ Liking
12.	↑ Information Seeking	Negative	↓ Intimacy Level of Communication
13.	↑ Information Seeking	Positive	↑ Reciprocity
14.	↑ Information Seeking	Negative	↓ Similarity
15.	↑ Information Seeking	Negative	↓ Liking
16.	↑ Intimacy Level	Negative	↓ Reciprocity
17.	↑ Intimacy Level	Positive	↑ Similarity
18.	↑ Intimacy Level	Positive	↑ Liking
19.	↑ Reciprocity	Negative	↓ Similarity
20.	↑ Reciprocity	Negative	↓ Liking
21.	↑ Similarity	Positive	↑ Liking

Expansions of Uncertainty Reduction Theory

Many researchers have tested URT and based their studies on the tenets of the theory. Furthermore, Berger and several colleagues refined and expanded the theory, taking into account new research findings. URT has been expanded and modified in several areas, including antecedent conditions, strategies, developed relationships, social media, and context. We'll address each of these areas briefly below.

Antecedent Conditions

Berger (1979) has suggested that three antecedent (prior) conditions will activate uncertainty reduction processes. The first condition occurs when the other person has the *potential* to reward or punish. If Gia is a very popular, charismatic figure on campus, her attention may be seen as a reward by Lucas. Likewise, Lucas might experience a rejection by her as punishing. If Lucas finds out that a friend thinks Gia is boring and unattractive or if he discovers she has a bad reputation on campus, he won't see her attention as rewarding or her rejection as punishing. Thus, according to Berger, Lucas will be more motivated to reduce his uncertainty the more he thinks that Gia has the ability to reward or punish him.

A second antecedent condition is created when the other person behaves contrary to expectations. In the case of Gia and Lucas, social norms suggest that staring is impolite. When Gia perceives Lucas staring intently at her for an extended period of time, her expectations for his behavior are violated, and Berger predicts that her desire to reduce her uncertainty increases.

The third and final condition exists when a person expects future interactions with another. Lucas knows that he will continue to see Gia in class for the rest of the semester. Yet, because he has discovered that she's an engineering major, he may feel that he can avoid her in the future. In the first case, Berger would expect Lucas's desire to increase predictability to be high—he knows he'll be seeing Gia often; in the second case, Lucas's desire level is lower because Gia has a different major, and they can avoid each other after this class ends.

Strategies

A second area of expansion pertains to strategies. Berger (2016) suggests that people, in attempting to reduce uncertainty, employ tactics from three categories of strategies: passive, active, and interactive. At the core of each is the goal of "desired information" (p. 2) from those with whom we are communicating. First, there are **passive strategies**, whereby an individual assumes the role of unobtrusive observer of another. **Active strategies** exist when an observer engages in some type of effort other than direct contact to find out about another person. For instance, a person might ask a third party for information about the other. Finally, **interactive strategies** occur when the observer and the other person engage in direct contact or face-to-face interaction—that is, conversation that may include self-disclosures, direct questioning, and other information-seeking tactics. Although these strategies are critical to reducing uncertainty, Berger believes that certain behaviors, such as asking inappropriately sensitive questions, may increase rather than decrease uncertainty, and people may need additional reduction strategies in that case. Kami Kosenko (2011) found support for all these strategies in a study examining transgender adults' communication about safer sex. In her study all of these strategies were illustrated in the participants' responses.

passive strategies reducing uncertainties by unobtrusive observation

active strategies reducing uncertainties by means requiring effort that isn't direct contact

interactive strategies reducing uncertainties by engaging in conversation

To further illustrate these strategies, consider Lucas and Gia once again. The time they spend in class covertly observing each other falls into the passive category. When Lucas observes how Gia reacts to jokes the professor tells in lecture, he is utilizing a particular passive tactic called **reactivity searching**, or observing Gia doing something specific. A different passive tactic, called **disinhibition searching**, would require Lucas to observe Gia in more informal settings outside the classroom to see how she behaves when her inhibitions are down. If either one of them engages friends to find

reactivity searching a tactic in the passive strategy involving watching a person doing something specific

disinhibition searching a passive strategy involving watching a person's natural or uninhibited behavior in an informal environment

out information about the other, they'll be using an active strategy. When they speak after class, they illustrate the interactive strategy to reduce their uncertainties.

Tara Emmers and Dan Canary (1996) argue that in established relationships an additional strategy is employed. They call this strategy *uncertainty acceptance*, and it includes tactics such as simply trusting your partner. Emmers and Canary suggest that accepting or trusting your partner even when you aren't completely certain about what's happening is a viable strategy for coping with uncertainty in developed relationships. This additional strategy leads us to a discussion of the third area of expansion for URT: developed relationships.

Developed Relationships

When Berger and Calabrese (1975) first wrote their theory, they were interested in describing initial encounters between strangers. They stated a clear and narrow boundary around their theoretical insights. In the intervening years, however, the theory has been expanded to include developed relationships, as the acceptance strategy just described indicates. Berger (1982, 1987) acknowledged this expansion with updates to the theory. First, he comments that uncertainties are ongoing in relationships, and thus the process of uncertainty reduction is relevant in *developed* relationships as well as in *initial* interactions. This conclusion broadens earlier claims by Berger and Calabrese that specifically limited URT to initial encounters.

The inclusion of the three antecedent conditions discussed previously (potential for reward or punishment, deviation from expectations, and anticipation of future interactions) points us toward an examination of uncertainty in developed relationships. Specifically, we will expect rewards from, be surprised by, and anticipate future interactions with those with whom we have ongoing relationships.

Students Talking: Sonia

I enjoyed the section of the chapter that talked about expanding the theory into developed relationships. I've been working for the same boss for four years, yet many times I feel a sense of uncertainty with her like the theory talks about. You'd think I'd know what to expect from Tara after working for her for several years, but she still keeps surprising me. Last week, she completely changed her attitude, and I wasn't quite sure how to take it. For several weeks she's been telling me that we have to finish a project much more quickly than the time we usually take with our work. So I have been killing myself to get it done. Then last week she said we could slow down, and she didn't explain herself at all. I immediately used the active strategy and started asking a couple of other people in the office to see if I could figure out what she was thinking. So, overall, I think the theory should be expanded to developed relationships, and I'm glad it's expanding to relationships in the workplace too. We have a ton of uncertainties at my job, and they were made even worse with all the uncertainty around COVID-19!

Uncertainty in developed relationships may be different than it is in initial encounters. It may function dialectically within relationships; that is, there may be a tension between reducing and increasing uncertainty in developed relationships. See our discussion of Relational Dialectics Theory in **Chapter 25** for a thorough discussion of this type of tension. Berger and Calabrese (1975) anticipated this tension when first presenting their theory. They noted that "while uncertainty reduction may be rewarding up to a point, the ability to completely predict another's behavior might lead to boredom. Boredom in an interpersonal relationship might well be a cost rather than a reward" (p. 101). Gerald R. Miller and Mark Steinberg (1975) mention a similar belief, noting that people have a greater desire for uncertainty when they feel secure than they do when they feel insecure. This suggests that as people begin to feel certain about their relationships and their partners, the excitement of uncertainty becomes desirable. Neuliep and Grohskopf (2000) agree, stating that the linear relationship between uncertainty and other communication variables may not hold beyond initial interaction.

Let's examine this contradiction between certainty and uncertainty a bit further through the example of Lucas and Gia. When they met for the first time after class, their uncertainty was high. If their subsequent conversations reduce their uncertainties and they begin a friendship, then their relationship will involve a level of predictability—that is, both will be able to predict certain things about the other because of the time they spend together. Yet this predictability (certainty) may get tedious after a time, and they may feel their relationship is in a rut. At this point, the need for uncertainty, or novelty, will increase, and they might try to build some variety into their routine to satisfy this need.

Uncertainty and uncertainty reduction processes may operate in developed relationships in somewhat the same ways that Berger and Calabrese theorize they do in initial interactions. Research conducted by Sally Planalp and her colleagues (Planalp, 1987; Planalp & Honeycutt, 1985; Planalp, Rutherford, & Honeycutt, 1988) discovered that dating couples found that at times their uncertainty increased. When this happened, the individuals were motivated to reduce it through their communication behaviors. In a study of 46 married couples, Lynn Turner (1990) reached similar conclusions. Therefore, according to these researchers, we must not assume that once relationships begin, uncertainty disappears. However, Leanne Knobloch and her colleagues (2007) observe that uncertainty may be undesirable in marriage because it leads the partners to be more negative in evaluating conversations with each other.

Another example of how URT has been extended into developed relationships is found in the research of Malcolm Parks and Mara Adelman (1983). Parks and Adelman studied the social networks (friends and family members) of an individual and indicate that these third-party networks can be quite important information sources about a romantic partner. They note that network "members may comment on the partner's past actions and behavioral tendencies. They may supply ready-made explanations for the partner's behavior or serve as sounding boards for the individual's own explanations" (p. 57). They conclude that the more partners communicate with their social networks, the less uncertainty they will experience. Furthermore, the researchers found that the less uncertainty people feel, the less likely they will be to dissolve a relationship with another.

Based on Parks and Adelman's (1983) research in established relationships and social networks, Berger and Gudykunst (1991) posited an eighth axiom and seven resulting new theorems. The new axiom asserted that romantic partners who interact with their partner's social network experience less uncertainty about their partner than do those who don't have this interaction. The more interaction with the social network, the less uncertainty there will be.

Some researchers who were interested in how URT is applied to established relationships suggested that in developed relationships, people experienced a different type of uncertainty than they did in initial encounters. This uncertainty was labeled **relational uncertainty**

> **relational uncertainty** a lack of certainty about the future and status of a relationship

and defined as lack of certainty about the future and the status of the relationship. Berger (1987) discussed this new uncertainty type and noted that it stands in the way of relational stability. Additional research (e.g., Ficara & Mongeau, 2000; Knobloch & Solomon, 2003) established that relational uncertainty is distinct from the individual uncertainty that Berger and Calabrese originally theorized about. Relational uncertainty is found to be different from individual uncertainty because it exists at a higher level of abstraction (Knobloch & Solomon, 2003). Relational uncertainty was found to be useful in predicting several other behaviors within relationships, such as how participants will handle face threats (Knobloch, Satterlee, & DiDomenico, 2010) and communicative responses to betrayal (Levine, Kim, & Ferrara, 2010).

Marianne Dainton and Brooks Aylor (2001) examined how relational uncertainty operated in three different types of relationships: long-distance relationships with no face-to-face interaction, long-distance relationships with some face-to-face interaction, and geographically close relationships. The researchers were interested to see how relational uncertainty, jealousy, maintenance, and trust interacted in these three types of relationships. This is a relevant investigation because, as they note, 25–40 percent of romantic relationships between college students are long distance.

They found overall, as URT would predict, that the more uncertainty existed in a relationship the more jealousy, the less trust, and the fewer maintenance behaviors also existed. So, if Lola and Greta have a relationship, and they feel uncertain about each other's affection and goals, for instance, they are likely to have a high degree of jealousy and a low degree of trust in each other and demonstrate only a few relational maintenance behaviors. On the other hand, if Lola and Greta feel confident in each other and their relationship, they will likely express a low level of jealousy and a high degree of trust in each other and exhibit a variety of maintenance behaviors such as texting often, using pet nicknames for each other, and so forth. Dainton and Aylor found that face-to-face contact is critical to reducing relational uncertainty. People in long-distance relationships with no face-to-face interaction suffered from significantly more relational uncertainty. However, those who were geographically close did not differ significantly from those in long-distance relationships with some face-to-face interaction, which was not exactly what URT would predict. The researchers conclude that this is a fruitful line for further research into the utility and heurism of URT.

Social Media and Computer-Mediated Communication

Another area where URT has been expanded is in social media and computer-mediated communication in general. For instance, Artemio Ramirez, Jr. and Joseph Walther (2009) and others (e.g., Flanagin, 2007; Tidwell & Walther, 2002) have noted that URT can be applied to computer-mediated communication. Some research (e.g., May & Tenzek, 2016) indicates that information seeking on the internet is similar to how URT describes it in a pre-online environment—involving the passive, active, and interactive strategies that we've discussed.

Ramirez and Walther (2009), however, observe that there are some significant differences in information seeking online versus face-to-face. According to these researchers, online information seeking allows communicators to "employ several approaches to information acquisition sequentially or simultaneously in order to reduce uncertainty" (p. 73). Further, the online environment means that information-seeking sources can begin to pile up without a person's consent and outside of their con-

> **extractive strategy** a special type of active strategy for reducing uncertainty. It involves gathering information about another online such as by Googling them or reading their Facebook page and so forth

trol. Googling someone (or using other online search engines to obtain information about a specific person) reveals a great deal of information about them that they might not even be aware is in the public domain. Artemio Ramirez and his colleagues (Ramirez, Walther, Burgoon, & Sunnafrank, 2002) have noted that Googling forms a special case of active information seeking; they label it the **extractive strategy**.

Students Talking Tech: Ryder

I am not sure about the conclusion that it's better to interact face-to-face for increased liking. I have a lot of friends that I've never met, but I am talking to them all the time online and during games we play together. I think I like them just as much, if not more, than friends I talk to IRL [in real life]. I sometimes think that I'm a better version of me online than I am IRL. I can take a break and think about what I'm going to say so I don't sound like a jerk, and sometimes, I have to admit I can blow up and not be so cool in person.

Thomas Wagner (2018) specifically examined the extractive strategy by testing how uncertainty reduction and liking were affected by looking at a person's Facebook page after meeting them in person compared to having a face-to-face encounter only. Wagner's approach was grounded in Axioms 1 and 7 of URT. He found that uncertainty was reduced no matter whether the two had interacted face-to-face only or whether the extractive strategy was used via Facebook viewing. Liking, on the other hand, was more affected by face-to-face encounters. People tended to express more liking for those they had only spoken to in person than for those they had investigated via Facebook after their interaction.

Cynthia Palmieri, Kristen Prestano, Rosalie Gandley, Emily Overton, and Qin Zhang (2012) found that self-disclosures on Facebook decreased uncertainty, in line with URT's predictions. And, in another study using URT principles, Jayeon Lee (2015) examined the role that social media play in perceptions of journalists. Lee discovered that the social media activities of journalists affected audience viewpoints. In particular, Lee concluded that journalists who self-disclosed information related to their personal life were viewed positively. Additionally, one study found that both participants in an encounter didn't even have to be human for the effect of self-disclosure to occur. Seo Young Lee and Junho Choi (2017) found that when personal service agents such as Siri and Alexa engaged in "self-disclosure," users felt less uncertain and were more pleased with the interaction than when there was no self-disclosure.

Online dating provides a fruitful arena for the application of URT, and its use is rapidly growing in the United States. From 2013 to 2017, the use of online dating sites in the United States grew 300 percent (Jung et al., 2017).

Researchers (e.g., Gibbs, Ellison, & Lai, 2011; Sharabi, 2021) have observed that in the online dating arena, uncertainty reduction is a salient concern. Liesel Sharabi's (2021) work shows that viewing a profile online (i.e., gathering more information) prior to the date, had a positive effect on liking and perceptions of similarity as URT predicts. Gibbs and her colleagues (2011) note that online disclosures (unlike those offline) don't need to occur in the symmetric fashion that Berger and Calabrese (1975) described (first I tell you where I am from and then you reciprocate by telling me where you grew up). Instead, online information-seeking behaviors may occur prior to hearing a person's disclosures, as well as simultaneously or after disclosures are made. For instance, Andrea can Google Jer at any time during a possible online connection and use extractive strategies. Further, Gibbs and her colleagues suggest that this asymmetric information exchange is important to online daters because the risk of deception online is so large that it contributes to concerns about personal safety. However, they found that online daters use a great many interactive strategies, such as direct questioning online, that confirmed URT's theoretical claims as well as those found in earlier research (Antheunis, Valkenburg, & Peter, 2010). But one-third of Gibbs's 562 respondents reported triangulating information (comparing what a person said to information found about them on websites or in public records), supporting the notion that the online environment adds different information-seeking strategies for uncertainty reduction.

Context

URT originally sought to explain communication in interpersonal contexts, but it has been used in additional contexts as well, including culture, family communication, mass communication, health care, and the workplace. As Michael Boyle and his colleagues (2004) argued, the basic logic of URT can be applied to contexts beyond the interpersonal. We'll discuss each of these contexts briefly below, and you'll note that many of the studies we cite are situated at the interface of two contexts, such as family and social media.

Culture   The expansion and adaptation of URT to culture is credited to William Gudykunst (1993, 2005). Gudykunst (1995) extended URT into a new theory that deals specifically with culture, which he identified as Anxiety-Uncertainty Management (AUM). According to James Neuliep (2016), uncertainty is a cognitive state while anxiety pertains to an affective (emotional) state. AUM suggests that "many people, regardless of culture, experience anxiety when communicating, or when they anticipate communicating, with persons from different cultures or ethnic groups, especially during initial encounters" (Neuliep, p. 7). Neuliep (2012) contends that uncertainty is an aversive state—a claim also identified as a basic assumption of URT.

Gudykunst and Tsukasa Nishida (1986a) discovered differences in low- and high-context cultures. According to Edward T. Hall (1977), **low-context cultures** are those in which meaning is found in the explicit code or message itself. Examples of low-context cultures are the United States, Germany, and Switzerland. In these cultures, plain, direct speaking is valued. Listeners are supposed to be able to understand most of the meaning based solely on the words a speaker uses. In **high-context cultures**, nonverbal messages play a more significant role, and much of the meaning of a message

low-context cultures cultures, as in the United States, where most of the meaning is in the code or message

high-context cultures cultures, like Japan, where the meaning of a message is in the context or internalized in listeners

is internalized by listeners or is induced by listeners from the situation. Japan, Korea, and China are examples of high-context cultures. These cultures value indirectness in speech because listeners are expected to ignore much of the explicit code in favor of understood meanings cued by nonverbals and context.

With respect to research on low- and high-context cultures, Gudykunst and Nishida (1986b) found that frequency of communication predicts uncertainty reduction in low-context cultures, but not in high-context cultures. The researchers also discovered that people use direct communication (asking questions) to reduce their uncertainty in individualistic cultures. In collectivistic cultures, more indirect communication is used with individuals who are not identified as members of the cultural in-group. Based on this research, then, people from different cultures engage in different kinds of communication to reduce their uncertainty.

A concept similar to uncertainty reduction is **uncertainty avoidance**, which is an attempt to shun or avoid ambiguous situations (Hofstede, 1991; Smith, 2015), and cultures may be characterized by

uncertainty avoidance an attempt to avoid ambiguous situations

their attitude toward, and tolerance for, uncertainty. Geert Hofstede believes that the perspective of people in high–uncertainty avoidance cultures is "What is different is dangerous," whereas people in low–uncertainty avoidance cultures believe "What is different is curious" (1991, p. 119). Gudykunst and Yuko Matsumoto (1996) point out that a number of cultures differ in their uncertainty avoidance (see **Figure 5.1**). Understanding these differences can help us understand communication behaviors in other countries.

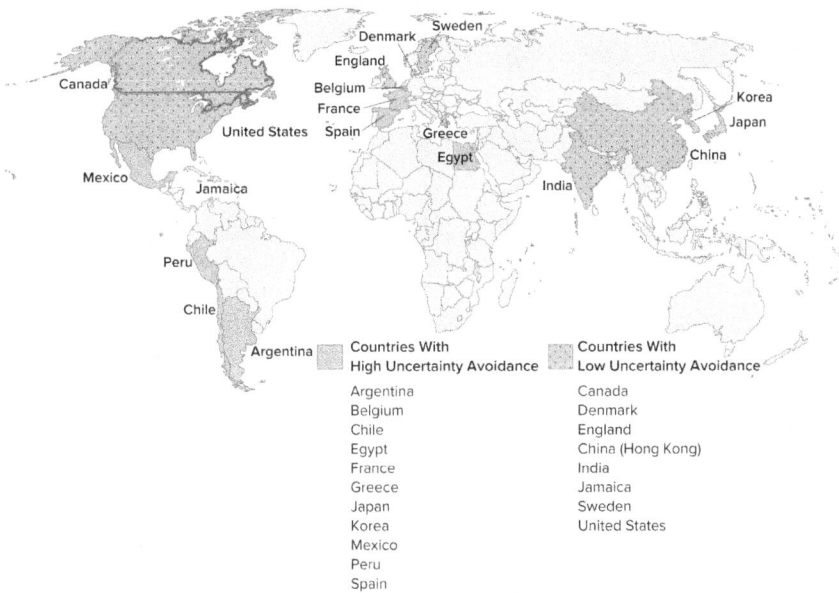

Figure 5.1 Countries and Uncertainty Avoidance

Family Communication   Margaret Stewart (2018) examines a specific type of family, military families, through the lens of URT. Stewart observes that "American military families cope with uncertainty, transiency, and geographic separation during deployments as a regular part of domestic life, and the contemporary military lifestyle is hallmarked by recurrent and lengthy deployments" (p. 1). Stewart investigated how military families used computer-mediated communication for relational maintenance in the hopes of reducing their uncertainties. She interviewed women whose husbands were military men currently deployed outside of the United States. She found that uncertainty was a dominant theme in all of their responses. The wives used social media to communicate with their husbands; however, communication via social media was experienced as a paradox for her respondents. Although it often did reduce the partners' uncertainties, it also could serve to increase uncertainties and spur jealousy.

A different application of URT to family communication comes from Amy May and Kelly Tenzek's (2016) work examining how gay couples wishing to have children via a surrogate try to reduce their uncertainties. The researchers, using URT as a framework, investigated what information gay men posted online to reduce their uncertainty about the surrogacy process. May and Tenzek found seven themes characterizing the content of what gay men posted. They conclude that their study provides a starting point for investigating how family members use online communication to reduce uncertainties in many of the ways predicted by URT.

Mass Communication   Michael Boyle and his coresearchers (2004) found that people used information-seeking behaviors after the terrorist attacks in the United States on September 11, 2001, in many of the ways predicted by URT. Some researchers (e.g., Perse & Rubin, 1989; Rourke & McGloin, 2019) have examined how viewers use two of the strategies advanced by URT (active and passive) to reduce uncertainties about characters seen on television shows, finding that passive strategies are emphasized. Specifically, Brenda Rourke and Rory McGloin discovered that when viewers with Asperger syndrome (AS) watch a show like *The Big Bang Theory*, which features a character with AS, they reduce uncertainty about themselves as well as the character.

Health Care   Some research has investigated doctor–patient communication using URT (e.g., Sheer & Cline, 1995). And URT has been tested in the context of health care reform. Specifically, Lindsay Neuberger and Kami Silk (2016) investigated what motivated the lay public (who express high uncertainty around the topic of health care reform) to engage in information-seeking behaviors. Their findings were not congruent with URT's predictions, however, because uncertainty alone wasn't the main motivating factor for people to engage in information seeking. Alternatively, Evan Perrault (2018) concluded that URT could be extended profitably into the health care context. His study found, as URT predicts, that when patients are able to view videos of a prospective medical care provider speaking about their background, uncertainties are reduced more than if they only read the provider's biography. At the intersection of health care and workplace communication, Li and Lee (2023) investigated employees' health care disclosures at work. The researchers found, consistent with URT, that employees were helped in their decisions to disclose by engaging in a wide variety of communication behaviors.

Workplace   Some research has set uncertainty reduction principles in the context of the workplace (e.g., Charoensukmongkol & Suthatorn, 2022; Hargie, Tourish, & Wilson, 2002; Morrison, 2002). Stevina Evuleocha and Steve Ugbah (2018) used URT and discovered that organizations can reduce uncertainty about job candidates when human resources engages in information seeking on social network sites such as Facebook and X for job screening. The authors do caution that there are legal and ethical considerations in this practice, however. In another approach to workplace communication, Shirish Srivastava and Shalini Chandra (2018) used URT to investigate how companies can create virtual global workplaces. They conclude that virtual world users experience several uncertainties about this context, and "URT can be a useful lens for examining such concerns" (p. 782). Peerayuth Charoensukmongkol and Tipruch Phungsoonthorn (2022) found URT was useful in investigating how two international universities in Thailand dealt with employees' uncertainties related to COVID-19.

Integration and Critique

The beginnings of our relationships are usually unpredictable, and URT helps us understand those challenges. Further, Berger (2016) posits that "uncertainty is 'big business'" in that governmental and nongovernmental organizations lose billions of dollars because "managers abhor uncertainty."

Over a decade after the publication of the original theory, Berger (1987) admitted that some of the propositions of URT were flawed. Other writers concur. Although URT has stimulated a great deal of discussion and research, it also has been criticized. As you think about how useful URT is, keep the following criteria in mind: utility and heurism.

Integration

Communication Tradition	Rhetorical	Semiotic	Phenomenological	Cybernetic	**Socio-Psychological**	Socio-Cultural	Critical
Communication Context	**Intrapersonal**	**Interpersonal**	Small Group/Team	Organizational	Public/Rhetorical	Mass/Media	Cultural
Approach to Knowing	**Positivistic/Empirical**	Interpretive/Hermeneutic	Critical

Critique

Evaluation Criteria	Scope	Logical Consistency	Parsimony	**Utility**	Testability	**Heurism**	Test of Time

Utility

Some researchers believe that the major assumptions of the theory are flawed. Michael Sunnafrank (1986) argues that reducing uncertainty about the self and another in an initial encounter is not an individual's primary concern. Instead, Sunnafrank argues, "a more primary goal is the maximization of relational outcomes" (p. 9). Sunnafrank calls for a reformulation of URT that takes into account the importance of predicted outcomes during initial interactions. Predicted Outcome Value (POV) would explain the case in our chapter's opening somewhat differently than URT would. Sunnafrank would contend that Lucas will be more concerned with maximizing rewards in a potential relationship with Gia than in figuring out what she might do and why she is doing it. Actually, Sunnafrank suggests that URT might kick in after Lucas decides what the predicted outcomes of talking with Gia will be. Jina Yoo's (2009) findings offered some empirical support for POV's claims as opposed to URT's.

Berger's (1986) response to Sunnafrank is that outcomes cannot be predicted without knowledge and reduced uncertainty about oneself, one's partner, and one's relationship. It is Berger's contention that uncertainty reduction is independent of, as well as necessary to, predicted outcome values. In fact, he believes that if one remains highly uncertain, there really are no predicted outcome values. Furthermore, Berger responds to Sunnafrank's critique by noting that the act of predicting an outcome serves as a means to reduce uncertainty. Thus, Berger concludes that Sunnafrank has simply expanded the scope of URT rather than offering an alternative to it.

Theory-Into-Action

 Sometimes people might think that it could be good to increase, rather than reduce, uncertainty. For example, imagine a dilemma when beginning a romantic relationship just before Valentine's Day. It's probably hard to figure out how to celebrate the holiday with a relatively new romantic partner: Should you play hard to get (increase uncertainty) or wear your heart on your sleeve (decrease uncertainty)? You could get some help from a study that showed women Facebook profiles of men who had supposedly viewed their Facebook profile. The women were divided into groups: One group was told that these men liked them the most of any profiles they'd viewed; one group was told the men found them to be average; and the third group was told they couldn't be sure about the men's feelings–they either liked them a lot or they thought they were average. The women responded most positively to the third group–the men they weren't sure about, not the ones they thought liked them the most. Further, the study also showed that being uncertain works best if you think the other person actually likes you. So that's good information to know when you first start a relationship. If you think the other person likes you, create uncertainty. If you think they don't like you yet, reduce uncertainty! This seems to represent a combination of Predicted Outcome Values and Uncertainty Reduction Theory.

A second problem with URT's utility has to do with its validity. Recall that even Berger (1987) has admitted some validity problems, yet he wasn't willing to give up on the theory. Some of his more skeptical colleagues, however, assert that given the tight logical structure of an axiomatic theory, if one building block is wrong, then much of the resulting theory is suspect. Kathy Kellermann and Rodney Reynolds (1990) point to problems with Axiom 3, which suggests that high uncertainty causes high levels of information-seeking behavior.

Their study of over a thousand students failed to find support for the third axiom. Instead, they found that "*wanting* knowledge rather than *lacking* knowledge is what promotes information-seeking in initial encounters with others" (p. 71 [emphasis added]). Kellermann and Reynolds point out that many times we may be uncertain about another, but because we have no interest in them, we are not motivated to reduce our uncertainties by information-seeking behaviors. People engage in communication, therefore, not to reduce uncertainty, but because they care about the other, are interested in the other, or both. In a different vein, Dale Brashers (2001) also questions the validity of Axiom 3. He notes with reference to post–September 11 anxieties that sometimes more information results in a greater sense of uncertainty. Interestingly, however, Dell McKinney and William Donaghy (1993) found some empirical support for Axiom 3, so the debate on this issue undoubtedly will continue concerning URT's usefulness.

Heurism

It's clear from our earlier discussion on expansions of the theory that URT has been expanded into many contexts, making it highly heuristic. For instance, in addition to the contexts we mentioned previously, the theory has also been integrated into research examining small groups (Booth-Butterfield, Booth-Butterfield, & Koester, 1988). And, as we noted, research in areas as diverse as doctor–patient communication (Perrault & Silk, 2015) and computer-mediated communication (Lundy & Drouin, 2016) has used URT as a framework.

Closing

URT has made a very important contribution to the field of communication, even as it's generated some theoretical disputes. It places communication and communication variables in a central position, and it marked the beginning of communication researchers focusing on their own discipline for theoretical explanations rather than borrowing theories from other disciplines. Further, URT is one of the top ten most frequently cited communication theories within the communication discipline (Chung, Barnett, Kim, & Lackaff, 2013).

Discussion Starters

 Case-In-Point: Why is examining initial interactions like that of Gia and Lucas' an important undertaking for communication theorists? Provide at least one example to support your view.

Try-It-Your-Selfie: How do you think this theory can be employed to explain the way dating sites like Hinge work? Interview someone who has used a dating site and argue whether URT is relevant in this context.

1. Are there times when asking questions in initial encounters with others only results in more uncertainty? Give examples. Has reducing your uncertainty about someone ever led to you liking the person less? Describe how this occurs.

2. Do you agree with Berger and Calabrese's assumption about the developmental process of interpersonal relationships? Give examples that support or contest the notion that relationships use communication to pass through entry, personal, and exit phases.

3. What additional factors or events exist—other than those presented in this chapter—when two people meet for the first time that might have an impact on their uncertainty levels? Be sure to be specific and provide appropriate examples.

4. How useful is URT when it comes to examining communication across cultures? Do you think it's appropriate to expand the theory to contexts other than initial encounters between strangers? Explain the reasons for your answer.

References

Antheunis, M. L., Valkenburg, P. M., & Peter, J. (2010). Getting acquainted through social network sites: Testing a model of online uncertainty reduction and social attraction. *Computers in Human Behavior, 26*(1), 100–109.

Berger, C. R. (1979). Beyond initial interaction: Uncertainty, understanding, and the development of interpersonal relationships. In H. Giles & R. St. Clair (Eds.), *Language and social psychology* (pp. 122–144). Blackwell.

Berger, C. R. (1982). *Social cognition and the development of interpersonal relationships: The quest for social knowledge.* Paper presented at the First International Conference on Personal Relationships, Madison, WI.

Berger, C. R. (1986). Uncertain outcome values in predicted relationships: Uncertainty reduction theory then and now. *Human Communication Research, 13*(1), 34–38.

Berger, C. R. (1987). Communicating under uncertainty. In M. E. Roloff & G. R. Miller (Eds.), *Interpersonal processes: New directions in communication research* (pp. 39–62). Sage.

Berger, C. R. (2011). From explanation to application. *Journal of Applied Communication Research, 39*(2), 214–222.

Berger, C. R. (2016). Uncertainty reduction strategies. In C. R. Berger & M. E. Roloff (Eds.), *The international encyclopedia of interpersonal communication* (pp. 1-9). John Wiley & Sons.

Berger, C. R., & Bradac, J. J. (1982). *Language and social knowledge: Uncertainty in interpersonal relations*. Arnold.

Berger, C. R., & Calabrese, R. J. (1975). Some explorations in initial interaction and beyond: Toward a developmental theory of interpersonal communication. *Human Communication Research, 1*(2), 99-112.

Berger, C. R., & Gudykunst, W. B. (1991). Uncertainty and communication. In. B. Dervin & M. J. Voight (Eds.), *Progress in communication sciences* (Vol. 10, pp. 21-66). Ablex.

Blalock, H. M. (1969). *Theory construction: From verbal to mathematical formulations*. Prentice Hall.

Booth-Butterfield, M., Booth-Butterfield, S., & Koester, J. (1988). The function of uncertainty reduction in alleviating primary tension in small groups. *Communication Research Reports, 5*(2), 146-153.

Boyle, M. P., Schmierbach, M., Armstrong, C. L., McLeod, D. M., Shah, D. V., & Pan, Z. (2004). Information seeking and emotional reactions to the September 11 terrorist attacks. *Journalism & Mass Communication Quarterly, 81*(1), 155-167.

Brashers, D. E. (2001). Communication and uncertainty management. *Journal of Communication, 51*(3), 477-497.

Charoensukmongkol, P., & Phungsoonthorn, T. (2022). The interaction effect of crisis communication and social support on the emotional exhaustion of university employees during the COVID-19 crisis. *International Journal of Business Communication, 59*(2), 269-286.

Charoensukmongkol, P., & Suthatorn, P. (2022). How managerial communication reduces perceived job insecurity of flight attendants during the COVID-19 pandemic. *Corporate Communications: An International Journal, 27*(2), 368-387.

Chung, C. J., Barnett, G. A., Kim, K., & Lackaff, D. (2013). An analysis on communication theory and discipline. *Scientometrics, 95*(3), 985-1002.

Dainton, M., & Aylor, B. (2001). A relational uncertainty analysis of jealousy, trust, and maintenance in long-distance versus geographically close relationships. *Communication Quarterly, 49*(2), 172-188.

Deyo, J., Walt, P., & Davis, L. (2011). Rapidly recognizing relationships: Observing speed dating in the South. *Qualitative Research Reports in Communication, 12*(1), 71-78.

Emmers, T. M., & Canary, D. J. (1996). The effect of uncertainty reduction strategies on young couples' relational repair and intimacy. *Communication Quarterly, 44*(2), 166-182.

Evuleocha, S. U., & Ugbah, S. D. (2018). Profiling: The efficacy of using social networking sites for job screening. *Journal of Employment Counseling, 55*(2), 48-57.

Ficara, L. C., & Mongeau, P. A. (2000, November). Relational uncertainty in long-distance college student dating relationships. Paper presented at the annual meeting of the National Communication Association, Seattle, WA.

Flanagin, A. J. (2007). Commercial markets as communication markets: Uncertainty reduction through mediated information exchange in online auctions. *New Media & Society, 9*(3), 401-423.

Gibbs, J. L., Ellison, N. B., & Lai, C. H. (2011). First comes love, then comes *Google*: An investigation of uncertainty reduction. *Communication Research, 38*(1), 70-100.

Gudykunst, W. B. (1993). Toward a theory of effective interpersonal and intergroup communication: An anxiety/uncertainty management (AUM) perspective. In R. L. Wiseman & J. Koester (Eds.), *International and intercultural communication annual* (pp. 33-71). Sage.

Gudykunst, W. B. (1995). Anxiety/uncertainty management (AUM) theory. In R. Wiseman (Ed.), *Intercultural communication theory* (pp. 8-58). Sage.

Gudykunst, W. B. (2005). An Anxiety/Uncertainty Management (AUM) theory of effective communication: Making the mesh of the net finer. In W. B. Gudykunst (Ed.), *Theorizing about intercultural communication* (pp. 281–322). Sage.

Gudykunst, W. B., & Matsumoto, Y. (1996). Cross-cultural variability of communication in personal relationships. In W. B. Gudykunst, S. Ting-Toomey, & T. Nishida (Eds.), *Communication in personal relationships across cultures* (pp. 19–56). Sage.

Gudykunst, W. B., & Nishida, T. (1986a). Attributional confidence in low- and high-context cultures. *Human Communication Research, 12*(4), 525–549.

Gudykunst, W. B., & Nishida, T. (1986b). Social penetration in close relationships in Japan and the United States. In R. Bostrom (Ed.), *Communication yearbook 7* (pp. 592–610). Sage.

Hall, E. T. (1977). *Beyond culture.* Anchor/Doubleday.

Hargie, O., Tourish, D., & Wilson, N. (2002). Communication audits and the effects of increased information: A follow-up study. *The Journal of Business Communication, 39*, 414–436.

Hofstede, G. (1991). *Cultures and organizations: Software of the mind.* McGraw Hill.

Jung, S., Roh, S., Yang, H., & Biocca, F. (2017). Location and modality effects in online dating: Rich modality profile and location-based information cues increase social presence, while moderating the impact of uncertainty reduction strategy. *Cyberpsychology, Behavior, and Social Networking, 20*(9), 553–560.

Kellermann, K., & Reynolds, R. (1990). When ignorance is bliss: The role of motivation to reduce uncertainty in uncertainty reduction theory. *Human Communication Research, 17*(1), 5–35.

Knobloch, L. K. (2008). Uncertainty reduction theory. In L. A. Baxter & D. O. Braithwaite (Eds.), *Engaging theories in interpersonal communication: Multiple perspectives* (pp. 133–144). Sage.

Knobloch, L. K., Miller, L. E., Bond, B. J., & Monnone, S. E. (2007). Relational uncertainty and message processing in marriage. *Communication Monographs, 74*(2), 154–180.

Knobloch, L. K., Satterlee, K. L., & DiDomenico, S. M. (2010). Relational uncertainty predicting appraisals of face threat in courtship: Integrating uncertainty reduction theory and politeness theory. *Communication Research, 37*(3), 303–334.

Knobloch, L. K., & Solomon, D. H. (2003). Responses to changes in relational uncertainty within dating relationships: Emotions and communication strategies. *Communication Studies, 54*(3), 282–305.

Kosenko, K. A. (2011). The safer sex communication of transgender adults: Processes and problems. *Journal of Communication, 61*(3), 476–495.

Lee, J. (2015). The double-edged sword: The effects of journalists' social media activities on audience perceptions of journalists and their news products. *Journal of Computer-Mediated Communication, 20*(3), 312–329.

Lee, S. Y., & Choi, J. (2017). Enhancing user experience with conversational agent for movie recommendation: Effects of self-disclosure and reciprocity. *International Journal of Human–Computer Studies, 103*, 95–105.

Levine, T. R., Sang-Yeon Kim, & Ferrara, M. (2010). Social exchange, uncertainty, and communication content as factors impacting the relational outcomes of betrayal. *Human Communication, 13*(4), 303–318.

Li, J-Y., & Lee, Y. (2023). To disclose or not? Understanding employees' uncertainty and behavior regarding health disclosure in the workplace: A modified socioecological approach. *International Journal of Business Communication, 60*(1), 173–201.

Lundy, B. L., & Drouin, M. (2016). From social anxiety to interpersonal connectedness: Relationship building within face-to-face, phone and instant messaging mediums. *Computers in Human Behavior, 54,* 271–277.

May, A., & Tenzek, K. (2016). "A gift we are unable to create ourselves": Uncertainty reduction in online classified ads posted by gay men pursuing surrogacy. *Journal of GLBT Family Studies, 12*(5), 430–450.

McKinney, D. H., & Donaghy, W. C. (1993). Dyad gender structure, uncertainty reduction, and self-disclosure during initial interaction. In P. Kalbfleish (Ed.), *Interpersonal communication: Evolving interpersonal relationships* (pp. 33–50). Erlbaum.

Miller, G. R. (1981). Tis the season to be jolly: A yule-tide 1980 assessment of communication research. *Human Communication Research, 7*(4), 371–377.

Miller, G. R., & Steinberg, M. (1975). *Between people: A new analysis of interpersonal communication.* Science Research Associates.

Mongeau, P. A., Jacobsen, J., & Donnerstein, C. (2007). Defining dates and first date goals: Generalizing from undergraduates to single adults. *Communication Research, 34*(5), 526–527.

Morrison, E. W. (2002). Information seeking within organizations. *Human Communication Research, 28*(2), 229–242.

Neuberger, L., & Silk, K. J. (2016). Uncertainty and information-seeking patterns: A test of competing hypotheses in the context of health care reform. *Health Communication, 31*(7), 892–902.

Neuliep, J. W. (2012). The relationship among intercultural communication apprehension, ethnocentrism, uncertainty reduction, and communication satisfaction during initial intercultural interaction: An extension of anxiety and uncertainty management (AUM) theory. *Journal of Intercultural Communication Research, 41*(1), 1–16.

Neuliep, J. W. (2016). Uncertainty and anxiety in intercultural encounters. In C. R. Berger & M. E. Roloff (Eds.), *The International Encyclopedia of Interpersonal Communication* (pp. 1–9). John Wiley & Sons.

Neuliep, J. W., & Grohskopf, E. L. (2000). Uncertainty reduction and communication satisfaction during initial interaction: An initial test and replication of a new axiom. *Communication Reports, 13*(2), 67–77.

Palmieri, C., Prestano, K., Gandley, R., Overton, E., & Zhang, Q. (2012). The Facebook phenomenon: Online self-disclosure and uncertainty reduction. *China Media Research, 8*(1), 48–53.

Parks, M. R., & Adelman, M. B. (1983). Communication networks and the development of romantic relationships: An expansion of uncertainty reduction theory. *Human Communication Research, 10*(1), 55–79.

Perrault, E. K. (2018). Adding multimedia cues to medical providers' online biographies: Do pictures, video, and B-roll matter? *Journal of Health Communication, 23*(5), 462–469.

Perrault, E. K., & Silk, K. J. (2015). Reducing communication apprehension for new patients through information found within physicians' biographies. *Journal of Health Communication, 30,* 743–750.

Perse, E. M., & Rubin, R. B. (1989). Attribution in social and parasocial relationships. *Communication Research, 16*(1), 59–77.

Planalp, S. (1987). Interplay between relational knowledge and events. In R. Burnett, P. McGhee, & D. Clarke (Eds.), *Accounting for relationships: Social representations of interpersonal links* (pp. 173–191). Methuen.

Planalp, S., & Honeycutt, J. M. (1985). Events that increase uncertainty reduction in personal relationships. *Human Communication Research, 11*(4), 593–604.

Planalp, S., Rutherford, D. K., & Honeycutt, J. M. (1988). Events that increase uncertainty in personal relationships II: Replication and extension. *Human Communication Research, 14*(4), 516–547.

Ramirez, A., Jr., & Walther, J. B. (2009). Information seeking and interpersonal outcomes using the internet. In T. D. Afifi & W. A. Afifi (Eds.), *Uncertainty, information management, and disclosure decisions: Theories and applications* (pp. 67–84). Routledge.

Ramirez, A., Jr., Walther, J. B., Burgoon, J. K., & Sunnafrank, M. (2002). Information seeking strategies, uncertainty, and computer-mediated communication: Toward a conceptual model. *Human Communication Research, 28*(2), 213–228.

Rourke, B., & McGloin, R. (2019). A different take on *The Big Bang Theory*: Examining the influence of Asperger traits on the perception and attributional confidence of a fictional TV character portraying characteristics of Asperger syndrome. *Atlantic Journal of Communication, 27*(2), 127–138.

Shannon C., & Weaver, W. (1949). *The mathematical theory of communication*. University of Illinois Press.

Sharabi, L. L. (2021). Online dating profiles, first-date interactions, and the enhancement of communication satisfaction and desires for future interaction. *Communication Monographs, 88*(2), 131–153.

Sheer, V. C., & Cline, R. J. (1995). Testing a model of perceived information adequacy and uncertainty reduction in physician–patient interactions. *Journal of Applied Communication Research, 23*(1), 44–59.

Smith, P. B. (2015). To lend helping hands in-group favoritism, uncertainty avoidance, and the national frequency of pro-social behaviors. *Journal of Cross-Cultural Psychology, 46*(6), 759–771.

Solomon, D. H. (2016). Uncertainty and relationship development. In C. R. Berger & M. E. Roloff (Eds.), *The international encyclopedia of interpersonal communication* (pp. 1–5). John Wiley & Sons.

Srivastava, S. C., & Chandra, S. (2018). Social presence in virtual world collaboration: An uncertainty reduction perspective using a mixed methods approach. *MIS Quarterly, 42*(3), 779–803.

Stewart, M. C. (2018). Uncertainty reduction and technologically mediated communication: Implications to marital communication during wartime deployment. *Ohio Communication Journal, 56,* 136–148.

Sunnafrank, M. (1986). Predicted outcome value during initial interactions: A reformulation of uncertainty reduction theory. *Human Communication Research, 13*(1), 191–210.

Tidwell, L. C., & Walther, J. B. (2002). Computer-mediated communication effects on disclosure, impressions, and interpersonal evaluations: Getting to know one another a bit at a time. *Human Communication Research, 28*(3), 317–348.

Turner, L. H. (1990). The relationship between communication and marital uncertainty: Is her marriage different from his marriage? *Women's Studies in Communication, 13*(2), 57–83.

Wagner, T. R. (2018). When off-line seeks information online: The effect of modality switching and time on attributional confidence and social attraction. *Communication Research Reports, 35*(4), 346–355.

Yoo, J. H. (2009). Uncertainty reduction and information valence: Tests of uncertainty reduction theory, predicted outcome value, and an alternative explanation? *Human Communication, 12*(2) 187–198.

CHAPTER 6
Social Exchange Theory

*Based on the research of **John Thibault** and **Harold Kelley***

"Every individual voluntarily enters and stays in any relationship only as long as it is adequately satisfactory in terms of his [sic] rewards and costs."

—John Thibault and Harold Kelley

Meredith Daniels and LaTasha Evans

Meredith Daniels and LaTasha Evans had been best friends since they served as hall monitors together in fourth grade. After elementary school they had both moved on to Collins High School, where they were now seniors. For the past three years, they'd suffered through homework, dating dilemmas, and other typical high school concerns. In addition, the two friends had also coped with racial issues because Meredith was white and LaTasha was Black. In their hometown of Biloxi, Mississippi, the heritage of racism formed a barrier to their friendship. Although Biloxi was now fairly progressive, Meredith's grandfather had been very uncomfortable about her friendship with LaTasha.

Also, LaTasha's Uncle Benjamin had participated in Freedom Marches in the 1960s and had formed some unfavorable opinions about whites. For a short time, her uncle had been a member of a Black separatist organization. He had some difficulty with LaTasha and Meredith's friendship, too. Both the young women had worked hard to maintain their relationship despite these family members' objections.

When LaTasha and Meredith were together, they tended to forget about all these problems. They seemed like sisters to each other, closer than many sisters they knew. They had the exact same sense of humor, and they could always cheer each other up with a goofy look or some silly jokes about their past. They enjoyed the same movies (horror/thrillers—they'd seen *Get Out* together three times) and the same subjects in

school (English and French) and had similar taste in clothes (1960s retro) and boyfriends (intellectual guys).

LaTasha and Meredith had weathered some threats to their friendship and come out stronger. When they saw the movie *Till*, it brought up a lot about their different heritages. The movie dramatized the story of Emmett Till, a Black 14-year-old from Chicago, who was tortured and killed by two white men in 1955 while he was visiting relatives in a small town in Mississippi. The white men said Emmett had disrespected a white woman in a grocery store. LaTasha said Meredith could never understand how that tragedy and the countless others like it still happening today affected her as a young Black woman. Meredith didn't disagree and the two talked through it, finally concluding that their commonalities were more important to them than their racial differences, even though racial violence in the U.S. did affect them differently.

But at home, race was a big deal to their families. Meredith's parents said they didn't object to her friendship with LaTasha, but they were unhappy when Meredith socialized with Black boys and went to parties where she might be one of only two or three white girls there. LaTasha's parents also had no problem with Meredith; they liked her and understood the friendship. But they drew the line when it came to dating white boys, and they didn't like her hanging out too much with Meredith's family, which included Will, Meredith's older brother, who had asked LaTasha out once. LaTasha's parents were very proud of their Black heritage, and they told all their children how important it was to maintain their traditions and way of life. For them this meant no dating or marriage with someone of another race. LaTasha's cousin had recently married a Japanese woman, and the whole family was having difficulty accepting the couple.

Now that LaTasha and Meredith were entering their senior year of high school, things had become even more difficult. LaTasha's parents were adamant that she attend a historically Black college after graduation. Meredith's family wanted her to go to a small college in southern California because both her parents had graduated from this school, and they had many relatives living near the college. Both LaTasha and Meredith, however, wanted to go to college together or at least be somewhat near each other. In addition, they resented how much time and energy all the discussions about college seemed to take up. It was almost ruining their senior year!

When they were together, they could usually forget about all the hassle and just have fun as usual, but the pressure was taking a toll on their friendship. Although they tried not to think about it, they both were concerned about the future. It was hard not to be able to tell each other everything as they always had in the past, but Meredith and LaTasha both found that talking to each other about college was stressful, so they mainly avoided the subject. Privately, each wondered what was going to happen and how she would get along next year without her best friend.

A Social Exchange theorist examining Meredith and LaTasha's relationship would predict that it might be heading for some trouble because it currently seems to be costing the two more than it's rewarding them. As Marianne Dainton (2015) asserts, interracial relationships always will require a great deal of energy to maintain, perhaps because there are not many role models in U.S. culture for such relationships even though evidence suggests that interracial friendships enhance personal well-being and can reduce cultural prejudices (Killen, Luken Raz, & Graham, 2022).

costs elements of relational life with negative value

rewards elements of relational life with positive value

Social Exchange Theory (SET) is based on the notion that people think about their relationships in economic terms. This means that people tally up the costs of being in a relationship and compare those to the rewards the relationship provides. **Costs** are the elements of relational life that have negative value to a person, such as the time and effort one has to put into maintaining a relationship or the negatives one has to put up with in their partner (such as the amount of time they spend with online games or how messy they are). In our chapter-opening scenario, the stress and tension that LaTasha and Meredith feel about the issue of college, as well as their families' lack of support and their differing racial backgrounds, all provide costs to their relationship. **Rewards** are the elements of a relationship that have positive value. In Meredith and LaTasha's case, the fun they have together, the loyalty they show for each other, their shared history, their ability to talk through problems, and the sense of understanding they share are all rewards.

All relationships provide us rewards, or positives. Families, friends, and loved ones generally give us a sense of acceptance, support, and companionship. Some friends provide us status just by being with us. Friends and families keep us from feeling lonely and isolated. Some friends teach us helpful lessons. Yet, all relationships incur costs too. Relationships require some time and effort on the part of their participants. When friends spend time with each other, which they must do to maintain the relationship, they are unable to do other things with that time, so in that sense the time spent is a cost. Friends might need attention at inopportune times, and then the cost is magnified. For instance, if you had to finish a paper for class, and your best friend just had a family crisis and wanted to talk to you, you can see how the friendship would cost you something in terms of time.

Communication researchers are interested in SET because communication is the means people have for negotiating an exchange, offering rewards, and creating, as well as potentially reducing, costs to their partners.

For instance, when LaTasha wants to engage in relational maintenance with Meredith, that means she wants to talk to her about what's going on in their friendship and perhaps discuss how to fix it. When Meredith's parents express, verbally or nonverbally, their displeasure with the parties Meredith attends with LaTasha, they're using communication to do this and, consequently, creating some costs to the friendship.

George Homans is generally credited with bringing an exchange perspective to human behavior. Homans examined the idea of exchange in small groups and it was later explored in interpersonal relationships (Trevino & Tilly, 2016). Although some of what we present in this chapter can be applied to group behavior, the essence of what we delve into relates to how individuals think about and communicate in their interpersonal relationships.

The Social Exchange perspective argues that people calculate the overall worth of a particular relationship by subtracting its costs from the rewards it provides (Monge & Contractor, 2003), using the following equation:

$$Worth = Rewards - Costs$$

Positive relationships are those whose worth is a positive number; that is, the rewards are greater than the costs. Relationships where the worth is a negative number (the costs exceed the rewards) tend to be negative for the participants. SET predicts that the worth of a relationship influences its **outcome,** or whether people will continue with a relationship or terminate it. Positive relationships are expected to endure, whereas negative relationships will probably terminate.

> **outcome** whether people continue in a relationship or terminate it

Although the situation is more complex than this simple equation suggests, as we'll discuss later in the chapter, the equation does state the essence of what many exchange theorists argue. Ronald Sabatelli and Constance Shehan (1993) note that the Social Exchange approach views relationships through the metaphor of the marketplace, where each person acts out of a self-oriented goal of profit taking. However, Laura Stafford (2008) qualifies that assertion by pointing out that economic exchanges and social exchanges have differences: social exchanges involve a connection with another person; social exchanges involve trust, not legal obligations; social exchanges are more flexible; and social exchanges rarely involve explicit bargaining.

Students Talking: Rosa

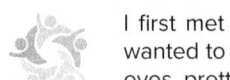

I first met my now fiancé, Tara, on Match. We had quite a discussion about whether we wanted to get married or continue to live together. The costs of getting married were, in my eyes, pretty limited because we'd have tax benefits, and it would just be easier to introduce each other and talk about our relationship. In Tara's eyes, there were several risks, such as whether we were both prepared to say 'til death do us part or whether we are ready to be monogamous. We both thought about these risks a lot and spent many days and nights talking about them. In the end, we agreed that the benefits of being married outweighed the costs, so we got engaged. It's funny to find that there's a theory about what we went through.

We have been talking in general about exchange theories and the perspective of Social Exchange; this is because (1) Social Exchange provides an overarching view of human beings that might be taken up by other specific theories like Social Penetration Theory, for instance, and (2) there are several different theories of Social Exchange. Michael Roloff (1981, 2016) discusses five specific Social Exchange theories. Roloff observes that these theories are tied together by a central argument that "the guiding force of interpersonal relationships is the advancement of both parties' self-interest" (1981, p. 14). Furthermore, Roloff notes that these theories don't assume that self-interest is a negative thing; rather, when self-interest is recognized, it will actually enhance a relationship. Roloff also argues that there are significant differences among these five theories—some of which exist because they were developed by researchers in different disciplines (e.g., psychology, social psychology, and sociology). It's beyond our purposes here to differentiate among all the theories of Social Exchange. We'll focus on what may be the most popular theory: John Thibault and Harold Kelley's (1959) Theory of Interdependence. Although Thibault and Kelley called their theory the Theory of Interdependence and actually explicitly argued that it wasn't a theory of Social Exchange (Kelley & Thibault, 1978), it is often referred to as SET because it fits into the exchange framework we've described (Roloff, 2016).

Theory at a Glance • Social Exchange Theory: Theory of Interdependence

Social Exchange Theory posits that the major force in interpersonal relationships is the satisfaction of both people's self-interest. Self-interest is not considered necessarily bad and can be used to enhance relationships. Interpersonal exchanges are thought to be somewhat analogous to economic exchanges where people are satisfied when they receive what they consider to be a fair return for their efforts in a relationship. Thibault and Kelley's Theory of Interdependence, a specific type of SET, argues that people's feelings of satisfaction or dissatisfaction in a relationship rest on the interdependence between the partners. This means you are able to be happy only depending on your partner's happiness and the actions they take relative to you.

Assumptions of Social Exchange Theory

All Social Exchange theories are built upon several assumptions about human nature and the nature of relationships. Some of these assumptions should be clear to you after our introductory comments. Because SET is based on a metaphor of economic exchange, many of these assumptions flow from the notion that people view life as a marketplace. In addition, Thibault and Kelley base their theory on two conceptualizations: one that focuses on the nature of individuals and one that describes relationships between two people. They look to drive reduction, an internal motivator, to understand individuals and to gaming principles to understand relationships between people. Thus, the assumptions they make fall into two categories.

The assumptions that SET makes about *human nature* include the following:

- Humans seek rewards and to avoid punishments.
- Humans are rational beings.
- The standards that humans use to evaluate costs and rewards vary over time and from person to person.

The assumptions SET makes about the *nature of relationships* include the following:

- Relationships are interdependent.
- Relational life is a process.

We'll discuss each of these assumptions in turn.

The notion that humans seek rewards and to avoid punishment is consistent with the psychological theory of drive reduction (Roloff, 1981). This approach assumes that people's behaviors are motivated by some internal drive mechanism. When people experience the drive mechanism (such as thirst), they are motivated to reduce it (by getting a drink, for instance), and the process of doing so is a pleasurable, rewarding one. This assumption helps Social Exchange theorists understand why LaTasha and Meredith enjoy each other's company: the drives they experience are loneliness and a desire for understanding and companionship. They reduce these drives by seeking each other's friendship and by spending time together.

The second assumption—that humans are rational—is critical to SET. The theory rests on the notion that within the limits of the information that's available to them, people calculate the costs and rewards of a given situation and guide their behaviors accordingly. This also includes the possibility that, faced with no rewarding choice, people choose the least costly alternative. In the case of LaTasha and Meredith, it's costly to continue their friendship in the face of all the stress and family objections they're experiencing. Yet both young women may believe that it's less costly than ending their friendship and trying to find someone else who could provide the level of affection and understanding that they've shared for the past 9 years. When we discuss evaluating a relationship later in the chapter, we'll explain this idea further.

The third assumption—that the standards people use to evaluate costs and rewards vary over time and from person to person—suggests that the theory must take diversity into consideration. No one standard can be applied to everyone to determine what is a cost and what is a reward. Thus, LaTasha may grow to see the relationship as more costly than Meredith does (or vice versa) as their standards change over time. However, SET is a lawlike theory, as we described in **Chapter 3,** because SET claims that although individuals may differ in their definition of rewards, the first assumption is still true for all people: we are motivated to maximize our profits and rewards while minimizing our losses and costs (Pascale & Primavera, 2016).

As we mentioned earlier in this chapter, Thibault and Kelley take those three assumptions about human nature from drive reduction principles. In their approach to relationships, they drew on a set of principles from *game theory* (Balliet, Tybur, & Van Lange, 2017). The classic game they developed to illustrate their first assumption about relationships is called the Prisoner's Dilemma (see **Figure 6.1**). This game supposes that two prisoners are being questioned about a crime they deny committing. They have been separated during questioning, and they are given two choices: they can confess to the crime, or they can persist in their denials. The situation is further complicated by the fact that the outcome for them is not completely in their own hands, individually. Instead, each prisoner's outcome is a result of the combination of their two responses, or to put it another way, their outcomes are interdependent. The configuration of their possible choices is called a 2 × 2 matrix because there are two of them and they each have two choices: confess or deny.

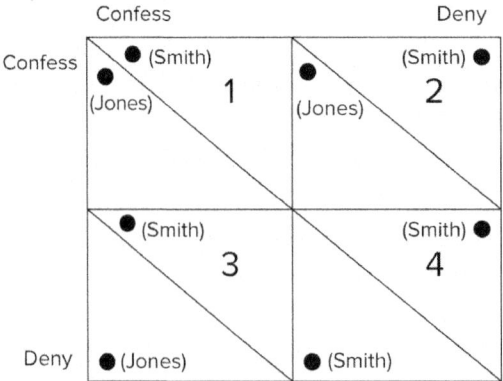

Figure 6.1 The Prisoner's Dilemma

If we call one of the prisoners Jones and the other Smith, we can see their choices and outcomes:

1. Jones confesses and Smith confesses = they both receive a life sentence.
2. Jones confesses and Smith denies = Jones goes free and Smith is jailed for life.
3. Jones denies and Smith confesses = Smith goes free and Jones is jailed for life.
4. Jones denies and Smith denies = they both serve a short jail term.

It's clear that the outcomes for Smith and Jones are interdependent. The outcome in each case depends on the relationship between Smith and Jones's answers, not on one answer alone. This concept is so central to Thibault and Kelley that they named their theory the Theory of Interdependence rather than Social Exchange or Game Theory. They wanted to avoid the notion of win–lose in Game Theory, and they wished to stress that Social Exchange is a function of interdependence.

Theory-Into-Action

 SET explains relationships between people through the metaphor of juggling relational costs and rewards. In thinking about online communication and privacy, people often adopt the same type of cost-versus-rewards type thinking. If you consider how much data you provide to Google and other search engines just by giving them access to your search history, you'll see what we're talking about. One technology director for a start-up company called the amount of data we allow others access to "data exhaust," saying that we emit data exhaust every time we use technology. But, as the editor-in-chief of *Wired* magazine stated, people have to decide what they want to give up in order to gain the benefits that social media and various forms of technology provide. If you think about having a relationship with social media, you can definitely see it in terms of SET. It's also the case that you leave traces of information for others to find on social media, and in that way, you're giving up some privacy too. For instance, Meredith might text LaTasha about how mad she is at her family and say some unkind things about her parents. If one of her parents picked up her phone before she deleted the texts, Meredith's privacy would be invaded, and she might find herself in a conflict with her parents. As you use technology, it's good to be mindful of this trade-off and think about whether the rewards you're getting outweigh the costs of the privacy you're losing. See, for example, Cloarec, Meyer-Waarden, & Munzel (2021). These authors talk about how Social Exchange Theory can be applied to the relationship people establish with the technology they use.

When we think of LaTasha and Meredith's situation, we can see that if Meredith decides to cut back on her friendship with LaTasha, LaTasha will inevitably be affected. Her own decisions about the costs and rewards of the relationship are contingent on Meredith's decisions. Thus, whenever any one member of a relationship acts, both their partner and the relationship as a whole are influenced. Human beings are interdependent in all aspects of their social, relational lives. Try to think of a relationship that you're in today in which interdependence, to some degree, does not exist—we think you'll find it impossible to do so.

The second assumption states that relational life is a process. This acknowledges the importance of time and change in relational life. Specifically, time affects exchanges because past experiences guide judgments about rewards and costs, and these judgments impact subsequent exchanges. For example, if Kathy dislikes school and has a very low opinion of teachers and then takes a class that exceeds her expectations, and she finds that she really likes this particular teacher, their relationship and Kathy's expectations about future relationships with teachers will be shaped by the process. Further, the notion of process allows us to see that relationships constantly change and evolve.

Given these assumptions about humans and relationships, we're ready to examine three important aspects related to the theory: dimensions of interdependence, evaluation of a relationship, and power and exchange patterns.

Dimensions of Interdependence

As we just discussed, Thibault and Kelley created the Prisoner's Dilemma to illustrate how people are interdependent in social situations. From this assumption, they induced four dimensions of interdependence that describe regularities across social situations: mutual dependence, power, conflict, and coordination (1978). Later, Kelley and his colleagues added two more dimensions: future interdependence and information certainty (Gerpott, Balliet, Columbus, Molho, & de Vries, 2018; Kelley et al., 2003). It's important to clarify that these dimensions are part of the situation, not the individuals. In fact, people often don't have a clear sense of their interdependence with others. In the case of the Prisoner's Dilemma, for example, the individual prisoners aren't aware of what the interrogators are saying to the other prisoner. But as the game shows, it's beneficial to be able to figure out the dimensions of interdependence in a given situation so that you'll figure out the best partner to pick, the best behavior to engage in yourself, and so forth (Gerpott et al., 2018).

We'll discuss each of these six dimensions briefly. First, to understand **mutual dependence,** or the extent to which an individual's outcome in a situation depends not just on their own behavior but on their behavior in combination with their partner's, we note that situations vary along a continuum of interdependence. If Meredith is leaving school for the day, for instance, and needs to decide which books she'll bring home, she's independent because her decision really has no effect on LaTasha's outcome. In other situations, such as the decisions related to where they'll go to college, their outcome is mutually dependent. Researchers have noted that in mutually dependent situations, people are usually attentive and committed to their partners as well as more cooperative, prosocial, and less aggressive (Gerpott et al., 2018).

> **mutual dependence** the extent to which a desired outcome is dependent on not just one's own behavior, but the behavior of a relational other
>
> **power** the degree of dependence a person has on another for outcomes
>
> **conflict** when two people's desired outcomes don't correspond with one another
>
> **coordination** working with a partner to achieve desired outcomes

As we discuss later, SET is concerned with how power is distributed between people. **Power** is defined in this context as *asymmetric dependence,* meaning a situation in which one partner is more dependent on the other for having their outcomes satisfied. So, if LaTasha has become convinced by her parents that she should attend a historically Black college, then she's less dependent on Meredith's college decisions and thus gains more social power in their relationship. **Conflict** refers to lack of correspondence in outcomes in a particular situation. In a workplace, there's often conflict between bosses who want the most productivity at the least cost and employees who want the highest salary without increasing their workload. Conflict exists in this situation but not in a situation where the partners' outcomes are more in line with each other's, such as when a couple wants to go on a vacation together. **Coordination** refers to situations in which people need to work together to achieve their desired outcomes. When a couple wants to go on vacation, they need to adjust and coordinate their behaviors to plan the vacation.

Future interdependence means that the outcomes of one interaction affect future interactions and outcomes between the same partners. When a boss turns down an employee's request for a raise without really listening to the employee, the employee might feel less committed to the organization and begin looking for work elsewhere.

> **future interdependence** when the outcomes of one interaction affect subsequent outcomes and interactions

The employee might avoid their boss in the future as a consequence of having had a negative interaction in the past. In contrast, a positive interaction might increase the likelihood of partners being more cooperative with each other in the future (Gerpott et al.,

information certainty when partners believe they know what each other wants for an outcome

2018). **Information certainty** means that each partner believes they know what the other wants for an outcome. Like most of these dimensions, information certainty is on a continuum, and in social situations people often are only relatively certain, or sometimes even completely uncertain, of what their partner wants. Further, information certainty influences a partner's behaviors. If partners possess (or think they possess) information certainty, they probably don't pay much attention to the other's behavior. That will change as a person becomes less certain. For example, if LaTasha is pretty sure she and Meredith are on the same page about their need to do whatever it takes to remain friends and go to college together, she won't scrutinize Meredith's every comment about college. But if Meredith does or says something that makes LaTasha less certain, she's likely to pay more attention to what Meredith's saying. If she becomes more uncertain over time, that might reduce cooperation between the two. See **Table 6.1** for a summary of the dimensions.

Table 6.1 Six Dimensions of Interdependence

DIMENSION	DEFINITION
Mutual dependence	How each person's outcomes are determined by the other's actions in combination with their own.
Power	How much an individual determines their own and others' outcomes, compared to how much their partner does.
Conflict	How much the outcomes for two partners correspond with one another. Conflict results from lack of correspondence.
Coordination	How the partners need to act together to achieve desired outcomes.
Future interdependence	How behavior in the present situation affects behavior and outcomes in future interactions.
Information certainty	How much a person believes they know their partner's preferred outcomes.

Adapted from Gerpott, F. H., Balliet, D., Columbus, S., Molho, C., & Vries, R. E. (2018).

Evaluation of a Relationship

As we mentioned previously, SET is more complex than the simple equation of worth that we initially presented. Social Exchange includes "both a notion of a relationship, and some notion of a shared obligation in which both parties perceive responsibilities to each other"

comparison level (CL) a standard for what a person thinks they should get in a relationship

(Lavelle, Rupp, & Brockner, 2007, p. 845). One of the most interesting parts of Thibault and Kelley's theory is their explanation of how people evaluate their relationships with reference to whether they'll stay in them or leave them. Thibault and Kelley claim that this evaluation depends on two types of comparisons: comparison level and comparison level for alternatives. The **comparison level (CL)** is a standard representing what people feel they should receive in the way of rewards and costs from a particular relationship and it provides a measure of relationship satisfaction. Thus, Meredith and LaTasha both have a subjective feeling about what they should give and get from a friendship. Their CL has been shaped by all their past friendships, by family members' advice, and by popular culture such as media representations of friendships that give them ideas about what is expected from this relationship.

CLs vary among individuals because they are subjective. Individuals base their CL, in large part, on past experiences with a specific type of relationship. Because individuals have very different past experiences with similar types of relationships, they develop different CLs. For example, if Suzanne has had many friendships that required her to do a great deal of listening and empathizing, her CL will include this. If Ali has not experienced friends requiring this listening behavior from her, she won't expect to encounter this cost in friendship. Because people often interact with people from their own culture, they share many relational expectations due to messages they have received from cultural norms and popular culture's depictions of relationships (Rawlins, 1992). Thus, people from the same culture will overlap somewhat in their CL.

Thibault and Kelley argue that our satisfaction with a current relationship derives from comparing the rewards and costs it involves to our CL. If our current relationship meets or exceeds our CL, the theory predicts we will be satisfied with the relationship. Yet people sometimes leave satisfactory relationships and stay in ones that are

> **comparison level for alternatives (CLalt)** how people evaluate a relationship based on what their alternatives to the relationship are

not so satisfying. Thibault and Kelley explain this seeming inconsistency with their second standard of comparison, the **comparison level for alternatives (CLalt).** This refers to "the lowest level of relational rewards a person is willing to accept given available rewards from alternative relationships or being alone" (Roloff, 1981, p. 48). In other words, the CLalt provides the threshold for evaluating a relationship compared to realistic alternatives to that relationship (Dillow, Malachowski, Brann, & Weber, 2011). For example, in a study of community theater participation, Michael Kramer (2005) notes that people who live in towns where there is only one community theater are likely to continue participating with fewer rewards than those who have other options for theater in their towns. He observes that this finding emphasizes the importance of CLalt.

CLalt provides a measure of stability rather than satisfaction. Thibault and Kelley stipulated that the CLalt suggests how likely it is that one friend would leave their relationship with another, even though it's mainly a satisfying one, for something they think would be better. As we mentioned previously, it's also the case that even if a relationship is somewhat problematic, CLalt predicts that the partners would remain in it if they believed it wasn't realistic that they'd find something better. If we further examine LaTasha and Meredith's relationship using CL and CLalt, we might speculate that Meredith finds her friendship with LaTasha very satisfying in general. She might give it an 8 out of a 10-point scale, where 1 is horrible and 10 is perfect. The 8 would be Meredith's outcome score for her relationship with LaTasha. If Meredith's expectations of friendship, or her CL, is only a 6, then Meredith should be satisfied with her friendship with LaTasha. Furthermore, if Meredith thinks that by being alone now as she decides about college she would rate herself at a 3, and she has few other friends that could even come close to her relationship with LaTasha, then her CLalt will be low. Her 8 outcome with LaTasha will exceed it, and SET predicts that Meredith will want to remain friends with LaTasha. **Table 6.2** summarizes six possible combinations among the outcome, the CL, and the CLalt and the resulting state of the relationship predicted by Thibault and Kelley's version of SET.

Table 6.2   How Outcome, CL, and CLalt Affect the State of a Relationship

RELATIVE VALUE OF OUTCOME, CL, CLalt	STATE OF THE RELATIONSHIP
Outcome > CL > CLalt	Satisfying and stable
Outcome > CLalt > CL	Satisfying and stable
CLalt > CL > Outcome	Unsatisfying and unstable
CLalt > Outcome > CL	Satisfying and unstable
CL > CLalt > Outcome	Unsatisfying and unstable
CL > Outcome > CLalt	Unsatisfying and stable

> means greater than
Adapted from Roloff, (1981).

If we calculate the numbers above for Meredith's friendship with LaTasha based on the format specified in **Table 6.2,** Meredith's outcome, 8, is greater than her CL (6), and her CL is greater than her CLalt, which is 3. Thus, her pattern fits the first example in the table and the theory predicts that the state of her relationship is satisfying and stable.

This kind of calculation—although perhaps a little unrealistic in turning a relationship into a single number—suggests why some people remain in relationships that are abusive. If people see no alternative and fear being alone more than being in the relationship, SET predicts they will stay. Some scholars have written about women in abusive relationships, using this theoretical reasoning to explain why women stay with violent men (e.g., Stafford & Kuiper, 2022; Walker, 1984). Other researchers (Cox & Kramer, 1995) note that managers in organizations use these calculations to help them make decisions about dismissing employees. And Bishop, Scott, Goldsby, and Cropanzano (2005) investigate how employees use SET thinking in forming commitments to their jobs.

Although Thibault and Kelley distinguished between CL and CLalt in terms of satisfaction and stability, not all researchers agree. In a study done in Korea, researchers Hyesoo Kim and Seongyeon Auh (2019) found that both CLalt and CL were related to dating relationship stability in college students. Perhaps this is unsurprising because satisfaction with a relationship is likely related to relationship stability.

Power and Exchange Patterns

In addition to studying how people calculate their relational outcomes, Thibault and Kelley were interested in how people adjust their behaviors when interacting with their relational partners. Thibault

behavioral sequences a series of actions designed to achieve a goal

and Kelley suggest that when people interact, they are goal directed. This is congruent with their assumption that human beings are rational. People, according to Thibault and Kelley, engage in **behavioral sequences,** or a series of actions designed to achieve their goal. These sequences are the heart of what Thibault and Kelley conceptualize as social exchange. In enacting these behavioral sequences people are interdependent, as we've mentioned. Their interdependence relates to the concept of *power*—the dependence a person has on another for outcomes. If Meredith depends more on LaTasha for rewards than vice versa, then LaTasha has greater power than Meredith in their relationship. Further, if Meredith cares more about their relationship than LaTasha does, again, LaTasha will have greater power. This example illustrates what Willard Waller

(1937) termed the principle of least interest. Waller's principle informs a great deal of how Thibault and Kelley thought about power in relationships (Carpenter, 2017).

There are two types of power in SET: fate control and behavior control. **Fate control** is the ability to affect a partner's outcomes. For example, if Meredith withholds her friendship from LaTasha, she affects LaTasha's outcome. If LaTasha cannot replace Meredith as a friend, Meredith's behavior gives her fate control over LaTasha. This presumes that Meredith does not care about the relationship. If she does care about it, then withholding her friendship is a punishment for her as well, which gives LaTasha a certain amount of fate control in the relationship, too.

> **fate control** the ability to affect a partner's outcomes

Behavior control is the power to cause another's behavior to change by changing one's own behavior. If Meredith texts LaTasha, it's likely that LaTasha will stop what she's doing to read the text and then text Meredith back. If LaTasha is talking with Meredith in person and suddenly falls silent, Meredith will probably change her behavior in response. She might stop talking too, or she might question LaTasha to find out if something is wrong.

> **behavior control** the power to change another's behavior

Thibault and Kelley state that people develop patterns of exchange to cope with power differentials and to deal with the costs associated with exercising power. These patterns describe behavioral rules or norms that indicate how people trade resources in an attempt to maximize rewards and minimize costs. Thibault and Kelley describe three different matrices in Social Exchange to illustrate the patterns people develop: the given matrix, the effective matrix, and the dispositional matrix.

The **given matrix** represents the behavioral choices and outcomes that are determined by a combination of external factors (the environment) and internal factors (the specific skills each interactant possesses). When two people engage in an exchange, the environment may make some options more difficult than others. LaTasha's and Meredith's families, for instance, are part of the environment that's making their friendship more difficult. Furthermore, the given matrix depends on the skills people bring to a social exchange. If people lack skills for ballroom dancing, for example, it's unlikely that they'll spend time dancing together. To a certain degree, the given matrix represents "the hand you are dealt."

> **given matrix** the constraints on your choices due to the environment and/or your own skill levels

People may be restricted by the given matrix, but they are not trapped by it. They can transform it into the **effective matrix**, "which represents an expansion of alternative behaviors and/or outcomes which ultimately determines the behavioral choices in social exchange" (Roloff, 1981, p. 51). If people don't know how to tango, they can take lessons and learn the dance, transforming the given matrix into the effective matrix. If Meredith and LaTasha think their families are bothering their friendship too much, they can engage in conflict with them until they change their families' minds. Or they can stop talking about each other at home and keep their friendship secret so that they can avoid their families' negative sanctions.

> **effective matrix** the transformations you are able to make to your given matrix, by learning a new skill, for example

The final matrix, the **dispositional matrix**, represents the way two people believe that rewards ought to be exchanged between them. If Meredith and LaTasha think that friends ought to stick together no matter how much outside interference they get, that will affect their dispositional matrix. If people view exchanges as competition or a game, these beliefs will be reflected in their dispositional matrix.

> **dispositional matrix** the beliefs you have about how rewards should be exchanged in a relationship

Thibault and Kelley assert that if we know the kinds of dispositions people have (the dispositional matrix) and the nature of the situation in which they are operating (the given matrix), then we'll know how to predict the transformations they'll make (the effective matrix) to impact the social exchange. Thus, if we understand that LaTasha expects unqualified loyalty from her friends, and from herself as a friend, and we know that her family's opposition to her friendship to Meredith is not too strong, we might predict that LaTasha will defend Meredith to her family and attempt to change their beliefs. If her family has a stronger opposition, we might predict that LaTasha will simply remain friends with Meredith without trying to change her family's beliefs. The dispositional matrix guides the transformations people make to their given matrix; these transformations lead to the effective matrix, which determines the social exchange.

Integration and Critique

SET has generated a great deal of research and has been called "one of the major theoretical perspectives in the field of social psychology" (Cook & Rice, 2003, p. 53). SET is grounded in quantitative research, making it an empirically driven framework from which to examine interpersonal relationships. As we evaluate SET, we'll address the criteria of scope, utility, testability, and heurism.

Integration

Communication Tradition	Rhetorical | Semiotic | Phenomenological | Cybernetic | **Socio-Psychological** | Socio-Cultural | Critical
Communication Context	**Intrapersonal** | **Interpersonal** | **Small Group/Team** | **Organizational** | Public/Rhetorical | Mass/Media | Cultural
Approach to Knowing	**Positivistic/Empirical** | Interpretive/Hermeneutic | Critical

Critique

Evaluation Criteria	**Scope** | Logical Consistency | Parsimony | **Utility** | **Testability** | **Heurism** | Test of Time

Scope

When examining SET on the basis of scope, some critics comment that the theory fails to explain the importance of group solidarity in its emphasis on individual need fulfillment (England, 1989). This critique argues that the exchange framework is too focused on the separate self, and it overemphasizes rationality and self-interest (Sabatelli & Shehan, 1993). This priority overlooks and undervalues the connected self. In some ways, this objection has ontological considerations as well, but it also suggests that the scope of the theory is too narrow. SET only considers the individual as a unique entity without focusing on the individual as a member of a group. Because of this, critics argue, SET cannot account for relationships in cultures that prioritize connection over individuality, for example.

Utility

The criterion of utility suggests that if the theory doesn't present an accurate picture of people, it isn't useful. The important question for SET becomes: Are humans really that calculating (Zafirovski, 2005)? SET has been criticized for the conceptualization of human beings it advances. In the theory, humans are seen as rational calculators, coming up with numerical equations to represent their relational life. Many people object to this understanding of humans, asking whether people really rationally calculate the costs and rewards to be realized in a relationship. Social Exchange assumes a great deal of cognitive awareness and activity, which several researchers have questioned (Berger & Roloff, 1980). Researchers haven't come to a definitive answer about how much people calculate their relational life, but this calculation probably ebbs and flows according to many factors. First, some contexts may make people more self-aware than others. As LaTasha and Meredith receive more pressure to decide about college, they may think about their relationship more than they did when they were younger. Second, some individual differences might affect how people process information. Some people are simply more self-aware than others (Snyder, 1979). As researchers continue to work with this theory, they must account for these and other factors relative to their calculations.

Also, SET doesn't completely explain how people do these calculations, and some research suggests that the process might not be conscious at all. Rather, people might simply automatically and unintentionally track positive and negative experiences involving their partners, and these affective, intuitive feelings "become automatically associated with the partner in memory to form and update one's automatic partner attitude" (Hicks & McNulty, 2019, p. 255). So, it's not that LaTasha and Meredith each sit down and calculate the worth of their relationship consciously. Instead, they simply have an automatic association of the worth of the relationship stored in memory that gets adjusted periodically as new situations arise.

In addition, critics wonder if people are really as self-interested as SET assumes. Steve Duck (1994) argues that applying a marketplace mentality to the understanding of relational life misrepresents what goes on in relationships. He suggests that it's wrong to think about personal relationships in the same way that we think about business transactions, like buying a house or a car. This suggestion relates to the approaches to knowing one brings to the theory, as we discussed in **Chapter 3**. For some people, the analogy of the marketplace is appropriate, but for others it's not and may be highly offensive.

Testability

A common criticism of SET is that it's not testable. As we discussed in **Chapter 3,** one important attribute of a theory is that it is testable and capable of being proven false. The difficulty with SET is that its central concepts—costs and rewards—are not clearly defined. As Sabatelli and Shehan (1993) note,

> It becomes impossible to make an operational distinction between what people value, what they perceive as rewarding, and how they behave. Rewards, values, and actions appear to be defined in terms of each other (Turner, 1978). Thus, it is impossible to find an instance when a person does not act in ways so as to obtain rewards. (p. 396)

When the theory argues that people do what they can to maximize rewards and then also argues that whatever people do is rewarding, it's difficult to disentangle the two concepts. As long as SET operates with these types of circular definitions, it will be untestable and, thus, unsatisfactory. However, Roloff (1981) observes that some work has been done to create lists of rewards in advance of simply observing what people do and labeling that as rewarding because people are doing it. For example, Uriel Foa and Edna Foa (1974) began clearly defining rewards apart from people's behavior 50 years ago.

Heurism

People who support SET point out that it has been heuristic. Studies in many diverse areas, from corporations (e.g., Davis-Sramek, Hopkins, Richey, & Morgan, 2022; Muthusamy & White, 2005) to foster care (Timmer, Sedlar, & Urquiza, 2004), have been framed using the tenets of Social Exchange. Researchers (e.g., DeHart, 2012) have suggested that Social Exchange might offer a useful framework for examining coaching people in positive communication. Researchers have also examined communication in romantic relationships (Carpenter, 2017; Frisby, Sidelinger, & Booth-Butterfield, 2015), theater groups (Kramer, 2005), microblogging (Liu, Min, Zhai, & Smyth, 2016), and between humans and service robots (Kim, So, & Wirtz, 2022) using SET. Research on microblogging and between humans and bots suggest that a theory conceptualized in 1959 still has relevance and still generates questions and explanations in an entirely different communication environment. Nusrat Jahan and Seung Woon Kim (2021) agree and found that SET explained people's interaction in online communities. Furthermore, the emphasis that Thibault and Kelley placed on interdependence is congruent with many researchers' notions of interpersonal relationships. As a theory that examines human relationships, SET resonates with a number of scholars.

Students Talking Tech: Graham

I'm kind of wondering how social media might affect SET. The chapter said some researchers looked at microblogging and online communities with SET. But I think that some social media might change the theory—or at least change what we think are costs and rewards. Before we had texting, for instance, I'm sure people would think it was a cost to be with a friend who didn't always pay attention to you. I know my mom thinks that's a huge cost, and she's always yelling at me and my sisters if we try to talk to her and also check texts or Snapchat. But I'm not sure that I think it's too big of a deal. Sure, a person can overdo it, but it's not an issue to me when a friend checks texts while also talking to me. We're the generation of multitasking, and everyone pretty much does this. Maybe the whole theory doesn't have to change, but I do think that some parts of it might have to take a new way of communicating into account.

Closing

Since its introduction decades ago, researchers have found SET relevant for framing studies in an ever-changing complex society. The theory emphasizes the need for people to assess the benefits and risks of their relationships and ultimately calculate their value. Given the importance and unpredictability of relational life, we're likely to see this theory incorporated into research for years to come.

Discussion Starters

 Case-In-Point: Discuss Meredith and LaTasha's given matrix, transformational matrix, and dispositional matrix.

Try-It-Your-Selfie: How do you think SET could (or could not) explain a relationship between gamers who interact frequently online, but never meet face-to-face? Does it weaken the theory if it cannot explain online relationships, or does that simply become a boundary condition of the theory? Explain your reasoning.

1. Explain the problem with SET regarding testability. Is there anything that Social Exchange theorists could do to make the theory more testable? Use examples in your response.

2. Choose a current relationship of yours and perform a cost–benefit analysis on it. Assess whether the relationship meets, fails to meet, or exceeds your comparison level and your comparison level for alternatives.

3. How does SET explain the unselfish things that people do that do not seem calculated to gain rewards for themselves?

4. Have you ever stayed in a relationship because you thought you did not have any other alternatives? Explain.

References

Balliet, D., Tybur, J. M., & Van Lange, P. A. M. (2017). Functional interdependence theory: An evolutionary account of social situations. *Personality and Social Psychology Review, 21*(4), 361–388.

Berger, C. R., & Roloff, M. E. (1980). Social cognition self-awareness and interpersonal communication. In B. Dervin & M. Voigt (Eds.), *Progress in communication sciences* (Vol. 2, pp. 158–172). Ablex.

Bishop, J. W., Scott, K. D., Goldsby, M. G., & Cropanzano, R. (2005). A construct validity study of commitment and perceived support variables: A multifoci approach across different team environments. *Group Organizational Management, 30*(2), 153–180.

Carpenter, C. J. (2017). A relative commitment approach to understanding power in romantic relationships. *Communication Studies, 68*(1), 115–130.

Cloarec, J., Meyer-Waarden, L., & Munzel, A. (2021). The personalization–privacy paradox at the nexus of social exchange and construal level theories. *Psychology & Marketing, 39*(3), 647–661.

Cook, K. S., & Rice, E. (2003). Social exchange theory. In J. Delamater (Ed.), *Handbook of social psychology* (pp. 53–76). Kluwer Academic/Plenum.

Cox, S. A., & Kramer, M. W. (1995). Communication during employee dismissals: Social exchange principles and group influences on employee exit. *Management Communication Quarterly, 9*, 119–161.

Dainton, M. (2015). An interdependence approach to relationship maintenance in interracial marriage. *Journal of Social Issues, 71*(4), 772–787.

Davis-Sramek, B., Hopkins, C. D., Richey, R. G., & Morgan, T. R. (2022). Leveraging supplier relationships for sustainable supply chain management: Insights from social exchange theory. *International Journal of Logistics Research and Applications, 25*(1), 101–118.

DeHart, P. (2012). *Laws common and unwritten: On the natural law and the possibility of civic cooperation for the common good.* APSA 2012 Annual Meeting Paper. Retrieved from http://ssrn.com/abstract=2110682.

Dillow, M. R., Malachowski, C. C., Brann, M., & Weber, K. D. (2011). An experimental examination of the effects of communicative infidelity motives on communication and relational outcomes in romantic relationships. *Western Journal of Communication, 75*(5), 473-499.

Duck, S. (1994). *Meaningful relationships.* Sage.

England, P. (1989). A feminist critique of rational-choice theories: Implications for Sociology. *The American Sociologist, 20,* 14-28.

Foa, U., & Foa, E. (1974). *Societal structures of the mind.* Thomas.

Frisby, B. N., Sidelinger, R. J., & Booth-Butterfield, M. (2015). No harm, No foul: A social exchange perspective on individual and relational outcomes associated with relational baggage. *Western Journal of Communication, 79*(5), 555-572.

Gerpott, F. H., Balliet, D., Columbus, S., Molho, C., & de Vries, R. E. (2018). How do people think about interdependence? A multidimensional model of subjective outcome interdependence. *Journal of Personality and Social Psychology, 115*(4), 716-742.

Hicks, L. L., & McNulty, J. K. (2019). The unbearable automaticity of being . . . in a close relationship. *Current Directions in Psychological Science, 28*(3), 254-259.

Jahan, N., & Kim, S. W. (2021). Understanding online community participation behavior and perceived benefits: A social exchange theory perspective. *PSU Research Review, 5*(2), 85-100.

Kelley, H. H., & Thibault, J. (1978). *Interpersonal relations: A theory of interdependence.* John Wiley & Sons.

Kelley, H. H., Holmes, J. G., Kerr, N. L., Reis, H. T., Rusbult, C. E., & Van Lange, P. A. M. (2003). *An atlas of interpersonal situations.* Cambridge University Press.

Killen, M., Luken Raz, K., & Graham, S. (2022). Reducing prejudice through promoting cross-group friendships. *Review of General Psychology, 26*(3), 361-376.

Kim, H., & Auh, S. (2019). The influences from comparison level and comparison level for alternatives on the dating relationship stability, and mediating effects from the commitment among the college students: An application of the interdependence theory. *Family and Environment Research, 57*(1), 127-142.

Kim, H., So, K. K. F., & Wirtz, J. (2022, October). Service robots: Applying social exchange theory to better understand human-robot interactions. *Tourism Management, 92.*

Kramer, M. W. (2005). Communication and social exchange processes in community theater groups. *Journal of Applied Communication Research, 33*(2), 159-182.

Lavelle, J. J., Rupp, D. E., & Brockner, J. (2007). Taking a multifoci approach to the study of justice, social exchange, and citizenship behavior: The target similarity model. *Journal of Management, 33*(6), 841-866.

Liu, Z., Min, Q., Zhai, Q., & Smyth, R. (2016). Self-disclosure in Chinese micro-blogging: A social exchange theory perspective. *Information & Management, 53*(1), 53-63.

Monge, P. R., & Contractor, N. (2003). *Theories of communication networks.* Oxford University Press.

Muthusamy, S. K., & White, M. A. (2005). Learning and knowledge transfer in strategic alliances: A social exchange view. *Organization Studies, 26*(3), 415-441.

Pascale, R., & Primavera, L. H. (2016). *Making marriage work: Avoiding the pitfalls and achieving success.* Roman & Littlefield.

Rawlins, W. K. (1992). *Friendship matters: Communication, dialectics, and the life course.* Aldine De Gruyter.

Roloff, M. E. (1981). *Interpersonal communication: The social exchange approach.* Sage.

Roloff, M. E. (2016). Social exchange theories. In C. R. Berger & M. E. Roloff (Eds.), *The international encyclopedia of interpersonal communication,* (pp. 1-19). John Wiley & Sons.

Sabatelli, R. M., & Shehan, C. L. (1993). Exchange and resource theories. In P. G. Boss, W. J. Doherty, R. LaRossa, W. R. Schumm, & S. K. Steinmetz (Eds.), *Sourcebook of family theories and methods: A contextual approach* (pp. 385–411). Springer Science & Business Media.

Snyder, M. (1979). Self-monitoring processes. In L. Berkowitz (Ed.), *Advances in experimental social psychology* (pp. 86–131). Academic Press.

Stafford, L. (2008). Social exchange theories. In L. A. Baxter & D. O. Braithwaite (Eds.), *Engaging theories in interpersonal communication: Multiple perspectives* (pp. 377–389). Sage.

Stafford, L., & Kuiper, K. (2022). Social exchange theories: Calculating the rewards and costs of personal relationships. In D. O. Braithwaite & P. Schrodt (Eds.), *Engaging theories in interpersonal communication: Multiple perspectives* (3rd ed., digital ed.).

Thibault, J. W., & Kelley, H. H. (1959). *The social psychology of groups*. John Wiley & Sons.

Timmer, S. G., Sedlar, G., & Urquiza, A. J. (2004). Challenging children in kin versus nonkin foster care: Perceived costs and benefits to caregivers. *Child Maltreatment, 9*(3), 251–262.

Trevino, A. J., & Tilly, C. (2016). *George C. Homans: History, theory, and method*. Routledge.

Walker, L. (1984). *The battered woman syndrome*. Springer.

Waller, W. (1937). The rating and dating complex. *American Sociological Review, 2*(5), 727–734.

Zafirovski, M. (2005). Social exchange theory under scrutiny: A positive critique of its economic-behaviorist formulations. *Electronic Journal of Sociology, 2*(2), 1–40.

CHAPTER 7
Social Penetration Theory

*Based on the research of **Irwin Altman** and **Dalmas Taylor***

> *"Communication and disclosure intimacy appear to be the sine qua non of developing satisfying interpersonal relationships."*
>
> —Irwin Altman and Dalmas A. Taylor

Jason LaSalle

About three years ago, Jason LaSalle's wife, Nikkya, died in a car accident, leaving Jason a single parent of 8-year-old twins. Since his wife's death, he has struggled both financially and emotionally. He has worried about making his rent and van payments and about meeting his children's needs. For the past three years, Jason has worked odd jobs around the neighborhood to supplement his modest income as custodian for a local cinema complex. In addition, Jason has been lonely. He is shy around others, especially women. Nikkya was the only woman he really felt comfortable with, and he misses her a great deal.

Jason's sister, Kayla, is always trying to get him out of the house. One night, she took it upon herself, hired a babysitter, and picked up her brother to go out. This evening was especially important to Kayla because she had also invited her friend Elise Porter, who was recently divorced. Kayla thought that Elise might be a good match for her brother. She was hoping that Elise's easygoing nature and her great sense of humor would appeal to Jason. Throughout the evening, Jason and Elise talked about a variety of things, including their experiences as single parents, her divorce, and the two children they were each raising. Much of their night was spent dancing or talking to each other. The evening ended with Jason and Elise promising to get together again soon.

As Jason drove home to his apartment, he only thought about Nikkya. He was lonely. It had been three years since he had shared any intimacy with an adult. When he arrived home, his heart raced as he caught sight of the family picture taken at Disney World shortly before her death. He wasn't sure if it was a good time to pursue an intimate relationship, and yet he wanted a chance to see what kind of person Elise was. He knew that future dates would inevitably require him to talk about Nikkya, and he felt that these conversations would be very tough. He would have to open up emotionally to Elise, and the idea of being placed in such a vulnerable position seemed challenging.

After he paid the babysitter and closed the door behind her, he walked into the twins' room and gave each a kiss on the forehead. Sitting drinking a beer in the living room, Jason felt unsure about embarking upon something new, exciting, and a bit frightening.

When we say we're close to someone, we often act as though others understand precisely what we mean. That's not always the case, however. Saying that you are close or intimate with someone may not be universally understood. Certainly, as we think about cultural influences upon the intimacy process, saying that "we're close" means different things to different people across the globe.

To understand the relational closeness between two people, Irwin Altman and Dalmas Taylor (1973) conceptualized Social Penetration Theory (SPT). The two conducted research into the area of social bonding among various types of couples. Their theory illustrates a

social penetration process of bonding that moves a relationship from superficial to more intimate

pattern of relationship development, an evolution that they identified as social penetration. First, so we have a common foundation, a definition is in order. **Social penetration** refers to a process of relationship bonding whereby individuals move from superficial communication to more intimate communication. According to Altman and Taylor, intimacy involves more than physical intimacy; other dimensions of intimacy include intellectual and emotional, and the extent to which a couple shares activities (West & Turner, 2023). The social penetration process, therefore, necessarily includes verbal behaviors (the words we use), nonverbal behaviors (our body posture, the extent to which we smile, etc.), and environmentally oriented behaviors (the space between communicators, the physical objects present in the environment, etc.).

Altman and Taylor (1973) believe that people's relationships vary tremendously in their social penetration. From husband–husband

trajectory pathway to closeness

to supervisor–employee to golf partners to physician–patient, the theorists conclude that relationships "involve different levels of intimacy of exchange or degree of social penetration" (p. 3). The authors note that relationships follow some particular **trajectory,** or pathway, to closeness. Furthermore, they contend that relationships are somewhat organized and predictable in their development. Because relationships are critical and are pivotal to our humanity (Ruell, 2023), Social Penetration theorists attempt to unravel the simultaneous nature of relational complexity and predictability. Finally although many individuals may have established online relationships, Altman and Taylor did not conceptualize this development in their writing. Still, we retain the discussion of technology in this chapter.

The opening story of Jason LaSalle and his arranged date illustrates a central feature of SPT. The only way for Jason and Elise to understand each other is for them to engage in personal conversations; such discussion requires each sharing personal bits of information. As the two become closer, they will move from a nonintimate relationship to an intimate one. In addition, each person's personality will influence the direction of the relationship. So Jason and Elise's relationship will be influenced by Jason's shyness and Elise's easygoing manner. The future of Jason's relationship with Elise is based on a multiplicity of factors— factors that we will explore throughout this chapter.

Early discussions of SPT began during the 1960s and 1970s, an era when opening up and talking candidly was especially valued as an important relational strategy. Through the years, however, communication researchers and practitioners acknowledged that cultures can vary tremendously in their endorsement of openness as a relational skill, and some scholars are now questioning the initial enthusiasm for so much relational openness in general. Therefore, as you read this chapter, keep in mind that we are discussing a theory that is rooted in a generation for which speaking freely and without boundary was a highly valued characteristic. Nevertheless, much of the theory remains relevant today as we live in a society where openness is still a valued personal characteristic. Simply read Ashish's comments in Students Talking for verification of openness!

Students Talking: Ashish

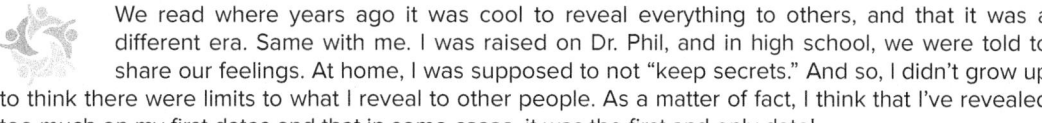

We read where years ago it was cool to reveal everything to others, and that it was a different era. Same with me. I was raised on Dr. Phil, and in high school, we were told to share our feelings. At home, I was supposed to not "keep secrets." And so, I didn't grow up to think there were limits to what I reveal to other people. As a matter of fact, I think that I've revealed too much on my first dates and that in some cases, it was the first and only date!

To begin, we outline several assumptions of SPT. We then identify the catalyst for the theory and provide a detailed example of the stages of the penetration process.

Theory at a Glance • Social Penetration Theory

Relational life is quite complex. Yet, interpersonal relationships evolve in some gradual and predictable fashion. Social Penetration theorists believe that self-disclosure is the primary way that superficial relationships progress to intimate relationships. Although self-disclosure can lead to more intimate relationships, it can also leave one or more persons vulnerable.

Assumptions of Social Penetration Theory

Social Penetration Theory (called a "stage theory" by Mongeau, Henningsen, & Oliver-Blackburn, 2021), has enjoyed widespread acceptance by a number of scholars in the communication discipline. Part of the reason for the theory's appeal is its straightforward approach to relationship development. Although we alluded to some assumptions earlier, we will explore the following assumptions that guide SPT:

- Relationships typically progress from nonintimate to intimate.
- Relational development is generally systematic and predictable.
- Relational development includes depenetration and dissolution.
- Self-disclosure is at the core of relationship development.

To begin, one fundamental assumption posits that relational communication between people begins at a rather superficial level (nonintimate) and moves along a continuum to a more intimate level. On their date, arranged by Kayla, Jason and Elise no doubt talked about trivial issues related to being single parents. They also likely shared how difficult it is to have enough time in the day to do everything, but they probably didn't express highly personal feelings about their family lives. These initial conversations at first may appear unimportant, but as Jason discovers, such conversations allow an individual to size up the other and provide the opportunity for the early stages of relational development. Jason feels awkward, but this awkwardness can pass. With time, relationships have the opportunity to become intimate.

Not all relationships fall into the extremes of nonintimate or intimate (Punyanunt-Carter, 2019). In fact, many of our relationships are somewhere in between these two poles. Often, we may want only a moderately close relationship. For instance, we may want a relationship with a coworker to remain sufficiently distant so that we do not know what goes on in her house each night or how much money she has in the bank. Yet we need to know enough personal information to have a sense of whether she can complete her part of a team project.

The second assumption of SPT pertains to predictability. Specifically, Social Penetration theorists argue that relationships progress fairly systematically and predictably. Some people may have difficulty with this claim. After all, relationships—like the communication process—are dynamic and ever changing, but even dynamic relationships follow some acceptable standard and pattern of development.

To better understand this assumption, again consider Jason LaSalle. Without knowing all the specifics of his situation, we could figure out that if he pursues a relationship with Elise, he will have to work through his emotions about Nikkya. In addition, he must inevitably reconcile how their families might merge if the relationship progresses into more intimacy. We could probably predict that the relationship will move slowly at first while both Jason and Elise work out their feelings and emotions.

These projections are grounded in the second assumption of the theory: relationships generally move in an organized and predictable manner. Although we may not know precisely the direction of a relationship or be able to predict its exact future, social penetration processes are rather organized and predictable. We can be fairly sure, for instance, that Jason and Elise will not introduce each other to important people in their families before they date a few more times. We would also expect that neither would declare their love for the other before they exchanged more intimate information. Of course, a number of other events and variables (time, personality, and so forth) affect the way relationships progress and what we can predict along the way. As Altman and Taylor (1973) conclude, "People seem to possess very sensitive tuning mechanisms which enable them to program carefully their interpersonal relationships" (p. 8).

The third assumption of SPT pertains to the notion that relational development includes depenetration and dissolution. At first, this may sound a bit peculiar. Thus far, we have explored the coming

depenetrate slow deterioration of relationship

together of a relationship. Yet relationships do fall apart, or **depenetrate,** and this depenetration can lead to relationship dissolution. Elise, for example, may be unprepared for Jason's past and may wish to depenetrate and ultimately dissolve the relationship.

Addressing depenetration and dissolution, Altman and Taylor liken the process to a film shown in reverse. Just as communication allows a relationship to move forward toward intimacy, communication could move a relationship back toward *nonintimacy*. If the communication is conflictual, for example, and this conflict continues to be destructive and unresolved, the relationship may take a step back and become less close. And Social Penetration theorists think that depenetration—like the penetration process—is often systematic.

If a relationship depenetrates, it does not mean that it will auto-matically dissolve or terminate. At times, relationships experience **transgressions,** or a violation of relational rules, practices, and expec-tations. These transgressions may seem unworkable and, at times,

transgression a violation of relational rules, practices, and expectations

they are. In fact, Laura Blackie and Kate McLean (2022) point out that transgressions can be quite influ-ential in relational life. In **Chapter 16,** we discuss unwanted repetitive patterns of conflict in couples, which, in part, represent these transgressions. We note that recurring conflicts characterize a number of different relationship types and that couples generally learn to live with these conflicts. You may believe conflict or relational transgressions will inevitably lead to dissolution, but depenetration does not necessarily mean that the relationship is doomed. In fact, it might enliven a dormant relationship or allow relational partners to frame the depenetration in positive ways (Masur, 2019).

The final assumption contends that self-disclosure is at the core of relationship development. **Self-disclosure** can be generally defined as the purposeful process of revealing significant information about yourself to others. Usually, the information that makes up self-

disclosure is of a significant nature. For instance, revealing that you like to play the piano may not be all that important; revealing a more personal piece of information, such as that you are a practicing Catholic or that you don't support COVID vaccinations may influence the evolution of a relationship.

According to Altman and Taylor (1973), nonintimate relationships progress to intimate relationships because of self-disclosure. This process allows people to get to know each other in a relationship. Further, Ann Weber (2019) notes that "healthy intimacy" is related to the appropriateness of time and place. In other words, sometimes revealing personal information can be untimely and done in the wrong environment. Imagine, for example, someone revealing that they have a prison record on their second date at a family reunion! Self-disclosure helps shape the present and future relationship between two people, and "making [the] self accessible to another person is intrinsically gratifying" (p. 50). Elise will understand the challenges that lie ahead for her in a relationship with Jason by hearing Jason reveal his feelings about his wife's death and his desire to begin dating again. In turn, because social penetration requires a "gradual overlapping and exploration of their mutual selves by parties to a relationship" (p. 15), Elise, too, would have to self-disclose her thoughts and feelings.

Theory Into Practice • Social Penetration Theory

Theoretical Claim: Reciprocity with similar levels of "risk" typically leads to increased intimacy between people.

Practical Implication: Lindsey reveals to Bradley that her father will not attend their wedding because he "disapproves" of a biracial marriage. Bradley states that his mother, too, had also felt uncomfortable about their upcoming cultural union, but after he talked with her, she was more welcoming. He stated that he would like to sit down with Lindsey and with her dad to communicate how much the couple love each other. Lindsey feels more connected to her fiancé than ever before because of his gracious overture.

Finally, we should note that self-disclosure can be strategic or nonstrategic. That is, in some relationships, we tend to plan out what we will say to another person. In other situations, our self-disclosure may be spontaneous. Spontaneous self-disclosure is widespread in our society. In fact, researchers have used the phrase

"**stranger-on-the-train** (or plane or bus) phenomenon" to refer to those times when people reveal information to complete strangers in public places. To be precise, stranger-on-the-train refers to those moments of "sharing with an anonymous seat-mate intimate details [of which] not even our closest friends are aware" (Bareket-Bojmel & Shahar, 2011, pp. 732–733). Think about how many times you have been seated next to a stranger on a trip, only to have that person disclose personal information throughout the journey. Or, what about the times that you did the same thing to someone else? Interpersonal communication researchers continue to investigate why people engage in this activity. One conclusion that has emerged relates to whether this "seatmate" talk results in more positive experiences. Julianna Schroeder, Donald Lyons, and Nicholas Epley

(2022), for instance, found that when commuters in London were randomly assigned to either talk with strangers or travel alone, those who talked with strangers reported more pleasant commuting experiences than those who were alone.

"Tearing Up" the Relationship: The Onion Analogy

Previously we discussed the importance of revealing information about oneself of which others are unaware. In their discussion of SPT, Altman and Taylor incorporate an onionskin structure (**Figure 7.1**).

public image outer layer of a person; what is available to others

They believe that a person like Jason LaSalle can be compared to an onion, with the layers (concentric circles) of the onion representing various aspects of a person's personality. The outer layer is an individual's **public image,** or that which is available to the naked eye. This is the "publicly observable self" where "private information that is stored at deeper layers must be discovered" (Fox, 2015, p. 6). Jason's public image is an African American male in his mid-40s who is slightly balding. Elise Porter is also an African American, but is significantly taller than Jason and has very short hair. A layer of the public image is removed, however, when Jason discloses to his date his frustrations with being a single father.

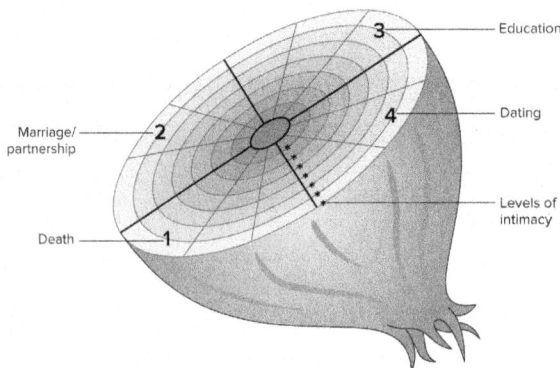

Figure 7.1   The Social Penetration Process of Jason LaSalle

As the evening evolves for the two of them and things are looking favorable, Jason and Elise no doubt begin to reveal additional layers of their personalities. And, only by developing their relationship can

reciprocity the return of openness from one person to another

their inner layers be revealed (Punyanunt-Carter, 2019). For instance, Elise may reveal that she, too, experiences single-parent anxieties. This **reciprocity,** or the process whereby one person's openness leads to the other's openness, is a primary component in SPT. In fact, researchers have shown consistently that when one person divulges personal information, the other person is likely to reciprocate similar levels of sensitive information (Acquisti, John, & Loewenstein, 2012). Reciprocity has been shown to be significant in both established and new relationships, such as Jason and Elise's. Further, in terms of online relationships, people are more inclined to self-disclose to others online than face-to-face (see Social Information Processing Theory, **Chapter 8**). One reason explaining this fact is that some scholars feel that even carrying a cell phone suggests that we are permanently online and always prepared to reveal something to others (Masur, 2019).

Before leaving the discussion of self-disclosure, we should point out that penetration is usually viewed along two dimensions: breadth and depth. **Breadth** refers to the number of various topics discussed in the relationship; **breadth time** pertains to the amount of time that relational partners spend communicating with each other about these various topics. **Depth** refers to the degree of intimacy that guides topic discussions. In the initial stages, relationships can be classified as having narrow breadth and shallow depth. For Jason LaSalle, it is feasible that his first date with Elise was characterized

breadth the number of topics discussed in a relationship

breadth time amount of time spent by relational partners discussing various topics

depth degree of intimacy guiding topic discussion

this way. Most likely, the two did not discuss many topics, and what they did discuss probably lacked intimate overtones. As relationships move toward intimacy, we can expect a wider range of topics to be discussed (more breadth), with several of those topics marked by depth.

Students Talking Tech: TJ

 Just thinking about public image and social media made me think about LinkedIn. My picture is really formal and I have posted everything that is related to my professional career choices. When I look at my Instagram page, my profile has nothing that would show I'm even in a career, and certainly I don't look "professional." I know the two types of media have different purposes and my public image on both shows that!

A few conclusions are important with respect to the breadth and depth of self-disclosure. First, shifts or changes in central layers (of the onion) have more of an impact than those in outer, or peripheral, layers. Because an individual's public image, or outer layer, represents those things that others can see, or what we can term the *superficial*, we expect that if there are changes in the outer layer, the consequence is minimal. For example, if Elise changed her hairstyle, her relationship with Jason would be less affected than if she changed her opinion about premarital sex.

Second, the greater the depth, the more opportunity for a person to feel vulnerable. Imagine that Jason reveals some inadequacy about himself to Elise—for instance, the fact that he was on public assistance for a few years after his wife's death. When he reveals this personal information to Elise, she can respond in several different ways. She can simply say, "Wow," and not venture further into the discussion. Or she can reply, "That must have been very hard for you," communicating compassion. A third possible response is "I don't see anything wrong with that. Millions of people need some help at some point in their lives." The latter response demonstrates even more compassion and an effort to diffuse the possible anxiety that Jason is feeling. How Elise responds influences how vulnerable Jason feels. As you can see, the first response may elicit a high degree of vulnerability, whereas the third response may invoke little vulnerability.

As you reflect on the topic of self-disclosure, keep in mind that an individual should be judicious in using self-disclosure. That is, although self-disclosure generally moves a relationship toward more closeness, if people disclose too much during the early stages of a relationship, they may actually end the relationship. Some partners may be ill equipped and underprepared to know another so intimately. Also note that trust is an inherent part of the disclosure and reciprocity processes. In fact, it's very difficult to disentangle the relationship between trust and self-disclosure as the two are inextricably linked (Espinoza, Garcia-Fornes, & Sierra, 2012). If we desire reciprocity in disclosure, we must try to gain the trust of the other person and, similarly, feel trustful of the other person. If a person reveals that they had COVID, it's likely this disclosure was preceded by a trusting bond between you and the other. Perhaps you previously disclosed that you've had COVID or even that you believe that the virus may eventually infect everyone. This disclosure is both thoughtful and appropriate. We have included other guidelines for self-disclosure in **Table 7.1.**

Table 7.1   Guidelines for Self-Disclosure

ASK YOURSELF	CONSIDERATIONS
Is the other person important to you?	Reveal significant pieces of information about yourself to those people with whom you have developed a close personal relationship.
Is the risk of disclosing reasonable?	Determine whether revealing significant information about yourself has great risk associated with it. Assess the risk potential of your disclosure.
Are the amount and type of disclosure appropriate?	Discern whether you are revealing too much or too little information. Examine the timing of the disclosure.
Is the disclosure relevant to the situation at hand?	Constant disclosure is not typically useful in a relationship. Don't share everything.
Is the disclosure reciprocated?	Unequal self-disclosure usually creates an imbalanced relationship. Wait for reciprocity.
Will the effect be constructive?	If not employed carefully, disclosure can be used in destructive ways. Use care in disclosing information that may be perceived as damaging.
Are cultural misunderstandings possible?	Work to maintain cultural sensitivity and empathy as people disclose to you and you disclose to others.

A Social Exchange: Relational Costs and Rewards

Social Penetration Theory is grounded in several principles from many theories related to relationship development. Chiefly among these is Social Exchange Theory (Thibault & Kelley, 1959). As we discussed in **Chapter 6,** this theory suggests that social exchanges that are shared between communicators are, in fact, mutual obligations which have a significant impact upon the relationship. Altman and Taylor based some of their work on social exchange processes; that is, an exchange of resources between individuals in a relationship. Specifically, rewards and costs relate to Social Exchange Theory.

Taylor and Altman (1987) argue that relationships can be conceptualized in terms of rewards and costs, two terms from Social Exchange Theory (**Chapter 6**). Rewards are those relational events or behaviors that stimulate satisfaction, pleasure, and contentment in a relational

reward–cost ratio balance between positive and negative relationship experiences

partner, whereas costs are those relational events or behaviors that stimulate negative feelings or experiences. Quite simply, if a relationship provides more rewards than costs, then individuals are more likely to stay in that relationship. However, if an individual believes that there are more costs to being in a relationship, then relationship dissolution may be probable. For instance, Jason LaSalle will most likely regulate the closeness of his relationship with Elise by assessing a **reward–cost ratio,** which is defined as the balance between positive and negative relationship experiences. If Jason believes he is deriving more pleasure (nurturance, supportive teasing, and so forth) than pain (frustration, insecurity, and so forth) from being in his relationship with Elise, then it is likely that he is fairly satisfied at the moment. His own expectations and experiences must also be taken into account in the reward–cost ratio. As Taylor and Altman point out, "Rewards and costs are consistently associated with mutual satisfaction of personal and social needs" (1987, p. 264).

Jon Carter/CartoonStock Ltd

To understand this a bit better, consider the following two conclusions observed by Taylor and Altman: (1) Rewards and costs have a greater impact early on in the relationship than later in the relationship, and (2) relationships with a reservoir of positive reward/cost experiences are better equipped to handle conflict effectively. We will examine each of these briefly.

The first conclusion suggests that in the early stages of a relationship, we are more inclined to look at the "bigger" issues in a relationship rather than focus on a singular cost or reward. So, for instance, it is probable that Jason will be impressed with Elise if she is willing to give Jason space during the early stages of their relationship; for Jason, rushing into a relationship may be a bit overwhelming, and Elise's patience may be viewed as an important relational reward. Elise, however, may view Jason's early ambivalence as an indicator of things to come. She may, therefore, decide that his uncertainty is simply too much of a cost to endure and want to dissolve the relationship sooner than Jason does. We can't ignore the complexity of relational life with the two.

With respect to the second conclusion regarding costs and rewards, Taylor and Altman note that some relationships are better able to manage conflict than others. As relational partners move on in a relationship, they may experience a number of disagreements. Over the years, couples become accustomed to managing conflict in various ways, creating a unique relational culture that allows them to work through future issues. There may be more trust in handling a conflict in established relationships. In addition, the relationship is not likely to be threatened by a single conflict because of the couple's stockpile of experiences in dealing with conflict.

In sum, then, relationships often depend on both parties strategically assessing the rewards and costs. If partners feel that there are more rewards than costs, chances are that the relationship will survive. If more costs are perceived than rewards, the relationship may depenetrate or dissolve. However, keep in mind that both partners may not see an issue similarly; a cost by one person may be viewed as a reward by the other. For one partner, an unexpected pregnancy may be a joyful event; for the other partner, this pregnancy may be viewed with dread because of the financial challenges a newborn will pose.

The Social Exchange perspective relies on both parties in a relationship to calculate the extent to which individuals view the relationship as negative (cost) or positive (reward). According to Social Exchange thinking, as relationships come together, partners ultimately assess the possibilities within a relationship as well as the perceived or real alternatives to a relationship (Carpenter & Greene, 2015). These evaluations are critical as communicators decide whether the process of social penetration is desirable. In the following section, we identify the stages of the social penetration process.

Stages of the Social Penetration Process

The decision about whether a potential relationship appears satisfying is not immediate. As we mentioned earlier, SPT is viewed as a "stage" theory. Further, relationship development occurs in a rather systematic manner, and decisions about whether people want to remain in a relationship are not usually made quickly. Not all relationships go through this process, and those that do are not always romantically based relationships. In fact, most of the close relationships in which we find ourselves in our lives are not romantic at all (Blackie & McLean, 2022). To demonstrate how each stage functions in relationships that are *not* romantic, we provide a scenario for you to think about. We then talk about each stage and refer back to the example. **Figure 7.2** outlines the four stages of the social penetration process.

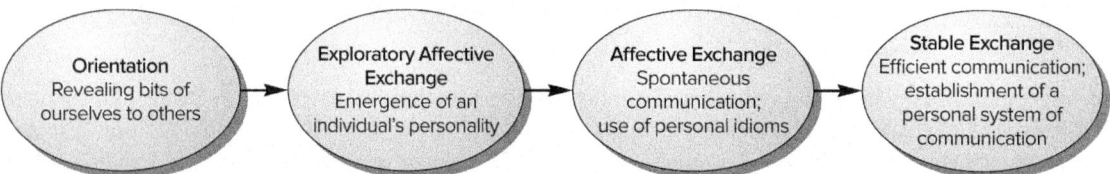

Figure 7.2 Stages of Social Penetration

Consider the relationship between Carmen and Brennan, two 21-year-olds who both decided to live off-campus for their junior year. The two did not know each other before meeting in their Communication course, but both found that they wanted to live away from campus and both wanted to pledge one of the sororities. The two come from different parts of the country: Carmen is from Los Angeles, and Brennan was raised in rural Pennsylvania. They also differ in family makeup in that Carmen is one of four siblings and Brennan is an only child. Finally, Carmen is the first in her family to attend college while Brennan is a third-generation college student. They have only met each other in class and are now about to have their first breakfast together.

Orientation: Revealing Bit by Bit

The earliest stage of interaction, called the **orientation stage,** occurs at the public level; only bits of ourselves are revealed to others. During this stage, comments are usually on the cliché level and reflect superficial aspects of individuals. People usually act in socially desirable ways and are cautious of disturbing any societal expectations. Individuals smile pleasantly and react politely in the orientation stage. Conflict is usually avoided at this stage (Carpenter & Greene, 2015).

orientation stage stage of social penetration that includes revealing small parts of ourselves

Taylor and Altman (1987) note that people tend not to evaluate or criticize during the orientation stage. This behavior would be perceived as inappropriate by the other and might jeopardize future interactions. If evaluation does occur, the theorists believe that it will be couched in soft overtones. In addition, both parties actively avoid any conflict so that they have further opportunity to size up each other.

The orientation stage can be understood by examining the dialogue between Carmen and Brennan during their breakfast:

> CARMEN: I have to admit that I am a bit weirded out about all of this. I mean I know you from class, but I am a tad awkward. I think this will work fine ... I'm rambling, right?!.
>
> BRENNAN: I agree. [awkward silence] I have to be upfront: I've read a lot of your posts and I agree with almost all of them! That's a good thing.
>
> CARMEN: But, look: We both love lacrosse, and maybe we'll both make the team. We both want to rush, we're now Insta friends, and I hope ... [Brennan interrupts]
>
> BRENNAN: I love to study near ... Sorry. You go ahead.
>
> CARMEN: No, you go.
>
> BRENNAN: I was going to say that I hope that we have some chances to get away and maybe even head to my family's lake house. I love to study near the water and if it's warm, I definitely love to swim. I haven't had time this past summer, though, because I was working too much.
>
> CARMEN: Believe it or not, I don't know how to swim! I tried to learn, but I just don't seem to be coordinated.
>
> BRENNAN: Hey! I'm a good swimmer. I'll teach you when we get some time.
>
> CARMEN: Ha. If I eat any more of this stuff, I'll definitely know how to float!

As you can see, both women speak in a rather superficial and very formalized conversation, and neither one is really judging the other. In fact, Brennan has an opportunity to tell Carmen how strange it is that she doesn't know how to swim, but she chooses to stay supportive. There is a modest level of self-disclosure about social media, but not enough to cause any conflict and tension.

Exploratory Affective Exchange: The Self Emerges

In the orientation stage, interactants are cautious about revealing too much about themselves to each other. The **exploratory affective exchange stage,** however, is an expansion of the public areas of the self and occurs when aspects of an individual's personality begin to emerge. Both people start to "explore" each other and small pieces

> **exploratory affective exchange stage** stage of social penetration that results in the emergence of our personality to others

of their private life become more public. The theorists note that this stage is comparable to the relationships we have with casual acquaintances and friendly neighbors. Like other stages, this stage includes both verbal and nonverbal behaviors. People might begin to use some catch phrases that are idiosyncratic to the relationship. There is a small amount of spontaneity in communication because individuals feel more relaxed with each other, and they are not as cautious about blurting out something that they may later regret. Also, more touch behavior and more affect displays (such as facial expressions) may become part of the communication repertoire with the other person. Taylor and Altman tell us that many relationships don't proceed beyond this stage, although with social media, some relationships may never have a clear ending at this stage. Retaining posts or screen-shotting posts may allow the "relationship" to continue.

To gain a clearer picture of the exploratory affective exchange stage, let's return to Carmen and Brennan. This time, however, consider that the two have been roommates for about eight weeks now, and each is getting a better idea about the personality of the other. Both are now pledging Delta Sigma and are sitting in their apartment talking about their pledging experience.

BRENNAN: Okay, Carmen. I don't know why we're doing this.

CARMEN: I know. At times, I think it's weird that we're doing all this just to have this "sister-hood" experience. But, okay, the girls do a lot of good and I mean they're just very cool and not stuck on themselves. It's also the only sorority that allows you to live outside the campus. And just like me, the "recruiters" told you that they liked our social media posts, right?

BRENNAN: Right, I know. But I'm not having any fun! I am so tired, too, and I ...

CARMEN: Please, Brennan, it's fine to chill! We both agreed to do this and if you want to back out now, do it. But, all of this shade is getting me down.

BRENNAN: ... Um ... you're the one who needs to chill ... and I'm not whining! Just because I say I'm having second thoughts on pledging doesn't mean I'm a whiner. I'm just not sure if we're just trying to conform or ...

CARMEN: Whatever. I know that I wanted to pledge because I wanted a group of friends I could trust and who would be there if I needed help. It's okay to take things as they come and not get too worked up before things happen.

Clearly, Brennan and Carmen are starting to feel more comfortable around each other! In fact, words like "shade" and "chill" reflect the sort of catchphrases to which Taylor and Altman refer. Furthermore, Carmen is slowly revealing more personal information about her family and her perceptions of Brennan as a whiner. Brennan, in turn, openly questions the value of the pledge process. Their exploratory affective exchange is generally one characterized by candor, engagement, and ease.

Affective Exchange: Commitment and Comfortability

This stage is characterized by close friendships and intimate partners. The **affective exchange stage** includes those interactions that are more "freewheeling and casual" (Taylor & Altman, 1987, p. 259) in that communication is frequently spontaneous and individuals make quick decisions, often with little regard for the relationship as a whole. This stage represents further commitment to the other individual; the interactants are comfortable with each other.

> **affective exchange stage** stage of social penetration that is spontaneous and quite comfortable for relational partners

The stage includes those nuances of a relationship that make it unique; a smile may substitute for the words "I understand," or a penetrating gaze may translate into "We'll talk about this later." We

> **personal idioms** private, intimate expressions stated in a relationship

might also find individuals using **personal idioms** (Hopper, Knapp, & Scott, 1981), which are private ways of expressing a relationship's intimacy through words, phrases, or behaviors. Idiomatic expressions—such as "honey" or "babe"—carry unique meaning for two people in a relationship. Personal idioms act as a couple's private communication system. These idioms are different from the catchphrases we discussed in the exploratory affective exchange in that idioms usually characterize more established relationships, whereas catchphrases may develop at any point in an initial interaction. We should add that this stage, like the exploratory affective exchange stage, may also include criticisms. As the theorists contend, these criticisms, hostilities, and disapprovals may exist "without any thought of threat to the relationship as a whole" (Altman & Taylor, 1973, p. 139). Consequently, barriers to closeness may be broken down, but many people still protect themselves from becoming too vulnerable.

Returning to our example, Carmen and Brennan have been together for a little more than 10 weeks. They have had ample opportunity to understand a number of idiosyncrasies about each other; living with someone usually does that to people. Their conversation centers around a date that Brennan had on Saturday night:

> BRENNAN: I simply refuse to believe that I even spent one hour with him! I mean I have dated a lot of guys, but this one was one of the weirdest.
>
> CARMEN: I've seen some of your dates, Roomie. Ah, remember Alex and his pet tarantula? Hello?
>
> BRENNAN: Word. I know. Hey, at least he didn't take his cat on a date!
>
> CARMEN: I knew you would say that. But, listen, that guy was so nice ... you never see guys act like gentlemen these days! I mean even his TikToks were about his cats! He was so sweet.
>
> BRENNAN: Yeah, and it's a wonder why his cats didn't claw you while you guys were together! I mean, THREE CATS?! Who has that many?
>
> CARMEN: I know the kind of guy you like, Brennan. And, yes, we're different from each other on guys. I just keep things to myself and won't wear my heart on my sleeve like you do! I'm much more discreet, shall we say?
>
> BRENNAN: There's a reason why I call you "Carmen Chameleon!"
>
> CARMEN: Funny, not funny.

As you can sense, there is noticeable comfort in the relationship right now. Altman and Taylor would argue that because it appears that the two know each other so well, they are willing to expose their relational values, their perceptions of the other's dating experiences, among other things. There is a slight tension embedded in their conversation, but nothing that seems to prompt the other to terminate the conversation. Their interpersonal "barriers" appear to be down and both are comfortable sharing very personal comments about each other. And, to be sure, affective exchanges may include both positive and negative exchanges.

Stable Exchange: Raw Honesty and Intimacy

The fourth and final stage, **stable exchange,** pertains to an open expression of thoughts, feelings, and behaviors that results in a high degree of spontaneity and relational uniqueness. During this stage, partners are highly intimate and synchronized; that is, behaviors between the two sometimes recur, and partners are able to assess and predict the behavior of the other fairly accurately. At times, the partners may tease each other about topics or people. This teasing, however, is done in a friendly manner.

stable exchange stage stage of social penetration that results in complete openness and spontaneity for relational partners

Social Penetration theorists believe that there are relatively few misinterpretations in communication meaning at this stage. The reason for this is simple: both partners have had numerous opportunities to clarify any previous ambiguities and have begun to establish their own personal system of communication. As a result, communication, according to Altman and Taylor, is efficient. At this stage, two independent people begin to emerge as "one" insofar as they have begun to peel away the levels that they have used to protect themselves (recall our onion analogy).

We return to our example of Carmen and Brennan. It's now finals week, and obviously the two are very tense. Yet they both realize that this week must be free of conflict, and both also realize that after this week, they will not see each other for a month because of the break. The stable exchange stage is very apparent when we listen to their conversation:

> CARMEN: So, okay, this week we *have* to be careful not to blow up at each other. I have three finals and a paper due.
>
> BRENNAN: Tell me about it. Did I ever tell you that if I don't get at least a B average, my parents told me that I'm on "probation," and they won't help with my tuition.
>
> CARMEN: Seriously? I wish I even had help with my tuition; even my fees for pledging almost put me over. Next term, I'm going to have to pick up another part-time job.
>
> BRENNAN: I'm lucky, I know.
>
> CARMEN: Yep. Not everyone has mommy and daddy to help them.
>
> BRENNAN: Hey, I'm not going to apologize because my parents work hard.
>
> CARMEN: I'm not asking you to apologize. I never told you this, either, but if I don't get one of the Delta scholarships, this year may be my last here.
>
> BRENNAN: That's crazy. Don't even say that.
>
> CARMEN: I'm gonna keep studying. I can't think about anything, but "right now."

The stable exchange stage includes individuals who are willing to expose intimate parts of themselves. Clearly, both Carmen and Brennan care about each other and manifest various levels of vul-

dyadic uniqueness distinctive relationship qualities

nerability. Although our earlier example suggested a conflicted relationship, there is now what Altman and Taylor (1973) call **dyadic uniqueness,** or distinctive relationship qualities such as humor and sarcasm.

As we mentioned earlier, this multistage approach to intimacy can get convoluted with periodic spurts and slowdowns along the way. In addition, the stages are not a complete picture of the intimacy process. There are a number of other influences, including a person's background and values and even the environment in which the relationship exists. The social penetration process is a give-and-take experience whereby both partners continue to work on balancing their individual needs with the needs of the relationship.

Finally, similar to most of the theories you will read about in this book, Altman and Taylor could never have envisioned the internet when conceptualizing this theory. As you review it, consider the technological influences inherent in the stages (we address the interplay between relationship development and online communication in **Chapter 8**). For instance, as we all already know, some relationships grow online and this online disclosure can influence the path of the relationship. Consider how the stages might be influenced if technology was introduced into the penetration process. Will sub-stages exist? How would texting enhance or detract from a couple's intimacy journey? These and many other questions will have to be considered once we unpack technology's influence upon SPT.

Integration and Critique

SPT has been appealing since its inception. Altman and Taylor have proposed an intriguing model by which to view relational development. The theory's empirically derived conclusions have drawn the attention of researchers over the years who also applied quantitative methods to understanding relational closeness. The theory had its beginnings during a time of great openness in society. Over the years, thanks to the media, among other institutions, Westernized cultures have grown even more comfortable with revealing personal information, resulting in a theory that continues to have resonance. As you think about SPT and its value, consider the following criteria: scope and heurism.

Integration

Communication Tradition	Rhetorical | Semiotic | Phenomenological | Cybernetic | **Socio-Psychological** | Socio-Cultural | Critical Intrapersonal
Communication Context	**Interpersonal** | Small Group/Team | Organizational | Public/Rhetorical | Mass/Media | Cultural
Approach to Knowing	**Empirical** | Interpretive | Critical

Critique

Evaluation Criteria	**Scope** | Logical Consistency | Parsimony | Utility | Testability | **Heurism** | Test of Time

Scope

The scope of SPT is of concern. In fact, the scope of the theory "makes it difficult to adequately test it as a whole" (Mongeau & Henningsen, 2008, p. 370). Some scholars contend, for instance, that self-disclosure—one of the key themes of the theory—may be too narrowly interpreted. For instance, some researchers (e.g. Liu, Wang, & Wang, 2019) believe that self-disclosure depends on a number of factors (e.g., timing, relationship intimacy, etc.), and not simply the need to reveal to people over time. Further, self-disclosure is a fairly complex and unpredictable process, sometimes catching people off guard (Howe & Betts, 2023).

Further, Mark Knapp, Anita Vangelisti, and John Coughlin (2014) reject the notion that relationship development is as linear as is suggested in SPT. They believe relationships are embedded in other relationships, and in turn, these relationships affect the communication between partners. Therefore, other people may influence the direction of a relationship. In addition, the linearity of the theory suggests that the reversal of relational engagement (recall that Altman and Taylor likened relationship disengagement to a film shown in reverse) is relational disengagement. Leslie Baxter and Erin Sahlstein (2000) assert that the concept of information openness and closedness cannot be understood in isolation; there is much more going on in a relationship than simple self-disclosure.

To be fair, Altman later revisited the social penetration processes and amended his original thinking with Taylor. Altman explained that being open and disclosive should be viewed in conjunction with being private and withdrawn (Altman, Vinsel, & Brown, 1981; Taylor & Altman, 1987). In a sense, Altman proposes what Baxter and Montgomery articulate in Relational Dialectics Theory (**Chapter 25**). C. Arthur VanLear (1991) underscored this thinking by concluding that there are two competing cycles of openness and closedness in both friendships and romantic relationships. Jason LaSalle and Elise Porter from our opening story will surely experience this push and pull of self-disclosure as their relationship progresses. It is likely that as both of them share pieces of information, each will also remain private about other issues.

Heurism

There can be no doubt that SPT and the concept of self-disclosure have yielded literally thousands of studies. Therefore, SPT is highly heuristic. Researchers have studied and written about the effects of self-disclosure, for example, on various types of relationships and across a variety of populations including relational development on dating sites such as Bumble (Halversen, King, & Silva, 2022). In addition, families (West & Turner, 2019), physicians (Trahan & Goodrich, 2015), alcoholics (Greenberg & Smith, 2015), and college instructors (Hannah & Meluch, 2022) have all been investigated. Furthermore, the effects of culture on the penetration process (e.g., Liu, Wang, & Liu 2019) have also been investigated. Finally, HIV status (Catona, Greene, Magsamen-Conrad, & Carpenter, 2016), blogging (Luo, Guo, Lu, & Chen, 2018), and COVID-19 pandemic stress (Gallegos, Zaring-Hinkle, & Bray, 2022) and their relationship to the self and to self-disclosure have also been studied.

Closing

Despite some criticism, SPT remains an integral theory pertaining to relationship development and has generated scholarly interest. In particular, the theory has resonated with interpersonal communication scholars and those interested in looking at how relational intimacy evolves. Relationship development can be exhausting, invigorating, and challenging at times, and SPT helps people understand those challenges.

Discussion Starters

 Case-In-Point: If their relationship develops further, what do you think Jason and Elise will talk about as the two get to know each other better? Will there be any risk involved as they disclose to each other? Explain with examples.

Try-It-Your-Selfie: Discuss the possible/probable interplay of SPT with Instagram, X, TikTok, and LinkedIn. When self-disclosing to another person, several things can go wrong. Explain the consequences of poorly planned or inappropriate self-disclosure. Provide examples along the way.

1. What similar patterns cut across escalating relationships? Discuss marital relationships, relationships between friends, and parent–child relationships as individuals move toward intimacy.

2. Some critics have charged that SPT focuses too much on self-disclosure. Others, however, contend that self-disclosure forms the basis of most intimate relationships. What do you think? Is there a compromise between the two views?

3. If you outlined the stages of a past romantic relationship of yours, would it follow the sequencing that Taylor and Altman suggest? What similarities are there to the social penetration process? What differences are there? Provide examples.

4. Other than an onion, what additional analogies or metaphors can you think of that would apply to SPT?

References

Acquisti, A., John, L. K., & Loewenstein, G. (2012). The impact of relative standards on the propensity to disclose. *Journal of Marketing Research, 49*, 160–174.

Altman, I., & Taylor, D. A. (1973). *Social penetration: The development of interpersonal relationships.* Holt, Rinehart & Winston.

Altman, I., Vinsel, A., & Brown, B. (1981). Dialectic conceptions in social psychology: An application to social penetration and privacy regulation. In L. Berkowitz (Ed.), *Advances in experimental social psychology* (pp. 130–145). Academic Press.

Bareket-Bojmel, L., & Shahar, G. (2011). Emotional and interpersonal consequences of self-disclosure in a lived, online interaction. *Journal of Social and Clinical Psychology, 30*(7), 732–759.

Baxter, L. A., & Sahlstein, E. M. (2000). Some possible directions for future research. In S. Petronio (Ed.), *Balancing the secrets of private disclosures* (pp. 289–300). Erlbaum.

Blackie, L. E., & McLean, K. C. (2022). Examining the longitudinal associations between repeated narration of recent transgressions within individuals' romantic relationships and character growth in empathy, humility, and compassion. *European Journal of Personality, 36*(4), 507–528.

Carpenter, A., & Greene, K. (2015). Social penetration theory. *The International Encyclopedia of Interpersonal Communication*, 1–4.

Catona, D., Greene, K., Magsamen-Conrad, K., & Carpenter, A. (2016). Perceived and experienced stigma among people living with HIV: Examining the role of prior stigmatization on reasons for and against future disclosures. *Journal of Applied Communication Research, 44*, 136–155.

Espinoza, A., Garcia-Fornes, A., & Sierra, C. (2012). Self-disclosure decision-making based on intimacy and privacy. *Information Sciences, 211*, 93–11.

Fox, J. (2015). Models of relationship development. *The International Encyclopedia of Interpersonal Communication*, 1–9.

Gallegos, M. I., Zaring-Hinkle, B., & Bray, J. H. (2022). COVID-19 pandemic stresses and relationships in college students. *Family Relations, 71*(1), 29–45.

Greenberg, M., & Smith, R. A. (2015). Support seeking or familial obligation: An investigation of motives for disclosing genetic test results. *Health Communication, 31*, 668–678.

Halversen, A., King, J., & Silva, L. (2022). Reciprocal self-disclosure and rejection strategies on Bumble. *Journal of Social and Personal Relationships, 39*(5), 1324–1343.

Hannah, M., & Meluch, A. L. (2022). The risks and benefits of disclosing to students: College instructors' perceptions of their disclosures in the classroom. *Texas Speech Communication Journal, 46,* 31–45.

Hopper, R., Knapp, M. L., & Scott, L. (1981). Couples' personal idioms: Exploring intimate talk. *Journal of Communication, 31*, 23–33.

Howe, D., & Betts, L. (2023). *Attachment across the lifecourse: A brief introduction.* Bloomsbury Academic.

Knapp, M. L., Vangelisti, A. L., & Caughlin, J. P. (2014). *Interpersonal communication and human relationships.* Pearson.

Liu, Z., Wang, X., & Liu, J. (2019). How digital natives make their self-disclosure decisions: A cross-cultural comparison. *Information Technology & People, 32*(3), 538–558.

Luo, N., Guo, X., Lu, B., & Chen, G. (2018). Can non-work-related social media use benefit the company? A study on corporate blogging and affective organizational commitment. *Computers in Human Behavior, 81*, 84–92.

Masur, P. K. (2019). Theories of self-disclosure. In P. Masur (Ed.), *Situational privacy and self-disclosure* (pp. 69–88). Springer.

Mongeau, P. A., & Henningsen, M. L. (2008). Stage theories of relationship development. In L. A. Baxter & D. O. Braithwaite (Eds.), *Engaging theories in interpersonal communication* (pp. 363–376). Sage.

Mongeau, P. A., Henningsen, M. L. M., & Oliver-Blackburn, B. M. (2022). Developmental theories of relationships: Uncertainty, stage model, and turning point approaches to relationships' lifespan. In D. Braithwaite & P. Schrodt (Eds.), *Engaging theories in interpersonal communication* (pp. 328–341).

Punyanunt-Carter, N. (2019). Persuasion and personal relationships. In S. Wahl & E. Morris (Eds.), *Persuasion in your life* (pp. 171–191). Routledge.

Ruell, S. (2023). *Making relationships great again.* Printed by author.

Schroeder, J., Lyons, D., & Epley, N. (2022). Hello, stranger? Pleasant conversations are preceded by concerns about starting one. *Journal of Experimental Psychology: General, 151*(5), 1141–1153.

Taylor, D. A., & Altman, I. (1987). Communication in interpersonal relationships: Social penetration processes. In M. E. Roloff & G. R. Miller (Eds.), *Interpersonal processes: New directions in communication research* (pp. 257–277). Sage.

Thibault, J. W., & Kelley, H. H. (1959). *The social psychology of groups.* John Wiley & Sons.

Trahan, D. P., & Goodrich, K. M. (2015). "You think you know me, but you have no idea": Dynamics in African American families following a son's or daughter's disclosure as LGBT. *The Family Journal, 23,* 147–157.

VanLear, C. A. (1991). Testing a cyclical model of communicative openness in relational development: Two longitudinal studies. *Communication Monographs, 58,* 337–361.

Weber, A. L. (2019). *Introduction to psychology.* HarperCollins.

West, R., & Turner, L. H. (2019). *An introduction to communication.* Cambridge University Press.

West, R., & Turner, L. H. (2023). *Interpersonal communication.* Sage.

CHAPTER 8
Social Information Processing Theory

Based on the research of *Joseph Walther*

> *"There are times when less interpersonal or socioemotional communication is beneficial."*
>
> —Joseph Walther

Corrina Abrams

Corrina Abrams was bored. She was studying for her midterm in business statistics and needed a break. She felt as though she would freak out if she continued to try any more practice problems. As she got up to get a drink, she thought about going online to check any email from the "**betogether.com**" website. The site—a dating location—was popular among college students and was often the topic of conversation whenever Corrina and her friends got together. That evening, as Corrina checked her email, she noticed one from another college student who wrote with the subject line "Salut. Je m'ennuie." Corrina Googled the French salutation ("*Hello, I'm bored*") and immediately responded to him. She was happy to read the email (written in English) from Marcus, a Canadian who, like Corrina, was also a college student. She figured a five-minute break chatting with a guy who was also in college would be pretty cool.

As Corrina sat and chatted with her Canadian tech companion, she soon realized that she was discussing some pretty personal information with him. She told him about helping her roommate deal with a recent assault, how she was the first kid in her family to go to college, and how she remained close friends with both of her parents after their divorce. Marcus, in turn, revealed that he was in a technical college because his parents couldn't help him with tuition and that he was paying for school by doing some amateur modeling. For some reason, she liked where the conversation with Marcus was going. He seemed very different from any other person and he never judged her; nor did he pry inappropriately.

Over the course of a month, Corrina and Marcus continued to email, FaceTime, and text each other. They were amazed how much they had in common. They both loved country music and had dozens of country music CDs. They both loved reading *Wired* magazine. The two also talked about each other's family and about the differences between online relationships and those that began in a bar or coffee shop. And, although Corrina was excited to talk, she didn't really want to tell Marcus about her low grades in college and the fact that she was on academic probation. It was too important for her to focus on the positive, especially since she liked the fact that she felt he was similar to her in a lot of ways. After several days of e-chats, Corrina thought that Marcus was simply too good to be true. And Marcus seemed to be quite enamored with her. He figured that because she was a business major, Corrina was smart and would graduate with a job in hand. The credentials of his new online interest impressed him. It was clear that the two had shared quite a bit of information over the week. And, they hadn't even met face-to-face yet.

Soon, the two were Snapchatting, and as a result, their friendship circles began to merge. They posted pictures of themselves on Instagram ("*He really is cute,*" she admitted; "*Wow, she could be in a magazine,*" he thought). These were only pictures, but the two developed a virtual world that reflected their personalities. The social linkage that started on a whim now turned into a full-blown technological relationship of six weeks.

Still, Corrina couldn't stop wondering whether all of this was simply a fantasy or whether it could develop into something bigger. Deep down, she wondered whether she was being "catfished." She knew there was a difference between talking online and seeing someone in person—she and Marcus talked and laughed a lot about that. And she realized she had never even seen anything other than Marcus's face on her computer screen!

After a while, Corrina was seriously thinking of meeting her newfound acquaintance. But, then again, she thought, maybe just continuing the online relationship would fulfill her needs right now—she just wasn't sure. She kept wondering whether it was too good to be true. Had she really found a soul mate nearly 400 miles away in another country? She still couldn't believe that all of this happened simply because she went online during a night of studying!

The development and maintenance of our relationships have, as you have read elsewhere in this book, fascinated researchers for decades. In fact, as we noted in the beginning of our book, one of the most researched areas in the communication field is the evolution of human relationships. The theories related to the Interpersonal Communication context, in fact, underscore the complexity and excitement of theories related to relationships. Still, with one exception—the focus of this chapter—no communication theory has focused exclusively on what is, arguably, the most important development in modern times: the internet.

Technology has always been important to interpersonal relationships, yet, interestingly, researchers only began investigating this interface a few decades ago because most scholars didn't see the influence of computer technology on communication. In fact, the process of meeting someone years ago followed a predictable trajectory. Although some variation existed, the relational journey went something like this: we meet someone in person, we "date" that person socially, we introduce that person to our friends and family, and we marry that person (see **Chapter 7** on Social Penetration Theory, for more information on predictable relational development). To some of you, this approach may still make sense to you. Yet to others of you, this may sound like a scene out of a 1950s movie! After all, you might ask, what about dating sites and other technologies that are influencing relationships? When can I swipe left or right?!

According to the Pew Research Center (2021), in 2005, just 5 percent of adults used a social media platform. In 2011, they reported that it was nearly 50 percent. Today, that number is over 70 percent! Facebook and YouTube are the most popular of all social networking sites, followed by Instagram, LinkedIn, and X. In all, the United States has over 186 million Facebook users, while in India, for instance, there are over 370 million (oberlo.com; retrieved October 18, 2023; **https://www.oberlo.com/statistics/facebook-users-by-country#:~:text=Number%20of%20Facebook%20users%20by,around%2022.1%25%20of%20the%20population**). In China, over 250 million people use Weibo (China's version of X), while in the United States, X has nearly 80 million active users. Finally, as most of you can confirm, Snapchat, Instagram, and TikTok have gained significant usage over the years, attesting to the immediate and visually stimulating option for social media users. Clearly, we've come a long way since the telephone!

The development of online relationships is at the heart of a theory known as Social Information Processing (SIP). The theory, introduced by Joseph Walther (1992, 1993, 2008, 2015) long before other scholars envisioned the enormous influence of the internet, rests upon the interplay among technology, relationships, and self-presentations. For Walther and other SIP advocates, the possibility of developing and cultivating online relationships is not only probable, it occurs with high degrees of success and relationship satisfaction. Online intimacy, during the 1990s, was something new and something that few scholars and practitioners felt was possible.

To imagine what SIP suggests, we almost have to suspend our traditional interpretation of what it means to have a "relationship." As you will read in other chapters that focus on relationship development, many interpersonal communication theorists prior to Walther argued that relational life is one that is shaped and sustained by face-to-face (FtF) communication. Walther tweaked this fundamental perspective and determined that the nonverbal cues to which we are accustomed in our FtF relationships, although important, were not essential to developing and securing an interpersonal relationship. He acknowledges that while both individuals in either FtF or online relationships have the same needs for uncertainty reduction and affinity (see **Chapter 5,** on Uncertainty Reduction Theory), online interactants "adapt their linguistic and textual behaviors" to how others communicate with them (Walther, Anderson, & Park, 1994, p. 65). That means that unlike FtF relationship development, in online relationship development, relational participants will engage each other personally through both the words they use, the frequency of words, and the meanings embedded in those words.

Theory at a Glance • Social Information Processing Theory

Individuals have the ability to establish online relationships and these relationships have intimacy equal to or greater than the intimacy achieved in face-to-face (FtF) relationships. Even without nonverbal cues, through various tech options (e.g., email, texting, etc.), SIP theory argues that relationships have the potential to be significant in people's lives. Individuals use computer-mediated communication (CMC) to get to know each other and use this information to form impressions of each other. Because messages travel via one primary channel, it takes longer for relationships to achieve the same level as those that are FtF. In some cases, online relationships may be viewed as more important than FtF relationships.

So, it's not only what Corrina and Marcus reveal about themselves that allows their relationship to develop and mature, but also the number of times that they communicate with each other. A difference exists, for instance, if Corrina emails Marcus once a week instead of once a day. And, after six weeks, Walther and others would suggest that this relationship has all the workings of an intimate experience.

SIP Theory has, at its core, impression management (Bernstein, 2023; Walther, 2016). For our purposes, we define **impression management** as either a strategic or unconscious effort to influence another's perception. Much of the earlier research on impression management focused on FtF communication and the nuances with meeting someone. A person's self-image was viewed as important in relational development. Later applications of impression management were undertaken once online communication began (Ellison, Heino, & Gibbs, 2006). Ellison and her colleagues, for example, found that various self-presentational behaviors exist in online dating environments. They report on three selves: the **actual self** (attributes of an individual), the **ideal self** (attributes an individual ideally possesses), and the **ought self** (attributes an individual should possess). These selves are manifested and are of particular importance in online conversations and relationships, allowing an individual to manage his or her online persona or identity.

impression management the strategic or unconscious effort to influence

actual self the attributes a person possesses

ideal self the attributes a person ideally possesses

ought self the attributes a person should possess

"Well, this is certainly awkward. It seems that we've *both* been a bit imaginative in creating our online profiles."

Eric Hodecker/CartoonStock Ltd

Identity management is directly associated with the social information process (Rosenberg & Egbert, 2011). As online participants, we necessarily forgo our reliance upon nonverbal cues in our dialogues. Therefore, the linguistic cues we send to others will be undertaken with a concern for self-presentation. Walther (2011) contends that we "accrue impressions" of others in our online relationships. He further states that "communicators are motivated to develop interpersonal impressions" (p. 458). Therefore, we can see that impressions are central to SIP and that the theory relies heavily on how individuals are interpreting (the words of) each other online.

Theoretical Turbulence: The Cues Filtered Out

Many of the communication theories (e.g., Social Penetration Theory) we address in this book are rooted in other theoretical perspectives from various fields of study. Yet unlike some theories (e.g., Communication Accommodation Theory), SIP was conceptualized, in part, by addressing the shortcomings of other theories that addressed communication mediums. These theories are termed **cues filtered-out theories** (Culnan & Markus, 1987; Walther, 2011), meaning that because of a lack of nonverbal cues, online relational development has little integrity. Walther's research critiqued past methodological and conceptual problems with this thinking. He subsequently worked toward establishing a communication theory that more accurately reflected the intersection among communication, online environments, the self, and relationships. He also believed that a medium can support or carry many cues for others and that ultimately, communicators will seek whatever medium is available in order to cultivate their relationships. For Walther and other SIP scholars, nonverbal cues have the capacity to let others know about a sender, their personality, likes, dislikes, and so forth. Let's briefly discuss two of those filtered-out theoretical perspectives that influenced Walther's theory: Social Presence Theory and Media Richness Theory. We provide a snapshot of each theoretical lens below.

> **cues filtered-out theories** theories that address the lack of nonverbal cues as being detrimental to online relationship development

Social Presence Theory (Short, Williams, & Christie, 1976) pertains to the extent that individuals are aware of each other via various communication mediums, that is, the awareness of individuals during an interaction—their characteristics, qualities, physicalities, and so forth. According to the theory, some mediums have a higher degree of social presence (e.g., video) and others possess lower degrees of social

> **Social Presence Theory** a theory that posits the extent to which people are aware of each other via various communication mediums

presence (e.g., audio and text based). High-degree mediums are viewed as opportunistic for relational growth and lower-degree mediums are perceived as being more impersonal and unemotional in nature. Social presence is a "subjective quality," meaning that individuals may not agree on whether or not there is a high or low degree of engagement in an interaction. With respect to online environments, social presence manifests in the manner in which messages are sent and received by others in the mediated environment (Kreijns, Xu, & Weidlich, 2022). To a large extent, according to the theory, a high degree of social presence results in a warmer (and more satisfying) interpersonal experience. In fact, in one study applying Social Presence Theory to online classroom perceptions, researchers discovered that instructors can enhance student learning by providing "warmer" teaching behaviors, including delivering feedback via email, frequently "visiting" online student groups/forums, and employing fun fonts, emojis, or other visual stimuli to students (Dixson, Greenwell, Rogers-Stacy, Weister, & Lauer, 2017).

Media Richness Theory (Daft & Lengel, 1986) also functions in the evolution of Social Information Processing Theory. The theory, first explored in the organizational context, suggests that while communicating online, the medium is too narrow to allow for relationship development. To this end, individuals experience uncertainty and equivocality in their communication with others. In the theory, "richness" is evaluated in several ways,

> **Media Richness Theory** a theory that advances the notion that communication can be classified according to message complexity

including the medium's capacity for immediate feedback, the number of cues and channels used, and the extent to which a message is personalized (Gu, Higa, & Moodie, 2011). Similar to Social Presence Theory, scholars studying Media Richness posit that mediums interpreted as "rich media" help to support both verbal and nonverbal cues. According to the theory, "richest" media are FtF communications, which prompt individuals to communicate faster; rich media are used for more complex conversations and tasks. The "leanest" media are telephones, memos, and letters, which frequently prompt others to avoid communicating in more complex ways.

Walther (2011) believes that while both theories have important theoretical notions, the online world is much more complex than what is inferred by Social Presence and Media Richness. For example, text-based communication between online participants–despite the absence of nonverbal cues–has relational value. To this end, both Social Presence Theory and Media Richness Theory suffer from a limited understanding of relational life online. Walther argues that if interactants communicate enough times and with sufficient breadth and depth, nonverbal communication does not remain critical in relationship development. In fact, he and his colleagues posit that SIP "focuses more on language content and style as more important conduits of interpersonal information" (Walther, Van Der Heide, Ramirez, Burgoon, & Peña, 2015, p. 8). Consider Corrina from our opening example. She clearly established a foundation with Marcus via email. Other platforms eventually provided more information, but the two relied upon a "lean" medium, with words used to communicate their feelings.

Assumptions of Social Information Processing Theory

By now, we hope that you already understand that online relationships have the potential to be expressive, intimate, and in many cases, long term. SIP researchers like Joseph Walther are intrigued by how identities are managed online and how relationships are able to move from superficiality to intimacy. To further unpack SIP, we offer the following three assumptions related to the model:

- Computer-mediated communication (CMC) provides unique opportunities to connect with people.
- Online communicators are motivated to communicate (favorable) impressions to others.
- Online interpersonal relationships require extended time and more accumulated messages to develop equivalent levels of intimacy seen in FtF interpersonal relationships.

Our first assumption rests on the premise that CMC is a unique opportunity to build interpersonal relationships with others. First, although we alluded to it earlier, let's define CMC. Those studying the connection between human communication and technology embrace **computer-mediated communication,** which is a process in which people perceive, interpret, and exchange information via large networked telecommunications systems. These CMC systems are vast, are almost always text based, and include, among others, instant messaging and email. CMC has been identified as "an organic setting" (Tong & Walther, 2015, p. 188) and it can be both synchronous and asynchronous. **Synchronous communication** occurs when both sender and receiver are online simultaneously. **Asynchronous communication** exists when time constraints influence the sending and receiving of various messages and responses. Asynchronous messages exist at various times and do not require the concurrent presence of a sender.

computer-mediated communication (CMC) process in which people perceive, interpret, and exchange information via large networked telecommunications systems

synchronous communication a process that occurs when both sender and receiver are online simultaneously

asynchronous communication a process that occurs when both sender and receiver are online at different times, owing to time constraints

The relationship between CMC and relational development is underscored by Walther (2011): "Computer-mediated communication (CMC) systems, in a variety of forms, have become integral to the initiation, development, and maintenance of interpersonal relationships" (p. 443). Consider the various systems to which Social Information Processing scholars subscribe: email, X, Facebook, blogs, Zoom, and others. Now, consider whether or not you have had an interpersonal relationship via any of these formats. Chances are that many or most of you will say that you've built a relationship via these technological venues. You will likely agree with Walther and others who claim that CMC systems provide for relational growth. In fact, the developers of eHarmony, Match, and Tinder would be out of business if they didn't subscribe to this claim!

Yet some of you may not fully embrace, for instance, this notion of relationship development over the internet. You may subscribe to the belief that for interpersonal communication and an interpersonal relationship to exist, face-to-face communication must inevitably occur (consider our discussion of "interpersonal communication" in **Chapter 2**). In fact, one research team asserts that computer technologies are not replacing the value of FtF conversations (Hu, Wood, Smith, & Westbrook, 2004). Yet, the researchers do accept that "individuals are able to form impressions, gain interpersonal knowledge, and develop relationships solely through textual interaction." Perhaps you think that although CMC is a legitimate way to cultivate an understanding of another, establishing a relationship requires the parties to actually meet. Perhaps you feel that despite an ongoing interaction, Corrina and Marcus from our chapter-opening story will never develop a "relationship."

Students Talking: Delia

I'm sure I'm the only one who disagrees with some parts of this theory. I'm sure it has been researched a lot and I know there's a lot of validity to it. But, I'm having a hard time understanding how we can have more online intimacy than in person. I think if we start believing this stuff, then we will never know that being with someone—face-to-face—is more important than being online. I know that this contradicts the theory, but I don't ever want people to think that we can get really close online and think that's the same as getting close in person. Just doesn't make sense to me.

Much of Joseph Walther's research (e.g., DeAndrea & Walther, 2011; Walther, 2011) undercuts this latter perception. He first argues that people will communicate personal information online and in many cases, people will disclose more online than they will in person. In fact, research shows that people will reciprocate high levels of intimate information while online (Blackie & McLean, 2022), a topic we revisit later in the chapter. Further, a study by Walther and Jeong-woo Jang (2012) demonstrates that even participatory websites (e.g., YouTube, etc.) and their online interactants can "both facilitate and complicate social influence" (p. 2), providing additional evidence that CMC remains opportunistic for online discussions, reactions, and relationships. Think of this claim in this way: as you visit a website, many times you're there for social support, information, or simply to read and/or listen. Relational life is similar. Many times, "visitors" enter our lives and we end up keeping them around for quite some time. In essence, they become part of our relational fabric. Consider the serendipity related to Corrina and Marcus. She was not anticipating meeting anyone. Yet CMC provided her the chance to chat with someone and based upon the levels of disclosure and the mutual exchange of personal information, the two seem to be on their way to an interpersonal relationship.

CMC is clearly different from face-to-face communication, but it offers an unparalleled opportunity to meet someone whom you would never meet FtF. Moreover, relationships established via CMC systems prompt the same emotions and feelings we find in all relationships, including social support, jealousy, and happiness (Green-Hamann, Eichhorn, & Sherblom, 2011). Additionally, since CMC systems are available around the globe, the uniqueness of being able to cultivate online relationships with someone who is, say, 5,000 miles away cannot be ignored. It's clear, too, that many people *prefer* this form of communication because, as of 2017, almost 40 percent of marriages and long-term heterosexual relationships in the U.S. began online, eclipsing any other source of meeting (such as being introduced by friends, etc.) (Rosenfeld, Thomas, & Hausen, 2019).

A second assumption of Social Information Processing Theory was alluded to earlier in the chapter. Online participants are motivated to present themselves in strategic ways. Impression management is essential in online relationships (Walther, 1993) and participants undertake efforts to ensure particular impressions. Individuals like Corrina Abrams, for example, obviously wish to avoid presenting their various "selves" to others in negative ways (recall that Corrina did not want to reveal her poor academic status in college).

Researchers have found that social networking sites (SNS) like Facebook are filled with people who wish to provide a number of different self-presentations to others (Bryant, Marmo, & Ramirez, 2011; Yau & Reich, 2018). Since the more Facebook friends one has, the more attractive the individual is viewed to be (Walther, Van Der Heide, Kim, Westerman, & Tong, 2008), managing one's online impressions is of importance to many users. In terms of romantic relationships, if someone is either single or partnered, a number of options are available to communicators that would communicate their relational image to others: (1) display relationship status (e.g., single, married, divorced, etc.); (2) utilize a picture that displays a partner; (3) provide dialogue about the user's partner; and (4) dialogue in chat rooms aimed at a person's relational goals (e.g., marriage and/or cohabitation, etc.) (Utz & Beukeboom, 2011; Winter, Haferkamp, Stock, & Kramer, 2011). Importantly, presenting oneself in an authentic way remains paramount to many (Darvin, 2022).

Students Talking Tech: Rachel

 As I was reading this chapter, I went back to my social media and saw that I have different pictures for each profile. I don't think that's a problem but now I'm wondering whether that means I have different identities based on what I'm using. Maybe there is something to the fact that my profile and my identity and related to each other. I know that some of my followers talk about my profile pic and some say I should change it because "it's not me." Hmm.

How others present and manage themselves online remains important on various SNS and numerous CMC system platforms. Because numerous sites include others monitoring a number of different verbal and nonverbal displays, people deliberately use different behaviors to project a particular online identity (Bullingham & Vasconcelos, 2013). For some, openness is a primary goal; for others, impression management is undertaken in more covert ways. Clearly, if others' opinions make a difference in relational goals, then a person will invest some time on managing and influencing those opinions (Toma & Hancock, 2012).

A third assumption of SIP Theory states that different rates of information exchange and information accrual affect relationship development. To understand this assumption, let's reflect upon Corrina Abrams and her online relational experiences with Marcus. She relies on several back-and-forth messages with Marcus and also undertook this technological exchange over the course of several weeks. In fact, she begins to really wonder about whether or not Marcus is her "soul mate," an intuition that many people in FtF relationships also consider after numerous significant interactions.

This assumption reflects Walther's contention that online relationships have the same capacity to become intimate as those that are established face-to-face. Recall our earlier discussion of filtered-out theories, which, as you will remember, render nonverbal cues as nearly absent in CMC platforms and systems. Despite this, a great deal of research has reported that given sufficient time and an accrual of messages, intimacy can be, and usually is, achieved.

Chronemic cues, or those cues related to how people perceive, use, or respond to time, and the accumulation of messages are two notions that are interrelated. Walther (2008) states that "when sufficient time elapses so that ample communicative exchanges are made, personal and relational information accrues and CMC is no less effective than FtF interaction at developing impressions and managing interpersonal relations" (p. 393). Walther and other SIP researchers believe that nonverbal cues we see in FtF encounters (e.g., smiling, touching, etc.) are important. Yet online relationships that rely upon text-based or email-based communication—despite the absence of nonverbal cues (except chronemic cues)—can be highly intimate. He posits that "language and writing are held to be highly interchangeable with nonverbal cues" (p. 393).

> **chronemic cues** cues related to how people perceive, use, or respond to time

Years ago, as we noted previously, relational life (and many interpersonal communication theories) placed a high premium on the value of nonverbal communication. Because researchers focused on FtF experiences, things like eye contact, touch, and tone of voice, among other characteristics, were important to study. And, today, we still continue to do so because, eventually, even a relationship that started online typically results in meeting that person. Think, for instance, of the initial impressions of meeting someone; clearly you're attuned to both what the person says and how they say it. Yet Walther is suggesting that when we have any discussion about online relationships, we need to suspend the notion that nonverbal cues are both paramount and mandatory in order for a relationship to exist. His research supports the claim that if messages are delivered over a period of time and these "verbal messages" are adapted and transferred into nonverbal codes, an online relationship can become quite intimate (Walther, 2012).

These claims may be counterintuitive to you and prior to Walther's theory, interpersonal researchers relied extensively on the value of nonverbal communication in relationship development. And, we agree that nonverbal behaviors *are* critical to interpersonal relationships and that people often present idealized versions of themselves online, rendering their online presence suspect at times (Venter, 2019). Nonetheless, Social Information Processing Theory is suggesting that although the messages are verbal, communicators "adapt" to the restrictions of the online medium, look for cues in the messages from others, and modify their language to the extent that the words compensate for the lack of nonverbal cues (Walther, DeAndrea, & Tong, 2010). In addition, Walther contends that although there is no precise comparison, it does generally take (at least) four times longer to exchange as much information via CMC as it does to do so face-to-face (even when communicating in "real time"). Further, these messages "build up" over time and provide online participants sufficient information from which to begin and develop interpersonal relationships. All of the aforementioned allows for a relationship resulting in a relatively high level of intimacy.

With these fundamental assumptions of SIP Theory identified, we now turn our attention to a few central features and concepts related to the theory. As you reflect on this theory and read the following, keep in mind that the theory had its beginnings in the early 1990s. To some researchers at the time, Walther's theory was considered to be both provocative and pioneering because he was the first communication theorist to discover that online relationships can be just as valuable and satisfying as those established FtF. Let's dig further into the theory by discussing two areas of the theory that have gained quite a bit of attention. First, we discuss the hyperpersonal perspective and next we address a principle that Walther has termed *warranting*.

Hyperpersonal Perspective: "I Like What I Read and I Want More"

As Walther conceptualized and clarified SIP Theory, he was struck by how vivid and poignant the online communication was between interactants. He conducted research that showed, among other things, that communication between online participants was viewed as more powerful than those undertaken FtF. Walther (1996, 2011) argues that the impressions we cultivate via CMC systems and those relationships we develop and maintain "exceed the desirability and intimacy that occur in parallel off-line interactions" (2011, p. 460).

Individuals online can take their time thinking about responses and can choose to do it in a synchronous manner (e.g., Zoom) or without the online partner around (asynchronous). Senders and receivers have the opportunity to "think before they speak." All of this, Walther advances, tends to lead to friendlier online relational environments. He terms this the **hyperpersonal perspective** (Walther has also called this the *hyperpersonal effect* and *hyperpersonal model*).

hyperpersonal perspective a dynamic associated with Social Information Processing Theory that suggest people have the potential to cultivate more intimate relationships online than those that are FtF

Walther's approach to understanding CMC relationships was a result of consulting and researching theories from a variety of fields, a belief we discussed previously. After investigating SIP Theory and its various trajectories, Walther (1996) discovered that despite the prevailing views of some regarding the limitation of online relationships, individuals actually took advantage of CMC and strategically positioned themselves to others in very favorable ways, in ways that an FtF context cannot. Walther believes that online participant/users "exploit" the technological aspects of the medium to manage their impressions and to help their relationships. Consequently, those relationships we establish online are often more intimate than those we establish FtF.

The hyperpersonal perspective takes all of this into consideration and much more. This perspective entails a number of different areas related to CMC and interpersonal relationships, including message tone, message complexity, personal language, editing behaviors, and composing time, among others (Walther, 2007). In other words, online communicators can be selective in how they present themselves and how they respond to another individual and an inordinate depth of intimacy can take place within online relationships.

The hyperpersonal perspective is more than saying that an online relationship is intimate. Walther, writing in a number of different scholarly venues, articulated its complexity. We elucidate four components he studied, thinking back to our beginning chapter when we introduced the various models of communication: (1) sender, (2) receiver, (3) channel, and (4) feedback. These four, you may recall, constitute many of the models of communication we explained in **Chapter 1**. For now, though, let's discuss each so that you can better understand this intriguing extension of SIP Theory.

Sender: Selective Self-Presentation

According to Walther, senders have the ability to present themselves in highly strategic and highly positive ways. This self-presentation is controlled and it serves as a foundation for how CMC users get to know one another. The fundamental underpinning of this component of the hyperpersonal perspective is affinity seeking. That is, senders provide information online that prompts affinity in others. Walther (1996), for instance, playfully states that when we send online messages to another, we don't need to hold in our waist and we can nod, smile, or feign interest without others knowing. Senders, then, may provide personal disclosures that represent an "idealized self." They may do more than self-disclose. Walther states that a sender's word choices and expressions of affinity ("*You sound so cool*") as well as the extent to which a sender agrees with a receiver are all part of a hyperpersonal experience. Earlier, we alluded to online self-presentations. According to some researchers (e.g., Venter, 2019), people optimize their self-presentations because they have time to "think it through" (p. 4).

Think of our earlier story of Corrina Abrams. It is clear that to impress Marcus, she is presenting herself in very positive ways. Think of the different feelings she triggers: she tells Marcus that she helped her roommate after an assault ("compassion"), how she is the first child in her family to attend college ("tenacity"), and that she is good friends with both of her parents following their divorce ("nonjudgmental"). As a sender, Corrina is communicating to Marcus an image that is overwhelmingly positive. She is transmitting preferable cues (Walther, 1992) to Marcus and is certainly managing her online image in a way that casts her as an other-centered and thoughtful person. Her social currency is apparently accelerated because she chooses to emphasize more flattering examples about herself. In other words, she uses CMC to edit her disclosive messages, making her more desirable in Marcus's eyes. In the end, she is engaging in preferential actions to elicit preferential reactions.

Receiver: Idealization of the Sender

At the core of the receiver component in the hyperpersonal perspective is attribution. **Attributions** are those evaluations and judgments we make based on the actions or behaviors of others. It's a concept that is perceptual in nature; we perceive others and come up with some conclusions based upon what we perceive. Walther (2011) believes that

attributions evaluations and judgments we make based on the actions or behaviors of others

receivers tend to "fill in the blanks" on perceptions that are either incomplete or missing altogether rather than forming an impression of someone. Walther et al. (2015) also contend that as receivers, we tend to exaggerate our perceptions of the message sender. We make attributions of another's behavior (e.g., the sender). Think of his claim this way: we already know that a sender is providing favorable information to a receiver online. Now, consider the fact that because a receiver is not privy to all the "typical" cues we find in FtF conversations, the receiver tends to attribute and, according to the theory, the receiver may "overattribute." A receiver, for instance, is likely to think that a sender has more similarities than differences. A receiver may tend to compare the sender to someone else they know, employing some sort of "perceptual personality" framework (*"You sound like my cousin Barry. He's my favorite relative."*). Finally, a receiver may experience an overreliance on the minimal cues available online and forget that the relationship they have with a sender is based on words—the misspellings, typographical errors, use of punctuation, and so forth (Lea & Spears, 1992; Walther, 1996). In fact, consider the attributions taking place in a pioneering study undertaken several years ago. Harry Witchel and his research team (2020) discovered that serious misgivings regarding online health information were a result of spelling errors and emphatic capitalization in those messages. In fact, trustworthiness of online health information was sacrificed because of a sender's grammatical errors.

In terms of Corrina and Marcus, as a receiver, Marcus overattributed Corrina's credentials. He perceived her major in business as indicative of her intellectual prowess. We know that impression is incorrect. Further, he assumed she would have a job before graduation, a debatable point again, given that Corrina was struggling academically and may even be on academic probation. This is not to suggest that the two will not or do not have an authentic relationship. Rather, the two are engaging in hyperpersonal behaviors that may render the relationship less profound than the two of them perceive it to be.

Channel Management

Earlier in the chapter, we talked about synchronous and asynchronous communication. You may recall that when both sender and receiver are online simultaneously, we have synchronous experiences. When either the sender or receiver sends a message without the presence of the other, we have asynchronous experiences.

CMC does not require both sender and receiver to be online at the same time. In fact, the asynchronous nature of CMC allows online participants to reflect upon, edit, and review their comments before hitting the "Send" button. Think about the times, for instance, where you were speaking with someone FtF in the earlier stages of meeting. While exchanging information with one another, you don't have a lot of time to think about what you say and at times, your anxiety may even prompt you to blurt out something inappropriate.

Now consider the previous example as one undertaken online. Upon meeting someone, you have the opportunity to think about texts or emails before sending them, a point we introduced earlier. Further, prior to sending messages, you can rewrite them for clarity, sense, and relevancy. We can edit, delete, rewrite, or otherwise modify our messages in order to "reflect intended effects before sending them" (Walther et al., 2015, p. 15). With some CMC systems, you can even "retract" emails before they reach the receiver. Online asynchronous experiences allow for "optimal and desirable" communication. Senders and receivers pay little attention to the physical cues, as we learned earlier, and so the energies and "cognitive resources" (Walther, 2011, p. 461) are placed into ensuring that the messages are of high quality. Walther contends that the more relational the affection or more desirable the other communicator is, the more editing in message composition.

Feedback

A fourth component embedded in a hyperpersonal perspective is feedback. You may remember from **Chapter 1** that feedback is essentially those words or behaviors that are communicated to us in an interaction.

Feedback's relationship to CMC and the hyperpersonal approach is a bit more complicated. In this case, Walther interprets feedback as behavioral confirmation, which is a "reciprocal influence that partners exert" (Walther, 1996, p. 27). In communication theory, we refer to this as self-fulfilling prophecy, a term we introduced in our discussion of Symbolic Interactionism (**Chapter 15**). This prophecy essentially is a tendency for an individual's expectation of a target person to evoke a response from that person which, in turn, reaffirms the original prediction. So, if Corrina believes, based on his emails and texts, that Marcus is a trusting person, she will treat him as such and he, in return, will be trusting, reaffirming her original supposition.

Walther's (2011) hyperpersonal perspective acknowledges a feedback system this way: "When a receiver gets a selectively self-presented message and idealizes its source, that individual may respond in a way that reciprocates and reinforces the partially modified personae, reproducing, enhancing, and potentially exaggerating them" (p. 463). For instance, once we get a message from a sender with whom we have affinity, our feedback to that person will be aligned with how they presented themselves to us. If their message is one of admiration for us, we, in turn, may be inclined to embrace that admiration. And, we may be inclined to even amplify that admiration! Because cues in an online environment are limited, the feedback that does occur is often exaggerated or magnified. A sender and receiver have "heightened expectations and idealized impressions" (Ramirez & Wang, 2008, p. 34).

The four components—sender, receiver, channel, and feedback—suggest that the hyperpersonal perspective is an ongoing, dynamic experience. As you consider this theoretical extension, keep in mind that online communication is usually complex when we are discussing human relationships. Walther (2008) concludes that SIP is a "process" theory because both information and interpersonal meaning is accumulated over time, providing online partners an opportunity to establish a relationship over time. The dynamics of being online, with individuals with whom we're establishing relationships, are often unpredictable. For instance, Stephanie Tom Tong and Walther (2015) discovered that perceptual *dis*confirmation can exist. And, recall that time is of importance here (Ramirez & Wang, 2008). That is, a hyperpersonal experience is not one that occurs immediately for everyone. Corrina and Marcus, for example, communicate over several weeks and their hyperpersonal relationship may be quite different from, say, that of another couple, which may require additional time for intimacy to be achieved. We now turn our focus to a second feature of SIP that has attracted some scholarly attention: warranting.

Warranting: Gaining Confidence Online

A conversation on SIP and online relationships is incomplete without understanding one significant element embraced in any long-standing relationship: trust. As we discussed the self-presentation of online participants, we noted that individuals may present themselves in less than truthful ways. It's social media, after all, complete with an online world filled with avatars and emojis! In fact, some people assert online identities and believe that others will simply subscribe to everything that is written or posted. What happens if the partners decide to initiate a FtF meeting? Will the self-presentations be deemed truthful? Will the impressions be accurate?

To remedy distorted and deceptive online presentations, Joseph Walther and his colleague Malcolm Parks (2002) articulated the need for "warranting" behavior by communicators. In the online universe, **warranting** is defined as "the perceived legitimacy and validity of information about another person that one may receive or observe online" (Walther, 2011, p. 466). Walther argues that online relationship development—like those that are FtF—can fall prey to misleading or manipulative overtures (the television show *Catfish* features these sorts of communication). To this end, in order for CMC users to feel more confident about online assertions, a warranting "value of information" will frequently take place that allows for more truthfulness and accuracy in presentations. Many online partners ensure that warranting exists in order to reduce uncertainty, to engender

warranting the perceived legitimacy and validity of information about another person that one may receive or observe online

more confidence about the truthfulness of comments and self-presentations, and ultimately, to grow their relationship. Warranting usually takes the form of a sender connecting the receiver with the sender's off-line network. Further, warranting varies in terms of acceptability and integrity. Personal webpages, for instance, will likely have less warranting value than, say, one that has been officiated by a third party. Or, a photo included by a CMC user may have less warranting value than a photo found in an official media outlet.

Other platforms are opportunistic for warranting. Walther contends that because we can control our impressions online, we have low and high warranting. If a profile owner is able to control content online, then that is a **low warranting** process. If others can control or manipulate it, it is considered a **high warranting** process. If you're on Instagram, your followers will comment on your pictures and phrases; they may either agree or disagree with your claims. If you're in "finsta" (a "fake" Instagram account typically reserved for close friends and/or family members), chances are that you have

> **low warranting** occurs when profile owners can modify their online content
>
> **high warranting** occurs when profile owners are unable or incapable of controlling or manipulating their online content

unfiltered posts to avoid, for instance, potential employers discovering the "real you." On LinkedIn, others can verify employment and educational history. On Facebook, others can tag you, or you can post status location and relationship updates. Clearly, the warranting process permits the introduction of the "trust but verify" premise we find in many discussions, including those related to nuclear disarmament!

Walther and his research team (Walther, Van Der Heide, Hamel, & Shulman, 2009) assert that efforts at corroborating information and visual cues will enhance the online relationship efficacy between people. In a study that included mock-up Facebook postings, the researchers found that self-proclaimed physical attractiveness by a Facebook user was found to be more believable when others confirmed the claim. What this suggests is that "one's own claims of attractiveness on Facebook are suspect" (p. 248), but when others can corroborate those claims, this self-presentation contains more credibility.

Related to the above, a central point of SIP Theory pertains to the self-presentation of information. This online impression management is opportunistic for dishonesty. According to Walther, CMC users will frequently try to ensure the accuracy of their profiles, photos, images, and words they convey. Individuals like Corrina and Marcus may one day meet if they continue to have a satisfying online relationship. Right now, neither may be skeptical, but if they each continue to reveal more information about themselves, both may want verification of that information. And, it's likely that both will desire to know what the other looks like— in person. So, if both Corrina and Marcus now take steps to provide information that has high warranting value, it's probable that this technological relational experience will one day become a real-world experience.

Theory Into Practice • Social Information Processing Theory

Theoretical Claim: Warranting allows an individual to validate the legitimacy of a person's identity via online/social media.

Practical Implication: Desiree and Jazz just met online a few weeks ago. Desiree told her new boyfriend that she is a homebody and doesn't like to go out a lot. But, Jazz "saw" an entirely different Desiree as he looked at her past pictures boozing it up on her Instagram account. Thus, Jazz's perception of Desiree's credibility is now vastly different than when they first started chatting.

Integration and Critique

Throughout this book, we have attempted to give you both theories that are rooted in other fields of study and theories that have their infancy in the communication discipline. SIP Theory is an attempt to bridge the two. While Walther clearly identified other theoretical frameworks in the conceptualization of SIP, his theory, nonetheless, centers on communication and human relationships. Further, the theory clearly has evolved from quantitative investigation, making it positivistic in nature. With his infusion of CMC, the theory is a unique and important development in communication theory. To evaluate the theory, we examine three areas: scope, utility, and testability.

Integration

Communication Tradition	Rhetorical | Semiotic | Phenomenological | Cybernetic | **Socio-Psychological** | Socio-Cultural | Critical
Communication Context	Intrapersonal | **Interpersonal** | Small Group/Team | Organizational | Public/Rhetorical | Mass/Media | Cultural
Approach to Knowing	**Empirical/Post-Positivistic** | Interpretive | Critical

Critique

Evaluation Criteria	Scope | Logical Consistency | Parsimony | **Utility** | **Testability** | Heurism | Test of Time

Scope

Any theory dealing with CMC will inevitably be criticized for its expansiveness and lack of nuance. Yet although Walther's theory addresses relationship development, his theoretical thinking over the years has silenced many critics who argue that it suffers from a lack of refinement. Walther spent several years ensuring that the "cumbersome" area known as CMC doesn't result in a theory that's too broad. Some early research (e.g., Walther & Tidwell, 1995) began by examining how CMC is affected by time. And, then Walther's theory evolved to reflect more specificity (e.g., warranting perspective) and address different types of online communicators. Therefore, the theory does not suffer from too much breadth. Rather, SIP has moved from generalized thinking about online relationships to more specific discussions, such as physical attraction and impression management, among other areas that have captured much research attention over the years. Although Walther has claimed that the theory has application to both individuals and small groups, his primary focus has remained on impression formation and online communication.

Utility

As SIP was discussed above, we're confident that many of you were nodding your heads in agreement with what Walther argues. In other words, the usefulness of the theory became apparent to you. Perhaps Walther (2011) best addresses the usefulness of his theory this way: "Newer theories have also arisen, some barely tested, the ultimate utility of which remains to be seen" (p. 444). Further, given that on average, over 150 billion text messages are sent every month in the United States (Statista, retrieved August 18, 2023, https://www.statista.com/statistics/185879/number-of-text-messages-in-the-united-states-since-2005/) Walther believes that SIP's utility will continue. So, while some may think the theory needs more refinement, it has already resonated deeply in both the scholarly and applied communities.

Testability

One area of SIP that has received some concern and criticism relates to its testability. In essence: how does one investigate a receiver in a CMC episode who may or may not be interested in online relationships? In other words, the theory assumes participation in an online development but testing that interest can be problematic. Further, the theory does not take into account those on the internet who may be "trolling," a common undertaking found all over online platforms (Furian & March, 2023).

Interestingly, Walther has been a self-reflective critic of his own theory pertaining to the issue of "time" in CMC relationships (Walther, 2011). Later writing by Walther (2016), however, suggested that while using CMC, messages take longer to process than those conveyed FtF. There may be even more of a need to consider time when different online users are using different online platforms.

Second, in discussing the hyperpersonal perspective, Walther admits that not all of the theoretical components of his hyperpersonal approach have been researched sufficiently (nor tested). He and his research colleagues (2011) relate that "despite its importance, feedback has received the least direct attention in research on the hyperpersonal model of CMC" (p. 5).

In addition, some writers have looked at the receiver's behavior in the hyperpersonal perspective. Specifically, the receiver's perceptions help contribute to online intimacy (Jiang, Bazarova, & Hancock, 2011). Yet the researchers conclude that the "receiver component has never been isolated from the sender's effect" (p. 61). In other words, although the hyperpersonal approach states that the sender and receiver are both unique and interrelated, researchers have not fully investigated whether there is a particular receiver behavior that leads to greater online intimacy. Jiang and her colleagues found that online intimacy is a result of a sender's self-disclosure *as well as* a receiver's perceptions of the disclosure, resulting in research that was among the first to provide this causal link. Walther (2016) later notes that researchers have established that self-disclosure is more frequent in CMC encounters than in FtF encounters because CMC users "must employ 'interactive' strategies [including] asking personal questions in order to prompt a partner's self-disclosure" (p. 9). In other words, the more information that is actively solicited will result in more self-disclosure taking place. Such a conclusion provides for a testable area related to the hyperpersonal perspective.

Third, in examining the warranting hypothesis, Walther and his colleagues (2009) accept the fact that high warranting value may exist on those matters that have strong social desirability. For instance, physical attractiveness is a highly desirable trait in the United States, making it socially desirable. So, as Walther accepts, online communicators would seek corroboration for those qualities that society deems important (or desirable). Whether or not other less socially desirable qualities are prone to warranting overtures is not fully explained. Testing these traits, then, may require alternative processes.

Closing

SIP Theory arrived in the communication discipline at the time when the rest of the research world was starting to examine the internet for its likely influence on interpersonal communication and human relationships. As we alluded to earlier, Joseph Walther is a scholarly prophet, forecasting the importance of looking at online relationships in the early 1990s.

We're sure that as you reviewed Walther's comparisons between CMC and FtF relationships, many of you were wondering whether or not we will continue to be engaged in CMC with little chance of FtF meetings. This will likely never be the case. Yet we cannot ignore the fact that Walther's theory remains a pivotal framework to consider as we envision future relationship development in an uncertain technological time.

Discussion Starters

Case-In-Point: Corrina and Marcus have begun to establish their relationship online after just a few conversations. Do you believe that their experiences reflect what others experience when going online in the same manner as Corrina? Do any demographic issues (e.g., age, biological sex, cultural background) make a difference in online relationship development?

Try-It-Your-Selfie: Walther's followers have begun to examine what happens when online communicators meet for the first time. What experiences have you had related to meeting another via your social media platforms?

1. Think of other situations in which a "cues filtered-out" approach exists. Use examples in your response.

2. Envision that you are a social media expert who is embarking upon an online relationship with someone who is not tech-savvy. Discuss the similarities and differences in this online relationship development.

3. React, with examples, to the concerns and cautions expressed by Delia (in the Students Talking box). Do you believe her views are appropriate? Why or why not?

4. Provide an example for each component of the hyperpersonal perspective and determine whether or not the approach merits modification.

References

Bernstein, L. (2023). *Social information processing theory (SIP).* Salem Press.

Blackie, L. E., & McLean, K. C. (2022). Examining the longitudinal associations between repeated narration of recent transgressions within individuals' romantic relationships and character growth in empathy, humility, and compassion. *European Journal of Personality, 36*(4), 507–528.

Bryant, E. M., Marmo, J., & Ramirez, A. (2011). A functional approach to social networking sites. In K. Wright & L. Webb (Eds.), *Computer-mediated communication in personal relationships* (pp. 3–20). Peter Lang.

Bullingham, L., & Vasconcelos, A. C. (2013). "The presentation of self in the online world": Goffman and the study of online identities. *Journal of Information Science, 39*, 1–29.

Culnan, M. J., & Markus, M. L. (1987). Information technologies. In F. M. Jablin, L. Putnam, K. H. Roberts, & L. W. Porter (Eds.), *Handbook of organizational communication: An interdisciplinary perspective* (pp. 420–443). Sage.

Daft, R. L., & Lengel, R. H. (1986). Organizational information requirements, media richness and structural design. *Management Science, 32*, 554–571.

Darvin, R. (2022). TikTok and the translingual practices of Filipino domestic workers in Hong Kong. *Discourse, Context & Media, 50*, 100655.

DeAndrea, D. C., & Walther, J. B. (2011). Attributions for inconsistencies between online and offline self-presentations. *Communication Research, 38*(6), 805–825.

Dixson, M. D., Greenwell, M. R., Rogers-Stacy, C., Weister, T., & Lauer, S. (2017). Nonverbal immediacy behaviors and online student engagement: Bringing past instructional research into the present virtual classroom. *Communication Education, 66*(1), 37–53.

Ellison, N., Heino, R., & Gibbs, J. (2006). Managing impressions online: Self-presentation processes in the online dating environment. *Journal of Computer-Mediated Communication, 11*(2), 415–441.

Furian, L., & March, E. (2023). Trolling, the Dark Tetrad, and the four-facet spectrum of narcissism. *Personality and Individual Differences, 208*, 112169.

Green-Hamann, S., Eichhorn, K. C., & Sherblom, J. C. (2011). An exploration of why people participate in Social Life support groups. *Journal of Computer-Mediated Communication, 16*, 465–491.

Gu, R., Higa, K., & Moodie, D. (2011). A study on communication media selection: Comparing the effectiveness of the media richness, social influence, and media fitness. *Journal of Service Science and Management, 4*(3), 291–299.

Hu, Y., Wood, J. F., Smith, V., & Westbrook, N. (2004). Friendships through IM: Examining the relationship between instant messaging and intimacy. *Journal of Computer-Mediated Communication, 10*(1).

Jiang, L. C., Bazarova, N. N., & Hancock, J. T. (2011). The disclosure-intimacy link in computer-mediated communication: An attributional extension of the hyperpersonal model. *Human Communication Research, 37*(1), 58–77.

Kreijns, K., Xu, K., & Weidlich, J. (2022). Social presence: Conceptualization and measurement. *Educational Psychology Review, 34*(1), 139–170.

Lea, M., & Spears R. (1992). Paralanguage and social perception in computer-mediated communication. *Journal of Organizational Computing, 2*, 321–341.

Pew Research Center. (2021, April). *Social media fact sheet.* Retrieved from https://www.pewinternet.org/fact-sheet/social-media/.

Ramirez A., Jr., & Wang, Z. (2008). When online meets offline: An Expectancy Violations Theory perspective on modality switching. *Journal of Communication, 58*, 20–39.

Rosenberg, J., & Egbert, N. (2011). Online impression management: Personality traits and concerns for secondary goals as predictors of self-presentation tactics on Facebook. *Journal of Computer-Mediated Communication, 17*, 1–18.

Rosenfeld, M. J., Thomas, R. J., & Hausen, S. (2019). Disintermediating your friends: How online dating in the United States displaces other ways of meeting. *Proceedings of the National Academy of Sciences, 116*(36), 17753–17758.

Short, J. A., Williams, E., & Christie, B. (1976). *The social psychology of telecommunications.* Wiley.

Statista. Total number of SMS and MMS sent in the United States from 2005–2021. Retrieved August 18, 2023 from https://www.statista.com/statistics/185879/number-of-text-messages-in-the-united-states-since-2005/.

Toma, C. L., & Hancock, J. T. (2012). What lies beneath: The linguistic traces of deception in online dating profiles. *Journal of Communication, 62*(1), 78–97.

Tong, S. T., & Walther, J. B. (2015). The confirmation and disconfirmation of expectancies in computer-mediated communication. *Communication Research, 42*(2), 186–212.

Utz, S., & Beukeboom, J. (2011). The role of social network sites in romantic relationships: Effects on jealousy and relationship happiness. *Journal of Computer-Mediated Communication, 16*, 511–527.

Venter, E. (2019). Challenges for meaningful interpersonal communication in a digital era. *HTS Theological Studies, 75*(1), a5339.

Walther, J. B. (1992). Interpersonal effects in computer-mediated interaction: A relational perspective. *Communication Research, 19*, 52–90.

Walther, J. B. (1993). Impression development in computer-mediated interaction. *Western Journal of Communication, 57*(4), 381–398.

Walther, J. B. (1996). Computer-mediated communication: Impersonal, interpersonal, and hyperpersonal interaction. *Communication Research, 23,* 3–43.

Walther, J. B. (2007). Selective self-presentation in computer-mediated communication: Hyperpersonal dimensions of technology, language, and cognition. *Computers in Human Behavior, 23,* 2538–2557.

Walther, J. B. (2008). The social information processing theory of computer-mediated communication. In L. Baxter & D. O. Brathwaite (Eds.), *Engaging theories in interpersonal communication* (pp. 391–404). Sage.

Walther, J. B. (2011). Theories of computer-mediated communication and interpersonal relations. In M. L. Knapp & J. A. Daly (Eds.), *The handbook of interpersonal communication* (4th ed.) (pp. 443–479). Sage.

Walther, J. B. (2012). Interaction through technological lenses: Computer-mediated communication and language. *Journal of Language and Social Psychology, 31,* 397–414.

Walther, J. B. (2015). Social information processing theory (CMC). *The international encyclopedia of international communication.* Wiley-Blackwell.

Walther, J. B., Anderson, J. E., & Park, D. W. (1994). Interpersonal antisocial communication. *Communication Research, 21,* 460–487.

Walther, J. B., & Jang, J. (2012). Communication processes in participatory websites. *Journal of Computer-Mediated Communication, 18,* 2–15.

Walther, J. B., & Parks, M. (2002). Cues filtered out, cues filtered in: Computer-mediated communication and relationships. In M. L. Knapp & J. A. Daly (Eds.), *Handbook of interpersonal communication* (pp. 529–563). Sage.

Walther, J. B., & Tidwell, L. C. (1995). Nonverbal cues in computer-mediated communication, and the effect of chronemics on relational communication. *Journal of Organizational Computing, 5*(4), 355–378.

Walther, J. B., DeAndrea, D. C., & Tong, S. T. (2010). Computer-mediated communication versus vocal communication in the attenuation of preinteraction impressions. *Media Psychology, 13,* 364–386.

Walther, J. B., Van Der Heide, B., Hamel, L., & Shulman, H. (2009). Self-generated versus other-generated statements and impression in computer-mediated communication: A test of warranting theory using Facebook. *Communication Research, 36,* 229–253.

Walther, J. B., Van Der Heide, B., Ramirez, A., Burgoon, J. K., & Peña, J. (2015). Interpersonal and hyperpersonal dimensions of computer-mediated communication. In S. Sundar (Ed.), *The handbook of the psychology of communication technology* (pp. 3–22). John Wiley & Sons.

Walther, J. B., Van Der Heide, B., Kim, S. Y., Westerman, D., & Tong, S. T. (2008). The role of friends' appearance and behavior on evaluations of individuals on Facebook: Are we known by the company we keep?. *Human Communication Research, 34*(1), 28–49.

Winter, S., Haferkamp, N., Stock, Y., & Kramer, N. C. (2011). The digital quest for love—The role of relationship status in self-presentation on social networking sites. *Cyberpsychology: Journal of Psychosocial Research on Cyberspace, 5*(2), Article 3.

Witchel, H. J., Thompson, G. A., Jones, C. I., Westling, C. E., Romero, J., Nicotra, A., Maag, B., & Critchley, H. D. (2020). Spelling errors and shouting capitalization lead to additive penalties to trustworthiness of online health information: Randomized experiment with laypersons. *Journal of Medical Internet Research, 22*(6), e15171.

Yau, J. C., & Reich, S. M. (2018). "It's just a lot of work": Adolescents' self-presentation norms and practices on Facebook and Instagram. *Journal of Research on Adolescence, 29*(1), 196–209.

CHAPTER 9
Structuration Theory

Based on the research of **Anthony Giddens, M. Scott Poole, David R. Seibold,** *and* **Robert D. McPhee**

"Democracy hence implies not just the right to free and equal self-development, but also the constitutional limitation of power."

—Anthony Giddens

Tim Clifford and Bayside City Tire Company

Bayside City Tire Company (BCT) is a multinational company with many employees. One of the employees, Tim Cliffords, is a new Production Division manager. Jeremy has been a production floor supervisor for 19 years. He has been assigned the task of showing Tim, his new boss, around for a couple of weeks.

On the first day, Jeremy took Tim around the production floor to introduce him to the workers in his division. When approached, all of the workers called Tim by his last name: "Good morning, Mr. Cliffords." "It's great having you onboard." Tim responded to the workers by saying, "Mr. Cliffords is my father. Please, call me Tim."

As Tim and Jeremy walked away, Tim asked, "Why is everyone so formal?"

Jeremy responded, "That's company policy. We believe that addressing one another in this way establishes a sense of respect for our supervisors at Bayside. I know it sounds weird, but corporate is corporate."

Tim dissented: "Well, I'm head of the division, and I would really prefer if we do things my way. First names are fine with me."

Jeremy paused: "Well, sure, this is your division, but this is the way they have been taught to communicate with one another. I mean, think of how confused they will be if they are allowed to address you by your first name, and other managers expect to be addressed with more formal titles."

Tim just didn't like what he was hearing. He thought that if they were "confused" by a title change, then they had bigger issues going on. But he didn't want to disrupt the policy that was already in place, despite how odd it was. He told Jeremy, "Okay, let's go with it. I guess I'll have to get used to the workers calling me 'MISTER Cliffords' until I can get this straightened out. Besides, I don't want them to think that I am going against company rules or that I want to change things." Tim's reluctance was palpable.

Later that day, Tim talked to Angela Griffith, the Human Relations head for BCT, about changing the policy to allow employees and staff to address one another less formally. Angela listened to Tim's ideas and tried to tell him that in the grand scheme of the BCT, the issue was not that critical. Tim disagreed, arguing that interpersonal relationships are crucial to productivity. After some discussion, Angela agreed to call a meeting of the Production Division supervisors to discuss the potential change. It seemed to be a "minor" matter, but the climate of his division in the company was a critical issue to Cliffords. Perhaps a small change in one part of the company could result in company changes overall.

At the meeting, the six supervisors all voiced their opinions about the change in policy. Janette, a 15-year veteran with BCT, pointed out that she liked the formality because women in that type of work environment rarely get respect, and a title is one way to get that.

Darnell, who had joined BCT only 3 years prior after graduating with his business administration degree, countered by stating that he thought the workers didn't feel comfortable coming to him to discuss problems because of the formality and rules that were in place at BCT. He wanted a more friendly environment than what they had.

Wayne, who is a third-generation employee of BCT, argued, "You don't mess around with tradition. This is the way it's always been, and it has worked just fine for the past 30 years."

Angela listened to all their opinions, but the supervisors ended up splitting equally on the policy change issue. Angela never envisioned that even a simple change in policy related to calling supervisors by a title would resonate as it did with the division. And yet, she also knew that these sorts of rules, while seemingly unimportant, can affect the morale and productivity, and therefore, needed to be clarified for everyone. Although some people may see these issues as petty, in some companies such as BCT, these sorts of things can escalate quickly.

Just as a contractor uses a blueprint as a guide in building a structure, members of an organization use rules to state expectations of behavior and communication within the organization. As we learned in **Chapter 5** (Uncertainty Reduction Theory), many people are uncomfortable with uncertainty and ambiguity.

Consider the need for structure in your own life. Usually, students want to know the structure or rules so they will understand what is expected of them in class. If a professor were simply to announce, "This term we'll have a couple of tests, a few quizzes, and you'll need to do a project," pretty much all of you would likely be uncomfortable because very few can succeed with such limited information. You would want more specific instructions, or rules, regarding page-length requirements of any papers, due dates, expectations for exams, and so on. Thus, the school and the instructor provide students with a course syllabus, a *blueprint* of the rules for the class, in which the structure is created and maintained. However, we can also change the structure of an institution by adapting rules or creating new ones. Feedback from student evaluations, for instance, may guide a professor to alter the syllabus or the exam criteria in the future.

Anthony Giddens (1994) believes that social institutions are produced, reproduced, and transformed through the use of rules. David Seibold and Karen Kroman Myers (2006) contend that social institutions are "organized around members' interactional processes and practices: disseminating information, allocating resources, accomplishing tasks, making choices, managing disagreements, and the like" (p. 143). The structure of these institutions is the focus of Structuration Theory. Ultimately, when interpreting "social structures," consider that they are defined as the overall pattern that evolves in a society (Sheposh, 2023). Social structures are influenced by a number of things, including hierarchies, norms, and various institutions in a culture.

Recall from our discussion in **Chapter 3** that rules indicate and prescribe how things are to be done; they suggest a course of action in a setting. In Structuration Theory, rules in a social institution go beyond telling employees what they can and cannot do. In interactions among people, the rules guiding these conversations allow employees to maintain or alter an organization. An instance in which this would apply would involve a company's talking about poor customer service to make changes in the customer service policy. The rules governing these conversations might pertain, for instance, to hierarchy (e.g., the extent to which an employee can share a candid opinion with superiors). We return to a more detailed examination of rules later in the chapter.

In the chapter-opening example, the employees involved in the managers' meeting could potentially influence the structure of the organization by decisions made during their debate. If they had voted to change the rule of formally addressing one another, the entire organizational structure may have necessitated change. Communication is what allows for an organization to change. And the social structures that exist are not random but are structured, or patterned, in distinct ways. Further, "there are regularities in the way we behave and in the relationships we have with one another. But social structure is not like the physical structure like a building, which exists independently of human actions" (Giddens & Sutton, 2017, p. 5). Embedded in this view are not only rules, but also the properties, resources, and relationships within the society's social structures (Olowa, Witt, Morganti, Teittinen, & Lill, 2022).

Giddens (1979, 1993) views social structures as a double-edged sword. Interestingly, the structures and the rules that we create usually restrict our behavior. However, these same rules also enable us to understand and interact with others. We need rules to guide our decisions about how we are expected to behave. These rules may be either explicitly stated (such as grievance procedures that are outlined in an employee manual) or implicitly learned (such as respecting one another by providing each member of the group an opportunity to voice their opinion).

Groups and organizations are coordinated around various social interactions—for example, socializing new members through new employee receptions, arriving at decisions during conference calls, conducting meetings in person or via videoconference, or teaching new skills in employee-training sessions. Giddens (1984, 2003) points out that the key to making sense of the communication that occurs in these groups and organizations is to examine the

> **system** a group or organization and the behaviors that the group engages in to pursue its goals
>
> **structure** the rules and resources used to sustain a group or organization

structures that serve as their foundation. He makes a distinction between the concepts of system (as we discussed in **Chapter 3**) and structure. The term **system,** in this sense, refers to the group or organization itself and the behaviors and practices that the group engages in to pursue its goals. Organizations may engage in many behaviors/practices, including new employee assimilation processes and performance appraisals (Seibold & Myers, 2006). The term **structure** refers to the rules and resources members use to create and sustain the system, as well as to guide individual behaviors related to such behaviors/practices.

In the opening scenario, both the BCT organization and the group meeting of the division managers can be viewed as a system. Their goal is to discuss the problem of using formal names to address one another, as well as to conduct the daily operations of the organization. To assist them in accomplishing their goals in an efficient manner, both of these systems have a formal set of rules that includes guidelines for how employees are expected to address one another at BCT as well as guidelines for allowing everyone to voice their opinion during the group meeting. These represent the structures of the group and organization. Tim Cliffords is frustrated by the structure that restricts his employees from communicating with him informally. The rules for interactions between supervisors and subordinates contradict what he has learned about building relationships and enhancing employee productivity. Angela employs rules for conducting open discussions about issues before any decisions about changes are made. Thus, BCT is created and guided by structures— rules that explain how to demonstrate respect for one another.

Although Giddens first drew attention to organizational structures, communication scholars were instrumental in ushering in an application of structuration principles to other contexts. For example, Marshall Scott Poole (1990) and his colleagues David Seibold and Robert McPhee (McPhee, 2015; McPhee & Canary, 2016; McPhee, Poole, & Iverson, 2014; Poole & McPhee, 2005; Poole, Seibold, & McPhee, 1986, 1996; Seibold & Myers, 2006) have studied the application of structuration principles to the communication field. Their theoretical approach has been called *Adaptive* Structuration Theory to explain how task groups use and strategically adapt information technology, rules, and resources to accomplish organizational/group goals (Poole & McPhee, 2005). We will address this adaptive nature throughout the chapter, yet, we will continue to emphasize the original thinking of Giddens and discuss Structuration Theory as a perspective with appeal to many areas in the study of communication. As Dennis Mumby (2011) suggests, there is an embedded relationship between communication and structuration principles. He notes that communication not only reflects organizational reality, but also "creates and maintains the meanings that guide organized life and motivate particular actions" (p. 194).

To begin, it's important to define structuration. In a general sense, structuration allows people to understand their patterns of behavior—the structures of their social system. Specifically, **structuration** is described as "the process by which systems are produced and reproduced through members' use of rules and resources" (Poole et al., 1996, p. 117). Structures are not considered permanent but are continuously evolving and created by the communication among the people in a group/organization. For structuration theorists, organizational structures are produced and replicated by people who interact on a daily basis, trying to accomplish personal and company goals (Miller, 2018).

> **structuration** the production, reproduction, and transformation of social environments through rules and resources in relationships

Structuration allows people to understand their patterns of behavior. With respect to small groups, Poole and colleagues conclude that the key to understanding groups is through an analysis of the structures that underlie them. Rules and resources for communicating and arriving at decisions are typically learned from the organization itself and from members' past experiences and personal rules. These same rules and resources are reaffirmed as a result of their application or use; the group may decide to keep them in their existing format or modify them to meet the changing needs of the group.

Let's return to our opening scenario. In arriving at a decision on whether BCT should change its rule about having employees address one another formally, Angela enacts a rule of decision making that requires the group to discuss member viewpoints before making changes. BCT may also have a rule for company meetings in which the majority vote decides the action that should be taken. If Angela fails to call a meeting of the managers to solicit their opinions, many of them might be displeased. Or, if Angela doesn't give everyone the opportunity to state an opinion on the rule while the rest of the group observes an unspoken rule to respect others' opinions (i.e., challenge ideas but do not attack people), some of the group's members will be dissatisfied with the outcome of the meeting. Janette, for one, depends on the rule to maintain her credibility and authority when interacting with her male employees. The fact that Angela follows the rule of calling a meeting to discuss the issue reaffirms the belief that this is a good guideline to follow in making decisions in the organization. However, some of the group's members may not be pleased with the decision to keep the rule in its current form, and they may invoke a new rule among their own teams in which they permit subordinates to address them in a less formal manner. Thus, the rule will be altered from its original state.

Theory at a Glance • Structuration Theory

 Organizations create structures, which can be interpreted as an organization's rules, and resources. These structures, in turn, create social systems in an organization. Organizations achieve a life of their own because of the way their members use their structures. Power structures guide the decision making that takes place in these organizations.

Structuration provides a useful foundation for examining the impact that rules, hierarchies, and resources have on group decisions and organizational communication. Structures produce a system and they also represent the outcomes of a system. In addition, it helps describe how these rules are altered or confirmed through interactions. Finally, structuration is communicative: "Talk is action. If structure is truly produced through interaction, then communication is more than just a precursor to action; it is action" (Modaff & Butler, 2021, p. 121).

Assumptions of Structuration Theory

Structuration Theory is rather thorny because it deals with people, technology, resources, behavior, tasks, norms, and organizational life (Iyamu, 2019). Therefore, to assist in unraveling this complexity, we first consider some of the basic assumptions that guide the theory:

- Groups and organizations are produced and reproduced through actions and behaviors.
- Communication rules serve as both the medium for, and an outcome of, interactions.
- Power structures are present in organizations and guide the decision-making process.

Underscoring the first assumption, Giddens proposes that every action or behavior results in the production of something new—a **fresh act.** Each of the actions or behaviors in which a group or orga- **fresh act** something new developed from action or behavior

nization engages is influenced and affected by the past. This history serves as a reference for understanding what rules and resources are required to operate within the system. Consider, for instance, when a group leader decides to conduct a vote using an anonymous ballot. If past voting using the ballot has proven to be effective for group members, history is influencing the rules for operation within that system.

Structuration advances the notion that every time we communicate with someone, we are establishing a new beginning by (1) creating a new rule or expectation, (2) altering an existing rule, *or* (3) reaffirming rules that have been used in the past. In establishing these new beginnings, we will still rely on past rules and expectations to guide our behaviors. Thus, we never escape our history—it continually influences our decisions for behavior in groups and organizations.

Recall the interaction between Tim and Jeremy from our opening story. According to Structuration Theory, each interaction that takes place between Tim and Jeremy is considered new and creates something unique in the structure of their future interactions. At the same time, Jeremy and Tim each bring to the interaction a set of rules (based on their history together) and expectations that will guide and shape their communication with each other. In other words, as we observed in **Chapter 1,** their fields of experience become critical in their communication with one another. The employees address Tim by his last name, Mr. Cliffords, the first time they meet. Tim responds by saying, "Please, call me Tim." Tim creates something new in the social structure by establishing a precedent (a new norm) of asking his subordinates to call him by his first name. This precedent is influenced by the past expectations that each man brings to the interaction. In other words, the experiences that Tim and his employees bring to the interaction affect their expectations for the structure that will guide their future communication. Tim believes that an open and relaxed communication style works, whereas Jeremy and others operate under a very different set of rules for interaction—formal address works best. In this case, Jeremy's rules for conversing serve the dual function of guiding his current behavior and setting expectations for future behavior.

All of our communicative actions exist in relationship to the past. Angela may have learned in past group meetings that members are not supportive of others' thoughts and opinions. She may have decided simply to state the problem and have group members vote on a potential solution without first soliciting the advice and opinions of group members. In this instance, historical rules would have been used as a reference for altering the group's rules in future interactions.

The second assumption of Structuration Theory refers to the notion that rules simultaneously provide guidelines for behavior as well as possible constraints on behavior. Similar to many other theories in this text, rules become front and center in the understanding of a particular communication theory. Within this theory, the structure of a group includes a network of rules and resources that are used by its members in making decisions about what communication behaviors are expected. Katherine Miller (2009) writes about organizational rules this way: "Rules act as recipes for social life, in that they are generalizable procedures about how to get things done" (p. 215). Some rules take precedence over others, and history influences the action that is enacted. That is, if a rule worked well in the past, it's likely to be retained; if not, it's likely to be modified or abandoned. Moreover, rules "make certain kinds of conduct possible, while precluding others" (Nicolini, 2013, p. 47). The challenge lies in determining which rule will be the most efficient or productive in achieving the goals of the group or organization. We return to a discussion of rules later in the chapter.

Let's take a moment to examine this issue a little more closely. Giddens (1979) asserts that a rule can only be truly understood "in the context of the historical development" as a whole (p. 65). So, Structuration Theory assumes that to understand the rules of a social system, individuals need to know some of the background resources that led to the rule. Perhaps if Tim from our chapter-opening scenario were to investigate the origins of the rule for addressing superiors, he would have a better understanding of why this rule is currently enforced. There may have been a past incident when employees showed a blatant disregard for their supervisors by substituting nicknames for the informal first names that they were permitted to use in addressing their superiors. For example, Tim may discover that Wayne was called "Wayne the Pain" by his employees years earlier when the rule was not in existence. The company subsequently developed more formal rules to help Wayne enhance his credibility and gain the respect of his employees. However, if after conducting his research, Tim does not discover a satisfactory reason for the structure or rules that are in place, he may try to alter this particular rule. Not only may a refusal to attend meetings demonstrate his personal rule, but it may also be an attempt to exert his power in the group.

The third assumption guiding Structuration Theory states that power is an influential force in arriving at decisions in organizations. In this theory, **power** is perceived as the ability to achieve results—it enables

power imposition of personal will on others

us to accomplish our goals. Giddens believes that power is a two-way street; any time two people are engaged in communication with each other, both sources have a certain level of power that they bring to the interaction. Even subordinates have some power over superiors (e.g., deciding to take a "sick day" on a very busy day, etc.). We all have power, but as we know from Media Ecology and Cultural Studies theorists, some have more than others. It's important to consider the role that rules play in this discussion of power. Based on the history of an organization, typically rules have been established to grant some members a particular form of power over other members. In BCT, a rule of the group is that Angela has the power to call a meeting of the members.

Tim wants his employees to call him "Tim" and not "Mr. Cliffords," but the structure—in this case a company norm—does not allow production floor workers to communicate informally. However, Tim's personal communication rule is that his workers should call him by his first name. His rule is based on his training, which taught him that in this situation it would be better for worker morale (and productivity) if subordinates were permitted to call him "Tim." Tim's decision to tell Jeremy to support keeping things the way they are for the time being is based not solely on one rule, but also on Tim's power as a supervisor in this organization. In convincing Angela to call a meeting of division managers to discuss the feasibility of adopting this rule at BCT, Tim is once again able to exert a sense of power that is granted to him as a member of the group and as a division manager. However, we see that all of the division managers bring power in one form or another to the group decision-making process. Darnell brings the power associated with his knowledge of managerial techniques as a result of his MBA. Wayne's extensive background and knowledge of the organization's history afford him a different type of power to bring to the discussion. Angela contributes the power associated with her position as Human Resources coordinator at BCT.

To summarize, when it comes to decision making, no one power structure in the web of organizational rules is more important than the other. We know that supervisors and subordinates can each exercise various "power grabs" at various times. Power, in Giddens' view, is a two-way street, and the fact that an individual is even invited to participate in discussions and decision making indicates that they have a certain amount of power over others.

Central Concepts of Structuration Theory

Like other communication theories, Structuration Theory is composed of a number of component parts. And, Beth Bonniwell Haslett (2015) asserts that these concepts "have profound implications for organizing and communicating" (p. 6). Let's look at the various elements that are central to understanding the complexity of rules and the influence they have on communication within systems. The concepts we will discuss include agency and reflexivity, duality of structure, and social integration.

Agency and Reflexivity

Structuration Theory is based on the notion that human activity is the source that creates and recreates the social environment in which we exist. Two key terms related to this perspective are agency and agent. **Agency** is defined as the specific behaviors or activities that humans engage in, guided by the rules and contexts in which interactions take place. Agency is normally undertaken by individuals, groups, or organizations themselves. **Agent** refers to the person who engages in these behaviors and who has the capacity to make a difference. For example, students serve as the agents who engage in the agency of attending classes at a college or university. The context of the classroom provides a template of rules that the agents (students) are expected to follow. If the context of the class is a large lecture setting, the rules might dictate that the particular agency (behaviors) to be performed in asking a question be formal in nature (e.g., raising one's hand to ask a question).

agency behaviors or activities used in social environments

agent a person engaging in behaviors or activities in social environments

Students Talking: Haley

I'm thinking about how different structuration themes are in my current job as a waitress. I respond to the company's "rules" and have to abide by them (clock in at a specific time or share tips with other servers). I also can see how the hierarchy affects my job. When the owner is around, we're all more stressed. When she's not, we are more relaxed. I guess the entire restaurant is like a social structure!

M. Scott Poole, David Seibold, and Robert McPhee (1986, 1996) apply the notion of agency to their examination of small groups by proposing that a group's members are aware of and knowledgeable about the events and activities that take place around them. This awareness guides their decision to engage in particular behaviors (agency). According to Structuration theorists, organizations (and groups) engage in a process known as reflexivity. **Reflexivity** refers to the actors' ability to monitor their actions and behaviors. In essence, members of an organization are able to look into the future and make changes in the organization's structure if it appears as if things are not going to work according to plan. An important

reflexivity a person's ability to monitor their actions or behaviors

element in agency and reflexivity is the ability of an individual to articulate the reasons for their choices of behavior (Drefs, 2023). When an instructor asks a student why they didn't ask the question during class time, the student may respond that the number of questions being asked was too overwhelming for them. Thus, the agent has a level of consciousness about their behavior and can explain why a particular behavior was chosen over another.

In employing agency and reflexivity, organizations reflect on the structures and systems that are in place, and members have the ability to explain the reasons for the behaviors as well as the ability to identify their goals. This awareness occurs on two levels. **Discursive consciousness** refers to the ability of a person to state their thoughts in a language that can be shared with other members of the organization. Put another way, this consciousness pertains to knowledge that can be expressed through words to others. **Practical consciousness**

> **discursive consciousness** a person's ability to articulate personal goals or behaviors
>
> **practical consciousness** a person's inability to articulate personal goals or behaviors

refers to those actions or feelings that cannot be put into words. Modaff and Butler (2021) state that "some activities and/or feelings are easily explained by individuals (discursive consciousness), while other experiences, behaviors, and feelings are not as easily put into words (practical consciousness)" (p. 118).

To understand these terms, let's look at our opening scenario. Tim may be able to explain the reasons he proposed the change in BCT's rule for supervisor–subordinate communication. He engages in reflexivity by describing his past supervisory experiences and the success that he had in relating to his former employees. Thus, he is using a level of discursive consciousness. What Tim may not be able to articulate is the practical consciousness, or the internal feelings, that he experiences in a formal work environment. Tim might find it difficult to explain the warm, familial feelings that emerge when his employees address him by first name. He also might not realize the fact that when they address him as "Mr. Cliffords," he feels as though they are addressing his father or grandfather—maybe he does not perceive himself as being old enough to be addressed as "Mr." The agency, Tim's decision to address the issue, is based on his ability to reflect on previous experience.

Reflexivity also enables Tim to make a prediction about how these behaviors will impact employer–employee relationships in the future. Tim's ability to articulate his reasons for his attitudes and actions provides others with an understanding of why he pursues these goals. Both personal and organizational rules are influencing the decision of how employees should address supervisors in the organization. The task Tim faces is in deciding whether to maintain the organizational rule or alter it to accommodate his personal rules.

Duality of Structure

Rules and resources fulfill dual functions in organizations. According to the principle of **duality of structure,** members of an organization depend on **rules** and resources to guide their decisions about the behaviors or actions that they will employ in their communication. When an individual chooses to either follow a rule or alter the rule, then the way that the rule is followed in future interactions changes. Following a rule makes it "valid." Sarah Tracy (2013) contends that duality of structure can be interpreted as follows: "Structure is created from the top down and from the bottom up" (p. 61).

> **duality of structure** rules and resources used to guide organizational decisions about behaviors or actions
>
> **rules** general routines that the organization or group follows in accomplishing goals

To better understand this relationship, let's explore the distinction between rules and resources. In Structuration Theory, rather than viewing rules as strictly guidelines for *why* something must be done, it is more useful to view them as an instruction manual for *how* a goal may be accomplished. As was stated earlier in this chapter, and as referred to in **Chapter 3,** these rules may be explicitly stated or implicitly learned. At BCT, Tim realizes that it would not be effective simply to change the communication expectations for his team without first consulting the other managers in the organization. He employs a personal rule that states that he should respect his colleagues and their reasons for organizational protocol. He accomplishes that goal by implementing this rule; he requests a meeting with the human resources coordinator. Next, a group meeting is called to give all division supervisors the opportunity to express their opinions. The group's rule of giving every member equal voice in arriving at decisions is employed. Finally, a vote is taken to decide the preferred action of the majority of the group's members. Tim could initially decide simply to invoke his own rules and disregard the rules of the organization, but he decides to discuss the issue with his colleagues and supervisors to arrive at a decision. A vital element in understanding why some rules are enforced over others lies in understanding the power that certain agents have in the decision-making process.

Resources refer to the power that individuals bring to the group or organization. Power in this regard can be both tangible (office, supplies, etc.) and intangible (decision-making style). Additional sources of power can be technological in nature. For instance, integrating ChatGPT and overseeing social media initiatives may give you more power than other colleagues without this tech know-how. Power is influential because it leads an individual to take action or initiate change.

> **resources** attributes or material goods that can be used to exert power in an organization

An organization can employ two types of resources: allocative and authoritative. **Allocative resources** refer to the material assistance generated by an organization to help the group in accomplishing its goal. Suppose a group of college students wants to have access to a facility where they can work out during their breaks between classes or during their "off" time. Some of the group's members decide to write a proposal to college administrators and provide them with a list of activities that they are willing to undertake to raise some of the money if matching funds are secured. Thus, a plan for providing allocative, or material, resources has been established.

> **allocative resources** material assistance used to help groups accomplish their goals

Authoritative resources pertain to the interpersonal characteristics that are employed during communication interactions. Recall from **Chapter 2** that interpersonal communication pertains to the dynamic between two people. In Structuration Theory, interpersonal communication is viewed as the ability to influence others. Put another way, authoritative resources allow a person to execute power in an organization. Every person possesses a degree of power and influence on the operations of an organization, a claim we identified earlier in the chapter.

> **authoritative resources** interpersonal assistance used to help groups accomplish their goals

One key point about power in this theory is how agents can employ power to get what they want in a social system. To understand better what we mean by power, consider one model of social power (French & Raven, 1959; Raven, 2004) that has had lasting appeal. These bases of power can be thought of as the various authoritative resources employed in organizations. As Giddens (1984) suggests, when individuals are oppressed, they still have resources (such as power) available to overcome the status quo. Each power type is explained with our chapter opening in mind (see **Table 9.1**).

Table 9.1   Power Type as a Resource in an Organization

TYPE OF POWER	DEFINITION
Reward	Luca has the ability to provide something of value to Jayden.
Coercive	Luca can deliver punishment to Jayden.
Referent	Luca achieves agreement or compliance because Jayden respects Luca and desires to be like Luca.
Legitimate	Luca exerts control over Jayden because of Luca's title or position.
Expert	Luca possesses special knowledge or expertise that Jayden needs.

Reward Power   **Reward power** is based on a person's perception that another has the ability to provide positive reinforcements. These rewards may come in the form of praise, material rewards, or simply removal of negative aspects of the system. Tim's employees may decide to accommodate his request to address him by his first name because they perceive him as having power to promote them or give them raises. If this is the case, reward power is a resource that is affecting communication in the organization.

reward power perception that another person has the ability to provide positive outcomes

coercive power perception that another person has the ability to punish you

referent power perception that another person has the ability to achieve compliance because of established personal relationships

Coercive Power   If Tim's employees fear that they will be demoted or fired as a result of failing to comply with his wishes to establish relationships on a first-name basis, coercive power may be influencing decisions and communication. **Coercive power** is based on the expectation that an individual has the ability to exact punishment. A person may comply with another's request simply to avoid negative consequences such as losing credibility in front of one's coworkers or getting a rotten work schedule.

Referent Power   Perhaps Tim's employees choose to address him by his first name primarily because he is a friendly, likable person who demonstrates a genuine interest in his workers. Then, the resource guiding communication decisions is **referent power,** or the ability of an individual to engage compliance based on the fact that personal relationships have been established between the two interactants.

Theory-Into-Action

 Structuration Theory posits that resources refer to the power potential that individuals bring to the groups in an organization. What might not have been considered, however, is that people are resources themselves. That is, not only is the power a resource, but the person is also a resource. Consider, for instance, the phrase invoked by so many companies: "Our best resource is our employee!" While some may cringe at this claim, the reality is that an employee's field of experience—their personality, likes, dislikes, technological competency, beliefs, patterns of communication—all come into play on the job. These characteristics can be considered organizational resources because a company never knows when a particular quality will be of importance. A coffee barista who is multilingual will be advantageous to a coffee company with a diverse client base. An employee who is from a large family can provide an immeasurable advantage if the target consumer is "the family." Bringing one's attributes and experiences into a job is part of what it means to be viewed as a company resource. Perhaps that's why one of the most central units in organizational life is called "Human Resources."

Legitimate Power   Recall Tim's comment to Jeremy: "I'm the head of the division, and I would really prefer if we do things my way." The influence a person exerts on the basis of their position or title is called **legitimate power.** If division managers decide to retain the current communication rules simply because they respect Wayne and his tenure with the company, legitimate power is a resource guiding their decision. This type of power is associated with one's right to exert influence.

legitimate power perception that another person has the ability to exert influence because of title or position

Expert Power   **Expert power** refers to one's ability to exert influence over others based on the knowledge or expertise that one possesses. In the opening scenario, we learned that Darnell holds an MBA and has extensive knowledge about effective managerial communication strategies in the workplace. If the division managers decide to base their decision to adopt a less-formal environment in the workplace on Darnell's knowledge, his expert power serves as a resource in the decision.

expert power perception that another person has the ability to exert influence because of special knowledge or expertise

In another example, many students have assigned expert power to a teacher based on their previous experiences in classes with knowledgeable instructors. However, if an instructor were to enter a classroom and use profanity or only provide examples based on what they read on Instagram, those same students might alter their personal rule that all teachers possess expert power; they might decide to employ a rule in the future that requires them to be tentative in their decision to assign power to an instructor. This is an example of duality of structure.

Students Talking Tech: Emily

 I'm sitting here re-reading this chapter because it's pretty tough to get through once. But, the one thing I did understand the first round was the issue of power. I think I can apply this to social media for sure. First, there's the "haves and have nots"—those who have access and those who don't. Then, there's the issue of who decides to block who when texting or on Facebook. And, there's also the issue of posting something that goes viral. When that happens, it's like the person has a lot of power because others have liked what they read or saw. So, the power—social media connection is clear to me.

Social Integration

In addition to agency and reflexivity and a duality of structure, an additional key concept functions prominently in Structuration Theory. **Social integration** refers to the reciprocity of communication behaviors between people in an interaction. This is an ongoing

> **social integration** reciprocity of communication behaviors in interaction

process whereby members of an organization become acquainted with one another and form expectations based on previous impressions or information that is learned. If members are interacting with one another for the first time, their knowledge of one another will be quite limited, and the process of social integration will be much more extensive. However, as members become acquainted with one another, the social integration process relies heavily on structures that are recalled from past interactions. This is similar to the self-disclosure process that we explained in **Chapter 7.**

We already know that each group or organizational member brings their own background, experiences, and expectations to a communication event. However, as a member of an organization, an individual brings knowledge to the situation that is subject to change based on the influence of both internal and external sources. As we gain a sense of how we and others fit into the group, we begin to communicate and act in ways that indicate the roles we expect each member to fulfill. Expectations for patterns of behavior are established, but these expectations could potentially change as the members of the group interact and evolve.

It's important to note that organizational scholars have found that social integration may be limited. That is, it may be nearly impossible to avoid the fact that people lose self-sufficiency and control (in organizations and in life). These are termed "dividuals," meaning that their individuality has been compromised. As Nicolas Bencherki and James Snack (2016) observe: "People are not whole, coherent entities that fit in a single organization, in the same way as the Matryoshka dolls nest within each other. Rather, they each constitute societies that hold together the many 'pre-individual' elements that constitute them" (p. 284). So although social integration is a profound communication process in Structuration Theory, it can be limited insofar as people may lose their full identity as an organizational member.

Recall the meeting of the division managers. Because Tim is the newest member of the group, the other division managers may be uncertain about what type of power to attribute to him as he presents his ideas. The same applies to Tim's interactions with his subordinates. As Tim and the others engage in subsequent interactions, his view of himself and others' views of Tim will evolve.

Application of Time and Space

We now examine the role of time and space in organizations. Structuration theorists believe that all social interaction in an organization is composed of temporal (time) and spatial (space) dimensions. This is called time–space distanciation (TSD) and is an important feature of the theory. The actual communication or interaction that takes place in the organization can be examined as existing in real time and as taking place in real locations (space).

We hear a message as it occurs in a context. For instance, think about a conversation that your supervisor has with you about company layoffs while the two of you are walking to your cars at the end of the workday. The supervisor's decision to talk about this topic, in this place (parking lot), and at this particular time would be of interest to Structuration theorists. However, researchers would likely need to look more deeply to determine the factors that motivated your boss to communicate this message to you. Furthermore, does your boss make reference to past experiences with company layoffs? Does she look ahead to potential consequences if such a layoff were to happen? And, especially relevant to many communication researchers: if a layoff were to be announced, how would it be communicated to employees? A mass email? Public posting? How would the conversation, as Poole and McPhee (2005) express it, "bind" (p. 180) space and time in the discussion of company layoffs?

Essentially, space is viewed as a contextual element that has meaning for the various members of a group or organization. The elements of time and space are factors that enable us to engage in communication. One's view of their time spent in a group spans both time and space and influences the decisions that are made. For example, in our opening scenario, it is unlikely that a single event or a single location has solely influenced Tim's action to request a change in the company's policy. Rather, it's a series of experiences and references that are the result of his own managerial history—and perhaps even his own experience as an employee in an organization—that have influenced his choice of action.

Integration and Critique

Organizations remain central to our lives. Company employees spend almost 90 percent of their time in group or team meetings. In fact, with COVID-19 in the rear-view mirror for many companies, employees are spending time in both FtF and Zoom meetings 3–5 hours per week (Flynn, 2023, **https://www.zippia.com/advice/meeting-statistics/**). Structuration Theory provides an important framework for understanding these communication opportunities. The theory has been historically linked to empiricism, making it primarily aligned with quantitative methods. Still, some organizational scholars have used case studies, resulting in research that employs qualitative methods as well. And yet, the studies related to the Structuration process in group decision making in particular tend to be qualitative in nature (Pilny, Poole, Reichelmann, & Klein, 2017, p. 415). Among the criteria relevant to evaluating theory, we identify two for our discussion: scope and parsimony.

Integration

Communication Tradition	Rhetorical | Semiotic | Phenomenological | **Cybernetic** | Socio-Psychological | **Socio-Cultural** | Critical
Communication Context	Intrapersonal | Interpersonal | **Small Group** | **Organizational** | Public/Rhetorical | Mass/Media | Cultural
Approach to Knowing	**Positivistic/Empirical** | **Interpretive/Hermeneutic** | Critical

Critique

Evaluation Criteria	**Scope** | Logical Consistency | **Parsimony** | Utility | Testability | Heurism | Test of Time

Scope

Structuration Theory can be applied to numerous social settings involving teams and small groups and scores of communication interactions, including those that are technologically based (Madsen & Matusitz, 2022). As noted previously, the areas of communication that have applied the theory with the most theoretical success are organizational communication and group/team communication. Our focus has been on the organizational environment, namely, how the structures created in organizations influence communication and decisions.

Still, some concerns related to the theory's breadth exists. At first glance, you may view the theory as too expansive, resulting in too broad a scope. With a discussion of rules, resources, power, and discourse, it would seem that the theory is trying to do too much at one time. Indeed, in his review of the theory, Rob Stones (2005) asserts that Giddens's theory should be "clearer, tighter, and more systematic than Giddens has been about Structuration theory's distinctive and defining characteristics" (p. 1).

Parsimony

Related to the theory's scope is the notion of the theory's clarity. Recall that the parsimony criterion responds to the following question: is the theory easy to understand or cumbersome? Some organizational communication scholars have argued the theory is laden with terminology that is challenging to understand. For example, Wafa Kort and Jamel Gharbi (2013) state that "there is a consensus toward the difficulty that the readers encounter when they try to understand Structuration theory" (p. 95). In other words, people agree that the theory is somewhat unnecessarily complicated.

Stephen Banks and Patricia Riley (1993) agree. They point out that Structuration Theory is difficult to understand: "Structuration lacks certain characteristics that communication researchers and other social science scientists often find appealing: It is not quickly read, immediately intuitive, or parsimonious" (p. 178). Banks and Riley present many concepts as they examine the intricate process of how organizations and groups structure their communication and arrive at decisions. Their advice to those who are researching this theory in an attempt to understand organizations and groups is to "begin at the beginning" (p. 181). This requires insight and understanding of the historical rules brought into an organization by each member—and this is an extremely difficult task to accomplish. Furthermore, Banks and Riley suggest that scholars resist the temptation to apply pre-established categories in explaining how organizations (and groups) are developed and how they experience change. Perhaps Daniel Broger's (2011) comments summarize the notion of the theory's potential ambiguity. He notes that the theory "requires considerable time and effort on the part of scholars not familiar with social theory in general and structuration theory in particular to come to grips with Giddens's comprehensive theory" (p. 3).

Closing

Structuration theorists Giddens, Poole, Seibold, and McPhee provide communication scholars a theory that considers the interplay between and among people and their resources, rules, hierarchies, and relationships in organizations (and small groups). The theory has drawn its critics, and Giddens has not helped his own theoretical reputation since he appears to have ignored many of the criticisms and failed to provide clarifications on the criticisms levied against him and his theory (Broger, 2011). Nonetheless, complexity and lack of deference are not tantamount to theoretical rejection. Structuration Theory, therefore, will continue to resonate to those interested in the rules, principles, and processes in organizational life.

Discussion Starters

 Case-In-Point: Recall a time when you were involved in a group similar to the Bayside City Tire Company—a time when a decision had to be made. What were some of the rules that influenced the process of decision making? Were the rules changed as a result of the decision that was reached or the process that was followed in arriving at that decision? If so, how?

Try-It-Your-Selfie: Reflecting upon your social media, identify how you and others have co-created various formal and informal rules across different platforms. Be aware of the different hierarchies that may have influenced the rules.

1. One of the assumptions of this theory is that power structures are present in groups and guide the decision-making process by providing us with information on how to best accomplish our goals. Discuss the potential positive and negative implications of power as an element of structuration.

2. Giddens proposes that structure (rules and resources) should be viewed not as a barrier to interaction, but as a necessary part of the creation of the interaction. Do you agree or disagree with this position? Defend your answer.

3. Structuration Theory proposes that structures themselves should be viewed as being nontemporal and nonspatial. Discuss the significance of this idea. Provide an example from your own experience of the influence of time and space in a group or organization.

4. How would you explain your family using any of the principles of Structuration Theory?

References

Banks, S. P., & Riley, P. (1993). Structuration theory as ontology for communication research. In S. A. Deetz (Ed.), *Communication yearbook 16* (pp. 167–196). Sage.

Bencherki, N., & Snack, J. P. (January 2016). Contributorship and partial inclusion: A communicative perspective. *Management Communication Quarterly, 30*(3), 279–304.

Broger, D. (2011). *Structuration theory and organization research*. [Unpublished doctoral dissertation, University Press of St. Gallen]. https://www1.unisg.ch/www/edis.nsf/SysLkpByIdentifier/3873/$FILE/dis3873.pdf.

Drefs, I. (2023). On-site actors' agency within international media development. *Global Media Journal, 12*(2).

Flynn, J. (2023, February 14). *28+ incredible meeting statistics (2023): Virtual, Zoom, in-person meetings and productivity* [Webinar]. https://www.zippia.com/advice/meeting-statistics/.

French, J. R., & Raven, B. (1959). *The bases of social power.* University of Michigan Press.

Giddens, A. (1979). *Central problems in social theory: Action, structure, and contradiction in social analysis.* University of California Press.

Giddens, A. (1984). *The constitution of society: Outline of the theory of structuration.* University of California Press.

Giddens, A. (1993). *New rules of sociological method: A positive critique of interpretive sociologies* (2nd ed.). Polity.

Giddens, A. (1994). *Beyond left and right: The future of radical politics.* Polity.

Giddens, A. (2003). *Runaway world: How globalization is reshaping our lives.* Taylor & Francis.

Giddens, A., & Sutton, P. W. (2017). *Sociology* (8th ed.). Polity.

Haslett, B. B. (2015). Structuration theory. *The International Encyclopedia of Interpersonal Communication,* 1-7.

Iyamu, T. (2019). Understanding the complexities of enterprise architecture through structuration theory. *Journal of Computer Information Systems, 59*(3), 287-295.

Kort, W., & Gharbi, J. E. (2013). Structuration theory amid negative and positive criticism. *International Journal of Business and Social Research, 3*(5), 92-104.

Madsen, H., & Matusitz, J. (2022). Benefits of Google technologies for organizations: Perspectives from adaptive structuration theory. *International Journal of Technology Management & Sustainable Development, 21*(1), 5-18.

McPhee, R. (2015). Agency and the four flows. *Management Communication Quarterly, 29,* 1-6.

McPhee, R. D., & Canary, H. E. (2016). Structuration theory. *The International Encyclopedia of Communication Theory and Philosophy,* 1-15.

McPhee, R., Poole, M. S., & Iverson, J. (2014). Structuration theory. In L. L. Putnam & D. Mumby (Eds.), *The SAGE handbook of organizational communication: Advances in theory, research, and methods* (pp. 75-100). Sage.

Miller, E. (2018). *The structuration of gendered cultures within state legislatures: A communication approach.* [Master's Thesis, Hollins University].

Miller, K. (2009). *Organizational communication: Approaches and processes.* Wadsworth/Cengage.

Modaff, D., & Butler, J. (2021). *Organizational communication: Foundations, challenges, and misunderstandings.* Cognella.

Mumby, D. (2011). *Reframing difference in organizational communication studies: Research, pedagogy, and practice.* Sage.

Nicolini, D. (2013). *Practice theory, work, and organization: An introduction.* Oxford University Press.

Olowa, T., Witt, E., Morganti, C., Teittinen, T., & Lill, I. (2022). Defining a BIM-enabled learning environment—An adaptive structuration theory perspective. *Buildings, 12*(3), 292.

Pilny, A., Poole, M. S., Reichelmann, A., & Klein, B. (2017). A structurational group decision-making perspective on the commons dilemma: Results from an online public goods game. *Journal of Applied Communication Research, 45*(4), 413-428.

Poole, M. S. (1990). Do we have any theories of group communication? *Communication Studies, 41,* 237–247.

Poole, M. S., & McPhee, R. D. (2005). Structuration theory. In S. May & D. K. Mumby (Eds.), *Engaging organizational theory & research* (pp. 171–195). Ablex.

Poole, M. S., Seibold, D. R., & McPhee, R. D. (1986). A structurational approach to theory-building in group decision-making research. In R. Y. Hirokawa & M. S. Poole (Eds.), *Communication and group decision-making* (pp. 237–264). Sage.

Poole, M. S., Seibold, D. R., & McPhee, R. D. (1996). The structuration of group decisions. *Communication and group decision making, 2,* 114–146.

Raven B. H. (2004). Six bases of power. In G. R. Goethals, G. J. Sorenson, & J. M. Burns, (Eds.). *Encyclopedia of leadership* (pp. 134–139). Sage.

Seibold, D. R., & Myers, K. K. (2006). Communication as structuring. In G. J. Shepherd, J. St. John, & T. Striphas (Eds.), *Communication as ... Perspectives on theory* (pp. 143–152). Sage.

Sheposh, R. (2023). *Structuration theory.* Salem Press.

Stones, R. (2005). *Structuration theory.* Palgrave Macmillan.

Tracy, S. J. (2013). *Qualitative research methods.* Wiley-Blackwell.

CHAPTER 10
Organizational Information Theory
Based on the research of *Karl Weick*

"Sensemaking is tested to the extreme when people encounter an event whose occurrence is so implausible that they hesitate to report it for fear that they will not be believed."

—Karl Weick

Dominique Martin

As she read about the increasing corporate scandals, Dominique Martin knew that it would be just a matter of time. Her instincts became real when the Vice President of Compliance and Standards at BankNG emailed her to come speak to him about the new federal regulations on oversight and financial disclosure. Dominique was an expert on banking compliance and she was the go-to person at BankNG whenever new regulations emerged. With the recent federal bill signed into law, Dominique's instructions were clear: "Get it done," her vice president told her.

Dominique Martin could feel the stress build as the months progressed and as she considered all the avenues related to the "FedReg" project. Her expertise was sought at every moment as various BankNG units involved in regulation contacted her for assistance. In fact, she became even busier as media coverage caused many in the media to question the ability of companies to accommodate the new law, causing even more pressure for Dominique. The task was not an easy one because the law involved new procedures for auditing, corporate governance, internal operations, and financial disclosure. Special project teams dealing with the challenge of bringing their bank and its systems up to date were immediately established. Whether it was face-to-face communication or Slack team meetings, Dominique's position as a facilitator was paramount.

As she approached the door of the conference room, Dominique thought about how happy and relieved she would be once the conversion was over. Should BankNG's standards be enacted efficiently and effectively, Dominique wondered whether or not she would be promoted at the bank. But, for now, she had to deal with BankNG's compliance and she needed a project management team that was responsive, competent, and immediate. The vice president had appointed her project manager, and today she would meet for the first time with members of various teams via Zoom. Several employees from the Boston, Dallas, Denver, and Seattle offices would take part. The project goal involved the conversion of all BankNG information systems so that they would be compatible for the required changes. The FedReg project team consisted of almost 80 people who would be responsible for various aspects of the conversion. Dominique was used to managing a team of 12 employees who were all located in the same city, so dealing with nearly seven times as many colleagues was a very new challenge.

As Dominique began the meeting, she asked each team leader in the various cities to introduce their team members and their assigned duties. As they went around to the various locations, Dominique became overwhelmed. Although a compliance checklist was available, BankNG was simply too large and too complex for a simple checklist to be effective. There were so many different areas to manage in this project: conducting proper homework on the law and its various components, coordinating information with the federal regulatory agencies, providing updates to and obtaining feedback from the bank board members and shareholders, establishing an audit committee, keeping the other divisions within the bank informed of changes that were being made, and how those changes would affect their departments ... and that was only the beginning! Each of the teams needed to be kept informed about the others' progress in order to meet their deadlines. Emails, texts, and internal social media announcements would be pivotal.

As the meeting progressed, Dominique realized that the success of this project would depend on effective communication with all the various teams in the different states. They developed a communication strategy to help keep the teams in the different offices across the country informed of the events associated with the project. Although the home office was located in Boston, the regional offices needed to be made aware of the plans and protocols so that everyone was on board. The plan included hiring an internal communication coordinator to manage all the messages that would be sent about compliance. Next, they decided that a company wiki would be established so that all project members could communicate about the information needed to

complete the project. Also, an electronic newsletter would be distributed periodically to update all members of BankNG about the progress that was being made. Finally, they decided that they would use software to record the goals, resources, deadlines, and accomplishments associated with the project. In the end, despite the pressures and stressors, Dominique's leadership prevailed and BankNG was ultimately compliant with little resistance and few problems.

Although several months have passed, Dominique continues to wonder whether she facilitated the regulatory project effectively. She knew that she was a good team leader, but even today, she thinks about whether her thoughts, actions, and activities were the most appropriate at the time. She felt confident that everyone had a voice in the matter and that their conversations helped move FedReg along. Still, she couldn't rid herself of a nagging doubt about whether she managed the events to the best of her ability. She knew that some team members were still upset about some tense moments in the compliance journey, but she didn't know the specifics. Yet, she had to move on to the next project, thinking about what she learned from her experience with FedReg.

The task of managing vast amounts of information is a typical challenge for nearly all organizations. As our options for new communication channels increase, the number of messages that we send and receive, as well as the speed at which we send them, increases as well. Not only are organizations faced with the task of decoding the messages that are received, but they are also challenged with determining which people need to receive the information to help achieve the organization's goals. New media are enabling companies to accomplish their goals in ways never before seen. Videoconferencing, teleconferencing, Slack, and webinars allow people like Dominique to provide teams with the opportunity to share and react to great amounts of information simultaneously. Each of the teams was given the opportunity to decide what information was essential to its tasks or to request additional information that would be needed in the future. Clearly, the introduction of new technologies is reshaping existing organizational functions and structures (Livingston, Cunningham, & Forbes, 2023).

Sometimes the information an organization receives is ambiguous. In the case of the FedReg project, each team depended on the others to provide information so that it could complete its portion of the project. Teams needed the information to be presented in a way they could understand. The "Customer/Shareholder Services" team may have little knowledge of highly technical jargon, so they depended on IT to clarify the information and present it to them in a way that they could then communicate to their customers and shareholders. Without this exchange and management of information—particularly given the federal directive—BankNG's regulation accommodation would probably fail.

Some writers of organizational communication have used the metaphor of a "living system" to describe organizations (De Kroon & Kaat, 2015). Just as living systems engage in a process of activities to maintain their functioning and existence, an organization must have a procedure for dealing with all the information it needs to send and receive to accomplish its goals. Much like systems, organizations are made up of people and teams that are interrelated. Each depends on the other to accomplish their goals.

To try to unpack the complexity of this theory, throughout this chapter, we will invoke an experience that we all understand: college life. Consider the process that colleges and universities go through to recruit new students. The marketing office conducts research to find out what criteria matter to potential students. Are graduation rates and job placement something high school students are looking for in deciding where to pursue their education? Is the diversity of the student body an important factor in making a decision? Does student financial aid remain paramount? Will the recent rating by *The Princeton Review* have an impact on student decisions? Or, are students attracted to a campus that has superior computer facilities, dorms, or cafeteria food? After collecting these data, the publications office will use this information to develop publicity materials that appeal to potential students. Recruitment fairs may be held in major cities to provide parents and students with the opportunity to ask admissions counselors about the school. While this process is taking place, the school collects feedback and monitors the reactions of potential students and their parents in order to make changes in their current recruitment strategies. The admissions office reduces uncertainty about what qualities are attractive to students while at the same time assisting students in making their decision by providing information about the school.

Karl Weick developed an approach to describe the process by which organizations (like a college or university) collect, manage, and use the information that they receive. Rather than focusing his attention on the structure of the organization in terms of the roles and rules that guide its members, Weick emphasizes the process of organizing. In doing so, the primary focus is on the exchange of information that takes place within the organization and how members take steps to understand this material. Weick (1995) believes that "organizations talk to themselves" (p. 281). To this end, organizational members are instrumental in the creation and maintenance of message meaning. Further, Weick and Kathleen Sutcliffe (2015) state that "organizing holds events together and reliable performance depends on sustained organizing" (p. viii).

Weick sees the organization as a system taking in confusing or ambiguous information from its environment and making sense out of it. Therefore, organizations will evolve as they try to make sense out of themselves and their environment. Going back to our opening story, rather than focusing on Dominique's role as Project Manager or on the specific communication rules for sending messages between and among superiors and subordinates, Weick's Organizational Information Theory (OIT) directs our attention to the steps that are necessary to manage and use the information for the BankNG project. As it becomes more difficult to interpret the information that is received, an organization needs to solicit input from others (often multiple sources) to make sense of the information and to provide a response to the appropriate people or departments.

Theory at a Glance • Organizational Information Theory

The main activity of organizations is the process of making sense of equivocal and ambiguous information. Organizational members accomplish this sensemaking process through enactment, selection, and retention of information. Organizations are successful to the extent that they are able to reduce equivocality through these means.

The Only Constant Is Change (in Organizations)

Weick's theory focuses on the process that organizations undergo in their attempt to make sense out of all the information that bombards them on a daily basis. His view of organizations is unique: they contain a high level of abstraction, "defined by knowledge-intensive, high-risk, and uncertain core technologies" (Johnson & Kruse, 2019, p. 4). Often the process results in changes in the organization and its members. In fact, Weick states that "organizations and their environments change so rapidly that it is unrealistic to show what they are like now, because that's not the way they're going to be later" (1969, p. 1). According to this approach, it would be unrealistic to try to depict what colleges or universities and their surrounding environment look like today because it's likely that they will change. The major fields of study chosen by students may change as organizations' needs for employees change. Consider, for instance, the early 1990s. To assist them in making changes in their computer systems, many companies were seeking graduates in the fields of computer technology and management information systems. Today, however, some of these same companies are more interested in Library Science or Communication Studies degree candidates because of their understanding of information literacy. As you can see, organizational demands are often influenced by cultural demands.

The focus of Organizational Information Theory is on the communication of information that is vital in determining the success of an organization. It's quite rare that one person or one department in an organization has all the information necessary to complete a project. This knowledge typically comes from a variety of sources. However, the task of information processing is not completed simply by attaining information; the difficult part is in deciphering and distributing the information that is gained. To understand this process better, we will discuss how two major perspectives influenced OIT: General Systems Theory and the Theory of Sociocultural Evolution.

General Systems Theory

To explain the influence of information from an organization's external environments and to understand the influence that an organization has on its external environments, Weick applied General Systems Theory in the development of his approach to studying how organizations manage information. As we discussed in **Chapter 3,** Ludwig von Bertalanffy is most frequently associated with the systems approach. Von Bertalanffy believed that patterns and wholes exist across different types of phenomena. To this end, he proposed that when there's a disruption in one part of a system, it affects the entire system. In sum, systems theorists argue that there are complex patterns of interaction among the parts of a system, and understanding these interactions will help us understand the entire system.

Systems thinking is especially useful in understanding the interrelationships that exist among various organizational units. Organizations are usually made up of different departments, teams, or groups. Although these units may focus on independent tasks, the goals of the organization as a whole typically require sharing and integrating the information that each of the teams has to arrive at a solution or conclusion. Organizations depend on combined information so that they can make any necessary adjustments in order to reach their goal. They may need additional information, they may need to email information to other departments or people within the organization, or they may need to hire outside consultants to make sense of the information. If one team fails to address the information needed to fulfill its obligation in the completion of the project, achieving the final goal will probably be delayed for the entire organization.

In the chapter-opening scenario, we learned that Dominique needed to consult with different teams from different states. Each state's team had a particular responsibility in order to complete the FedReg project. Yet, the interrelationships among the teams could not be ignored. That is, each had a dependency on the other; one needed the other for information and could not act without it. If such information were not received, enacting BankNG's changes would be delayed.

An important component of General Systems Theory, and one that is essential to making sense of information in an organization, is feedback, a term we introduced in **Chapter 1** of this book. Feedback, you will recall, is the information received, and in this case, the information by an organization and its members. It's important to remember that this information can be either positive or negative. The organization and its members can then choose to use the information to maintain the current state of the organization or an organization can decide to initiate changes as they relate to the system's (organization's) goals. Through feedback, units are able to determine if the information that is being transmitted is clear and sufficient to achieve the desired goals.

The decision of the organization to request or provide feedback reflects a selective choice made by the group in an effort to accomplish its goals. If an organization hopes to survive and accomplish its goals, it will continue to engage in cycles of feedback to obtain necessary information and reduce its uncertainty about the best way to accomplish its goals. This process reflects a Darwinian approach to how organizations manage information, as we discuss next.

Darwin's Theory of Sociocultural Evolution

A second perspective that has been used to describe the process by which organizations collect and make sense out of information is the **theory of sociocultural evolution** (Stuart-Fox, 2023). You may have heard of the "survival of the fittest," which is an apt phrase to describe the theory. The eventual goal of any organization is survival,

theory of sociocultural evolution
Darwin's belief that only the fittest can survive challenging surroundings

and, like humans, it works to discover the best strategies for getting by. Although this approach is used to describe the social interactions that take place in an organization with regard to making sense out of information, its origins are in the field of biology.

The theory of evolution was originally developed to describe the adaptation processes that living organisms undergo in order to thrive in a challenging ecological environment. Charles Darwin (1948) explained these adaptations in terms of mutations that allow organisms to cope with their various surroundings. Some organisms could not adapt and died, whereas others made changes and prospered. Taking the example of Dominique Martin and the FedReg challenge, corporations that took steps to accommodate the law quickly have a greater chance to prosper and thrive. Those who did not make the attempt to adapt to the changing legal landscape will likely be faced with severe consequences including financial penalties.

Marion Blute (2010) explains the processes by which organizations and their members adapt to their social surroundings. The sociocultural theory of evolution examines the changes people make in their social behaviors and expectations to adapt to changes in their social surroundings. Many times, these changes involve team-member efforts to be creative (Sawyer, 2012).

Consider the scenario at the beginning of this chapter. Suppose that Dominique had a team of representatives from BankNG's offices in Japan involved in this project. If she wanted to make a quick decision on how to address the FedReg changes in the Japanese offices, her first tendency might be to employ a communication style recognized as appropriate by the U.S. employees. However, she would soon realize that the Japanese approach to business is usually different from that practiced in the United States. Instead of quickly presenting the facts and determining financial processes, Japanese bank employees prefer an approach that emphasizes the development of rapport. Only after the relationship has been developed would the issue of signing an agreement be presented. Dominique must engage in the process of sociocultural evolution to adapt to the norms and expectations of her overseas team.

Weick adapts sociocultural evolution to explain the process that organizations undergo in adjusting to various information pressures. These pressures may be the result of information overload or ambiguity. Although the evolutionary approach is useful in describing the adaptations that are necessary to process information, General Systems Theory is also an essential piece in this puzzle because it highlights the interrelatedness among organizational teams, departments, and employees in the processing of information. We now continue our discussion of OIT and identify its underlying assumptions.

Assumptions of Organizational Information Theory

OIT is one way of explaining how organizations make sense out of information that is confusing or ambiguous. It focuses on the process of organizing members of an organization to manage information rather than on the structure of the organization itself. A number of assumptions underlie this theory:

- Human organizations exist in an information environment.
- The information an organization receives differs in terms of equivocality.
- Human organizations engage in information processing to reduce equivocality of information.

The first assumption states that organizations depend on information in order to function effectively and accomplish their goals. Weick (2020) views the concept of information environment as distinct from the physical surroundings in which an organization is housed. He proposes that these information environments are created by the members of the organization. Members establish goals that require them to obtain information from both internal and external sources. However, these inputs differ in terms of their level of understandability.

Consider the university Admissions and Enrollment office example. A school can use numerous channels to gain information about student needs: it may develop a website to answer prospective students' questions and to solicit student feedback; it may conduct surveys at high school academic fairs to gain more information about student desires; it may host focus group interviews with current students to discover needs and concerns; or it may ask alumni to provide examples from their educational experiences to attract future students. Once the school has received messages from all of these external sources, the university must decide how to communicate messages internally to establish and accomplish its goals for current and future students. The possibilities for information are endless, and the university must decide how to manage all the available potential messages.

The second assumption proposed by Weick focuses on the ambiguity that exists in information (Boivin & Brummans, 2022). Messages differ in terms of their understandability. An organization needs to determine which of its members are most knowledgeable or experienced in dealing with particular information that is obtained. A plan to make sense of the information needs to be established. In fact, Karl Weick (2020) argues that when things are ambiguous, people will do their best to work with it. That is, he believes that at times, employees have no choice but to deal with ambiguous messages. The key is to simply accept it and try to make sense of it, an idea we explore a bit later in the chapter.

At BankNG, each member of each team must be able to interpret and understand the messages accurately. Given the federal law requirements, however, and the complexities related to the law, many messages are not always clear because individuals often assume understanding and often communicate in ways that are less than clear. Further, each office has had little experience working with such a massive undertaking.

equivocality the extent to which organizational messages are uncertain, ambiguous, and/or unpredictable

Messages in BankNG, then, according to Weick's theory, may be frequently equivocal. **Equivocality** refers to the extent to which messages are complicated, uncertain, and unpredictable. Equivocal messages are often sent in organizations. Because these messages are not clearly understood, people need to develop a framework or plan for reducing their ambiguity about the message. And, it's not one person's responsibility to reduce organizational equivocality, but everyone's, because "it leads to a safer and more adaptive system" (Mumby & Kuhn, 2019, p. 116). Yet, given that so many Western cultures prefer clarity rather than ambiguity (Liu, Volcic, & Gallois, 2019), we can see that nearly all U.S. companies would eschew equivocal messaging in their offices.

You may be tempted to think that equivocality is ineffectual in an organization. Yet, as Eric Eisenberg (2007) reminds us, equivocality is not necessarily problematic. He states that rather than viewing equivocality as difficult, "Weick turned this idea on its head, arguing instead that equivocality is the engine that motivates people to organize" (p. 274). Eisenberg further clarifies that equivocality may "make coordinated action possible" (p. 274). When individuals in an organization reduce equivocality, they engage in a process that tries to make sense out of excessive information received by the organization. We delve further into equivocation a few more times later in the chapter.

In an attempt to reduce the ambiguity of information, the third assumption of the theory proposes that organizations engage in joint activity to make information that is received more understandable. Weick (2020) sees the process of reducing equivocality as a joint activity among members of an organization. It is not the sole responsibility of one person to reduce equivocality. Rather, this is a process that may involve several members of the organization. Consider the BankNG example. Each department needs to use information from other units, but they also need to provide information to these same units to accomplish the tasks necessary to meet the organization's federal compliance. This illustrates the extent to which departments in an organization may depend on one another to reduce their ambiguity. An ongoing cycle of communicating feedback takes place in which there is a mutual give-and-take of information.

Key Concepts and Conceptualizing Information

Weick's theory of Organizational Information contains a number of key concepts that are critical to an understanding of the theory. They include information environment, rules, and cycles. We now explore each in detail.

Information Environment: The Sum Total

Information environment is an integral part of Weick's theory. Information environment is a core concept in understanding how organizations are formed as well as how they process information. Every day, we are faced with literally thousands of stimuli that we could potentially process and interpret. However, it's unrealistic to think that an organization or its members could or would possibly process all the information that is available. Thus, we are faced with the tasks of selecting information that is meaningful or important and focusing our senses on processing those cues.

Think, for instance, about where you are sitting while you're reading this. Are you in your apartment? Dorm room? Kitchen? Library? Outdoors? Now, look at all the various stimuli, including people, noise, color, physical surroundings, and so forth. It's fair to say that

information environment the availability of all stimuli in an organization

you would only process those stimuli that have "value" to you. The availability of all stimuli is considered to be the **information environment.** Organizations are composed and are sustained because of information that is vital to their formation and continues to be essential to their existence. For example, the FedReg project team was formed because new federal laws required publicly held companies to revamp their financial protocols. If the bank is to survive during a competitive era, it must comply or else face dire consequences on the horizon. So, the information environment not only comprises internal communications between and among different units and offices, but there are external stimuli (e.g., governmental mandates) that also must be considered.

Essentially, organizations have two primary tasks to perform in order to successfully manage these multiple sources of information: (1) They must interpret the external information that exists in their information environment, and (2) they must coordinate that information to make it meaningful for the members of the organization and its goals. These interpretation processes require the organization to reduce the equivocality of the information to make it meaningful.

Students Talking Tech: Jason

I keep trying to find how this theory relates to technology, and I think it's just assumed by Weick. He talks about the "information environment" (IE), and although he doesn't say it directly, today, that environment can also include social media. It may seem crazy to say that Twitter and Facebook affect an office, but they do, and I can easily see that with so many more options available on Facebook than the other platforms, there's a higher chance for equivocality to happen.

Rules: Guidelines to Analyze

Weick (2020) proposes two communication strategies that are essential if the organization hopes to reduce message equivocality. The first strategy requires that organizations determine the rules for reducing the level of equivocality of the message inputs as well as for choosing the appropriate response to the information received. In a sense, then, organizations self-govern with respect to getting things accomplished. In OIT, **rules** refer to the guidelines that an organization has established for analyzing the equivocality of a message as well as for guiding responses to information. For example, if any of the FedReg team leaders experience ambiguity over the information that is provided to them from the computer technicians, they are instructed to contact the technician team leader for clarification. If they were to contact Dominique first, an additional step would be added to the process of reducing equivocality because she would have to contact the technicians and then report back to the team leaders. Recall the example of the university Admissions and Enrollment office which conducted research in order to design materials that appeal to students. As a result, when it receives information from students inquiring about the university, the rule that it applies in dealing with the information environment is to respond to that message by sending out their informational brochure. Both organizations have rules for determining the equivocality of the information and for identifying the appropriate way in which they should respond to the messages.

> **rules** guidelines in organizations as they review responses to equivocal information

Weick provides examples of rules that might cause an organization to choose one cycle of information or feedback over another for reducing the equivocality of messages. These rules include duration, personnel, success, and effort. Examples of duration and personnel rules guiding communication can be seen in our chapter-opening story. **Duration** refers to a choice made by an organization to engage in communication that can be completed in the least amount of time. For example, BankNG has rules that state who should be contacted to clarify

> **duration** organizational rule stating that decisions regarding equivocality should be made in the least amount of time

technical information. These rules prevent people from asking those who are not knowledgeable about the topic. In establishing these rules, BankNG increases its efficiency by having employees go directly to the person who can provide the necessary information,

> **rule of personnel** organizational rule stating the most knowledgeable workers should resolve equivocality

thus eliminating delays that might result from having to channel questions through several different people. In doing so, BankNG is also guided by the **rule of personnel.** This rule states that people who are the most knowledgeable should emerge as key resources to reduce equivocality. Computer or information technicians, not human resources personnel, are consulted to reduce equivocality of technical information associated with the project.

Self-Governance in an Age of Rules

When an organization chooses to employ a plan of communication that has been proven effective in the past at reducing equivocality of information, **success** is the influential rule that is being applied. A university knows that many of its potential students' questions and concerns can be answered via a well-researched brochure or website.

> **success** organizational rule stating that a successful plan of the past will be used to reduce current equivocality

They have proven to be a successful recruiting tool in the past, as the enrollment figures of the school have increased by 4 percent per year over the past 5 years. Thus, the university knows that this is a successful way of reducing information ambiguity for students.

The **rule of effort** influences the choice to use a brochure to promote the university. This rule guides organizations in choosing an information strategy that requires the least amount of effort to reduce equivocality. Rather than fielding numerous phone calls from potential students asking the same questions about the school, the university's

> **rule of effort** organizational rule stating that decisions regarding equivocality should be made with the least amount of work

decision to print a brochure that answers frequently asked questions is the most efficient means of communicating all the university has to offer. The decision made by many companies to implement automated telephone customer service is another example of how organizations reduce equivocality of information. Rather than requiring a customer to explain the reason for the call multiple times as the customer is connected to various employees, the customer chooses a number that best matches the problem or concern. This directs the customer to the appropriate department.

Students Talking: Nicolas

One area of this theory that I can relate to is the issue of equivocality. I'm a waiter at a "national chain" restaurant and we have all sorts of confusion when it comes to our supervisors. They're all supposed to follow the "corporate rules," but each one seems to have their own way of supervising. One says to be super kind to the kitchen staff and another says that we should be kinder to the customers. The third supervisor basically leaves us alone and seems to always blame the customer for the complaints. This is the ultimate of an equivocal company climate and many of us simply do our best without worrying about these different leadership styles.

Cycles: Act, Respond, Adjust

If the information that is received is highly equivocal, the organization may engage in a series of communication behaviors in an attempt to decrease the level of ambiguity. Weick labels these *systems of behavior cycles*. The more equivocal the message, the more **cycles** needed to reduce equivocation (Herrmann, 2007). The cycle of communication behaviors used to reduce equivocality includes three stages: act, response, and adjustment. An **act** refers to the communication statements and behaviors used to indicate

cycles series of communication behaviors that serve to reduce equivocality

act communication behaviors indicating a person's ambiguity in receiving a message

one's determination to reduce ambiguity. For example, as a team leader, Dominique may say to a member of the auditor independence team, "The financial disclosure research team wants your group's input on the ethics related to the stock transactions of the company officers." In deciding to solicit information and subsequently reduce any ambiguity, Dominique is employing the rule of personnel.

A second step in the communication cycle is response. **Response** is defined as a reaction to the act. That is, a response that seeks clar-

response reaction to equivocality

ification in the equivocal message is provided as a result of the act. The auditing team leader might reply, "It is essential that all possible avenues examining conflict of interest and officer stock transactions be explored. This is absolutely of primary importance."

As a result of the response, the organization formulates a response in return as a result of any **adjustment** that has been made to the information that was originally received. If the response to the act

adjustment organizational responses to equivocality

has reduced the equivocality of the message, an adjustment is made to indicate that the information is now understood. If the information is still equivocal, the adjustment might come in the form of additional questions designed to clarify the information further. In other words, if Dominique is still uncertain, she will likely continue to ask questions, thereby reducing message equivocality.

Feedback is an essential step in the process of making sense of the information that is received. Weick uses the term *double-interact loops* to describe the cycles of act, response, and adjustment in information exchanges. **Double-interact loops** refer to multiple communi-

double-interact loops cycles of an organization (e.g., interviews, meetings) to reduce equivocality

cation cycles that are used to assist the organization's members in reducing the equivocality of information. Employees may have working lunches, informal chats in the employee break room, interviews, conference calls, among other undertakings to continue to reduce information equivocality. Imagine the double interact as you imagine shoveling snow. The goal is to get the snow removed so that no accidents occur. Double interacts in companies are necessary to "clean up" the ambiguities and uncertainties. Failure to do so will not only prohibit goals from being attained, but also can have consequences. Because these cycles require members in the organization to communicate with one another to reduce the level of ambiguity, Weick suggests that the relationships among individuals in the organization are more important to the process of organizing than the talent or knowledge that any one individual brings to the team. Hence, the General Systems Theory philosophy—"the whole is greater than the sum of its parts"—is apparent.

The Principles Related to Equivocality

By now, we're sure that you're able to see that equivocality remains one important feature of OIT. Indeed, it's a critical theme woven throughout the theory, even as the theory has been revisited and refined by scholars (Turbanti, 2023). In this section, we discuss equivocality by looking at three guiding principles of equivocality and also at ways to reduce this communication process in organizational life.

Organizations use several principles when dealing with equivocality. An organization must analyze a) relationship among the equivocality of information, b) rules the organization has for removing the equivocality, and c) cycles of communication that should be used. When analyzing the relationship among these three variables, a few conclusions are possible. If a message is highly equivocal, chances are that the organization has few rules for dealing with the ambiguity. As a result, the organization has to employ a greater number of cycles of communication to reduce the level of equivocality of the information. No self-sufficient company can survive in an environment of sustained equivocality. The organization will examine the degree of equivocality of the information (inputs) it receives and determine if it has enough rules that will assist in guiding the cycle of communication that should be employed to reduce the ambiguity.

To exemplify, a technical question is highly equivocal to those who are not trained in the field of expertise. Thus, the organization may not have a set of rules to guide communication responses and may rely solely on one rule (e.g., personnel) to guide the cycles of communication. If inputs are easily understood by many members of the organization (e.g., a customer may ask, *"Is BankNG doing anything to prepare for federal regulations?"*), more rules (i.e., success, effort, duration) can be employed in reducing the equivocality. The more equivocal the message, the fewer rules that are available to guide cycles of communication; the less equivocal the message, the more rules that are available to assist the organization in reducing equivocality, thus reducing the number of cycles that are needed to interpret the information.

A second principle proposed by Weick (2020) deals with the association between the number of *rules* that are needed and the number of *cycles* that can be used to reduce the equivocality. If the organization has only a few rules available to assist it in reducing equivocality, a greater number of cycles will be needed to filter out the ambiguity. In our chapter-opening example, a majority of the members in the organization view technical information related to federal regulation as highly equivocal. The only rule that many of them have for dealing with this equivocality is to communicate with a person (personnel) who is knowledgeable in the appropriate area (e.g., external auditing, financial disclosure, etc.). Thus, more cycles of information exchange will be employed between the technician and the organization's employees to reduce the ambiguity of the information.

The third principle proposed by Weick relates to a direct relationship between the number of cycles used and the amount of equivocality that remains. The more cycles that are used to obtain additional information and make adjustments, the more equivocality is removed. Weick proposes that if a larger number of cycles is used, it is more likely that equivocality can be decreased than if only a few cycles are employed.

Although BankNG has increased the potential for equivocality of information by providing its employees with multiple resources for obtaining information about the project's status (videoconferences, telephone conferences, wikis), it has also provided its employees with a larger number of potential cycles that can be employed to reduce this equivocality. Employees can engage in online dialogues in an attempt to answer their questions and concerns about the FedReg project. They can download database files that track the progress of the project and inquire if other teams will be meeting various deadlines that are crucial to their success in the project's completion. And they have the benefit of having an internal communication team to manage all the messages they receive from various sources. Thus, there is the potential for a greater number of cycles that can be used to reduce the equivocality of information.

Reducing Equivocality: Trying to Use the Information

Reducing equivocality is both necessary and complex. According to Weick (1995) and Phil Salem (2023), organizations evolve through stages in an attempt to integrate the rules and cycles so the information can be easily understood and is meaningful. The process of equivocality reduction is essentially an interpersonal process and occurs through the following three stages: enactment, selection, and retention.

Enactment: Assigning Message Importance

Enactment refers to how information will be received and interpreted by the organization. Andrew Herrmann (2007) states, "Enactment starts with the bracketing or framing of a message in the environment by an individual" (p. 18). During this stage, the organization must

enactment interpretation of the information received by the organization

analyze the inputs it receives to determine the amount of equivocality that is present and to assign meaning to the information. Existing rules are reviewed in making decisions about how the organization will deal with the ambiguity. If the organization determines that it does not have a sufficient number of rules for reducing the equivocality, various cycles of communication must be analyzed to determine their effectiveness in assisting the organization in understanding the information. Weick believes that this action stage is vital to the success of an organization. If a university made no effort to interpret information from potential students, it would not be able to address their concerns and desires in choosing a college to attend. Eric Eisenberg and H. L. Goodall (2013) observe that enactment may be Weick's most "revolutionary" concept (p. 109).

Weick believes that one affiliate of enactment is **sensemaking,** or the attempt to create understanding in situations that are complex and uncertain (Turbanti, 2023). For Weick (1995), sensemaking includes "the placement of items into frameworks, comprehending, redress-

sensemaking creating awareness and understanding in situations that are complex or uncertain

ing surprise, constructing meaning, interacting in pursuit of mutual understanding, and patterning" (p. 6). Sensemaking "starts with chaos" (Weick, Sutcliffe, & Obstfeld, 2009) and covers many forms of communication, including routines, arguments, symbols, commitments, and other actions and behaviors. Weick (2012) also believes that sensemaking involves storytelling because organizational stories have the potential to create meaning between and among employees that otherwise would have been more challenging. Further, Weick contends that sensemaking can be difficult to achieve because of the following information problems found in most organizations: (1) no information, (2) too much information, (3) not enough information, and (4) equivocal information insofar as information has multiple meanings (Johnson & Kruse, 2019).

Weick (2012) also argues that decisions and sensemaking are related but are not synonymous. In fact, Weick quotes a firefighter/crew chief as he "shifts" from decisions to sensemaking:

> If I make a decision it is a possession, I take pride in it, I tend to defend it and not to listen to those who question it. If I make sense, then this is more dynamic and I listen and I can change it. A decision is something you polish. Sensemaking is a direction for the next period. (cited in Weick, 2012, p. 22)

Interestingly, some researchers believe that a prospective *sensemaking* happens at times. In their study of surgical teams, Ragnar Rosness, Tor Evjemo, Torgeir Haavik, and Irene Wærø (2016) found that when surgical teams considered plausible projections of issues that might occur during the surgery and how these situations might be handled, the surgery was viewed as safer and more efficient. Therefore, some sort of decisions take place but these decisions are considered in light of the consequences and implications of the decision.

Selection: Interpreting the Inputs

Once the organization has employed various rules and cycles to interpret its information environment, it must analyze what it knows and choose the best method for obtaining additional information to further reduce equivocality. This stage is referred to as **selection.** In this stage of organizing, the group is required to make a decision about the rules and cycles that will be used. If the information is still ambiguous, the organization has to look into the resources that it has available and determine if it has any additional rules that could help in reducing the ambiguity.

> **selection** choosing the best method for obtaining information

For instance, imagine Kelvin is overwhelmed with a task that was assigned to him by his boss. As he considers all the inputs required of him, Kelvin becomes visibly agitated and anxious. He has one week to report back to his supervisor, and he is entering a time period at work during which he is constantly interrupted by colleagues seeking his assistance. His perception of being inundated has resulted in his dealing with a project that has a high degree of equivocality. As he oversees the project, he has to attend several meetings, which, interestingly, result in additional information and additional ambiguity. If Kelvin is to reduce his feeling of being overwhelmed, he will have to develop a plan to obtain information and disseminate it appropriately in order to finish the task. Reviewing organizational rules regarding communication channels may assist Kelvin as he manages the inputs. In this case, his selection process will be instrumental in his reduction of equivocality. As an organizational member, his expertise will also be quite important in the equivocality decline (Weick & Sutcliffe, 2015).

Retention: Remembering the Small Stuff

Once the organization has reviewed its ability to deal with ambiguity, it analyzes the effectiveness of the rules and cycles of communication and engages in retention. In the **retention** stage, the organization stores information for later use. This stage requires organizations to look at what to deal with and what to ignore or leave alone. If a particular rule or cycle was beneficial in assisting the organization in reducing the equivocality of information, it's likely that it will be used to guide the organization in future decisions of a similar nature. Suppose that Dominique found the videoconferences to result in even more confusion among the team members because they were bombarded with more information than was either desired or required to complete various tasks in the project. She will remember this and likely refrain from using the technology as a means of sharing project information in the future. Instead, she may choose to use online discussions to allow team members to choose the information essential to their portion of the task and skip over information that is of little relevance to them. If strategies for dealing with equivocality are deemed to be useful, they will be retained for future use.

> **retention** collective memory allowing people to accomplish goals

So, through enactment, selection, and retention, organizations can begin the process of reducing and eliminating their ambiguity. These three processes are not always going to occur at once; indeed, one may take much longer than the other. As we noted earlier, these stages are primarily interpersonal in nature and dealing with human relationships—like organizational decisions—can be unpredictable.

Integration and Critique

Karl Weick's Organizational Information Theory has been accepted as an influential theoretical framework for explaining how organizations make sense of the information they receive for their existence. Like several other communication-centered theories, OIT draws from other theoretical perspectives that explain the processes that organizations undergo to receive input from others. Further, the theory has been investigated using the scientific method, making it an empirically driven process. And, the theory is clearly communicative in nature. As Eric Eisenberg (2007) states, "Weick's most valuable contributions have been his insistence on the centrality of language and communication in the construction of organizational reality, and his sustained focus on communication as a site for improving our understanding of cognition, culture, and social interaction" (p. 284). Moreover, Weick's contribution to the organizational communication literature is highly recognized. As Bob Johnson and Sharon Kruse (2019) state, "few contemporary theorists have exerted a more potent and expansive influence on the study of organizations than Karl Weick" (p. 1). Clearly, Karl Weick's influence upon scholarship is meritorious.

Students Talking Tech: Kathryn

 During high school, I worked as a deli clerk at a convenience store. I hated the job but needed the money to save for college. I do remember a time when equivocality was present at the workplace. The manager had a habit of coming into work and videotaping employees without any notice. We couldn't stand it, and we looked at her actions as equivocal. The employees described the work environment as equivocal because there were different views of the manager's behavior. One of us thought she (the manager) was keeping us on our toes. Another thought she watched us on video because she didn't trust us to be by ourselves. Personally I thought she was taping without telling anyone because she wanted to show her power. I never did find out why she surprised us with her "tech visits," but I'm long gone now.

Integration

Communication Tradition	Rhetorical	Semiotic	Phenomenological	**Cybernetic**	Socio-Psychological	Socio-Cultural	Critical
Communication Context	Intrapersonal	Interpersonal	Small Group/Team	**Organizational**	Public/Rhetorical	Mass/Media	Cultural
Approach to Knowing	**Positivistic/Empirical**	Interpretive/Hermeneutic	Critical

Critique

Evaluation Criteria	Scope	**Logical Consistency**	Parsimony	**Utility**	Testability	**Heurism**	Test of Time

Logical Consistency

Recall that theories must make sense and make clear the concepts under discussion. Weick's theory seems to fail the test of logical consistency. One prevailing criticism pertains to the belief that people are guided by rules in an organization. Yet organizational scholars note that "we puzzle and mull over, fret and stew over, and generally select, manipulate, and transform meaning to come up with an interpretation of a situation" (Papa & Daniels, 2014, p. 114). Therefore some organizational members may have little interest in the communication rules in place at work. Individuals are not always so conscious or precise in their selection procedures, and their actions may have more to do with their intuition than with organizational rules. As employees become more immersed in the organizational milieu, they may be guided more by instinct if that instinct is accurate, ethical, and thoughtful.

An additional criticism underscoring the problems of logical consistency is that OIT views organizations as static units in society (Taylor & Van Every, 2000). These researchers challenge Weick's view by noting that "at no point are inherent contradictions in organizational structure and process even remotely evoked" (p. 275) in his research. Organizations have ongoing tensions, and these need to be identified and examined in light of Weick's claims. Furthermore, given the dynamic changes in organizations due to corporate mergers, downsizing, offshore outsourcing of employee work, and the continuing changes taking place with technology, static or frozen assessments of organizations are shortsighted.

Utility

The theory's utility is underscored by its focus on the communication process. OIT focuses on the process of communication rather than on the role of communicators themselves. This is of great benefit in understanding how members of an organization engage in collaborative efforts with both internal and external environments to understand the information they receive. Rather than attempt to understand the people in an organization—and their unpredictability—Weick decides to unravel the complexities of information processing, which makes this a rather useful theoretical undertaking.

Heurism

OIT is heuristic and has prompted considerable scholarly discussion. The theory has inspired thinking in research on a variety of topics, including active shooters on college campuses (Long, 2022), nomadic work (Bean & Eisenberg, 2006), surgical procedures (Rosness et al., 2016), organizational humor (Heiss & Carmack, 2012), leadership in high schools (Carraway & Young, 2015), and cohesion in the U.S. Army (Van Epps, 2013). The theory has also been applied to such diverse populations as zoo volunteers (Kramer & Danielson, 2016) and Ghanaians living with HIV/AIDS (Latzoo, 2015). Weick clearly was influential in the work of organizational communication scholars.

Closing

Karl Weick has centralized the notion of organizational ambiguity and his theory resonates today. OIT continues to attract theorists and practitioners interested in the intersection of organizations, messages, and people. It has been called "the sensemaking model" (Abu-Shaqra & Luppicini, 2016, p. 62) and has prompted researchers from across a number of disciplines to study the vagaries of organizational life. To that end, the theory will continue to resonate in cultures where corporations have to deal with uncertain and challenging ebbs and flows.

Discussion Starters

 Case-In-Point: Based on your understanding of the concepts of Organizational Information Theory, what additional avenues could Dominique engage in to facilitate a solution to the FedReg project in BankNG? Respond using at least two concepts discussed in the chapter.

Try-It-Your-Selfie: Considering all of the social media available to you, consider which platform (e.g., X, Facebook, Instagram, LinkedIn, etc.) would be the most appropriate for reducing equivocality in both a for-profit (e.g., Disney) and nonprofit (e.g., Doctors Without Borders) organization.

1. Recall an organization in which you are or were a member. Can you remember an incident when you or the organization received ambiguous information? If so, what rules did you or the organization use in dealing with this equivocality?

2. Weick describes the process of enactment, selection, and retention to understand how organizations deal with information inputs. Provide an example of how your school has employed these strategies in making sense of information on your campus.

3. Do you think that highly equivocal information requires more complex communication processes in order to make sense of the input? Defend your answer.

4. Discover if your college or university solicits feedback from students regarding its promotion of the school. Discuss the methods that your school uses.

References

Abu-Shaqra, B., & Luppicini, R. (2016). Technoethical inquiry into ethical hacking at a Canadian university. *International Journal of Technoethics, 7,* 62–76.

Bean, C. J., & Eisenberg, E. (2006). Employee sensemaking in the transition to nomadic work. *Journal of Organizational Change Management, 19*(2), 210–222.

Blute, M. (2010). *Darwinian sociocultural evolution: Solutions to dilemmas in cultural and social theory.* Cambridge University Press.

Boivin, G., & Brummans, B. H. (2022). What's pragmatic about ambiguity in the communicative constitution of organizations?: The Case of CCO scholarship's establishment. In J. Basque, N. Bencherki, & T. Kuhn (Eds.), *The Routledge handbook of the communicative constitution of organization* (pp. 47–59). Routledge.

Carraway, J. H., & Young, T. (2015). Implementation of a districtwide policy to improve principals' instructional leadership principals' sensemaking of the skillful observation and coaching laboratory. *Educational Policy, 29,* 230–256.

Darwin, C. (1948). *The origin of species.* Random House.

De Kroon, A., & Kaat, S. (2015). *Systemic consulting: The organisation as a living system.* Het Noorderlicht.

Eisenberg, E. M. (2007). *Strategic ambiguities: Essays on communication, organization, and identity.* Sage.

Eisenberg, E., Goodall, H. L., & Tretheway, A. (2013). *Organizational communication: Balancing creativity and constraint.* Bedford/St. Martin's.

Heiss, S. N., & Carmack, H. J. (2012). Knock, knock; who's there? Making sense of organizational entrance through humor. *Management Communication Quarterly, 25,* 106–132.

Herrmann, A. (2007). *Narrative and ethnography as existential phenomenological approaches to organizational sensemaking.* Paper presented at the annual meeting of the International Communication Association, San Francisco, CA.

Johnson, B. L., Jr., & Kruse, S. D. (Eds.). (2019). *Educational leadership, organizational learning, and the ideas of Karl Weick: Perspectives on theory and practice.* Routledge.

Kramer, M. W., & Danielson, M. A. (2016). Developing and re-developing volunteer roles: The case of ongoing assimilation of docent zoo volunteers. *Management Communication Quarterly, 30,* 103–120.

Latzoo, C. (2015). Questioning HIV/AIDS-related stigma: Lived disempowerment and potential empowerment among Ghanaians living with HIV/AIDS. *Journal of Communication, 41,* 238–258.

Liu, S., Volcic, Z., & Gallois, C. (2019). *Introducing intercultural communication: Global cultures and contexts.* Sage.

Livingston, L. A., Cunningham, I., & Forbes, S. L. (2023). Using technological innovation to manage and develop sport officials. *Managing Sport and Leisure,* 1–3.

Long, C. J. (2022). New media in times of crisis: New agendas in communication. *International Journal of Communication, 16,* 4.

Mumby, D., & Kuhn, T. A. (2019). *Organizational communication: A critical introduction* (2nd ed.). Sage.

Papa, M. J., & Daniels, T. (2014). *Organizational communication: Perspectives and trends.* Sage.

Rosness, R., Evjemo, E., Haavik, T., & Wærø, I. (2015). Prospective sensemaking in the operating theatre. *Cognition, Technology & Work, 18,* 53–69.

Salem, P. J. (2023). *Organizational communication dynamics and higher education.* Routledge.

Sawyer, K. (2012). Extending sociocultural theory to group creativity. *Vocations and Learning, 5,* 59–75.

Stuart-Fox, M. (2023). Major transitions in human evolutionary history. *World Futures, 79*(1), 29–68.

Taylor, J. R., & Van Every, E. J. (2000). *The emergent organization: Communication as its site and surface.* Lawrence Erlbaum.

Turbanti, G. (2023). *Philosophy of communication.* Palgrave Macmillan.

Van Epps, G. (2013). *Relooking unit cohesion: A sensemaking approach.* BiblioGov.

Weick, K. E. (1979). *The social psychology of organizing.* Addison-Wesley.

Weick, K. E. (1995). *Sensemaking in organizations.* Sage.

Weick, K. E. (January 2012). Organized sensemaking: A commentary on processes of interpretive work. *Journal of Human Relations, 65*(1), 141–153.

Weick, K. E. (2015). Ambiguity as grasp: The reworking of sense. *Journal of Contingencies and Crisis Management, 23,* 117–123.

Weick, K. E. (2020). Sensemaking, organizing, and surpassing: A handoff. *Journal of Management Studies, 57*(7), 1420–1431.

Weick, K. E., & Sutcliffe, K. M. (2015). *Managing the unexpected: Sustained performance in a complex world* (3rd ed.). John Wiley & Sons.

Weick, K. E., Sutcliffe, K. M., & Obstfeld, D. (2009). Organizing and the process of sensemaking. In K. E. Weick (Ed.), *Making sense of the organization* (pp. 129–152). John Wiley & Sons.

CHAPTER 11
Agenda Setting Theory

*Based on the research of **Maxwell McCombs** and **Donald Shaw***

> *"In choosing and displaying news, editors, newsroom staff, and broadcasters play an important part in shaping political reality. Readers learn not only about a given issue, but also how much importance to attach to that issue from the amount of information in a news story and its position. In reflecting what candidates are saying during a campaign, the mass media may well determine the important issues—that is, the media may set the 'agenda' of the campaign."*
>
> —Maxwell E. McCombs and Donald Shaw

Sally D'Amato

While Sally was fixing dinner for her extended family, she had the TV playing in the background. She was only half listening, but she stopped peeling carrots when the news came on and she heard a story about how Prince Harry probably would never reconcile with his brother William because of things Harry's wife Meghan had done and said. Sally hadn't thought a lot about Harry and Meghan before, but she remembered that they'd done an interview with Oprah Winfrey that she'd seen parts of and Meghan had accused the royal family of being racist and cruel to Harry had agreed and supported Meghan. She also remembered that South Park had done a really funny parody of Harry and Meghan asking for privacy when they were giving interviews and making Netflix documentaries. Sally laughed to herself about the problems of the rich and entitled. All during dinner, she talked to her family about the royal family. Her family all agreed that Meghan had probably ruined the great relationship it seemed like William and Harry had after the tragic death of their mother, Princess Diana.

A few weeks later, Sally was at the nursing home, where she worked as an aide. She pulled out her cell to check her texts, and her Apple News alerts popped up. The first one was about Harry and Meghan again. Sally had to laugh when she read that Harry was going to be on Stephen Colbert's show. It was certainly hard to believe that they really moved to California from England to have more privacy. And, it did seem like Meghan's influence had a lot to do with it. The article noted that Meghan, as an actress, loved the spotlight and Harry was just going along for the ride. Sally wondered how long their marriage would last and she felt sorry for their two kids. Just then, Sally looked up and Michah, one of the other aides, came into the break room. Sally told Michah the story, and the two of them spent the rest of their break discussing Prince Harry and Meghan Markle's relationship. Neither of them thought it would last.

The following week, Sally was at her computer at home working on a paper for her class at the local college in Contemporary Social Problems. Sally was trying to finish the BA she had started 15 years before so she could go on to become a registered nurse. This class satisfied a social–cultural core requirement, and she was actually enjoying it even though it didn't seem to have anything to do with nursing. The professor had asked the students to write about what they thought had been one of the most compelling social problems of the previous 10 years. Sally typed in "social problems in the United States" in a Google search, and she was amazed at the results. She stopped counting the various problems after she got to 100. Sally realized there were a lot of issues happening in the United States that she hadn't thought about much and wondered a bit about why that was the case. She thought back over the previous few weeks and considered some of the issues she had been thinking and talking about. Upon reflection, she had to admit most of them weren't too significant. Sally doubted that Meghan and Harry's relationship would even make the list for the top 1,000 social problems in the United States! And yet the royal family and Meghan and Harry had been in the media, sometimes more than the other problems she'd discovered in her Google search. Sally sighed and got started on her paper. She knew she had a lot of work ahead of her.

Over time scholars have had various ideas about how influential the media are in people's lives. In the early days of mass media, people were seen as helpless victims of the powerful mass media. This notion was eventually discredited and replaced by what is called a *limited effects model* of the mass media, which acknowledges that media influence people, but that media's influence is minimized or limited by certain aspects of individual audience members' personal and social lives. In addition, researchers argue, audience members, themselves, play a role in the mass communication process (see our more complete discussion of this history of media effects in our chapter on Uses and Gratifications Theory, **Chapter 13**). Agenda Setting research began with a belief in the powerful effects of the media, but later refinements put it into the camp of limited effects (McCombs, 2004; McCombs & Bell, 1996; McCombs & Shaw, 1993; Min & McCombs, 2016). When Agenda Setting Theory adopted the limited effects perspective, its practitioners began to ask questions such as why some voters expose themselves to certain messages more than other voters do and how politicians control the agenda through their rhetorical strategies (e.g., Montiegel & Robinson, 2019). In the next section, we discuss the evolution of agenda setting research in more detail.

History of Agenda Setting Research

The history of agenda setting research can be conceptualized in three stages: (1) pretheoretical conceptualizing, (2) establishing the theory, and (3) elaborating the theory. We will discuss each of these stages briefly.

Pretheoretical Conceptualizing

Most researchers (e.g., Dearing & Rogers, 1996) talk about the first stage of agenda setting research as consisting of the conceptualizations of several scholars in different fields beginning to think and write about the relationships among the media, the audience, and the policymakers in the United States. The first person to contribute to this line of thought according to James Dearing and Everett Rogers was Robert E. Park. Park was a sociologist at the University of Chicago (1915–1935), and he is thought to be the first scholar of mass communication. He devised the notion of media gatekeeping and began to discuss some of the issues that are now incorporated into Agenda Setting Theory. Park noted that editors are *gatekeepers* because they have the power to "kill" stories and to promote other stories that are submitted to them by correspondents, reporters, and news agencies. This statement related to later developments in Agenda Setting Theory because Park distinguished between issues that become public and those that don't come to the public's attention.

After Park, Walter Lippmann was a pioneer of the pretheoretical stage. Walter Lippmann was a scholar of propaganda and public opinion as well as an influential newspaper columnist and presidential adviser. In 1922, he wrote a book called *Public Opinion,* and he titled the first chapter "The World Outside and the Pictures in Our Heads." He made the argument that the mass media connect the two. According to Lippmann, the events that happen in the world are brought to people by the mass media and the way these events are reported shape how people structure the images of these events in their minds (Fahmy, Bock, & Wanta, 2014). Lippmann did not use the term *agenda setting*, but his writing was very influential on the development of the theory.

In 1948, Harold D. Lasswell, a political scientist at the University of Chicago, contributed an important chapter to an anthology about communication that had far-reaching implications for Agenda Setting Theory. In this chapter, Lasswell talked about two important functions of mass media: surveillance and correlation. **Surveillance** is the process of newspeople scanning the information that's in the environment and deciding which of the many events that are occurring deserve attention in their news outlets. In discussing this function, Lasswell was echoing Park's notion of gatekeeping. Lasswell argued that news reporters, editors, and so forth decide which of the multitude of possible stories will be the ones to reach the public via their papers or other outlets. Obviously, in this process, the media do exert powerful effects—they are in charge of what the public gets information about and how that information is presented. People often complain that the news is all bad and that the good things that happen don't get reported (the old adage "if it bleeds, it leads" refers to the likelihood that bad news will be featured in the media). This speaks to the surveillance process that is controlled by newspeople.

> **surveillance** the process of newspeople scanning the information that is in the environment and deciding which of the many events that are occurring deserve attention in their news outlets

Lasswell (1948) describes the function of **correlation** as the way that media direct our attention to certain issues through communicating them to the public and policymakers. In this function, media synchronize the various groups in society so that they're paying attention to the same things at the same time. Lasswell spoke of the "correlation of the parts of society in responding to the environment" (p. 38). The result of the media orchestration of our attention was "a correlation of attention on certain issues at the same time by the media, the public, and policymakers" (Dearing & Rogers, 1996, pp. 11–12). This function is illustrated when there is a national/international event (like a presidential inauguration, the Super Bowl, or the Olympics) or a natural catastrophe in the United States (like wildfires in Maui, flooding in Southern California, or historically sweltering hot temperatures in the Southwest). The media correlate our attention to these things in real time, so we might hear about storm damage at the same time we hear or read about the results of the Women's World Cup in soccer, for example.

> **correlation** the way that media direct our attention to certain issues through communicating them to the public and to policymakers at the same time

Establishing the Theory

All of these ideas by early researchers came together in the second stage of agenda setting research. This stage was marked by a study Maxwell McCombs and Donald Shaw (1972) conducted which took these early concepts and put them to an empirical test. This landmark study examined the public's and the media's agendas during the 1968 U.S. presidential election. Agenda Setting Theory was focused on issues of politics in its beginnings. McCombs and Shaw were interested to test the hypothesis, derived from the ideas of scholars like Lasswell, Park, and Lippmann, that the mass media create an agenda through their selection of what to include in the news, and this agenda influences public perception of what is important. In this 1972 study, McCombs and Shaw hypothesized a causal relationship between the media and the public agendas, which stated that the media agenda would, over time, become the agenda for the public.

To test their hypothesis, they interviewed 100 undecided voters during the three weeks just prior to the presidential election in November of 1968. Although elements of their hypotheses changed in subsequent studies, one of the enduring contributions of this early work was the way they measured the two variables of interest:

the public agenda and the media agenda. The public agenda for these undecided voters was measured by their responses to a survey question: "What are you *most* concerned about these days? That is, regardless of what politicians say, what are the two or three *main* things that you think the government *should* concentrate on doing something about?" (McCombs & Shaw, 1972, p. 178). They ranked the issues based on the frequency with which they were mentioned and found five main issues—foreign policy, law and order, fiscal policy, public welfare, and civil rights—which were mentioned most frequently by the respondents. These five issues formed the public agenda.

Theory at a Glance • Agenda Setting Theory

In choosing and displaying news, editors, newsroom staff, webcasters, and anchors play an important part in shaping social and political reality. When readers and viewers consume news, they not only learn about a given issue, but they also learn how much importance to attach to that issue by the amount of coverage and position relative to other stories it's given by the media. When examining what candidates say during a campaign, Agenda Setting Theory suggests that the media may well determine the important issues—that is, the media may set the "agenda" of the campaign. How influential the media are in this agenda setting function depends on several factors, including media credibility, the extent of conflicting evidence, shared values, and the audience's need for guidance.

They measured the media agenda by counting the number of news articles, editorials, and broadcast stories in the nine main media outlets that served the area where the undecided voters lived. These media sources included television, newspapers, and news magazines. McCombs and Shaw found an almost perfect correlation between the rank order of the five issues on the media agenda as measured by their content analysis of media coverage of the election campaign, and the five issues on the public agenda as determined by their survey of the 100 undecided voters.

At this point in the history of the theory, the ideas articulated by Lippmann and others now had a name: "McCombs and Shaw named this transfer of salience from the media agenda to the public agenda the agenda setting influence of mass communication" (McCombs & Bell, 1996, p. 96). The theory was launched, and hundreds of articles, books, and monographs have followed.

Elaborating the Theory

Further research (Weaver, Graber, McCombs, & Eyal, 1981) expanded Agenda Setting Theory beyond the public issues McCombs and Shaw had begun exploring in 1972. Other key political elements were added to the agenda such as candidate image and voter interest in campaigns. More recent research added the question: who sets the media agenda? This is a complicated question, and research has suggested a variety of answers. Steven Littlejohn, Karen Foss, and John Oetzel (2021) suggest that there are four types of power relations between the media and other sources that might provide an answer: (1) high-power source and high-power media; (2) high-power source and low-power media; (3) low-power source and high-power media; and (4) both media and source are low power.

In the first case, a popular president could be a source to a well-funded media outlet with a decent reputation, such as MSNBC. In this situation, the two would be equals in setting the agenda, which will work well for them if they see things similarly but will result in struggles if they are not on the same side of the important issues of the day. In the second scenario, the source (an influential politician) has more power than the media (a local paper), so then the source will be able to set the agenda for the media, think about the relationship between former President Trump and Fox News, for instance. In the third relationship, the media are able to set their own agenda because the source is not considered to have much of a voice. In this case, the media may marginalize the source, and the source will have trouble getting access to the public to discuss their issues. A welfare group who wants to have their agenda broadcast on national television might be in this situation. Finally, in the last relationship, Littlejohn, Foss, and Oetzel suggest that events will probably set the public agenda because neither the source nor the media have much power (i.e., a local official and a small town website).

In addition, researchers have expanded the theory to examine what they call intermedia influence on the agenda setting process; they've noted that news organizations affect each other's agendas. For example, some research (Lim, 2011) has investigated the influence major news websites have on each other's agendas. Lim's 2011 study, set in South Korea, found that the major news websites there did influence the agendas of online newspapers as well as, to a certain extent, each other. Other studies (e.g., Johnson, 2011; Maier, 2010; Meraz, 2011; Ragas & Kiousis, 2010) found some support for the influence various media have on each other. In addition, Maxwell McCombs and Tamara Bell (1996) observed that the intermedia effect can come from individual newspeople as well as news organizations. As they note, journalists live in "an ambiguous social world," so they often rely on each other for confirmation and as a source of ideas. McCombs and Bell mention several studies of cases in which journalists followed each other in reporting about specific issues. They refer to this agenda influence as pack journalism. To some extent, pack journalism is similar to groupthink, a theoretical model we discuss in **Chapter 18.** McCombs and colleagues (e.g., McCombs & Funk, 2011) have suggested that intermedia influence is the wave of the future for agenda setting research.

Finally, some research examines how well Agenda Setting Theory predicts and explains agendas in an era of social media (e.g., Lee & Xu, 2018). We will discuss this stage in the evolution of the theory more when we examine some of the critiques of the theory.

Next, we present the major assumptions of Agenda Setting, which allow us to understand the foundation of the theory.

Assumptions of Agenda Setting Theory

Agenda Setting Theory rests on three basic assumptions:

- The media establish an agenda and in so doing are not simply reflecting reality but actually shaping and filtering reality for the public.
- The media's concentration on the issues that comprise their agenda influences the public's agenda, and these together influence the policymakers' agenda.
- The public and policymakers have the possibility to influence the media's agenda as well.

These three assumptions are woven into Agenda Setting Theory and suggest the interaction the theory specified among the media, the public, and policymakers. First, the media both shape and filter the reality in which we live. We may not be aware of it, but media are constantly providing us a lens both to understand and reflect our social reality. Along the way, the media also establish an agenda for us to consider. Imagine, for instance, all of the information around us. Now, imagine the various media (e.g., radio, television, the internet, social media, etc.). Somehow all of this information has to be sifted through, some pieces chosen, and some disregarded, and then those that remain have to be packaged in some way. The media do all this for us, forming the way we receive information and the extent to which the information reaches us. In other words, media shape what we hear, read, or attend to and we are usually unwittingly compliant in that process. Consider, for instance, our opening story. Michah and Sally's conversation focused on Meghan Markle and

Prince Harry's relationship rather than on any of the other social issues Sally discovered in her Google search. It's likely that the media influenced the conversation by presenting the story about Meghan and Harry and filtering out other (arguably more important) information related to stories of other social issues such as the war in Ukraine, the several indictments against former President Donald Trump, homelessness, immigration, or the association between climate change and natural disasters.

A second assumption of Agenda Setting Theory relates to the gatekeeping function of the media. The theory assumes that the media focus on the issues that comprise an agenda and, in turn, influence the public's agenda and subsequently the agenda of decision makers. Let's think about this a bit further with an example. Over the past several years, there has been a concerted effort to deal with bullying episodes taking place in schools. Although the number of kids being bullied is high, it was the media's sustained attention to the topic that inevitably prompted parents to talk to school boards to demand local policy change and ultimately prompted several state legislatures to enact anti-bullying laws. The media became gatekeepers of the bullying topic in that they highlighted the kids who were bullied, their stories of survival, and the ways in which schools could and should reduce bullying. As a result, this issue made it to the policymakers' agenda and changes were enacted.

The third assumption of the theory states that the policymakers and the public can affect the media's agenda. Because an interrelationship exists among the three elements (the media, the public, and those in prominent decision-making roles), it's possible that the media will initiate an agenda because of the influence (or pressure) brought about by the other two elements. For example, consider the increase in legislation restricting gun access following the school shooting at Marjory Stoneman Douglas High School in Parkland, Florida. Since the shooting in 2018, where 17 people lost their lives, 11 states have enacted legislation restricting access to guns in some way. Additionally, the federal government passed a law banning the sale of bump stocks for guns. Although people in the U.S. continue to die at unacceptable rates because of gun violence, few would argue against the fact that legal changes were undertaken because of student activism, led by survivors of the Stoneman Douglas school shooting. Clearly, although some media outlets were reporting on this topic for years before the shooting at Stoneman Douglas, it was the public, spearheaded by student activists and policymakers who prompted the issue to come to the forefront of the media's agenda.

With these assumptions in mind, we turn to the three-part process that comprises Agenda Setting Theory.

Three-Part Process of Agenda Setting

The agenda setting process consists of three parts: setting the media agenda, setting the public agenda, and setting the policy agenda. The **media agenda** refers to the priority of issues to be discussed in mediated sources. The **public agenda** is the result of the media agenda interacting with what the public thinks. And, finally, the public agenda interacts with what is considered important by policymakers to create the **policy agenda.** In a simple format, the theory states that the media agenda affects the public agenda, which in turn affects the policy agenda, a process we suggested in the second assumption of the theory (see **Figure 11.1**).

media agenda the priority placed on issues discussed in mediated sources

public agenda the result of the media agenda interacting with what the public thinks

policy agenda the result of the public agenda interacting with what policymakers think

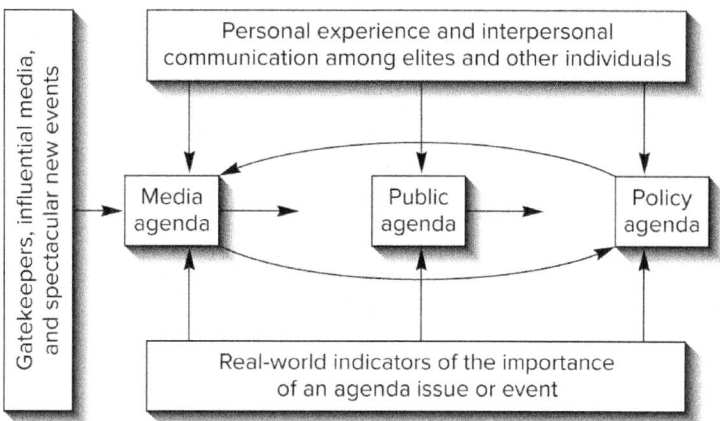

Figure 11.1   The Impact of Public and Media Agenda on Policy Agenda

Rogers & Dearing, 1988.

In addition, this simple formulation is complicated by a few other factors. First, agenda setting is concerned with **salience,** or the degree to which one issue is perceived as important relative to the other issues on the agenda (McCombs, 2004). Agenda setting researchers are more interested in salience than in the usual concerns of public opinion researchers such as positive and negative attitudes toward an issue. Salience allows agenda setting researchers to capture what the media agenda is and what the media are telling the public to think about. Thus, how salient or important an issue is perceived to be by the audience will have an effect on the degree of influence felt. Although you might disagree with Sally and Michah talking about Prince Harry and Meghan Markle rather than climate change or more important issues, the Westernized focus on "celebrity culture" seems to be one reason for their attention to the topic. Further, Agenda Setting became a limited effects model in part because the power of the media agenda is dependent on a variety of factors, including media credibility, the extent of conflicting evidence available to the consumer, the extent to which people share the values of the media, and the public's need for guidance or orientation (Sevenans, Walgrave, & Epping, 2016).

> **salience** the degree to which an agenda issue is perceived as important relative to the other issues on the agenda

If an audience member doesn't believe a media source is credible, they will likely dismiss the agenda promoted on that source. If Mandy's a political liberal, for instance, she would be unlikely to believe much of what she might hear from a source such as Tucker Carlson. And, if Samuel is politically conservative, he'll probably dismiss most of the agenda promoted on Pod Save America or on any political podcast perceived as liberal or progressive. Further, if Randy listens to talk shows that promote the value of the sanctity of marriage, he'll likely be convinced by their agenda because he shares that value. However, if Marianne believes that marriage is an oppressive institution, she probably won't accept the agenda presented on those shows.

However, people don't always accept the media agenda and the theory has to explain why that happens. Agenda Setting Theory attempts to do this by explicating the media consumer's need for guidance or orientation (McCombs, 2004). This factor explains that people sometimes don't adopt the media agenda because of two key variables: relevance and uncertainty. **Relevance** is defined as the perception of personal importance the issue holds for someone. If people believe they aren't at all involved in an issue, such as greenhouse gases, for instance, they won't look to the media for guidance on the issue and thus will not be affected by the media agenda.

> **relevance** a factor explaining why people seek guidance from the media agenda. It refers to how personally affected they feel by an issue
>
> **uncertainty** a factor explaining why people seek guidance from the media agenda. It refers to how much information a person believes they already possess about an issue

Uncertainty refers to how much information people think they have about an issue. If they believe they have a great deal of information about the candidates in a presidential election, for instance, their uncertainty is low, and thus they won't have a need for guidance from the media. If, however, they are unsure if they have enough information, then they do need guidance from the media agenda. These two variables work together to explain deviations from the general principles of Agenda Setting Theory. If relevance and uncertainty are both high, then Agenda Setting should be predictive. If relevance and uncertainty are both low, then the theory's assertions won't be as predictive. To think about these contingency conditions, you might reflect on what sort of relevance and uncertainty Sally possessed in the opening vignette for this chapter.

In addition to this three-part process, Agenda Setting exists on three separate levels. These levels evolved as the theory developed over time. We address these levels next.

Three Levels of Agenda Setting

Agenda Setting Theory proposes that the agenda setting function has three levels (Rogerson & Roselle, 2016). The original conception of the theory identified only the first level of agenda setting. This level focuses on the list of important issues that constitute the **agenda** as

> **agenda** a list of what issues people should think about as decided by an entity, such as the media

decided by some entity such as the media (i.e., what issues to think about). This first level was perhaps best reflected by the often quoted aphorism that the press "may not be successful much of the time in telling its readers what to think, but it is stunningly successful in telling its readers what to think about" (Cohen, 1963, p. 13). Later, a second level, sometimes called *attribute agenda setting*, was added to the theory. This level asserts that the media do guide their consumers in how to think. The second level focuses on which parts (or attributes) of the issues are most important (i.e., how to think about these issues). Most recently, a third level has been proposed that acknowledges the interplay between news and reality (i.e., how to connect issues to create a sense of social reality).

Students Talking: Fatima

 This theory is spot on when we talk about international issues. The media like to portray all Arabs as terrorists and extremists. And when COVID-19 hit, everyone in the U.S. blamed the Chinese, when Chinese Americans had nothing to do with it! But the media kept "telling me" what to think, and even on some social media, I was told how to vote. The idea that the media inform is not as truthful as the fact that the media try to persuade. I feel as if the media try to "brainwash" me in some way or another. Kind of doesn't give you a chance to think for yourself about any of it. So, I disagree with that Cohen quote. I think the media do try to tell you what to think, not just what to think about. Although I guess my responses show that people can argue against the media agenda too.

In discussing the second level of agenda setting, Todd Gitlin (1980) introduced the concept of *framing* in his examination of how CBS television coverage of the 1960s student movement made it seem less important than it actually was. Researchers noted that framing could be accomplished in many ways. In newspapers, things like the size of headlines, photographs included with the story, a story's overall length and placement allow the editors to frame its importance and highlight the aspects of it that they deemed most important. On television, the visuals accompanying the story add to the ability of newspeople to frame a story. Some research (e.g., Miller & Roberts, 2010) has examined what they call visual agenda setting, which is concerned exclusively with visuals. This study asked 466 Louisiana State University students to respond to imagery about Hurricane Katrina six weeks after the storm. They found that most people chose the compelling, repetitive imagery shown in the dominant media. So the researchers concluded that the principles of visual agenda setting were supported. However, they also found that the result was qualified by how close the respondent was to the news event. Students who were more personally affected by Katrina chose more personal images and disregarded the images that the media selected, supporting the notion of relevancy that we discussed previously.

Other researchers expanded the notion of framing to include affect (e.g., Coleman & Wu, 2010; Entman, 1993), and also talked about a related process they call priming. **Priming** (Clawson & Oxley, 2017) is a cognitive process whereby what the media present temporarily, at least, influences what people think about afterwards in processing additional information. For example, if you watch or hear news reports about

priming a cognitive process whereby what the media present temporarily, at least, influences what people think about afterwards in processing additional information

ongoing nuclear tests conducted by North Korea, you might be primed to have more anxiety about that country's ability to bomb the United States than if you hadn't paid attention to those news reports.

In conceptualizing the third level of agenda setting, Lei Guo (2016) developed a new model he calls Network Agenda Setting (NAS). Guo used NAS to address the question of whether the media still exercise significant effects on the public in an era of citizen journalists, in which both the media and the public are fragmented and anyone can be a media producer. Guo concludes that the media do still set the agenda, stating "that the news not only tells us what to think and how to think, but also determines how we associate different messages to conceptualize social reality" (p. 3). By this Guo means that the media tell us how to associate disparate ideas in our minds. Guo provides the following example: if the U.S. media create a link between unemployment in the United States and trade with China, this connection becomes salient to media consumers. Guo calls links like this information networks and argues that media network agendas are formed when these links are reported on in the media, which, in turn, affects the public network agendas. In one application of NAS (Guo, Mays, & Wang, 2019), the researchers took an international question concerning the dispute in the South China Sea and found that NAS was useful in discovering how the U.S. narrative (and agenda setting effect) transcended national borders via X (formerly known as Twitter).

Theory-Into-Action

 Imagine being a person who experiences mental health issues or having a person with these issues in your family. If you consulted media regularly, you'd find a skewed picture of mental health issues that probably wouldn't ring true to your own experiences or those of your family member. The media agenda tends to either stigmatize or trivialize mental health issues. For instance, after a traumatic public experience such as a school shooting or a bombing, the media often try to explain the alleged perpetrator's behavior by labeling them "crazy" or "autistic." This occurred in the media coverage of the shooting at Sandy Hook Elementary School in Newtown, Connecticut, in 2012. Media descriptions of Adam Lanza, the shooter responsible for the horrific event, first included words such as "odd," "aloof," and "a loner." Later, media reports included terms such as "lacked empathy" and finally concluded Lanza was "on the autism spectrum" and suffering from "a mental illness like Asperger's." This type of reporting fosters stigmatization by insinuating that people with a mental illness are (1) all the same—violent, and (2) should be isolated from society because they're dangerous. This ignores that not all people with autism or depression, for instance, manifest their condition in the same ways, and that people with mental health issues are more likely to be the victims of violence (i.e., bullying and so forth) than to be the perpetrators.

Alternatively, the media agenda can dismiss some mental health issues, such as anorexia for example, as not being as serious as they actually are. The media might also present some issues like obsessive compulsive disorder (OCD) as being beneficial to the person experiencing it, such as in the television show *Monk*, in which the lead character is shown as using his OCD to help him solve crimes. Further, sometimes mental health issues are oversimplified in the media, such as on X (formerly known as Twitter), where #OCD is used to mean a strong desire for cleanliness and order.

If you had any of these issues or had a loved one who does, these media agendas would probably be disturbing and unsettling. In some cases, it might lead you to organize with others to combat these characterizations in the media.

Integration and Critique

Agenda Setting Theory is a venerable theory of mass communication; it has a history spanning back to the beginning of the 20th century, and it is still being employed today in studies of media and public communication. It is clearly a theory that has stood the test of time. Although it has its detractors, and some of its central concepts may need adaptation for an era of social media, fragmented publics, and the capacity for everyday people to be gatekeepers and "editors" of information, Agenda Setting Theory still has many adherents. As we think about the value of Agenda Settings's theoretical claims, we'll pay special attention to the following evaluative criteria: scope, utility, and heurism.

Integration

Communication Tradition	Rhetorical | Semiotic | Phenomenological | Cybernetic | **Socio-Psychological** | Socio-Cultural | Critical
Communication Context	Intrapersonal | Interpersonal | Small Group | Organizational | Public/Rhetorical | **Mass/Media** | Cultural
Approach to Knowing	**Positivistic/Empirical** | Interpretive/Hermeneutic | Critical

Critique

Evaluation Criteria	**Scope** | Logical Consistency | Parsimony | **Utility** | Testability | **Heurism** | Test of Time

Scope

Some researchers have critiqued the scope of Agenda Setting Theory. Occasionally, the complaint is that the scope is too large and sometimes the opposite is discussed. Some of the problems around this issue have to do with the concept of framing. First, some research (e.g., Takeshita, 2006) asserts that framing is a separate theory from Agenda Setting and could supersede it altogether. This is labeled an identity problem with Agenda Setting. Toshio Takeshita comments that some researchers believe Agenda Setting Theory, in adding the second level of attribute agenda setting (or framing), is actually colonizing other theories and overreaching the appropriate scope for the theory. Takeshita concludes that the two theories can coexist and more empirical work will determine which one provides the better explanation for media's influence on the public. Further, Takeshita notes that Agenda Setting Theory has the advantage over framing theory in methodological terms because scholars have developed clear operational definitions of the media and the public agendas that work well in quantitative studies.

Utility

Douglas Cannon and Laura Cannon (2019) found that Agenda Setting Theory was useful in examining how historians have interpreted a significant labor strike in Texas in 1938. Historians had discussed the strike in the context of its leader being a Communist. But Cannon and Cannon argue that subjecting contemporaneous newspaper accounts to analysis guided by Agenda Setting Theory shows that historians have given communism more salience than people did at the time. The researchers show how Agenda Setting Theory remains useful for questions interrogating historians' interpretations. Further, Yunjuan Luo and colleagues (Luo, Burley, Moe, & Sui, 2019) pointed out, in a rigorous meta-analysis, that studies using Agenda Setting Theory from 1972 to 2015 showed consistency across the findings in support of media's agenda-setting effects.

However, one study examining election data from Germany (Geiß, 2022) argued that Agenda Setting Theory doesn't sufficiently account for the context around agendas. For instance, if the public believes an issue is "played out" or has been discussed too much, the theory won't be as predictive as it will be if the issue is highly important to the public.

Further, some questions have been raised about whether the theory is useful in testing questions relevant to our current media environment. When people have so much freedom in their quests for information and media sources are multiple and fragmented, perhaps the tenets of Agenda Setting Theory will not be supported. Some studies have been conducted to test this question and the results have been mixed. On the whole, however, they suggest that Agenda Setting Theory can still be applied in a media environment that's anything but monolithic. In a study exploring age-related differences in agenda setting (Coleman & McCombs, 2007), the researchers found that even though media use differentiated the generations (the youngest generation used newspapers and television significantly less than the older two generations and used the internet significantly more), the agenda setting effect was still apparent regardless of which type of media was used. Additionally, Natalia Aruguete (2017) in a thorough review of Agenda Setting research concludes that although the fragmentation of media sources is a threat to Agenda Setting Theory, as long as there is a hierarchy among bloggers, for instance, the premises of the theory still seem relevant. And, one study (Gilardi, Gessler, Kubli & Müller, 2022) specifically examined the relationship among the agendas of social media, traditional media, and the social media agenda of politicians. Congruent with the predictions of Agenda Setting Theory, the researchers found that the three agendas all had reciprocal influence on one another and no one agenda dominated the others, except in the case of environmental advocacy where the social media agenda was more influential than traditional media's agenda.

Jennifer Brubaker (2008), however, found that television viewers and internet users ranked a series of issues differently than the general media did. Brubaker concluded that Agenda Setting is not useful as a theoretical framework when people have so much freedom in their media choices. Still other research (e.g., Ragas & Kiousis, 2010) did find evidence of first- and second-level agenda setting relationships. This study examined the agenda setting effects among explicitly partisan news media coverage, political activist groups, citizen activists, and official campaign advertisements on YouTube, all in support of the same candidate, Barack Obama, in 2008. The authors concluded that Agenda Setting Theory is applicable across a variety of media.

Sharon Meraz (2011) found a slightly more complicated result in that her study revealed traditional media outlets were unable to set the agenda for political blogs. Yet, she also found that ideologically diverse political blog networks were able to have an impact on traditional media's online news agenda, and, to a lesser extent, their newsroom blog agenda. She concluded that there was a reduction of traditional media's agenda setting influence. But, she also argued that Agenda Setting Theory could work to explain the greater interdependence between traditional media and political blogs, and in fact, noted that some blogs like the Huffington Post were operating like the traditional media and may exert an agenda setting function of their own. Other studies (e.g., Johnson, 2011; Maier, 2010) concurred noting that although traditional news media have less of an ability to set the public agenda than in the past, they still perform an agenda setting function, and what is found on news websites and in citizen-journalists' postings often correlates strongly with what appears on mainstream media's agenda.

Finally, a study in 2010 (Weeks & Southwell) used a novel approach to test the agenda setting effects of the traditional media (television and newspapers) on the use of newer media (Google searches). This study examined the relationship of television and newspaper coverage of the rumor circulating during the 2008 presidential campaign that Barack Obama was Muslim. The results showed, as Agenda Setting Theory would predict, that the more this rumor was covered in traditional media the more Google searches there were on the topic.

Students Talking Tech: Christian

I like parts of this theory. It makes sense to me that what I talk about with my friends and family is influenced by what I hear and see in media. If that weren't the case, I'd just be talking about stuff that actually happens in front of me here at school. So, that part of the theory seems right to me. But, I just can't see how there can be a public agenda anymore. We all read and listen to such different sources, how can there be topics that occupy everyone at the same time? If I'm reading a blog about aviation and my sister is reading *Motherlode* (because she has a new baby), I can't see us having the same agenda. I think this theory was right when it was first developed in the 1970s, but it has to be changed to keep up with the times.

Heurism

With respect to heurism, Agenda Setting Theory certainly has been successful. It has supported hundreds of studies since its inception in 1972. From 1972 to 2015, more than 400 studies have used Agenda Setting Theory (Zhou, Kim, & Kim, 2015) in some form. Additionally, researchers (Kim, Kim, & Zhou, 2017) found that the number of Agenda Setting Theory studies has kept increasing over time. A Google search of the theory in 2019 yielded 46 million results. Clearly, the theory has resonated with both scholars and laypeople.

Further, although the theory initially studied political agendas, subsequent studies have been situated in a wide variety of fields and topics. Many Agenda Setting studies still focus on politics, but they are not confined to that topic, and they aren't confined to political issues in the United States as evidenced by studies that have examined and situated the theory in Iraq, Taiwan, Mainland China, and Poland (Guo et al., 2015), for example. Further, the theory's application to such important topics as human trafficking (Papadouka, Evangelopoulos, & Ignatow, 2016), physician time management and patient satisfaction (Robinson, Tate, & Heritage, 2016), and consumption of organic food (Danner, Hagerer, Pan, & Groh, 2022) suggests multiple valuable lines of scholarship. As we have discussed throughout the chapter, Agenda Setting research examining new media, traditional media, political issues, and responses to visual stimuli among other issues, attests to the robust, heuristic quality of this theory.

Closing

Agenda Setting is a theoretical model that, despite being many decades old, continues to be relevant today. All around the globe we have seen media transform and evolve and there can be no doubt that media resonate with millions of people. We have seen the agenda setting function of the media front and center in a number of different areas. Moreover, media consumers are often unaware of the influence that the mediated agenda has upon them. As we move further into new forms of media and myriad ways to communicate a message using media, we are likely to see Agenda Setting Theory, with some modifications, take on more prominence than ever.

Discussion Starters

 Case-In-Point: Does Sally's situation, from our chapter-opening scenario, ring true for you? Are the topics that she hears about through media actually charting an agenda for her to think about? Have you ever had an experience like Sally's, where you realize that there are a lot of important issues in the world that you haven't thought about before? Do you think it's because the media haven't directed you to these issues or is it due to some other reasons? If there are other reasons, what are they?

Try-It-Your-Selfie: Ask a variety of people what is on their minds in terms of the important issues of the day. Ask these same people what media they use to obtain information. Try to determine if people's agendas are different based on the media they turn to for news and entertainment. Reflect on what your findings imply for Agenda Setting Theory.

1. (Why) do you think it's important to know the history of Agenda Setting Theory? (How) does it help you to understand and/or apply the theory to know about its evolution over time?
2. Do you agree that framing is a part of Agenda Setting Theory or do you think that it is a competing theory that suggests Agenda Setting is no longer useful?
3. Discuss a news event that has gone through the three-stage process suggested by Agenda Setting Theory: first it is placed on the media agenda, then it arrives at the public's agenda, and finally it reaches the policymakers' agenda where actual policy is made that relates to the news event.
4. Who do you think sets the agenda for the media? Do you agree with the material in the chapter about how the media agenda gets established? Explain your answer.

References

Aruguete, N. (2017). The agenda setting hypothesis in the new media environment. *Comunicación y Sociedad, 28,* 35–58.

Brubaker, J. (2008). The freedom to choose a personal agenda: Removing our reliance on the media agenda. *American Communication Journal, 10*(3), 1–14.

Cannon, D. F., & Cannon, L. E. (2019). Headlines vs. history: A case study comparing agenda-setting results to historical interpretations of the 1938 San Antonio pecan shellers' strike. *Media History, 25*(2), 208–224.

Clawson, R. A., & Oxley, Z. M. (2017). *Public opinion: Democratic ideals democratic practice* (3rd ed.). Sage.

Cohen, B. (1963). *The press and foreign policy.* Princeton University Press.

Coleman, R., & McCombs, M. (2007). The young and agenda-less? Exploring age-related differences in agenda setting on the youngest generation, baby boomers, and the civic generation. *Journalism & Mass Communication Quarterly, 84*(3), 495–508.

Coleman, R., & Wu, H. D. (2010). Proposing emotion as a dimension of affective agenda setting: Separating affect into two components and comparing their second-level effects. *Journalism & Mass Communication Quarterly, 87*(2), 315–327.

Danner, H., Hagerer, G., Pan, Y., & Groh, G. (2022). The news media and its audience: Agenda setting on organic food in the United States and Germany. *Journal of Cleaner Production, 354*(20), 13150.

Dearing, J. W., & Rogers, E. M. (1996). *Agenda-setting.* Sage.

Entman, R. M. (1993). Framing: Toward clarification of a fractured paradigm. *Journal of Communication, 43*(4), 51–58.

Fahmy, S., Bock, M. A., & Wanta, W. (2014). *Visual communication theory and research: A mass communication perspective.* Palgrave Macmillan.

Geiß, S. (2022). The media's conditional agenda-setting power: How baselines and spikes of issue salience affect likelihood and strength of agenda-setting. *Communication Research, 49*(2), 296-323.

Gilardi, F., Gessler, T., Kubli, M., & Müller, S. (2022). Social media and political agenda setting. *Political Communication, 39*(1), 39-60.

Gitlin, T. (1980). *The whole world is watching: Mass media in the making and unmaking of the New Left.* University of California Press.

Guo, L. (2016). A theoretical explication of the network agenda setting model: Current status and future directions. In L. Guo & M. McCombs (Eds.), *The power of information networks* (pp. 3-18). Routledge.

Guo, L., Chen, Y. N. K., Vu, H., Wang, Q., Aksamit, R., Guzek, D., ... McCombs, M. (2015). Coverage of the Iraq War in the United States, Mainland China, Taiwan and Poland: A transnational network agenda-setting study. *Journalism Studies, 16*, 343-362.

Guo, L., Mays, K., & Wang, J. (2019). Whose story wins on Twitter? *Journalism Studies, 20*(4), 563-584.

Johnson, K. A. (2011). Citizen journalism, agenda-setting and the 2008 presidential election. *Web Journal of Mass Communication Research, 28*, 23-33.

Kim, Y., Kim, Y., & Zhou, S. (2017). Theoretical and methodological trends of agenda-setting theory. *The Agenda Setting Journal. Theory, Practice, Critique, 1*(1), 5-22.

Lasswell, H. D. (1948). The structure and function of communication in society. In L. Bryson (Ed.), *The communication of ideas: A series of addresses* (pp. 215-228). Harper.

Lee, J., & Xu, W. (2018). The more attacks, the more retweets: Trump's and Clinton's agenda setting on Twitter. *Public Relations Review, 44*(2), 201-213.

Lim, J. (2011). First-level and second-level intermedia agenda-setting among major news websites. *Asian Journal of Communication, 21*(2), 167-185.

Littlejohn, S. W., Foss, K. A., & Oetzel, J. G. (2021). *Theories of human communication* (12th ed.). Waveland Press.

Luo, Y., Burley, H., Moe, A., & Sui, M. (2019). A meta-analysis of Coleman news media's public agenda-setting effects, 1972-2015. *Journalism & Mass Communication Quarterly, 96*(1), 150-172.

Maier, S. (2010). All the news fit to post? Comparing news content on the web to newspapers, television, and radio. *Journalism & Mass Communication Quarterly, 87*(3), 548-562.

McCombs, M., & Funk, M. (2011). Shaping the agenda of local daily newspapers: A methodology merging the agenda setting and community structure perspectives. *Mass Communication & Society, 14*(6), 905-919.

McCombs, M. E. (2004). *Setting the agenda: The mass media and public opinion.* Polity.

McCombs, M. E., & Bell, T. (1996). The agenda-setting role of mass communication. In M. B. Salwen & D. W. Stacks (Eds.), *An integrated approach to communication theory and research* (pp. 93-110). Informa UK Limited.

McCombs, M. E., & Shaw, D. L. (1972). The agenda-setting function of mass media. *The Public Opinion Quarterly, 36*(2), 176-187.

McCombs, M. E., & Shaw, D. L. (1993). The evolution of agenda-setting research: Twenty-five years in the marketplace of ideas. *Journal of Communication, 43*(2), 58-67.

Meraz, S. (2011). Using time series analysis to measure intermedia agenda-setting influence in traditional media and political blog networks. *Journalism & Mass Communication Quarterly, 88*(1), 176-194.

Miller, A., & Roberts, S. (2010). Visual agenda-setting and proximity after Hurricane Katrina: A study of those closest to the event. *Visual Communication Quarterly, 17*(1), 31-46.

Min, Y., & McCombs, M. E. (2016). Agenda setting in American political campaigning. In W. L. Benoit (Ed.), *Praeger handbook of political campaigns in the United States* (pp. 45-62). Praeger.

Montiegel, K., & Robinson, J. D. (2019). "First" matters: A qualitative examination of a strategy for controlling the agenda when answering questions in the 2016 U.S. Republican primary election debates. *Communication Monographs, 86*(1), 23-45.

Papadouka, M. E., Evangelopoulos, N., & Ignatow, G. (2016). Agenda setting and active audiences in online coverage of human trafficking. *Information, Communication & Society, 19,* 655-672.

Ragas, M. W., & Kiousis, S. (2010). Intermedia agenda-setting and political activism: MoveOn.org and the 2008 presidential election. *Mass Communication & Society, 13*(5), 560-583.

Robinson, J. D., Tate, A., & Heritage, J. (2016). Agenda-setting revisited: When and how do primary-care physicians solicit patients' additional concerns? *Patient Education & Counseling, 99,* 718-723.

Rogers, E. M., & Dearing, J. W. (1988). Agenda-setting research: Where has it been? Where is it going? In J. A. Anderson (Ed.), *Communication yearbook 11* (pp. 555-594). Sage.

Rogerson, K., & Roselle, L. (Eds.). (2016). *Routledge studies in global information, politics, and society.* Routledge.

Sevenans, J., Walgrave, S., & Epping, G. J. (2016). How political elites process information from the news: The cognitive mechanisms behind behavioral political agenda-setting effects. *Political Communication, 33*(4), 605-627.

Takeshita, T. (2006). Current critical problems in agenda-setting research. *International Journal of Public Opinion Research, 18*(3), 11-32.

Weaver, D. H., Graber, D. A., McCombs, M. E., & Eyal, C. H. (1981). *Media agenda-setting in a presidential campaign. Issues, images and interest.* Praeger.

Weeks, B., & Southwell, B. (2010). The symbiosis of news coverage and aggregate online search behavior: Obama, rumors, and presidential politics. *Mass Communication and Society, 13*(4), 341-360.

Zhou, S., Kim, Y., & Kim, Y. (2015). Theoretical and methodological trends of agenda setting theory: A thematic meta-analysis of the last five decades. Paper presented at the annual meeting of the Association for the Education in Journalism and Mass Communication, San Francisco, CA.

CHAPTER 12
Spiral of Silence Theory

*Based on the research of **Elisabeth Noelle-Neumann***

"I have not written anything in my life that I did not believe to be true."

—Elisabeth Noelle-Neumann

Carol Fahey

Since her husband died, Carol Fahey finds herself going to the "Elder-berry Cafe" each morning. There, she encounters a rambunctious cast of characters each breakfast, including GK, a Vietnam War veteran who sings Broadway songs; Nancy, an ER nurse who tells lively stories about former patients; and Nick, a tech-savvy New England lobsterman who is an avid newspaper reader and blogger. This morning's breakfast was especially interesting because the conversation quickly turned to an article on spanking children that Nick read online.

After reading the brief article to the group, Nick offered his opinion on the topic: "I agree with this writer. I don't see anything wrong with spanking a kid. Look at this survey in the paper. Over 60 percent of the state believes it's okay to spank, but only 40 percent of the country does. Nowadays, you can't lay a hand on a kid. They're ready to sue you, or you'll get some state worker to come into your own house and take your kid away. It's not right."

"I agree," said Nancy. "I can tell you that my neighbor's daughter is almost eight and a holy terror. But, noooo, her mother won't touch her! I don't get it. If that was my child, I wouldn't mind putting her over my knee and giving her a good wallop! The girl's parents don't want to send 'the wrong message' to her so she gets away with a lot. It's a lot of 'timeout' crap!"

GK became more interested in the subject as Nancy spoke. Like the others, GK had a strong opinion on the subject: "Look. How many people

at this table were spanked when they were little?" All raised their hands. "And how many of you think that you're violent people?" None showed any response. "There. That's my point. Today, they tell you that if you spank your own kid, then that kid is going to end up violent. But look at us. We aren't violent. We don't hurt anyone. There's just too much of this political correctness out there, and too many parents simply have no rights anymore."

Carol continued staring out over the heads of her breakfast companions. She, like the others, had an opinion on the subject. But her thoughts differed from those of the others. She did not believe in spanking a child at all. She had been spanked like the rest of her friends, but her dad didn't know when to stop. Carol had often been physically abused. She thought about the number of parents who are not able to stop at just one slap on the behind. She also thought about what hitting accomplished. Children can be taught right and wrong, she thought, without being hit. Are parents that naive, she thought, to think that there is only one way to punish a child?

"Hey, Carol," Nick interrupted, "You're pretty quiet. What's your take on all this?"

Carol thought for a quick moment. Should she disagree with the rest of them? What about all the people in her community who also agree with spanking? Carol quickly recalled seeing a news program on the topic about a month ago, and the reporter had interviewed several adult children who had been abused. She wondered how many of them were spanked when they were young and yet, all of them stated that they spank their children and they don't feel it's abuse in any way.

Carol knew that she disagreed with her breakfast colleagues, but how could she begin to explain all of her thoughts? They wouldn't understand. It's probably better simply to go with the flow, she surmised.

"Oh, I don't know. I can see how some kids need special attention. But sometimes, parents get too angry."

"C'mon Carol," Nancy interrupted. "There are a lot of …"

"Well, I guess I agree with it. I hope that it's not done that often, though." As the server arrived at the table to pour more coffee, the conversation quickly turned to other news. Privately, Carol thought about why she had deferred to the group's will. She didn't want to be alone in her viewpoint, nor did she want to explain the personal and sordid details of her past. As Nick began to talk about last night's city council meeting, Carol wondered whether she would ever speak up on the subject again.

Our opinions of events, people, and topics change periodically in our lives. Consider, for example, your opinions about dating when you were a 15-year-old and your opinions about dating now. Or consider the opinions you held of your family members during your adolescence and those you hold today. Your opinions on a wide range of topics—such as cybertalk and raising children—have likely evolved over the years. Opinions are not static and frequently change over the years.

One important influence on our opinions is the media. Media help to shape who we are today. Often, this influence is subtle; at other times, it is more direct. The media's influence on public opinion is what Elisabeth Noelle-Neumann studied, dating back to the 1930s and 1940s. It was in the early 1970s, however, that she conceptualized the Spiral of Silence Theory. And, yet, as we will learn later, today, scholars continue to discuss this theory in many ways, particularly during a time when (social) media are front and center in the everyday lives of people across the globe.

Originally, the theory focused on traditional types of media such as newspapers and television. Yet, as the theory attracted new attention over the years, the application of its concepts to emerging media emerged (Sohn, 2022). Later in the chapter, we will include a brief discussion of the newer research integrating online communication and the Spiral of Silence Theory. You will discover that this theory—with origins in the early 20th century—has enjoyed sustained scholarly interest over the past 70 years!

Noelle-Neumann's Spiral of Silence Theory is important to address for several reasons. The theory directly relates to the freedom of speech, which is the cornerstone of U.S. democracy (Rosenberg, 2023). Further, it is a theory that weaves communication and public opinion, two critical areas in virtually any democracy around the globe (Donsbach, Salmon, & Tsfati, 2013). Third, it has intellectual roots in six of the seven contexts of communication, which we identified in **Chapter 2:** intrapersonal, interpersonal, public, mass/media, small group/team, and culture. In fact, Spiral of Silence scholars have demonstrated the cultural influence of the theory around the globe (Eilders & Porten-Cheé, 2023), a topic we examine later.

Noelle-Neumann focuses on what happens when people provide their opinions on a variety of issues that the media have defined for the public. Yet, she contends issues should be moralistic in nature or value laden (i.e., those have definitive points of view) (Gearhart & Zhang, 2018). The Spiral of Silence Theory suggests that people who believe that they hold a minority viewpoint on a public issue will remain in the background where their communication will be restrained; those who believe that they hold a majority viewpoint will be more encouraged to speak. Noelle-Neumann (1983) contends that the media focus more on the majority views, underestimating or even ignoring minority views. Those in the minority typically will be less assertive in communicating their opinions, thereby leading to a downward spiral of communication. Interestingly, those in the majority will overestimate their influence and may become emboldened in their communication. Subsequently, the media will report on their opinions and activities. This theory, then, adheres to a premise we underscored in **Chapter 2:** groups and teams are highly influential in our lives.

Theory at a Glance • Spiral of Silence Theory

Because of their enormous power, media have a lasting and profound effect on public opinion. Spiral of Silence theorists believe that mass (and social) media work simultaneously with majority opinion to silence minority beliefs on cultural and social issues in particular. A fear of isolation prompts those with minority views to examine the beliefs of others. Individuals who fear being socially isolated are prone to conform to what they perceive to be the majority view. Every so often, however, the silent majority raises its voice in activist ways.

The minority views of Carol Fahey and the behavior of her breakfast friends underscore the gist of the Spiral of Silence Theory. Listening to her colleagues' opinions on spanking, Carol feels that she is alone in thinking that spanking is wrong. The theory suggests that Carol is likely influenced by media reports of over 60 percent of the state supporting spanking for discipline and also by her own recollection of a television news show featuring abused children who as grown-ups spanked their own children and who did not believe it was abusive. Carol perceives her opinion to be a minority view, and consequently she speaks less. Conversely, those in our story who agree with spanking as discipline (Nick, Nancy, and GK) appear to be inspired by the state survey responses; this prompts even more assertive communication on their part.

The difference between these majority and minority views at the senior center is further clarified by Noelle-Neumann (1991). She believes that those in the majority have the confidence to speak out. They may even display their convictions by wearing buttons, brandishing bumper stickers, and emblazoning their opinions on the clothes they wear (today, these individuals will likely be posting or blogging about their convictions). Holders of minority views, however, are sometimes cautious and silent, which reinforces the public's perceptions of their weakness. Nick, Nancy, and GK are clearly confident in their opinions, whereas Carol fosters a sense of nonassertiveness by choosing to engage in an opinion different from those of the people who have spoken.

The Spiral of Silence Theory uniquely intersects public opinion and media. To understand this interface better, we first unravel the notion of public opinion, a key component of the theory. We then examine three assumptions of the theory.

The Court of Public Opinion

As a researcher, Noelle-Neumann was interested in clarifying terms that may have multiple meanings. At the core of the Spiral of Silence Theory is a term that is commonly accepted, but one that she felt was misconstrued: public opinion. As a founder and director of the Allensbach Institute, a polling agency in Germany, Noelle-Neumann contended that interpretations of public opinion have been misguided. In fact, although she identified more than 50 definitions of the term since the theory's inception, none satisfied her and she further lamented that most researchers erroneously link public opinion with government, resulting in a limited understanding of the term (Noelle-Neumann, 1993).

Although many years have passed since the theory's original expression, and it's viewed as a pre-internet-era study of communication, the concept of public opinion "is [still] particularly encumbered by the thicket of confusion, misunderstandings, and communication problems" (Noelle-Neumann & Petersen, 2004, pp. 339–340).

Further, writers continue to state that public opinion is more important than ever (Claussen & Oxley, 2016). Attempting to provide some understanding of this key term in the theory, Noelle-Neumann has provided some clarity. She appropriately separates public opinion into two discrete terms: public and opinion.

She notes that there are three meanings of *public*. First, there is a legal association with the term. **Public** suggests that it is open to everyone, as in "public lands" or "public place." Second, public pertains to the concerns or issues of people, as in "the public responsibility of journalists." Finally, public represents the social-psychological side of people. That is, people not only think inwardly but also think about their relationships to others. The phrase "public eye" is relevant here. Noelle-Neumann (1993) concludes that individuals know whether they are exposed to or sheltered from public view, and they adjust themselves accordingly. She claims that the social-psychological side of public has been neglected in previous interpretations of public opinion, and yet, "this is the meaning felt by people in their sensitive social skin" (p. 62).

> **public** legal, social, and social-psychological concerns of people
>
> **opinion** expression of attitude

An **opinion** is an expression of an attitude. Opinions may vary in both intensity and stability. Invoking the early French and English interpretation of opinions, Noelle-Neumann notes that opinion is a level of agreement of a particular population. In the spiral of silence process, opinion is synonymous with something regarded as acceptable.

Putting all of this together, Noelle-Neumann (1993) defines **public opinion** as the "attitudes or behaviors one must express in public if one is not to isolate oneself; in areas of controversy or change, public opinions are those attitudes one can express without running the danger of isolating oneself" (p. 178). So, for Carol Fahey, her opinion on spanking would not be regarded as acceptable by her breakfast club. Because she fears being isolated from her particular early-morning community, she silences her opinions.

> **public opinion** attitudes and behaviors expressed in public in order to avoid isolation

Noelle-Neumann and Petersen (2004) argue that public opinion is a dynamic process and limited by time and place. To that end, they note that "a spiral of silence only holds sway over a society for a limited period of time" (p. 350). So, there are both short- and long-term components to public opinion. For instance, the public's opinion on legalizing marijuana has changed dramatically over the years. Suppose, in 1969, we were asked to answer the question: should marijuana use be legalized? At that time, per the Gallup poll, only about 12 percent of respondents would have answered affirmatively. Today, national polling shows that nearly 90 percent of respondents support legalizing marijuana use (van Green, 2022, **https://www.pewresearch.org/fact-tank/2022/11/22/americans-overwhelmingly-say-marijuana-should-be-legal-for-medical-or-recreational-use/**). Think about the dynamics related to the changes in response to this issue. Two reasons for changing public opinion seem reasonable. First, an increase in support is likely due to the belief that medical marijuana use can have lasting benefits for sick patients. Second, this upward trend in support may also be because some people think that criminalizing usage is a waste of time and resources. These and other reasons help explain the dynamism surrounding public opinion (look at **Figure 12.1** for an illustration of this topic as it relates to the theory).

PUBLIC OPINION AS COMMUNICATED BY THE MEDIA

Majority View [as reported by the media]
"People should be able to smoke pot for medicinal reasons."

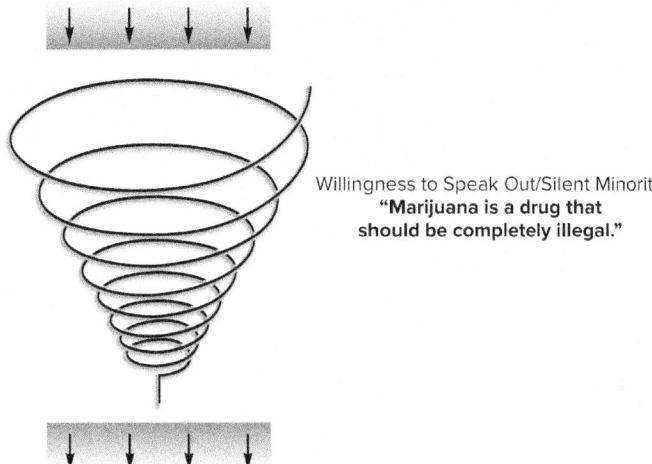

Willingness to Speak Out/Silent Minority
**"Marijuana is a drug that
should be completely illegal."**

[but because of a fear of isolation...many remain silent]
**"I don't want to be perceived as lacking
compassion for the sick and suffering."**

Figure 12.1   The Spiral of Silence: Medicinal Marijuana

Essentially, "public opinion" refers to the collective sentiments of a population on a particular subject. Most often, the media determine what subjects will be of interest to people, and the media often make a subject controversial. For example, the drug Viagra, used to treat impotence, was considered a medical marvel until the media discovered that many health plans covered this drug but did not cover female contraceptives. Many media outlets subsequently reported that this was an overtly sexist practice (Chen, 2016).

Noelle-Neumann (1991) notes that public opinion may be influenced by who approves or disapproves of our views. In 2011, for instance, President Obama instructed the Department of Justice not to defend the Defense of Marriage Act, a law, at the time, that prohibited same-sex partners from marrying. Your opinion on whether you support this congressional action will likely be shaped by spokespeople on both sides of the issue as well as by friends and family members. In addition, as we will learn later, social media platforms such as Facebook and Snapchat will also affect your views. In the end, therefore, a spiral of silence is the response to the shifting opinions of others.

Assumptions of Spiral of Silence Theory

With public opinion as our backdrop to the theory, we now explore three assumptions of the Spiral of Silence Theory. Noelle-Neumann (1991, 1993) has previously addressed these assertions:

- Society threatens deviant individuals with isolation; fear of isolation is pervasive.
- A fear of isolation causes individuals to try to assess the climate of opinion nearly all the time.
- Public behavior is affected by public opinion assessment.

The first assumption asserts that society holds power over those who do not conform through threat of isolation. Noelle-Neumann believes that the very essence of our society depends on people commonly recognizing and endorsing a set of values. And it's public opinion that determines whether these values have equal conviction across the populations. When people agree on a common set of values, then their fear of isolation decreases. When there is a difference in values, fear of isolation sets in.

Like many theorists, Noelle-Neumann is concerned with the testability of this assumption. After all, she notes, are members of a society really threatened with isolation? How could this be? She believes that simple polling could not accurately assess this area (e.g., How much do you fear isolation?). Questions such as these ask respondents to think too abstractly, because it's likely that few respondents have ever thought about isolation.

Noelle-Neumann employs the research values of Solomon Asch (1951), a social psychologist in the 1950s. Asch conducted the following laboratory experiment more than 50 times with eight to ten research subjects:

Which of the following lines on the right is equal to the line on the left?

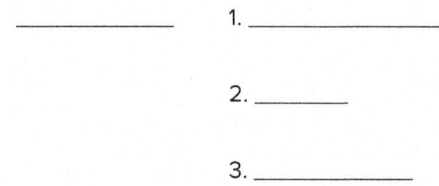

You're probably quick to say that line 3 is equal to the line provided in this line judgment task/question. The group of research subjects, however, disagreed. After going around the room, the experimenter's assistants (who were in on the experiment) all named line 1 as the one that was equal to the line on the left. The unsuspecting subjects began to name line 1 as the correct response. In fact, Asch discovered that several times around, the unsuspecting subjects named the incorrect response. Asch believed that individuals frequently feel great pressure to agree with others, even though the others are incorrect. Borrowing from the theory, there is a very real fear of isolation.

Elizabeth Blakeslee (2005) of *The New York Times* notes that Asch's research conclusions on social conformity still exist. She reports on the implications of following a group in many areas of society, including jury decisions and elections. She notes that "the unpleasantness of standing alone can make a majority opinion seem more appealing than sticking to one's own beliefs" (p. D3). Interestingly, Xin Qin and their associates (2022) noted that today, social robots are inducing normative conformity, further reaffirming the thinking of the theory.

Responding to primary criticisms of the Asch studies—that people did not have a real fear of isolation but rather a lack of confidence in their own judgment—Noelle-Neumann engaged in a more realistic threat-of-isolation test. She believed that requiring subjects to assess a moral or aesthetic conviction was more realistic than any laboratory experiments conducted by Asch. In fact, Noelle-Neumann felt that there had to be "topical controversy" on a "morally charged" topic that prompted people to be involved (Eilders & Porten-Cheé, 2023) in order for a spiral of silence to be present. Indeed, these moral issues should be contemporary (in the public spotlight) and issues on which the public is divided. Think about such topics as marriage equality, abortion rights, climate change, and other topics on which divergent points of view clearly exist.

For Noelle-Neumann, freedom to smoke was (and continues to be) an issue "in the spotlight." During interviews with smokers, she showed them a picture with a person angrily saying, "*It seems to me that smokers are terribly inconsiderate. They force others to inhale their health-endangering smoke.*" Respondents were asked to phrase responses to the statement. The results indicated that in the presence of nonsmokers, many smokers were less willing to support smokers' rights overtly.

The second assumption of the theory identifies people as constant assessors of the climate of public opinion. Noelle-Neumann contends that individuals receive information about public opinion from two sources: personal observation and the media. First, let's discuss how people are able to personally observe public opinion and then examine the role of the media.

Students Talking: Yudie

I've only participated in one protest in my life and that was one of the most exciting times in my life! It was a climate change rally and I surely used my "quasi-statistical sense" because I had to forecast how much resistance the 30 of us would have in front of a congressperson. I was really surprised at the fact that there was no one—not one person—who was opposed to our protests. We even got pizza delivered to us by the office staff! My "sense" was clearly wrong.

Noelle-Neumann (1991) states that people engage in a quasi-statistical ability to appraise public opinion. A **quasi-statistical sense** means that people are able to estimate the strength of opposing sides in a public debate. They are able to do this by listening to the views of others and incorporating that knowledge into their own viewpoints. For instance, Carol Fahey's quasi-statistical sense makes her believe that she is the only person at her breakfast table who opposes spanking. She can see that she is vastly outnumbered on the topic and therefore is able to assess the "local" public opinion on the subject. Noelle-Neumann calls this a quasi-statistical frequency "organ" in that she believes that people like Carol are able to numerically estimate where others fall on the topic. The theorist states that this organ is on "high alert" during periods of instability. So our quasi-statistical sense works overtime when we see that our opinions on a subject are different from those of the majority around us. This sense is, as a rule, an unconscious process.

> **quasi-statistical sense** personal estimation of the strength of opposing sides on a public issue

Personal observations of public opinion can often be distorted and inaccurate. Noelle-Neumann (1993) calls the mistaken observations about how most people feel **pluralistic ignorance.** She notes that people "mix their own direct perceptions and the perceptions filtered through the eyes of the media into an indivisible whole that seems to derive from their own thoughts and experiences" (p. 169). Consider Carol's assessment of the opinions on spanking. With the vast majority of people around her supporting this type of discipline, she may believe that she is clearly in the minority. One or both sides in the debate, however, can overestimate their ability to estimate opinion. Especially with such lopsided support on a topic (as with the group at the cafe), Noelle-Neumann believes that people can become disillusioned.

> **pluralistic ignorance** mistaken observation of how most people feel

People not only employ their personal observations of public opinion, but as we've argued earlier in this chapter and in others, they also rely on the media. Yet, Noelle-Neumann insists that the media's effects are frequently indirect. Because people in many societies are inherently social in nature, they talk about their observations to others. And, people seek out the media to confirm or disconfirm their observations and then interpret their own observations through the media. This can be illustrated through Carol's future behaviors. First, if she returns home from the cafe and reveals her beliefs on spanking to others, she may encounter several neighbors who share her opinion. Next, if she watches the evening news or reads her Facebook posts on the subject and learns that the majority of the country oppose spanking, this will likely resonate deeply with her. She will also be affected by any media reports that disproportionately publicize opposition to spanking. Finally, later discussions that Carol might have on the subject may invoke the media. She may tell others that even the YouTube videos she views online tend to support her point of view.

The final assumption of the theory is that the public's behavior is influenced by evaluations of public opinion. Noelle-Neumann (1991) proposes that public behavior takes the form of either speaking out on a subject or keeping silent. If individuals sense support for a topic, then they are likely to communicate about it; if they feel that others do not support a topic, then they maintain silence. She continues, "The strength of one camp's signals, or the weakness of the other's, is the driving force setting the spiral in motion" (p. 271). In sum, Spiral of Silence scholars would argue that people will generally act according to how other people feel (Codington-Lacerte, 2018).

Noelle-Neumann believes that human beings have an aversion to discussing topics that do not have the support of the majority. To test this assumption, consider interviewing people on your campus about a controversial issue such as physician-assisted suicide. If straw polls in your campus newspaper show that almost 70 percent of the campus opposes this, then according to the theory, students, faculty, and staff may be less inclined to speak out in favor of the practice. A willingness to speak out may have more to do with one's convictions and an assessment of overall trends in society. That is, if there is a liberal climate on your campus, there may be more willingness to speak out; if a conservative climate exists, people may feel less inclined to offer their opposition. And, as we continually note throughout this book, the fields of experience of the communicators will likely influence the willingness.

These three assumptions are important to consider as we further delineate Noelle-Neumann's theory. In **Figure 12.1,** we illustrate several concepts and themes emerging from the theory's assumptions.

Personal opinions, a fear of being alone in those opinions, and public sentiment lay the groundwork for discussing the remainder of the theory. Each of these areas is influenced by a significant part of U.S. society: the media. Let's now overview the powerful influence of the media in the Spiral of Silence Theory.

The Media's Influence

As we have discussed, the Spiral of Silence Theory rests on public opinion. Noelle-Neumann (1993) cautions, however, that "much of the population adjusts its attitudes to the tenor of the media" (p. 272). Nancy Eckstein and Paul Turman (2002) agree. They claim that "the media may provide the force behind the spiral of silence because it is considered a one-sided conversation, an indirect public form of communication where people feel helpless to respond" (p. 173). Further, as Francis Delisay (2012) concludes, "Media can influence the public's perceptions of opinion climates" (p. 485). Finally, some authors assert that the media frequently inform the public about which policy positions to support and the extent to which they should form opinions about them (Hong & Li, 2022).

A willingness to speak out depends greatly on the media. Without support from others for divergent views, people will remain consonant with the views offered in the media. In fact, Noelle-Neumann (1993) believes that the media may even provide biased words and phrases so people can confidently speak about a subject. And, if certain words or phrases are favored by the media, then many people will fall silent (Consider the difference, for instance, if the media used "abuse" rather than the word "spanking"). The extent to which Carol Fahey, in our chapter-opening scenario, will offer her views about spanking, then, will likely rest on what position the various media have taken on the subject. And, although many of us rely on the internet, George Gerbner (Cultivation Analysis, **Chapter 26**) reminds us that television remains among the most influential of all media forms.

In explaining why the media have such influence, Noelle-Neumann believes that the public is not offered a broad and balanced interpretation of news events. Consequently, the public is given a limited view of reality. This restrictive approach to covering cultural events and activities narrows an individual's perception. Certainly, many audiences are active and critical and not all will be so passive in believing everything the media say. Yet, as W. James Potter (2016) acknowledges, most people in the United States seek out media that are aligned with their values and practices. Inevitably, Potter concludes, many suffer from media and information illiteracy.

Consider the theorist's three characteristics of the news media: ubiquity, cumulativeness, and consonance. **Ubiquity** refers to the fact that the media are pervasive sources of information. Because media

> **ubiquity** the belief that media are everywhere

are everywhere, they are relied on when people seek out information. The morning television talk shows, the internet, office gossip, streaming services, and online work spaces all point to the media's ubiquity. Nick, a member of Carol Fahey's morning group, is quick to talk about the recent surveys done in the state about perceptions of spanking. He has the source immediately at hand. Even Carol recalls a television program as she thinks about spanking.

The **cumulativeness** of the media refers to the process of the media repeating themselves across programs and across time. This conclusion might resonate this way: you read about an event on your TikTok

> **cumulativeness** the belief that media repeat themselves

feed when you awaken, listen to the same story on the radio as you drive to work, see posts about the story on X later in the day, repost it on Instagram, then watch the story on the evening news before going to bed. Noelle-Neumann calls this a "reciprocal influence in building up frames of reference" (1993, p. 71). It can become problematic when the original source is left unquestioned, and yet, three media (social media, radio, and television) rely on that source. The theory suggests that conformity of voice influences what information gets released to the public to help them develop an opinion.

Finally, **consonance** pertains to the similarities of beliefs, attitudes, and values held by the media. In fact, events or news items are frequently shared by multiple news agencies (e.g., the Associated Press, etc.). Noelle-Neumann states that consonance is produced from a tendency for newspeople to confirm their own thoughts and opinions, making it look as if those opinions were emanating from the public.

> **consonance** the belief that all media are similar in attitudes, beliefs, and values

Each of these three qualities—ubiquity, cumulativeness, consonance—allows for majority opinions to be heard. Those wishing to avoid isolation will usually remain silent.

It is not surprising that the media are influential in public opinion. Many surveys have demonstrated that people consider the media to have too much power in U.S. society. Consider also that information is frequently filtered through news reporters, social media influencers, and other outlets. As a result, what is presented—or in the case of this theory, what is perceived—may not be an accurate picture of reality. Imagine, for instance, the frustration of many *unemployed* disabled individuals as they read or listen to reports about the success of the Americans with Disabilities Act. Or, maybe you have watched stories about the many people who have been forced off of public assistance, but you probably haven't seen many stories describing the dire circumstances of many families as a result of funding cuts. And, although there have been news reports on employers hiring military veterans, this is hardly helpful to the many vets still looking for long-term employment. If the media report these "success" stories often enough, Noelle-Neumann says they are identifying what should be noticed, deciding what questions should be asked, and determining whether various social policies and programs are effective, a point we noted earlier. In other words, people experience the climate of public opinion through the prism of mass media.

Theory-Into-Action

 A symbiotic relationship exists between professional athletes and the media that cover them. That is, sports figures need media for promotion or marketing and because they are key figures in advocating for cultural or societal values. The media, concurrently, need athletes because, frankly, their photographs and words sell. So, what happened when athletes like NFL players Colin Kaepernick, Patrick Mahones, and Lamar Jackson "took a knee" during the National Anthem in silent protest? With so few athletes and club owners joining their effort, it was as if these three and a handful of others were part of the "hard core." And yet, it's fair to say that conservative media, in particular, were influential in terms of telling their viewers and readers that NFL players should get off their knees and start a dialogue. Could the media have finally begun silencing an athlete, or will an athlete always have the upper hand by getting media more viewers and readers?

As you can see, then, when people look to media for a glimpse into the perceptions and beliefs of the population, they are likely to receive anything but an impartial representation. **Dual climates of opinion** often exist—that is, a climate that the population perceives directly and the climate the media report. For instance, Carol Fahey may compare her personal perceptions of spanking with those surveyed perceptions published in the newspaper. What is remarkable is that despite the differences in opinion, many people decide to remain silent. To understand what motivates people to speak out, Noelle-Neumann developed the train test.

> **dual climates of opinion** difference between the population's perception of a public issue and the way the media report on the issue

The Train Test

For Spiral of Silence theorists, examining whether or not people will speak out requires a methodology that is clear, testable, representative, and replicable. To support her claims, Noelle-Neumann conceptualized the train test (or plane or bus as well). The **train test** is an assessment of the extent to which people will speak out with their own opinion. According to the Spiral of Silence Theory, people on two different sides of an issue will vary in their willingness to express views in public. To study this, the researchers gave respondents sketches showing two people in conversation. The researcher asked a respondent, "Which of the two would you agree with, Person A or Person B?" This question would then be followed up with a more pivotal question—for example, one that might test opinions pertaining to food safety. Essentially, the train test asks people a question such as the following:

> **train test** an experiment used to assess the extent to which people will speak out

> **Suppose that you have a five-hour train ride ahead of you and a person sits next to you and starts to discuss the problems of food safety. Would you talk or not talk about the topic to the person?**

This question was repeated several times with various subjects. It focused on a number of topics, ranging from nuclear power plants to abortion to racial segregation. The test revealed a number of factors that help determine whether a person will voice an opinion. They include the following:

- Supporters of a dominant opinion are more willing to voice an opinion than those in the minority opinion.
- Because of a fear of isolation, people tend to refrain from publicly stating their position if they perceive that this perception will attract laughter, mockery, or similar threats of isolation.
- There are various ways of speaking out—for example, hanging posters, displaying bumper stickers, and distributing flyers.
- Men (ages 45–59) from large cities are more likely to speak out.
- People are more likely to voice an opinion if it agrees with their own convictions as well as fits within current trends and the spirit of the era.
- People will voice an opinion if it aligns with societal views.
- People tend to share their opinions with those who agree with them more than with those who disagree.
- People draw the strength of their convictions from a variety of sources, including family, friends, and acquaintances.
- People may engage in **last-minute swing,** or jumping on the bandwagon of the popular opinion during the final moments of conversation.

last-minute swing jumping on the bandwagon of popular opinion after opinions have been expressed

The train test proved to be an interesting approach to studying public opinion. The method simulates public behavior when two schools of thought exist on a subject. For those who are willing to speak out, there are opportunities to sway others. And there are times when the minority opinion speaks out loudly. We now examine this group.

The Hard Core

Every now and then, the silent minority rises up. This group, called the **hard core,** "remains at the end of a spiral of silence process in defiance of the threats of isolation" (Noelle-Neumann, 1993, p. 170). The hard core represents a group of individuals who know that there

hard core group(s) at the end of the spiral willing to speak out at any cost

is a price to pay for their assertiveness. They try to buck the dominant way of thinking and are prepared to directly confront anyone who gets in their way or who refuses to allow their voices to be heard (**Figure 12.2**).

Noelle-Neumann invokes the work of social psychologist Gary Shulman in attempting to understand the hard core better. Shulman argues that if the hard core becomes large enough, the majority voice becomes less powerful. Several decades ago, for example, when AIDS first appeared in the United States, it was common for many to believe that those diagnosed with the disease should be quarantined (majority opinion). It didn't take long, however, for people's opinions to reject this narrow-minded view, primarily as a result of the hard core's efforts to educate the public. It was not long before this silent hard core discovered that others had adopted their view. In this situation, the hard core was instrumental in changing public opinion, although it is true that the hard core frequently engaged in both rational and irrational acts to make its point.

Figure 12.2   Examples of the "Hard Core" in the United States

For further evidence of the hard core, let's discuss an example pertaining to religion and religious opinion. Although we realize that not all people believe in God, God pervades our intellectual, political, and popular culture (people say "*God bless you*" when others sneeze, we have "*In God We Trust*" on our dollar bills, politicians end their speeches with "*God Bless America,*" and so forth).

Despite pervasive references to God, many people do not believe in God. Moreover, some of these individuals contend that the country's Constitution requires a separation between church and state, and therefore any religious references in tax-supported venues should be eliminated. However, this opinion may not be shared by others who identify with an organized religious group such as Islam or Christianity. Over 80 percent of the world identifies with an organized religious group. Whether the media report on visits by the Pope, present video clips of politicians leaving religious services each week, or solicit quotes for news stories from the clergy, they continue to imply that religion is an integral part of people's lives.

The minority—civil libertarians who advocate extracting religion from public-supported activities—has been vocal regarding their opinions; these hard-core dissenters have not blended into the background. For instance, over the past several years, atheist families, in particular, have sued in courts either to remove "under God" from the Pledge of Allegiance or to allow young children to abstain from the recitation. Although courts have almost always dismissed such cases, they have brought this issue to the forefront of cultural discourse. The hard core might also claim victory as they witness cities removing religious icons (nativity sets, crucifixes, etc.) from city parks during holidays. And, interestingly, with media outlets covering such legal victories, the hard core may be reconfiguring majority opinion.

If we think about more contemporary examples of the hard core, five groups seem particularly relevant: the Tea Party, Occupy Wall Street, Black Lives Matter, #MeToo, and former President Trump's supporters—all movements that were influential at some point from 2011 to 2024. These groups decided to defy expectations and cast themselves in unique, innovative, and controversial ways. The Tea Party—an offshoot of the Republican Party—worked aggressively to ensure that politicians they supported were aligned with their conservative values and practices. Occupy Wall Street, a protest movement that addressed social and economic inequality, greed, and corruption, advocated on behalf of the "99 percent," or the people they claimed were paying for the errors of those in power—the 1 percent. Black Lives Matter remains a social activist group fighting for equality, peace, and justice for all Blacks and a group dedicated to promoting the dignity and integrity of Black people everywhere. #MeToo was founded to weed out the "rich and powerful" who have been able to get away with various levels of sexual harassment and sexual assault. And, the Keep America Great Again (Donald Trump supporters) movement continue to deny the 2020 election results and unabashedly support the former president in all of his endeavors. All five groups—despite having very diverse missions—clearly championed their goals and causes in the media and, despite not representing the majority of the population, they received great media attention and continue to do so in various ways. Now consider the relationship between the hard core and any topic related to COVID-19 (e.g., masks, vaccinations, etc.).

Speaking Out About Various Issues

One important clarification that scholars have noted regarding the Spiral of Silence theory relates to the issues about which individuals may speak out. It's important to point out that not all issues or topics are of equal nature. Expanding the work of other scholars (e.g., Yeric & Todd, 1996), Sherice Gearhart and Weiwu Zhang (2018) articulated and differentiated among various typologies or issue types. In particular, they noted that the types of issues that will prompt others to speak out (and subsequently enhancing the validity of the theory) are enduring issues, emerging issues, and transitory issues.

Enduring issues are concerned with those topics that, historically, have elicited divergent points of view. For instance, one of the most lasting controversial issues in Western societies is the topic

enduring issues concerned with topics that have a historical legacy

of reproductive rights for women and in particular, abortion. The issue culminated with the 1973 U.S. Supreme Court decision (*Roe v. Wade*) protecting a woman's right to choose whether to have an abortion. In 2022, decades later, the Court reversed its landmark decision and the consequences continue to be felt at the local, state, and federal levels.

The second type of issue related to speaking out refers to those that are emerging. Specifically, **emerging issues** are concerned with those topics that have little to no history but that have arisen as a result of various cultural events. So, for instance, while marriage equality was rarely talked about in the 1980s, today, it is legal in the U.S. and in

emerging issues concerned with topics that have little to no history, but which have arisen because of cultural events

many countries around the globe. However, once there is universal buy-in on the issue, then it is no longer viewed as emerging. Think, for instance, about sexual harassment. In the 1980s, it was only discussed infrequently. Today, there are laws against the behavior, and these laws have been around so long that the topic is no longer considered to be evolving.

The word *transitory* generally means "brief, or passing quickly." In some ways, Gearhart and Zhang (2018) would agree with this view. Yet, for them and other Spiral of Silence theorists, **transitory issues** are concerned with topics that appear periodically, such as during election times or during global events such as the Olympics. For

transitory issues concerned with topics that periodically occur in a particular culture or around the globe

example, every few years, with the Olympics set in a country across the globe, a discussion occurs related to doping and drug enhancements. Or, think about the topics that take place each time a presidential election is held in the United States: immigration, national debt, and civil rights, among other topics, are forefront, on many voters' minds.

The Spiral of Silence and Social Media

As we alluded to earlier in the chapter, the Spiral of Silence Theory had its infancy during a time when social media did not exist. As we noted, this is a "pre-internet" theory. Yet, over the years, scores of researchers have employed the theory in their studies on social media. We present a few of the many findings to demonstrate the new ways of thinking about the theory.

Jesse Fox and Katie Warber (2015) looked at Facebook under the rubric of "queer identity management." In particular, they found that Facebook has "social costs" (p. 92), and this site affects how LGBT+ individuals hide their identities. For some in their study, Facebook is viewed as a hostile social media platform, and it was for some of the research subjects more appropriate to stay silent rather than face the consequences of being "outed." Study participants felt that if they voiced their opinions or identity online, they would be viewed as social outcasts, perpetuating the fear of isolation. Kim Bartel Sheehan (2015) examined three different "hot topics": immigration, a national gas tax, and an invasion of North Korea by the South. She found that online social capital did influence a willingness to speak up. In particular, it was discovered that on sites such as Facebook and X, speaking up was related to "bridging" social capital. That is, people were most comfortable speaking up on these topics when they felt a connection with others who held similar views and values. Further, Bartel Sheehan found that "the majority of people holding the majority opinion *believed* that they held the majority opinion."

In addition to Bartel Sheehan's research, others have examined the theory by intersecting gender, culture, and X. For example, Ali Dashti, Hamed Al-Abdullah, and Hasan Johar (2015) explored online political participation of Kuwaiti women. Up until 2006, Kuwaiti women were the only female population in the Arab states who were not afforded full political rights, including the right to run for office and vote. Technology—particularly X—was instrumental as women (and others) began to express their views and talk about their political activities. This study was motivated out of this backdrop.

Students Talking Tech: Caroline

Talk about feelings of "social isolation"! I posted an article on Facebook awhile back on homeless people. The article talked about how a large percentage of people who are homeless are that way because they "wanted" to be. I should've known better. I not only felt like I was at the end of the spiral, but that the spiral was jabbing me in the heart! I had people threaten me, bully my sister online, and even one said that he'd follow me to make sure that I ended up homeless! I had not one person (!) who supported me. Eventually, after a week, I took the post down, but it shows that there really isn't any "free speech" in the United States.

Dashti, Al-Abdullah, and Johar (2015) hypothesized that while women may not be willing to share their views face-to-face, social networks such as X facilitated these sorts of views. Their study was conducted with 323 female students in media and political science courses. The researchers discovered that gender and who a woman/participant knows influenced the willingness to talk about an issue. With men that women do not know, the women/participants acted according to the basic premises of the theory (e.g., silent when dominant voices were present). However, when these women/participants interacted with women they did not know, the women/participants had little problem challenging the majority. When discussing matters related to local political issues, there were no differences in those speaking face-to-face or via X. The research team concluded that X is one way to ease the obstacles that women often experience while discussing politics because the fear of isolation is reduced.

In a study examining netizens in China and their perceptions of speaking out regarding genetically modified food and food additives, one research team (Xu, Liu, Kim, & Chon, 2021) discovered that people's self-perception about their ability to influence others affected their willingness to speak out. In addition, when others noted that they didn't know a lot about food safety, they relied on the opinions of others who claimed to be "closely identified" with the topic. So, for instance, if JT worked in a manufacturing plant that produced food additives, JT would be viewed in high regard by those who knew little about additives.

A fifth study is quite pioneering in nature and reflects a more modern approach to Noelle-Neumann's "train test" we discussed earlier in the chapter. Looking at Edward Snowden's revelations of widespread government surveillance of U.S. citizens' phone and email records, researchers (Hampton et al., 2014) at the Pew Research Center surveyed over 1,800 adults about several issues related to the National Security Agency (NSA). They specifically were interested in their opinions about the leaks, their willingness to speak out in both face-to-face and online settings, and their perceptions of the different opinions of the topic. (Snowden is now a Russian citizen.)

Like Noelle-Neumann, several conclusions were drawn from this "test." Overall, the Pew Research team concluded that if people felt that their social media friends and followers disagreed with their opinions, they were less likely to state their views in public gatherings (e.g., restaurants, etc.). In addition, five additional conclusions were derived from the survey:

1. People were less likely to discuss the Snowden/NSA story via social media than in person.
2. Social media did not provide an alternative discussion platform for those who were not willing to discuss the Snowden/NSA story.
3. In both personal and online settings, people were more willing to share their views if they thought their audience agreed with them.
4. Previous theoretical findings as to people's willingness to speak up in various settings also applies to social media users.
5. Facebook and X users were less likely to share their opinions in many face-to-face settings.

Contributing to the theory's evolution, Hampton et al. (2014) identified three influential factors related to speaking out: a confidence in how much they knew, the intensity of their opinions, and the extent of their interest.

The previous studies are just a handful of the current surplus of research using the Spiral of Silence Theory. As more social media platforms come and go, the theory will have ongoing modifications.

Integration and Critique

The Spiral of Silence Theory is one of the few theories in communication that focuses on public opinion. Indeed, the theory has been identified as an important foundation for examining the human condition (Csikszentmihalyi, 1991). Further, it has deep roots in quantitative methods. Noelle-Neumann, as a scholar, is credited for introducing survey research in Germany and co-founded the reputable *International Journal of Public Opinion Research*. Scholars have identified Noelle-Neumann as almost spiritual-like, noting that she was "a contributing and ethical colleague, always ready to share her knowledge, her findings, and most of all her wisdom" and that she was "confidant of Chancellors and Editors, known by generations of policy makers and opinion leaders throughout the world, [and] she always was helpful to colleagues, and by her example, influenced thousands" (Worcester, 2010, p. 153). To evaluate the theory, we draw attention to two areas: logical consistency and heurism.

Integration

Communication Tradition	Rhetorical | Semiotic | Phenomenological | **Cybernetic** | **Socio-Psychological** | Socio-Cultural | Critical
Communication Context	Intrapersonal | Interpersonal | Small Group | Organizational | Public/Rhetorical | **Mass/Media** | Cultural
Approach to Knowing	**Positivistic/Empirical** | Interpretive/Hermeneutic | Critical

Critique

Evaluation Criteria	Scope | **Logical Consistency** | Parsimony | Utility | Testability | **Heurism** | Test of Time

Logical Consistency

Despite the accolades, Noelle-Neumann's theory has not avoided substantial criticism. And much of that criticism pertains to the lack of logical consistency in several of the terms and concepts. Charles Salmon and Gerald Kline (1985) state that the Spiral of Silence fails to acknowledge a person's ego involvement in an issue. At times, people may be willing to speak because their ego is involved in the topic (e.g., if a promotion at work depends on assertiveness). Carroll Glynn, Andrew Hayes, and James Shanahan (1997) raise the issue of various selectivity processes, meaning that individuals will often avoid a topic that conflicts with their own views.

Carroll Glynn and Jack McLeod (1985) note two additional shortcomings pertaining to the logical consistency of the theory. First, they believe that the fear of isolation may not motivate people to express their opinions. They claim that Noelle-Neumann did not empirically test her belief that fear of isolation prompts people to speak out, relying instead on something called the "train test." Yet, some scholars have found a direct relationship between fear of isolation and an examination of opinion climates and opinion expression (Kim, 2012). Second, Glynn and McLeod were troubled by how the theory was developed by, and relied extensively upon, (West) German media. They doubt whether the characteristics of the media then and there (ubiquitous, cumulative, and consonant) apply to the media in the United States today.

Noelle-Neumann has responded to several of her critics, notably in defending her emphasis on the media. She remains convinced that the media are instrumental in public opinion. She writes that "by using words and arguments taken from the media to discuss a topic, people cause the point of view to be heard in public and give it visibility, thus creating a situation in which the danger of isolation is reduced" (Noelle-Neumann, 1985, p. 80). She continues by noting that not once did the spiral of silence process contradict the media's position on a topic (Noelle-Neumann, 1993). In terms of application across cultures, Noelle-Neumann agrees that any theory of public opinion must have cross-cultural applicability. However, she posits, it is important to note that most U.S. researchers desire a rational explanation for human behavior, but not all behavior can be explained sensibly.

Still, she does accept that the train test may be limited in cross-cultural adaptation. As a result, Noelle-Neumann (1993) updated the version to read:

> Assume you are on a five-hour bus trip, and the bus makes a rest stop and everyone gets out for a long break. In a group of passengers, someone starts talking about whether we should support [insert topic] or not. Would you like to talk to this person, to get to know his or her point of view better, or would you prefer not to? (p. 217)

Of course, you may doubt whether simply changing a train test to a bus test broadens the cross-cultural application of the theory.

Heurism

The theory has attracted writers and scholars who have discussed its merits in a variety of ways. First, some writers (Simpson, 1996) have attempted to discredit the theory because of its lack of application beyond one culture; other scholars, however, have supported the cross-cultural application of the theory (Kim, 2012). This sort of scholarly dialogue enhances the heuristic appeal of the theory. Perhaps Donald Shaw (2014), while reviewing the corpus of research using the Spiral of Silence Theory, said it most succinctly: "her theories have taken hold worldwide" (p. 841).

Spanning many decades, researchers have employed the theory and many of its central concepts in their studies, including the following topics: a declaration to make English the official language of the United States (Lin & Salwen, 1997), religion in the classroom (Eckstein & Turman, 2002), X usage and speaking out (Miyata, Yamamoto, & Ogawa, 2015), college student sexual values (Chia & Lee, 2008), online surveillance and willingness to voice opinions (Stoycheff, 2016), and the theory's influence on both Europe's refugee crisis (Johansson, 2018) and whether people will speak out about the use of force by law enforcement (Italiano, 2018). Finally, although much research has been undertaken to defend and support the Spiral of Silence, one study has found that a fear of isolation related to a minority viewpoint didn't exist. Specifically, Irfan Chaudhry and Anatoliy Gruzd (2020) examined nearly 52,000 public comments to over 100 stories on race posted on the Facebook page of the Canadian Broadcasting Corporation. They discovered that contrary to expectations, the vocal minority were quite comfortable expressing unpopular racist views. It's important to point out that 25 percent of coded comments illustrated a counter-discursive message, meaning that they either challenged or undercut the racist perceptions.

Regardless whether one is supporting, defending, or challenging the theory, it's clear that in the eyes of many, the Spiral of Silence theory is worth studying.

Closing

The Spiral of Silence will continue to generate discussion among media scholars. The theory has sustained considerable criticism, and with a central emphasis on political discussion, researchers will continue to assess the theory's vitality. Still, "But there is certainly accordance, even among the most ferocious critics of this theory, that it has been one of the most influential of all theories developed in communication research and political communication over the last half century" (Donsbach, Tsfati, & Salmon, 2014, p. 1).

Whether people openly express majority or minority viewpoints on an issue may not be directly proportional to the media's involvement on the issue, but it's clear that the public has come to rely on the media in global society. The theory, therefore, will likely have lasting effects that have not yet been imagined.

Discussion Starters

 Case-In-Point: Carol Fahey feels embarrassed about offering her opinions to a group that does not share her beliefs. Consider a similar time in your life. Did you speak out, or did you decide to remain quiet? What motivated your decision?

Try-It-Your-Selfie: Thinking about one's willingness to speak out, would posting your views on such controversial topics as COVID-19 vaccination or gun control be influenced by any social media platforms (e.g., TikTok, YouTube, Instagram, etc.)?

1. Discuss the times that you have been part of the hard-core minority. How did you behave? How did your confidence and self-esteem influence your behavior?
2. Does it make a difference to you to learn that Noelle-Neumann was once a newspaper journalist for Nazi publications? Why or why not?
3. Do you believe that given all the different mediated sources available today, the U.S. media are ubiquitous, consonant, or cumulative? Exemplify your responses.
4. Noelle-Neumann believes that the media help to influence minority views. Based on your observations of the media over the past several years, do you agree or disagree with this claim? What examples can you provide to defend your position?

References

Asch, S. E. (1951). Effects of group pressure upon the modification and distortion of judgments. In J. Guetzkow (Ed.), *Groups, leadership, and men: Research in human relations* (pp. 177–190). Russell & Russell.

Blakeslee, S. (2005, June 28). What other people say may change what you see. *New York Times,* 11.

Chaudhry, I., & Gruzd, A. (2020). Expressing and challenging racist discourse on Facebook: How social media weaken the 'spiral of silence' theory. *Policy & Internet, 12*(1), 88–108.

Chen, A. (2016). Covering Viagra but not birth control. *JSTOR Daily.* Retrieved from https://daily.jstor.org/cover-viagra-but-not-birthcontrol/.

Chia, S. C., & Lee, W. (2008). Pluralistic ignorance about sex: The direct and the indirect effects of media consumption on college students' misperception of sex-related peer norms. *International Journal of Public Opinion Research, 20,* 53–73.

Claussen, R. A., & Oxley, Z. (2016). *Public opinion: Democratic practice.* CQ Press.

Codington-Lacerte, C. (2018). Spiral of silence. *Salem Press.*

Csikszentmihalyi, M. (1991). Reflections on the "Spiral of Silence." In J. A. Anderson (Ed.), *Communication yearbook 14* (pp. 294–298). Sage.

Dashti, A. A., Al-Abdullah, H. H., & Johar, H. A. (2015). Social media and the spiral of silence: The case of Kuwaiti female students' political discourse on Twitter. *Journal of International Women's Studies, 16*(3), 42–53.

Delisay, F. (2012). The spiral of silence and conflict avoidance: Examining antecedents of opinion expression concerning the U.S. military buildup in the Pacific Island of Guam. *Communication Quarterly, 60,* 481–503.

Donsbach, W., Salmon, C. T., & Tsfati, Y. (Eds.). (2013). *The spiral of silence: New perspectives on communication and public opinion.* Routledge.

Donsbach, W., Tsfati, Y., & Salmon, C. T. (2014). The legacy of spiral of silence theory. An introduction. In W. Donsbach, C. T. Salmon, & Y. Tsfati (Eds.), *The spiral of silence: New perspectives on communication and public opinion* (pp. 1-18). Routledge.

Eckstein, N. J., & Turman, P. D. (September 2002). "Children are to be seen and not heard": Silencing students' religious voices in the university classroom. *Journal of Communication and Religion, 25*(2), 166-192.

Eilders, C., & Porten-Cheé, P. (2023). Effects of online user comments on public opinion perception, personal opinion, and willingness to speak out: A cross-cultural comparison between Germany and South Korea. *Journal of Information Technology & Politics, 20*(3), 323-337.

Fox, J., & Warber, K. M. (2015). Queer identity management and political self-expression on social networking sites: A co-cultural approach to the spiral of silence. *Journal of Communication, 65,* 79-100.

Gearhart, S., & Zhang, W. (2018). Same spiral, different day? Testing the spiral of silence across issue types. *Communication Research, 45*(1), 34-54.

Glynn, C. J., Hayes, A. F., & Shanahan, J. (1997). Perceived support for one's opinions and willingness to speak out: A meta-analysis of survey studies on the spiral of silence. *Public Opinion Quarterly, 61,* 452-463.

Glynn, C. J., & McLeod, J. M. (1985). Implications of the spiral of silence for communication and public opinion research. In K. R. Sanders, L. L. Kaid, & D. Nimmo (Eds.), *Political communication yearbook 1984* (pp. 43-65). Southern Illinois University Press.

Hampton, K. N., Rainie, L., Lu, W., Dwyer, M., Shin, I., & Purcell, K. (2014). *Social media and the "spiral of silence."* Pew Research Center.

Hong, C., & Li, C. (2022). Spiral of Silence or social loafing? A parallel mechanism to explain why people defend their stances on controversial sociopolitical issues. *International Journal of Communication, 16,* 21.

Italiano, R. (2018). *Willingness to speak in the context of police who use force.* [Master's Thesis, Marquette University Press].

Johansson, B. (2018). Expressing opinions about the refugee crisis in Europe: The spiral of silence and crisis communication. *Journal of International Crisis and Risk Communication Research, 1*(1), 57-82.

Kim, S-H. (2012). Testing fear of isolation as a causal mechanism: Spiral of silence and genetically modified (GM) foods in South Korea. *International Journal of Public Opinion Research, 24,* 306-324.

Lin, C. A., & Salwen, M. B. (1997). Predicting the spiral of silence on a controversial issue. *Howard Journal of Communications, 8,* 129-141.

Miyata, K., Yamamoto, H., & Ogawa, Y. (2015). What affects the Spiral of Silence and the hard core on Twitter? An analysis of the nuclear power issue in Japan. *American Behavioral Scientist, 59,* 1129-1141.

Noelle-Neumann, E. (1983). The effect of media on media effects research. *Journal of Communication, 33,* 157-165.

Noelle-Neumann, E. (1985). The spiral of silence—A response. In K. R. Sanders, L. L. Kaid, & D. D. Nimmo (Eds.), *Political communication yearbook, 1984* (pp. 43-65). Southern Illinois University Press.

Noelle-Neumann, E. (1991). The theory of public opinion: The concept of the spiral of silence. In J. A. Anderson (Ed.), *Communication yearbook 14* (pp. 256-287). Taylor & Francis.

Noelle-Neumann, E. (1993). *The spiral of silence: Public opinion—our social skin.* University of Chicago Press.

Noelle-Neumann, E., & Petersen, T. (2004). The spiral of silence and the social nature of man. In L. L. Kaid (Ed.), *Handbook of political communication* (pp. 339-356). Taylor & Francis Group.

Potter, W. J. (2016). *Media literacy* (8th ed.). Sage.

Qin, X., Chen, C., Yam, K. C., Cao, L., Li, W., Guan, J., & Lin, Y. (2022). Adults still can't resist: A social robot can induce normative conformity. *Computers in Human Behavior, 127,* 107041.

Rosenberg, I. (2023). *The fight for free speech. Ten cases that define our first amendment freedoms.* New York University Press.

Salmon, C. T., & Kline, F. G. (1985). The spiral of silence ten years later: An examination and evaluation. In K. Sanders, L. L. Kaid, & D. Nimmo (Eds.), *Political communication yearbook 1984* (pp. 3–29). Southern Illinois University Press.

Shaw, D. (2014). Book review: The Spiral of Silence: New perspectives on communication and public opinion. *Journalism & Mass Communication Quarterly,* 841–842.

Sheehan, K. B. (2015). The many shades of greenwashing: using consumer input for policy decisions regarding green advertisements. *Communicating Sustainability for the Green Economy,* 53–65.

Simpson, C. (1996). Elisabeth Noelle-Neumann's "spiral of silence" and the historical context of communication theory. *Journal of Communication, 46,* 149–173.

Sohn, D. (2022). Spiral of silence in the social media era: A simulation approach to the interplay between social networks and mass media. *Communication Research, 49*(1), 139–166.

Stoycheff, E. (2016). Under surveillance examining Facebook's Spiral of Silence effects in the wake of NSA internet monitoring. *Journalism & Mass Communication Quarterly, 93,* 296–311.

van Green, T. (2022, November 22). Americans overwhelmingly say marijuana should be legal for medical or recreational use. https://www.pewresearch.org/fact-tank/2022/11/22/americans-overwhelmingly-say-marijuana-should-be-legal-for-medical-or-recreational-use/.

Worcester, R. (2010). A special Tribute to Elisabeth Noelle-Neumann. *International Journal of Public Opinion Research, 22,* 153.

Yeric, J. L., & Todd, J. R. (1996). *Public opinion: The visible politics* (3rd ed.). F. E. Peacock Publishers.

Xu, L., Liu, J., Kim, J., & Chon, M. G. (2021). Are Chinese netizens willing to speak out? The spiral of silence in public reactions to controversial food safety issues on social media. *International Journal of Environmental Research and Public Health, 18*(24), 13114.

CHAPTER 13
Uses and Gratifications Theory

*Based on the research of **Elihu Katz, Jay G. Blumler,** and **Michael Gurevitch***

"Books and cinema have been found to cater to needs concerned with self-fulfillment and self-gratification: they help to connect individuals to themselves. Newspapers, radio, and television all seem to connect individuals to society."

—Elihu Katz, Jay G. Blumler, and Michael Gurevitch

Ryan Grant

It was a dreary Friday night and 20-year-old Ryan Grant was trying to figure out what he wanted to do. He was going to school at Middleton State University and he worked for his father part-time in the family's hardware store. This had been a rough week. He had to put in a lot of extra hours because they were doing inventory and his dad needed help cataloging the merchandise. This morning he took an exam in his communication class at the university. Because he hadn't studied last week, Ryan had crammed until 2 A.M. Now the week was over, and it was time to escape both work and school.

Ryan felt exhausted and burned out. He wouldn't mind doing something social, but he knew he wouldn't be the best company. He considered two choices: he could stay home and watch something on Hulu or Netflix or he could try to get some friends together to go to the movies. It could be good to go out with people who would help him loosen up. But, he also liked the idea of being at home and not exerting so much energy.

Ryan was faced with two opposing arguments. Here was the argument for staying home: first, he wouldn't have to spend anything extra, and he'd be justifying the money he spent monthly to subscribe to the streaming services. He could dress and look how he wanted, and he could watch what he wanted when he wanted. There were a couple of shows that friends had told Ryan were

good that he'd not seen yet. At home, he could also command the best seat in the house. If he wanted to, he could invite over friends, making the night more social.

But, the argument for going out to the movies seemed just as strong. He could see that new action/adventure movie he'd been waiting to see. Also, because Ryan was a "techie" who appreciated all technology, he knew he'd enjoy the movie theater's excellent sound system and huge screen, which were far better than what he had at home. Finally, he could easily have fun going out. Although he'd never really thought about it too much, he enjoyed sitting with his friends in the darkened theater, sharing the same experience, and he loved movie popcorn!

Still, he was torn, so Ryan weighed the positives and negatives of each option. If he chose a streaming service, he'd have to deal with watching on a small screen. Also, if he did have friends over, he might have to endure fighting over what to watch. On the other hand, he could turn off the television if he didn't like the show and simply hang out with his friends. The arguments for and against going out to the movies seemed equivalent. On the plus side for going out, he knew that watching a movie was a great escape from his real life, and he would be able to talk about the movie on Monday with his friends. But, it was also true that he'd have to drive to the multiplex and try to find a parking space.

To top it off, he would have to pay more than $20 for the movie ticket, popcorn, and a drink. Ryan's decision about what to do came down to a simple question: what does a movie theater offer versus what streaming services offer? As Ryan considered this question, other alternatives occurred to him: he could play Minecraft, or, if all else failed, he could always just go to bed early!

Ryan is doing what we all do when dealing with the media: he's thinking about different media and making choices. Consider how many times you have found yourself in a situation similar to Ryan's. You may have decided that you needed some relaxation and thought about all the options before making up your mind. The process may not have taken very long, but it was a process that required thinking about what was available.

In the early days of mass media (the era of the penny newspaper, radio, and silent movies), **Mass Society Theory**—the idea that average people are helpless victims of powerful mass media—defined the relationship between audiences and the media they consumed (see our discussion of the Spiral of Silence Theory in **Chapter 12**). This notion was eventually discredited, in large part because social science—and simple observation—couldn't confirm the idea of media and media messages being so powerful. Not only were most people not directly affected by the media's messages, but when they were influenced, they weren't all influenced in exactly the same way.

> **Mass Society Theory** the idea that average people are the victims of the powerful forces of mass media
>
> **limited effects theories** the perspective replacing Mass Society Theory; holds that media effects are limited by aspects of the audience's personal and social lives
>
> **Individual Differences Perspective** a specific approach to the idea of limited effects; concentrates on the limits posed by personal characteristics
>
> **Social Categories Model** a specific approach to the idea of limited effects; concentrates on the limits posed by group membership

In time, Mass Society Theory was replaced by what we now call the limited effects theories (we introduced the notion of limited effects in **Chapter 11**). **Limited effects theories** conceptualize media influence as limited or minimized by some aspects of individual audience members' personal or social lives. Researchers have taken two approaches to the limited effects orientation. First, the **Individual Differences Perspective** sees media's power as restrained by an audience member's personal factors such as intelligence and self-esteem. For example, smart people and more secure people are better able to resist unwanted media impact. A second limited effects approach, the **Social Categories Model,** views media's power as limited by audience members' associations and group affiliations. For example, Republicans tend to spend time with other Republicans, who help them interpret media messages in a consistent, Republican-friendly manner. This effectively limits any influence media messages alone might have.

You may have noticed that neither of these views affords audience members much credibility. The first (Mass Society Theory) suggests that people simply aren't smart enough or strong-minded enough to protect themselves against unwanted media effects. The second (limited effects theories) suggests that people have relatively little personal choice in interpreting the meaning of the messages they consume and in determining the level of impact those messages will have on them because these effects are limited only by group membership or some individual factors that not everyone possesses. Eventually, in response to these unflattering views of typical audience members, theorists Elihu Katz, Jay G. Blumler, and Michael Gurevitch (1974) presented a systematic and comprehensive articulation of audience members' role in the mass communication process. Their thinking was formalized as Uses and Gratifications Theory (UGT). While UGT is still considered a limited effects model, researchers believe its value lies in its ability to clarify how effects can and do happen and in its focus on how audience members are active participants in the achievement of those effects. Additionally, Marianne Dainton and Alexandra Stokes (2015) note that although UGT was initially proposed to study one-way media (think about the linear model we explained in **Chapter 1**), it's now being applied to forms of media, like social media, where audiences are interactive (e.g., Gamage, Tajeddini, & Tajeddini, 2022; Rathnayake & Winter, 2018).

Further, while granting the media some effects, UGT gives the audience credibility and holds that people actively seek out specific media and specific content to generate specific gratifications (or results). Researchers using UGT view people as active because they're able to examine and evaluate various types of media to accomplish communication goals. As we saw in our opening, Ryan not only identified the specific media he was willing to consider, but was also able to determine for himself the uses he could and would make of each, and the personal value of those uses.

Assumptions of Uses and Gratifications Theory

UGT provides a framework for understanding when and how individual media consumers become more or less active and the consequences of that increased or decreased involvement. Many of the assumptions of UGT were clearly articulated by the founders of the approach (Katz et al., 1974). There are five basic assumptions of UGT:

- The audience is active and its media use is goal oriented.
- The initiative in linking need gratification to a specific medium choice rests with the audience member.
- The media compete with other sources for need satisfaction.
- People have enough self-awareness of their media use, interests, and motives to be able to provide researchers with an accurate picture of that use.
- Value judgments of media content can only be formed by the audience.

The theory's first assumption, about an active audience with goal-oriented media use, is fairly straightforward. Individual audience members pick and choose actively among different media and are making these choices based on goals they wish to accomplish via the media. Researchers have created several ways of classifying audience needs, or the goals that audience members have for seeking out specific media that we discuss later in the chapter.

In our chapter opening, we saw Ryan choosing between two competing media: a streaming service at home, and a movie in a theater. All of us have our favorite content within a given medium, and we all have reasons for selecting a particular medium. At the movies, for instance, many of us like love stories rather than war films; some of us prefer to be entertained at the end of a long day rather than to be educated about a historical event. Although it's getting more difficult to do because of state laws, some drivers prefer to talk on their cell phones over long trips; it not only passes the time but also allows people to stay connected with their family and friends. Other drivers may prefer to listen to podcasts rather than talking on the phone during a long drive. There are people who enjoy watching home improvement or cooking shows on cable and others who prefer dramatic fare. Audience members choose among various media for different gratifications. UGT researchers have disagreed about the extent to which an audience is active, however, and we'll discuss their various stances on this topic later in the chapter.

UGTs' second assumption links need gratification to a specific medium choice that rests with the audience member. Because people are active agents, they take initiative. We may choose shows such as the *Drew Barrymore Show* when we want to be entertained and a podcast such as the *TED Radio Hour* when we want to be informed. Although it's also possible that we listen to the podcast because we find it entertaining rather than informative, and perhaps we find sources of information in Drew Barrymore's talk show. The implication here is that audience members have a great deal of autonomy in the mass communication process and make their choices for their own idiosyncratic purposes.

The third assumption—that media compete with other sources for need satisfaction—means that there are many activities for an audience member to choose among. On a first date, for example, the partners could go out to the movies, stay home and watch something on a streaming video service, go to a club and hang out with friends, or any one of many other options. Additionally, cultural context affects media selection. The media and their audiences don't exist in a vacuum. Both are part of the larger society, and the relationship between media and audiences is influenced by that society. For instance, someone who is an infrequent consumer of media—who, for example, finds more gratification in conversations with friends and family—may turn to the media with greater frequency when seeking information during a national political election or in wartime (Kozman & Melki, 2018).

The fourth assumption of UGT relates to a methodological issue that has to do with researchers' ability to collect reliable and accurate information from media consumers. To argue that people are aware enough of their own media use, interests, and motives to be able to provide researchers with an accurate picture of that use reaffirms the belief in an active audience, and it also implies that people are cognizant of that activity. In fact, the early research in Uses and Gratifications included questioning respondents about why they consumed particular media. This qualitative approach included interviewing respondents and directly observing their reactions during conversations about media. The thinking surrounding this data collection technique was that people are in the best position to explain what they do and why they do it. Interestingly, as the theory evolved, the methodology also changed. Researchers began to abandon their qualitative analysis in favor of more quantitative procedures. Yet, the quantitative questionnaires drew on the information from the interviews and observations collected in the qualitative period.

The fifth assumption is also less about the audience than it is about those who study it. It asserts that researchers should suspend value judgments linking the audience's needs to specific media or content. Uses and Gratifications theorists argue that because it's individual audience members who decide to use certain content for certain ends, the value of media content can be assessed only by the audience. Even if a researcher might be inclined to label content in shows such as *Love Is Blind,* where people date and possibly marry without ever seeing one another, as tacky, UGT assumes that the content is functional if it provides gratification for the audience.

UGT and its assumptions gained acceptance from researchers for a number of reasons. First, the limited effects researchers began to run out of things to study. After all the variables that limited media influence were chronicled, what was left to say about the process of mass communication? Second, the limited effects perspective failed to explain why advertisers spend billions of dollars a year to place their ads in the media or why so many people spend so much time consuming the media. Third, some observers speculate that people often decide whether specific media effects are desirable and intentionally set out to achieve those effects. If this is so, researchers ask, what does this say about limited effects? Finally, although many negative effects were documented by limited effects researchers, such as the relationship between mediated violence and aggressive behavior, positive uses of media were left unexamined in limited effects research.

These factors produced a subtle shift in the focus of those researchers working within the limited effects paradigm. Their attention moved from the things media do to people to the things people do with media. If effects occur at all, either positive or negative, it's because audience members want them to happen or at least let them happen. Thus, scholars applying Uses and Gratifications Theory are interested in the following question: what do consumers do with media?

As you can see, UGT presumes an active media consumer. Considering that this overarching principle contradicts the views offered by other media theorists and other theoretical perspectives, it's important to trace the theory's development, which we do in the next section.

Theory at a Glance • Uses and Gratifications Theory

People are active in choosing and using particular media to satisfy specific needs. Although Uses and Gratifications takes a limited effects position, it does so by positing that the media have that limited effect on audiences because users are able to exercise choice and control. People are self-aware, and they're able to understand and articulate the reasons they use media. They see media use as a good way to gratify the needs they have. Uses and Gratifications Theory is primarily concerned with the following question: What do people do with media?

History of Uses and Gratifications Research

Uses and Gratifications theorists have investigated the uses/gratification perspective in stages over time. The first stage of uses and gratifications research (prior to the formulation of the theory itself) consisted of acknowledging that people can and do actively participate in the mass communication process. The pioneering work of Herta Herzog (1944) was instrumental in establishing this perspective. Herzog sought to classify the reasons people engage in different forms of media behavior, such as newspaper reading and radio listening. Wanting to understand why so many women were attracted to radio soap operas, Herzog interviewed dozens of soap opera fans and identified three major types of gratification. First, some people enjoyed the dramas because of the emotional release they found in listening to the problems of others. Second, some listeners seemed to engage in wishful thinking—that is, they gained a vicarious satisfaction from listening to the experiences of others. Finally, some people felt that they could learn from these programs because "if you listen to these programs and something turns up in your life, you would know what to do about it" (p. 25). Herzog's work was critical to developing UGT because she was the first published researcher to provide an in-depth examination of media gratifications. She's sometimes credited with having originated UGT (although its label was to come much later).

Continuing the first stage of uses and gratifications research, Wilbur Schramm (1954) developed a means of determining "which offerings of mass communication will be selected by a given individual" (p. 19). His **fraction of selection** represents precisely the process that Ryan goes through when he thinks about his choices of going out to a movie or watching a show at home on Hulu or Netflix. The fraction of selection is the expectation of reward divided by the effort required to attain the reward.

fraction of selection Schramm's idea of how media choices are made; the expectation of reward divided by the effort required

$$\frac{\text{Expectation of reward}}{\text{Effort required}}$$

Schramm sought to make clear that audience members judge the level of reward (gratification) they expect from a given medium or message against how much effort they must make to secure that reward—an important component of what would later become known as UGT.

The second stage of uses and gratifications research began when researchers created typologies representing all the reasons people had for media use.

For instance, Jay Blumler and another colleague, Denis McQuail (1969), examined the reasons that people watch political programs. They found a number of motives for watching political broadcasts, and this work formed an important foundation for researchers developing and applying Uses and Gratifications Theory. These researchers found that people had a need either to connect with or to disconnect from others. Researchers also found categories of needs associated with acquiring information or knowledge, achieving pleasure, obtaining status, strengthening relationships, and relaxing. As you will recall, Ryan Grant was trying to work through two simultaneous needs: the need for being social and the need to relax.

Another typology of reasons for media use came from the origina-tors of the theory and includes the following needs: **cognitive needs,** or using the media as a source of information; **affective needs,** or using the media for a pleasant experience; **personal needs,** or using the media as a source for enhancing personal credibility; **social needs,** or using media to enhance connections with others; and **tension release needs,** or using media for escape or diversion (Katz, Gurevitch, & Haas, 1973). In **Table 13.1** we summarize these need types and present media sources used to satisfy them.

cognitive needs using the media as a source of information

affective needs using the media for a pleasant experience

personal needs using the media as a source for enhancing personal credibility

social needs using media to enhance connections with others

tension release needs using media for escape or diversion

Alan Rubin (1981) found that motivations for television use clus-tered into the following categories: to pass time, for companionship, excitement, escape, enjoyment, social interaction, relaxation, infor-mation, and to learn about a specific content. Other researchers (McQuail, Blumler, & Brown, 1972) asserted that media use could be categorized with only four basic divisions: diversion, personal relationships, personal identity, and surveillance.

Table 13.1 Needs Gratified by the Media

NEED TYPE	DESCRIPTION	MEDIA EXAMPLES
Cognitive	Acquiring information, knowledge, comprehension	Television (news), YouTube videos (e.g., "How to Install Ceramic Tile"), movies (documentaries or films based on history, e.g., *The Deepest Breath*)
Affective	Emotional, pleasant, or aesthetic experience	Music, movies based on art or dance
Personal	Enhancing credibility, confidence, and status	YouTube videos (e.g., "Speaking With Conviction")
Social	Enhancing connection with family and friends	Social media such as Instagram, Facebook, or Snapchat
Tension release	Escape and diversion	Movies: action/adventure, rom-com, etc.; TV: late night talk shows, reality shows, etc.; video games

Adapted from Katz, Gurevitch, & Haas, 1973.

In the third stage of uses and gratifications research, researchers began to link specific reasons for media use with variables such as needs, goals, benefits, the consequences of media use, and individual factors. In this effort, researchers worked to make the theory more explanatory and predictive. Alan Rubin and Mary Step (2000) conducted a study that exemplifies this stage of the research. Rubin and Step examined the relationship of motivation, interpersonal attraction, and **parasocial interaction** (the relationship we feel we have with people we know only through the media) to listening to public affairs talk radio. They found that motivations for entertainment and information acquisition interacted with perceptions of the parasocial relationship to explain why listeners tuned in to talk radio and why they found a host credible.

parasocial interaction the relationship we feel we have with people we know only through the media

The next stage of UGT research is ongoing, and now researchers are interested in how the theory operates with respect to newer media (e.g., Meier, 2018; Rathnayake & Winter, 2018). We'll discuss this research in some detail later in the chapter. In addition, research using UGT continues to generate lists of gratifications that audiences use traditional media to satisfy. For example, Darrin Brown, Sharon Lauricella, Aziz Douai, and Arshida Zaidi (2012) were interested in the motivations people had for watching crime dramas on TV. They found some support for the idea that people satisfied their curiosity through watching these shows.

Changing Positions on Media Effects

The history of UGT has much to do with how researchers shifted and changed their positions on media effects. As we explained previously, researchers moved from a position where they saw the media as very powerful to one where media effects were seen as more limited. UGT moved even further to the position of the active audience and less powerful media. However, there was controversy about how much control audience members actually possessed. Jay Blumler (1979), one of the originators of UGT, believed that some scholars had gone too far in describing the active audience. He argued that the theory meant to assert that even when audience members are active—even when they determine for themselves the uses they wish to make of mass media and the gratifications they seek from those uses—media effects can and do occur because the media are powerful too. The failure of researchers applying UGT to consider the possibility of important media effects led Jay Blumler (1985) to assert that it wasn't the original theorists' intention to imply that audience members are always totally free in either the uses they make of media or the gratifications they seek from them; the world in which media consumers live shapes them just as surely as they shape it, and content does have intended meaning.

Students Talking: Andre

It's interesting to think about how much choice we have in picking media. There are a lot of options, but sometimes I wonder how much difference there really is among these choices. What's actually different about *House Hunters* and *House Hunters International* and *Beach House Hunters*? I've watched these shows with my girlfriend, and I don't see much difference among them. They all basically tell the same story—people want a new house, search for it, and then buy one. And when a movie has multiple parts, like the Jason Bourne movies, or *Die Hard 1, 2*, etc., and all the Marvel movies—how different are they? Or for that matter, what's the difference between *Die Hard* and Jason Bourne? And doesn't one cat video seem pretty much like all cat videos? It seems like there are a lot of shows out there, but they're all really similar. When the theory says we're active choice makers, I have to question that. I don't know if we have that much choice when we're restricted to what the media offer us.

Blumler and his colleagues point to the original set of premises of UGT, which makes clear their position on limited effects and the relationship between audience and media. Elihu Katz, Jay G. Blumler, and Michael Gurevitch (1974) wrote in developing UGT that "social situations" in which people find themselves can be "involved in the generation of media-related needs" (p. 27) in five ways. First, social situations can produce tensions and conflicts, leading to pressure for their easement through the consumption of media. That is, we live in the world, and events in it can compel us to seek specific media and content. For instance, in early 2023 there was a great deal of information and controversy about what Fox newscasters said privately about the results of the 2020 U.S. presidential election compared to what they said publicly. In private, several of the newscasters indicated they believed Joe Biden had won and there had been no fraud. But, on the air, they often pushed what was known as the Big Lie, that the election had been stolen from Donald Trump and he was the rightful winner. This controversy was fed by legal documents filed by Dominion Voting Systems, who sued Fox News for defamation. The case was based on Fox's on-air accusations that Dominion's tabulation machines had flipped votes for Trump over to Biden. This was a social situation fraught with tension and conflict. Where did you go to ease the pressure? Did you seek out more information through TV, radio, or the internet? Did you use social media sites to share your opinion of the case, Trump, Biden, Fox News, and so forth?

Second, social situations can create an awareness of problems that demand attention, information about which may be sought in the media. Simply stated, the world in which we live contains information that makes us aware of things that are of interest to us, and we can find out more about those interests through the media. In 2023, everyone everywhere—at work, at school, in virtually every social situation you entered—was talking about multiple weather-related disasters that brought wildfires to Maui, a heat dome to most parts of the mainland U.S., and rare flooding to Southern California. This situation demanded your attention. You probably turned to the media for information, perspective, and analysis.

Third, social situations can restrict real-life opportunities to satisfy certain needs, and the media can serve as substitutes or supplements. In other words, sometimes the situations in which you find yourself make the media the best, if not the only, source possible. Your social situation as a college student made it unlikely that you could directly get information about Fox News or weather extremes yourself. You probably couldn't interview the Fox newscasters for yourself. Nor could you easily travel to Los Angeles and Maui to check out the weather yourself. You wanted to know what was going on in these situations, but the reality of your position in society meant that you had little choice but to rely on the media to meet that need.

Fourth, social situations often elicit specific values, and their affirmation and reinforcement can be facilitated by the consumption of related media materials. Again, you're a college student. You are an educated person. The media offer an appropriate location for the affirmation and reinforcement of the knowledge and awareness that you value.

Finally, social situations demand familiarity with media; these demands must be met to sustain membership in specific social groups. As a student, you're viewed as a future leader. Not only should you have had an opinion about important events in the culture, but also you should have had something to say about the media's performance in reporting on these events. Lacking those opinions, you may have been regarded as out of it or uninformed.

Katz and his colleagues (1974) note that we should ask three things about these five issues. As we do this, we'll be able to see the ways media and audiences together create effects. First, are the mass media instrumental in creating this social situation? What role did various media outlets play? Based on what information did we form our opinions? Second, are the mass media instrumental in making the satisfaction of this situation's related needs so crucial? Why, for instance, was it important to have an opinion at all? Third, who put this issue on the public's agenda? Who determined that it was more important than any of the myriad events that were happening in the world? In raising these three questions, UGT overlaps a bit with Agenda Setting Theory, which we discussed in **Chapter 11.**

Key Concepts

By now, you know that UGT is an approach that takes into consideration the importance of the active audience. And the active audience is the first key concept of UGT and constitutes the foundation for a discussion of the other key terms. The interpretation of an **active audience** is an important part of UGT. For our purposes, we draw upon the words of Mark Levy and Sven Windahl (1985), who define the active audience as follows:

> **active audience** the idea that audiences have a voluntaristic and selective orientation toward media use

> As commonly understood by gratifications researchers, the term "audience activity" postulates a voluntaristic and selective orientation by audiences toward the communication process. In brief, it suggests that media use is motivated by needs and goals that are defined by audience members themselves, and that active participation in the communication process may facilitate, limit, or otherwise influence the gratifications and effects associated with exposure. Current thinking also suggests that audience activity is best conceptualized as a variable construct, with audiences exhibiting varying kinds and degrees of activity. (p. 110)

In addition to the active audience, activity and activeness also are key concepts in the theory. UGT distinguishes between activity and activeness to better understand the variability of audience activity that Levy and Windahl (1985) speak about. Although the terms are related, **activity** refers more to what the media consumer does (e.g., they choose to go online for news rather than read it in the newspaper). **Activeness** refers to the audience's freedom and autonomy in any media situation.

> **activity** refers to what the media consumer does

> **activeness** refers to how much freedom the audience really has in the face of mass media

Jay G. Blumler (1979) offers several kinds of *activity* in which media consumers could engage. These include utility, intentionality, selectivity, and imperviousness to influence, which are four additional key concepts of UGT. First, the media have uses for people, and people can put media to those uses. This is termed **utility.** For example, people listen to the car radio or check an app (e.g., Waze) to find out about traffic on their route somewhere. They go online to download music. They read *Consumer Reports* to find out the latest ratings on new car options. **Intentionality** occurs when people's prior motivations determine their consumption of media content. When people want to be entertained, they usually find a comedy. When they want greater detail about a news story,

> **utility** the ability to utilize the media for the specific purpose they represent, such as using a traffic app to navigate a trip

> **intentionality** a cognitive behavior that occurs when people's prior motives determine use of media

> **selectivity** audience members' use of media reflects their existing interests

> **imperviousness to influence** refers to audience members constructing their own meaning from media content

they might go to news-centered networks such as CNN or MSNBC because they have found these needs gratified by these media previously. A third type of audience activity is termed **selectivity,** which means that audience members' use of media can reflect their existing interests and preferences. If you like jazz, you might listen to the jazz program on the local radio station. If you're interested in tech trends, you are likely a reader of *Wired.* If you're interested in politics, you probably subscribe to *Politico* or read political blogs. Finally, an **imperviousness to influence** suggests that audience members construct their own meaning from content and that meaning influences what they think and do. People who enjoy action/adventure films and television shows might select them for catharsis and as a way to release aggression without becoming aggressive in their real lives. Further, people often actively avoid certain types of media influence. For example, some people try to buy products on the basis of quality and value and avoid listening to ads.

Activeness is relative. Some people are active participants in the mass communication process; others are more passive. We all know people who live their lives through TV or who follow every fad and fashion presented in the media. Terms like *couch potato* and *sofa spud* developed from the idea that many folks simply sit back and take in whatever is presented on the TV set in front of them. On the other hand, we also know that people can be quite deliberate in their reasons for consuming media. Your friend may listen to rap because of the beat. You may listen to rap not only for its rhythms, but also for its social commentary. For you, watching a documentary on climate change may prompt you to think about the amazing photography. For your sister, however, the movie may have been interpreted as a social commentary about the lack of U.S. action on global warming.

Activeness is also variable. A person can be inactive at times (*"I'll just turn on the television for background noise"*) and then become quite active (*"The news is on! I'd better watch"*). Our level of activeness often varies by time of day and type of content. We can be active users of the internet by day, using it to obtain news or to do research for a class paper, and then become passive consumers of Jimmy Fallon's show at night, trying to kick back and relax. Further, activeness is limited by what is presented by the media. Audience members can only seek gratification from media that are accessible to them (Gitlin, 2002, 2007).

Contemporary Applications of Uses and Gratifications Theory

Throughout this book, we try to provide you the most recent research related to a particular theory, even when the theory originated many years ago. UGT originated in the 1970s, and media in the 21st century has certainly changed in the intervening years. To this end, we wish to provide you an overview of how the emergence of new communication technologies has actually revived the theory. Let's examine a few areas where UGT has been applied to social media.

Almost all researchers expect that changes in media will continue to change our communication practices. And, many researchers believe that UGT will be able to explain the ways that people use the internet and SMSs, as well as cell phone technology and other media. As James Shanahan and Michael Morgan (1999) observed, there's an "underlying consistency of the content of the messages we consume and the nature of the symbolic environment in which we live" (p. 199) even if the delivery technology changes. They assert that new technologies have always developed by adopting the message content from the technology that was previously dominant. They argue, for example, that films took their content from serialized literature, radio did the same, and television simply repackaged radio programming. Marshall McLuhan (see **Chapter 23**) noted that new media merely provide new bottles for old wine. The question for Uses and Gratifications researchers is whether the motivations people brought to their use of "old" media still apply to "new" media. In 2022, Grace Falgoust, Emma Winterlind, Prachi Moon, Alden Parker, Heidi Zinzow, and Kapil Chalil Madathil found that UGT provided a useful framework for understanding why young people participated in potentially dangerous TikTok challenges such as ingesting large quantities of salt or other substances. Further, these researchers argued that using UGT can point to ways to discourage such harmful activities.

Theorists are interested in finding out whether social media so alters the message and the experience that UGT no longer applies or has to be radically modified. Access to new technologies has changed and extended our abilities for entertainment and information gathering, and media researchers require greater understanding of the personal and social reasons people have for using media. For instance, some studies that examine social media through the lens of UGT have found the gratifications are different from those asserted by earlier research exploring traditional media. Devadas Menon (2022) studied why people post photographs on Instagram and noted that the gratifications derived from this activity included some that differed from the typologies we reviewed earlier and may be unique to the type of media under investigation. Menon found seven gratifications: disclosure, peer influence, trend influence, self-promotion, diversion, habitual pastime, and social interaction, some of which paralleled those found earlier (e.g., diversion) and others which were new (e.g., self-promotion). Researchers need to reflect on how finding these new gratifications might or might not change the tenets of UGT.

Theory Into Practice • Uses and Gratifications Theory

Theoretical Claim: People are goal-oriented in their use of social media.

Practical Implication: As a social media manager for a startup, Amanda knows she has to provide content that offers opportunities for her target market to satisfy specific gratifications. So, when Amanda puts together content, she's mindful of providing information about the products, but she's also adopted the strategy of soliciting user-generated content (i.e., asking users to send in pictures of themselves with the product) that she can post. User-generated content provides information but also satisfies social needs. In addition, Amanda has become good at adding aesthetically pleasing pictures on Instagram that show the products in a beautiful light, so she can satisfy that need for her market as well. Further, she's occasionally experimented with posting some humorous videos that barely mention the product but provide escape and diversion for viewers with just a short reminder about her product. Amanda has been successful in diversifying her social media use so that a wide variety of gratifications is offered to consumers.

Researchers examining the following topics have found UGT to be a valuable theoretical framework: cell phone use (e.g., Lauricella, Cingel, Blackwell, Wartella, & Conway, 2014; Leung & Wei, 2000); social media use and privacy concerns (e.g., Quinn, 2016; Tan, Kim, & Qi, 2018); digital game playing (e.g., DeSchutter & Malliet, 2014; Lucas & Sherry, 2004; Meier, 2018); social networking websites such as Facebook (e.g., Dainton & Stokes, 2015; Dhir, Chen, & Chen, 2017; Gruzd, Haythornthwaite, Paulin, Gilbert, & del Valle, 2018); and fitness and dating apps (e.g., Lee & Cho, 2017; Sumter, Vandenbosch, & Ligtenberg, 2017). However, some research (e.g., Anderson, 2011; Zeng, 2011) suggests that while the basic logic of UGT holds in studies of a variety of media, the exact list of gratifications will change based on the specific medium. For instance, Isolde Anderson (2011) found that people utilizing CaringBridge, an online resource offering personalized websites for those in need of care (e.g., those undergoing a serious illness, military deployment, adoption, etc.), reported at least two new kinds of gratifications they obtain from using the CaringBridge site. These include spiritual support from a higher power and social presence, or feeling tied to the greater community.

Robert LaRose and Matthew Eastin (2004) suggest that when applying UGT to social media, the theory can be enhanced by the addition of some new variables such as expected activity outcomes and social outcomes. Expected activity outcomes concern what people think they will obtain from the medium. LaRose and Eastin found that people expect that using the internet will improve their lot in life. Social outcomes involve social status and identity. LaRose and Eastin speculate that people may enhance their social status by finding like-minded others through the internet and expressing their ideas to them. The authors also suggest that "perhaps the internet is a means of constantly exploring and trying out new, improved versions of our selves" (p. 373).

A few studies have examined how new media satisfy users' gratifications compared to traditional media (e.g., Ha & Fang, 2012; Min & Kim, 2012). These studies found that media such as email and websites are perceived as superior to traditional media and provide more gratifications for both getting news and mobilizing people to action than do traditional media.

Another study examined the same dilemma Ryan faced in our chapter opening story—whether to see a show at a movie theater or to watch something at home on a streaming service (Tefertiller, 2017). In this study, Alec Tefertiller found that audiences are increasingly selective about which films to see in a theater, and their reasoning didn't depend as much on the factors Ryan considered as it did on the type of film. Viewers wanted to watch visually exciting, involving films on the big screen and not at home, according to the 331 respondents in the study.

Based on the finding that online behaviors reflect the offline cultures in which they're situated (Abeele & Roe, 2011), some researchers have applied UGT to questions of social networking use across cultures. One study found that participants from cultures deemed individualistic were more engaged and more likely to use social networking sites than were those from collectivist cultures (Ifinedo, 2016). Wenzhen Xu, Jiro Takai, and Li Liu (2018) explored the motives people in Japan and China gave for using social networking sites and then compared those to past studies using Western participants. They found that the reasons their respondents gave for seeking out social media sites were generally similar to those given by Western samples, with one specific difference: East Asians expressed motives more concerned with social connections than did Western respondents.

Integration and Critique

UGT had its greatest influence in the 1970s and 1980s. The theory has been investigated largely using quantitative approaches, although some research has been qualitative in nature (DeSchutter & Malliet, 2014).

UGT is especially relevant when discussing how people use newspapers (newspapers are made up of discrete sections, each aimed at a specific type of reader seeking specific types of information) or magazines (publications with very specific, demographically targeted readers) for some definite goals such as making a decision or judgment. Yet, the theory remains relevant today. As Elizabeth Perse (2014) notes, UGT remains one of the most used theories in mass communication research. As we've just discussed, the most recent research using the theory relates to online communication. Thus, the theory is experiencing a resurgence as researchers work to test its premises with new media. However, there are some critiques of UGT, and they mainly focus on the following criteria: logical consistency and heurism.

Integration

Communication Tradition	Rhetorical | Semiotic | Phenomenological | Cybernetic | Socio-Psychological | **Socio-Cultural** | Critical
Communication Context	Intrapersonal | Interpersonal | Small Group | Organizational | Public/Rhetorical | **Mass/Media** | Cultural
Approach to Knowing	**Positivistic/Empirical** | Interpretive/Hermeneutic | Critical

Critique

Scope | **Logical Consistency** | Parsimony | Utility | Testability | **Heurism** | Test of Time

Logical Consistency

Denis McQuail (1984) believes that the theory suffers from a lack of theoretical coherence, in large part because some of the theory's terminology lacks clear definition. For instance, the term *audience* can be difficult to define, especially in the 21st century (e.g., Livingstone, 2015). As Kim Christian Schrøder (2018) observes, in a digital society there's often no "audience" per se but rather multiple actors who participate in the message in various ways. Thus, a singular interpretation of "the audience" can be fraught with challenges because an audience member might be a reader, a viewer, a listener, or a participant with a variety of different responsibilities and engagement levels. Because *audience* forms the foundation for the key concepts of UGT, it's important that the theory provide a consistent definition for it; critics believe the theory does not do so. And, you'll note that we didn't define *audience* as one of the key concepts of the theory, but rather began with the term *active audience* because that's what the theory does.

Students Talking Tech: Briana

Wow, that part in the chapter about how journalism can be irresponsible kind of got to me. It's creepy to think about another country meddling in the U.S. elections and maybe even changing the things we think. I heard a podcast that said that people in the United Kingdom voted to leave the European Union in large part because they believed things they saw on Facebook that weren't even true—like how little the EU had done for them or how dangerous immigrants from Turkey were. The woman on the podcast said the town where people told her about the dangerous immigrants didn't even have any Turkish immigrants, but people believed it did because they saw it on Facebook. I think that's a problem for the theory—if the media can't be trusted, our freedom as choice-making audience members is in trouble.

McQuail (1984) also notes that the theory relies too heavily on the functional use of media, because there are times when the media can be reckless. For example, sometimes sloppy, inaccurate, or unethical journalism makes its way to the audience. There are several well-known examples of reporters who abuse their positions and operate without regard to professional ethics. For instance, there's the infamous case of Janet Cook, a *Washington Post* reporter in the early 1980s who was awarded a Pulitzer Prize for an article she wrote about an 8-year-old African American boy, "Jimmy," who became addicted to heroin when he was 5 years old because his mother and her boyfriend, who were both addicts, gave him drugs. After the Pulitzer was awarded to Cook, she was exposed for completely fabricating the story. Another case involved Jayson Blair, a 27-year-old prolific *New York Times* reporter in the early 2000s who published stories about Iraq war veterans' families, among other topics, that were eventually revealed to be untrue in many respects. He claimed to have interviewed people who actually didn't exist, and he said he traveled to a variety of places when he was, in fact, in New York. These reporters were immediately fired after their unethical behaviors became known, but social media allow for citizen journalism where anyone with an internet connection can publish ideas, allegations, and photos without the checks and balances or the training that accompany professional journalism. This provides for dysfunctional journalism at times, and this is what critics of UGT are concerned about. When we think about how Russia interfered in the 2016 U.S. presidential election by putting untrue messages on Facebook and X, we can see the very real problems associated with dysfunctional media.

Heurism

UGT rates very highly on the criterion of heurism. UGT research has spanned several decades, and the theory has framed a number of research studies. In addition to the early pioneers Katz, Blumler, Gurevitch, and their colleagues, others have employed the theory and its thinking in their research on gay and bisexual men (Miller, 2015); information sources related to college students (Parker & Plank, 2015); texting and public health emergency messaging (Karasz, Li-Vollmer, Bogan, & Offenbecher, 2014), online group buying motivations (Xiao, 2018); young people's participation in TikTok challenges (Falgoust, Winterlind, Moon, Parker, Zinzow, & Madathil, 2022); and digital photo sharing on Facebook (Malik, Dhir, & Nieminen, 2016).

Closing

The value of UGT today rests on its ability to provide a framework for considering the audiences and individual media consumers in contemporary mass communication research and theory. Uses and Gratifications may not be the defining theory in the field of media today, but it serves the discipline well as we seek to understand the intersection between media offerings and consumer choice.

Discussion Starters

 Case-In-Point: Are there choices other than those identified for Ryan Grant to consider in his decision to do something on Friday night? How do these alternatives relate to UGT? Use examples in your response.

Try-It-Your-Selfie: UGT assumes that media present content and consumers choose when and how to consume it. How does the internet threaten to disrupt this model? How might UGT adapt to allow for the transformation of traditional media consumers into online media users and producers? Reflect on your own practices with social media. How good an explanation does UGT provide for them?

1. How active a media consumer are you? Are you always thoughtful in your choice of media content? Do you bring different levels of activeness to different media—newspapers versus radio, for example?

2. UGT has been criticized for failing to consider when media are dysfunctional. Do you agree with this critique? If you do agree, how do you think the theory can adapt to this problem? Is it a fatal flaw in the theory? Explain your answer. If you do not agree, explain why not.

3. What's your position on the effects of the media on media consumers? Do you agree with limited effects or do you think the media's effects are greater than what UGT and other limited effects models suggest? Explain your answer with examples.

4. What difference does genre make to UGT? If Ryan wanted to see an action/adventure film or a quirky romantic comedy, how would that affect his choice between Netflix or Hulu and the local movie multiplex? How does UGT account for media genre and content?

References

Abeele, M. V., & Roe, K. (2011). New life, old friends: A cross-cultural comparison of the use of communication technologies in the social life of college freshmen. *Young, 19*(2), 219-240.

Anderson, I. K. (2011). The uses and gratifications of online care pages: A study of CaringBridge. *Health Communication, 26*(6), 546-559.

Blumler, J. G. (1979). The role of theory in uses and gratifications studies. *Communication Research, 6*(1), 9-36.

Blumler, J. G. (1985). The social character of media gratifications. In K. E. Rosengren, L. A. Wenner, & P. Palmgreen (Eds.), *Media gratifications research: Current perspectives* (pp. 41-60). Sage.

Blumler, J. G., & McQuail, D. (1969). *Television in politics: Its uses and influence.* University of Chicago Press.

Brown, D., Lauricella, S., Douai, A., & Zaidi, A. (2012). Consuming television crime drama: A uses and gratifications approach. *American Communication Journal, 14*(1), 47-61.

Dainton, M., & Stokes, A. (2015). College students' romantic relationships on Facebook: Linking the gratification for maintenance to Facebook maintenance activity and the experience of jealousy. *Communication Quarterly, 63*(4), 365-383.

DeSchutter, B., & Malliet, S. (2014). The older player of digital games: A classification based on perceived need satisfaction. *The European Journal of Communication Research, 39*(1), 67-88.

Dhir, A., Chen, G. M., & Chen, S. (2017). Why do we tag photographs on Facebook? Proposing a new gratifications scale. *New Media & Society, 19*(4), 502-521.

Falgoust, G., Winterlind, E., Moon, P., Parker, A., Zinzow, H., & Madathil, K. C. (2022). Applying the uses and gratifications theory to identify motivational factors behind young adult's participation in viral social media challenges on TikTok. *Human Factors in Healthcare, 2,* 100014.

Gamage, T. C., Tajeddini, K., & Tajeddini, O. (2022). Why Chinese travelers use WeChat to make hotel choice decisions: A uses and gratifications theory perspective. *Journal of Global Scholars of Marketing Science, 32*(2), 285-312.

Gitlin, T. (2002). *Media unlimited: How the torrent of images and sounds overwhelms our lives.* Henry Holt.

Gitlin, T. (2007). *The bulldozer and the big tent: Blind republicans, lame democrats, and the recovery of American ideals.* John Wiley & Sons.

Gruzd, A., Haythornthwaite, C., Paulin, D., Gilbert, S., & del Valle, M. E. (2018). Uses and gratifications factors for social media use in teaching: Instructors' perspectives. *New Media & Society, 20*(2), 475-494.

Ha, L., & Fang, L. (2012). Internet experience and time displacement of traditional news media use: An application of the theory of the niche. *Telematics & Informatics, 29*(2), 177-186.

Herzog, H. (1944). What do we really know about daytime serial listeners? In P. F. Lazarsfeld & F. N. Stanton (Eds.), *Radio research, 1942-1943* (pp. 3-33). Duell, Sloan and Pearce.

Ifinedo, P. (2016). Applying uses and gratifications theory and social influence processes to understand students' pervasive adoption of social networking sites: Perspectives from the Americas. *International Journal of Information Management, 36*(2), 192-206.

Karasz, H. N., Li-Vollmer, M., Bogan, S., & Offenbecher, W. (2014). Targeting young adult texters for public health emergency messages: A Q study of uses and gratifications. In R. Ahmed (Ed.), *Health communication and mass media: An integrated approach to policy and practice* (pp. 115-131). Gower, Ltd.

Katz, E., Blumler, J. G., & Gurevitch, M. (1974). Utilization of mass communication by the individual. In J. G. Blumler & E. Katz (Eds.), *The uses of mass communication: Current perspectives on gratifications research* (pp. 19–32). Sage.

Katz, E., Gurevitch, M., & Haas, H. (1973). On the use of the mass media for important things. *American Sociological Review, 38*(2), 164–181.

Kozman, C., & Melki, J. (2018). News media uses during war. *Journalism Studies, 19*(10), 1466–1488.

LaRose, R., & Eastin, M. S. (2004). A social cognitive theory of internet uses and gratifications: Toward a new model of media attendance. *Journal of Broadcasting & Electronic Media, 48*(3), 358–377.

Lauricella, A. L., Cingel, D., Blackwell, C., Wartella, E., & Conway, A. (2014). The mobile generation: Youth and adolescent ownership and use of new media. *Communication Research Reports, 31*(4), 357–364.

Lee, H. E., & Cho, J. (2017). What motivates users to continue using diet and fitness apps? Application of the uses and gratifications approach. *Health Communication, 32*(12), 1445–1453.

Leung, L., & Wei, R. (2000). More than just talk on the move: Uses and gratifications of the cellular phone. *Journalism & Mass Communication Quarterly, 77*(2), 308–320.

Levy, M., & Windahl, S. (1985). The concept of audience activity. In P. L. Palmgreen, L. A. Wenner, & K. E. Rosengren (Eds.), *Media gratifications research: Current perspectives* (pp. 109–122). Sage.

Livingstone, S. (2015). Audiences and publics: Reflections on the growing importance of mediated participation. In S. Coleman, G. Moss, & K. Perry (Eds.), *Can the media serve democracy: Essays in honour of Jay G. Blumler* (pp. 132–140). Palgrave Macmillan.

Lucas, K., & Sherry, J. L. (2004). Sex differences in video game play: A communication-based explanation. *Communication Research, 31*(5), 499–523.

Malik, A., Dhir, A., & Nieminen M. (2016). Uses and gratifications of digital photo sharing on Facebook. *Telematics and Informatics, 33*(1), 129–138.

McQuail, D. (1984). With the benefits of hindsight: Reflections on uses and gratifications research. *Critical Studies in Mass Communication, 1*(2), 177–193.

McQuail, D., Blumler, J. G., & Brown, J. (1972). The television audience audience: A revised perspective. In D. McQuail (Ed.), *Sociology of mass communication* (pp. 135–165). Penguin Books.

Meier, K. (2018). Journalism meets games: News-games as a new digital genre. Theory, boundaries, utilization. *Journal of Applied Journalism & Media Studies, 7*(2), 429–444.

Menon, D. (2022). Uses and gratifications of photo sharing on Instagram. *International Journal of Human-Computer Studies, 168,* 102917.

Miller, B. (2015). They're the modern-day gay bar: Exploring the uses and gratifications of social networks for men who have sex with men. *Computers in Human Behavior, 51*(PtA), 476–482.

Min, S., & Kim, Y. M. (2012). Choosing the right media for mobilization: Issue advocacy groups' media niches in the competitive media environment. *Mass Communication and Society, 15*(2), 225–244.

Parker, B. J., & Plank, R. E. (2015). The information sources of college students: A uses and gratifications perspective. In A. Manrai & H. Meadow (Eds.), *Global perspectives in marketing for the 21st century* (pp. 69–72). Springer.

Perse, E. (2014). *Uses and gratifications.* Oxford Bibliographies Online.

Quinn, K. (2016). Why we share: A uses and gratifications approach to privacy regulation in social media use. *Journal of Broadcasting & Electronic Media, 60*(1), 61–86.

Rathnayake, C., & Winter, J. S. (2018). Carrying forward the uses and grats 2.0 agenda: An affordance-driven measure of social media uses and gratifications. *Journal of Broadcasting & Electronic Media, 62*(3), 371–389.

Rubin, A. M. (1981). An examination of television viewing motives. *Communication Research, 8*(2), 141–165.

Rubin, A. M., & Step, M. M. (2000). Impact of motivation, attraction, and parasocial interaction on talk radio listening. *Journal of Broadcasting & Electronic Media, 44*(4), 635–654.

Schramm, W. L. (1954). *The process and effects of mass communication.* University of Illinois Press.

Schrøder, K. C. (2018). Audience reception. In P. M. Napoli (Ed.), *Mediated communication* (pp. 105–128). Walter de Gruyter.

Shanahan, J., & Morgan, M. (1999). *Television and its viewers: Cultivation theory and research.* Cambridge University Press.

Sumter, S. R., Vandenbosch, L., & Ligtenberg, L. (2017). Love me Tinder: Untangling emerging adults' motivations for using the dating application Tinder. *Telematics and Informatics, 34*(1), 67–78.

Tan, X., Kim, Y., & Qin, L. (2018). Impact of privacy concern on using mobile social networking apps: An empirical study. *International Journal of Mobile Communication, 16*(3), 286–306.

Tefertiller, A. (2017). Moviegoing in the Netflix age: Gratifications, planned behavior, and theatrical attendance. *Communication & Society, 30*(4), 27–44.

Xiao, L. (2018). Analyzing consumer online group buying motivations: An interpretive structural modeling approach. *Telematics and Informatics, 35*(4), 629–642.

Xu, W., Takai, J., Liu, L. (2018). Constructing the social media uses and gratifications scale on Japanese and Chinese samples: Comparing content to Western conceived scales. *Intercultural Communication Studies, 27*(1), 125–144.

Zeng, L. (2011). More than audio on the go: Uses and gratifications of MP3 players. *Communication Research Reports, 28*(1), 97–108.

CHAPTER 14
Face-Negotiation Theory

*Based on the research of **Stella Ting-Toomey***

"The manner in which individuals conceive of their self-images should have a profound influence on how they construct meanings, interpret speech codes, form relationships, and infer underlying speech rules and premises from their unique identity lens."

—Stella Ting-Toomey

Professor Jie Yang and Kevin Bruner

The first 10 weeks of her academic term in the United States had gone quite well. As a faculty member from China, Professor Jie Yang felt that the communication courses she taught were well attended by students who were eager to participate. Students frequently asked her questions about what Chinese life was like, often concentrating on college life in particular. Jie was more than willing to answer their questions. She, too, had asked students about life in the United States and about what students thought of Chinese–U.S. relations. Although there were a few intercultural difficulties in translation, overall Yang felt that despite what was going on at the national level, excellent interpersonal relationships had been cultivated in such a short period of time. She had every reason to think that things would continue to unfold comfortably.

The professor's instincts, however, proved incorrect. As her intercultural communication class began to prepare for their individual presentations, class members were showing signs of tension. In addition to a written final project on a research topic of their choice, the professor had asked that each student provide a brief oral presentation of what they had studied. Throughout the term, she had listened to student complaints about how certain journal articles were not available online and a lot of concerns about the time crunch. Still, she felt that her assignments were important and despite the complaining, didn't back down.

One evening after class, Kevin Bruner, a graduating senior, challenged Professor Yang directly. He stated that there simply wasn't going to be enough time to complete his final presentation. He believed that the professor was asking too much of him, considering that he had missed 2 weeks of class early in the term because of pneumonia. "It's completely unfair," Bruner lamented. "This is way too much for me right now."

Professor Yang was sympathetic to Kevin's concerns. She agreed that he was under some pressure and reassured him that he was capable of finishing his project despite his past attendance record. "Look, I believe that you're likely an excellent student, Kevin," the professor related. "And obviously I can tell that you're very upset, as I'm sure I would be. But, as a hard worker, I know you want to do your best. You surely don't want me to treat you any differently than the others. I mean, what kind of ethics would that be? Everyone is under pressure here."

Kevin wouldn't hear of it and even surprised himself when he replied: "I don't want to be disrespectful, but this is too much! I refuse to believe that you won't let a good student have some extra time. It's not like I'm not going to do it." Kevin continued by outlining his plan. If she would give him an "incomplete," then he would turn in the final paper 2 weeks after the class ended. "I'll try to give my presentation on what I've written," he continued, "but I'm not sure it's going to be that good."

Professor Yang grew weary with the discussion. "Kevin, you underestimate yourself. You have a few weeks left. I trust you'll be able to finish it thoroughly."

The tone of the conversation quickly changed. "Well, let me see," Kevin interrupted, "I don't think you know the American system yet. I'm being up front and honest about the fact that I won't be able to finish the project. You keep telling me that I will. I know that you've only been in the States a short time and I don't think you know how our system works. You've got to give students a chance. The shakiness in Kevin's voice seemed to now match the passion of his words."

Professor Yang maintained her composure, despite the offending words: "Kevin, in this class, we all have time constraints and outside responsibilities. I, like everyone else, have a lot to finish before the end of the term. But we can't simply shrug off our responsibilities. I can tell that you're a reasonable student and again, don't want to give you a special accommodation that I don't give to other students. Still I know that you were out sick, so how about this: email me a detailed outline of your paper by Monday. You have the weekend to get it into good shape. I will look it over on Monday and then you take my comments and rework the outline into a paper. Instead of the 8–10 page requirement, you can have a 6–8 page requirement. That is the best that I can and will do! But, I will tell you that if you're even 1 minute late with the paper, I won't accept it."

Kevin followed her lead. "Of course, I'll do my best but it'll help a lot if you look over the outline a bit before I turn it in. I know I was pretty rude, so sorry about that."

"I may be from China, but I can say that almost anywhere, professors simply want their students to do their best and have high standards," Professor Yang replied, wondering if she'd made the right decision. "And you should know that I do expect you to do your best. I'm sure you will get everything done on time." As Kevin walked away from his conversation, he couldn't help but think about his confrontational approach to his conflict. He knew that he had work to do and that the boundaries had been clearly set for him by Professor Yang.

Working out conflicts like Kevin's is not easy. In the United States particularly, individuals try to manage their conflicts in a solution-oriented manner, frequently disregarding the other's cultural values or norms. Although people from a number of different cultures share Kevin's approach, some cultures would not endorse his strategies for conflict resolution. In fact, many professors and students would be aghast at the manner in which Kevin attempted to "resolve" his conflict with a classroom "supervisor."

Kevin finds himself in a situation common to many students. Yet, unlike many students, Kevin overtly confronts Professor Yang about her expectations. Although he was given the assignment in the early part of the term, he knows that the remaining time will not be sufficient to complete the task. He then tries to negotiate a different result with his professor. His dominating approach does little to rattle Professor Yang as she calmly tries to gain closure on the conflict.

Kevin's conflict with Professor Yang underscores much of the thinking behind Face-Negotiation Theory (FNT) by Stella Ting-Toomey. The theory is multifaceted, incorporating research from intercultural communication, conflict, politeness, and "facework," a topic we explore later in the chapter. FNT has cross-cultural appeal and application because Ting-Toomey has focused on a number of different cultural populations, including those in Japan, South Korea, Taiwan, China, and the United States. As Ting-Toomey (1988) comments: "Culture provides the larger interpretive frame in which 'face' and 'conflict style' can be meaningfully expressed and maintained" (p. 213). Ting-Toomey asserts that members from different cultural backgrounds have various concerns for the "face" of others. This concern leads them to handle conflict in different ways. These comments form the backdrop to FNT.

Our chapter-opening example of Kevin Bruner and Professor Yang represents the heart of FNT. Representing two different cultural backgrounds, Kevin and his professor seem to have two different interpretations of how to manage the difficulty that Kevin is having with his final project. Kevin clearly desires to turn his work in late, whereas Professor Yang wants him to turn it in with the rest of the class. She tries to ease the conflict between the two by highlighting Kevin's qualities, and clearly she does not want to embarrass Kevin which she easily could have, given her credentials. Rather, she encourages him to work out this conflict by focusing on his ability to get things done, not the remaining time left in the term.

Professor Yang and Kevin Bruner engage in behavior that researchers have termed *face*. Because face is an extension of one's self-concept (Donohue & Druckman, 2022), it has become the focus of much research in a number of fields of study, including management, international diplomacy, anthropology, sociology, and linguistics, among others. In fact, because FNT rests primarily on this concept, let's first interpret the meaning behind the term and then examine some central assumptions of the theory.

About Face

Ting-Toomey bases much of her theory on face and facework. **Face** is clearly an important feature of life, an "exhilarating metaphor" (Ting-Toomey & Dorjee, 2019, p. 332) for a self-image that pervades

face a metaphor for the public image people display

all aspects of social life. More specifically, face is a "claimed source of desired social self-image in a relational or international setting" (p. 319). The concept of face has evolved in interpretation over the years. It originates with the Chinese who have two conceptualizations of face: *lien* and *mien-tzu*, two terms describing identity and ego (Hu, 1944).

Erving Goffman (1967) is generally credited with situating face in contemporary Western research. He noted that face is the image of the self that people display in their conversations with others. Ting-Toomey and her colleagues (Oetzel, Ting-Toomey, Yokochi, Masumoto, & Takai, 2000) observe that face pertains to a favorable self-worth and/or projected other worth in interpersonal situations. People do not "see" another's face; rather, face is a metaphor for the boundaries that people have in their relationships with others. In essence, then, face is the desirable self-image that a person wishes to convey to another based upon society's interpretation of what is appropriate and successful.

Goffman (1967) described face as something that is maintained, lost, or strengthened. At the time of his writing, Goffman did not envision that the term would be applied to close relationships. As a sociologist, he believed that face and all that it entailed was more applicable to the study of social groups. Over time, however, the study of face has been applied to a number of contexts, including close relationships and small groups.

Ting-Toomey incorporates thinking from research on politeness that concludes that the desire for face is a universal concern (Brown & Levinson, 1978). Hongyan Lan (2016) asserts: "Face and facework are regarded as universal phenomena and people of every culture are always negotiating face" (p. 41). Ting-Toomey (1988; 2004) and her colleagues (1991) expands on Goffman's thinking and argues that face is a projected image of one's self and the claim of self-respect in a relationship. She believes that face "entails the presentation of a civilized front to another individual" (Ting-Toomey, 1994a, p. 1) and that face is an identity that two people conjointly define in a relational episode. Further, face is a "socially approved self-image and other-image con-sideration issues" (Ting-Toomey & Chung, 2005, p. 268). Ting-Toomey and her colleague Beth Ann Cocroft (1994) identify face as a "pancultural phenomenon" (p. 310), meaning that individuals in all cultures share and manage face; face cuts across all cultures.

Going back to our opening story, Ting-Toomey and other Face-Negotiation theorists would likely be interested in knowing that Professor Yang is from China and Kevin Bruner was born in the United States. Their cultural backgrounds influence the way that they relate to each other and the way face is enacted. That is, Ting-Toomey believes that although face is a universal concept, there are various representations of it in various cultures. Face needs exist in all cultures, but all cultures do not manage the needs similarly. Ting-Toomey contends that face can be interpreted in two primary ways: face concern and face need. **Face concern** may relate to either one's own face or the face of another. In other words, there is a self-concern and an other-concern. Face concern answers the question, "Do I want attention drawn toward myself or toward another?" **Face need** refers to an inclusion–autonomy dichotomy. That is, "Do I want to be associated with others (inclusion) or do I want dissociation (autonomy)?"

face concern interest in maintaining one's face or the face of others

face need desire to be associated or disassociated with others

Face and Politeness Theory

As we noted earlier, Ting-Toomey was influenced by research on politeness. In a general sense, politeness is concerned with appropriateness of behavior and procedures as they relate to establishing and maintaining harmony in relationships (Kerbrat-Orecchioni, 2012). In particular, politeness theorists (Brown & Levinson, 1978, 1987) contend that people will use a politeness strategy based on the perception of face threat. Politeness theory suggests that a single message can provoke more than one face threat and can both support and threaten face needs simultaneously, and that politeness and face threats influence subsequent messages. Drawing on over a dozen different cultures around the world, and based on field work of at least three languages (Feng, 2015), politeness researchers discovered that two types of universal needs exist: positive face needs and negative face needs. **Positive face** is the desire to be liked and admired by significant others in our lives; **negative face** refers to the desire to be autonomous and unconstrained. Karen Tracy and Sheryl Baratz (1994) note that these "face wants" are necessarily a part of relationships. They support their claim as follows:

positive face desire to be liked and admired by others

negative face desire to be autonomous and free from others

> Recognition of existing face wants explains why a college student who wanted to borrow a classmate's notes typically would not ask for them boldly ("*Lend me your notes, would you?*"), but more frequently would ask in a manner that paid attention to a person's negative face wants ("*Would it be at all possible for me to borrow your notes for just an hour? I'll get them back to you right away*") (p. 288).

Theory at a Glance • Face-Negotiation Theory

How do people in individualistic and collectivistic cultures negotiate face in conflicts? Face-Negotiation Theory is based on face management, which describes how people from different cultures manage conflict negotiation to maintain face. Self-face and other-face concerns explain the conflict negotiation between and among people from various cultures and cultural backgrounds.

Brown and Levinson's research illustrates a dilemma for individuals who try to meet both types of face needs in a conversation. Trying to satisfy one's face need usually affects the other face need. For instance, our opening example shows that Professor Yang wants Kevin to work at achieving his full potential. Her positive face needs, however, are met with the challenges of putting in more time with Kevin, thereby costing her negative face needs. To dig a bit deeper into positive face, we draw upon the work of Sally Hastings and Gina Castle Bell (2018) (who themselves have drawn upon the work of Lim and Bowers, 1991). They identify a few additions related to positive face, including fellowship face.

Facework

When communicators' positive or negative face is threatened, they tend to seek some recourse or way to restore their or their partner's face. Ting-Toomey (1994a), following Brown and Levinson, defines

> **facework** actions used to deal with face needs/wants of self and others

this as facework, or the "actions taken to deal with the face wants of one and/or the other" (p. 8). Ting-Toomey and Leeva Chung (2005) also comment that facework is "about the verbal and nonverbal strategies that we use to maintain, defend, or upgrade our own social self-image and attack or defend (or 'save') the social image of others" (p. 268). In other words, **facework** pertains to communication strategies people use to make whatever they're doing consistent with their face. Ting-Toomey equates facework with a "communication dance that tiptoes" between respect for one's face and the face of another. Catherine Kerbrat-Orecchioni (2012) interprets facework as a way to smooth over the actions that may damage or threaten conversations and relationships (we return to this issue a bit later in the chapter).

Tae-Seop Lim and his co-authors (Lim & Ahn, 2016; Lim & Bowers, 1991) extend the discussion by identifying three types of facework: tact, solidarity, and approbation. First, **tact facework** refers to the extent to which one respects another's autonomy. This allows a per-

> **tact facework** extent to which a person respects another's autonomy

son freedom to act as they please while minimizing any impositions that may restrict this freedom. For instance, Professor Yang engages in tact facework with Kevin while he relates his problems with the course assignments. She could, of course, respond to him by stating that he should just keep quiet and work, but instead she uses tact facework strategies—she asks him for suggestions while avoiding directives.

The second type of facework, **solidarity facework,** pertains to a person accepting the other as a member of an in-group. Solidarity enhances the connection between two speakers. That is, differences

> **solidarity facework** accepting another as a member of an in-group

are minimized and commonalities are highlighted through informal language and shared experiences. For instance, Professor Yang notes that, like Kevin, she has responsibilities, and people cannot simply go back on their duties because of a time crunch. Her conversational style reflects an approachable professor, not one who uses language reflecting a status difference.

The final type of facework is **approbation facework,** which involves minimizing blame and maximizing praise of another. Approbation facework exists when an individual focuses less on the negative aspects of another and more on the positive aspects. Despite her real

> **approbation facework** focusing less on the negative aspects and more on the positive aspects of another

feelings about Kevin's experiences, Professor Yang employs approbation facework by noting that he is a hard worker and an excellent student. She also explains that he has the ability to get everything accomplished. In other words, she recognizes Kevin's positive attributes while avoiding blame.

Our introductory comments on face and facework form an important backdrop to the understanding of FNT. The theory proposed by Ting-Toomey also rests on a number of other issues needing attention and clarification. We begin to unravel the theory in more detail by presenting three key assumptions of FNT.

Assumptions of Face-Negotiation Theory

Several assumptions of Face-Negotiation Theory take into consideration the key components of the theory: face, conflict, and culture. With that in mind, the following guide the thinking of Ting-Toomey's theory:

- Self-identity is important in interpersonal interactions, with individuals negotiating their identities differently across cultures.
- The management of conflict is mediated by face and culture.
- Certain acts threaten one's projected self-image (face).

Facework is closely linked with identity (Ting-Toomey & Dorjee, 2019). The first assumption highlights self-identity, in particular, which we describe as the personal features or character attributes

self-identity personal attributes of an individual

of an individual. In their discussion of face, William Cupach and Sandra Metts (1994) observe that when people meet, they present an image of who they are in the interaction. This image is "an identity that [someone] wants to assume and wants others to accept" (p. 3). **Self-identity** includes a person's collective experiences, thoughts, ideas, memories, and plans. People's self-identities do not remain stagnant, but rather are negotiated in their interactions with others. People have a concern with both their own identity or face (self-face) and the identity or face of another (other-face), a process that is associated with impression management (West & Turner, 2020).

Just as culture and ethnicity influence self-identity, the manner in which individuals project their self-identities varies across cultures. Mary Jane Collier (1998) relates that cultural identity is enacted and "contested in particular historical, political, economic, and social contexts" (p. 132). Delores Tanno and Alberto Gonzalez (1998) note that there are "sites of identity," which they define as "the physical, intellectual, social, and political locations where identity develops its dimensions" (p. 4). Self-identity, therefore, is influenced by time and experience. Consider, for example, the self-identity of a politician as she begins her new term in office. She most likely will be overwhelmed and likely challenged by her new position and the responsibilities that go along with it. Yet, over time and with experience, that frustration will be replaced with confidence and a new perspective on her identity as a representative of others.

Inherent in this first assumption is the belief that individuals in nearly all cultures hold a number of different self-images and that they negotiate these images continuously. Ting-Toomey (1993) states that a person's sense of self is both conscious and unconscious. That is, in scores of different cultures, people carry images that they habitually or strategically present to others. Ting-Toomey believes that how we perceive our sense of self and how we wish others to perceive us are paramount to our communication experiences.

The second assumption of FNT relates to conflict, which is a central component of the theory. Conflict in this theory, however, works in tandem with face and culture. For Ting-Toomey (1994b), conflict can damage the social face of individuals and can serve to reduce the relational closeness between two people. As she relates, conflict is a "forum" for face loss and face humiliation. Conflict threatens both partners' faces, and when there is an incompatible negotiation over how to resolve the conflict (such as insulting the other, imposing one's will, etc.), the conflict can exacerbate the situation. Ting-Toomey states that the way humans are socialized into their culture influences how they will manage conflict. That is, some cultures, like the

United States, value the open airing of differences between two people; other cultures believe conflict should be handled discreetly. We return to these cultural orientations a bit later in the chapter.

This confluence of conflict, face, and culture can be exemplified in our story of Professor Yang and Kevin Bruner. It's apparent that Kevin's conflict with Professor Yang centers on his desire to receive an "incomplete" and her determination to have him complete the project. Because Professor Yang does not acquiesce to his wishes, Kevin tries to maintain his face by agreeing to his professor's compromise. In other words, he expresses a need to preserve his face with his professor.

A third assumption of FNT pertains to the effects that various acts have on one's face. Incorporating politeness research, Ting-Toomey (1988) asserts that face-threatening acts (FTAs) threaten either the positive or the negative face of the interactants. FTAs can be either direct or indirect and occur when people's desired identity is challenged. Direct FTAs are more threatening to the face of others, whereas indirect FTAs are less so.

Ting-Toomey and Mark Cole (1990) note that two actions make up the face-threatening process: face-saving and face restoration. **Face-saving** involves efforts to prevent events that either elicit vulnerability or impair one's image. Face-saving often prevents embarrassment. For instance, French is the primary language of a friend of one of your authors, Rich. Although he does speak fluent English, he periodically splices French phrases into his conversations with others. Because others are usually not prepared for this, Rich introduces him as someone whose primary language is French. In this example, Rich is using a face-saving technique.

> **face-saving** efforts to avoid embarrassment or vulnerability

Face restoration occurs after the loss of face has happened. Ting-Toomey and Cole (1990) observe that people attempt to restore face in response to the events. For instance, people's excuses are face-restoration techniques when embarrassing events occur (Cupach & Metts, 1994). Excuses (*"I thought it was her job"*) and justifications (*"I'm not a morning person"*) are commonplace in face restoration. These face-maintenance strategies and their relationship to each other are represented in **Figure 14.1.**

> **face restoration** strategy used to preserve autonomy and avoid loss of face

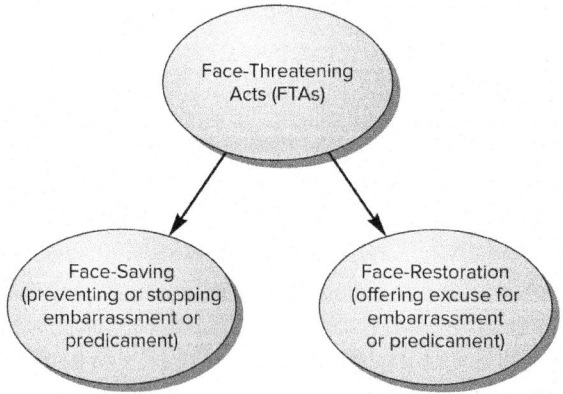

Figure 14.1 Face-Maintenance Framework

Source: Adapted from Ting-Toomey, S., & Cole, M. (1990).

So far, we have detailed face and facework as well as three primary assumptions of FNT. We now turn our attention to a discussion of additional areas of the theory. First, we explore individualism and collectivism, and then we examine how conflict—a central feature of the theory—functions in the theory.

Individualistic and Collectivistic Cultures

Students Talking: Luis

The idea of politeness and facework is the part of the theory that I found most interesting. If I apply it to my parents' different backgrounds and when they get into arguments, then it's easier for me to understand. My mother is a first-generation American, and she came from Japan, which is a country known for being polite. When they argue, she is very concerned that my dad not lose his face and a lot of times, she uses face-restoration tactics. My dad, though, is prone to being more concerned with being "right," and he seems to want to always protect his way of communicating—even though he's usually wrong!

Culture is not a static variable. It's interpreted along many dimensions. To this end, Ting-Toomey examines culture and how it interrelates with face and conflict by employing thinking derived from both Harry Triandis (1972, 1988) and Geert Hofstede (1980, 2003). Both were instrumental in helping us understand cultural variability. In FNT, culture can be organized around two ends of a bipolar continuum: individualism and collectivism (Ting-Toomey, 2010). At one end is a culture that places a premium on the value of individual identity; at the other end is a culture that values group identity. Individualistic cultures are "independent self" cultures, and collectivistic cultures are "interdependent self" cultures. It's important to note that cultures across the world vary in individualism and collectivism (**Table 14.1**). These two dimensions play a prominent role in the way that facework and conflict are managed. But Hofstede reminds us that cultural comparisons can change over time, like changes in population demographics and health opportunities, for instance. Jung-Soo Yi (2021, p. 1) eloquently underscores this notion of cultural change:

> Rapid industrialization and westernization in many Asian countries have forced those populations to adjust to new cultural phenomena in their societies. Latin American nations have struggled through social, political, and financial turbulences. Many incidents of ethical and racial conflicts have made a huge impact to the culture of the United States and have developed new perspectives of cultural evolutions.

Ting-Toomey (2010) along with her colleagues (Ting-Toomey, Gao, Trubisky, Yang, Soo-Kim, Lin, & Nishida, 1991) clarify that individualism and collectivism apply not only to national cultures, but also to co-cultures within national cultures. That is, different racial and ethnic groups within the United States may vary in their individualism and collectivism. For example, Ting-Toomey and her research team observe that whereas many European Americans in the United States identify with individualistic values and beliefs, when first-generation immigrant groups from Mexico or Japan arrive, they tend to retain their collectivistic orientation. Let's explore the concepts of individualism and collectivism a bit further.

Table 14.1   Rankings of Individualism and Collectivism Around the World*

RANK	COUNTRY	RANK	COUNTRY
1	United States	28	Turkey
2	Australia	29	Uruguay
3	Great Britain	30	Greece
4/5	Canada/Netherlands	31	Philippines
4/5	Netherlands	32	Mexico
6	New Zealand	33/35	Yugoslavia
7	Italy	33/35	Portugal
8	Belgium	33/35	East Africa
9	Denmark	36	Malaysia
10/11	Sweden	37	Hong Kong
10/11	France	38	Chile
12	Ireland	39/41	Singapore
13	Norway	39/41	Thailand
14	Switzerland	39/41	West Africa
15	Germany	42	Salvador
16	South Africa	43	South Korea
17	Finland	44	Taiwan
18	Austria	45	Peru
19	Israel	46	Costa Rica
20	Spain	47/48	Pakistan
21	India	47/48	Indonesia
22/23	Japan	49	Colombia
22/23	Argentina	50	Venezuela
24	Iran	51	Panama
25	Jamaica	52	Ecuador
26/27	Brazil	53	Guatemala
26/27	Arab countries		

*The lower the number, the more individualistic the country is rated; the higher the number, the more collectivistic the country is rated.
Source: Hofstede, G. (2003).

In **Chapter 16**, we continue to discuss the parameters and overall meaning of both individualism and collectivism. But, as you may recall, although a theorist may use the same word as another theorist, there may be slight variations in meaning. Thus, we briefly address

individualism a cultural value that places emphasis on the individual over the group

these two terms within an FNT framework. In a general sense, when people emphasize the individual over the group, they are articulating an individualistic perspective. **Individualism** refers to the tendency of people to highlight individual identity over group identity, individual rights over group rights, and individual needs over group needs (Ting-Toomey, 1994b). Individualism is the "I" identity (*"What's in it for me?"*). Individualism was likely the first value that developed in the early formation of the United States because it promotes independence, initiative, and individual achievement. Values that are individualistic highlight freedom, honesty, comfort, and personal equality, among others (Ting-Toomey & Chung, 2005).

Theory-Into-Action

So much is said about the United States being a country where rugged individualism is of paramount importance. After all, for centuries, one of the most famous phrases coming out of the United States was "pull yourself up by your bootstraps," which roughly means that you should get yourself out of your own predicaments and challenges. In other words, people should improve their own situation by their own efforts. Clearly, that's individualist. But it's not that simple. In fact, millions in the United States rely heavily on the culture/federal government for public aid, food stamps, and nutrition assistance, among others. Maybe a proverb normally attributed to African cultures precisely captures the collectivistic spirit of the United States: "It takes a village." Can it be that the reputation of the United States as an individualistic-only culture is faulty or out of date?

As you see, individualism involves self-motivation, autonomy, and independent thinking. Individualism also suggests direct communication with another. Think about Kevin Bruner's comments expressing his desire to complete his project on his own time. He is clearly projecting an individualistic approach to Professor Yang.

According to intercultural communication scholars, individualism is esteemed in the United States (Jandt, 2018). In addition to the United States, a number of other cultures are viewed as individualistic. Australia, Great Britain, Canada, the Netherlands, and New Zealand are examples of individualistic cultures. Italy, Belgium, and Denmark are also considered individualistic. These cultures stress individual achievement and value independence.

Whereas individualism focuses on one's personal identity, collectivism looks outside the self. Ting-Toomey (1994b) comments that collectivism is the emphasis of group goals over individual goals, group obligations over individual rights, and in-group needs over

collectivism a cultural value that places emphasis on the group over the individual

individual wants. **Collectivism** is the "we" identity (*"We can do this"*). People in a collectivistic culture value working together and viewing themselves as part of a larger group. People are obligated to the group, and the self is defined in relationship to others. Collectivistic societies, consequently, value inclusion. Collectivistic values emphasize harmony, respecting parents' wishes, and fulfillment of another's needs, among others. Examples of collectivistic cultures include Indonesia, Venezuela, Panama, Mexico, Ecuador, and Guatemala.

Face Management and Culture

So, how do individualism and collectivism relate to Ting-Toomey's theory? Ting-Toomey and Chung (2005) argue that "members [of cultures] who subscribe to individualistic values tend to be more

face management the protection of one's face

self-face-oriented and members who subscribe to group-oriented values tend to be more other- or mutual-face oriented in conflict" (p. 274). If you are a citizen of an individualistic society, you are more likely to be concerned with controlling your own autonomy and boundaries for behavior. You would also want choices to satisfy self-face needs. Returning back to our opening story: dissatisfied with the assignment and its deadline, Kevin Bruner is seeking autonomy and wants another choice from his professor. Ting-Toomey believes that in individualistic cultures, **face management** is overt in that it involves protecting one's face, even if it comes to bargaining. Kevin Bruner engages in face negotiation that promotes confrontation, and as Ting-Toomey notes, members of individualistic cultures like Kevin will tend to use more autonomy-preserving face strategies in managing their conflict than will members of collectivistic societies.

Collectivistic cultures "are concerned with the adaptability of self-presentation image" (Ting-Toomey, 1988, p. 224). Adaptability, then, allows for interdependent bonds with others (positive face). What this means is that members of collectivistic communities consider their relationship to others when discussing matters and feel that a conversation requires ongoing maintenance by both communicators. For instance, Professor Yang makes efforts to demonstrate her connection to Kevin by empathizing with his conflict ("*I can tell that you're very upset, as I'm sure I would be*"). She also demonstrates a collectivistic orientation by asking whether it would be fair to grant him an extension and not offer the same alternative to other students. Professor Yang, then, as a member of a collectivistic-centered culture, seeks both self-face and other-face needs.

Ting-Toomey believes that conflict is opportunistic when members from two different cultures—individualistic and collectivistic—come together with a disregard for how the other handles conflictual situations. In fact, nearly all communication scholars believe that if managed effectively, interpersonal conflict can result in positive changes in any relationship. To this end, FNT takes into consideration the influence culture has on the way conflict is handled within relationships across the globe.

Students Talking Tech: Cassie

Thinking through this theory, I can't help but think of how many times I've dealt with Tweeters around the world who want to get into an argument with me. I have my strong political opinions about the current American president, and I don't mind putting them online. But I have these people who want to get into an argument with me saying that I should never be a "traitor" to my country and criticize the president. I say, "No," and I can criticize anyone in my country as long as I own it. And, I do. I own my words and don't feel I need to "connect" to anyone else if they disagree with me.

Managing Conflict Across Cultures

The individualistic–collectivistic cultural dimension influences the selection of conflict styles. These styles refer to patterned responses, or typical ways of handling conflict across a variety of communication encounters (Ting-Toomey & Chung, 2005; Ting-Toomey & Oetzel, 2001). The styles include avoiding (AV), obliging (OB), compromising (CO), dominating (DO), and integrating (IN). In **avoiding,** people will try to stay away from disagreements and dodge unpleasant exchanges with others (*"I'm busy"* or *"I don't want to talk about that"*). They pretty much assume the conflict doesn't exist or even use a third party to manage the conflict. The **obliging** style includes a passive accommodation that tries to satisfy the needs of others or goes along with the suggestions of others (*"Whatever you want to do is fine"*). Individuals employing this style value their relationship even more than their personal goal in conflict. In **compromising,** individuals try to find a middle road to resolve impasses and use give-and-take so that a compromise can be reached (*"Why don't I just give up one week of my vacation and you give up a week of yours?"*).

avoiding staying away from disagreements

obliging satisfying the needs of others

compromising a behavior that employs give-and-take to achieve a middle-road resolution

The **dominating** style includes those behaviors that involve using influence, authority, or expertise to get ideas across or to make decisions (*"I'm in the best position to talk about this issue"*). At times, a person who is using the dominating style may be verbally aggressive. Finally, the **integrating** style is used by people to find a solution to a problem (*"I think we need to work this out together"*).

dominating using influence or authority to make decisions

integrating collaborating with others to find solutions

Integrating approaches involve remaining calm during conflictual episodes. As opposed to compromising, integrating requires a high degree of concern for yourself and for others. In compromising, a moderate degree exists.

Ting-Toomey believes that the decision to use one or more of these styles will depend on the cultural variability of communicators. Yet, conflict management necessarily takes into consideration the concern for self-face and other-face. We illustrate this relationship in **Figure 14.2.** From our chapter-opening story, Kevin Bruner, using a dominating style of conflict management, apparently has little concern for the face of his professor (self-face). Professor Yang, however, is more compromising in the way that she handles the conflict with her student (other-face).

Let's continue to clarify Ting-Toomey's discussion of conflict management by identifying the relationship of conflict style to facework. Ting-Toomey notes several relationships between conflict styles and face concern/face need. First, both AV and OB styles of conflict management reflect a passive approach to handling conflicts. A CO style represents a mutual-face need by finding middle-ground solutions to a conflict. Finally, a DO style reflects a high self-face need and a need for control of the conflict, whereas the IN conflict style indicates a high self-face/other-face need for conflict resolution.

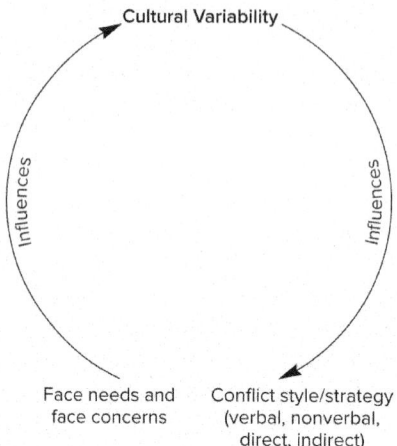

Figure 14.2   Cultural Variability, Face, and Conflict

With respect to comparisons across five cultures (Japan, China, South Korea, Taiwan, and the United States), Ting-Toomey and her research colleagues (1991) discovered the following:

- Members of U.S. cultures use significantly more dominating styles of conflict management (than Japanese and Koreans).
- The Taiwanese report using significantly more integrating styles of conflict management.
- The Chinese and Taiwanese use significantly more obliging conflict styles (than those in the U.S.).
- The Chinese use higher degrees of avoidance as a conflict style than other cultural groups.
- The Chinese use a higher degree of compromising than other cultures.

Additional research also showed that collectivistic cultures (China, Korea, and Taiwan) had a higher degree of other-face concern. Additional research has found that Arab participants preferred more integrating and avoiding styles while U.S. participants identified more with a dominating style (Khakimova, Zhang, & Hall, 2012). Further, in a study examining Ecuador and the United States, James Neuliep and Morgan Johnson (2016) discovered that Ecuadorians preferred avoiding and compromising much more than U.S. respondents. One additional study Croucher, Bruno, McGrath, Adams, McGahan, Suits, and Huckins (2012) employing Ting-Toomey's intercultural conflict styles concluded that among Indians, Hindus prefer integrating and dominating styles, with little use of avoiding and obliging styles. Muslims prefer integrating and compromising styles, with little preference toward dominating and avoiding styles.

As always, we caution you if you're ready to cast a broad cultural net and assume all cultures act one way or another. Yet, it's clear from the research on face and conflict that cultural variability influences the way conflict is managed. Let's revisit our opening discussion of Kevin Bruner and Professor Yang. According to FNT, because Professor Yang is Chinese—interpreted as a collectivistic society—she is likely to compromise with Kevin in their conflict. As you saw in our opening, she does try to compromise by asking him to write up a detailed outline and then alters the page parameters of the assignment. She also has a high degree of other-face concern. Kevin, on the other hand, is very dominating in his conflict style and possesses a great deal of self-face concern.

Finally, a word about culture and negotiation, two primary themes in FNT. While you may be tempted to ask about the differences between and among cultures with respect to negotiation skills, keep in mind that just as conflict styles are culturally based, so are negotiation styles. And, to be sure, negotiation is a "Westernized" concept that is not easily translatable to other cultures, namely, Eastern Societies (Schoen, 2022). What all of this means is that we all need to be cautious and sensible—as we have in this chapter's discussion—when applying terminology across the globe.

Integration and Critique

FNT assumes that people of various cultures are concerned with the presentation of their face. FNT artfully infuses conflict into its framework, explaining why members of two different cultures, for instance, manage conflict differently. Ting-Toomey asserts that different cultural values exist in dealing with conflict, and these conflictual episodes, in turn, are influenced by the face concerns and face needs of communicators. FNT has a rich history, and a research agenda that has been nearly all quantitative in nature, although, depending upon the research questions, Ting-Toomey and Dorjee (2019) believe that the theory can have both interpretive and critical value. Let's close this chapter by identifying two criteria related to the theory's effectiveness: logical consistency and heurism.

Integration

Communication Tradition	Rhetorical | Semiotic | Phenomenological | Cybernetic | **Socio-Psychological** | Socio-Cultural | Critical
Communication Context	Intrapersonal | Interpersonal | Small Group/Team | Organizational | Public/Rhetorical | Mass/Media | **Cultural**
Approach to Knowing	**Positivistic/Empirical** | Interpretive | Critical

Critique

Evaluation Criteria	Scope | **Logical Consistency** | Parsimony | Utility | Testability | **Heurism** | Test of Time

Logical Consistency

Interestingly, FNT has received some clarification from Ting-Toomey herself, prompting refinement of the theory. Recall that the theory rests on the differing experiences and perceptions of individualistic and collectivist cultures. Ting-Toomey uses this foundation to lay out the essence of her theory. At times, however, this cultural dimension may not fully explain cultural differences. In her own research, Ting-Toomey and colleagues (1991) discovered some discrepancies. She found that Japanese respondents showed more concern for self-face than U.S. respondents. In addition, although Ting-Toomey proposes that individualistic cultures are not usually compromising in their conflict styles, the highly individualistic U.S. respondents used a significantly higher degree of compromising when faced with a conflict. In this study, then, the "I" identity of the U.S. respondents was displaced.

Ting-Toomey and Cocroft (1994) respond to these differences in expectations by noting that looking at facework from the individualistic and collectivistic orientation "is a necessary starting point for facework behavior research" (p. 314). The researchers also state that many of the facework category systems in research reflect individualism–collectivism thinking, and therefore, FNT must necessarily begin from this vantage point.

Additional issues surrounding the logical consistency of FNT remain. As we have mentioned, Ting-Toomey (1988) has positioned the theory within the politeness perspective of Brown and Levinson (1987). She incorporates a number of the components of their thinking, including positive face and negative face. Ting-Toomey seems to be aligning herself with the notion that people invoke politeness strategies to reduce face threats (Feng, 2015). Yet, Tracy and Baratz (1994) believe that such labeling in Brown and Levinson's framework "may be too general to capture the face-concern most central to an interactant" (p. 290). That is, other issues pertaining to face concerns exist that are not identified by the researchers. Still, Susan Shimanoff (2009) states that politeness is, indeed, a practice people use to redress FTAs. Ting-Toomey and Cocroft (1994) agree with the fact that Brown and Levinson have presented an original template from which to draw, but report data that demonstrates several problems with their research.

Heurism

Ting-Toomey's FNT continues to spark interest among intercultural communication researchers, making it highly heuristic. Several of the key concepts and features of the theory have been studied. For instance, conflict management examined within the framework of "face" has been investigated with juvenile delinquents and youth-at-risk (Lim, Vadrevu, Chan, & Basnyat, 2012). The theory has been woven into research looking at family problems in Japan (Child, Pearson, & Nagao, 2006) and family secrets (Cho & Sillars, 2015) and it has been expanded beyond intercultural issues and has been applied to forgiveness (Zhang, Ting-Toomey, & Oetzel, 2014) and to crisis communication in Nigeria (George, 2016). In addition, face-threatening acts on Facebook have been investigated (Maíz-Arévalo, 2019) and also been integrated into research that has addressed depression in graduate students (Wiesenthal, Gin, & Cooper, 2023).

Closing

FNT will continue to intrigue communication researchers most importantly, because culture continues to be a critical issue in many societies. In addition, because Ting-Toomey has articulated "conflict" as the central feature of the theory, those interested in the explanation of conflict patterns will find the theory particularly appealing. When people from various cultures have a conflict, understanding how they maintain and negotiate face will have implications beyond the encounter. Ting-Toomey has given us an opportunity to think about how we can mediate the potential difficulties in communication among cultures, and she elegantly presents important information on a world dependent on communication.

Discussion Starters

 Case-In-Point: If he had to do it over again, what communication strategies would you recommend for Kevin Bruner in his conflict with Professor Yang? How might he save his own face and the face of his professor?

Try-It-Your-Selfie: Have you been to one of the countries categorized here as collectivistic? If so, what communication differences did you notice between that culture and U.S. culture?

1. Do you believe that Face-Negotiation Theory relies on people being reasonable agents who are capable of handling conflict? How can conflict become unreasonable? Explain with examples.
2. Interpret the following statement by Ting-Toomey through description and example: "Collectivists need to work on their ethnocentric biases as much as the individualists need to work out their sense of egocentric superiority." Do you agree or disagree with her view?
3. What evidence do you have that face maintenance is a critical part of U.S. society? Use examples in your response.
4. Consider the rapid changes of conflict both in your life and surrounding you. What elements of FNT are most applicable and why?

References

Brown, P., & Levinson, S. (1978). Universals in language usage: Politeness phenomenon. In E. Goody (Ed.), *Questions and politeness* (pp. 56–89). Cambridge University Press.

Brown, P., & Levinson, S. C. (1987). *Politeness: Some universals in language use.* Cambridge University Press.

Child, J. T., Pearson, J. C., & Nagao, M. (2006). Elusive family problems reported by Japanese college students: An examination of theme variation and methodology. *Journal of Intercultural Communication Research, 35,* 45–59.

Cho, M. K., & Sillars, A. (2015). Face threat and facework strategies when family (health) secrets are revealed: A comparison of South Korea and the United States. *Journal of Communication, 65,* 535–557.

Collier, M. J. (1998). Researching cultural identity: Reconciling interpretive and postcolonial perspectives. In D. V. Tanno & A. Gonzalez (Eds.), *Communication and identity across cultures* (pp. 122–147). Sage.

Croucher, S. M., Bruno, A., McGrath, P., Adams, C., McGahan, C., Suits, A., & Huckins, A. (2012). Conflict styles and high–low context cultures: A cross-cultural extension. *Communication Research Reports, 29*(1), 64–73.

Cupach, W. R., & Metts, S. (1994). *Facework.* Sage

Donohue, W. A., & Druckman, D. (2022). Perceptions of face threats in conflict. *International Journal of Conflict Management, 33*(3), 408–424.

Feng, H. (2015). Politeness theory. *The International Encyclopedia of Interpersonal Communication,* 1–13.

George, A. M. (2016). Crisis communication in Nigeria. In A. Swarz & M. Seeger (Eds.), *The handbook of international crisis communication* (pp. 259–268). Wiley & Sons.

Goffman, E. (1967). *Interaction rituals: Essays on face-to-face interaction.* Doubleday.

Hastings, S. O., & Castle-Bell, G. (2018). Facing our heuristic limits: Expanding the terminology for types of positive face. *Communication Quarterly, 66*(1), 96–110.

Hofstede, G. (1980). *Culture's consequences: International differences in work-related values.* Sage.

Hofstede, G. (2003). *Culture's consequences: Comparing values, behaviors, institutions, and organizations across nations.* Sage.

Hu, H. C. (1944). The Chinese concepts of "face." *American Anthropologist, 46*(1), 45–64.

Jandt, F. (2018). *An introduction to intercultural communication.* Sage.

Kerbrat-Orecchioni, C. (2012). From good manners to facework: Politeness variations and constants in France from the classic age to today. In M. Bax & D. Kadar (Eds.), *Understanding relational (im)politeness: Relational linguistic practices over time and across cultures* (pp. 131–153). John Benjamins.

Khakimova, L., Zhang, Y. B., & Hall, J. A. (2012). Conflict management styles: The role of ethnic identity and self-construal among young male Arabs and Americans. *Journal of Intercultural Communication Research, 41*(1), 37-57.

Lan, H. (2016). Conflict management between Japanese and Mexican friendship. In J. G. Oetzel Ting-Toomey & S. Ting-Toomey (Eds.), *The SAGE handbook of conflict communication: Integrating theory, research, and practice* (pp. 41-53). Sage.

Lim, S. S., Vadrevu, S., Chan, Y. H., & Basnyat, I. (2012). Facework on Facebook: The online publicness of juvenile delinquents and youths-at-risk. *Journal of Broadcasting & Electronic Media, 56*, 346-361.

Lim, T., & Ahn, S. (2016). Politeness and social influence. In C.R. Berger & M. E. Roloff (Eds.), *The International Encyclopedia of Interpersonal Communication*, (pp. 1-9). John Wiley & Sons.

Lim, T., & Bowers, J. W. (1991). Facework, solidarity, approbation, and tact. *Human Communication Research, 176*, 415-450.

Maíz-Arévalo, C. (2019). Losing face on Facebook: Linguistic strategies to repair face in a Spanish common interest group. In P. Bou-Franch & P. Blitvich (Eds.), *Analyzing digital discourse* (pp. 283-309). Palgrave Macmillan.

Neuliep, J. W., & Johnson, M. (2016). A cross-cultural comparison of Ecuadorian and United States face, facework, and conflict styles during interpersonal conflict: An application of face-negotiation theory. *Journal of International and Intercultural Communication, 9,* 1-19.

Oetzel, J. G., Ting-Toomey, S., Yokochi, Y., Masumoto, T., & Takai, J. (2000). A typology of facework behaviors in conflicts with best friends and relative strangers. *Communication Quarterly, 48*, 397-419.

Schoen, R. (2022). *Getting to yes* in the cross-cultural-context: "one size doesn't fit all"—a critical review of principled negotiations across borders. *International Journal of Conflict Management, 33*(1), 22-46.

Shimanoff, S. (2009). Facework theories. In S. W. Littlejohn & K. A. Foss (Eds.), *Encyclopedia of communication theory* (pp. 374-377). Sage.

Tanno, D. V., & Gonzalez, A. (1998). Sites of identity in communication and culture. In D. V. Tanno & A. Gonzalez (Eds.), *Communication and identity across cultures* (pp. 3-7). Sage.

Ting-Toomey, S. (1988). Intercultural conflict styles: A face negotiation theory. In Y. Y. Kim & W. B. Gudykunst (Eds.), *Theories in intercultural communication* (pp. 213-238). Sage.

Ting-Toomey, S. (1991). Intimacy expression in three cultures: France, Japan, and the United States. *International Journal of Intercultural Relations, 15*, 29-46.

Ting-Toomey, S. (1993). Communicative resourcefulness: An identity negotiation perspective. In R. Wiseman & J. Koester (Eds.), *Intercultural communication competence* (pp. 72-111). Sage.

Ting-Toomey, S. (1994a). Face and facework: An introduction. In S. Ting-Toomey (Ed.), *The challenge of facework* (pp. 1-14). SUNY Press.

Ting-Toomey, S. (1994b). Managing intercultural conflicts effectively. In L. A. Samovar & R. E. Porter (Eds.), *Intercultural communication: A reader* (pp. 360-372). Wadsworth.

Ting-Toomey, S. (2004). Translating conflict face-negotiation theory into practice. In D. Landis, J. Bennett, & M. Bennett (Eds.), *Handbook of intercultural training* (pp. 215-240). Sage.

Ting-Toomey, S. (2010). Applying dimensional values in understanding intercultural communication. *Communication Monographs, 77*, 169-180.

Ting-Toomey, S., & Chung, L. C. (2005). *Understanding intercultural communication*. Roxbury Publishing Company.

Ting-Toomey, S., & Cocroft, B. A. (1994). Face and facework: Theoretical and research interests. In S. Ting-Toomey (Ed.), *The challenge of facework* (pp. 307-340). SUNY.

Ting-Toomey, S., & Cole, M. (1990). Intergroup diplomatic communication: A face-negotiation perspective. In F. Korzenny & S. Ting-Toomey (Eds.), *Communication for peace: Diplomacy and negotiation* (pp. 77–95). Sage.

Ting-Toomey, S., & Dorjee, T. (2019). *Communicating across cultures*. Guilford Publications.

Ting-Toomey, S., & Oetzel, J. (2001). *Managing intercultural conflict effectively*. Sage.

Ting-Toomey, S., Gao, G., Trubisky, P., Yang, Z., Kim, H. S., Lin, S., & Nishida, T. (December 1991). Culture, face maintenance, and styles of handling interpersonal conflict: A study in five cultures. *The International Journal of Conflict Management, 2*(4), 275–296.

Tracy, K., & Baratz, S. (1994). The case for case studies of facework. In S. Ting-Toomey (Ed.), *The challenge of facework* (pp. 287–306). SUNY Press.

Triandis, H. C. (1972). *The analysis of subjective cultures*. Wiley.

Triandis, H. C. (1988). Collectivism vs. individualism: A reconceptualization of a basic concept in cross-cultural psychology. In G. Verma & C. Bagley (Eds.), *Cross-cultural studies of personality, attitudes, and psychology* (pp. 60–95). Macmillan.

West, R., & Turner, L. H. (2020). *Interpersonal communication* (4th ed.). Sage.

Wiesenthal, N. J., Gin, L. E., & Cooper, K. M. (2023). Face negotiation in graduate school: The decision to conceal or reveal depression among life sciences Ph. D. students in the United States. *International Journal of STEM Education, 10*(1), 35.

Yi, J. S. (2021). Revisiting Hofstede's uncertainty-avoidance dimension: A cross-cultural comparison of organizational employees in four countries. *Journal of Intercultural Communication, 21*(1), 46–61.

Zhang, Q., Ting-Toomey, S., & Oetzel, J. G. (2014). Linking emotion to the conflict face-negotiation theory: A U.S.-China investigation of the mediating effects of anger, compassion, and guilt in interpersonal conflict. *Human Communication Research, 40,* 373–395.

CHAPTER 15
Symbolic Interaction Theory

*Based on the research of **George Herbert Mead***

> *"The self has the characteristic that it is an object to itself, and that characteristic distinguishes it from other objects and from the body."*
>
> —George Herbert Mead

Roger Thomas

Roger Thomas stared at his reflection in the mirror and straightened his tie. He gave himself a last glance and decided that he looked pretty professional. He was a little apprehensive about the new job, but he was excited, too. He had just graduated from Carlton Tech in Omaha, Nebraska, with a degree in engineering, and had landed a terrific job with a great firm in Houston, Texas. This made for a lot of changes in his life. It was a bit overwhelming. He had been born and raised in central Nebraska, and he had never really been in a city bigger than Omaha until he went on his job interviews. Now he was living in Houston! It had all happened so quickly that Roger could almost feel his head spin.

Roger was the first person in his family to graduate from college. As far back as he could recall, his family had been farmers, and although he knew that engineering was something he loved and excelled at, he felt a little confused about how to behave off the farm and in a completely new life. It also didn't help that he was so far from home. Whenever he had felt stressed at Carlton, he had gone home to see his family, and they always made him feel better. He remembered one day during his first year at Carlton when he felt impossibly out of place and uncomfortable. He felt like he didn't know how to act as a college student. He went home for the weekend, and being in a familiar place with his family instantly gave him confidence. When he returned to Carlton on Monday, he felt much more self-assured.

Even though his parents had not attended college themselves, they respected education and communicated this to Roger. They expressed pride in him and his accomplishments. They also told him how his younger brothers looked up to him. This gave Roger confidence in himself, and he liked the idea that he was blazing a new trail for his family. Also, whenever he visited, he appreciated his parents' qualities; they were so calm and steady. As they went about their tasks on the farm, they demonstrated the peace and harmony that Roger wanted to find in his own life and work. After seeing them, he always had a renewed sense of self.

Now Roger decided he would have to get on Face-Time or Zoom soon to see and talk with his family, but today he'd have to face his new job alone. Yet even thinking about his family made him feel a little stronger. He was smiling when he got to the office. He was greeted warmly by the office assistant, who showed him into the conference room. He waited there for the other new hires to join him. By 9:05 A.M., they were all gathered, and their boss came in to give them an orientation speech. While the boss was talking, Roger looked around at his colleagues.

There were 10 new employees in all, and they could not have been more different. Roger was the youngest person in the room by at least 5 years. He was a bit alarmed when he realized that he must be the one with the least experience. He tried to calm himself down. He thought of his parents' pride in him and how his brothers looked up to him. He remembered his favorite teacher telling him that he was one of the best engineering students ever to go through Carlton. Thinking about this helped Roger, and after the boss was finished speaking, he felt prepared to face the challenges of the job. During the break, he even had the confidence to begin talking to one of his new colleagues. He introduced himself and discovered that he didn't have less experience than she did. Helen Underwood explained that she had grown up on a farm outside a small Texas town. After she had graduated from high school, she'd gotten a clerical job with an engineering firm in town. After working for a couple of years, she'd decided to go to school and get an engineering degree. Roger was amazed to meet someone else who came from a farming background. Helen told Roger she was really impressed that he had graduated from Carlton. She knew it had a wonderful reputation, and its internship program was supposed to be the best in the country. Roger replied that he had been really lucky to go there and had loved working at his internship, where he'd learned a great deal. Helen said she was nervous about starting out at this firm, and Roger smiled and nodded.

This conversation made him feel much better about the challenges ahead of him. Even though Helen was in her 40s, they had a lot in common, and they were in the same situation at the firm. Roger thought they would be friends.

As Roger prepares for his first day of work and then listens his boss and meets his new colleague, he's engaging in the dynamic exchange of symbols. George Herbert Mead, who is credited with originating the Theory of Symbolic Interaction, was fascinated with humans' ability to use symbols; Mead proposed that people act based on the symbolic meanings that are communicated in a given situation. People use language and symbols in their everyday communication with others. Symbolic Interaction (SI) Theory centers on the relationship between symbols (or verbal and nonverbal codes) and interactions between people using these symbols. As Lorie Sicafuse and Monica Miller (2010) observe, generally in social encounters people don't respond to stimuli directly but rather to symbolic representations of stimuli negotiated through interactions with others. Implicit in this assertion is Mead's belief that meaning is created in interactions between people, and there is no such thing as objective meaning (Blumer, 1936; Fynbo, 2018; Resch, Knapp, & Schrittesser, 2022).

Ralph LaRossa and Donald C. Reitzes (2009) suggest that Symbolic Interaction Theory is "essentially ... a frame of reference for understanding how humans, in concert with one another, create symbolic worlds and how these worlds, in turn, shape human behavior" (p. 136). LaRossa and Reitzes reflect Mead's contention about the interdependency between the individual and society. In fact, SI forms a bridge between theories focusing attention on individuals and theories attending to social forces. This is the case because of Mead's belief that individuals are active, reflective participants in their social contexts. Many researchers concur with this picture of the relationship between people and social contexts. For example, Tal Laor (2022) investigated the use of social media sites by members of disadvantaged social groups in Israel. Laor found people chose a specific SMS based on their belief that it would enable them to affiliate with a large group of others sharing their cultural codes. Also, showing the relationship between the individual and society, Dana Berkowitz and Linda Liska Belgrave (2010) used SI to frame their study examining how drag queens in South Florida take their marginalized status and turn it into something empowering. In the same vein, Michelle Stoner and her colleagues (Stoner, Aamlid, Hilliard, & Knox, 2019) use SI to interpret how international students in U.S. colleges make sense of conflicting messages from their culture and those from the United States about intercultural romantic relationships. Additionally, Spencer Olmstead, Kathryn Conrad, and Kristin Anders (2018) found that SI was a helpful framework to analyze how first-year college students become socialized to a hook-up culture.

Although Mead was a sociologist, the ideas of SI have been very influential in communication studies because of Mead's emphasis on the importance of messages and communication within social interactions (e.g., Harrigan, Dieter, Leinwohl, & Marrin, 2015; Lucas & Steimel, 2009; Poe, 2012). For instance, Pamela Poe (2012) explored how older people perceive health messages using SI. Julia Moore and Jenna Abetz (2019) used SI to examine threads on Reddit where parents discussed their regrets about having children. Abou Bakar, Hiba Khan, Noor Hazarina Hashim, and Richard Lee (2023) found the theory useful in understanding marketing communication. Jan Fernback (2007) studied online social networking through the lens of SI, and Matthew Loveland and Delia Popescu (2011) extended the theory to an examination of web-based deliberations about democracy.

Several researchers observe, however, that SI is a community of theories, rather than simply one theory. Many theorists refer to the Chicago School and the Iowa School as the two main branches of the theory. We can understand these branches and the overall contentions of SI by a brief examination of the history of the theory's development.

History of Symbolic Interaction Theory

The intellectual ancestors of Symbolic Interaction Theory were the early 20th-century pragmatists, such as John Dewey and William James. The pragmatists believed that reality is dynamic, which was not a popular idea at that time. In other words, they had different ontological beliefs than many other leading intellectuals at the time. The pragmatists advanced the notion of an emerging social structure and insisted that meanings were created in social interaction. They were activists, or critical theorists, who saw science as a way to advance knowledge and improve society.

Two different universities, the University of Iowa and the University of Chicago, employed scholars who subscribed to the ideas of the pragmatists, and at both schools, SI was advanced. At Iowa, Manford Kuhn and his students were instrumental in affirming the original ideas of the theory as well as expanding it over time. The Iowa group advocated some new ways of looking at the self, but their approach was viewed as eccentric, and not universally embraced by researchers. Thus, most of the principles and developments of Symbolic Interaction stemmed from the Chicago School.

Both George Herbert Mead and his friend John Dewey were on the faculty at the University of Chicago. Mead had studied both philosophy and social science, and he lectured on the ideas that form the core of the Chicago School. As a popular teacher who was widely respected, Mead played a critical role in establishing the perspective of the Chicago School, emphasizing the importance of communication to life and social encounters. Interestingly, Mead published very little during his academic career, but after he died, his students collaborated on a book based on his lectures. They titled the book *Mind, Self, and Society* (Mead, 1934), and it contains the foundations of Symbolic Interaction Theory. The name Symbolic Interaction wasn't Mead's idea. One of his students, Herbert Blumer, actually coined the term, but it was clearly Mead's work that began the Chicago branch of the theoretical movement.

The Iowa and Chicago schools of SI diverged on methodological grounds (or epistemology). Mead and his student Herbert Blumer contended that the study of human beings couldn't be conducted using the same methods as the study of other things. They advocated the use of case studies and histories and nondirective interviews. The Iowa School adopted a more quantitative approach to their studies. Kuhn believed that the concepts of SI could be operationalized, quantified, and tested. To this end, Kuhn developed a technique called the Twenty-Statements Self-Attitudes questionnaire. A research respondent taking the twenty-statements test is asked to fill in 20 blank spaces in answer to the question, Who am I? Some of Kuhn's colleagues at Iowa became disenchanted with this way of understanding the self and broke away to form the "new" Iowa School. Carl Couch was one of the leaders of this new school. Couch and his associates began studying interaction behavior through videotapes of conversations, rather than simply examining the twenty-statements test, although they still subscribed to a quantitative approach. Couch's innovative research studies led to a division in the Iowa School between the "old" Iowa school, influenced by Kuhn, and the "new" Iowa school, shaped by Couch's ideas (Carter & Fuller, 2015).

In addition to the Iowa and Chicago branches of SI, several other variations of the theory exist. Further, many theoretical approaches, such as Goffman's dramaturgical analysis (1958) and various theories of identity (e.g., McCall & Simmons, 1978; Stryker, 1980) as well as Berger and Luckman's Social Construction of Reality approach (1966), owe some debt to the central concepts of SI. Despite all the various branches of Symbolic Interaction, Mead's central concepts remain relatively constant in most interpretations. Consequently, we'll examine the basic assumptions and the key concepts that Mead outlined and Blumer later elaborated.

Theory at a Glance • Symbolic Interaction Theory

People are motivated to act based on the meanings they assign to people, things, and events. These meanings are created in the language that people use both in communicating with others (interpersonal context) and in self-talk or their own private thoughts (intrapersonal context). Language allows people to develop a sense of self and to interact with others in the community. As people interact with others they work together co-constructing a sense of reality.

Assumptions of Symbolic Interaction Theory

Symbolic Interaction Theory is based on ideas about language, the self, and the relationship of self to society. To explicate this statement more specifically, we'll spend some time detailing the three major assumptions of SI as well as several basic ideas each assumption supports.

The assumptions framing Symbolic Interactionism (Carter & Fuller, 2015; LaRossa & Reitzes, 2009) include:

- Individuals construct meaning via the communication process.
- Self-concept is a motivation for behavior.
- A unique relationship exists between the individual and society.

Individuals Construct Meaning via the Communication Process

First, Symbolic Interaction Theory assumes that individuals construct meaning through the communication process because meaning is not intrinsic to a thing or idea. It takes people to make meaning. In fact, the goal of interaction is to create shared meaning between the interactors. This is the case because without shared meaning communication is extremely difficult, if not impossible. Imagine trying to talk to a friend if you had to explain your own idiosyncratic meaning for every word you used, and your friend had to do the same. Of course, sometimes we assume that we agree with our conversational partner on a meaning only to discover that we are mistaken (Mom: *"I said get ready as fast as you can."* Joe: *"One hour was as fast as I could get ready."* Mom: *"But I meant for you to be ready in 15 minutes."* Joe: *"You didn't say that!"*), but, frequently, we can count on people having common meanings in a conversation.

This first assumption supports three additional ideas (Blumer, 1969; LaRossa & Reitzes, 2009). They include:

- Humans act toward others on the basis of the meanings those others have for them.
- Meaning is created in interaction between people.
- Meaning is modified through an interpretive process.

Humans Act Toward Others on the Basis of the Meanings Those Others Have for Them

This statement explains behavior as a loop between stimuli and the responses people exhibit toward those stimuli. SI theorists such as Herbert Blumer were concerned with the meaning behind behavior. They looked for meaning by examining psychological and sociological explanations for behavior. Thus, as researchers study the behaviors of Roger Thomas (from our beginning scenario), they see him making meanings that are congruent with the social forces that shape him. For instance, Roger assigns meaning to his new work experience by applying commonly agreed-upon interpretations to the things he sees. When he sees the age of his coworkers, he believes that they have more experience than he does because in the United States, we often equate age with experience. As Roger approaches Helen, he expects to find that she's more experienced than he is. Furthermore, SI researchers are interested in the meaning that Roger attaches to his encounter with Helen (e.g., he is cheered up and believes they will become friends, and thus, his future actions toward her will reflect this belief).

Meaning Is Created in Interaction Between People

As we've noted, SI stresses the intersubjective basis of meaning. Meaning can exist, according to Mead, only when people share common interpretations for the symbols they exchange in interaction. Blumer (1969) clarifies this idea by comparing three ways of accounting for the origin of meaning. One approach regards meaning as being intrinsic to the thing. Blumer states, "Thus, a chair is clearly a chair in itself ... the meaning emanates, so to speak, from the thing and as such there is no process involved in its formation; all that is necessary is to recognize the meaning that is there in the thing" (pp. 3–4). This approach indicates that most people would understand the meaning of a thing in the same way.

A second approach to the origin of meaning sees it as "brought to the thing by the person for whom the thing has meaning" (Blumer, 1969, p. 4). This position supports the popular notion that meanings are in people, not in things. In this perspective, meaning is explained by isolating the psychological elements within an individual that produce meaning. This approach focuses on the connotative meanings that individuals develop based on their idiosyncratic experiences with something. For instance, when Roger hears that Helen was raised on a farm, that information has a different meaning for him than it would have for another coworker who was raised in an urban environment.

Mead's theory takes a third approach to meaning, one that is congruent with many communication researchers' perspectives. SI advances that meaning occurs between people. Meanings are "social products" or "creations that are formed in and through the defining activities of people as they interact" (Blumer, 1969, p. 5). Therefore, if Roger and Helen didn't share a common language and didn't agree on denotations and connotations of the symbols they exchanged, no meaning would result from their conversation. Furthermore, the meanings created by Helen and Roger are unique to them and their relationship. A study of how police officers symbolically construct the meaning of their jobs (Innes, 2002) illustrates this assumption by showing how the police talk differently about murders when they talk to the public and when they talk among themselves.

Meaning Is Modified Through an Interpretive Process

Blumer notes that the interpretive process has two steps. First, interactors point out the things that have meaning. Blumer argues that this part of the process is different from a psychological approach and consists of people engaging in communication with themselves. Thus, as Roger gets ready for work in the morning, he communicates with himself about things that are meaningful to him. The second step involves communicators selecting, checking, and transforming the meanings in the context in which they find themselves. When Roger talks with Helen, he listens for her remarks that are relevant to the areas he's decided are meaningful. Further, in his interpretation process, Roger depends on shared social meanings that are culturally accepted. Roger and Helen are able to converse relatively easily because they both come from similar co-cultures.

A study by Sidharth Muralidharan, Carrie La Ferle, and Sanjukta Pookulangara (2018) examined how meanings are influenced by cultural values. The authors found that religious symbols in ads encouraging intervention in domestic violence in India were effective. These ads tended to positively influence willingness to report abuse for the Hindu adults participating in the study, especially those who were devout. In a different social context, the ads probably wouldn't be as powerful.

Self-Concept Is a Motivation for Behavior

The second assumption of SI focuses on the importance of the **self-concept,** or the relatively stable set of perceptions that people hold of themselves. When Roger (or any social actor) asks the question "Who am I?" the answer relates to self-concept. The characteristics

self-concept a relatively stable set of perceptions people hold about themselves

Roger acknowledges about his physical features, roles, talents, emotional states, values, social skills and limits, intellect, and so forth make up his self-concept. This notion is critical to Symbolic Interaction Theory. Furthermore, SI describes the ways in which people develop self-concepts. The theory pictures individuals with active selves, grounded in social interactions with others (see **Figure 15.1**). This assumption suggests two conclusions, according to LaRossa and Reitzes (2009):

- Individuals develop self-concepts through interaction with others.
- Self-concepts provide an important motive for behavior.

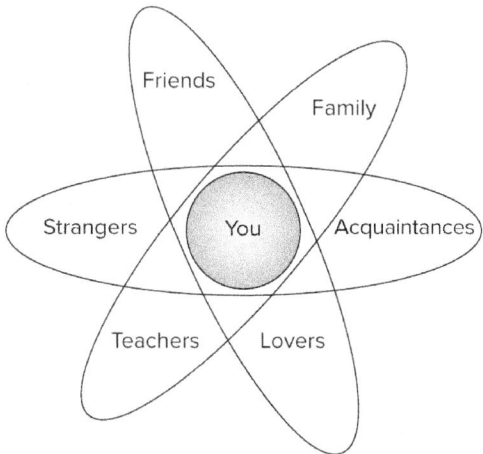

Figure 15.1 How the Self-Concept Develops

Individuals Develop Self-Concepts Through Interactions with Others

This conclusion asserts that it's only through contact with others that we develop a sense of self. People are not born with self-concepts; they learn them through interactions. According to SI, infants have no sense of an individuated self. During the first year of life, children begin to differentiate themselves from their surroundings. This is the earliest development of the self-concept. Mead's theory contends that this process continues through the child's acquisition of language and the ability to respond to others and internalize the feedback they receive. Roger has a sense of self because of his contacts with his family, his teachers, his colleagues, as well as friends and others he interacts with. Their interactions with him tell him who he is. Granberg (2011) found that, in the context of identity change following weight loss, it wasn't enough to successfully lose weight and sustain the weight loss over time. The identity change was not really solidified until interactions with others validated the new identity.

Students Talking Tech: Danielle

The chapter didn't talk much about how the self could develop through feedback you get online, but the comment about weight loss kind of got to me. I'm in an online weight loss support group, and I've really come to see myself as a normal-sized person through that group. I lost about 80 pounds before I came to college, and I needed help to see myself as others were seeing me at school. I considered myself a formerly fat person, but no one here on campus knew that about me. If I hadn't had my online group, I think I would have struggled more. They helped me get a good sense of my identity without ever meeting me IRL.

Self-Concepts Provide an Important Motive for Behavior

The notion that beliefs, values, feelings, and assessments about the self affect behavior is a central tenet of SI. Mead argues that because human beings possess a self, they are provided with a mechanism for self-interaction (i.e., intrapersonal communication). This mechanism is used to guide behavior and conduct. It's also important to note

self-fulfilling prophecy a prediction about yourself causing you to behave in such a way that it comes true

that Mead sees the self as a process, not as a structure. Having a self forces people to construct their actions and responses, rather than simply expressing them. So, for instance, if you feel great about your abilities in your communication theory course, then it's likely that you will do well in the course because your behaviors will be motivated by this self-concept. In fact, it's likely that you'll feel confident in all of your courses. This process is often called the **self-fulfilling prophecy,** or the self-expectations that cause a person to behave in such a way that their expectations are realized. When Roger believes that engineering is something he excels in, he's setting himself up to make a self-fulfilling prophecy about his performance at his new job. When he behaves in ways congruent with that self-fulfilling prophecy and does excellent work, the prophecy comes true and reinforces success in future behavior.

Ellis Rosen/CartoonStock Ltd

A Unique Relationship Exists Between the Individual and Society

The final assumption of SI pertains to the relationship between complete individual freedoms and absolute social constraint on the individual. Mead and Blumer took a middle position on this issue. They tried to account for both order and change in social processes. Symbolic Interaction Theory recognizes that both social structural and personal factors influence behavior (Andersen, 2020). Individuals are influenced in their thoughts and actions by social forces and processes, by shared meanings and symbols, and by individual agency and self-motives (Mead, 1934). Donald Reitzes and his colleagues (Reitzes, DePadilla, Sterk, & Elifson, 2010) noted that because of this middle position, acknowledging the importance of both the individual and society, SI allowed them to examine how people quit smoking by providing an integrated framework of social factors and individual factors contributing to the "smoker identity." Don Haviland, Nathan Alleman, and Cara Cliburn Allen (2017) used this framework to examine how full-time but non-tenure-track faculty at U.S. colleges and universities experience collegiality with their peers and with tenure-track faculty members. They were guided both by personal factors relating to congeniality and by structural, social factors relating to their low status vis-à-vis those on the tenure track.

Conclusions related to this assumption include the following:

- People and groups are influenced by cultural and social processes.
- Social structure is worked out through social interaction.

People and Groups Are Influenced by Cultural and Social Processes

This means that social norms constrain individual behavior. For instance, when Roger gets ready for his first day at his new job, he selects a navy suit, a white oxford shirt, and a burgundy and blue striped tie. His preferred mode of dress would be jeans and a flannel shirt, but he chooses clothing that he feels will be socially appropriate for this job. Furthermore, culture strongly influences the behaviors and attitudes that we value in our self-concepts. For instance, in the United States, people who see themselves as assertive are likely to be proud of this attribute and reflect favorably on their self-concept. This is the case because the United States is an individualistic culture that values assertiveness and individuality. In some Asian cultures, cooperation and community are highly valued, and the collective is more important than the individual. Thus, Asians who see themselves as assertive might feel ashamed of having assertiveness be part of their self-concept.

Mary Roffers (2002) notes that a college assignment to design a personal website was very difficult for a Hmong student in her class. The student explained that talking about oneself is not approved of in his culture and putting information about himself on the website felt inappropriate. In contrast, Jenny Davis (2010) found that people in the United States enjoyed creating a personal website for the control it gave them in negotiating their self-presentation.

Social Structure Is Worked Out Through Social Interaction

The second conclusion mediates the position taken by the first conclusion. SI acknowledges that individuals can modify social situations. For example, many U.S. workplaces have instituted "casual Fridays," when employees wear casual clothing rather than the typical office clothing that Roger selected. When COVID-19 forced employees to work remotely many enjoyed working in pajamas, often not turning on their cameras for meetings. In this way, the participants in the interaction modify the structure and aren't completely constrained by it. In other words, SI theorists believe that humans are choice makers. In our opening scenario, Roger chooses to introduce himself to Helen; he is not bound to do so by forces outside his control. In making choices, Roger exerts his individuality and demonstrates that he isn't completely constrained by culture or situation.

Students Talking: Francisco

I liked the example about casual Fridays in the workplace. When I went to work at my current job there were a lot of rules that I found difficult to follow. At first, it seemed like I was going to be constrained by the structure I found on the job. But then I started to talk to people at work, and we were able to make a couple of changes that makes working there a lot better for me. So, I think that idea in SI is true. People have some constraints on them because of social rules, but people also have free will and can change some of the rules. If I hadn't been able to change things, I probably would have quit—which would have been my choice, too.

Key Concepts

Previously we stated that the book outlining Mead's thinking was titled *Mind, Self, and Society.* The title of the book reflects the three key concepts of SI. We describe each here, noting how other important issues relate to these basic three. You'll notice that the three key concepts overlap to some extent.

Mind

Mead defines **mind** as the ability to use symbols that have common social meanings, and he believes that humans develop minds through interaction with others. However, people cannot really interact with others until they learn **language,** or a shared system of verbal and nonverbal symbols organized in patterns to express thoughts and feelings. Language depends on what Mead calls **significant symbols,** or those symbols that evoke basically the same meaning for many people. For instance, most people in the United States probably would hold similar meanings for words like *democ-*

mind the ability to use symbols with common social meanings

language a shared system of verbal and nonverbal symbols

significant symbols symbols whose meaning is generally agreed upon by many people

racy, 9/11, and *individuality.* When U.S. citizens use these words, the terms act as significant symbols. Speakers using significant symbols are relatively confident that others understand what they mean and they are also confident they can anticipate how their listeners will respond when they hear them. When Marti tells her friend Becca that she's afraid a candidate for president is a threat to democracy she expects Becca to understand what she's worried about without further explanation, for example.

By using language and interacting with others, we develop what Mead calls *mind,* and this enables us to create an interior setting for the society that we see operating outside us. Thus, mind can be viewed as the way people internalize society. Yet mind doesn't just depend on society. Mead suggests that there's a reciprocal relationship: mind both reflects and creates the social world. As people learn language, they learn the social norms and cultural values that constrain them. But they also learn ways to shape and change that social world through interaction. When children in the United States learn to talk, they may learn to say "please" and "thank you" as cultural indicators of politeness. But people also are free to create unique, personal ways of expressing politeness or anything else they wish to communicate. So, in one family, for instance, saying "mayberry" means the speaker is making an effort to be especially nice to the listener. The family created this special meaning because of a TV show they laughed at together showing characters acting overly polite and kind.

Closely related to the concept of mind is the notion of **thought,** which Mead conceives of as an inner conversation. While Roger, in

thought an inner conversation

our opening story, prepares for his new job, he reviews all the experiences that brought him to that time and place. He thinks about his family's example and support, he remembers a favorite teacher, and he tells himself that he will be successful at this challenge. Through this intrapersonal conversation, Roger sorts out the meaning of his new situation. Society is still important, though, because Mead holds that without social stimulation and interaction with others, people wouldn't be capable of holding inner conversations or sustaining thought.

According to Mead, one of the most critical activities that people accomplish through thought is **role taking,** or the ability to symbol- ically think of oneself as the imagined self of another person. This

role taking the ability to put oneself in another's place

process is also called perspective taking because it requires that one suspend one's own perspective on an experience and instead view it from the imagined perspective of another. For example, if Helen thought about Roger after their meeting and reflected on how he must have felt to be younger than the other new employees, then she would be role taking. Whenever we try to imagine how another person might view something or when we try to behave as we think another would, we are role taking. Mead suggests that role taking is a symbolic act that can help clarify our own sense of self, even as it allows us to develop the capacity for empathy with others. Christopher Johnson, Jordan Kelch, and Roxanna Johnson (2017) use this concept in exploring techniques for connection between caregivers and people living with dementia. SI has also been used in conjunction with the systems approach (discussed in **Chapter 3**) to illustrate how non-Indigenous researchers can utilize a kind of role-taking to "discover valuable insights into Indigenous perspectives and interpretations that might otherwise be ignored or neglected" (Ali et al., 2022, p. 197).

Self

As Mead theorizes about self, he observes that through language, people have the ability to be both subject and object to themselves. As subject, we act, and as object, we observe ourselves acting. Mead calls the subject, or acting self, the **I** and the object, or observing self, the **Me.** The I is spontaneous, impulsive, and creative, whereas the Me is more reflective and socially aware. The I might want to go

> **I** the acting self who is spontaneous, impulsive, and creative
>
> **me** the object that we observe who is more reflective and socially aware

out and party all night, whereas the Me might exercise caution and acknowledge the homework assignment that should be done instead of partying. Mead sees the self as a process that integrates the I and the Me.

Further, when we reflect on ourselves, we do so from the perspective of others, and Mead defines **self** as the ability to see ourselves as others do. From this, you can see that Mead doesn't believe that self comes from introspection or simply thinking on one's own. For Mead, the self develops from a particular kind of role taking—that is, imagining how we look to another person. Borrowing a concept originated by the sociologist Charles Cooley in 1912, Mead refers

> **self** imagining how we look to another person
>
> **looking-glass self** another term for "self"; our ability to see ourselves as another sees us

to this as the **looking-glass self,** or our ability to see ourselves in the reflection of another's gaze. Cooley (1972) believes that three principles of development are associated with the looking-glass self: (1) we imagine how we appear to others, (2) we imagine their judgment of us, and (3) we feel hurt or pride based on these self-feelings. We learn about ourselves from the way others treat us, view us, and label us. For example, when Rachel participated in the Great Bike Ride Across Iowa, she rode a three-speed bike 523 miles, from one end of the state to the other. The ride took 1 week, and after about 3 days, she felt she could not pedal a minute longer. But just as she was about to give up, a man biked up beside her and said, "You're amazing for going on this ride on a three-speed bike. You are just great. Keep it up." As he pedaled off, she straightened up and thought to herself, "Well, I guess I *am* amazing. I can finish this ride!" The label the man gave her actually changed her feelings of exhaustion and made her see her accomplishments and herself differently and more positively; and that gave her renewed strength to finish the ride.

Theory Into Practice • Symbolic Interaction Theory

Theoretical Claim: Individuals develop self-concepts through their interactions with others.

Practical Implication: Because Brianna has received ongoing supportive messages about her abilities in class, on the lacrosse field, and in the church choir, she evaluates herself as able to meet nearly any challenge presented to her.

Some researchers (e.g., Gilovich & Keltner, 2015; Ishida, 2016; Zhao, 2017) refer to the looking-glass self as *reflected appraisals*, or people's perceptions of how others see them. Joanne Kaufman and Cathryn Johnson (2004) used the concept of reflected appraisals to examine how gay men and lesbians develop and manage their identities. They argue that SI is a much better framework for understanding identity construction than the stage models often applied to gay and lesbian identity development. Kaufman and Johnson found that gay men and lesbians who experienced positive reflected appraisals (in regard to their sexual identity) had an easier time developing their identity than did those people who experienced negative reflected appraisals.

Reflected appraisals relate to another type of prophecy that affects people's self-concept and behavior. Earlier in the chapter, we spoke of self-fulfilling prophecies as being self-expectations that affect behaviors. For example, Roger tells himself repeatedly that he will succeed at his job and then engages in behaviors that are congruent with his expectations of success. In turn, these behaviors likely ensure that he will succeed. By the same token, negative self-talk can create situations where predictions of failure come true. The second type of prophecy is called the **Pygmalion effect,** and it refers to the expectations of others influencing one's actions, and self-concept, as in our example of Rachel on the bike ride across Iowa.

> **Pygmalion effect** living up to or down to another's expectations of us

The name for this prophecy comes from the myth of Pygmalion, on which a play of the same name and the musical *My Fair Lady* were based. In the play and the musical, the main character, Eliza, states that the difference between an upper-class lady and a poor flower girl is not in her behavior, but in how others treat her. This phenomenon was tested in a classic study by Robert Rosenthal and Lenore Jacobson (1968). Rosenthal and Jacobson told elementary-school teachers that 20 percent of their students were gifted. But the names of these "gifted" students were simply drawn at random. Eight months later, these students showed significantly greater gains in IQ compared to the rest of the children in the class. Rosenthal and Jacobson concluded that this was the result of teachers' expectations (and behaviors based on these expectations) toward the "gifted" children. These children responded to the way the teachers acted toward them, and actually did better in the classroom as a result.

Society

Mead argues that interaction takes place within a dynamic social structure that we call culture or society. Mead defines **society** as the web of social relationships that humans create. Individuals engage in society through behaviors that they choose actively and voluntarily. Society thus features an interlocking set of behaviors that individuals continually adjust. Society exists prior to any given individual, but is also created and shaped by individuals, acting in concert with others (Bern-Klug, 2009).

> **society** the web of social relationships humans create and respond to

Society, then, is made up of individuals, and Mead talks about two specific parts of society that affect the mind and the self. Mead's notion of **particular others** refers to the individuals in society who are significant to us. These people are usually family members, friends, work colleagues, and supervisors. We look to particular others to get a sense of social acceptability and a sense of self. When Roger thinks of his parents' opinion of him, he is deriving a sense of self from particular others. The identity of the particular others and the context influence our sense of social acceptability and our sense of self. Often, the expectations of some particular others conflict with those of others. For example, if Roger's family wants him to work hard and be successful, whereas his friends want him to party and ignore work, he's likely to experience conflict.

> **particular others** individuals who are significant to us

The **generalized other** refers to the viewpoint of a social group or the culture as a whole. It's given to us by society, and "the attitude of the generalized other is the attitude of the whole community" (Mead, 1934, p. 154). The generalized other provides information about roles, rules, and attitudes shared by the community. The generalized other also gives us a sense of how other people react to us and of general social expectations. This sense is influential in developing a social conscience. The generalized other may help

> **generalized other** the attitude of the whole community

mediate conflicts generated by conflicting groups of particular others. If your friends want you to party and your family wants you to study, for instance, your knowledge of the importance of a college degree in your culture may lead you to decide in favor of your parents' position.

Integration and Critique

Symbolic Interaction Theory has been a powerful theoretical framework for over 90 years. The theory is mainly derived from qualitative work, although many recent applications have been quantitative in nature. The theory provides striking insights about human communication behavior in a wide variety of contexts. Without a doubt, SI is a clear and coherent theory because it begins with the role of the self and progresses to an examination of the self in society. Yet the theory is not without its critics. As you think about Mead's theory, consider three areas of evaluation: scope, utility, and testability.

Integration

Communication Tradition	Rhetorical	**Semiotic**	**Phenomenological**	Cybernetic	Socio-Psychological	Socio-Cultural	Critical
Communication Context	**Intrapersonal**	**Interpersonal**	Small Group/Team	Organizational	Public/Rhetorical	Mass/Media	Cultural
Approach to Knowing	Positivistic/Empirical	**Interpretive/Hermeneutic**	Critical

Critique

Evaluation Criteria	**Scope**	Logical Consistency	Parsimony	**Utility**	**Testability**	Heurism	Test of Time

Scope

Some critics (e.g., Van Krieken et al., 2014) complain that Symbolic Interaction Theory is too broad to be useful. The theory covers too much ground, these critics assert, to fully explain specific meaning-making processes and communication behaviors. It's helpful to have a theory that explains a wide variety of human behavior, but when a theory purports to explain everything, it will be vague and difficult to apply. Critics say SI tries to do too much, and its scope needs to be refined. In response to this criticism, proponents explain that SI isn't one unified theory; rather, it's a framework that can support many specific theories. On the other hand, SI could be faulted for being too narrow to apply to new developments in communication such as social media. Relatively few studies have utilized SI to frame studies examining text messages Instagram, Facebook, and so forth.

Utility

The second area of criticism concerns the theory's utility. SI has been faulted as not useful for two reasons (Charmaz, 2014). First, although Mead lectured about both the individual (mind) and society, critics think SI focuses too much on the individual, and second, the theory ignores some important concepts that are needed to make its explanation complete. In the first case, critics observe that the theory's focus on the individual's power to create reality ignores the extent to which people live in a world not of their own making. SI theorists regard a situation as real if the actors define it as real. But Erving Goffman (1974) comments that this notion, although not completely untrue, ignores physical reality. For instance, if Roger and his parents agreed that he was an excellent engineer and that he was doing a wonderful job at his new firm, that would be reality for them. Yet it wouldn't acknowledge the fact that Roger's boss perceived his skills as inadequate and fired him. Others counter by citing that the theory treads a middle ground between freedom of choice and external constraint (Sicafuse & Miller, 2010). SI recognizes the validity of external constraint, but it also emphasizes the importance of shared meanings for creating reality. Further, as we observed previously, SI provides a bridge connecting the individual with society. Thomas Billard and Erique Zhang (2022) examine how this bridge offered by SI helps forge new theoretical ground in their study of how transgender people are represented in the media.

Secondly, some critics argue that the theory ignores important concepts such as emotion and self-esteem, stating that SI doesn't explain the emotional dimension of human interaction. Further, critics note that SI discusses how we develop a self-concept, but it doesn't have much to say about how we evaluate ourselves. These deficiencies render the theory less useful than it should be in explaining the self. Regarding self-esteem, Symbolic Interactionists agree that it's not a focus of the theory. But they point out that this isn't a flaw in the theory; it's simply beyond the bounds of what Mead chose to investigate. With reference to the lack of attention to the emotional aspects of human life, proponents respond that although Mead does not emphasize these aspects, the theory can accommodate emotions. In fact, some researchers have applied the theory to emotions with success. For instance, Paul Maciejewski, Francesca Falzarano, Wan Jou She, Wendy G. Lichtenthal, and Holly G. Prigerson (2022) used an SI perspective to examine bereavement and found that SI was a useful model for framing their study. Julia Moore and Jenna Abetz (2019) used an SI framework to study the emotion of regret, and they argue that the theory has a great deal to offer in understanding emotion, stating:

> Symbolic interactionists challenge the conventional view that emotions are experienced in reaction to external stimuli. Instead, social conventions and norms shape individuals' experiences of emotions (Fields et al., 2006). Moreover, feeling an emotion is different from expressing an emotion outwardly, and individuals often self-regulate their emotional displays to align with feeling rules, or latent guidelines about what a person should feel in a particular situation that are shared within a particular culture or group (Hochschild, 1979). Emotions are also linked to group membership, where the self is located in relation to others, and help individuals organize through *role-taking emotions*. When experiencing these emotions, a person imagines the reactions of generalized others and then feels and regulates their emotions in response. Role-taking emotions of guilt, shame, and embarrassment are experienced in reaction to an imagined untoward response from the generalized other. These emotions foster self-control, which, in turn, accomplishes social control (Fields et al., 2006; Shott, 1979). (p. 394)

Testability

With regard to testability, critics comment that the theory's broad scope renders its concepts vague. When so many core concepts are nominal, meaning they're not directly observable, it's difficult to test a theory. Again, the response to this critique argues that Mead's theory is a general framework, not a single theory. In more specific theories derived from SI, such as Role Theory, for example, the concepts are more clearly defined and are capable of falsification, satisfying the criterion of testability.

Closing

SI has critics, but it still remains an enduring theory. It supports research in multiple contexts, and it is constantly being refined and extended. Further, it's one of the leading conceptual tools for interpreting social interactions, and its core concepts provide the foundation for many other theories communication researchers find useful. Thus, because Symbolic Interaction Theory has stimulated so much conceptualizing, it has accomplished a great deal of what theories aim to do.

Discussion Starters

 Case-In-Point: Discuss Roger Thomas's initial reactions to his new job in Houston. How do they specifically relate to his sense of self? How useful is SI in understanding what's going on with Roger as he prepares for his new job?

Try-It-Your-Selfie: The theory hasn't been applied to social media too much. Construct a hypothesis that you could test or a study you could conduct that would utilize SI in examining social media use. Discuss how you might proceed with the study and how SI would be useful (or not) in framing your study.

1. Do you believe Mead's argument that one cannot have a self without social interaction? Would a person raised in relative isolation, for example, have little to no sense of self? Explain your answer.
2. Has there been a time in your life when your sense of self changed dramatically? If so, what contributed to the change? Did it have anything to do with others in your life?
3. Do you agree with the emphasis that Mead places on language as a shared symbol system? Is it possible to interact with someone who speaks a completely different language? Explain your position.
4. One of the criticisms of SI is that it puts too much emphasis on individual action and not enough emphasis on the constraints on individuals put upon them by society and their social position. What do you think about this criticism? Give examples to support your position.

References

Ali, T., Buergelt, P. T., Maypilama, E. L., Paton, D., Smith, J. A., & Jehan, N. (2022). Synergy of systems theory and symbolic interactionism: A passageway for non-Indigenous researchers that facilitates better understanding Indigenous worldviews and knowledges. *International Journal of Social Research Methodology, 25*(2), 197–212.

Andersen, M. L. (2020). *Sociology: The essentials,* 10th ed. Cengage.

Bakar, A., Khan, H., Hazarina Hashim, N., & Lee, R. (2023). The strange bedfellows of packaging cues and religiosity. *Journal of Global Scholars of Marketing Science, 33*(1), 31–44.

Berger, P. L., & Luckman, T. (1966). *The social construction of reality: A treatise in the sociology of knowledge.* Anchor Books.

Berkowitz, D., & Belgrave, L. L. (2010). "She works hard for the money": Drag queens and the management of their contradictory status of celebrity and marginality. *Journal of Contemporary Ethnography, 39*(2), 159–186.

Bern-Klug, M. (2009). A framework for categorizing social interactions related to end of life care in nursing homes. *The Gerontologist, 49*(4), 495–507.

Billard, T. J., & Zhang, E. (2022). Toward a transgender critique of media representation. *Journal of Cinema and Media Studies, 61*(2), 194–199.

Blumer, H. (1936). Social attitudes and nonsymbolic interaction. *Journal of Educational Sociology, 9,* 515–523.

Blumer, H. (1969). *Symbolic interactionism: Perspective and method.* Prentice Hall.

Carter, M. J., & Fuller, C. (2015). Symbolic interactionism. *Sociopedia.isa.*

Charmaz, K. (2014). *Constructing grounded theory.* Sage.

Cooley, C. H. (1972). *Human nature and social order.* Free Press.

Davis, J. (2010). Architecture of the personal interactive homepage: Constructing the self through MySpace. *New Media & Society, 12*(7), 1103–1119.

Fernback, J. (2007). Beyond the diluted community concept: A symbolic interactionist perspective on online social relations. *New Media & Society, 9*(1), 49–69.

Fynbo, L. (2018). The uncommon ground: Drunk drivers' self-presentations and accountings of drunk driving. *The Qualitative Report, 23*(11), 2634–2647.

Gilovich, T., & Keltner, D. (2015). *Social psychology.* W. W. Norton.

Goffman, E. (1958). *The presentation of self in everyday life.* University Press of Edinburgh Social Sciences Research Centre.

Goffman, E. (1974). *Frame analysis: An essay on the organization of experience.* Harper & Row.

Granberg, E. M. (2011). "Now my old self is thin": Stigma exits after weight loss. *Social Psychology Quarterly, 74*(1), 29–52.

Harrigan, M. H., Dieter, S., Leinwohl, J., & Marrin, L. (2015). "It's just who I am ... I have brown hair. I have a mysterious father": An exploration of donor-conceived offspring's identity construction. *Journal of Family Communication, 15*(1), 75–93.

Haviland, D., Alleman, N. F., & Cliburn Allen, C. (2017). "Separate but not quite equal": Collegiality experiences of full-time non-tenure-track faculty members. *The Journal of Higher Education, 88*(4), 505–528.

Innes, M. (2002). Organizational communication and the symbolic construction of police murder investigations. *British Journal of Sociology, 53*(1), 67–68.

Ishida, J. (2016). Optimal promotion policies with the looking-glass effect. In S. Ikeda, H. K. Kato, F. Ohtake, & Y. Tsutsui (Eds.), *Behavioral interactions, markets, and economic dynamics* (pp. 543–563). Springer.

Johnson, C., Kelch, J., & Johnson, R. (2017). Dementia at the end of life and family partners: A symbolic interactionist perspective on communication. *Behavioral Sciences, 7*(4), 42–51.

Kaufman, J. M., & Johnson, C. (2004). Stigmatized individuals and the process of identity. *The Sociological Quarterly, 45*(4), 807–833.

Laor, T. (2022). My social network: Group differences in frequency of use, active use, and interactive use on Facebook, Instagram and Twitter. *Technology in Society, 68,* 101922.

LaRossa, R., & Reitzes, D. C. (2009). Symbolic interactionism and family studies. In P. G. Boss, W. J. Doherty, R. LaRossa, W. R. Schumm, & S. K. Steinmetz (Eds.), *Sourcebook of family theories and methods: A contextual approach* (pp. 135–162). Springer Science & Business Media.

Loveland, M. T., & Popescu, D. (2011). Democracy on the web. *Information, Communication & Society, 14*(5), 684–703.

Lucas, K., & Steimel, S. J. (2009). Creating and responding to the gen(d)eralized other: Women miners' community-constructed identities. *Women's Studies in Communication, 32*(3), 320–347.

Maciejewski, P. K., Falzarano, F. B., She, W. J., Lichtenthal, W. G., & Prigerson, H. G. (2022). A micro-sociological theory of adjustment to loss. *Current Opinion in Psychology, 43,* 96–101.

McCall, G. J., & Simmons, J. L. (1978). *Identities and interactions.* Free Press.

Mead, G. H. (1934). *Mind, self and society: From the standpoint of a social behaviorist.* University of Chicago Press.

Moore, J., & Abetz, J. S. (2019). What do parents regret about having children? Communicating regrets online. *Journal of Family Issues, 40*(3), 390–412.

Muralidharan, S., La Ferle, C., & Pookulangara, S. (2018). Can divine intervention aid in domestic violence prevention? An analysis of bystanders' advertising attitudes and reporting intentions in India. *Journal of Promotion Management, 24*(1), 1–24.

Olmstead, S. B., Conrad, K. A., & Anders, K. M. (2018). First semester college students' definitions of and expectations for engaging in hookups. *Journal of Adolescent Research, 33*(3), 275–305.

Poe, P. Z. (2012). Direct-to-consumer drug advertising and "health media filters": A qualitative study of older adult women's responses to DTC ads. *Atlantic Journal of Communication, 20*(3), 185–199.

Reitzes, D. C., DePadilla, L., Sterk, C. E., & Elifson, K. W. (2010). A symbolic interaction approach to cigarette smoking: Smoking frequency and the desire to quit smoking. *Social Focus, 43*(3), 193–213.

Resch, K., Knapp, M., & Schrittesser, I. (2022). How do universities recognise student volunteering? A symbolic interactionist perspective on the recognition of student engagement in higher education. *European Journal of Higher Education, 12*(2), 194–210.

Roffers, M. (2002). Personal communication, Milwaukee, WI.

Rosenthal, R., & Jacobson, L. (1968). *Pygmalion in the classroom: Teacher expectation and pupils' intellectual development.* Holt, Rinehart & Winston.

Sicafuse, L. L., & Miller, M. K. (2010). Social psychological influences on the popularity of amber alerts. *Criminal Justice and Behavior, 37*(11), 1237–1254.

Stoner, M., Aamlid, C., Hilliard, T., & Knox, D. (2019). What parents don't know: International students' romantic relationships in the United States. *College Student Journal, 53*(1), 42–46.

Stryker, S. (1980). *Symbolic interactionism: A social structure version.* Benjamin Cummings.

Van Krieken, R., Habibis, D., Smith, P., Hutchins, B., Martin, G., & Maton, K. (2014). *Sociology: Themes and perspectives.* Pearson Education.

Zhao, S. (2017). Self as second-order object: Reinterpreting the Jamesian "me." *New Ideas in Psychology, 42,* 8–16.

CHAPTER 16

Coordinated Management of Meaning

*Based on the research of **W. Barnett Pearce** and **Vernon Cronen***

> *"I am prepared to argue that the quality of our personal lives and of our social worlds is directly related to the quality of communication in which we engage."*
>
> —Barnett Pearce

The Taylor–Murphys

About 4 years ago, Jessie Taylor decided that she could not stay in her abusive marriage and left her husband, taking her two children—Megan, 8, and Melissa, 6—with her. They currently live in a small apartment, and Jessie knows that the place is too cramped for the three of them. Yet, with her upcoming marriage to Ben Murphy, Jessie realizes that her living situation will change very soon. Her work hours as a relatively new law clerk, however, are quite long; at times, she must be in the office for 12-hour days. As a result, Jessie's children frequently require adult supervision in the early evening. Although Jessie would rather be at home with her children, she realizes that she cannot count on her ex-husband's child support payments, and she must keep her job. She hopes that her approaching marriage to Ben will help ease the financial and familial challenges.

Ben Murphy's wife died a little over a year ago. He parents his 3-year-old son, Patrick, but gets a great deal of help from both of his two sisters. He feels bad about leaving Patrick for the day but is grateful that his family is there, because his job as a state trooper is frequently unpredictable. He never really knows when he is going to be called out for an emergency or when he will be asked to work overtime. Lately, however, his sisters have made some comments, and Ben worries that the babysitting may be turning into a burden for them both. Ben is hoping that his marriage to Jessie will ease his reliance upon his sisters.

One evening, as Ben and Jessie are discussing final wedding plans, the two begin to talk about their future family. Ben is very excited about raising three children and looks forward to his son, Patrick, having new siblings. Jessie, however, is nervous about the logistics of bringing new people into her children's lives. Megan and Melissa are not pleased about their impending "blended" family. They have already had disagreements with Ben about a number of issues, including computer use and after-school activities.

As Ben and Jessie sit in front of the fireplace, they openly talk about the challenges, obstacles, and frustrations that they know they will experience in just a few months. Ben admits, "First, I need to tell you that I love you and that should be the most important thing right now. And I really think that the kids will come around in time—probably after we've been together for a while. A lot of families in our situation start out like this. There's a lot of chaos and then things begin to settle down. Hey, we're not all that unusual. Just look at any Netflix show about families!"

Jessie agrees. "Yeah, I know we're not the first family arranged this way. And I know things will work out. But when? And how are we all going to keep from screaming at each other? I mean, there are a lot of different personalities ... I think about how we're going to adjust, and I think about. ..." Her words trailed off and ended as Ben moved closer to her, hugged her, and told her that everything would work out.

Most people take their conversations for granted. When individuals speak to one another, they often fall into predictable patterns of talk and rely on prescribed social norms. The once-accepted norms for our face-to-face dialogues have been supplanted with less clear norms related to online communication. Yet, the complexity related to both the actual and virtual worlds cannot be ignored. To understand what takes place during a conversation, Barnett Pearce and Vernon Cronen developed Coordinated Management of Meaning (CMM). For Pearce and Cronen, people communicate on the basis of rules. Although we will address this later in the chapter, rules help us not only in our communication with others, but also in our interpretation of what others are communicating to us. CMM helps explain how individuals co-create the meaning in a conversation. Jessie Taylor and Ben Murphy, for instance, are in the early stages of developing rules and communication patterns that will govern their newly configured family's interaction.

For our purposes, CMM generally refers to how individuals establish rules for creating and interpreting meaning and how those rules are enmeshed in a conversation where meaning is constantly being coordinated. Human communication, therefore, is guided by rules. Cronen, Pearce, and Harris's (1982) summary of CMM is informative here: "CMM theory describes human actors as attempting to achieve coordination by managing the ways messages take on meaning" (p. 68). And it is important to note that Pearce (2007) succinctly underscores the transactional nature of communication we identified in **Chapter 1** by noting: "CMM invites you to ask, of the passing moment, 'What are we making together?'" (p. xi). Clearly, the theory requires an understanding of the co-creation of "social worlds" that Pearce believes exist. Finally, Pearce (2012) notes, "CMM does things to us; it changes in constructive ways the way we think and relate to others" (p. 17). As we discuss this theory, we will underscore a number of issues associated with it.

All the World's a Stage

To describe life experiences, Pearce and Cronen (1980) use the metaphor "undirected theater" (p. 120). They believe that in life, as in theater, a number of actors are following some sort of dramatic action and other actors are producing "a cacophonous bedlam with isolated points of coherence" (p. 121). Pearce (1989) describes this metaphor in eloquent detail:

> Imagine a very special kind of theater. There is no audience: everyone is 'on stage' and is a participant. There are many props, but they are not neatly organized: In some portions of the stage are jumbles of costumes and furniture; in others, properties have been arranged as a set for a contemporary office; in yet another, they depict a medieval castle.... Actors move about the stage, encountering sets, would-be directors, and other actors who might provide a supporting cast for a production of some play (p. 48).

The theorists believe that in this theatrical world, there is no one grand director, but rather a number of self-appointed directors who manage to keep the chaos in check.

Conversational flow is essentially a theater production. Interactants direct their own dramas, and, at times, the plots thicken without any script. For many people, how they produce meaning is equivalent to their effectiveness as communicators. To continue the metaphor, when actors enter a conversation, they rely on their past acting (fields of) experiences to achieve meaning. How they perceive the "play" is their reality, but the roles they play in the production are not known until the production begins. To this end, the actors are constantly coordinating their scripts with one another.

The theatrical metaphor was later reconsidered by Pearce (2007). He noted that it was somewhat incomplete as he considered the ebb and flow of conversations and the unpredictability of dialogue. As we noted earlier, however, conversations are frequently chaotic. Pearce and Cronen indicate that the actors who are able to read another's script will likely attain conversational coherence. Those who do not will likely need to coordinate their meaning. Of course, even agreeing upon what conversational script to follow can be difficult. Jessie and Ben, for example, may agree that achieving family harmony is essential, but may not agree on how to achieve that harmony. As Pearce (2007) observes, people may battle it out with respect to what script they will enact and then continue to argue about it.

This notion of a creative theatrical production was in stark contrast to the perspective held by other researchers when CMM was conceptualized. Early discussions of CMM centered on the need to break away from the empirical tradition that characterized much theory building at that time. To shape their theory, Pearce and Cronen looked to a number of different disciplines, including philosophy (Wittgenstein), psychology (James), and education (Dewey). Further, at the heart of CMM, according to Catherine Creede, Beth Fisher-Yoshida, and Placida Gallegos (2012), is the following question: what are we making together? Before delving into the theory's central features, we first consider three assumptions of the CMM.

Theory at a Glance • Coordinated Management of Meaning

In conversations and through the messages we send and receive, people co-create meaning. As we create our social worlds, we understand that rules exist, and we employ them to construct and coordinate meaning. That is, rules guide communication between people. CMM focuses on the relationship between an individual and their society. And, people play a significant role in the co-construction of their own social realities. Through a hierarchical structure, people come to organize meaning of literally hundreds of messages received throughout the day.

Assumptions of Coordinated Management of Meaning

CMM focuses on the self and its relationship to others; it examines how an individual assigns meaning to a message. The theory is especially important because it centers on the relationship between an individual and their society. Referring back to the theater metaphor, consider the fact that all actors must be able to improvise—using their personal repertoire of acting experiences—as well as reference the scripts that they bring into the drama.

Human beings, therefore, are capable of creating and interpreting meaning. In addition to this, there are a few assumptions related to this theory as well:

- Human beings live in communication.
- Human beings co-create a social reality.
- Information transactions depend on personal and interpersonal meaning.

The first assumption of CMM points to the centrality of communication. That is, human beings live *in* communication. Communication is, as noted in **Chapter 1,** a dynamic process that is more than talk; communication, according to CMM, is also a way of creating and doing things (Pearce, 2007). Pearce (1989) claims that "communication is, and always has been, far more central to whatever it means to be a human being than had ever been supposed" (p. 3). Further underscoring the interplay between communication and CMM, Paige Marrs (2012) advances that communication is the "substance of human community" and that CMM is an essential part of the discourses embedded in these communities. Finally, in CMM, Christine Oliver (2014) contends that communication can be identified as a "performative phenomenon" (p. 273).

We live in communication. In adopting this claim, Pearce rejects traditional models of the communication process such as the linear model to which we referred in **Chapter 1.** Rather, CMM theorists propose a counterintuitive orientation: they believe that social situations are created by interactions. Because individuals create their conversational reality, each interaction has the potential to be unique. We might think we know how a conversation may end, but so many times, the results are far from what we expected. This perspective requires CMM adherents to cast aside their pre-existing views of what it means to be a communicator. Further, CMM theorists call for a re-examination of how individuals view communication because "Western intellectual history has tended to use communication as if it were an odorless, colorless, tasteless vehicle of thought and expression" (Pearce, 1989, p. 17). Pearce and Cronen contend that communication must be reconfigured and contextualized in order to begin to understand human behavior. When researchers begin this journey of redefinition, they start investigating the consequences of communication, not the behaviors or variables that accompany the communication process (Cronen, 1995). All of this will inevitably require us to place communication front and center in nearly all of our social encounters.

To illustrate this assumption, consider our opening story. Although Jessie and Ben may believe that they have covered most of the details associated with merging their families, many more issues will appear as the families come together. Family members will create new realities for themselves, and these realities will be based on communication. Conversations will frequently be determined by what the family knows as well as what the family does not know. That is, parents and children will work through unexpected as well as expected joys and sorrows. Like many families in their situation, they may stumble upon areas they never considered. The two families will be working from different sets of conversation rules and therefore may arrive at very different conclusions as they discuss important issues.

A second assumption of CMM is that human beings co-create a social reality. Although we implied this assumption earlier, it merits delineation. The belief that people in conversations co-construct their social reality is called **social constructionism.** Generally speaking, social constructionists look at the norms that we take for granted and believe that these are both socially and culturally constructed (Mercadel, 2019). Pearce (2007) is clear in his social construction advocacy. He observes: "Rather than 'What did you

social constructionism belief that people co-construct their social reality in conversations

social reality a person's beliefs about how meaning and action fit within an interpersonal interaction

mean by that?' the relevant questions are 'What are we making together?' 'How are we making it?' and 'How can we make better social worlds?'" (pp. 30–31). These social worlds require an understanding of **social reality,** which refers to a person's beliefs about how meaning and action fit within their interpersonal encounters. When two people engage in a conversation, they each come with a host of past conversational experiences from previous social realities. Recall the discussion of "field of experience" when we introduced the interactional model of communication in **Chapter 1.** Current conversations, however, elicit new realities because two people are arriving at the conversation from different vantage points. In this way, two people co-create a new social reality.

Sometimes, these communication experiences are smooth; at other times they are cumbersome. As we alluded to earlier and as Gerry Philipsen (1995) concludes, "Many interactions are more messy than clean and more awkward than elegant" (p. 19). Our opening example of Jessie and Ben illustrates this assumption. Although Jessie and Ben have been dating for some time and are preparing for their wedding, CMM theorists believe that they will continue to co-create a new social reality. For instance, the two will have to manage the issue of Jessie's daughters' hesitation to support the marriage. As Jessie and Ben discuss the matter in front of the fireplace, regardless of how they have previously discussed it, they create a new social reality. Perhaps some new issues will emerge—child support, Jessie's job, the age of the children, Jessie's ex-spouse, and so forth—or perhaps Jessie, Ben, or both will adopt new perspectives on their future family makeup. In any event, the social reality that the two experience will be a shared reality.

The third assumption guiding CMM relates to the manner in which people control conversations. Specifically, Pearce (2012) concluded that "communication is about meaning" (p. 4) and that meanings are constantly changing from interaction to interaction.

Transactions depend on personal and interpersonal meaning, as distinguished many years ago by Donald Cushman and Gordon Whiting (1972). **Personal meaning** is defined as the meaning achieved when a person interacts with another and brings into the interaction their unique experiences. Cushman and Whiting suggest that personal meaning is derived from the experiences people have with one another, and yet "it is improbable that two individuals will interpret the same experience in a similar manner ... and equally improbable that they would select the same symbolic patterns to represent the experience" (p. 220). Personal meaning helps people in discovery; that is, it not only allows us to discover information about ourselves, but also aids in our discovery about other people.

> **personal meaning** the meaning achieved when a person brings their unique experiences to an interaction

When two people agree on each other's interpretation, they are said to achieve **interpersonal meaning.** Interpersonal meaning can be understood within a variety of contexts, including families, small groups, and organizations. This type of meaning is co-constructed by the participants. Achieving interpersonal meaning may take some time because relationships are complex and deal with multiple communication issues. A family, for example, may be challenged with financial problems one day, child-raising concerns the next, and elderly care the next. Or, there may be a high degree of variation in how family individuals contact each other; one may use Twitch, and the other may not even be fully comfortable with a laptop! Each of these scenarios will require family members to engage in unique communication pertaining to that particular family episode.

> **interpersonal meaning** the result when two people agree on each other's interpretations of an interaction

Personal and interpersonal meanings are achieved in conversations, frequently without much thought. Perceptive individuals recognize that they cannot engage in specialized personal meaning without explaining themselves to others. Interpersonal meaning must often be negotiated so that rules of meanings move from "in-house usage" to "standardized usage." Sharing meaning for particular symbols, however, is complicated by the fact that the meaning of many symbols is left unstated. For instance, consider a physician specializing in HIV who discusses recent drug therapies with a group of college students. As the physician discusses HIV, she must talk in nonmedical terms so that audience members will understand. However, despite honest efforts at avoiding jargon, it may be nearly impossible to avoid it completely. That is, as much as the physician tries, time may not permit her to fully explain the meaning of all specialized terms.

These three assumptions form a backdrop for discussing CMM. As these assumptions indicate, the theory rests primarily on the concepts of communication, social reality, and meaning. In addition, we can better understand the theory by examining a number of other issues in detail. Among these issues is the manner in which meaning is categorized.

The Hierarchy of Organized Meaning

By reading this chapter and the others in the book thus far, you probably realize already that the essence of communication is meaning. Further, "we live our lives filled with meanings, and one of our life challenges is to *manage* those meanings" (Pearce, 2012, p. 4). Therefore, according to CMM theorists, human beings organize meanings in a hierarchical manner. This is one of the core features of CMM, so we will discuss this at length. First we examine the meaning behind this claim, and then we look at the framework associated with the assumption. We have exemplified the hierarchy in **Figure 16.1.**

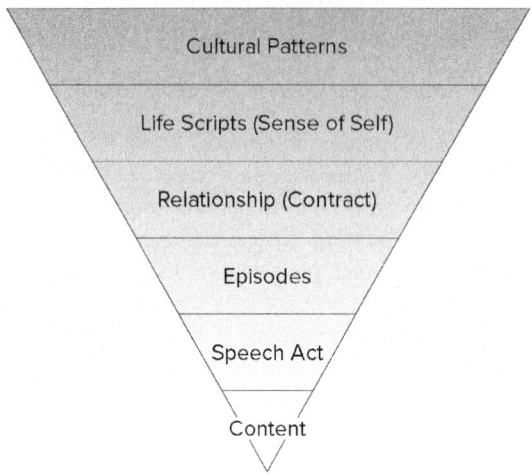

Figure 16.1 Hierarchy of Meaning

Suggesting that people organize meaning implies that they are able to determine how much weight to give to a particular message. It's true that people are constantly being bombarded with stimuli and that they must be able to organize the stimuli for communication to occur. This thinking is relevant to CMM. Imagine, for instance, Ian arriving at his new job on Monday morning. Throughout the day he—Ian identifies as a guy—will be exposed to a number of messages. From understanding company policy on medical leaves to overtime pay to computer safety to responding to email, texts, and internal memos, Ian must manage countless messages. As he returns home at the end of the day, he must organize the messages. In some way, Ian must try to coherently frame literally hundreds of messages from the day, sometimes using technology to organize the message load.

The process of organizing for Ian is similar to what many people experience when they speak to others. When people come together, they must try to handle not only the messages that are sent to them from others, but also the messages that they send to others. This helps people understand the full meaning of the messages. Let's consider how Pearce and Cronen (1980) illustrate the management of meaning.

CMM theorists propose six levels of meaning: content, speech act, episodes, relationship, life scripts, and cultural patterns. As you read about the levels, keep in mind that higher levels help us to interpret lower-level meanings. That is, each type is embedded in the other. Pearce and Cronen prefer to use this hierarchy as a model rather than as a true ordering system. They believe that no true ordering is appropriate because people differ in their interpretation of meaning at various levels. CMM theorists, then, propose a hierarchy to help us understand the sequencing of meaning in different people.

Content

The **content** level specifies the first step of converting raw sensory data into some meaning. For instance, during your break at work, you may convert the symbols being observed or sent into some sort

content the conversion of raw data into meaning

of meaning by their content. You may group information you hear about the boss in one category, information about the workplace environment in a second category, and information about the pay scale in a third category. For Ben Murphy in our chapter's opening, the words "I love you" convey information about Ben's reaction to Jessie, but the content of his words requires this additional level of meaning. Imagine the content level as a message without a context (Pearce, Cronen, & Conklin, 1979).

Students Talking: Juan Carlo

I can't think of a more specific example of coordinating meaning than when we were all consumed with COVID-19! The medical people were using words like "flattening the curve" and "droplet transmission." Those who thought that the virus was a joke were called "anti-vaxxers." Then the majority of us were using words like "super spreader" when a lot of people were together. We all eventually had to "coordinate" our meanings, but there was a lot of conflict while that was being done!

Speech Act

In discussing the second level of meaning, Pearce (2007) describes **speech acts** as a "class of very familiar things, such as *promises, threats, insults, speculations, guesses,* and *compliments*" (p. 105). These are actions we perform by speaking. Speech acts communicate the intention of the speaker and indicate how a particular communication should be taken. Using our earlier example of Ben Murphy, when Ben states "I love you" to Jessie, the phrase communicates more than an assertion. The phrase carries an affectional tone because of the speech act (Austin, 1975).

> **speech act** action we perform by speaking (e.g., questioning, complimenting, or threatening)

Furthermore, Pearce (1994) notes that "speech acts are not things; they are configurations in the logic of meaning and action of conversations, and these configurations are co-constructed" (p. 119). Therefore, we should stay aware that two people co-create the meaning of the speech act, a belief we talked about in our earlier assumption of CMM. Frequently, the speech act is defined both by the sender and by the response to what others have said or done. As Pearce concludes, "You cannot be a 'victim' unless there is a 'victimizer'" (p. 119). In addition, the relational history/field of experience must be taken into consideration when interpreting a speech act. It's difficult to figure out what a message means unless we have a sense of the dynamics between the participants. And speech acts don't always involve speech; Pearce acknowledges the importance of nonverbal communication, too.

Episodes

To interpret speech acts, Pearce and Cronen (1980) discuss **episodes,** or communication routines that have definable beginnings, middles, and endings. In a sense, episodes describe contexts in which people act. Irene Stein (2012) states that episodes can be small, such as discernible parts of a conversation, or large, such as the entire discussion between people. At this level, we begin to see the influence of context on meaning. Episodes vary tremendously—

> **episodes** communication routines that have recognized beginnings, middles, and endings
>
> **punctuation** process of identifying when an episode begins or ends

from picking up a hitchhiker to talking to your doctor about a diagnosis to having an affair with a coworker to fighting with your colleague over the direction of a project. Pearce (2007) states that episodes are sequences of speech acts that are "linked together as a story" (p. 132) (We return to the notion of stories a bit later in the chapter.). Individuals in a communication exchange may differ in how they punctuate an episode. **Punctuation** pertains to the process of identifying when an episode begins or ends. Episodes show how interactions are organized into a meaningful pattern. Pearce and Conklin (1979) clearly note that "coherent conversation requires some degree of coordinated punctuation" (p. 78). Different punctuation, however, may elicit different impressions of the episode, thereby creating "inside" and "outside" perspectives of the same episode. For example, Ben Murphy and Jessie Taylor may have punctuated differently their previous discussions about their future together. Ben may believe that dealing with the children will be better left until after their wedding, whereas Jessie may believe that the issue should be dealt with beforehand. Their subsequent episodes, therefore, will be partly determined by the way they handle their punctuation differences. Pearce (2005) believes that episodes are fairly imprecise because the actors in social situations find themselves in episodes that vary tremendously. He notes that episodes are also culturally based in that people bring to their interactions cultural expectations for how episodes should be executed. Finally, episodes can be highly structured within a social unit (e.g., family) that contains rituals, roles, or prescribed behavior.

Relationship

The fourth level of meaning is the **relationship,** whereby two people recognize their potential and limitations as relational partners. Pearce (2007) expands our traditional understanding of relationship.

relationship agreement and understanding between two people

In addition to partnerships between and among friends, spouses, and family members, he also believes that relationships can include "larger, less personal relationships such as corporations, cities, religions, and tennis clubs" (p. 200). He asserts that, in a sense, we're all "related" because of genes, language, and the planet's forces on each of us. These all coalesce to connect us, willingly or unwillingly.

Relationships are like contracts, which set guidelines and often prescribe behavior. In addition, relationships suggest a future. Few people take the time to outline relational issues unless they are concerned about their future together.

The relationship level suggests relational boundaries in that parameters are established for attitudes and behavior—for instance, how partners should speak to each other or what topics are considered taboo in their relationship. Pearce and Cronen (1980) note that

enmeshment the extent to which partners identify themselves as part of a system

boundaries distinguish between "we" and "they," or those people who are included in the contract and those who are not. The theorists use the term **enmeshment** to describe the extent to which people identify themselves as part of the relational system. Pearce (2012) further contends that enmeshment is the extent to which individuals get "caught up" in their patterns of communication. He asserts that people are enmeshed in a variety of different social systems, each comprised of various logics and meanings.

A relationship may prove invaluable as two people discuss issues that are especially challenging. For instance, there will certainly be some difficult discussions once the Taylor-Murphy family all resides together. Knowing the relational boundaries and the expectations the family members hold for themselves as well as for one another will be important as they become a stepfamily. Although upcoming events will be trying for this family, they will be managed more effectively with an understanding of the contract.

Life Scripts

Clusters of past and present episodes are defined as **life scripts.** Think of life scripts as autobiographies that communicate with your sense of self. You are who you are because of the life scripts in which you have engaged. And how you view yourself over your lifetime affects how you communicate with others. Imagine the differences

life scripts clusters of past or present episodes that create a system of manageable meanings with others

between Ben Murphy's and Jessie Taylor's life scripts. Their past fields of experience will be very informative as they try to deal with their future plans together. Ben comes from a very supportive and loving family, so he may expect conversations with Jessie to be characterized by these nurturing episodes. In fact, it is likely that his experiences as a single father and his perceptions of parenthood are less troubled because of past episodes co-created with his biological family. Jessie, however, did not have an affirming relationship with her ex-husband. Consequently, her past debilitating episodes influence how she now communicates with Ben. Moreover, in her interactions with Ben, she may expect something very different from what Ben expects from her. We should point out, though, that life scripts include those episodes that two people construct together. So, once Ben and Jessie begin to co-create their social world, they will simultaneously co-create a life script.

Cultural Patterns

When discussing cultural patterns, Pearce and Cronen (1980) contend that people identify with particular groups in particular cultures. Also, each of us behaves according to the actual values of our society. These values pertain to sex, race, class, and spirituality, among others. **Cultural patterns,** or archetypes, can be described as "very broad images of world order and [a person's] relationship to that order" (Cronen & Pearce, 1981, p. 21). That is, an individual's relationship to the larger culture is relevant when interpreting meaning. Speech acts, episode relationships, and life scripts are all understood within the cultural level. This is even more paramount when two people from two different cultures try to understand the meaning of each other's words. For instance, Kathryn Sorrells (2021) points out that the U.S. culture puts a premium on elevating the individual interests over a group's interest. This type of culture focuses on independence and initiative. Other cultures (such as Colombia, Peru, and Taiwan) emphasize the interests of the group, and these interests are put before the interests of the individual. Difficulty may arise when two people representing two different orientations interpret meaning from their particular vantage point. Culture, therefore, requires shared meanings and values (for more on this topic, read our discussion in **Chapter 14**).

cultural patterns images of the world and a person's relationship to it

Glenn McCoy/Gary/CartoonStock Ltd

The hierarchy of meaning is an important framework in helping us understand how meaning is coordinated and managed. The levels of meaning espoused by Pearce and Cronen are critical to consider when conversing with another. However, keep in mind that the theorists contend that their purpose is to model the way people process information, not to establish a true ordering. Also, remember that individuals vary in their past and present interactions. Finally, recall that your FtF dialogues are different from your Zoom dialogues which are different from your FaceTime dialogues. Creating meaning in each of these can be challenging and complex. Therefore, some people will have highly complex hierarchies and others will have simplified hierarchies. In addition, some people are able to interpret complex meaning, while others are not as proficient. Fisher-Yoshida (2012) succinctly summarizes the utility of the hierarchy by stating the model "reflects that communication occurs at more than one level at the same time, and the meaning we make is embedded in the contexts that went before and the contexts that follow" (p. 228).

Students Talking Tech: Jenny

I think applying the hierarchy to social media seems easy to me. I see a few levels. I see Content all the time because I have so many social media accounts—Instagram, Pinterest, Twitter, and Snapchat. I have so many pictures and messages I respond to, and I have to make sure I know what platform I'm on all the time. There are Life Scripts because how I respond—or even IF I respond—is based on who I am. I would respond to conservatives who complain about immigration because my mom is an immigrant. I wouldn't respond, though, to someone who was a racist. You can't change them. And, then I think that Cultural Patterns are involved all the time. I have some Instagram friends from different countries who talk about issues from their cultural viewpoints. I need to make sure we all understand each other.

Charmed and Strange Loops

The hierarchy of meaning presented earlier suggests that some lower levels can reflect back and affect the meaning of higher levels. Pearce and Cronen (1980) have termed this process of reflexivity a **loop.** Because the hierarchy cannot go on forever, the theorists propose that some levels reflect back. This supports their view that communication is an ongoing, dynamic, and ever-changing process, a topic we discussed in **Chapter 1.**

loop the reflexiveness of levels in the hierarchy of meaning

charmed loop rules of meaning are consistent throughout the loop

When loops are consistent throughout the hierarchy, Pearce and Cronen identify them as a **charmed loop.** Charmed loops occur when one part of the hierarchy confirms or supports another level. Furthermore, the rules of meaning are consistent and agreed upon throughout the loop. **Figure 16.2** illustrates a charmed loop. Note that there is consistency, or confirmation, in the loop. To illustrate, consider the following example. You hired a painter to repaint the first floor of your home. You agreed to pay the painter by the hour, and when you receive the bill you are shocked at the price. You confront the painter, stating that you observed a lot of "wasted" hours and that you are not going to pay the entire bill. The painter is prepared to challenge your judgment.

This encounter between painter and client exemplifies a charmed loop. In this case, the episode, or event, is the disagreement you both have over the costs of painting your home. Because your life script is your sense of self, your life script in this encounter is consistent with your passion to stand up for what you believe. In this example, the cultural pattern suggests that the two of you—painter and client—entered the conflict with two different views of the situation. The painter wants to get paid for a service and the client wants to feel satisfied with the service. In this brief example, the loop is charmed

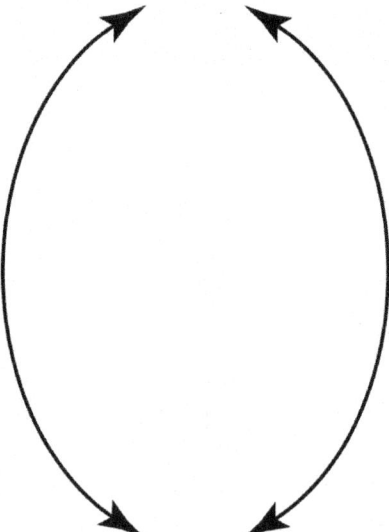

Figure 16.2 Charmed Loop

in that there is consistency among the levels (episode, life script, cultural pattern) in the hierarchy. In other words, the encounter makes sense in that your willingness to challenge the bill is *consistent* with who you are.

And the way you communicate with the painter is *consistent* with the cultural expectations of this type of relationship (i.e., you expect efficiency). The rules of meaning in this interaction are confirmed throughout the loop.

At times, however, some episodes are inconsistent with levels higher up in the hierarchy. Pearce and Cronen have called this a **strange loop.** Strange loops usually align with *intra*personal communication

> **strange loop** rules of meaning change within the loop

in that individuals engage in a sort of internal dialogue about their self-destructive behaviors. In a basic sense, Creede et al. (2012) contend that strange loops refer to some sort of a "stuck experience" that people have when they cycle through patterns that are rather oppositional. We illustrate strange loops in **Figure 16.3.**

To exemplify strange loops, the theorists offer the case of an alcoholic whose bouts with sobriety are followed by bouts of drinking. We present this strange loop in **Figure 16.4.** In this strange loop, confusion sets in. The life script of an alcoholic, for instance, suggests that drinking is out of control, so the alcoholic refuses to drink. But once the drinking stops, an alcoholic may feel a sense of control, so drinking begins again. In this example, the life script is the person's alcoholism, which manages meaning (or drinking) in particular episodes. The episodes are part of the alcoholic's life script. It's apparent, then, that the strange loop will continue to repeat itself. We call this a vicious cycle.

Charmed and strange loops are important to acknowledge because they explain times when there are both congruity and incongruity within the conversational hierarchy.

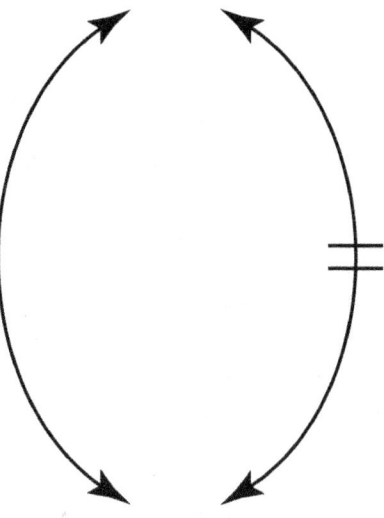

Figure 16.3 Strange Loop

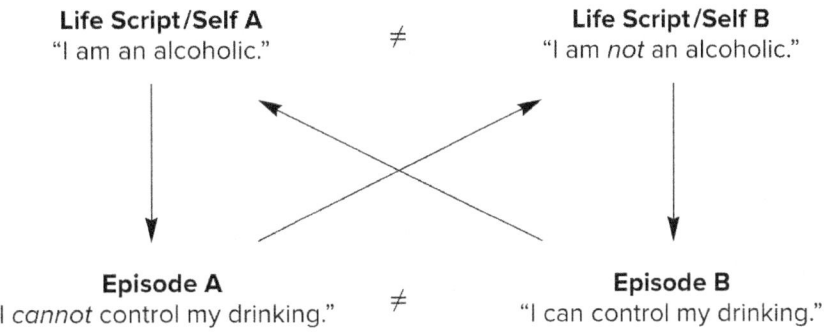

Figure 16.4 Example of a Strange Loop

With this background, we now turn our attention to the issue of coordination and examine how it functions in CMM.

The Coordination of Meaning: Making Sense of the Sequence

In his discussion of coordination, Pearce believes that coordination is best understood by watching people interact on a daily basis. Underscoring the first assumption of CMM, Pearce (2007) concludes that "'coordination' is probably not the first word that leaps to your mind when someone says 'communication.' Perhaps it should be" (p. 83).

Because people enter conversations with a variety of abilities and competencies, achieving coordination can be difficult at times. Coordination with others is challenging, in part because others try to coordinate their actions with ours. **Coordination** exists when two people try to make sense out of the sequencing of messages in their conversation. According to Pearce (2012), meaning is not achieved in isolation, but rather in coordination with other people. Three outcomes are possible when two people converse: they achieve coordination, they do not achieve coordination, or they achieve some degree of coordination (Philipsen, 1995). Gerry Philipsen reminds us that social reality is not perfectly coordinated, so the most likely outcome is partially achieved coordination.

> **coordination** trying to make sense of message sequencing

As noted above, there are three possible outcomes in coordination: people achieve it, they don't achieve it, or they partially achieve it. Let's look at some examples of each by borrowing the experiences of Jessie Taylor and Ben Murphy. They are discussing an upcoming camping trip after the family has been together for just a few months:

BEN: We need to be straightforward with the girls now. I don't see any problem with the two of us sleeping in the same tent and letting the kids sleep in their tent. Patrick can sleep with us if he gets upset.

JESSIE: I think you're right. It shouldn't be a problem.

BEN: Also, they really love him, which makes it easier for all of us.

JESSIE: The weekend should be a great bonding time for the families.

The two have coordinated their meanings of what to do with their children on the first family trip. Although each is entering the conversation with different experiences, Ben and Jessie have created a completed episode of meaning. We could say that conversational coordination is attained and we can say that they have shared meaning.

However, things are not usually this smooth when discussing such complex issues. For instance, think about what happens when a few words are added to a dialogue:

BEN: We need to be straightforward with the girls now. Like at home, we need our privacy. I don't see any problem with sleeping in the same tent and letting the kids sleep in their tent. Patrick can sleep with us if he gets upset. They just need to deal with it.

JESSIE: My girls won't like that. They're not the types who want a kid talking or crying ... even though they like Patrick a lot! And, by the way, I'm hoping you're not saying that our feelings are more important than our kids' feelings?

BEN: I didn't say that. ...

JESSIE: That's what I'm getting out of our conversation! Our first trip together shouldn't be spent trying to explain why Mom and Dad want their privacy.

BEN: All I was trying to do was to let the kids know that their Mom and Dad need to have their own space.

JESSIE: And that they need to deal with their own problems?! Look, I may be overreacting here, but. ...

BEN: Okay. Okay. Jess, I'm sorry that I ever brought up the issue.

As you can see, this episode is quite different from the previous one. In fact, it appears that Ben and Jessie have different interpretations of what the other is saying, clearly, conversational coordination is quite difficult, and their shared meaning is minimal.

Now consider the following dialogue between the two, depicting partial conversational coordination of meaning:

BEN: We need to be straightforward with the girls now. I don't see any problem with sleeping in the same tent and letting the kids sleep in their tent. Patrick can sleep with us if he gets upset.

JESSIE: It's more complicated than that. This is the first time that we're all going to be together for a weekend.

BEN: Honey, if we don't try this right away, it'll be more difficult when we all move in together.

JESSIE: But the kids are still trying to get this situation all figured out. Separating things out this way may be more of a pain than we thought.

BEN: How about if we bought a larger tent and let the kids sleep on one side and we'll be on the other?

JESSIE: That should work.

This partially coordinated dialogue closes the episode between the two parents. Both appear to be satisfied with the compromised sleeping arrangements. Although both parents had differing viewpoints at the outset, they arrived at a compatible conclusion about the sleeping. They each didn't get what they wanted at first, but through a very brief dialogue, managed to secure some ("partial") coordination (i.e., the kids are still sleeping together, but they'll all be in one tent). In your conversations with others—whether in person or online—coordination typically doesn't happen organically. We must work at it to achieve meaning.

Influences on the Coordination Process

Coordination is influenced by several issues, including a sense of morality and the availability of resources. Our social worlds are essentially moral places that prescribe ways we should act and the resources we should draw upon to live our lives with dignity and honor. Pearce (2014) claims that "we only know the social world from 'inside'" (p. 9). In this section, we examine both morality and resources.

First, coordination requires that individuals be concerned with a higher moral order (Pearce, 1989). Many CMM theorists, like Pearce, explain morality as honor, dignity, and character. Moral order involves ethics, and this order is essentially an opportunity for individuals to assert an ethical stance in a conversation. CMM theorists contend that ethics is an inherent part of the conversational flow.

Each person brings various moral orders into a conversation to create and complete an episode. Pearce contends that people simultaneously perform various roles, such as sister, mother, lover, student, employee, friend, or citizen. He believes that each of these roles carries various rights and responsibilities that differ from one person to another. Difficulty arises, however, if inconsistent moral obligations exist in conversations. For instance, in some cultures, men are identified as the sole decision makers and chief protectors of their family. This may conflict with women's obligations and may affect the coordination process in their conversations. These value differences may play themselves out throughout a relationship.

In addition to morality, coordination can be influenced by the resources available to an individual. When CMM theorists discuss **resources,** they refer to "the stories, images, symbols, and institutions that persons use to make their world meaningful" (Pearce, 1989,

> **resources** stories, symbols, and images that people use to make sense of their world

p. 23). Resources also include perceptions, memories, and concepts that help people achieve coherence in their social realities. CMM theorists embrace stories, in particular, and the theory "explores how we create our social reality through the interplay between stories lived and stories told" (Goering & Krause, 2018, p. 269). Further, our social worlds have more meaning and coherence because of the stories we share.

Let's focus on an example to illustrate how resources in a conversation vary from one person to another, causing coordination to be difficult. Consider the experiences of Loran and Wil. As a 19-year employee of a local textile manufacturer, Loran is respected by the seven employees he supervises. Although he does not have a college degree, he relishes his seniority in a company that almost shut down a few years ago. He has listened to the stories of his work colleagues, and over the years, his colleagues have heard his narratives, nearly all of them about his love for his job and his family! Wil, however, has a college degree and an MBA. He is not really into sharing any stories about himself. Because the company is interested in energizing its employees with fresh perspectives on management, Wil was hired as the new supervisor in Loran's department. These changes were marked by heated discussions and a great deal of disagreement between Loran and Wil. Loran's full range of resources—his historical understanding of the factory, his relationships with his employees, the perceptions he has of the company's goals, and the stories he shared with others—seem to be secondary to Wil's resources. Wil's resources are limited. He has little understanding of the company's history, he is not into offering any information about himself or his family, yet he does bring his college education to his job, a credential shared by few at the factory. Because an incompatibility between resources exists, Loran and Wil may have difficulty coordinating their meaning. Two important, but incompatible, sets of resources exist between Loran and Wil, and each may threaten conversation coordination.

Coordinating conversations is critical to communication. At times, coordinating with others is simple, but at other times it's quite challenging. People bring different resources into conversations, prompting individuals to respond to others based on their own management of meaning. In addition to resources, coordination relies on the rules that conversationalists follow. We now explore rules and their relationship to CMM.

Rules and Unwanted Repetitive Patterns

One way that individuals manage and coordinate meaning is through the use of rules. Earlier in the chapter we mentioned that CMM follows a rules perspective, meaning that people's dialogues include templates for how to behave. For CMM theorists, rules provide people opportunities to choose between and among alternatives. Once rules are established in a dialogue, interactants will have a sufficiently common symbolic framework for communication (Cushman & Whiting, 1972). CMM theorists argue that rule usage in a conversation is more than an ability to use a rule. Interactants must understand the social reality and then incorporate rules as they decide how to act in a given situation. In other words, in a conversation, a person needs to explore the context before, during, and after a communication encounter.

Rules are necessarily linked to time, place, relationship, self-concept, episode, culture, and other elements in a context (Harris & Sherblom, 2018). As you understand the notion of rules and their placement in CMM, keep in mind that rules are always dependent on context and that context is a multifaceted environment.

Two types of rules exist in CMM: constitutive and regulative. **Constitutive rules** refer to how behavior should be interpreted within a given context. In other words, constitutive rules tell us what types of behavior mean. We are able to understand another person's intention

> **constitutive rules** rules that organize behavior and help us to understand how meaning should be interpreted

because of the constitutive rules in place. For example, texting "I love you" has different implications when you send it to a roommate, a lover, a family member, or even a coworker. In each of these relationships, we adopt a rule that suggests that the relationship type (contract) and the episode will determine how the statement should be received.

A second type of rule assisting coordination is the regulative rule. Constitutive rules assist people in their interpretation of meaning, but they do not provide people with guidelines for behavior. That

> **regulative rules** guidelines for people's behavior

is the function of regulative rules. **Regulative rules** refer to some sequence of action that an individual undertakes, and they communicate what happens next in a conversation. There are, for instance, regulative rules for meeting a new coworker. You generally introduce yourself, welcome your colleague to the workplace, and indicate that you're available to answer appropriate questions.

To understand the interfacing of constitutive and regulative rules, consider the following situation. A heterosexual couple married for 20 years is having a crisis. The wife has discovered her husband's extramarital affair and must now decide what constitutive and regulative rules to follow. She decides on a venting session because the constitutive rule tells her that such an affair is wrong in their marriage. In turn, her husband must determine how to interpret the venting (constitutive rule) and must construct some sequence of response (regulative rule). As the two engage in their discussion (and co-create their social reality), they will ultimately discover each other's rules systems. Although they have been married for two decades, this situation has not surfaced before, and consequently they may not know what the possibilities are in terms of constitutive and regulative rules. As this discovery takes place, it is likely that the coordination of the conversation will be jeopardized. That is, each person may enact different episodes along the way. They may not always agree on the rules enacted, but at least they are able to make sense out of their conversational experiences.

Guy & Rodd/CartoonStock Ltd

If the couple continue to have sustained conflict, they may engage in what Vincent Cronen, Barnett Pearce, and Linda Snavely (1979) identify as unwanted repetitive patterns (URPs). **Unwanted repetitive patterns** are sequential and recurring interpersonal conflictual

> **unwanted repetitive patterns (URPs)** recurring, undesirable conflicts in a relationship

episodes that are considered unwanted by the individuals in the conflict. The individuals in a conflictual pattern may find themselves feeling helpless or powerless to cease the conflict (Fisher-Yoshida & Yoshida, 2022). Pearce and his associates studied hostile workplace relationships and noted that although privately each worker communicated a genuine desire to be amicable and efficient, publicly their discussions were hostile and ego bruising. This verbal sparring seemed to be ripe for explanation using CMM. The researchers explained that URPs arise because two people with particular rules systems follow a structure that obligates them to perform specific behaviors, regardless of their consequences. It may seem peculiar that people believe they are obligated to act a certain way. Yet as we alluded to earlier, Pearce (1989) states that "persons find themselves in URPs ... in which they report, with all sincerity, that 'I had no choice; I had to act this way'" (p. 39). Consider the notion of URPs this way: communication between two people has fallen into a pattern that is neither productive nor beneficial.

Why do two people continue to engage in URPs? First, they may see no other option. That is, the couple may not possess the skills to remove themselves from the conflict. Second, the couple may be comfortable with the recurring conflict. They know each other and know how the other generally communicates in conflictual situations. Third, a couple may unwittingly fall into this pattern, almost instinctively and by default. Finally, a couple may simply be too exhausted to work toward conflict resolution.

Theory-Into-Action

 The holidays in the United States are known for family gatherings. But the Norman Rockwell image of a turkey, a beautifully set table, and a smiling family is sometimes anything but idyllic. Family conflicts continue to be revved up, despite (or in spite of) our lived experiences with and shared stories about these people who grew up together in the same house. The recurring pattern can go something like this: Even after emailing or Zooming them, we see a sibling about once a year, thinking that they haven't changed much in a year (or even since they were little kids). We openly express our values about issues, such as politics, and we find out that our sibling vehemently disagrees with our views. We argue incessantly about the topic until dinner is done, dishes are cleaned, and people leave. Ultimately the conflict is unresolved, and the same pattern typically occurs at the next gathering unless ground rules regarding topics and conversations are established. It's a URP that revisits millions of families each year.

Integration and Critique

The CMM is one of the few theories to place communication explicitly as a cornerstone in its foundation. It has both quantitative and qualitative roots. Because communication is central to the theory, many scholars have employed the theory in their writings. Among the criteria for evaluating a theory, four seem especially relevant for discussion: scope, parsimony, utility, and heurism.

Integration

Communication Tradition	Rhetorical | Semiotic | **Phenomenological** | **Cybernetic** | Socio-Psychological | **Socio-Cultural** | Critical
Communication Context	**Intrapersonal** | **Interpersonal** | Small Group/Team | Organizational | Public/Rhetorical | Mass/Media | Cultural
Approach to Knowing	**Positivistic/Empirical (early research)** | **Interpretive (recent research)** | Critical

Critique

Evaluation Criteria	**Scope** | Logical Consistency | **Parsimony** | **Utility** | Testability | **Heurism** | Test of Time

Scope

It is unclear whether CMM is too broad in scope. Some communication scholars (e.g., Brenders, 1987) suggest that the theory is too abstract and that imprecise definitions exist. Further, Brenders believes that some of the ideas espoused by Pearce and Cronen lack parameters and beg clarification. The introduction of personal language systems, argues Brenders, is problematic and "leaves unexplained the social nature of meaning" (p. 342). Finally, Scott Poole (1983), in his review of CMM, notes that the theory may be problematic in that it is difficult to "paint with broad strokes and at the same time give difficult areas the attention they deserve" (p. 224).

Yet CMM theorists contend that such criticisms do not take into account the evolution of the theory and its refinement over the years (Barge & Pearce, 2004). Pearce (1995) candidly admits that during "the first phase of the CMM project, [our writings] were confused because we *could not* say what we were doing in the language of social science" (pp. 109–110). Therefore, critics should interpret the theory within the spirit of change; even theorists change as they clarify the goals of their theory. Furthermore, Cronen (1995) admits some early problems with the conceptualization of CMM by indicating that the way he and Pearce discussed the creation of meaning was originally confusing and "wrong-headed." Pearce, Cronen, and other CMM theorists believe that those who levy indictments regarding the scope of the theory should understand the time period in which the theory was developed. And yet, the breadth of the theory has been lauded as relevant to a variety of professions (Yim, 2022).

Parsimony

The scope may be broad, but do not think that the theory has not undergone tests of parsimony. At first glance, one may conclude that the theory is too cumbersome because it "is better understood as a worldview and open-ended set of concepts and model" (Barge & Pearce, 2004, p. 25). However, illustrating difficult concepts with a visual and succinct way of looking at conversations (vis-à-vis the hierarchy of meaning), for instance, has resulted in making CMM an approach to conversation coherency that is efficient and available to theoretical consumers. The theorists posit that "there is no reason why any research method could not be used in CMM research" (p. 25). One could argue, then, that Pearce and Cronen's illustration of the hierarchy of meaning is a visual and succinct way of looking at conversations. Therefore, even cumbersome and complex conversations can be analyzed and understood.

Utility

The application of this theory to individuals and their conversations is quite apparent. The practicality of looking at how people achieve meaning, their potential recurring conflicts, and the influence of the self on the communication process is admirable. CMM is one of the few communication theories that has been identified by both theorists and scholars as a "practical theory" (Creede et al., 2012). In fact, Pearce was always concerned with the utility of his theory since its inception (Pearce, 2012). As well, he is direct in calling CMM a practical theory and notes that the one question guiding CMM is "How can we make better social worlds?" (Pearce, 2007, p. 45). Finally, Pearce observes that the theory has evolved along three stages: interpretive, critical, and *"practical"* [emphasis added] (p. 52). Clearly, there is a pragmatic legacy with CMM. Perhaps Alan Holmgren (2004) stated it succinctly: CMM is like a Swiss Army knife that is useful in many situations.

Heurism

CMM is a very heuristic theory, spanning a number of different content areas, including examining sustainability farming (Hoffman, 2018), death and dying (Omilion-Hodges, Manning, & Orbe, 2019), student–professor relationships (Murray, 2014), refugee communities such as the Congolese (Hughes & Bisimwa, 2015), Muslim tourism (Oktadiana, Pearce, & Chon, 2016), and relations between Pakistan and India (Hussain, ur Rahman, & Iqbal, 2021). Furthermore, researchers have incorporated the theory and its tenets to understand living with cancer (Goering & Krause, 2017), high school hazing (DeWitt & DeWitt, 2012), the Presbyterian Church (Hutcheson, 2012), volunteer work (Creede, 2012), and families who have been tortured or persecuted (Montgomery, 2004). Some newer research (e.g., Hernandez, Punyanunt-Carter, Morris, & McKinin, 2021) has investigated the interplay among families, texting, and emotional communication.

Closing

Thanks in large part to CMM, we have a deeper understanding of how individuals co-create meaning in conversations. In fact, Ronald Arnett (2013) contends that Pearce provided a "communicative roadmap for navigating an era" (p. 6) in communication research. Further, the thinking and research of Pearce have been collated and are archived at Fitchburg State University, where a certificate in CMM in the Coordinated Management of Meaning Institute is available. The theory has aided us in understanding the importance of rules in social situations. CMM, and Pearce in particular, are credited for moving the theory into areas that have allowed for a "porous boundary" between his scholarship and his work toward social change (Lannaman, 2014, p. 257). It is clear that few can deny that CMM positioned communication at the core of human experience and remains a theory that has anchored communication as central to discourse.

Discussion Starters

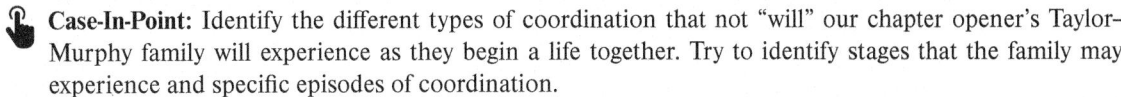

Case-In-Point: Identify the different types of coordination that not "will" our chapter opener's Taylor–Murphy family will experience as they begin a life together. Try to identify stages that the family may experience and specific episodes of coordination.

Try-It-Your-Selfie: Your social media accounts provide a number of different opportunities for the co-creation of meaning. Yet, communication noise is also possible. Using various examples, describe how platforms such as X, Slack, or Snap can both enhance and diminish meaning.

1. How might coordination be influenced by differing cultural backgrounds of communicators, including race, age, sexual identity, and geographical location, among others?
2. Chess and poker are games requiring coordination. Discuss some of life's "games" and how they require coordination. Be creative and give specific examples.
3. What types of meaning breakdowns have you experienced in your conversations with others? Be specific and identify the context, the situation, the individual, and the conversation topic.
4. Describe an episode in your life where telling a story enhanced the meaning between you and another person. What was the story about?

References

Arnett, R. C. (2013). Philosophy of communication as carrier of meaning: Adieu to W. Barnett Pearce. *Qualitative Research Reports in Communication, 14*(1), 1–9.

Austin, J. L. (1975). *How to do things with words.* Harvard University Press.

Barge, J. K., & Pearce, W. B. (2004). A reconnaissance of CMM research. *Human Systems, 15,* 13–32.

Brenders, D. A. (1987). Fallacies in the Coordinated Management of Meaning: A philosophy of language critique of the hierarchical organization of coherent conversation and related theory. *Quarterly Journal of Speech, 73,* 329–348.

Creede, C. (2012). Expanding meaning of my volunteer work in Uganda using circular questioning as a self-reflective journaling practice. In C. Creede, B. Fisher-Yoshida, & P. Gallegos (Eds.), *The reflective, facilitative, and interpretive practice of the Coordinated Management of Meaning* (pp. 123–137). Rowman & Littlefield.

Creede, C., Fisher-Yoshida, B., & Gallegos, P. V. (Eds.). (2012). *The reflective, facilitative, and interpretive practice of the Coordinated Management of Meaning: Making lives and making meaning.* Farleigh Dickinson University Press.

Cronen, V. E. (1995). Practical theory and the tasks ahead for social approaches to communication. In W. Leeds-Hurwitz (Ed.), *Social approaches to communication* (pp. 217–242). Guilford Press.

Cronen, V. E., & Pearce, W. B. (1981). Logical force in interpersonal communication: A new concept of the "necessity" in social behaviors. *Communication, 6,* 5–67.

Cronen, V. E., Pearce, W. B., & Harris, L. M. (1982). The Coordinated Management of Meaning: A theory of communication. In F. E. X. Dance (Ed.), *Human communication theory* (pp. 67–89). Harper & Row.

Cronen, V. E., Pearce, W. B., & Snavely, L. M. (1979). A theory of rule-structure and types of episodes, and a study of perceived enmeshment in undesired repetitive patterns (URPs). In D. Nimmo (Ed.), *Communication yearbook 3* (pp. 225–239). Transaction Books.

Cushman, D., & Whiting, D. C. (September 1972). An approach to communication theory: Toward consensus on rules. *The Journal of Communication, 22*(3), 217–238.

DeWitt, D. M., & DeWitt, L. J. (2012). Case of high school hazing: Applying restorative justice to promote organizational learning. *NASSP Bulletin, 96*, 13-22.

Fisher-Yoshida, B. (2012). Coordinated Management of Meaning (CMM) as reflective practice. In C. Creede, B. Fisher-Yoshida, & P. V. Gallegos (Eds.), *The reflective, facilitative, and interpretive practice of the Coordinated Management of Meaning.* Lexington Books.

Fisher-Yoshida, B., & Yoshida, R. (2022). Transformative learning and its relevance to coaching. In S. Greif, H. Moller, W. Scholl, J. Passmore, & F. Muller (Eds.), *International Handbook of Evidence-Based Coaching: Theory, Research and Practice* (pp. 935-948). Springer International Publishing.

Goering, E. M., & Krause, A. (2017). From sense making to decision making when living with cancer. *Communication & Medicine, 14*(3), 268-273.

Goering, E. M., & Krause, A. (2018). From sense making to decision making when living with cancer. *Communication & Medicine, 14*(3), 268-273.

Harris, T. E., & Sherblom, J. C. (2018). *Small group and team communication.* Waveland Press.

Hernandez, E. S., Punyanunt-Carter, N. M., Morris, A., & McKinin, M. (2021). I'd rather just text you my feelings: The effects of new technologies and social media on college students and emotional support. *Florida Communication Journal, 49*(1), 77-93.

Hoffmann, J. A. (2018). "Organic is more of an American term ... we are traditional farmers": Discourses of place-based organic farming, community, heritage, and sustainability. *Environmental Communication, 12*(6), 807-824.

Holmgren, A. (2004). Saying, doing and making: teaching CMM theory. *Human Systems: The Journal of Systemic Consultation and Management, 15*(2), 89-100.

Hughes, G., & Bisimwa, N. (2015). Hard to reach services? Liberating ourselves from the constraints of our practice. In T. Afuape & G. Hughes (Eds.), *Liberation practices: Toward emotional well-being through dialogue* (pp. 151-161). Routledge.

Hussain, R., ur Rahman, A., & Iqbal, M. (2021). Creating violence together: A study of Pakistan and India's national days celebrations through the lens of the CMM. *Journal of Languages, Culture and Civilization, 3*(4), 253-258.

Hutcheson, J. (2012). Achieving a transcendent episode. In C. Creede, B. Fisher-Yoshida, & P. V. Gallegos (Eds.), *The reflective, facilitative, and interpretive practice of the Coordinated Management of Meaning* (pp. 111-123). Farleigh Dickinson University Press.

Lannaman, J. (2014). Mixing metaphors and metamorphosis: CMM and the discourse of making better social worlds. In S. Littlejohn & S. McNamee (Eds.), *The Coordinated Management of Meaning: A Festschrift in honor of W. Barnett Pearce* (pp. 253-270). Farleigh Dickinson University Press.

Marrs, P. (2012). Taming the lizard: Transforming conversations-gone-bad at work. In C. Creede, B. Fisher-Yoshida, & P. V. Gallegos (Eds.), *The reflective, facilitative, and interpretive practice of the Coordinated Management of Meaning* (pp. 77-94). Farleigh Dickinson University Press.

Mercadel, T. (2019). Social Constructionism. *Salem Press.*

Montgomery, E. (2004). Tortured families: A Coordinated Management of Meaning analysis. *Family Process, 43*, 349-372.

Murray, D. S. (2014). Navigating toward andragogy: Coordination and management of student-professor conversations. *Western Journal of Communication, 78*, 310-336.

Oktadiana, H., Pearce, P. L., & Chon, K. (2016). Muslim travellers' needs: What don't we know? *Tourism Management Perspectives, 20*, 124-130.

Oliver, C. (2014). Coordinating logics of meaning and action: Developing a vocabulary of (un)consciousness. In S. Littlejohn & S. McNamee (Eds.), *The Coordinated Management of Meaning: A Festschrift in honor of W. Barnett Pearce* (pp. 271-290). Farleigh Dickinson University Press.

Omilion-Hodges, L. M., Manning, B. L., & Orbe, M. P. (2019). "Context matters:" An exploration of young adult social constructions of meaning about death and dying. *Health Communication, 34*(2), 139-148.

Pearce, K. (2012). Living into very bad news: The use of CMM as spiritual practice. In C. Creede, B. Fisher-Yoshida, & P. V. Gallegos (Eds.), *The reflective, facilitative, and interpretive practice of the Coordinated Management of Meaning* (pp. 277-294). Farleigh Dickinson University Press.

Pearce, W. B. (1989). *Communication and the human condition.* Southern Illinois University Press.

Pearce, W. B. (1994). *Interpersonal communication: Making social worlds.* Harper Collins Publishers.

Pearce, W. B. (1995). A sailing guide for social constructionists. In W. Leeds-Hurwitz (Ed.), *Social approaches to communication* (pp. 88-112). Guilford Press.

Pearce, W. B. (2005). The Coordinated Management of Meaning (CMM). In W. B. Gudykunst (Ed.), *Theorizing about intercultural communication* (pp. 35-54). Sage.

Pearce, W. B. (2007). *Making social worlds: A communication perspective.* Blackwell.

Pearce, W. B. (2012). Evolution and transformation: A brief history of CMM and a meditation on what using does to us. In C. Creede, B. Fisher-Yoshida, & P. V. Gallegos (Eds.), *The reflective, facilitative, and interpretive practice of the Coordinated Management of Meaning* (pp. 1-22). Farleigh Dickinson University Press.

Pearce, W. B., & Conklin, F. (1979). A model of hierarchical meanings in coherent conversation and a study of indirect responses. *Communication Monographs, 46,* 75-87.

Pearce, W. B., Cronen, V. E., & Conklin, F. (1979). On what to look at when analyzing communication: A hierarchical model of actors' meanings. *Communication, 4,* 195-220.

Pearce, W. B., & Cronen, V. E. (1980). *Communication, action, and meaning: The creation of social realities.* Praeger.

Philipsen, G. (1995). The Coordinated Management of Meaning theory of Pearce, Cronen, and associates. In D. Cushman & B. Kovacic (Eds.), *Watershed traditions in human communication theory* (pp. 13-43). SUNY Press.

Poole, M. S. (1983). Review of Communication, action and meaning: The creation of social realities. *Quarterly Journal of Speech, 69,* 223-224.

Sorrells, K. (2021). *Intercultural communication: Globalization and social justice.* Sage.

Stein, I. F. (2012). Levels of contact in professional coach-client communication. In C. Creede, B. Fisher-Yoshida, & P. V. Gallegos (Eds.), *The reflective, facilitative, and interpretive practice of the Coordinated Management of Meaning* (pp. 65-76). Farleigh Dickinson University Press.

Yim, S. H. (2022). A critique of coordinated management of meaning and circularity in relation to countering oppressive practice: Reflections from a trainee therapist. *Australian and New Zealand Journal of Family Therapy, 43*(3), 346-355.

CHAPTER 17
Communication Privacy Management Theory

*Based on the research of **Sandra Petronio***

> **"Underpinning [Communication Privacy Management Theory]
> is attentiveness to the variability of privacy choices and
> awareness of the dialectical nature found in private disclosures."**
>
> —Sandra Petronio

Lisa Sanders

Lisa Sanders knew it would be hard making it through her workday without getting sidetracked by someone. She worked in a social office and somebody always wanted to talk. She enjoyed her work and loved her coworkers, but the downside was that somebody was always around, ready to tell her something, and she simply had no time to waste today. She'd already used too much precious work time going through her emails. She had deleted about 65 pieces of junk, which was really getting to be a problem. The worst was that somehow she was on a porn list, and she kept getting solicitations for all sorts of nasty things. She hoped that the office didn't have some type of surveillance to observe all those messages. The whole thing made her feel vaguely uneasy, and she wondered if she should report it to someone in HR. Right now, though, she didn't have time, and she rushed around the office, keeping her eyes down, trying to avoid getting sucked into a conversation with anyone.

She went into the break room for a cup of coffee. That was a mistake. Yolanda and Michael were there, and they wanted to ask her opinion about a change the front office was making in how they did the billing. One thing led to another, and soon the three were talking about Yolanda's concerns about going on maternity leave next month. Lisa found herself enjoying the conversation even though she knew she had to get back to work.

After work, she went to Central University to take a night class so that she could finish her B.A. She was hopeful that she'd finish within the year, and she had her eye on a management position that she expected would be vacant by then. During the break, she visited with her friend Doug Banda, who was also finishing up his degree. She confided in Doug that she didn't think she was doing a very good job juggling work, school, and her personal life. It was a continual challenge, and today, her anxieties were getting the better of her. She surprised herself with how emotional she got while talking to Doug. Doug was a good friend; he just listened and offered a friendly hug of support. She felt better after talking with him.

Finally, class was over and Lisa went home. As soon as she got in the door, her phone rang. It was her mother calling with the usual gossip about the family. Most of what her mom had to say was about Lisa's sister-in-law, Margo. Lisa's mom didn't get along very well with Margo, and she often called Lisa to let off steam about something her daughter-in-law had done (or hadn't done). Today, it was the fact that Margo was taking her kids on a vacation, and Lisa's mom wouldn't get to see them for two weeks. Lisa listened and thought to herself that Margo surely had the right to take her own kids on a trip, but she just said, "I'm sorry, Mom, but you'll see the kids when they get back." As soon as she shut off her phone, her roommate, Adina Torelli, came home.

Lisa liked Adina, and she certainly couldn't afford the apartment without a roommate, but sometimes it felt like she didn't have a moment to herself. Adina had an overpowering personality, and she filled the apartment to the extent that Lisa often felt crowded. Adina always had something to say, and tonight was no exception. Adina wanted to tell Lisa all about her most recent fight with her boyfriend, Joel. Adina and Joel were always fighting, and Lisa privately thought Adina should dump the guy. But then she realized that Adina liked playing the drama queen role, so she stopped telling her any of her true reactions. Lisa grabbed a glass of wine and some crackers as she settled on the couch to listen to Adina's latest tale of woe.

As Lisa encounters the various people in her life—coworkers, classmates, family members, roommates, and so forth—she engages in a complex negotiation between what she keeps private and what she chooses to disclose. This chapter presents the theory of Communication Privacy Management (CPM), which helps us sort through and explain the complexities of this process. Sandra Petronio (2002), the theory's creator, states that CPM is a practical theory designed to explain the "everyday" issues described in Lisa's activities. As many researchers have observed, the question of whether to tell someone something we are thinking is a complicated one, yet it's one we face frequently in our daily lives (e.g., Venetis et al., 2012; Zhang, 2022). Petronio (2016a) notes that CPM is a "road map" that helps to explain "how people make judgments about managing their private information with other people" (p. 1).

Examining Lisa's day, we can see at least five instances when Lisa is occupied with questions of disclosure: (1) she worries about whether she should keep those unwanted pornographic emails a secret from people at work, (2) she engages in conversation with coworkers without telling them that she's feeling harried and needs to get back to work, (3) she confides in a friend about feeling she's not doing a good job managing everything at her work, at school, and in her personal life, (4) she listens to her mother's complaints without telling her what she thinks about Margo's vacation plans, and (5) she hears her roommate's disclosures without sharing what she really thinks about Adina's boyfriend.

All these examples illustrate CPM's contention that deciding what to reveal and what to keep confidential is not a straightforward decision but rather a continual balancing act. Both disclosure and privacy have potential risks and rewards for Lisa in all of the situations she encounters. There is always a tension, or a dialectic, involved in this balancing act. We'll talk about the Relational Dialectic Theory in **Chapter 25,** which also focuses on the tensions that make decisions between revealing or concealing information a challenge for most of us.

Further, Lisa also has to think about the risks and rewards her decisions may create for those with whom she interacts. How would Adina feel if Lisa told her that she thought her boyfriend was a loser, for instance? Finally, the act of revealing or withholding personal information has effects on relationships as well as on individuals (e.g., Helens-Hart, 2017). What happens to the relationship between Lisa and Doug as she confides in him about her feelings? All these concerns, relational and individual, create the complicated process of balance that Petronio addresses with CPM Theory. As Petronio (2002) observes:

> We try to weigh the demands of the situation with our needs and those of others around us. Privacy has importance for us because it lets us feel separate from others. It gives us a sense that we are the rightful owners of information about us. There are risks that include making private disclosures to the wrong people, disclosing at a bad time, telling too much about ourselves, or compromising others. On the other hand, disclosure can give enormous benefits.... [We may] increase social control, validate our perspectives, and become more intimate with our relational partners when we disclose.... The balance of privacy and disclosure has meaning because it is vital to the way we manage our relationships (pp. 1-2).

CPM attempts to do what few other theories have done: explain the process that people use to manage the relationship between concealing and revealing private information.

Students Talking: Leah

I know I like to keep a lot of information private, and I do think I own the information about me and can choose when and where to share—or not to share it at all. Topic avoidance is really how I get along with my father. We don't agree on much of anything, and if he knew all my personal beliefs, he would be arguing with me all the time. He doesn't need to hear how much I hate Trump when he supports him. Neither of us will change the other's mind, and I'd rather not get into it with him. It works pretty well to keep those boundaries in place, and my dad and I can talk about the weather, sports, and our dog—we get along just great that way.

CPM Theory is a relatively recent theory. Its recency and communication heritage are notable for two reasons. First, being a recent theory, it indicates the contemporary thinking going on in the communication discipline. It shows that fresh, new thinking continues to illuminate questions of communication behaviors. Having new theories illustrates the vibrancy of communication as a field. In case you were tempted to dismiss theory as something from musty tomes written by dead Greek men, Petronio's CPM Theory shows that theorizing is contemporary and alive.

Second, the fact that CPM grows specifically from a focus on communication shows the maturing and growth of the communication discipline. Some of the other theories in this text, you'll remember, originated in other disciplines. Communication researchers found them useful and borrowed them for their own work. For example, Symbolic Interaction Theory (**Chapter 15**) comes from sociology. Social Exchange Theory (**Chapter 6**) originated in psychology. Communication researchers have found these and other theories applicable for framing their studies examining communication behaviors. However, it's even more helpful to have theories that place communication concepts at the center of the explanation framework. CPM Theory does just that and allows researchers to focus on examinations of the communication process and specific communication practices.

Assumptions of Communication Privacy Management

CPM adheres to aspects of both the rules and systems approaches that we discussed in **Chapter 3,** and the theory is grounded in three assumptions about human nature that are congruent with these two approaches:

- Humans are choice makers.
- Humans are rule makers and rule followers.
- Humans' choices and rules are based on a consideration of others as well as the self.

With respect to the first assumption, Petronio believes that CPM Theory helps people better understand the choices they make and how these choices assist them in their relationships with others. Petronio and Reierson (2009) state that you have the "right to control" (p. 366) personal information, and if you decide to reveal it, it becomes "co-owned" (p. 366) by you and the person you've told. There are risks and rewards to sharing and withholding personal information. Think, for instance, about how you decide whether to tell a colleague about a poor job performance review you received. Your decision whether or not to reveal this personal information may have (lasting) relational consequences. Your colleague might draw closer to you and reveal they were also unhappy with their recent performance review. Or, your colleague might use this information against you and try to compete with you for scarce resources at work. If you don't tell, you risk the loss of support the colleague potentially might give and perhaps the knowledge that the bosses are unfairly evaluating all the employees. You must choose how to proceed in the face of those risks and rewards.

The second assumption means that CPM theorists acknowledge that rules play an instrumental part in our relational lives (refer back to **Chapter 3** for a more detailed discussion of this topic). Rules tell us what to reveal and what to withhold from others based on a "mental calculus." This calculus is grounded in a number of different areas, including culture, gender, and context. Keep in mind that "since people assume the right to retain jurisdiction over their private information" (Petronio, 2016b, p. 3), they will necessarily invoke rules to maintain that right.

The third assumption, that humans make choices and rules based on a consideration of the self and others, makes CPM a theory in the dialectic tradition. CPM focuses on the tensions inherent in being open to others while also maintaining privacy. Petronio (2016a) comments on the intersection between her theory and Relational Dialectics Theory (which we discuss in **Chapter 25**) by stating: "Grounding CPM theory within a dialectical framework allows more insights into privacy management by capturing the underlying logic and management of private information people actually use in their daily lives" (p. 2). That is, our needs for autonomy (self) and sociability (others) influence our decisions about what private information to reveal and what to conceal.

These assumptions, taken together, represent a picture of human beings who are active and engaged in relational life to the extent that self and other are intertwined. The notion of being intertwined is important to CPM. Not only are the self and the other in an engaged relationship, but disclosure is also intertwined with the concept of privacy. As Petronio (2002) has argued, privacy is only understood in a dialectical tension with disclosure. If we disclosed everything, we wouldn't have a concept of privacy. Conversely, if all information were kept private, the idea of disclosure wouldn't make any sense. It's only by pairing them that each concept can be defined. (See **Table 17.1** for a summary of the assumptions underlying all dialectic theories and those that are specific to CPM.)

Table 17.1 Assumptions of CPM

CPM THEORY	ALL DIALECTIC THEORIES
Human choice	Relational life is characterized by change
Humans make and follow rules	Contradiction is the fundamental fact of relational life
Social concerns intertwine with concerns for self	

Evolution of Communication Privacy Management Theory

Although the formal theory currently in use is relatively recent, the concepts and principles that form it show historical evolution. Around 40 years ago, Petronio and her colleagues published some studies outlining principles that would eventually become part of CPM (e.g., Petronio & Martin, 1986; Petronio, Martin, & Littlefield, 1984). In these studies, the researchers were interested in how people decided on the rules guiding their disclosure behavior. They noted that men and women have different criteria for judging when to be open and when to stay silent. These criteria lead to differing disclosure rules for men and women. The understanding of disclosure as rule governed is now a central part of CPM Theory.

In 1991, Petronio published her first attempt to codify all the principles of the theory. Her work then differed from her later (Petronio, 2002) conceptualization in two ways. First, the theory had more limited boundaries in 1991. At that time, Petronio referred to it as a **microtheory** because its boundaries were confined to privacy management within a marital dyad. Now, as we explain in this chapter, the theory is less restricted and attempts to explain privacy and disclosure in many other relationships and contexts. Petronio now refers to CPM as a **macrotheory** because its boundaries include a large variety of interpersonal relationships, such as students and faculty (e.g., Henningsen, Valde, Entzminger, Dick, & Wilcher, 2019), coworkers and clients (e.g., Helens-Hart, 2017), and families (e.g., Child, 2023), as well as a focus on online communication (e.g., Millham & Atkin, 2018) in addition to face-to-face interactions (e.g., Hays & Butauski, 2018).

microtheory a theory with limited boundaries

macrotheory a theory with expansive boundaries

Theory at a Glance • Communication Privacy Management Theory

People believe that they own private information about themselves and their families and loved ones. Because they own this information, they can choose whether and with whom they will share it. And, people experience enduring tension over the decision of whether to share private information or to withhold it; this ongoing tension between opposing positions makes CPM a dialectic theory. When someone tells another something private, the other becomes an authorized co-owner of the information. Even after sharing, however, the original owner believes they should still retain control over the information. People create rules for sharing or withholding their private information based on a variety of criteria that can be classified as either core (consistent to the person) or catalyst (variable based on the situation) criteria. When these rules break down for some reason (either intentionally or through a mistake), privacy turbulence occurs. When this happens, people try to address the turbulence in some way in order to restore the system.

The second difference in the theory was a name change. In 1991, Petronio called the theory Communication Boundary Management. When she published the fuller statement of the theory (Petronio, 2002), she renamed it Communication Privacy Management Theory. Petronio explained the new name as better "reflecting the focus on private disclosures. Though the theory uses a boundary metaphor to explain the management process, the name change underscores that the main focus of the theory is on private disclosures" (2002, p. 2).

The theory then underwent additional, more extensive, changes that Petronio reported on in 2013. She noted that she was responding to critiques that suggested the theory should be more accessible. To accomplish that she streamlined the basic framework of the theory from five principles to three key components: privacy ownership, privacy control, and privacy turbulence (see **Figure 17.1**). Additionally, by 2013, there was a sizable body of research using CPM, so based on those findings, eight axioms were established predicting how people manage private information.

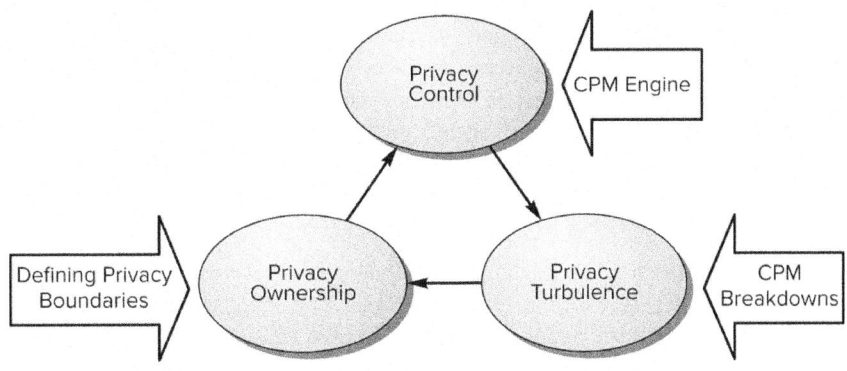

Figure 17.1 Key Components of CPM

Source: Petronio, 2013.

Petronio (2004, 2010, 2016a) expects (and hopes) that the theory will continue to grow and evolve as it's applied to practical questions of disclosure in relationships. Her 2004 article was subtitled "Narrative in Progress, Please Stand By," indicating that the theory continues to evolve and her ideas keep sharpening as empirical data allow her and other scholars using the theory to see and correct any weaknesses in CPM's explanatory power.

We'll discuss these three components and the eight axioms related to them next.

Key Concepts, Components, and Axioms of CPM

As we mentioned, CPM is constructed around three key components and eight axioms. But, before discussing them, we first need
to define three central concepts of the theory: private information,
private disclosures, and private boundaries. Understanding these
three terms is foundational to understanding the components and
axioms of CPM. Petronio (2000) commented that people define
private information as information about things that matter deeply
to them. The process of communicating private information in
relationships with others becomes **private disclosures.**

private information information about
things that matter deeply to a person

private disclosures the process of
communicating private information
to another

Petronio's emphasis away from the term *self-disclosure* marks a distinction between CPM's definition of disclosure and how traditional research on openness (e.g., Jourard, 1971) and other theories such as Social Penetration Theory, for example (see **Chapter 7**) view openness. CPM conceptualizes openness differently in three ways. First, private disclosure puts more emphasis on the personal content of the disclosure than does traditional self-disclosure literature. In so doing, CPM gives more credence to the substance of disclosures, or what is considered private. In addition, CPM examines how people disclose through a rule-based system. Thus, CPM focuses on rule structures created for sharing and withholding private disclosures. Finally, CPM does not consider that private disclosures are only about the self. As Petronio (2002) observes, "To fully understand the depth and breadth of a disclosure, CPM does not restrict the process to only the self, but extends it to embrace multiple levels of disclosures including self and group" (p. 3). She emphasizes that thinking about private disclosures this way makes them more dynamic than they'd be if they referred only to information about the self.

The third foundational concept in CPM is **private boundaries.** CPM relies on the boundary metaphor to make the point that there's a line between being public and being private. On one side of the boundary, people keep private information to themselves; on the other side, peo-

private boundaries the demarcation between private information and public information

ple reveal some private information to others in social relationships with them. Some research indicates that deciding to disclose private information can be a healthful choice (Joseph & Afifi, 2010); however, it is the discloser's choice. The boundaries around personal information are within people's control to set and change.

Boundaries might be set differently related to a variety of factors, however, some of which aren't completely within a person's control. For instance, age might have an influence on how boundaries are established. Children in the United States maintain relatively small privacy boundaries. The boundaries increase as children grow into adolescence and adulthood and cultivate a more developed sense of privacy. As people enter old age and need more help and support for managing their daily activities, their boundaries begin to shrink again (see **Figure 17.2**). In addition, gender and culture might play a role in how boundaries are drawn. One study (Cho, Rivera-Sánchez, & Lim, 2009) showed that age, gender, and nationality influenced people's conceptions of privacy online. Older females from individualistic cultures were the most concerned about online privacy and did the most to establish privacy boundaries online. We'll discuss the notion of boundaries further when we address the key component of Privacy Control.

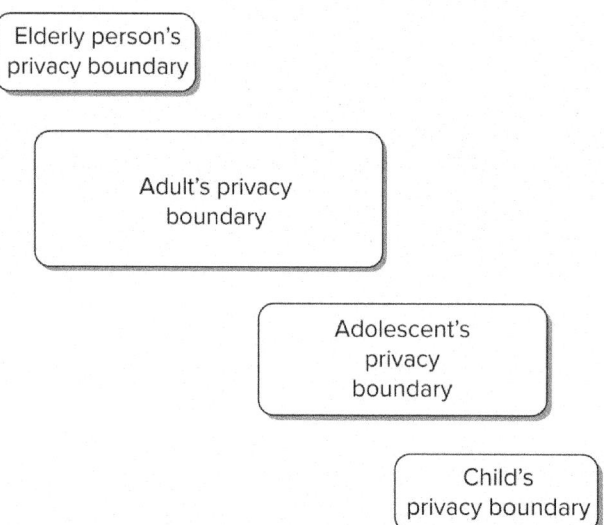

Figure 17.2   Boundaries and the Life Span
Source: Petronio, 2002.

With these foundational concepts in mind, we'll now discuss the three key components of CPM (refer to **Figure 17.1**), and the eight axioms that relate to them (Petronio, 2013). (A summary of this information is found in **Table 17.2.**)

Table 17.2 Key Concepts and Axioms of CPM Theory

Component 1:	**Privacy Ownership**	Axiom 1: People believe they own information about themselves and their significant others. Axiom 2: When original owners share their private information with others, those others become authorized co-owners of the information.
Component 2:	**Privacy Control**	Axiom 3: Even if an original owner reveals private information to an authorized co-owner, the original owner believes they still own the rights to the information. Axiom 4: Original owners develop privacy rules that govern how and when private information is shared based on a variety of core and catalyst criteria. Axiom 5: In order for original owners to continue to control their information after it's shared, they must coordinate and negotiate rules with authorized co-owners about whether or not they can tell anyone else, or whom they may or may not tell. Axiom 6: Information that is co-owned leads to collective privacy boundaries, with all the co-owners contributing private information. Axiom 7: Collective privacy boundaries are governed by privacy rule decisions in much the same way that a single owner's private information is governed.
Component 3:	**Privacy Turbulence**	Axiom 8: Privacy regulation doesn't always go smoothly, and problems can result ranging from relatively minor disruptions in the system to complete breakdowns where people believe their privacy has been betrayed.

Component 1: Privacy Ownership

Related Axioms: 1 & 2

As represented in **Figure 17.1,** one of the three key components of CPM is **privacy ownership,** which refers to how people feel about their private information. Two axioms flow from this component.

> **privacy ownership** the belief that a person owns all information about themselves

Axiom #1 predicts that people believe they "own" personal information about themselves and can manage it however they want. That is, people think they can keep their private information to themselves or share it with others as they wish (e.g., Brownlie, 2011; Bute, Brann, & Hernandez, 2019). As Petronio (2010) notes, however, this belief in ownership is a *perception*, and as such, might not always be factual. Petronio points to research she has done with Jeff Child and Judy Pearson (Child, Pearson, & Petronio, 2009) that focused on student bloggers, illustrating this discrepancy. While people claim they have privacy, or complete ownership of their private information, that claim is often contradicted online. However, Mary Helen Millham and David Atkin's (2018) study, grounded in CPM, found that if people were highly motivated to keep information private, then they'd carefully communicate online to be sure their privacy was maintained. Millham and Atkin concluded that people's *attitudes* about privacy were more important than whether they were communicating online or offline in terms of how well they were able to keep information private. Jeffrey Child and Shawn Starcher's (2016) study had a similar finding in that a person's concerns about privacy invasions were the most important issues in determining their efforts to preserve their private information on Facebook.

Axiom #2 states that when the "original owner" of private information discloses it to others, those others are considered "authorized co-owners" of the information. The original owner then perceives that the authorized co-owner has the responsibility to keep the information private. There might be more than one authorized co-owner, and if so, then they all share the responsibility to keep the information private. That responsibility might last forever, or it might be temporary.

Students Talking Tech: Tomás

It was interesting to read about those studies on Facebook and blogs, but I have to say I don't really agree that people hold much back on social media. Maybe I just follow a bunch of people who have no filters, but if they're holding something back I can't imagine what it would be. It seems to me that there aren't many limits on what people post from a lot of drinking stuff to sex stuff with a bunch of different people. I can't read too much of it at one time. To be honest, I'm not sure why I don't just get off Facebook. It's a time suck, and some of what these people post is disgusting. If they have any idea about privacy, I'd like to know what it is.

Component 2: Privacy Control

Related Axioms: 3 & 7

The second key component of CPM is **privacy control,** which refers to how the original owner of the information decides when (or whether) to reveal or conceal it. There are five axioms that come from this component, which we discuss below.

> **privacy control** how the owner of their information decides about revealing and concealing it

Axiom #3 asserts that when an information owner decides to disclose, they believe they still own the rights to their private information, they still control their privacy, and they are the ones to determine who might become an authorized co-owner in the future.

Axiom #4 predicts that people develop privacy rules to govern how their private information is shared. The rules have two elements: privacy rule development and privacy rule attributes. **Privacy rule development** is the process of creating rules for disclosure and it's guided by a variety of criteria such as cultural values, personal attitudes and beliefs, and the needs of the situation. Petronio (2013) argues that the criteria may be classified into two types: *core criteria* and *catalyst criteria*. Petronio (2016b) defines **core criteria** as a set of somewhat stable criteria that a person brings consistently

> **privacy rule development** the process of creating rules for disclosing private information based on criteria
>
> **core criteria** a set of stable criteria that a person uses consistently across situations in making privacy disclosure decisions

to these types of decisions across a variety of situations. CPM theorists assert that decisions to reveal or conceal can be dependent on issues such as culture. Think about how culture (e.g., gender, cultural community, sexual identity, etc.) guides an individual's expectations for privacy. Thus, we might understand Adina's desire to tell a lot of private information to Lisa in our chapter opening by noting that she is an Italian American woman, a cultural group that generally values openness of expression.

Catalyst criteria exist when "privacy rules have to be responsive to needed change" (Petronio, 2016b, p. 3). These criteria reflect situational influences that motivate people to change their core criteria. For example, in a study of bloggers (Child, Petronio, Agyeman-Budu, & Westermann, 2011), the researchers found that they used core criteria in writing their blogs, but after the blogs were posted, if they saw that what they'd written had the potential to hurt either themselves or others, they used that catalyst criterion (of avoiding hurt) to scrub out the posted message.

> **catalyst criteria** changes in core criteria based on responses to a specific situation where the core criteria don't seem to work for some reason

In another example, Petronio (2016b) shows how catalyst criteria might come into play in a family context. Suppose a couple who plans to divorce decides to reveal a great deal of information about the impending divorce to their children based on a core criterion rooted in the value of openness. However, after the disclosure, the couple might see that it's been harmful to their children to know all these details. The couple then will develop catalyst criteria based on this specific situation and will likely curtail future disclosures. Petronio notes that the motivation to disclose, the risk of the disclosure, the context of the situation, and the relationships involved can all prompt the privacy rule changes that will occur. Relatedly, Haley Kranstuber Horstman and her colleagues (Horstman, Butauski, Johnsen, & Colaner, 2017) found that a risk–benefit analysis was a critical part of the catalyst criteria adopted individuals used to decide whether or not to disclose their adoptive status to people outside their family.

The second element of privacy rules concerns **privacy rule attributes,** which refer to the ways people learn rules and their properties. CPM suggests that people generally learn rules through socialization processes or by negotiation with others to create new rules. For example, frequently new employees are socialized through informal networks in their company as to the behaviors the company values. Nate learned that his company valued nurturing relationships with coworkers. He learned this because the organization often invoked the metaphor of a family, hosted many social events for the workers, and tried in several other ways to impress on employees that close relationships were an important part of the organizational culture. When learned rules are inadequate or need modifying, however, then people collaborate to forge new rules. For instance, if Nate's company determined that its workers were socializing with each other to the detriment of productivity, it might work to renegotiate the rules.

> **privacy rule attributes** one of the features of privacy rules; they refer to the ways people learn about rules and their properties

Axiom #5 predicts that in order for original owners to continue to control their information after it's shared, they must coordinate and negotiate rules with authorized co-owners about whether or not they can tell anyone else, or whom they may or may not tell. This axiom predicts, for example, that now that Doug, from our opening story, is an authorized co-owner of Lisa's information concerning her stress and anxiety, Lisa needs to coordinate with him and set up rules governing how he's supposed to treat that information. Can he tell other students in the program, for instance? In order for Lisa to feel satisfied that she still maintains control of her information, Doug needs to know the parameters of what's allowable for him to share as an authorized co-owner of Lisa's private information.

Axiom #6 states that co-owned information leads to **collective privacy boundaries** where the information is not only about the self; it belongs to the relationship. When Lisa and her mother talk about Lisa's brother's wife, that information belongs to them as a dyad, and neither one of them reveals it to Margo or Lisa's brother, creating a collective boundary. The co-owners form a social network

> **collective privacy boundary** when private information belongs to a relationship between or among people

within these boundaries, and they might all share private information with one another. However, they might not always share everything within the social network. When they maintain privacy, they create a **personal boundary.** When Lisa doesn't tell Adina her opinion of Joel, that information has a personal boundary.

personal boundary a boundary around private information that includes just one person

Axiom #7 predicts that these collective privacy boundaries are regulated in much the same way the original owner's private boundaries are. Decision rules are made that govern who else, outside the social network, can know the information and who has the right to disclose it.

An important element of Axioms 5, 6, and 7 is **boundary coordination,** which refers to how we manage information that is co-owned. For example, when Lisa and her mother talk about Margo, Lisa's sister-in-law, it becomes clear that the information is to be kept from Margo and Lisa's brother. Boundary coordination is the process through which that decision is made and through which Lisa and her

boundary coordination one of the processes in the privacy rule management system; describes how we manage private information that is co-owned

mother become co-owners of private information. Boundaries are dynamic and can be redrawn over time. For example, when Rachel Norton was 18, her brother, aged 16, had an accident with the family car. Together, she and her brother had the car repaired and paid for the repairs without telling their parents. Now that they are both adults, they've told their parents about the incident. Thus, the boundary broadened to include the whole family.

Theory-Into-Action

 Given the internet's ubiquity, it might become difficult to own your own information. This is especially true for children. Some have coined the term "sharenting," which refers to the way parents post a variety of milestone events as well as the daily activities of their children on social media platforms to share with their followers. Children are concerned about how this violates their online privacy. Imagine if your parent(s) posted images of you that you thought were extremely unflattering, or if they posted about the most dramatic and thoughtless things you did or said to them. How would that make you feel? Are you concerned about what your parent(s) post about you? Today's children are members of the first generation whose whole lives might be displayed on social media. Aside from concerns about being seen in a poor light or being pictured when they were doing something especially childish that they now regret, kids might have other more significant concerns about "sharenting." Parents might be unwittingly exposing their kids to hacking, facial recognition tracking, pedophilia, identity fraud, and other serious security and privacy threats.

In an entry on a parenting blog, one poster discusses how lucky it was that there was no blogging when she was a rebellious teenager, so her mother wasn't able to write publicly about the things she did. She says when she was 15 and having a heated argument with her mother, she pulled out a knife and said, "This is how much pain I'm in. What do I have to do to make you understand that?" She goes on to say that she cut her arm slightly and waved the bloodied knife tip at her mom. Now, as an adult, she looks back at that melodramatic action and other mistakes she made growing up and reflects on how she was able to own that information about herself. She isn't so sure that's true of teens today. She acknowledges that parents might need an outlet for their own frustrations and concerns about their children, but she concludes that if parents post about their children's pain and frustrations, it eliminates their privacy and could harm their future in ways the parents can't even imagine.

Component 3: Privacy Turbulence

Related Axiom: 8

The third component in CPM Theory is privacy turbulence. **Privacy turbulence** exists when the rules of boundary coordination are unclear or when people's expectations for privacy management come into conflict with each other. One axiom flows from this component, which we discuss below.

> **privacy turbulence** the disruption that occurs when the rules of boundary coordination are unclear or conflicting

Axiom #8 predicts that regulating the boundaries around private information isn't always a smooth process. The problems that arise during the process can range from relatively minor disruptions in the system to complete breakdowns. When these problems occur, the people involved can experience clashes, which Petronio labels turbulence. In some cases, the turbulence results from what Petronio (2010) calls "fuzzy boundaries," by which she means that the boundaries are ambiguous. One partner could claim that they didn't know they weren't supposed to reveal the information. For instance, if your friend tells you private information, how do you know if you're allowed to share it with someone else? If your friend has told you explicitly not to tell anyone their secret, then the boundary is clear and unambiguous. However, if they just told you the secret without specifying whether you could tell anyone else, you might feel unsure about your responsibilities. You could suspect you aren't supposed to tell, but without instructions you feel uncertain. Or, partners might define private information differently. A romantic partner could think they're entitled to know all about the partner's past, while the other person might believe that the past is private and doesn't need to be revealed. In any event, turbulence will ensue when "expectations for privacy management are unfulfilled" (Petronio, 2010, p. 182).

CPM Theory asserts that when individuals experience turbulence, they'll try to make adjustments so that they can achieve coordination again. Lindsey Aloia's (2018) study supported this assertion generally but added the idea that the emotions experienced by the original owner will affect the communication behaviors they engage in after turbulence. When original owners feel a great deal of anger toward co-owner(s), they might not initially work with them to achieve coordination, and instead they might engage in yelling and accusations to express their anger at what they perceive as a betrayal.

Integration and Critique

CPM Theory has enjoyed significant attention from scholars in communication as well as from those in other disciplines. The theory appears to resonate because its focus on private information and the rules that guide its disclosure and concealment apply to a wide variety of communication behaviors in all types of relationships. Moreover, researchers have studied the theory using both quantitative and qualitative lenses. As you think about the merits of CPM, three criteria for theory evaluation are relevant: logical consistency, utility, and heurism.

Integration

Communication Tradition	Rhetorical | Semiotic | Phenomenological | Cybernetic | Socio-Psychological | **Socio-Cultural** | Critical
Communication Context	Intrapersonal | **Interpersonal** | Small Group/Team | Organizational | Public/Rhetorical | Mass/Media | Cultural
Approach to Knowing	**Positivistic/Empirical** | **Interpretive/Hermeneutic** | Critical

Critique

Evaluation Criteria	Scope | **Logical Consistency** | Parsimony | **Utility** | Testability | **Heurism** | Test of Time

Logical Consistency

One criticism of the theory that Petronio (2002) has discussed relates to its logical consistency. Some critics have observed that CPM uses the term *dialectic* inaccurately, claiming to be dialectic in nature when it's really based on dualistic thinking. The basis for the criticism stems from Leslie Baxter and Barbara Montgomery's (1996) distinctions among monologic, dualistic, and dialectic approaches (see **Chapter 25**). Using these distinctions, Baxter and Montgomery have argued that CPM takes a dualistic approach, treating privacy and disclosure as independent of one another and able to coexist in tandem.

Petronio (2002) responds to this criticism by noting that perhaps the accusation of dualistic thinking comes from the use of the terms *balance* and *equilibrium* in the early versions of CPM Theory. Petronio argues that CPM is not focused on balance in the psychological sense. She states: "Instead, [CPM] argues for coordination with others that does not advocate an optimum balance between disclosure and privacy. As an alternative, the theory claims there are shifting forces with a range of privacy and disclosure that people handle by making judgments about the *degrees* of privacy and publicness they wish to experience in any given interaction" (pp. 12–13 [emphasis in original]). Thus, Petronio argues that it is legitimate to call CPM Theory dialectical in nature.

Utility

CPM Theory is a useful theory. It offers an explanation for the delicate process of disclosing and concealing private information that people perform continually in their relationships with others. Furthermore, CPM may provide insights as that process becomes even more complex in an online environment. CPM is useful for explaining those daily intrusions into our lives from technological advances. As technology moves more and more of what we have considered private information into the public realm, we will need to understand the rule-based management system underlying this trend. As Petronio (2016b) aptly notes, "potential applications of communication privacy management continue to grow" (p. 7). Further, as researchers continue working with CPM, they uncover nuances that make the theory even more useful for future research. For example, disclosures aren't usually all-or-nothing propositions, and often people disclose only parts of their secrets or they find themselves using a sliding scale of disclosure (e.g., Rubinsky, 2018). As these insights are incorporated into CPM, its utility continues to increase.

Heurism

CPM Theory demonstrates heurism because it has been utilized as a framework in a variety of situations, including romantic relationships (e.g., Nichols, 2012), military families (e.g., Owlett, Richards, Wilson, DeFreese, & Roberts, 2015), academic advising (e.g., Thompson, Petronio, & Braithwaite, 2012), and health care (e.g., Petronio & Sargent, 2011; Romo, 2012). In 2020, Sandra Petronio and Jeff Child noted that the theory had over 1,000 citations and had been applied in studies spanning 44 different countries and 340 different contexts.

As we've mentioned, online communication is another exciting area where CPM has been applied. Jeff Child and his colleagues (e.g., Beam, Child, Hutchens, & Hmielowski, 2018; Child & Petronio, 2011; Child, Petronio, Agyeman-Budu, & Westermann, 2011; Child & Starcher, 2016; Child & Westermann, 2013) have been in the forefront of scholars who have explored notions of privacy and how they relate to online experiences. Others have examined online disclosures as well illustrating the heuristic nature of CPM (e.g., Baruh, Secinti, & Cemalcilar, 2017, in a meta-analytic review of online privacy concerns; Herrman & Tenzek, 2017, with reference to disclosing eating disorders on a pro-anorexia website; Romo, Thompson, & Donovan, 2017, with reference to how college students make privacy rules and respond to boundary turbulence about alcohol-related disclosures on social network sites). The evaluation of CPM Theory on these three criteria demonstrates CPM's success as a theoretical framework for communication questions.

Closing

As we continue to explore decisions to reveal private information to others or conceal it from them, CPM Theory will be an important and valuable explanatory framework. Our conversations and relationships with others are complex. CPM provides us a useful structure for unravelling the many and varied communication practices we engage in with relational partners where concealing or disclosing personal information is at issue.

Discussion Starters

 Case-In-Point: Apply CPM Theory to Lisa Sanders's case. How might the theory help Lisa to understand her situation?

Try-It-Your-Selfie: Look over the last week of your online activities. Note what you have posted on Instagram, TikTok, Facebook, or any of the SMSs you used over the past week. What did you reveal about yourself and what did you conceal? How does CPM theory help you understand your online disclosures?

1. What approach to theory building does Petronio take in CPM Theory (laws, rules, systems, or some combination)? Explain your answer.

2. Do you agree with Petronio's defense of CPM against the critique that it uses the concept of dialectics but really takes a dualistic approach to privacy management? Explain your reasoning.

3. Explain how you create privacy boundaries and disclosure rules. Do the axioms of CPM hold true in your personal experience? If they do, explain how you use core and catalyst criteria in making decisions about private information. If you don't think the components and axioms are useful in explaining your behaviors, give examples to illustrate why not.

4. Provide an example of how privacy turbulence affects CPM processes.

References

Aloia, L. S. (2018). The emotional, behavioral, and cognitive experience of boundary turbulence. *Communication Studies, 69*(2), 180–195.

Baruh, L., Secinti, E., & Cemalcilar, Z. (2017). Online privacy concerns and privacy management: A meta-analytical review. *Journal of Communication, 67*(1), 26–53.

Baxter, L. A., & Montgomery, B. M. (1996). *Relating: Dialogues and dialectics*. Guilford Press.

Beam, M. A., Child, J. T., Hutchens, M. J., & Hmielowski, J. D. (2018). Context collapse and privacy management: Diversity in Facebook friends increases online news reading and sharing. *New Media & Society, 20*(7), 2296–2314.

Brownlie, J. (2011). Not "going there": Limits to the professionalisation of our emotional lives. *Sociology of Health & Illness, 33*(1), 130–144.

Bute, J. J., Brann, M., & Hernandez, R. (2019). Exploring societal-level privacy rules for talking about miscarriage. *Journal of Social and Personal Relationships, 36*(2), 379–399.

Child, J. T. (2023). Family communication as boundary. In J. Manning, J. Allen, & K. D. J. Denker (Eds.), Family communication as…: Exploring metaphors for family communication (pp. 122–129). John Wiley & Sons.

Child, J. T., Pearson, J. C., & Petronio, S. (2009). Blogging, communication, and privacy management: Development of the blogging privacy management measure. *Journal of the American Society for Information Science & Technology, 60*(10), 2079–2094.

Child, J. T., & Petronio, S. (2011). Unpacking the paradoxes of privacy in CMC relationships: The challenges of blogging and relational communication on the internet. In K. B. Wright & L. M. Webb (Eds.), *Computer-mediated communication in personal relationships* (pp. 21–40). Peter Lang.

Child, J. T., Petronio, S., Agyeman-Budu, E. A., & Westermann, D. A. (2011). Blog scrubbing: Exploring triggers that change privacy rules. *Computers in Human Behavior, 27*(5), 2017–2027.

Child, J. T., & Starcher, S. (2016). Fuzzy Facebook privacy boundaries: Exploring mediated lurking, vague-booking, and Facebook privacy management. *Computers in Human Behavior, 54*, 483–490.

Child, J. T., & Westermann, D. A. (2013). Let's be Facebook friends: Exploring parental Facebook friend requests from a communication privacy management (CPM) perspective. *Journal of Family Communication, 13*(1), 46–59.

Cho, H., Rivera-Sánchez, M., & Lim, S. S. (2009). A multinational study on online privacy: Global concerns and local responses. *New Media & Society, 11*(3), 395–416.

Hays, A., & Butauski, M. (2018). Privacy, disability, and family: Exploring the privacy management behaviors of parents with a child with autism. *Western Journal of Communication, 82*(3), 376–391.

Helens-Hart, R. (2017). Females' (non)disclosure of minority sexual identities in the workplace from a communication privacy perspective. *Communication Studies, 68*(5), 607–623.

Henningsen, M. L. M., Valde, K. S., Entzminger, M. J., Dick, D. T., & Wilcher, L. B. (2019). Student disclosures about academic information: Student privacy rules and boundaries. *Communication Reports, 32*(1), 29–42.

Herrman, A. R., & Tenzek, K. E. (2017). Communication privacy management: A thematic analysis of revealing and concealing eating disorders in an online community. *Qualitative Research Reports in Communication, 18*(1), 54–63.

Horstman, H. K., Butauski, M., Johnsen, L. J., & Colaner, C. W. (2017). The communication privacy management of adopted individuals in their social networks: Disclosure decisions in light of the discourse of biological normativity. *Communication Studies, 68*(3), 296–313.

Joseph, A. L., & Afifi, T. D. (2010). Military wives' stressful disclosures to their deployed husbands: The role of protective buffering. *Journal of Applied Communication Research, 38*(4), 412–434.

Jourard S. M. (1971). *The transparent self* (rev. ed.). Van Nostrand.

Millham, M. H., & Atkin, D. (2018). Managing the virtual boundaries: Online social networks, disclosure, and privacy behaviors. *New Media & Society, 20*(1), 50–67.

Nichols, W. L. (2012). Deception versus privacy management in discussions of sexual history. *Atlantic Journal of Communication, 20*(2), 101–115.

Owlett, J. S., Richards, K., Wilson, S. R., DeFreese, J. D., & Roberts, F. (2015). Privacy management in the military family during deployment: Adolescents' perspectives. *Journal of Family Communication, 15*(2), 141–158.

Petronio, S. (2000). The boundaries of privacy: Praxis of everyday life. In S. Petronio (Ed.), *Balancing the secrets of private disclosures* (pp. 37–49). Erlbaum.

Petronio, S. (2002). *Boundaries of privacy: Dialectics of disclosure*. SUNY Press.

Petronio, S. (2004). Road to developing communication privacy management theory: Narrative in progress, please stand by. *Journal of Family Communication, 4*(3–4), 193–207.

Petronio, S. (2010). Communication privacy management theory: What do we know about family privacy regulation? *Journal of Family Theory & Review, 2*(3), 175–196.

Petronio, S. (2013). Brief status report on communication privacy management theory. *Journal of Family Communication, 13*(1), 6–14.

Petronio, S. (2016a). *Balancing the secrets of private disclosures*. Routledge.

Petronio, S. (2016b). Communication privacy management theory. In C. R. Berger & M. E. Roloff (Eds.), *The international encyclopedia of interpersonal communication* (pp. 1–9). John Wiley & Sons.

Petronio, S., & Child, J. T. (2020). Conceptualization and operationalization: Utility of communication privacy management theory. *Current Opinion in Psychology, 31*, 76–82.

Petronio, S., & Martin, J. N. (1986). Ramifications of revealing private information: A gender gap. *Journal of Clinical Psychology, 42*(3), 499–506.

Petronio, S., Martin, J. N., & Littlefield, R. (1984). Prerequisite conditions for self-disclosing: A gender issue. *Communication Monographs, 51*(3), 268–273.

Petronio, S., & Reierson, J. L. (2009). Regulating the privacy of confidentiality: Grasping the complexities through communication privacy management theory. In W. Afifi & T. Afifi (Eds.), *Uncertainty and information regulation in interpersonal contexts: Theories and applications* (pp. 365–383). Routledge.

Petronio, S., & Sargent, J. (2011). Disclosure predicaments arising during the course of patient care: Nurses' privacy management. *Health Communication, 26*(3), 255–266.

Romo, L. K. (2012). "Above the influence": How college students communicate about the healthy deviance of alcohol abstinence. *Health Communication, 27*(7), 672–681.

Romo, L. K., Thompson, C. M., & Donovan, E. E. (2017). College drinkers' privacy management of alcohol content on social-networking sites. *Communication Studies, 68*(2), 173–189.

Rubinsky, V. (2018). Revealing or concealing polyamory in the family: Cultural rules for communication polyamory to family members. *Women & Language, 41*(1), 16–38.

Thompson, J., Petronio, S., & Braithwaite, D. O. (2012). An examination of privacy rules for academic advisors and college student-athletes: A communication privacy management perspective. *Communication Studies, 63*(1), 54–76.

Venetis, M. K., Greene, K., Magsamen-Conrad, K., Banerjee, S. C., Checton, M. G., & Bagdasarov, Z. (2012). "You can't tell anyone but…": Exploring the use of privacy rules and revealing behaviors. *Communication Monographs, 79*(3), 344–365.

Zhang, X. (2022). Expression avoidance and privacy management as dissonance reduction in the face of online disagreement. *Telematics and Informatics, 75,* 101894.

CHAPTER 18
Groupthink

*Based on the research of **Irving Janis***

"Groups, like individuals, have shortcomings. Groups can bring out the worst as well as the best."

—Irving Janis

Melton Publishing Board of Directors

As they sat in the seminar room at Melton Publishing for the first time since COVID-19 vaccines were available, the seven men and women privately wondered when they would discover their next superstar author. Last week the board had heard the bad news: e-books were soaring in popularity, and print books—the bread and butter of the publisher—were sinking in numbers. As a result, Melton's profits were down significantly, and they either needed to overhaul the company right away or do what the major publishers often do: find a best-seller to counter some of the company's financial loss (the company had tried working on an e-book unit, but it had struggled against the mega-presses). Elizabeth Hansen, the board's director, suggested that the group review at least two books they thought could reverse the company's losses. From early reviews of one of the books—*Red Warnings,* a science-fiction novel—Hansen believed that here was an opportunity to turn around the downward spiral and give the small press a promising future. She knew that employee and board morale was down and felt that something had to be done quickly.

As director for the past few years, Elizabeth also knew that, in a way, her own credibility was on the line. She was chairing a board of literary people who, for the most part, had no life experience sitting on for-profit boards. And, she knew that the similarity among the members might result in a quick and reasonable solution to the company's financial circumstances. Although no clear solution was obvious at this difficult time for the company, Elizabeth was sure something had to be done.

Instead of Zooming, the group gathered together to talk about finances. Elizabeth opened the discussion with an upbeat message: "We'll beat this downward spiral," she said. She reminded the group that the 43-year-old company had a history of surviving rough times. "This is just another bump in the road," she reported.

The discussion soon turned to the book *Red Warnings*, which had been reviewed quite well in a national book magazine. Since the review's publication, orders for the book had skyrocketed. Elizabeth thought that the best way to generate more money for the small company, thereby resulting in more publicity, was to market *Red Warnings* aggressively: through social media; promotions on college campuses and at sci-fi conventions across the country; and, hopefully, with a national book club selection.

Randy Miles, another board member, disagreed with Elizabeth's plans. He thought that there was simply too much risk in concentrating so much time and money on one book. Randy's way to improve the financial situation of the company could be summed up in one word: cutbacks. After looking over the company's financial records, he had discovered that far too many people had been hired, resulting in overspending on salary, benefits, and book production. He knew that cutting back was not a popular strategy but reasoned that this wasn't the time for worrying about being popular.

Their meeting dragged on from an anticipated 2 hours to nearly 4 hours. Elizabeth and Randy continued to explain and defend their viewpoints. At times there was heated arguing; the two would frequently raise their voices to make their points. During their arguments, the remaining board members would try to be agreeable. The others knew they weren't all that knowledgeable about how to turn a company around, and, for the most part, they sat silently while the two leaders squared off against each other.

Finally, as the afternoon sun began to set, Tina, a writer from the Boston area, said, "Look, we've all been very patient while the two of you decide what's best for this company. And maybe we've learned for the future that we need to get more financial expertise on this board. But, for now, I have to speak. We need to make a decision—a hard decision. Whether a decision can even be made today is iffy. I recommend that we postpone any more talk until our next meeting so that we can all think about our options."

Randy interrupted, "Look, Tina, I appreciate your honesty, but this is not about whether we should build another bathroom! This is the company's future. I recommend that we hash this out now."

Elizabeth agreed, as did the other board members. As evening came, it was clear that Elizabeth and Randy were getting tired. "Listen, Randy, we're running low on energy and time. I think we should just get the two motions out and have the board vote." The other members agreed; it was obvious that they were all tired.

Randy reminded the group that they shouldn't rush their decision, but Tina reminded him that he was the one who hadn't wanted to postpone a decision. Although Randy tried to attack Tina's tone, the other group members began clamoring for a vote. Each member openly expressed support for Elizabeth's aggressive marketing plan. Because the group was small and he could sense the overwhelming support for Elizabeth, Randy decided to give in.

Participating in groups and teams is a fact of life. Whether people are at school, work, a volunteer agency, a spirituality meeting, jury rooms, combat squads or other venues, they frequently spend significant working hours with others in small groups. And, group and team communication remains paramount in cultures around the globe. To understand the nature of decision making in small groups, Irving Janis, in his book *Victims of Groupthink* (1972), explains what takes place in groups where group members are highly agreeable with one another. The original work in Groupthink focused on foreign-policy decision making, although the theory was later applied to several other areas, a topic we return to later. Janis contends that when group members share a common fate, there is great pressure toward conformity. He labels this pressure *Groupthink,* a term cleverly patterned after vocabulary found in George Orwell's *1984* (e.g., *doublethink*). We should point out that the word *Groupthink* was coined in 1952 by business author and editor William Whyte in *Fortune* magazine. About two decades later, Janis provided the communication context of this term and group experience, and since that time, scores of academic and popular press writers have invoked the term to illustrate its pervasiveness.

Groupthink is defined as a way of deliberating that group members use when their desire for unanimity overrides their motivation to assess all available plans of action. Janis contends group members frequently engage in a style of deliberating in which consensus seeking (need for

> **Groupthink** a way of group deliberation that minimizes conflict and emphasizes the need for unanimity

everyone to agree) outweighs good sense. Team members suppress any doubts and criticisms about their actions, resulting in outcomes that are much more risky. You may have participated in groups where the desire to achieve a goal or task was more important than developing reasonable solutions to problems. Janis believes that when highly similar and agreeable groups fail to consider fully any dissenting opinions, when they squash conflict just so they get along, or when the group members do not consider all solutions, they are prone to groupthink. He argues that when groups are "in" groupthink, they move to preserve the group. To this end, making peace is more important than making clear and appropriate decisions.

Theory at a Glance • Groupthink

An intrapersonal drive for getting along and for consensus drives groups to become highly cohesive. Highly cohesive groups frequently fail to consider alternatives to their course of action. When group members think similarly and do not entertain contrary views, they are also unlikely to share unpopular or dissimilar ideas with others. Groupthink suggests that these groups make premature decisions, some of which have lasting and tragic consequences.

Our example of the board of directors at Melton Publishing exemplifies the groupthink phenomenon. The board members are a group of people who apparently get along. The members are essentially connected by their literary backgrounds, making them more predisposed to groupthink. The group is under a deadline to come up with a thoughtful and financially prudent decision about the financial future of the company, especially during an era of e-book dominance. As the chapter unfolds, we will tell you more about how this group becomes susceptible to groupthink.

Before moving on, we first need to point out that Janis's theory emanates from his concern about historical decisions made by seemingly very smart people. In Janis's earlier writings, his focus on several high-profile foreign-policy decisions resulted in looking at "the inner circle" of those in charge, principally the U.S. President. Using principles from small group/team research, Janis (1982) explained why several foreign-policy decisions are flawed.

For instance, in the development of Groupthink, Janis analyzed five notorious matters of national importance: (1) the preparedness policies of the U.S. Navy at Pearl Harbor in 1941, (2) the decision to advance into North Korea by General McArthur, (3) the decision by President Kennedy to back the CIA's recommendation to invade Cuba at the Bay of Pigs shortly after Fidel Castro established a Communist government, (4) the decision by President Johnson to escalate the Vietnam War, and (5) the Watergate cover-up by President Nixon. Janis argues that each of these policy decisions was made by both the president and his team of advisors and that, because each group was under some degree of stress, they made hasty and what amounted to inaccurate decisions. He interviewed a number of people who were part of these teams and concluded that each of these policy fiascoes occurred because of groupthink. It was discovered that in each of these cases, the presidential advisors did not thoroughly test information before making their decisions. In other words, there was mutual agreement to avoid critical conversations and conflict. According to Janis, the group members failed to consider forewarnings, and their biases and desire for harmony overshadowed critical assessments of their own decisions.

The decision-making process underscoring Groupthink is historical. Randy Hirokawa, Dennis Gouran, and Amy Martz (1988), for instance, came to the conclusion that faulty decision making characterized the disaster of the space shuttle *Challenger.* Furthermore, John Schwartz and Matthew Wald (2003) comment that Groupthink principles were at work when the *Columbia,* another shuttle, broke up. A "quick analysis" by aviation giant Boeing and their engineers showed that "smart people working collectively can be dumber than the sum of their brains" (p. 4). The decision to invade Iraq in 2003 was also ripe with naive and uninformed advisers of President George W. Bush, insofar as they anticipated that it would be a "cakewalk," and "last six days [or] six weeks," but the "cost" totaled the loss of nearly 5,000 U.S. troops, the wounding of more than 32,000 others, and an estimated $806 billion in financial expenditures (Henderson, 2018).

And, more recently, Groupthink processes were prominent during the early days of COVID-19. Using the 1918 Spanish flu as a model, governments across the globe were concerned that another worldwide pandemic could exist. Yet, significant disagreements existed between being "overly confident" in predictions of how rapid the coronavirus would spread and the "over reliance" on science to make projections. Refusing to be "transparent" and a lack of contrarian perceptions regarding the lockdown, for instance, were seen as ever-present during the beginnings of COVID (Joffe, 2021), leading to groupthink among the "experts."

In sum, then, groupthink is often present because those in positions of authority or decision-making power surround themselves with those who are willing to "go along to get along." And, while most of the research related to Groupthink has, indeed, focused on major policy-oriented decisions, you will later understand that the theory's concepts have been applied to other types of decisions.

Assumptions of Groupthink

At its core, Groupthink is a theory associated with small group/team communication. In **Chapter 2,** we noted that teams are a part of virtually every segment of U.S. society. Janis focuses his work on **problem-solving groups** and **task-oriented groups,** whose main purpose is to make decisions and give policy recommendations. Decision making is a necessary part of these small groups. Other activities of small groups include information sharing, socializing, relating to people and groups external to the group, educating new members, defining roles, and telling stories (Rothwell, 2021). With that in mind, let's examine three critical assumptions that guide the theory:

problem-solving groups sets of individuals whose main task is to make decisions and provide policy recommendations

task-oriented groups sets of individuals whose main goal is to work toward completing jobs assigned to them

- Conditions in groups promote high cohesiveness.
- Group problem solving is primarily a unified process.
- Groups and group decision making are frequently complex.

The first assumption of Groupthink pertains to a characteristic of group life: cohesiveness. Conditions exist in groups that promote high cohesiveness. Ernest Bormann (1996) observes that group members frequently have a common sentiment or emotional investment, and, as a result, they tend to maintain a group identity. This collective thinking usually guarantees that a group will be agreeable and perhaps highly cohesive.

What is cohesiveness? You probably have heard of groups sticking together or having a high *esprit de corps*. This phrase essentially means that the group is cohesive. **Cohesiveness** is defined as the extent to which group members are willing to work together. It is a group's sense of togetherness. Cohesion arises from a group's attitudes, values, and patterns of behavior; those members who are highly attracted to other members' attitudes, values, and behaviors are more likely to be called *cohesive*.

cohesiveness a cultural value that promotes unity, togetherness, and mutual support in a small group or team

Cohesion is the glue that keeps a group intact. You may have been a member of a cohesive group, although it can be difficult to measure cohesiveness. For instance, is a group cohesive if all members attend all meetings? If all members communicate at each meeting? If everyone seems amiable and supportive? If group members use the word *we* instead of the word *I*? All of these? You know if you have been in a cohesive group, but you may not be able to tell others precisely why the group is cohesive.

Our second assumption examines the process of problem solving in small groups: it is usually a unified undertaking. By this, we mean that people are *not* predisposed to disrupting decision making in small groups. Members essentially strive to get along. Dennis Gouran (1998) notes that groups are susceptible to **affiliative constraints,** which means that group members hold their input rather than risk rejection. According to Gouran, when group members do participate, fearing rejection, they are likely "to attach greater importance to preservation of the group than to the issues under consideration" (p. 100). Group members, then, seem more inclined to follow the leader when decision-making time arrives. Taking these comments into consideration, the board of directors at Melton may simply be a group who recognizes the urgency associated with their financial dilemma. Listening to two board members, Elizabeth and Randy, therefore, is much easier than listening to seven. The two become leaders, and the group members allow them to set the agenda for discussion.

> **affiliative constraints** when members withhold their input rather than face rejection from the group

The third assumption underscores the nature of most problem-solving and task-oriented groups to which people belong: they are usually complex. In discussing this assumption, let's first look at the complexity of small groups and then at the decisions emerging from these groups. First, small group/team members must continue to understand the many alternatives available to them and be able to distinguish among these alternatives. In addition, members must not only understand the task at hand but also the people who provide input into the task. This contention was supported many decades ago and continues today. For instance, social psychologist Robert Zajonc (1965) studied what many people have figured out for themselves: the mere presence of others has an effect on us. He offered a very simple principle regarding groups: when others are around us, we become innately aroused, which helps or hinders the performance of tasks. Nickolas Cottrell and his research team (Cottrell, Wack, Sekerak, & Rittle, 1968) later clarified the findings of Zajonc and argued that what leads people to task accomplishment is knowing that an individual will be evaluated by other individuals. Cottrell and his colleagues believe that group members may be apprehensive or anxious about the consequences that other group members bring to the group. In our opening, for instance, the board members are eager to listen to others with ideas because they do not offer ideas themselves. If any of the five nonspeakers were to openly challenge the ideas of either Elizabeth or Randy, they would be asked for their own ideas. Accordingly, they simply yield the floor to those who have a specific plan.

Marvin Shaw (1981) discusses additional issues pertaining to groups and teams. He notes that a wide range of influences exist in a small group/team—age of group members, competitive nature of group members, size of the group, intelligence of group members, gender composition of the group, and leadership styles that emerge in the group. Further, the cultural backgrounds of individual group members may influence group processes. For instance, because many cultures do not place a premium on overt and expressive communication, some group members may refrain from debate or dialogue, to the surprise or disappointment of other group members. This may influence the perceptions of both participative and nonparticipative group members.

If group dynamics are both complex and challenging, why are people so frequently assigned to group work? Clearly, the answer rests in the maxim "Two heads are better than one." Groups are better at problem solving than individuals alone because of their availability to more information. Further, once a group or team participates in a decision, a higher level of commitment toward the group exists. Groups and team decisions, therefore, may be difficult and challenging, but through group work, people are able to achieve their goals more expeditiously and efficiently.

The relationship of this assumption to Groupthink should not escape you. Two issues merit attention. First, groups whose members are similar to one another are groups that are more conducive to groupthink (Myers & Anderson, 2008). We term this group similarity **homogeneity.** So, as we mentioned earlier, the board of directors at Melton is homogeneous in its backgrounds—they are all part of literary inner circles. This similarity is one characteristic that can foster groupthink.

> **homogeneity** group similarity

Second, group decisions that are not thoughtfully considered by everyone may facilitate groupthink. Quality of effort and the quality of thinking are essential in group decision making. For example, in our chapter-opening story, Elizabeth and Randy clearly offer opinions on what they think is the best course of action for the company. Their charisma, their ability to communicate their vision, and their willingness to openly share their ideas with the Melton team may be intoxicating to a board that is under pressure to resolve a financial dilemma. It's important to note that the two leaders have clear ideas about how to proceed next, failing to identify alternative and additional ways of looking at the problem.

What Comes Before: Antecedent Conditions of Groupthink

Students Talking Tech: Campbell

I know in class it was mentioned that this theory doesn't have a lot to do with social media because it's primarily about teams. But, I know of a time when I was working on a class project with my group on Zoom. We were rushed because we all had some "excuse" for not getting our part done. So, we "forgot" to think about any negative reactions to our ideas—which there were a lot of! Groupthink was, in a way, present in our Zooms because we were all too confident that we'd get an "A."

Janis believes that three conditions promote groupthink: (1) high cohesiveness of the decision-making group, (2) specific structural characteristics of the environment in which the group functions, and (3) stressful internal and external characteristics of the situation.

Group Cohesiveness

We have already discussed cohesiveness and its effects in relation to the three assumptions that guide Groupthink. Cohesiveness is also an antecedent condition. You may be wondering how cohesiveness can lead to groupthink. One reason this may be perplexing is that cohesion differs from one group to another, and different levels of cohesion produce different results. In some groups, cohesion can lead to positive feelings about the group experience and the other group members. Highly cohesive groups also may be more enthusiastic about their tasks and feel empowered to take on additional tasks. In sum, greater satisfaction is associated with increasing cohesiveness.

Despite the apparent advantages, highly cohesive teams may also bring about a troubling occurrence: groupthink. Janis (1982) argues that highly cohesive groups exert great pressure on their members to conform to group standards. Janis believes that as groups reach high degrees of cohesiveness, this euphoria tends to stifle other opinions and alternatives. Group members may be unwilling to express any reservations about solutions. And members even censor their own comments without being provoked. High-risk decisions, therefore, may be made without thinking through consequences. The risks involved in decisions pertaining to war, for instance, are extremely high. A decision to increase military troop presence in another country entails high risks for the soldiers, for the refugees who may flee the country, and for those who remain. Obviously, though, the risks associated with the decision pertaining to where a homecoming committee should hold its meetings are minimal in comparison to the increasing troop presence decision. You can see that risk varies across settings, and assessing risk is critical to the groupthink phenomenon.

Although people may feel confident that they will recognize groupthink when they see it, often they don't. Too much cohesion may be seen as a virtue, not a shortcoming. Imagine sitting around a conference table where everyone is smiling, affirming one another, and wanting to wrap things up (consider our opening story of Melton Publishing). Would you be willing to stop the head nodding and slaps on the back and ask, "But is this the best way to approach this?" To paraphrase the words from a famous fairy tale, who wants to tell the emperor that he has no clothes? Cohesiveness, therefore, frequently leads to conformity, and conformity is a primary route to groupthink.

Before moving on, let's make sure we're clear. We are not suggesting that cohesion automatically leads to groupthink. To be sure, cohesiveness is a necessary ingredient if groups or teams are to arrive at thoughtful, inclusive, and informed decisions. Still, when the effectiveness or consequences of a group's decision remain secondary to a group's cohesion, Janis contends that the group is *prone* to groupthink.

Structural Factors

Janis notes that specific structural characteristics, or faults, promote groupthink. They include (1) insulation of the group, (2) lack of impartial leadership, (3) lack of clear procedures for decisions, and (4) homogeneity of group members' backgrounds. **Group insulation** refers to a group's ability to be unaffected by the outside world and its happenings. Further, it is very easy for insulation to occur if group members do not solicit the opinions or views of those external to the group. In fact, they may be discussing issues that have relevance in the outside world, and yet the members are insulated from its influence. People outside the group who could help with the decision may even be present in the organization but not asked to participate.

> **group insulation** a group's ability to remain unaffected by outside influences

A **lack of impartial leadership** means that group members are led by people who have a personal interest in the outcome. An example of this point can be found in Janis's appraisal of President Kennedy's Bay of Pigs episode. When the president presided over the meetings on the Cuban invasion, Janis observes the following:

> **lack of impartial leadership** groups led by individuals who put their personal agendas first

> [At] each meeting, instead of opening up the agenda to permit a full airing of the opposing considerations, he allowed the CIA representatives to dominate the entire discussion. The president permitted them to refute immediately each tentative doubt that one of the others might express, instead of asking whether anyone else had the same doubt or wanted to pursue the implications of the new worrisome issue that had been raised. (1982, p. 42)

The deference of team members to their leader can be observed in the words of Arthur Schlesinger, Jr., a member of Kennedy's policy group: "I can only explain my failure to do more than raise a few timid questions by reporting that one's impulse to blow the whistle on this nonsense was simply undone by the circumstances of the discussion" (cited in Janis, 1982, p. 39). It is apparent that Kennedy considered other opinions to be detrimental to his plan, and alternative leadership was suppressed. We could probably sum up this perception with one word: *hubris*.

A final structural fault that can lead to groupthink is a **lack of decision-making procedures** and similarity of group members. First, some groups have few, if any, procedures for decision making; failing to have previously established norms for evaluation of problems

> **lack of decision-making procedures**
> failure to provide norms for solving group issues

can foster groupthink. Dennis Gouran and Randy Hirokawa (1996) suggest that even if groups recognize that a problem exists, they still must figure out the cause and extent of the problem. Groups, therefore, may be influenced by dominant voices and go along with those who choose to speak up. Other groups may simply follow what they have observed and experienced in previous groups. In fact, when independent analyses of the *Columbia* space shuttle disaster were undertaken after the craft's breakup, John Schwartz (2005) reports that NASA management was influential in decisions to launch on the fateful day in 2003. Schwartz quotes a former shuttle commander at the Johnson Space Center in Houston: "Managers 'ask for dissenting opinion because they know they are supposed to ask for them,' but the managers 'are either defending their own position or arguing against the dissenting opinion' without seriously trying to understand it" (p. A17). In other words, group members may try to challenge management, but their words are either muted or dismissed.

A second structural fault is the homogeneity of members' backgrounds. Janis (1982) notes that "lack of disparity in social background and ideology among the members of a cohesive group makes it easier for them to concur on whatever proposals are put forth by the leader" (p. 250). We alluded to this fault earlier in the chapter. Without diversity of background and experience, it may be difficult to debate critical issues.

Group Stress

The final antecedent condition of Groupthink pertains to the stress on the group—that is, stress on the team may evoke groupthink. Stress occurs when group members are influenced by issues, resources, or events both within and external to the group. When stress is high, group members may not see any reasonable solution and, therefore, rally around their leader. When decision makers are under great stress imposed by forces outside the group, faulty decision making happens. Stress, as explained in this theory, is a comprehensive concept in that it includes both internal and external stress. For instance, in a work environment, if there are deadlines of an assignment and you're unable to meet those deadlines, there is internal stress in that your supervisor established a time frame and you were unable to accommodate. However, there might also be external stress in that the reason you were unable to meet the deadline is because you have other obligations that you saw as more important (e.g., attending to a sick parent, etc.).

As you can see, teams frequently insulate themselves from outside criticism, form what ostensibly are close bonds, seek consensus, and ultimately develop groupthink. To get a clearer picture of what groupthink looks like, Janis (1982) identifies eight symptoms that can be assigned to three categories. We turn to these symptoms next.

Symptoms of Groupthink

Pre-existing conditions lead groups to concurrence seeking. **Concurrence seeking** occurs when groups try to reach consensus in their final decision. Consider the interpretation of consensus seeking

> **concurrence seeking** efforts to search out group consensus

from Andrea Hollingshead and her colleagues (2005): "Groupthink teams place such a high priority on supporting each other emotionally that they choose not to challenge one another" (p. 30). When concurrence seeking goes too far, Janis contends, it produces symptoms of groupthink. Janis (1982) observes three categories of symptoms: overestimation of the group, closed-mindedness, and pressures toward uniformity.

To illustrate these symptoms and to give a sense of how Janis conceptualized Groupthink, we examine a policy decision that because of the ongoing post-traumatic stress that military veterans frequently experience, continues to resonate today: the Vietnam War. During the height of the war, polls showed that most people felt that the objectives for going to Vietnam were ambiguous and ill-founded. War protestors were prominent in the 1960s and early 1970s, arguing that the country had no clear reason (i.e., policy) for entering the war. The controversy associated with Vietnam parallels the controversies that surrounded U.S. involvement in Iraq and Afghanistan and the war between Israel and Hamas. As with other foreign-policy efforts, there appeared to be no hurry for leaders to modify their style of governing, apparently having little regard for the impact of their decision making (Grube & Killick, 2023).

We present Janis's (1982) interpretation of what took place when President Lyndon Johnson's foreign-policy advisors examined decisions about bombing North Vietnam. Janis's comments are based on extensive discussions with individuals who were presidential advisors, and many of their comments are contained in the following analysis.

Overestimation of the Group

An **overestimation of the group** includes those behaviors that suggest the group believes it is more than it is. Two specific symptoms exist in this category (Katopol, 2016): illusion of invulnerability and a belief in the inherent morality of the group.

> **overestimation of the group** erroneous belief that the group is more than it is

Illusion of Invulnerability The **illusion of invulnerability** can be defined as a group's belief that they are special enough to overcome any obstacles or setbacks. The group believes it is invincible. With respect to the Vietnam War, Janis explains that President Johnson's

> **illusion of invulnerability** belief that the group is special enough to overcome obstacles

foreign-policy group wanted to avoid peace negotiations because they did not want to be viewed as having little bargaining power. The group, therefore, was willing to take a risk and believed that selecting bombing targets in North Vietnam was a wise course of action. The group members based their decision to bomb on four issues: the military advantage, the risk to U.S. aircraft, the potential to widen the conflict to other countries, and civilian casualties. Janis asks whether the team members shared the illusion that they were invulnerable when they confined their attacks to bombing targets.

Theory Into Practice • Groupthink

Theoretical Claim: The Groupthink symptom, illusion of unanimity, can be unconsciously undertaken by those who are shy, inexperienced, or reticent to speak up.

Practical Implication: As the majority of the school board determined that eliminating the philosophy program was necessary because of state budget cuts to education, Mel sat quietly, listening to everyone speak about the need to save money/reduce costs. When a straw vote was conducted, they agreed to the program cut because they were just recently elected to the board and had limited understanding of how the budget process worked.

Belief in the Inherent Morality of the Group   When group members have a **belief in the inherent morality of the group,** they are said to adopt the position that "we are a good and wise group" (Janis, 1982, p. 256). Because the group perceives itself to be good, they believe that their decision making must, therefore, be good. By embracing this belief, group members purge themselves of any shame or guilt, although they ignore any ethical or moral implications of their deci-

> **belief in the inherent morality of the group** assumption that the group members are thoughtful and good; therefore the decisions they make will be good

sion. Janis found it interesting that Johnson and his advisors were not concerned with bombing villages in North Vietnam; for them, the moral consequences did not outweigh the perception that the United States might be weak or fearful. In fact, the foreign-policy council continued to encourage bombing of Hanoi (North Vietnam), even though peace initiatives were concurrently under way in Poland. A sense of moral certitude prevailed because the president and his advisors believed that North Vietnam would not negotiate a surrender.

Closed-Mindedness

When a group is **closed-minded,** it ignores outside influences on the group. The two symptoms discussed by Janis in this category are stereotypes of out-groups and collective rationalization.

> **closed-mindedness** a group's willingness to ignore differences in people and warnings about poor group decisions

Out-Group Stereotypes   Groups in crisis frequently engage in **out-group stereotypes,** which are stereotyped perceptions of rivals or enemies. These stereotypes underscore the fact that any adversaries are either too weak or too stupid to counter offensive tactics. For Johnson's advisors, enemy meant *Communist*. It was this stereo-

> **out-group stereotypes** stereotyped perceptions of group enemies or competitors

type that kept the advisors from seeing the enemy as people. Janis (1982) reasons that because the North Vietnamese were considered to be Communist enemies, this embodiment of evil justified the "destruction of countless human lives and the burning of villages" (p. 111).

Collective Rationalization   The fourth symptom of Groupthink, **collective rationalization,** refers to the situation in which group members ignore warnings that might prompt them to reconsider their thoughts and actions before they reach a final decision. In

> **collective rationalization** situation in which group members ignore warnings about their decisions

most circumstances, this symptom can also be interpreted to be a rationalization of "bad news." President Johnson and his staff were given a number of forewarnings—from intelligence agencies, among others—about the implications of bombing North Vietnam. Some in the intelligence community believed that bombing sites such as oil facilities would do nothing to erode Communist operations. Nonetheless, Johnson's group maintained its unified position on escalating the bombing. Janis (1982) observed that the men in Johnson's inner circle were very convinced of the importance of the Vietnam War to the United States. Therefore, revisiting any strategies to exit the war was out of the question.

Pressures Toward Uniformity

The **pressure toward uniformity** can be enormous for some groups. Janis believed that some groups who go along to get along may be setting themselves up for groupthink. The four symptoms in this category are self-censorship, an illusion of unanimity, the presence of self-appointed mindguards, and direct pressure on dissenters.

> **pressure toward uniformity** occurs when group members go along to get along
>
> **self-censorship** group members minimize personal doubts and counter-arguments

Self-Censorship Earlier we discussed self-censorship. **Self-censorship** refers to group members' tendency to minimize their doubts and counterarguments. They begin to second-guess their own ideas. For those in President Johnson's foreign-policy group, self-censorship was exhibited when group members dehumanized the war experience. These advisors would not allow themselves to think of the civilian people being killed because such thinking would only personalize the war. Janis argues that silencing one's own opposing views and using in-group rhetoric further bolster the decisions of the group.

Illusion of Unanimity Another symptom of Groupthink is an **illusion of unanimity,** which suggests that silence means consent. Although some advisors in Johnson's inner circle felt differently about the Vietnam intervention, they were silent. This silence prompted others around the table to believe that there was consensus in planning and execution.

Self-Appointed Mindguards Groups in crisis may include **self-appointed mindguards**—group members who shield the group from adverse information. Mindguards believe that they act in the group's best interest. Walt Rostow, White House assistant, effectively played the role of mindguard in Johnson's group. Janis relates the following: "Rostow cleverly screened the inflow of information and used his power to keep dissident experts away from the White House. This had the intended effect of preventing the President and some of his advisers from becoming fully aware of the extent of disaffection with the war and the grounds for it" (1982, p. 119). Ironically, within the group, keeping the peace at the White House was more important than maintaining it in Vietnam.

> **illusion of unanimity** belief that silence equals agreement
>
> **self-appointed mindguards** individuals who protect the group from adverse information
>
> **pressures on dissenters** direct influence on group members who provide thoughts contrary to the group's

Pressures on Dissenters The final symptom involves pressuring any team member who expresses opinions, viewpoints, or commitments that are contrary to the majority opinion. Janis calls this **pressures on dissenters.** In his interviews with those within President Johnson's inner circle, Janis discovered that the group members formed a gentlemen's club, where mutual respect and congenial talk pervaded. Janis also later realized that these men often turned to one another for support, and they were loyal to one another. Is it any wonder, then, that group members believed that everyone agreed to the course of action? This attitude would prevail, of course, considering that those who openly disagreed frequently resigned or were replaced (Janis, 1982). Robert McNamara, Secretary of Defense under President Johnson, was a dissenter who believed that bombing North Vietnam was the wrong way to get the enemy to the negotiation table. However, he was under much pressure to avoid disagreeing with members of the gentlemen's club.

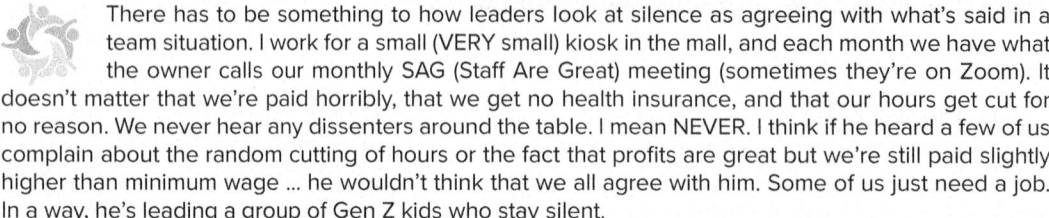

Students Talking: Angie

There has to be something to how leaders look at silence as agreeing with what's said in a team situation. I work for a small (VERY small) kiosk in the mall, and each month we have what the owner calls our monthly SAG (Staff Are Great) meeting (sometimes they're on Zoom). It doesn't matter that we're paid horribly, that we get no health insurance, and that our hours get cut for no reason. We never hear any dissenters around the table. I mean NEVER. I think if he heard a few of us complain about the random cutting of hours or the fact that profits are great but we're still paid slightly higher than minimum wage ... he wouldn't think that we all agree with him. Some of us just need a job. In a way, he's leading a group of Gen Z kids who stay silent.

Think Before You Act: Ways to Prevent Groupthink

It bears repeating: our discussion of Groupthink and the issues that accompany this theory may have given you the impression that all cohesive groups and teams will be prone to groupthink. This isn't true. Janis (1982) notes that cohesiveness is a necessary but not sufficient condition of groupthink. Nonetheless, frequently, when groups find themselves in highly cohesive situations and when decision makers are under great stress, groupthink may materialize.

How can group members learn to avoid groupthink, or at least work toward more healthy interactions? Janis (1989; Herek, Janis, & Huth, 1987) suggests that groups engage in vigilant decision making, which involves (1) looking at the range of objectives group members wish to achieve, (2) developing and reviewing action plans and alternatives, (3) exploring the consequences of each alternative, (4) analyzing previously rejected action plans when new information emerges, and (5) having a contingency plan for failed suggestions. Janis offers additional recommendations, but critics such as Paul 't Hart (1998) question whether Janis's recommendations inadvertently erode collegiality and foster group factionalism.

To avoid oversimplifying the groupthink problem, 't Hart (1990) has proposed four general recommendations for teams who may be prone to groupthink: (1) require oversight and control, (2) embrace whistle-blowing in the group, (3) allow for objection, and (4) balance consensus and majority rule. We will look at each recommendation (**Table 18.1**).

First, 't Hart believes that one way to enhance group decision making is to impose some external oversight and control. He argues that groups need to hold key decision makers accountable for their actions; this should be done *before* groups begin their deliberations about issues. Accountability may take the form of a committee that serves to enforce control (vis-à-vis rules, governance procedures, decorum, etc.). 't Hart theorizes that such committees prompt group members to challenge collective rationalizations and inaccurate perceptions. Reflecting on our example of President Johnson and his foreign-policy group, 't Hart proposes that the inner circle of advisors was insulated from external oversight. As well, there were inadequate intragroup measures to improve the group's decisions on North Vietnam.

Table 18.1 Preventing Groupthink

RECOMMENDATION	ACTION
Require Oversight and Control	*Establish a parliamentary committee:* develop resources to proactively monitor ongoing policy ventures; establish incentives to intervene; link personal fate to fate of group members.
Embrace Whistle-Blowing	*Voice doubts:* avoid suppressing concerns about group processes; continue to disagree and debate when no satisfactory answers are given; question assumptions.
Allow for Objection	*Protect conscientious objectors:* provide for group members' exits; do not play down the moral implications of a course of action; acknowledge private concerns about ethical issues in the group.
Balance Consensus and Majority Rule	*Alter rules governing choice:* relieve pressure on groups in minority positions; dissuade the development of subgroups; introduce a multiple advocacy approach to decisions.

Source: Adapted from 't Hart, P. (1990).

In addition to accountability, 't Hart proposes that **whistle-blowing** be embraced in a group's culture. That is, group members "should be encouraged to voice concerns rather than to voluntarily suppress them, to question assumptions rather than to accept them at face value, and to continue to disagree and debate when no satisfactory answers to their concerns are given by the rest of the group"

whistle-blowing process in which individuals report unethical or illegal behaviors or practices to others

(p. 385). Anna Mulrine (2008) writes that the 2003 Iraq invasion brought about great chaos in the planning and execution of decision making by the Army. To that end, the Army has developed a "cadre of devil's advocates" (p. 30) who are charged with "questioning prevailing assumptions to avoid getting sucked into that Groupthink" (p. 30) which can happen in stressful times. Ultimately, the Military Whistleblower Protection Act was amended to expand the rights of whistle-blowers protecting disclosures in legal cases and closing loopholes. President Obama eventually signed this overhaul of the Act and it became law in 2013. Embracing whistle-blowing in this way results in looking at a decision from alternate points of view. Further, scholars such as Eva Tsahuridu and Wim Vandekerckhove (2008) advocate that groups protect whistle-blowers because groups usually need dissenting voices when decisions have lasting and significant consequences.

A third suggestion by 't Hart is that groups allow **conscientious objectors,** or group members who refuse to participate in the decision-making process because it would violate their conscience. He reasons that groupthink causes groups to downplay the moral implications of their decisions, and if conscientious objectors know that they can

conscientious objectors group members who refuse to participate because it would violate personal conscience

exit a conversation based on moral or ethical grounds, then they may be more likely to speak up. Thus, these objectors may be able to raise doubts about the decision or even to protest it. President Johnson's decision to continue bombing Vietnam, therefore, may have been different if conscientious objectors had been allowed to introduce the moral implications of killing innocent people.

Finally, 't Hart advocates that teams not require consensus but work instead toward a majority of support. Because consensus demands that every group member agree on a decision, group members often feel pressured to consent (illusion of unanimity). 't Hart believes that groups should strive toward consensus but be prepared for majority support. If groups adopt this orientation, 't Hart believes, they will function more like teams. Janis (1982) discusses the fact that several members of President Johnson's inner circle of advisors (McNamara, Rusk, Bundy) wanted to temporarily halt the Hanoi bombings; we can only speculate how the passion for consensus affected both the United States and the Vietnamese people.

Polythink: Moving Beyond the Groupthink Phenomenon

There have been several academic discussions regarding groupthink, its effects, and ways to resolve it. And one view of team decision making employs a *continuum of cohesion* (Barr & Mintz, 2023; Mintz & Wayne, 2016). On one side of this continuum is the sort of group decisions that are rooted in uniformity; on the other side, decisions are rooted in divergent points of view. This latter point is called *polythink*, a decision-making dynamic involving group members who advocate for their individual opinions. Alex Mintz and Carly Wayne introduced this new way of looking at groups and admit that polythink relates to groups other than those charged with a task. Further, they note that polythink "is essentially the opposite of groupthink on the continuum of decision-making from 'completely cohesive' (Groupthink) to 'completely fragmented' (Polythink)" (p. 4).

Similar to groupthink, polythink occurs under various conditions. First, too much diversity of opinion occurs, rather than too little. This typically is illustrated by several team members having several ways of looking at an issue. Second, team members are rather relentless in pushing for their viewpoints to be entertained by the group. Third, group members do not have strong interpersonal commitments to each other, resulting in less understanding of the fields of experience of others. In fact, in discussions between various organizational units, even departmental rivalries can negatively affect the decisions. Fourth, polythink is opportunistic when there are *disjointed* decisions taking place; that is, decisions with major variation in dissension (e.g., vehemently agreeing with arguments, passionately rebuffing other arguments, such as those articulated in our opening story of Tina and Randy). Such a fragmented group will only result in fragmented decisions. Finally, leadership style can influence the development of polythink. Strategic leaders, for instance, may cultivate opinions of the more seasoned team members, alienating the newcomers who, despite their tenure, may be more creative and adept in decision making.

Clearly, polythink is different from groupthink, and yet both involve the complex interplay among communication, groups and teams, decision making, and individual input patterns. We will continue to see additional research into how each differ and overlap as we move into a future where face-to-face meetings are substituted with Teams and Zoom. Further, social media platforms will become prominent sites of research as both polythink and groupthink are unpacked to assess their relevancy.

Integration and Critique

Groupthink is a theory dedicated to understanding the decision-making process in small groups and teams. The theory has been tested and expanded using primarily experimental methods, making it aligned with a quantitative approach. Yet, some research (Rajakumar, 2019) notes that many of these experimental studies have "yet to show full or significant support" (p. 1) for Groupthink, but qualitative work—including case study analyses—has "provided substantial support for the model" (p. 1). Janis believes that groups frequently make decisions with profound consequences, and although he focused his efforts on foreign-policy groups, as we learned previously, the application of Groupthink terminology resonates in many other decision-making teams. Among the criteria for evaluating a theory, four are especially relevant for discussion: scope, testability, heurism, and test of time.

Integration

Communication Tradition	Rhetorical | Semiotic | Phenomenological | Cybernetic | **Socio-Psychological** | **Socio-Cultural** | Critical
Communication Context	Intrapersonal | Interpersonal | **Small Group/Team** | **Organizational** | Public/Rhetorical | Mass/Media | Cultural
Approach to Knowing	**Positivistic/Empirical** | **Interpretive/Hermeneutic** | Critical

Critique

Evaluation Criteria	**Scope** | Logical Consistency | Parsimony | Utility | **Testability** | **Heurism** | **Test of Time**

Scope

Despite the fact that many Groupthink principles can be applied to several types of groups, Janis was clear in his original conceptualization in applying Groupthink solely to decision-making groups in crisis periods; he does not readily apply his thinking to every group type. Although the theory has been applied to groups as diverse as presidential advisors and ice hockey teams, the focus has been in those groups that are *decision-making* groups. Nonetheless, Jennifer Kretchmar (2021) concluded that the theory became popular immediately, primarily because the scope was relevant to a variety of disciplines and various societal problems. Janis may not have intended the theory to "advance" so considerably in scope, but its wide range cannot be understated.

Testability

Group and team communication scholars have pointed to some validity problems with the theory, calling into question its testability. For instance, Jeanne Longley and Dean Pruitt (1980) criticize the validity of the theory. They argue that half of the symptoms of Groupthink are not associated with concurrence seeking—a key feature of the theory. They charge that "a theory should be a logical progression of ideas, not a grab-bag of phenomena that were correlated with each other in a sample of six cases" (p. 80). Further, James Rose (2011), in an exhaustive literature review on Groupthink, found that with respect to testing the entire "model" of Groupthink, problems abound. Finally, other researchers (e.g., Akhmad, Chang, & Deguchi, 2021) identified a number of contradictory conclusions regarding the testability of the theory—some supporting the testing protocols and others criticizing them. Because many of Janis's variables are not well-defined, the theory is prone to lapses in validity and reliability.

Heurism

The theory of Groupthink is a heuristic undertaking; the theory and many of its elements have been employed in a number of studies and have enjoyed the attention of many communication and social psychology scholars. In addition to foreign-policy decisions, writers have studied Groupthink and applied its concepts and tenets to student killings at Kent State (Hensley & Lewis, 2010), the 1989 Tiananmen Square disaster (Lee, 2020), the sexual abuse cover-up at Penn State University (Cohen & DeBenedet, 2012), Hurricane Katrina (Garnett & Kouzmin, 2009), physician attitudes toward the donor heart program (Farr & Colvin, 2019), and heterosexism in the academy (*"Don't Say Gay"*) (Goodrich, 2022). More recent developments have integrated Groupthink principles and practices with TikTok (Booth, 2022). The theory has also generated a number of assumptions about group behavior, and Groupthink remains an important part of the literature on group decision making. In addition, as we noted earlier, it has spawned new ways of looking at teams via polythink, which have been applied to the wars in Syria, Afghanistan, Iraq, and Iran (Barr & Mintz, 2023; Mintz & Wayne, 2016).

Test of Time

The theory of Groupthink has withstood the test of time. Although they have examined the theory from various scholarly angles, researchers continue to investigate many of features of the theory and the theory has gained both academic and popular attention. On the 30th anniversary of Groupthink, Schwartz and Wald (2003) called Janis a "pioneer in the study of social dynamics" (p. 4). Finally, given that government policy decisions will always exist and given that many governmental leaders surround themselves with individuals who are typically conflict-avoidant, the likelihood of future instances of groupthink remains rather high.

Closing

Some people believe that Groupthink may be more intuitively appealing than theoretically driven. The theory, however, continues to receive attention. In fact, Janis's thinking related to Groupthink has been quite influential in several fields of study, including communication, cognitive and social psychology, anthropology, and political science. Few would debate the failure of the foreign-policy fiascoes and historical blunders outlined by Janis: massive violence and casualties, loss of confidence in governmental decisions, and policymaking gone wrong. For these reasons alone, Janis is credited with helping us identify and examine one type of group decision-making problem. His theory continues to resonate, particularly today.

Discussion Starters

👆 **Case-In-Point:** What advice would you give the Melton Publishing board of directors before they meet to discuss their financial situation? Frame your advice with Groupthink concepts in mind.

⌨ **Try-It-Your-Selfie:** If you had a chance to create a YouTube channel about Groupthink and social media, what themes, practices, and cautionary tales would you include? Who would be your guests? What would be the target audience and how would you persuade them to subscribe to your channel?

1. Have you ever been in a small group/team with too much cohesiveness? If so, did groupthink develop? If so, how did you know? If not, what prevented it from occurring?
2. In his book, *Groupthink,* Janis asks if a little knowledge of groupthink is a dangerous thing. Why do you think Janis asks this question, and what are the consequences of knowing about Groupthink? Incorporate examples into your response.

3. Apply principles of Groupthink to recent domestic and foreign-policy decisions made by the United States.

4. Discuss the delicate balance between sufficient group cohesion and exceedingly high amounts of group cohesion. Illustrate your response with examples from your own group experiences.

References

Akhmad, M., Chang, S., & Deguchi, H. (2021). Closed-mindedness and insulation in groupthink: Their effects and the devil's advocacy as a preventive measure. *Journal of Computational Social Science, 4*, 455–478.

Barr, K., & Mintz, A. (2023). Groupthink, polythink, and con-div. In P. Mello & F. Ostermann (Eds.), *Routledge handbook of foreign policy analysis methods* (pp. 269–288). Routledge.

Booth, J. (2022). *Have we all gone mad? Why groupthink is rising and how to stop it.* Biteback Publishing.

Bormann, E. G. (1996). Symbolic convergence theory and communication in group decision making. In R. Y. Hirokawa & M. S. Poole (Eds.), *Communication and group decision making* (pp. 81–113). Sage.

Cohen, L. J., & DeBenedet, A. T. (2012, July 17). Penn State cover-up: Groupthink in action. *Time*, 180.

Cottrell, N. B., Wack, D. L., Sekerak, G. J., & Rittle, H. (1968). Social facilitation of dominant responses by the presence of an audience and the mere presence of others. *Journal of Personality and Social Psychology, 9*, 245–250.

Farr, M., & Colvin, M. M. (2019). What number are we? Groupthink and the donor heart offer freefall. *Circulation: Heart Failure, 12*.

Garnett, J., & Kouzmin, A. (2009). Crisis communication post-Katrina: What are we learning. *Public Organizational Review, 9*, 385–398.

Goodrich, K. M. (2022). Groupthink in counselor education: "Don't say gay" and the visibly invisible heterosexism of the academy. *The Journal for Specialists in Group Work, 47*(2), 83–89.

Gouran, D. S. (1998). The signs of cognitive, affiliative, and egocentric constraints in patterns of interaction in decision-making and problem-solving groups and their potential effects on outcomes. In J. Trent (Ed.), *Communication: Views from the helm for the 21st century* (pp. 98–102). Allyn & Bacon.

Gouran, D. S., & Hirokawa, R. Y. (1996). Functional theory and communication in decision-making and problem-solving groups. In R. Y. Hirokawa & M. S. Poole (Eds.), *Communication and group decision making* (pp. 55–80). Sage.

Grube, D. C., & Killick, A. (2023). Groupthink, polythink and the challenges of decision-making in cabinet government. *Parliamentary Affairs, 76*(1), 211–231.

Henderson, P. G. (2018). Anatomy of a national security fiasco: The George W. Bush administration, Iraq, and groupthink. *Humanitas, 31,* 46–80.

Hensley, T. R., & Lewis, J. M. (2010). *Kent state and May 4th–A social science perspective.* Kent State University Press.

Herek, G. M., Janis, I. L., & Huth, P. (1987). Decision making during international crises: Is quality of process related to outcome? *Journal of Conflict Resolution, 31*, 203–226.

Hirokawa, R. Y., Gouran, D. S., & Martz, A. E. (1988). Understanding the sources of faulty group decision making: A lesson from the *Challenger* disaster. *Small Group Behavior, 19*, 411–433.

Hollingshead, A. B., Wittenbaum, G. M., Paulus, P. B., Hirokawa, R. Y., Ancona, D. G., Peterson, R. S., & Yoon, K. (2005). A look at groups from the functional perspective. *Theories of small groups: Interdisciplinary perspectives, 21*–62.

Janis, I. L. (1972). *Victims of groupthink: A psychological study of foreign-policy decisions and fiascoes.* Houghton Mifflin Company.

Janis, I. L. (1982). *Groupthink: Psychological studies of policy decisions and fiascoes.* Houghton Mifflin Company.

Janis, I. L. (1989). *Crucial decisions: Leadership in policymaking and crisis management.* Free Press.

Joffe, A. R. (2021). COVID-19: Rethinking the lockdown groupthink. *Frontiers in Public Health, 9.*

Katopol, P. F. (2016). Groupthink: Group dynamics and the decision-making process. *Library Leadership & Management, 30,* 1-6.

Kretchmar, J. (2021). *Groupthink.* Salem Press.

Lee, T. C. (2020). Groupthink, qualitative comparative analysis, and the 1989 Tiananmen Square disaster. *Small Group Research, 51*(4), 435-463.

Longley, J., & Pruitt, D. G. (1980). Groupthink: A critique of Janis's theory. *Review of personality and social psychology, 1*(1980), 74-93.

Mintz, A., & Wayne, C. (2016). *The polythink syndrome: US foreign policy decisions on 9/11, Afghanistan, Iraq, Iran, Syria, and ISIS.* Stanford University Press.

Mulrine, A. (2008, June 2). To battle groupthink, the Army trains a skeptics corps. *U.S. News & World Report, 30,* 32.

Myers, S. A., & Anderson, C. M. (2008). *The fundamentals of small group communication.* Sage.

Rajakumar, N. (2019). *Why empirical studies of the Groupthink model have failed.* [CMC Senior Theses, Claremont McKenna College]. Retrieved from https://scholarship.claremont.edu/cmc_theses/2080.

Rose, J. (2011). Diverse perspectives on the Groupthink theory—A literary review. *Emerging Leadership Journeys, 4,* 37-57.

Rothwell, J. D. (2021). *In mixed company: Communicating in small groups and teams.* Cengage.

Schwartz, J. (April 4, 2005). Some at NASA say its culture is changing, but others say problems still run deep. *The New York Times.* A19.

Schwartz, J., & Wald, M. L. (2003, March 9). 'Group-think' is 30 years old, and still going strong. *New York Times,* p. 9.

Shaw, M. E. (1981). *Group dynamics: The psychology of small group behavior.* McGraw Hill.

Tsahuridu, E. E., & Vandekerckhove, W. (2008). Organisational whistleblowing policies: Making employees responsible or liable?. *Journal of Business Ethics, 82,* 107-118.

't Hart, P. (1990). *Groupthink in government: A study of small groups and policy failure.* Taylor & Francis.

't Hart, P. (1998). Preventing groupthink revisited: Evaluating and reforming groups in government. *Organizational Behavior and Human Decision Processes, 73,* 306-313.

Zajonc, R. B. (1965). Social facilitation. *Science, 149*(3681), 269-274.

CHAPTER 19

Organizational Culture Theory

*Based on the research of **Clifford Geertz, Michael Pacanowsky**, and **Nick O'Donnell-Trujillo***

> **"The particular potential of the organizational culture metaphor lies in its ability to liberate our thinking about both organizations and communication."**
>
> —Michael Pacanowsky and Nick O'Donnell-Trujillo

Amelia Callahan

As an employee of Grace's Jewelers, Amelia Callahan had always known that her job was unlike those of her friends. The company employed about 180 people throughout its 20 stores in the southeastern United States and targeted primarily teenage girls who frequented malls. Its founder, Grace Talmage, had always made it a point to bring in homemade cookies to employees like Amelia each month, helping them to feel good about working in such a small company.

Amelia's relationship with Grace had always been quite cordial. Why wouldn't it be? Despite the fact that the "big box" mall tenants were slowly leaving because of the online craze, Grace's location had been doing quite well ... as had Amelia. She received excellent commission rates and a reasonable health care package (including dental care) and got along superbly with her supervisor. In addition, Amelia and the other employees were able to wear casual clothing to work, which made them the envy of other workers in the mall. All of this may explain why Amelia had worked for the company for almost nine years and why she had no plans to leave—until now.

After 24 years in the business, Grace had decided it was time to sell the business and retire. Because of the profits over the years, Jewelry Plus, a large retail jewelry store, had decided to bid for the small chain. Although Grace hadn't wanted to sell to such a retail giant, its bid had been simply too good to pass up. In the end, she had decided

to sell her business, much to the disappointment of her employees. Amelia had been especially concerned after reading online gossip about the larger company's treatment of its employees and its way of handling day-to-day operations. She had privately wondered how much would change once Grace sold the stores. She needed the job, though, and had decided to stay on.

Amelia's instincts were right. Once the company's transition was completed, she had to undergo a "new employee" webinar orientation, which meant standing in front of all of the new employees and informally talking about why she had applied to the company. Among the new company policies were new dress codes and a new policy for store exchanges. Amelia could no longer wear casual clothing; instead, she had to wear a company uniform.

With respect to store exchanges, company policy changed from "complete satisfaction or 100 percent money-back guarantee" to "product must be returned within 10 days with register receipt." Although Amelia thought that this new policy would turn many customers away, the success of Jewelry Plus was proof enough that it had worked in the past.

Finally, with the new company, her health benefits no longer included dental or vision coverage. This lack of coverage was grist for the gossip mill. The story that Amelia heard during her "orientation" was that an employee actually lost a back tooth because she couldn't afford the dental bills! With all the changes in store policies, dress codes, and company philosophies, Amelia and a number of her coworkers felt overwhelmed. In fact, many of Amelia's coworkers, with whom she had worked over the past nine years, quit their jobs. As a single parent, however, Amelia felt that she couldn't resign just yet.

On top of all that, her new boss was a disaster! Amelia and her coworkers nicknamed him "The Shadow" because he stood right behind them as they waited on and rang up their customers. Having her supervisor watch everything she did was annoying and seemed pointless to Amelia, especially considering the fact that most of her customers were teenagers, who were fickle in their purchasing behaviors. She had no idea how other stores were handled, but in Amelia's store the manager was obsessed with policy. There was even a rumor that he scoured Instagram accounts of his employees!

Despite these concerns, Amelia went to the company's first picnic. She didn't really want to, but she thought that she should give the company a chance. As she and her coworkers drank iced tea and ate hot dogs, they seemed to bond. Former employees of Grace's Jewelers told the workers of the retail giant about the way things used to be. They in turn, seemed to be genuinely interested in hearing about people like Gabby, the 70-year-old retiree who told every customer she met about the time she got caught in a rainstorm while fishing. There was a lot of laughing about the old times.

The day ended quite differently from the way Amelia had envisioned it. She made a few friends, reminisced about the past, and felt a bit more comfortable with her future. Although she knew that her boss would be difficult to deal with, and as hard as she tried to reject the company gossip, Amelia decided that she would try to make the most out of her job. At the very least, she thought, she had some people in whom to confide.

Once you graduate from college, it's likely that you will work for an organization. Organizational life is characterized as much by change as it is by anything else. And, change is frequently marked by such feelings as excitement, anxiety, uncertainty, frustration, and disbelief. These emotions are especially acute during stressful times, for example, during a recession, company layoffs, and even pandemics!

Look online or go to any bookstore and you're sure to see a number of books on organizational life. This pop culture approach to the world of corporate USA is everywhere. Some authors tell us that there are *Ten Easy Ways to Get a Raise* or that there are *Eight Safe Steps to Being Promoted*. Other authors have made millions touting the importance of *Communicating with Difficult People* and *Working to Live and Living to Work*. Most of these books center on what people can do to make their lives easier in the workplace. The problem, however, is that organizational life is very complex. Few "easy ways" to anything exist in organizations and clearly, one difficult person one day may turn into a pleasant person the next day. And one social media post can influence a number of company decisions.

To understand organizational life beyond pop culture—including an organization's values, stories, goals, practices (including technological), and philosophies—Michael Pacanowsky and Nick O'Donnell-Trujillo (1982, 1983, 1990) conceptualized Organizational Culture Theory (OCT). Pacanowsky and O'Donnell-Trujillo believe that organizations can best be understood using a cultural lens, an idea originally proposed by anthropologist Clifford Geertz. They believe that researchers are limited in their understanding of organizations when they follow the scientific method; the quantitative approach, they believed, is constrained by its task of measuring, rather than discovering. Pacanowsky and O'Donnell-Trujillo (1982) argue that OCT invites all researchers "to observe, record, and make sense of the communicative behavior of organizational members" (p. 129). They embrace the "totality of lived experience within organizations" (Pacanowsky, 1989, p. 250). The theorists paint a broad stroke in their understanding of organizations by stating that "culture is not something an organization has; a culture is something an organization is" (Pacanowsky & O'Donnell-Trujillo, 1982, p. 146). In other words, culture is usually an intangible experience.

Culture is communicatively constructed by organizational practices, and culture is distinct to each organization. Culture within an organization is usually interpreted to be visible, that is, that which others can observe (e.g., furnishings, company website, etc.). Yet, the invisible component of culture (e.g., stories, morale, productivity) is typically the type of human capital that is the most influential upon company dynamics (Ince, 2023). For the theorists, understanding individual organizations is more important than generalizing from a set of behaviors or values across organizations. Further, it is important to accept the notion that organizational culture—or organizational climate—is created over a period of time, something Amelia figured out after her transition to Jewelry Plus. These thoughts form the backdrop of the theory.

"That's our mission statement."

Dave Carpenter/CartoonStock Ltd

Think about the types of organizations to which you now belong. Some of you may have found someone significant on Meetup and others you affiliate with because you have little choice (e.g., an HMO). Organizations vary in scope and size and contain a number of unique practices. For instance, one organization that we all have in common is an academic one—a college or university. You've heard and perhaps shared stories about certain professors and classes to take or to avoid. There are various rites of passage in college, such as attending new-student orientation, pledging during fraternity or sorority rush, establishing social media groups, and yes, eating cafeteria food! Practices such as advising and internships also characterize institutions of higher education, as do discussions related to student debt, child care, and personal health. Years ago, students visited professors to get information and clarification on different subjects; today, much of that is done via email. The academic organization clearly has a culture that is quite unique from others.

Students Talking Tech: Paco

Since COVID, I've heard nothing but bad things about how the culture of different companies has changed. My sister works in a hospital and they used to be "patient-oriented." Now, they just clock in their hours and move on. My partner works in a large restaurant and before COVID, people were very friendly. He loved going to work and a lot of times, he and his co-workers would go out afterwards. Now, they just go to work and leave. COVID has changed things in that people are not as sociable at work as they used to be.

It's clear that the essence of organizational life is found in its culture. In this sense of the word, *culture* does not refer to the variety of races, ethnicities, and backgrounds of individuals, a perspective we discussed in **Chapter 2.** Rather, according to organizational culture theorists, culture is a way of living in an organization. Organizational culture (and climate) includes connecting an organization with the individuals, which entails the emotional and psychological climate or atmosphere. Some authors call this "organizational genes" (Small, 2019, p. 11), and it usually involves those "invisible" components we identified earlier—employee morale, attitudes, productivity, competition, autonomy, and cooperation. Organizational culture also includes all the symbols (actions, routines, conversations, etc.) and the meanings that people attach to these symbols. Cultural meaning and understanding, in turn, are achieved through the interactions employees and management have with one another. Robert Lussier & Christopher Achua (2022) states that although organizational culture is difficult to describe, nearly all employees know what kind of culture an organization possesses. With all of this in mind, we begin our discussion of OCT by first interpreting culture and then presenting three assumptions of the theory.

Theory at a Glance • Organizational Culture Theory

 People are like spiders who are suspended in webs that they create and cultivate at work. An organization's culture is composed of shared symbols, each of which has a unique meaning. Organizational stories, rituals, websites, office decor, values, and rites of passage are examples of the culture of an organization and the essence of Organizational Culture theory. When both the visible and invisible components of an organization come together, company stakeholders are part of the organizational culture.

The Cultural Metaphor: Of Spider Webs and Organizations

As noted earlier in the book, culture is among the most complex terms to describe. The origin of the word *culture* is interesting. Culture originally referred to preparing the ground for tending crops

organizational culture the essence of organizational life

and animals. It was interpreted as fostering growth. When we contextualize "culture" within an organization, we need to understand that there is more than "what meets the eye." **Organizational culture** includes what is visible (e.g., office wall art, employee uniforms, etc.), but also what we can't "see," including, as we alluded to earlier, employee convictions, priorities, among other similar things.

Pacanowsky and O'Donnell-Trujillo (1982) believe that organizational culture is the essence of organizational life. As we mentioned earlier, they apply anthropological principles to construct their theory. Specifically, in their model, they adopt the Symbolic-Interpretive approach articulated by Clifford Geertz (1973). Geertz remarks that people are animals "suspended in webs of significance" (p. 5). He adds that people spin webs themselves. Pacanowsky and O'Donnell-Trujillo (1982) comment on Geertz's metaphor:

> The web not only exists, it is spun. It is spun when people go about the business of construing their world as sensible—that is, when they communicate. When they talk, write a play, sing, dance, fake an illness, they are communicating, and they are constructing their culture. The web is the residue of the communication process (p. 147).

A primary goal of researchers, then, should be to think about all possible weblike configurations (features) in organizations. Geertz invokes the image of a spider web deliberately. He believes that culture is like the webs spun by a spider. That is, webs are intricate designs, and each web is different from all others. Furthermore, webs "represent strength, life, and cohesion, but they are also things that need constant maintenance" (Modaff, Butler, & DeWine, 2012, p. 96). For Geertz, cultures are like this as well. Basing his conclusions on various cultures around the world, Geertz argues that cultural communities are all different and that their uniqueness should be celebrated. To understand a culture, Geertz believes that researchers should begin to focus on the meaning shared within it. We examine more of Geertz's beliefs later.

Pacanowsky and O'Donnell-Trujillo (1983) apply these basic principles to organizations. Employees and managers alike "spin their webs." People are critical in the organization, and therefore, it is important to study their behaviors in conjunction with the overall organization. Pacanowsky and O'Donnell-Trujillo claim that members of organizations engage in a number of communication behaviors that contribute to the culture of the company. They may do this through a number of ways, including gossiping, joking, backstabbing, or becoming romantically involved with others. Further, with 70 percent of companies using some form of social media (Wong, 2023), we need to accept that much of this communication will take place via technology.

As you think about these issues, think, too, about the various organizations and their various identities and values. Maybe you're thinking about Google, with founders Larry Page and Sergey Brin wanting to create a culture similar to the feeling of what it means to be a part of a great project in graduate school. Or perhaps you immediately thought of Amazon, with founder Jeff Bezos advocating a "gladiator culture," suggesting aggressive creation and execution of goals. Others of you might have considered the Disney corporate culture, which is leader-driven insofar as the actions and behaviors of the company's leaders dictate the type of culture in which employees find themselves (Disney Institute, 2023).

The organizational culture at Jewelry Plus will be inevitably revealed in a number of ways. You will recall that Amelia learned of the new owner through gossip and that the company picnic was a way for her to learn more about the new company culture. No doubt she will experience an organizational climate with her new job that is very different from what she experienced with Grace's Jewelers. The company has changed, the faces are new, and the rules reflect new ownership. Amelia also contributes to the spinning of the organizational web by both responding to company stories and passing them on to others. In sum, the web of organizational culture has been spun. This broad perspective underscores why Pacanowsky and O'Donnell-Trujillo (1983) argue that organizational culture "is not just another piece of the puzzle; it **is** the puzzle" (p. 146).

Theory Into Practice • Organizational Culture Theory

Theoretical Claim: Culture is an intangible and crucial component of all organizations.

Practical Claim: As the HR director of *BGone,* a start-up focusing on European travel, Akira knows that the company's mission is related to (a) how employees are treated and their opportunities (e.g., monthly social events, child care, stock options, etc.) and (b) community commitments (e.g., mentoring, etc.). Taken together, these efforts constitute the organizational culture that Akira must communicate to all employees.

Assumptions of Organizational Cultural Theory

The study of organizations and workplace settings began in earnest in the late 1960s, especially because of a 1967 NASA conference on organizational communication (Rocci & de Saussure, 2016). Since the NASA gathering, organizational culture scholars have tried to understand the various challenges, goals, behaviors, and people in the workplace. In fact, Trudy Mercadel (2018) notes that labor activists were instrumental over the years in focusing on worker job tedium, worker depression, and worker productivity, prompting investigations into the culture of various companies.

Three assumptions guide Organizational Culture Theory. As you work through these assumptions, keep in mind the diversity and complexity of organizational life. Also, understand that these assumptions emphasize the process view of organizations that Pacanowsky and O'Donnell-Trujillo advocate:

- Organizational members create and maintain a shared sense of organizational reality.
- The use and interpretation of symbols are critical to an organization's culture.
- Cultures vary across organizations, and the interpretations of actions within these cultures are diverse.

The first assumption pertains to the importance of people in organizational life. Specifically, individuals share in creating and maintaining their reality, underscoring the transactional model of communication, which we articulated in **Chapter 1.** These individuals include employees, supervisors, employers, and others. At the core of this assumption are an organization's values. Values are the standards and principles within a culture that have intrinsic worth to the culture. Values inform organizational members about what is important. Pacanowsky (1989) notes that values derive from moral knowledge and that people display their moral knowledge through narratives or stories. The stories that Amelia hears and shares, for example, will result in her understanding the organization's values. Further, if she shares those stories over Facetime, she will also be communicating her values about being an employee. If she reads and responds to various social media posts about the company, she is perpetuating that story.

People share in the process of discovering an organization's values. Being a member of an organization requires active participation in that organization. The meanings of particular symbols—for instance, why a company continues to interview prospective employees when massive layoffs are under way—are communicated by both employees and management. The symbolic meaning of hiring new people when others are being fired will not escape savvy workers; why dedicate money to new personnel when others are losing their jobs? Pacanowsky and O'Donnell-Trujillo (1982) believe that employees contribute to the shaping of organizational culture. Their behaviors are instrumental in creating and ultimately maintaining organizational reality.

The reality (and culture) of an organization are also determined in part by the symbols, the second assumption of the theory. Earlier we noted that Pacanowsky and O'Donnell-Trujillo adopted the Symbolic-Interpretive perspective of Geertz. This perspective underscores the use of symbols in organizations, and, as we mentioned in **Chapter 1,** symbols are representations for meaning. Organizational members create, use, and interpret symbols every day. These symbols, therefore, are important to the company's culture. Mary Jo Hatch (2018) extends the notion of symbols in her discussion of the categories of symbolic meaning (**Table 19.1**).

Table 19.1 Symbols of an Organizational Culture

GENERAL CATEGORY	SPECIFIC TYPES/EXAMPLES
Physical Symbols	art/design/logo
	buildings/decor
	dress/appearance
	material objects
Behavioral Symbols	ceremonies/rituals
	traditions/customs
	rewards/punishments
Verbal Symbols	anecdotes/jokes
	jargon/names/nicknames
	explanations
	stories/myths/history
	metaphors

Symbols include the verbal and nonverbal communication in an organization. Frequently, these symbols communicate an organization's values. Symbols may take the form of slogans that carry meaning. For example, several companies—past and present—have slogans that symbolize their values, including Walmart (*"Save Money–Live Better"*), GEICO (*"15 minutes could save you 15% or more"*), State Farm Insurance (*"Like a good neighbor, State Farm is there"*), *The New York Times* (*"All the News That's Fit to Print"*), and the *Washington Post* (*"Democracy Dies in Darkness"*). Finally, consider Disney's infamous slogan: "The most magical place on earth." Imagine trying to sustain that perception every day at work! The extent to which these symbols are effective relies not only on the media but also on how the company's employees enact them. Some employees will ignore the various company symbols, while others may post their reactions and experiences on X or other social media platforms.

For evidence of verbal symbols in an organization, consider this story. A supervisor named Willy communicates a great deal about values in casual conversation with his employees. Willy frequently tells long stories about how he handled a particular issue at a previous workplace. He often launches into detailed accounts of how, for instance, he managed to get his employees a bonus at the end of the year. His stories inevitably begin with a short vignette about his upbringing in Arkansas and end with a moral. At first, employees were unsure how to handle this type of communication. As time went on, however, they soon realized that Willy was trying to demonstrate a connection with his employees and to indicate that although problems may seem insurmountable, he knows ways to handle them. Through many of his stories, he communicates that he cares about the issues of the company and the workers. He also communicates a new view of what he thinks the organizational culture should be.

Our third assumption of OCT pertains to the variety of organizational cultures. Simply put, organizational cultures vary tremendously. The perceptions of the actions and activities within these cultures are just as diverse as the cultures themselves. Consider what it is like for Amelia as she transitions from Grace's Jewelers to Jewelry Plus. We have already provided a number of examples that underscore the various cultural issues within each company. Her perceptions, however, and her participation in the culture may differ from those of others. Some people might appreciate a cultural change after working so many years for the same small company.

As an employee in a small jewelry store, Amelia knew that a store's problems could readily be resolved and that any suggestions for changes were welcomed and enacted. The culture was such that employees were empowered to make quick decisions, often without supervisor approval. Exceptions to the store return policy, for instance, were handled by all employees. The store's founder felt that employees were in the best position to deal with difficult problems needing quick resolutions. In addition, employee rewards for customer service were routine, and conflict mediation and anger management programs were available for both employees and management. These organizational practices communicated the importance of a shared sense of organizational reality among employees. Employees at Grace's Jewelers got together regularly for F.A.C.—Friday Afternoon Club—at a local restaurant. These activities communicated the *esprit de corps* in the company. The employees of Grace's were members of an organizational culture who "constitute[d] and reveal[ed] their culture to themselves and to others" (Pacanowsky & O'Donnell-Trujillo, 1982, p. 131).

The organizational culture of Jewelry Plus is very different from that of Grace's, and Amelia's experiences with Jewelry Plus are very different from hers with Grace's Jewelers. The corporate giant has no exception to its policy on store returns, and any suggestions for store improvement must be placed in the employee suggestion box or emailed to the national headquarters. A sense of community is not encouraged at Jewelry Plus because tasks clearly promote autonomy. There are some efforts to ensure that employees have time together—through breaks, lunch, or holiday gatherings—but these opportunities are too limited to foster collegiality. In addition, email is closely monitored by supervisors. Clearly, without collegiality, stories, rituals, and rites of passage are restricted. It should be obvious that significant differences exist in the climates of Grace's and Jewelry Plus.

We have presented three assumptions of Organizational Culture Theory. Each is grounded in the belief that when researchers study organizational cultures, they will uncover a complex and intricate *web*. Pacanowsky and O'Donnell-Trujillo believe that the Symbolic-Interpretive perspective provides a realistic picture of the culture of a company. To gain a better sense of how they went about studying organizations, we turn our attention to the primary methodology employed in their work and the work of their predecessor, Clifford Geertz: ethnography.

Ethnographic Understanding: Laying It On Thick

Communication and performance studies scholar Dwight Conquergood (1992, 1994) studied one of the most intriguing of all research topics in communication: street gangs. In an effort to understand gang communication, Conquergood moved into a run-down building in Chicago known at the time as "Big Red." He lived in the building for nearly two years, observing and participating in virtually all parts of life occupied by gang members. Through observing, participating, and taking notes, Conquergood's research offered a view of gang communication virtually ignored in the media. He uncovered many private rituals and symbols, and his work enabled the gang population to have a "voice" never written about in the communication discipline. His efforts in revealing gang-related stories to others is part of ethnography, the underlying methodology of Organizational Culture Theory.

You will recall that Pacanowsky and O'Donnell-Trujillo based much of their work on Geertz's. Because Geertz's work was ethnographic in nature, let's briefly discuss the ethnographic orientation of Geertz and explain its relationship to the theory.

Geertz (1973) argues that to understand a culture, one must see it from the members' points of view. In order to do this, Geertz believes researchers should become ethnographers. Ethnography is a qualitative methodology that uncovers and interprets artifacts, stories, rituals, and practices to reveal meaning in a culture. Paul Atkinson (2022) notes that ethnographers frequently refer to their investigations as naturalistic research in that they believe that the manner in which they study cultures is much more natural than, say, that of quantitative researchers. In this spirit, Geertz remarked that ethnography is not an experimental science but rather a methodology that uncovers meaning. Discovering meaning, then, is paramount to ethnographers. Geertz, and later Pacanowsky and O'Donnell-Trujillo, primarily subscribe to direct observation, interviews, and participant observation in finding meaning in culture.

Students Talking: Meredith

I will be doing my ethnographic study on humor in my workplace. I work part-time for a startup company, and we laugh all the time! I mean we laugh the minute we start "work" (if you could call it that!), and we laugh at lunch and during breaks and after work. It sounds like we don't get work done, but just the opposite happens: We're having a great time, and that kind of culture gets work done. I tell my friends, and they either think I'm lying or they hate that they don't have that kind of company culture. I want to know why we have such a great time (it's definitely not the jobs because it's a software company) and whether the people at the top of the hierarchy wanted that when they started the company.

As an ethnographer, Geertz spent many years studying various cultures. His writings have addressed a number of diverse subjects, from Zen Buddhism to island life in Indonesia. During his stay in some of these places, he relied heavily on field notes and kept a **field journal,** recording his feelings and ideas about his interactions with members of a specific culture. In his writings, Geertz (1973) concludes that ethnography is a kind of **thick description,** or an explanation of the intricate layers of meaning underlying a culture.

field journal personal log to record feelings about communicating with people in a different culture from one's own

thick description explanation of the layers of meaning in a culture

Ethnographers, therefore, strive to understand the thick description of a culture and "to ferret out the unapparent import of things" (p. 26). Interestingly, Geertz believes that any cultural analysis is incomplete because the deeper one goes, the more complex the culture becomes. Therefore, it's not possible to be completely certain of a culture and its norms or values.

Geertz (1983) points out that this qualitative methodology is not equivalent to walking a mile in the shoes of those studied. This thinking only perpetuates "the myth of the chameleon fieldworker, perfectly self-tuned to [their] exotic surroundings, a walking miracle of empathy, tact, patience, and cosmopolitanism" (p. 56). Geertz suggests that a balance must be struck between naturally observing and recording behavior and integrating a researcher's values into the process. He states that "the trick is to figure out what the devil they think they are up to" (p. 58). This, as you might imagine, can be quite difficult for ethnographers, especially since Pacanowsky and his colleagues (2019) argue that organizational cultures both enable things to occur and constrain things from occurring.

Pacanowsky and O'Donnell-Trujillo were drawn to Geertz's ethnographic experiences and his articulation of the importance of observation, analysis, and interpretation. Their own research experiences with different co-cultures proved invaluable. For instance, Pacanowsky (1983) observed police in the Salt Lake [Utah] valley, and Trujillo (1983) studied a new- and used-car dealership. The diversity of their experiences in these smaller cultures in the United States prompted them to acknowledge that cultural performances, or what we call storytelling, are instrumental in communicating about an organization's culture. We return to the topic of performance a bit later in this chapter.

OCT is rooted in ethnography, and organizational culture is often understood with ethnographic principles in mind. Let's explore this method by using our example of Amelia Callahan. If ethnographers were interested in studying the culture of her new job at Jewelry Plus, they might begin by examining several areas: for instance, what sort of new corporate rules are in place? What do new employees like Amelia think about them? What types of strategies are used to ease the transition for employees like Amelia? Are there any corporate philosophies or ideologies? Are there morale problems? How are they resolved? Has the company responded to employee complaints? If so, how? If not, why not? What online comments about the store have been posted? Were there subsequent negative or positive threads? These and a host of other questions would begin the ethnographic process of understanding the organizational climate of Jewelry Plus.

We could never fully capture the excitement of ethnography in this limited space. It is a method that is not experimental; in fact, ethnographers believe that if we were to statistically study human behavior, we would be sacrificing the understanding of "voice" and behavior. We're confident that you have some foundational understanding of the basic processes associated with ethnography and an understanding of why Pacanowsky and O'Donnell-Trujillo embrace such a methodology in their work on organizational culture. Let's now expand on the topic of performance, which is a key component in OCT.

The Communicative Performance

Pacanowsky and O'Donnell-Trujillo (1982) contend that organizational members act out certain communication performances, which result in a unique organizational culture. **Performance** is a metaphor that suggests a symbolic process of understanding human behavior in an organization. Organizational performances frequently mimic the theatre, in that both supervisor and employees choose to take on various roles, or parts, in their organization. To some extent, the workplace can be viewed as a stage and each "actor" enacts different scripted and unscripted dialogues. Let's explore their system of performances.

> **performance** metaphor suggesting that organizational life is like a theatrical presentation

The theorists outline five cultural performances: ritual, passion, social, political, and enculturation. In **Table 19.2,** we identify these performances. As we discuss each, we will provide various examples from different contexts to communicate the diversity embedded in each performance. Although the category system is not necessarily exclusive, you will get an idea about the extent to which organizations vary in terms of how human behavior can be understood. Further, recall our discussion of organizational symbols. As you read about performance, keep in mind this is one way that an organization and its people interpret their environment.

Table 19.2   Cultural Performances in Organizations

Ritual Performances	Personal rituals—checking Snaps and email; task rituals—issuing tickets, collecting fees; social rituals—happy-hour gatherings; organizational rituals—department meetings, company picnics
Passion Performances	Storytelling, metaphors, and exaggerated speech—"this is the most unappreciative company," "follow the chain of command or it'll get wrapped around your neck"
Social Performances	Acts of civility and politeness; extensions of etiquette—customer thank-yous, water cooler chat, supporting another's "face"
Political Performances	Exercising control, power, and influence—"barking" bosses, intimidation rituals, use of informants, bargaining
Enculturation Performances	Acquired competencies over organizational career—learning/teaching roles, orientations, interviews

Ritual Performances

Communication performances that occur on a regular and recurring basis are termed **ritual performances.** Rituals include four types: personal, task, social, and organizational. **Personal rituals** include things that you routinely do each day at the workplace. For instance, many organizational members regularly check their voicemail or email when they get to work each day. **Task rituals** are routinized behaviors associated with a person's job. Task rituals get the job done. For instance, task rituals of employees of the Department of Motor Vehicles include overseeing eye and written examinations, taking pictures of prospective drivers, administering driving tests, verifying car insurance, and listening to the stories that employees share with one another collecting fees. **Social rituals** are the verbal and nonverbal

> **ritual performances** regular and recurring presentations in the workplace
>
> **personal rituals** routines done at the workplace each day
>
> **task rituals** routines associated with a particular job in the workplace
>
> **social rituals** routines that involve relationships with others in the workplace

routines that normally take into consideration the interactions with others. For instance, some organizational members get together for a happy hour in bars on Fridays, celebrating the work week's end.

With respect to your own social rituals, consider the social routines you experience in your classes. Many of you arrive early to catch up with your classmates on what has happened since the last time you spoke and continue the social ritual either during a class break or after class. Social rituals may also include nonverbal behaviors in an organization, including casual Fridays and employee-of-the-month awards. Finally, **organizational rituals** include frequently occurring company events such as division meetings, faculty meetings, and even company picnics like the one Amelia Callahan attended.

> **organizational rituals** routines that pertain to the organization overall

Passion Performances

The organizational stories that members enthusiastically relate to others are termed **passion performances.** Many times, people in organizations become fervent in their storytelling. Consider, for instance, the experience of Adam, who works at a national retail store. Adam and his coworkers hear and retell stories about their department supervisor. The story goes that the boss walks the perimeter of their department every 30 minutes to get an expanded view of the workers and customers. If the supervisor sees something that they think is peculiar, they call the employee into the back room, review a videotape of the event, and ask the employee what they will do to improve any future problems. Adam relates that all of his coworkers passionately tell this story over and over to both new and seasoned employees. In fact, even after six years, Adam's passion for sharing the story is the same as when he told it for the first time. Passion performances can occur online, too. In fact, one group of researchers found that social media stories can significantly influence a company's brand and branding practices (Dutta, Traymbak, Sharma, and Deka, 2023).

> **passion performances** organizational stories that employees share with one another

Social Performances

Social performances are the common extensions of civility, politeness, and courtesy used to encourage cooperation among organizational members. The adage that "a little goes a long way" relates directly to this performance. Whether with a smile or a "good morning" greeting, establishing some sense of collegiality is frequently part of an organization's culture. Performing these "niceties" may be viewed as trivial and yet, it is often difficult to be polite. When the mood is tense, it is both trying and somewhat insincere to smile or to wish another a "good morning." Most organizations wish to maintain a professional decorum, even in difficult times, and these social performances help to accomplish this.

> **social performances** organizational behaviors intended to demonstrate cooperation and politeness with others

Political Performances

When organizational cultures communicate **political performances,** they are exercising power or control. Acquiring and maintaining power and control is a hallmark of U.S. corporate life. In fact, some might argue that power and control pervade organizational life. Nonetheless, because by their nature most organizations are hierarchical, there must be someone with the power to accomplish things and with enough control to maintain the bottom line.

> **political performances** organizational behaviors that demonstrate power or control

As organizational members engage in political performances, they essentially communicate a desire to influence others. That is not necessarily a bad thing. Let's consider an extended example here. Consider the experiences of a group of nurses, for instance, at Spring Valley Hospital. For years, the nurses were quiet regarding their second-class status relative to the hospital's physicians. Recently, however, the nurses decided to speak out about their treatment. They talked to physicians, other medical staff, and patients. They posted YouTube videos expressing their anger about their experiences, particularly after the onslaught of COVID-19 and the burnout that so many of the nurses experienced. In this instance, they were exercising more control over their jobs. Their cultural political performances centered on being recognized for their competency as medical professionals and for their commitment to the mission of the hospital. Their goal was to be legitimized in the hospital by the physicians, their coworkers, and the patients. Their performances, no doubt, were critical in establishing a modified organizational climate.

Enculturation Performances

The fifth type of performance identified by Pacanowsky and O'Donnell-Trujillo is termed **enculturation performance.** Enculturation performances refer to how members obtain the knowledge and skills in order to become contributing members of the organization. These performances may be bold or subtle, and they demonstrate a member's competency within an organization. In the chapter-opening scenario, for example, a number of performances will be enacted to enculturate Amelia into her

> **enculturation performances** organizational behaviors that assist employees in discovering what it means to be a member of an organization

new position. She will watch and listen to her colleagues "perform" their thoughts and feelings on a number of issues: work hours, employee discounts, and the company newsletter, among others. In sum, Amelia will begin to know the organization's culture.

As we mentioned previously, these performances may overlap. It's possible, therefore, to have social performances considered ritual performances. Think about, for instance, greeting one coworker with "Good morning" or fetching coffee for another each day. In this example, the acts of politeness are considered to be a personal (and even task) ritual. Therefore, the performance may be both a social and a ritual performance.

Furthermore, performances may arise from a conscious decision to act out thoughts and feelings about an issue, as in our example of the nurses at Spring Valley Hospital. Or the performances may be more intuitive, as in our example of Amelia Callahan. It is clear that Pacanowsky and O'Donnell-Trujillo believe that communicative performances are critical to an organization's culture and are important symbols of communication.

Integration and Critique

Organizational Culture Theory, as articulated by Pacanowsky and O'Donnell-Trujillo, remains an important influence on organizational communication theory and research. The vast majority of research conducted using OCT has been qualitative in nature since, as we noted earlier, the theorists contend that honoring organizational "voice" is essential to understanding an organization. To evaluate the effectiveness of the theory, we discuss three criteria: logical consistency, utility, and heurism.

Integration

Communication Tradition	Rhetorical | Semiotic | Phenomenological | Cybernetic | Socio-Psychological | **Socio-Cultural** | Critical
Communication Context	Intrapersonal | Interpersonal | Small Group/Team | **Organizational** | Public/Rhetorical | Mass/Media | Cultural
Approach to Knowing	Positivistic/Empirical | **Interpretive** | Critical

Critique

Evaluation Criteria	Scope | **Logical Consistency** | Parsimony | **Utility** | Testability | **Heurism** | Test of Time

Logical Consistency

The logical consistency of the model should not go unnoticed. Recall that logical consistency refers to the notion that theories should follow a logical arrangement and remain coherent. From the outset, Pacanowsky and O'Donnell-Trujillo tried to remain true to their belief that the organization's culture is rich and diverse. They felt that listening to the communicative performances of organizational members was where we must begin in understanding "corporate culture." This is the basis from which much of the theory gained momentum.

Still, some believe consistency is lacking. Eric Eisenberg, Angela Treadway, Marianne LeGreco, and H.L. Goodall (2017), for instance, observe that Organizational Culture Theory relies heavily on shared meaning among organizational members. They comment that stories, for example, are not shared similarly across employees because different organizational stories are told by different organizational narrators. That is, although the theory advances that stories are told and retold and contribute to the culture of an organization, the stories may not have shared meaning. Think about, for instance, how one employee may find joy in sharing a story about a manager's cancer remission while another employee may find sharing such a story as inappropriate. Other scholars (e.g., Kiaos, 2022) agree. Although employees may have the same or similar experience in a company, a multitude of subjective meanings of those experiences exist.

Utility

The theory has utility because the information is applicable to nearly every employee in an organization. The approach is useful because much of the information from the theory (e.g., symbols, stories, rituals) has direct relationship to how employees work and their identification with their work environment (Schein, 2016). Because the theorists' work is based on real organizations with real employees, the researchers have made the theory more useful and practical. What is still unknown, however, is the effect that the virtual office will have upon the utility of OCT. That is, after the advent of COVID-19, many companies initiated remote work. Some, indeed, have returned to FtF office settings. Consequently, research is still underway regarding whether the assumptions and concepts of the theory will resonate as useful.

Heurism

The appeal of Organizational Culture Theory has been far and wide, resulting in a heuristic effort. Researchers, and practitioners have focused on several of the theory's fundamental themes in research examining engineering initiatives (Templin, 2012), farmers' markets (Hoelscher, Zanin, & Kramer, 2016), and nursing homes (Johnston & Womack, 2015). Organizational culture has framed research examining sustainability in the auto industry (Moslehpour, Chau, Tu, Nguyen, Barry, & Reddy, 2022), academic achievement (Özsoy & Uslu, 2019), and "pediatric cardiac specialists" (McMahon, Hickey, Nolke, & Penny, 2023). The original incarnation of OCT has been expanded insofar as many disciplines (e.g., marketing, consumer behavior, psychology, etc.) now incorporate many of the theory's underlying tenets.

Closing

Pacanowsky and O'Donnell-Trujillo were among the first communication scholars to examine organizational life by looking at both employees and their behaviors. Exploring organizational culture in this way enables researchers to appreciate the importance of connecting with the people and their performances at work. The theory will continue to be an important one to understand as organizational life becomes more layered and more complex and as technologies such as webinars, Slack, and Zoom continue to dominate the workplace.

Discussion Starters

👆 **Case-In-Point:** How can employees like Amelia Callahan ease into a new and different organizational culture? What advice would you give her as she begins her new job with Jewelry Plus?

⌨ **Try-It-Your-Selfie:** Pick two of the five organizational performances and provide examples of how the two might be illustrated via (a) X, (b) Instagram, (c) TikTok, and (d) LinkedIn.

1. Consider some of the organizations to which you belong. Identify the cultural performances that you have either observed or shared. How could you use these performances in your work?

2. Geertz has compared culture to a spider web. Now that you've completed the chapter, what other metaphors can you think of that could represent organizational cultures?

3. Discuss your preferences and experiences with various organizational cultures (including schools). What differences and similarities exist? Which type of culture did you prefer?

4. Imagine that you're an ethnographer who has been assigned to study your school's culture. How might you go about studying it? What visible and invisible cultural components would you find?

References

Alkhazraji, K. M. (1997). The acculturation of immigrants to U.S. organizations: The case of Muslim employees. *Management Communication Quarterly*, *11*, 217–265.

Atkinson, P. (2022). *Crafting ethnography*. Sage.

Carmack, H. J. (2017). Crafting a culture of patient safety: Structuring physicians' medical error disclosure and apology. *Ohio Communication Journal*, *55*, 29–42.

Conquergood, D. (1992). Life in Big Red: Struggles and accommodations in a Chicago polyethnic tenement. In L. Lamphere (Ed.), *Structuring diversity: Ethnographic perspectives on the new immigration* (pp. 95–144). University of Chicago Press.

Conquergood, D. (1994). Homeboys and hoods: Gang communication and cultural space. In
L. R. Frey (Ed.), *Group communication in context: Studies of natural groups* (pp. 23-55). Erlbaum.

Disney Institute (2023). *Disney institute training manual.* Walt Disney.

Dutta, M., Traymbak, S., Sharma, M., & Deka, J. (2023). Role of employees' social media stories in
employer branding: A qualitative study. In P. Mandan, S. Tripathi, F. Khalique, & G. Peri (Eds.),
Re-envisioning organizations through transformational change (pp. 53-74). Productivity Press.

Eisenberg, E. M., Treadway, A., LeGreco, M., & Goodall, H. L. (2017). *Organizational communication:
Balancing creativity and constraint* (8th ed.). Bedford/St. Martin's.

Gaudine, A., & Thorne, L. (2012). Nurses' ethical conflict with hospitals: A longitudinal study of outcomes.
Nursing Ethics, 19, 727-737.

Geertz, C. (1973). *The interpretation of cultures.* Basic Books.

Geertz, C. (1983). *Local knowledge.* Hachette Books.

Hatch, M. J. (2018). *Organization theory: Modern, symbolic, and postmodern perspectives.* Oxford University Press.

Hoelscher, C. S., Zanin, A. C., & Kramer, M. W. (2016). Identifying with values: Examining organizational
culture in farmers markets. *Western Journal of Communication, 80*(4), 481-501.

Ince, F. (2023). Transformational leadership in a diverse and inclusive organizational culture. In
R. Perez-Uribe, D. Ocampo-Guzman, N. Moreno-Monsalvo (Eds.), *Handbook of research on promoting
an inclusive organizational culture for entrepreneurial sustainability* (pp. 188-201). IGI Global.

Johnston, L. M., & Womack, D. F. (2015). Best practices in communication with older adults. *China Media
Research, 11,* 54-64.

Kiaos, T. (2022). An interpretative framework for analysing managerial ideology, normative control,
organizational culture and the self. *Cogent Business & Management, 10*(1).

Lussier, R. N., & Achua, C. F. (2022). *Leadership: Theory, application, & skill development.* Sage.

McMahon, C. J., Hickey, E. J., Nolke, L., & Penny, D. J. (2023). Organizational culture as a determinant of
outcome in teams: Implications for the pediatric cardiac specialist. *Pediatric Cardiology, 44*(3), 530-539.

Mercadel, T. (2018). *Organizational culture.* Salem Press.

Modaff, D., Butler, J., & DeWine, S. (2012). *Organizational communication: Foundations, challenges, and
misunderstandings.* Pearson.

Moslehpour, M., Chau, K. Y., Tu, Y. T., Nguyen, K. L., Barry, M., & Reddy, K. D. (2022). Impact of
corporate sustainable practices, government initiative, technology usage, and organizational culture on
automobile industry sustainable performance. *Environmental Science and Pollution Research, 29*(55),
83907-83920.

Özsoy, E., & Uslu, O. (2019). Examining the effects of sustainable organizational culture on academic
achievement. *Discourse and Communication for Sustainable Education, 10*(1), 37-46.

Pacanowsky, M. E. (1983). A small town cop. In L. L. Putnam & M. E. Pacanowsky (Eds.), *Communication
and organizations: An interpretive approach* (pp. 261-282). Sage.

Pacanowsky, M. E. (1989). Creating and narrating organizational realities. In B. Dervin, L. Grossberg, B. J.
O'Keefe, & E. Wartella (Eds.), *Rethinking communication* (pp. 250-257). Sage.

Pacanowsky, M. E., & O'Donnell-Trujillo, N. (1982). Communication and organizational cultures. *Western
Journal of Speech Communication, 46*(2), 115-130.

Pacanowsky, M. E., & O'Donnell-Trujillo, N. (1983). Organizational communication as cultural
performance. *Communication Monographs, 50,* 127-147.

Pacanowsky, M. E., & O'Donnell-Trujillo, N. (1990). Communication and organizational cultures. In S. R. Corman, S. P. Banks, C. R. Bantz, & M. E. Mayer (Eds.), *Foundations of organizational communication: A reader* (pp. 142-153). Longman.

Pacanowsky, M., Arsht, S. S., Mackey, A., Baxter, B. K., Banks, L., Henage, R. T., & Scott, A. (2019). Employee voice: Foundation to the scaffolding of CHG Healthcare's culture journey. *Organizational Dynamics, 48*, 16-27.

Rocci, A., & de Saussure, L. (Eds.). (2016). *Verbal communication*. Walter de Gruyter.

Schein, E. H. (2016). *Organizational culture and leadership*. Wiley & Sons.

Small, K. (2019). Defining company culture: It's about organizational genes, not blue jeans. *New Hampshire Business Review*.

Templin, P. (2012). A failure of culture. *Industrial Engineer, 44*, 1.

Trujillo, N. (1983). Performing Mintzberg's roles: The nature of managerial communication. In L. Putnam & M. E. Pacanowsky (Eds.), *Communication and organizations: An interpretive approach* (pp. 73-98). Sage.

Wong, B. (2023). *Top social media statistics and trends of 2023*. Forbes Advisor.

CHAPTER 20
The Rhetoric

*Based on the writings of **Aristotle***

"Character may almost be called the most effective means of persuasion."

—Aristotle

Camille Ramirez

Camille Ramirez knew that she would have to take a public speaking course for her major. Although she had been active in high school, serving as treasurer for the student council and playing on the lacrosse team, she had never done any formal public speaking. Now that she was in her second semester, she wanted to get the required course out of the way, so she had enrolled in Public Speaking 101.

The class seemed to be going quite well. She felt fairly confident about her speaking abilities; she had received two As and one B on her speeches so far. The final speech, however, would be her most challenging. It was a persuasive speech, and she decided that she would speak on the dangers of drinking and driving. The topic was a personal one for Camille because she would talk about her Uncle Jake, a wonderful man who had died the previous year in an accident with a drunk driver. As she prepared for the speech, she thought that she would blend both her feelings and her evidence into her presentation. Camille also thought that she would have to identify both sides of the drinking issue—the desire to "relax" and the need to be responsible.

On the day of her speech, as recommended to her, Camille took several deep breaths before the class started; it was a strategy that had worked before her previous speeches. As she approached the lectern, she could feel the butterflies well up in her stomach. She reminded herself about the topic and its personal meaning. So, as planned, she began with a short story about her favorite time with Uncle Jake—the time they went to Philadelphia to see the Liberty Bell. Camille then talked about the night—two weeks after her trip—that her uncle died; he was driving home from his daughter's soccer game, and a drunk driver slammed his car from behind, forcing Jake to the embankment. His car then slid into a pond where he drowned.

The room was silent as Camille finished the story. She proceeded to identify why she was speaking on the topic. She told the group that considering their classroom was filled with people under the age of 25, it was important that they understand how fragile life is. No social media account, she thought, could ever capture their reactions as her words resonated with the group. For the next five minutes, Camille mentioned her Uncle Jake several times as she repeated the importance of not drinking and getting behind the wheel of a car.

Camille felt relieved as she finished her speech. She thought that she had done a decent job with her assignment, and she also felt that it helped her to be able to talk about her uncle again. After class, several of Camille's classmates came up and congratulated her. Many of them posted words of sympathy on her Instagram page. They told her that it took a lot of guts to talk about such a personal subject in front of so many people. A few of them also commented that they felt her topic was perfect for the audience; in fact, one of her classmates said that he wanted Camille to give the same speech to his fraternity brothers. As Camille walked to her dorm, she couldn't help but think that she had made a difference and that the speech was both a personal and a professional success.

Contemporary life provides us scores of opportunities to speak in front of others. Politicians, spiritual leaders, physicians, building managers, and investment brokers are just a few of the many types of people who spend much of their time speaking to others—in both formal and informal ways. Especially as members of the academy, whether by choice or by accident, we find ourselves speaking in the classroom, in our organizations, with our professors (and other), on our dorm floors or apartments, among many other locations.

Studying public speaking and communication in general is important in U.S. society for several reasons. First, for nearly three decades, the National Association of Colleges and Employers has identified "communication skills" as paramount to securing and maintaining a job and essential to highlight on resumes. Second, public speaking, by definition, suggests that as a society people are receptive to listening to views of others, even views that may conflict with our own. Deliberation and debate in the United States and in other parts of the world are hallmarks of a democracy, especially when a democracy continues to rely heavily on social media (e.g., the "Arab Spring" in the 2010s). Third, when one speaks before a group, the information resonates beyond that group of people. For instance, when a politician speaks to a small group or team of constituents in southwest Missouri, what she says frequently gets told and retold to others. When a minister consoles a congregation after a fatal shooting at a local middle school, the words reverberate even into the living rooms of those who were not present at the service.

Finally, effective communication is identified as paramount in communication among individuals from around the globe and whom we find in our workplace (Guillén-Yparrea & Ramirez Montoya, 2023). It's not just a topic that resonates solely in the United States. Clearly, effective public speaking has the ability to affect individuals beyond the listening audience, and it is a critical skill for citizens of a democratic society.

Despite the importance of public speaking in our lives, it remains a dreaded activity. In fact, some opinion polls state that people fear public speaking more than they fear death! Comedian Jerry Seinfeld reflects on this dilemma: "According to most studies, people's number one fear is public speaking. Number two is death. Death is number two. Does that seem right? This means to the average person, if you have to go to a funeral, you're better off in the casket than doing the eulogy."

Public speaking is not so funny to people like Camille Ramirez. She must work through not only her anxiety about speaking before a group, but also her anxiety about discussing a very personal topic. For Camille, having a sense of what to speak about and what strategies to adopt is foremost in her mind. Based on her classmates' reactions, her speech remains effective. Camille may not know that the reasons for her success may lie in the writings of Aristotle, published more than 25 centuries ago.

Aristotle is generally credited with explaining the dynamics of public speaking. Aristotle's Rhetoric consists of three books: one primarily concerned with public speakers, the second focusing on the audience, and the third attending to the speech itself. His Rhetoric is considered by historians, philosophers, and communication experts to be one of the most influential pieces of writing in the Western world. In addition, many still consider Aristotle's works to be the most significant writing on speech preparation and speech making. In a sense, Aristotle was the first to provide the "how to" for public speaking. Lane Cooper (1932) agrees. Nearly a century ago, Cooper observed that "the rhetoric of Aristotle is a practical psychology, and the most helpful book extant for writers of prose and for speakers of every sort" (p. vi). According to Cooper, people in all walks of life—attorneys, legislators, clergy, teachers, and media writers—can benefit in some way when they read Aristotle's writings. That is some accolade for a man who has been dead for thousands of years!

Theory at a Glance • The Rhetoric

Aristotle's theory centers on the notion of rhetoric, which he calls the available means of persuasion. That is, a speaker who is interested in persuading their audience should consider three rhetorical proofs: logic (logos), emotion (pathos), and ethics/credibility (ethos). In the Rhetoric, audiences are key to effective persuasiveness, and the employment of rhetorical syllogisms, requiring audiences to supply missing pieces of a speech, are instrumental to persuading others.

To understand the power behind Aristotle's words, it's important first to understand the nature of the Rhetoric. In doing so, we will be able to present the simple eloquence of Rhetorical Theory. First, in order for you to understand the historical context of the theory, we present a brief history of life in Aristotle's day followed by a discussion of his definition of rhetoric.

The Rhetorical Tradition

The son of a physician, Aristotle was encouraged to be a thinker about the world around him. He went to study with his mentor, Plato, at the age of 17. Although he often imitated Plato, the two frequently had conflicting worldviews; therefore, their philosophies differed as well (Devender, 2023). Plato was always in search of absolute truths about the world. He didn't care whether these truths had practical value. Plato felt that as long as people could agree on matters of importance, society would survive. Aristotle, however, was more interested in dealing with the here and now. He wasn't as interested in achieving absolute truth as he was in attaining a logical, realistic, and rational view of society. In other words, we could argue that Aristotle was much more grounded than Plato, trying to understand the various types of people in Athenian society.

Because he taught diverse groups of people in Greek society, Aristotle became known as a man committed to helping the ordinary citizen—at the time, a land-owning male. During the day, common citizens (men) were asked to judge murder trials, oversee city boundaries, travel as emissaries, and defend their property against would-be land collectors (Golden et al., 2020). Because there were no professional attorneys at that time, many citizens hired **Sophists,** teachers of public speaking, to instruct them in basic principles of persuasion. These teachers established small schools where they taught students about the public speaking process and where they produced public speaking handbooks discussing practical ways to become more effective public speakers. Aristotle, however, believed that many of these handbooks were problematic in that they focused on the judicial system to the neglect of other contexts. Also, he thought that authors spent too much time on ways to arouse judges and juries: "It is not right to pervert the judge by moving him to anger or envy or pity—one might as well warp a carpenter's rule before using it," Aristotle observes (cited in Rhys & Bywater, 1954, p. 20). Aristotle reminds speakers not to forget the importance of logic in their presentations.

> **Sophists** teachers of public speaking (rhetoric) in ancient Greece

The Rhetoric could be considered Aristotle's way of responding to the problems he saw in these handbooks. Although he challenges a number of prevailing assumptions about what constitutes an effective presentation, what remains relevant to our discussion is Aristotle's definition of **rhetoric:** the available means of persuasion. In some ways, rhetoric is employed to modify or alter another's perspective and convince them of something. For Aristotle, however, availing oneself of all means of persuasion and trying to alter another do not translate into bribery or torture, common practices in ancient Greece. What Aristotle envisions and

> **rhetoric** a speaker's available means of persuasion

recommends is for speakers to work beyond their first instincts when they want to persuade others. They need to consider all aspects of speech making, including their audience members. When Camille prepared for her speech by assessing both her words and her audience's needs, she was adhering to Aristotle's suggestions for effective public speaking.

For some of you, interpreting rhetoric in this way may be unfamiliar. After all, the word has been tossed around by so many different types of people that it has lost Aristotle's original intent. For instance, Jasper Neel (1994) comments that "the term rhetoric has taken on such warm and cuddly connotations in the postmodern era" (p. 15) that we tend to forget that its meaning is very specific. For people like Neel, we must return to Aristotle's interpretation of rhetoric or we will miss the essence of his theory. For instance, politicians often indict their opponents by stating that their "rhetoric is empty" or that they're all "rhetoric, with little action." These sorts of criticisms only trivialize the active and dynamic process of rhetoric and its role in the public speaking process. Indeed, rhetoric is an art of using language and therefore, simply because someone talks or chats aimlessly does not mean that person is using rhetorical discourse. Consider this important caveat as you review this chapter.

Assumptions of the Rhetoric

Hundreds of various interpretations of Aristotle have been cataloged over the years. We understand the difficulty in distilling this information in a chapter such as this. Nonetheless, we have found that much of the Rhetoric can be understood with two primary assumptions in mind. We examine the following:

- Effective public speakers first consider their audience.
- Effective public speakers employ a number of proofs in their presentations.

The first assumption underscores the interpretation of communication that we presented in **Chapter 1:** Communication is a transactional process. Within a public speaking context, Aristotle suggests that the

audience analysis an assessment and evaluation of listeners

speaker–audience relationship must be acknowledged and even primary in the speaking process. Speakers should not construct or deliver their speeches without considering their audiences. Speakers need to be audience-centered, and they need to consider their audience members before sitting down to construct their speeches. They should think about the audience as a group of individuals with diverse motivations, decisions, and choices and not as some undifferentiated mass of homogeneous people. The effectiveness of Camille's speech on drinking and driving derived from her ability to understand her audience. She knew that students, primarily under the age of 25, rarely think about death, and, therefore, her speech prompted them to think about something that they normally would not consider. Camille, like many other public speakers, engaged in **audience analysis,** which is the process of evaluating an audience and its background (e.g., age, sex, educational level, etc.) and tailoring one's speech so that listeners respond as the speaker hopes they will. For some speakers, audience analysis may simply be nothing but interviewing a few people; for others, a full-blown survey posted on LinkedIn or Facebook may be able to capture large swaths of audience members.

It's apparent that Aristotle believed that audiences are crucial to a speaker's ultimate effectiveness. He believed that of the three components constituting speech making—speaker, subject, and audience—it's the audience/listener that determines whether or not a speech has been effective (i.e., meaning is achieved) (Bartlett, 2021). Each listener, however, is unique, and what works with one listener may fail with another. Expanding on this notion, Christopher Tindale (2015) observes that audiences are not always open to rational argument.

Consider Camille's speech on drinking and driving. Her speech may have worked wonderfully in the public speaking classroom, but she might have different results with a group of alcohol distributors. As you can see, understanding the audience is critical before a speaker begins constructing their speech. Further, as James Herrick (2016) asserts: in the end, rhetoric is about gaining compliance and in order to do that, the audience must be considered.

Theory Into Practice • The Rhetoric

Theoretical Claim: Audience analysis influences speaking effectiveness.

Practical Implication: When a politician speaks to senior citizens, they should consider talking about topics that resonate with this group (e.g., telemarketing scams, Medicare, Social Security, etc.). In doing so, it's likely their audience will be more apt to listen attentively and act accordingly. This is the essence of what it means to persuade another.

The second assumption underlying Aristotle's theory pertains to what speakers do in their speech preparation and their speech making. Aristotle's proofs refer to the means of persuasion, and, for

ethos the perceived character, intelligence, and goodwill of a speaker

Aristotle, three proofs exist: ethos, pathos, and logos. **Ethos** refers to the perceived character, intelligence, and goodwill of a speaker as they become revealed through their speech. Eugene Ryan (1984) notes that ethos is a broad term that refers to the mutual influence that speakers and listeners have on each other. Ryan contends that Aristotle believed that the speaker can be influenced by the audience in much the same way that audiences can be influenced by the speaker. Interviewing Kenneth Andersen, a communication ethicist, Pat Arneson (2007) relates Andersen's thoughts about Aristotle and ethos. Ethos, according to Andersen, is "something you create on the occasion" (p. 131). To that end, a speaker's ethos is not simply something that is brought into a speaking experience; it *is* the speaking experience. Melissa Waresh (2012) contends that ethos must necessarily take into consideration the relationship between speaker and audience. She states: "Ethos is character. Character implicates trust. Trust is based on relationship. Relationship persuades" (p. 229). In the end, ethics relates to the general norms of how choices are made by individuals.

Students Talking Tech: Desi

The three proofs of Aristotle relate to my social media accounts. First, I have to have credibility on my LinkedIn pages because potential employers will be reading my profile. Second, on my Instagram account, I have pictures of my mom and sisters hugging each other, with hundreds of "likes" from my followers (pathos). And my Twitter account is loaded with stats about how we've destroyed the planet. These are "mini speeches" to me and they're found all over my social media!

Aristotle felt that a speech by a trustworthy individual was more persuasive than a speech by an individual whose trust was in question. Michael Hyde (2004) contends that Aristotle felt that ethos is part of the virtue of another and, therefore, "can be trained and made habitual" (p. xvi). **Logos** is the logical proof that speakers employ—their arguments, evidence, and rationalizations. For Aristotle, logos involves using a number of practices, including using logical claims and clear language. To speak in "poetic" phrases results in a lack of clarity and naturalness. **Pathos** pertains to the emotions that are drawn out of listeners. Aristotle argues that listeners become the instruments of proof when emotion is stirred in them; listeners judge differently when they are influenced by joy, pain, hatred, or fear. Let's return to our example of Camille to illustrate these three Aristotelian proofs.

> **logos** logical proof; the use of arguments and evidence in a speech
>
> **pathos** emotional proof; emotions drawn from audience members

The ethos that Camille evokes during her presentation is important. Relating a personal account of her relationship with her Uncle Jake and describing his subsequent death at the hands of a drunk driver bolster perceptions of her credibility. Undoubtedly, her audience feels that she is a credible speaker by virtue of her relationship with Jake and her knowledge of the consequences of drinking and driving. Logos is evident in Camille's speech when she decides to logically argue that although drinking is a part of recreation, mixing it with driving can be deadly. Using examples to support her claims underscores Camille's use of logical proof. The pathos inherent in the speech should be apparent from the subject matter. She chooses a topic that appeals to her college listeners. They most likely will feel for Camille and reflect on how many times they or their friends have gotten behind the wheel after having a few drinks. The proofs of Aristotle's theory, therefore, guide Camille's effectiveness.

Each of these three—ethos, logos, and pathos—is critical to speech effectiveness. But each, alone, may not be sufficient. Keep in mind Kenneth Burke's belief that according to Aristotle, "an audience's confidence in the speaker is the most convincing proof of all" (Burke, 2007, p. 335). For Aristotle, logos is much more than offering evidence in a speech. He delineates this proof in more detail in his writings. In his discussion, he notes that speakers who consider logos must necessarily consider syllogisms. We now turn our attention to this critical Aristotelian principle.

The Syllogism: A Three-Tiered Argument

We noted that logos is one of the three proofs that, according to Aristotle, create a more effective message. Nestled in these logical proofs is something called syllogisms. The term requires clarification because there is some debate among scholars on its precise meaning.

Communication scholars have studied the Rhetoric and its meaning for centuries and have attempted to disentangle some of Aristotle's words. We look here at the term *syllogism,* defined as a set of propositions that are related to one another and draw a conclusion from the major and minor premises. Of all of the concepts related to the Rhetoric, scholars believe that the syllogism is Aristotle's "chief accomplishment in logic" (Darty, 2020, p. 73). Typically, syllogisms contain two premises and a conclusion. A **syllogism** is nothing more than a deductive argument, a group of statements (premises) that lead to another group of statements (conclusions). In other words, premises are starting points or beginnings used by speakers. They establish

> **syllogism** a set of propositions that are related to one another and draw a conclusion from the major and minor premises

justification for a conclusion. In a syllogism, both major and minor premises exist. Symbolically, a syllogism looks like this:

$$A \rightarrow B$$
$$B \rightarrow C$$
$$\text{Therefore,} \quad A \rightarrow C$$

In other words, if A and B are true, then C must be true. Consider this classic example of a syllogism:

Major Premise:	All people are mortal.
Minor Premise:	Aristotle is a person.
Conclusion:	Therefore, Aristotle is mortal.

Let's use the beginning story about Camille and construct a syllogism that she might employ in her speech:

Major Premise:	Drunk driving can kill people.
Minor Premise:	College students drink and drive.
Conclusion:	Therefore, college students can kill others by drinking and driving.

As a speaker, you might (unwittingly) incorporate syllogisms to persuade your audience. However, in an often complex and convoluted society, drawing such a clear conclusion from preliminary premises may not be appropriate. Syllogistic reasoning may also undercut a point you're making. For example, it is usually difficult to draw a clear conclusion when dealing with the behaviors of close friends or family members. Personality, fields of experience, and timing all intersect to make drawing a simple conclusion quite difficult. Further, syllogistic reasoning is, like many issues in this book, impacted by culture. Audience members do not always share a speaker's logical progression of ideas. Therefore, speakers need to be cautious in expecting audience members to draw conclusions in similar ways. This is where an audience analysis becomes critical in your speechmaking process.

Syllogisms are important in the speaking process for Aristotle. Speakers use them to enhance effectiveness in their speeches. In addition, speakers also incorporate other techniques that are labeled *canons.*

Canons of Rhetoric

Aristotle was convinced that, for a persuasive speech to be effective, speakers must follow certain guidelines or principles, which he called *canons.* Canons are recommendations for making a speech more compelling. Classical rhetoricians have maintained Aristotle's observations, and to this day, most writers of public speaking texts in communication adhere to these canons for effective speaking that early Greeks and Romans integrated.

Although his writings in the Rhetoric focused on persuasion, these canons have been applied in a number of speaking situations (Brummett, 2022). Five prescriptions for effective oratory exist and we now discuss these canons, which are highlighted in **Table 20.1.**

Table 20.1 Aristotle's Canons of the Rhetoric

CANON	DEFINITION	DESCRIPTION
Invention	Integration of reasoning and arguments in speech	Using logic and evidence in speech makes a speech more powerful and more persuasive.
Arrangement	Organization of speech	Maintaining a speech structure—Introduction, Body, Conclusion—bolsters speaker credibility, enhances persuasiveness, and reduces listener frustration.
Style	Use of language in speech	Incorporating style ensures that a speech is memorable and that a speaker's ideas are clarified.
Delivery	Presentation of speech	Delivering an effective speech complements a speaker's words and helps to reduce speaker anxiety.
Memory	Storing information in speaker's mind	Knowing what to say and when to say it eases speaker anxiety and allows a speaker to respond to unanticipated events.

Invention

The first canon is invention. This term can be confusing because invention of a speech does not mean invention in a scientific sense. **Invention** is defined as the construction or development of an argument that is relevant to the purpose of a speech. Invention is discovering all the proofs a speaker plans to use. Invention is broadly interpreted as the body of information and knowledge that a speaker brings to the speaking situation. This stockpile of information can help a speaker in their persuasive approaches.

invention a canon of rhetoric that pertains to the construction or development of an argument related to a particular speech

Suppose, for instance, you are presenting a speech on DNA testing. Invention associated with this speech would include appeals woven throughout your speech (e.g., "DNA helps living organisms pass along information to their offspring," "DNA is the fundamental blueprint for all life," or "DNA testing has proven to be instrumental in capturing criminals"). In constructing your arguments, you may draw on all these examples.

Aids to invention are identified as topics. **Topics,** in this sense, refer to the lines of argument or modes of reasoning a speaker uses in a speech. Speakers may draw on these invention aids as they decide which speaking strategy will persuade their audiences. Topics, therefore, help speakers enhance their persuasiveness.

topics an aid to invention that refers to the arguments a speaker uses

Speakers look to what are called **civic spaces,** or the metaphorical locations where rhetoric has the opportunity to effect change, and locations that may be opportunities to persuade others. Recall, for instance, Camille's decision to talk about drinking and driving in her public speaking class. As she speaks, she defines her terms, looks at

civic spaces a metaphor suggesting that speakers have "locations" where the opportunity to persuade others exists

opposing arguments, and considers ideas similar to her own. That is, she identifies a "location" in her speech where she is able to adapt to an audience that may be losing attention. Camille does whatever it takes to ensure that she has the chance to persuade her audience. And, to be sure, the online era has ushered in "digital spaces," allowing for this persuasion to take place in various virtual environments.

Arrangement

A second canon identified by Aristotle is called arrangement. **Arrangement** pertains to a speaker's ability to organize a speech. Aristotle felt that speakers should seek out organizational patterns for their speeches to enhance the speech's effectiveness. Artistic

> **arrangement** a canon of rhetoric that pertains to a speaker's ability to organize a speech

unity among different thoughts should be foremost in a speaker's mind. Simplicity should also be a priority because Aristotle believed that there are essentially two parts to a speech: stating the subject and finding the proof. At the time, he felt that speakers were either unaware of the significance of arrangement or felt that speeches were organized haphazardly, making speakers less effective.

Aristotle is very clear in his organizational strategy. If you've enrolled in or are currently enrolled in a public speaking class, you already know the template. Speeches generally follow a threefold approach: introduction, body, and conclusion. The **introduction** should first gain the audience's attention, then suggest a connection with the audience, and finally provide an overview of the speech's purpose.

> **introduction** part of an organizational strategy in a speech that includes gaining the audience's attention, connecting with the audience, and providing an overview of the speaker's purpose

Introductions can be quite effective in speeches that are intended to arouse emotionally. Gaining attention by incorporating emotional wording is an effective persuasive technique. Consider Camille's introductory words. She obviously captures the audience's attention by personalizing a very difficult subject. She then suggests her relationship with the topic, followed by an overview of her speaking purpose:

> Jake McCain was killed by someone he didn't know. He was a wonderful man and the person who killed him never knew that. Yet, Jake's death could have been prevented. You see, Jake McCain was killed by someone who was drunk. The driver will have a future, but Jake will never have a chance to see his grandchild grow up or see his niece get married. I know about Jake McCain: he's my uncle. Yes, my uncle. Today, I wish to discuss the dangers of drunk driving and identify how you can avoid becoming one of the many thousands who get behind a wheel after drinking too much.

Arrangement also includes the body and conclusion of the speech. The **body** includes all of the arguments, supporting details, and necessary examples to make a point. In addition to the entire speech being organized, the body of the speech also follows some sort of organizational structure. Aristotle states that audiences need to be

> **body** part of an organizational strategy in a speech that includes arguments, examples, and important details to make a point

led from one point to another. Although he never viewed the audience as passive, he felt that they needed to have some direction as speeches were delivered. Consider, for instance, the difficulty you'd have finding an address in a large city without a map; you'd be rightfully confused.

Finally, the **conclusion** or epilogue of a speech is aimed at summarizing the speaker's points and arousing emotions in the audience. Conclusions should be arrived at logically and should also attempt to reconnect with listeners. Camille's conclusion clearly demonstrates her desire to leave her listeners with a message:

> **conclusion** part of an organizational strategy in a speech that is aimed at summarizing a speaker's main points and arousing emotions in an audience

> So I leave you today after talking about the prevalence of drunk driving, the current laws associated with this behavior, and what you and I can personally do to help rid our society of this terrible and overlooked part of being a college student. The next time you go and have a drink, don't forget to give your keys to a friend. Or catch an Uber. I'm sure your family will thank you. Do it for me. Do it for my Uncle Jake.

We can feel Camille's passion for a topic that is both personal and personally difficult.

Style

The use of language to express ideas in a certain manner is called **style.** In his discussion of style, Aristotle includes word choice, word imagery, and word appropriateness. He believes that each type of rhetoric has its own style, yet style is often overlooked. He notes that strange words or **glosses** (e.g., antiquated/inappropriate words and phrases, such as "spaz" or "girl Friday") should be avoided. Speaking in terms that are too simplistic will also turn off an audience. To bridge this gap between the unfamiliar and the too familiar, Aristotle introduces the notion of **metaphor,** or a figure of speech that helps to make the unclear more understandable. When we use a metaphor, we have two thoughts merging together, resulting in one word or phrase. Metaphors are critical devices to employ in speeches, according to Aristotle, because they have the capacity to change the perceptions and the minds of listeners.

Mike Gruhn/CartoonStock Ltd

Style can be better understood through an example from Camille's speech on drunk driving. If Camille were concentrating on style, her speech would have the following passage:

> **style** a canon of rhetoric that includes the use of language to express ideas in a speech
>
> **glosses** outdated words in a speech
>
> **metaphor** a figure of speech that helps to make the unclear more understandable

> Drinking is often viewed as a means to release. After a very long day at work or at school, there may be nothing better than having a cold beer. So they say. Yet, too often, one beer turns into two, which by the end of a few hours, has turned into a six-pack. And the result can be tragic: how many times have you watched your friend or family member get into a car after a six-pack? This person can be as dangerous as a bullet, unleashed from a gun that is randomly pointed at someone. If you must drink, it's not only your business; it's my business, too.

Camille's words evoke some strong imagery; mentally, we can recreate the scene that she has laid out. Her word choice is unmistakable in that she uses familiar words. Finally, she uses the compelling metaphor of a bullet.

Delivery

Thus far we have concentrated on how a speech is constructed. Aristotle, however, was also interested in how a speech is delivered. In this case, **delivery** refers to the nonverbal presentation of a speaker's ideas. Delivery normally includes a host of behaviors, including eye contact, vocal cues, pronunciation, enunciation, dialect, body movement, and physical appearance. For Aristotle, delivery specifically pertains to the manipulation of the voice. He especially encouraged speakers to use appropriate levels of pitch, rhythm, volume, and emotion. He believed that the way in which something is said affects its intelligibility. Today, with such delivery annoyances as high-rising terminal ("upspeak") and gap fillers (e.g., "um," "eh"), delivery effectiveness has diminished.

delivery a canon of rhetoric that refers to the nonverbal presentation of a speaker's ideas

Aristotle believed that delivery could not be easily taught, yet it is crucial for a speaker to consider. He also taught that speakers should strive to be natural in their delivery. Speakers should not use any vocal techniques that may detract from the words and should strive to capture a comfortable presence in front of an audience. In other words, speakers should avoid being "gimmicky" in their presentations and strive for authenticity.

Memory

Storing invention, arrangement, and style in a speaker's mind is **memory.** In contrast to the previous four canons, Aristotle does not spend significant time delineating the importance of memory in speech presentation. Rather, he alludes to memory in his writings. Some scholars have called this "the lost canon" because in early Greek days, memorizing a speech was critical. Today, as Barry Brummett (2022) notes, with the advent of teleprompters and PowerPoint, "such a concern seems irrelevant" (p. 39). Throughout the Rhetoric, for instance, Aristotle reminds us to consider a number of issues prior to the presentation (e.g., examples, signs, metaphors, delivery techniques, etc.). He further notes that to speak persuasively, a speaker has to have a basic understanding of many of these devices when constructing and presenting a speech. In other words, speakers need to have memorized a great deal before getting up to speak.

memory a canon of rhetoric that refers to a speaker's effort in storing information for a speech

Today, people interpret memory in speech-making differently from Aristotle. Memorizing a speech often means having a basic understanding of material and techniques. Although other rhetoricians like Quintilian made specific recommendations on memorizing, Aristotle felt that familiarizing oneself with the speech's content was most appropriate. In some situations, a reading every word, which includes specific statistics, will be necessary—such as those which includes specific statistics will be necessary, such as the speeches undertaken by presidents of various countries. At other times, however, if you are to make an impression (i.e., create meaning), memorizing certain passages may be highly effective (Gregory, 2021). When Camille presents her speech on drinking and driving, for example, she has some parts of her speech committed to memory and other parts overviewed on notes.

The canons of rhetoric are incorporated into a number of different persuasive speeches. Our exploration of Aristotle's Rhetorical Theory concludes with a discussion of the three types of rhetoric.

Types of Rhetoric

You will recall that during Aristotle's time, citizens were asked to take part in a number of speaking activities—from judge to attorney to legislator. It was in this spirit that Aristotle identified different speaking situations for citizens to consider when conversing on trade, finance, national defense, and war. He denoted three types of rhetoric, or what he called three types of oratory, or speaking: forensic, epideictic, and deliberative. Before delving into these types, let's review an historical and cultural note related to this topic. Over the years, many Native American tribes and communities have employed oratory in both ceremonial and non-ceremonial ways. Kenneth McAllister (2023) contends that before the invasion of North America by Europeans, native populations had no written language. Consequently, speaking powerfully and persuasively was highly regarded when communicating about such issues as family histories, craft techniques, and rituals. As McAllister writes, oratorical effectiveness was instrumental across most of the tribal nations. **Forensic rhetoric** pertains to establishing a fact; at the core of forensic rhetoric is justice. **Epideictic rhetoric** is discourse related to praise or blame. **Deliberative rhetoric** concerns speakers who must determine a course of action—something should or should not be done. The three types refer to three different time periods: forensic to the past, epideictic to the present, and deliberative to the future. We discuss these three rhetorical types next and illustrate them in **Figure 20.1**.

> **forensic rhetoric** a type of rhetoric that pertains to speakers prompting feelings of guilt or innocence from an audience
>
> **epideictic rhetoric** a type of rhetoric that pertains to praising or blaming
>
> **deliberative rhetoric** a type of rhetoric that determines an audience's course of action

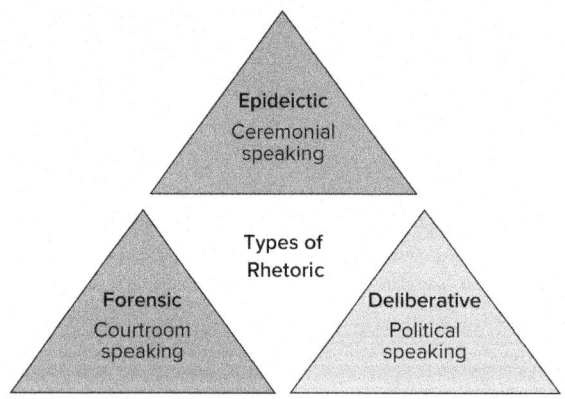

Figure 20.1 **Types of Rhetoric**

Forensic oratory, or judicial rhetoric, specifically refers to speaking in courtrooms. Its intent is to establish guilt or innocence. In Aristotle's day, forensic speakers directed their presentation to courtroom judges. Aristotle examined forensic rhetoric within a legal framework, and thus many of his beliefs on the law are found in the Rhetoric. Peter O'Connell (2017) notes that forensic speaking requires focusing on arguments that tap into judges' psyches, including their beliefs about why certain criminals act the way they do and which types of circumstances tempt people to break the law. O'Connell refers to this as "a law of performance" (p. 18). Because past actions are frequently indicative of a person's current behavior, forensics orators rely on previous behaviors.

Aristotle recognized that a person's character is critical in forensic rhetoric. He interprets character as both status (i.e., whether a person is young or old, rich or poor, fortunate or unfortunate) and morality (i.e., whether a person is just or unjust, reasonable or unreasonable). If people act voluntarily, Aristotle argued, the choices they make have consequences. To establish guilt, the forensic speaker needs to establish motivation for doing wrong. In speaking before an audience, then, speakers will invoke what Aristotle called the "moral habits" of a person.

Examples of forensic rhetoric abound in our society. Forensic speakers have played prominent roles in U.S. courtrooms. Attorneys, in particular, have effectively and persuasively used their forensic rhetoric over the years. One of the most memorable forensic presentations in history took place in the closely watched trial of football legend O. J. Simpson. Prosecutors tried to implicate Simpson's morals by playing a tape recording of a 911 call in which Simpson could be heard yelling at his wife and by showing pictures of her beaten body to the jury. More recent forensic efforts by prosecutors include the legal proceedings related to the attack on the U.S. Capitol in 2020. The numbers of "pro-Trump" supporters numbered in the thousands and scores of them were indicted on various criminal charges, including assaulting law enforcement, trespassing, and disrupting Congress, among other charges. Generally speaking, forensic speaking is employed to undercut the moral integrity of the defendant and to establish guilt.

Students Talking: Charlotte

There are so many parts of this theory that I can relate to. One part that really caught my eye was the forensic oration. I love the Supreme Court and read everything about it. But I love television that shows real-life courtrooms even more because I can watch prosecution in action. I wish they'd televise Supreme Court hearings so I'd see lawyers trying to belittle the character of others, and I see defendants trying to keep it together while it's being done to them. The theory makes sense, even today, to someone like me.

The second type of rhetoric, epideictic, is also called ceremonial speaking. Speeches during Aristotle's time were given in public arenas with the goal of praising, honoring, blaming, or shaming. Epideictic rhetors include people, events, organizations, or nations in their speeches. These speeches usually focus on social issues because, according to Aristotle, people are interested in the here and now. Epideictic speaking cannot be separated from ethos, Aristotle stated. He believed that by understanding the need to praise or blame, epideictic speakers understand the importance of their own character. For instance, a speech criticizing prison conditions may not resonate deeply with an audience if the speaker is on death row for murder.

At times, there are speeches that simply are generic in nature and yet, the target of blame is implied. Across the country, for instance, with nearly 650 mass shootings in the U.S. in 2022, it's not uncommon for speeches to advocate for more gun control. From the Lewiston (ME) bowling alley shootings with 18 people killed to the Stoneman Douglas (FL) High School killings of 17 students and staff to the 10 people of color gunned down in a Buffalo (NY) grocery store, many speeches at the time blamed the National Rifle Association (NRA) and others who advocate for more guns. With an average mass shooting taking place about two times a day, epideictic speeches will continue well into the future. It is clear that many speakers blame the NRA for its influence on restricting gun legislation and openly opposing any curtailment on gun purchases.

Epideictic rhetoric is also exemplified in funeral practices in our country. Eulogies, which are commonplace at many funerals, usually laud the life of the deceased. Commenting on contemporary values, the epideictic speaker at a funeral frequently compares the virtues of the dead person with those of society. For instance, after the death of his grandmother, Rich, one of your authors, was asked to give the eulogy at the funeral. During his speech, he talked about his grandmother's uplifting spirit and how she rarely complained about her ailments or about her financial situation. He evoked images of contemporary society in his speech, noting how unusual it is today for someone to refrain from self-centered complaining, especially in social media. His speech centered on a prevailing virtue of his grandmother—her selflessness—and also commented on society as a whole.

The third type of rhetoric is deliberative rhetoric, also called political rhetoric, and it was the focus of many of Aristotle's comments on rhetorical discourse. As we mentioned earlier, deliberative rhetoric is associated with the future—what an audience will do or think as a result of a speaker's efforts. Deliberative speaking, then, requires the speaker to be adept at understanding how their thoughts are aligned with those of the audience. The deliberative speaker should be prepared to consider subjects that are relevant to the audience and to which the speaker can personally relate. Aristotle identified five subjects on which people deliberated in his day: revenue, war and peace, the defense of the country, commerce, and legislation. Today's list of deliberative topics might include health insurance, climate change, relationships, student loan debt, and social justice. Deliberative speakers might try to raise interest in these topics, and once interest is piqued, they might find that listeners are more prone to being persuaded.

Larry Arnhart (1981) comments that the deliberative rhetorician needs to know not only the actual subject of deliberation but also the elements of human nature that influence deliberation. There are a number of topics, therefore, which are suited for deliberation and others that are not. Aristotle focused on what deliberative speakers can say to an assembly (e.g., a body of legislators), and today this deliberative oratory continues. Consider the following example. When asked to give a short presentation to her state's legislative committee on health insurance, Beverly, a 64-year-old mother of four, spoke about the health insurance of elderly people. As the caretaker of her 90-year-old mother-in-law, a patient in a local nursing home, Beverly knew precisely the kinds of persuasive strategies to use with the group of politicians. Her speech focused on the difficulties of being old and how these problems are amplified by not having enough insurance. She asked the legislators to consider their own aging parents in their discussions. She outlined five points of action for the committee to follow. Three of the points could be undertaken immediately: establishing a task force, interviewing elderly citizens, and setting up a toll-free number to solicit citizen concerns and complaints. The remaining two required funding from the legislature. At the conclusion of her brief speech, Beverly was satisfied that her suggestions would not be ignored. How might Beverly use her Facebook account to communicate the same message?

Aristotle would have approved of Beverly's rhetoric. Her recommendations were doable (the committee enacted three of the five), and she made her experiences relevant to her audience by asking the group to think about their own parents. This approach elicited personal identification, which is an important tactic in deliberative speaking. By eliciting these feelings, Beverly knew that she would be able to get her audience to agree with her thinking.

Integration and Critique

Aristotle's Rhetoric remains among the most influential theoretical foundations in communication studies. You can pick up any public speaking text and find discussions on delivery, organization, and style. Students of public speaking have benefited greatly from the words and values of Aristotle, and for this reason the theory will resonate deeply for years to come. With a few exceptions, his theory has drawn the attention of scholars who primarily identify as interpretive and critical, making the theory more aligned with the qualitative approach. However, keep in mind that some inferred aspects of the theory (e.g., speaking anxiety) are studied with an experimental lens. The evaluative criteria for communication we wish to discuss center on three primary areas: logical consistency, heurism, and test of time.

Integration

Communication Tradition	**Rhetorical** | Semiotic | Phenomenological | Cybernetic | Socio-Psychological | Socio-Cultural | Critical
Communication Context	Intrapersonal | Interpersonal | Small Group/Team | Organizational | **Public/Rhetorical** | Mass/Media | Cultural
Approach to Knowing	Positivistic/Empirical | **Interpretive** | **Critical**

Critique

Evaluation Criteria	Scope | **Logical Consistency** | Parsimony | Utility | Testability | **Heurism** | **Test of Time**

Logical Consistency

Critics of Aristotle's theory have taken issue with some tenets of the theory because some scholars believe that Aristotle's conclusions were primarily "scattered remarks and examples" (Curzer, 2015, p. 129). For instance, Aristotle has been criticized for contradiction and incoherence. Lord (1994) contends that in developing his theory, Aristotle blasts his contemporaries for focusing too much on the audience's emotions. Although Aristotle encourages speakers to avoid focusing on emotions while making their points, he proceeds to do just that when he stresses the importance of presenting emotions and invoking audience passions (pathos) during a speech. This makes the theory somewhat inconsistent.

Yet, John Cooper (1996) challenges Lord's critique. He argues that Aristotle was simply responding to the Sophists' messages of the day. Because most of the speeches in ancient Greece were directed to judges and rulers, Aristotle indicated that speakers should try to elicit feelings of pity in the courtroom. To do that, Aristotle felt that speakers should try to view judges in congenial ways (Bartlett, 2021).

Further criticism of the logical consistency of the theory has been offered. First, as we alluded to a bit earlier, scholars agree that the Rhetoric is a rather unorganized undertaking; In fact, the theory is assembled from Aristotle's lecture notes. It's not surprising, then, that Aristotle seems to discuss topics in a random and arbitrary manner. At times, Aristotle introduces a topic and then drops it, only to return to it later. His terminology is especially problematic for some scholars. Some writers contend that his work is nothing but an amalgam of quotations (Devender, 2023). You may not find this too earth-shattering, but recall that researchers need clear foundations of terms before they can embark upon testing or clarifying theory. Larry Arnhart (1981) concludes that Aristotle defined his terms in less than precise ways so that audiences (readers) would have a broader understanding of his words and ideas. Arnhart believes that this conscious decision to remain unclear does not mean that Aristotle's thoughts should be discarded.

Finally, the logical consistency is further challenged by an examination of how Aristotle views the audience. Critics charge Aristotle with ignoring the critical nature of listeners. For instance, Jasper Neel (1994) states,

"Aristotle makes clear that the introduction [of a speech] has nothing to do with the 'speech itself.' It exists only because of the hearer's weak-minded tendency to listen to what is beside the point" (p. 156). Eugene Ryan (1984) is more blunt: "Aristotle is thinking of listeners who have some difficulty keeping their minds on the speaker's business, are easily distracted, tend to forget what has gone on before, [and] are not absorbed with abstract ideas" (p. 47). From these writers, we get the impression that Aristotle perceived audiences to be incapable of being discriminating listeners or critical thinkers. It's important to note, though, that Aristotle was writing at a time when people were rather passive listeners; they had no Google news, social media feeds, or other access to information about world events accustomed to openly challenging their mentors. Aristotle's view of the audience is not so implausible.

Heurism

Few would argue that Aristotle's Rhetoric is one of the most heuristic theories found in communication. Scholars in a variety of fields of study, including political science, medicine, English composition, and philosophy, have studied Rhetorical Theory and incorporated Aristotelian thinking in their research. The theory has spawned a number of subareas in the communication discipline, such as communication apprehension, and has generated research. In fact, much of the writing in public speaking is based on the writings of Aristotle. Some interesting research has discussed the deliberative rhetoric of an African American activist (McClish, 2007) while others have specifically discussed the application of the syllogism to geography, technology, and engineering (Goldin, 2022). Much of the discussion related to preaching in churches can be directly attributed to Aristotle's thinking (Broadus, 2012). Aristotelian proofs have been employed to understand "Brexit" using YouTube (Finlayson, 2022), in the analyses of environmental reports (Higgins & Walker, 2012), and in the words of Abraham Lincoln (Horrocks, 2014). In addition, Aristotle has been invoked in a wide range of studies examining such diverse topics as military digital games (Sparrow, Harrison, Oakley, & Keogh, 2015), the Scottish vote on independence (Mackay, 2015), and hip-hop music (Sciullo, 2014). With an entire book dedicated to "quantifying Aristotle," too (Di Liscia & Sylla, 2022), the theory resonates in ways unimagined by the philosopher–theorist!

Test of Time

No other theory in the communication discipline has withstood the test of time as well as Aristotle's Rhetoric. With centuries behind it and public speaking textbooks, teachers, and researchers communicating Aristotelian principles, no other theory in the field of communication will ever achieve such longevity!

Closing

As the 21st century continues, we are in an informed position to reflect on some of the greatest written works of all time. The Rhetoric is clearly such a work. Even today, Aristotle's words have value to a global society immersed in mediated forms of communication. Some people may reject his thoughts as outdated in an age in which multiple ways of knowing are embraced. Yet, some scholars are more direct in their acclaims of this thinker: "The rational, scientific, and technological culture that pervades much of the Western world owes more to him than anyone else" (Woodfin & Groves, 2012, p. 3). Clearly, a theory focusing on how speakers use and engender emotions, logic, and trustworthiness cannot be ignored. With communication remaining a premium in nearly every society, the lessons that Aristotle has provided should be considered each time we approach an audience.

Discussion Starters

 Case-In-Point: In the chapter-opening scene, Camille Ramirez relied on Aristotle's view of public speaking. Do you believe that she could have been more effective? In what way? Use examples in your response.

⌨ **Try-It-Your-Selfie:** Construct a syllogism related to how you use your various social media accounts. Remember, you are discussing the "use" of your accounts, including how you use them, the frequency of use, who "uses" them with you, among others.

1. Aristotle's critics have focused on the fact that his theory is simply a collection of lecture notes that are contradictory, vague, and often narrow. Does this make a difference to you? Why or why not?

2. Employ syllogistic reasoning for and against each of the following topics: physician-assisted suicide, marriage equality, and climate change.

3. If Aristotle were alive today and you were his student, what additional suggestions would you offer him for a new edition of the Rhetoric? Why do you believe your suggestions are important to address in public speaking? Incorporate examples in your response.

4. Aristotle spent a great deal of time discussing the role of the audience. If you were giving a speech on safety to a group of convenience store employees, what sort of audience analysis would you undertake?

References

Arneson, P. (2007). A conversation about communication ethics with Kenneth A. Andersen. In P. Arneson (Ed.), *Exploring communication ethics* (pp. 131–142). Peter Lang.

Arnhart, L. (1981). *Aristotle on political reasoning: A commentary on rhetoric.* Northern Illinois University Press.

Bartlett, R. C. (2021). *Aristotle's art of rhetoric.* University of Chicago Press.

Broadus, J. (2012). *On the preparation and delivery of sermons.* Southern Grace Publishers.

Brummett, B. (2022). *Rhetoric in popular culture.* Sage.

Burke, K. (2007). On persuasion, identification, and dialectical symbols. *Philosophy and Rhetoric, 39,* 333–339.

Cooper, J. M. (1996). An Aristotelian theory of the emotions. In A. O. Rorty (Ed.), *Essays on Aristotle's Rhetoric* (pp. 238–257). University of California Press.

Cooper, L. (1932). *The rhetoric of Aristotle.* Appleton-Century Company.

Curzer, H. J. (2015). Aristotle's practical syllogisms. *The Philosophical Forum, 46,* 129–153.

Darty, D. (2020). Categorical syllogism. In U. Etuk & C. Ijiomah (Eds.), *Argument and evidence* (pp. 73–88). Inela Ventures and Publishers.

Devender, G. W. (2023). *Aristotle.* Salem Press.

Di Liscia, D. A., & Sylla, E. D. (2022). *Quantifying Aristotle: The impact, spread and decline of the Calculatores tradition.* Brill.

Finlayson, A. (2022). Brexit, YouTube and the populist rhetorical ethos. In C. Kock & L. Villadsen (Eds.), *Populist rhetorics: Case studies and a minimalist definition* (pp. 81–106). Springer International Publishing.

Golden, J. L., Coleman, W., Sproule, J. M., Golden, R., McHendry, G. F., Thorpe, E., Kurr, J. E. (2020). *The rhetoric of western thought: From the Mediterranean world to the global setting.* Kendall Hunt.

Goldin, O. (2022). Review of the enthymeme: Syllogism, reasoning, and narrative in ancient Greek rhetoric. *The Classical Review, 72*(1), 79–81.

Gregory, H. (2021). *Public speaking for college and career.* McGraw Hill.

Guillén-Yparrea, N., & Ramírez-Montoya, M. S. (2023). Intercultural competencies in higher education: A systematic review from 2016 to 2021. *Cogent Education, 10*(1), 2167360.

Herrick, J. A. (2016). *The history and theory of rhetoric: An introduction.* Routledge.

Higgins, C., & Walker, R. (2012). Ethos, logos, pathos: Strategies of persuasion in social/environmental reports. *Accounting Forum, 36,* 194–208.

Horrocks, T. A. (2014). *Lincoln's campaign biographies.* Southern Illinois University Press.

Hyde, M. (2004). Introduction: Rhetorically we dwell. In M. Hyde (Ed.), *The ethos of rhetoric* (pp. 12–27). University of South Carolina Press.

Lord, C. (1994). The intention of Aristotle's Rhetoric. In E. Schiappa (Ed.), *Landmark essays on classical Greek rhetoric* (pp. 157–168). Hermagoras Press.

Mackay, R. R. (2015). Multimodal legitimation: Selling Scottish independence. *Discourse & Society, 26,* 323–348.

McAllister, K. (2023). *Oratory.* Salem Press.

McClish, G. (2007). A man of feeling, a man of colour: James Forten and the rise of African American deliberative rhetoric. *Rhetorica, 25,* 297–328.

Neel, J. (1994). *Aristotle's voice: Rhetoric, theory and writing in America.* Southern Illinois University Press.

O'Connell, P. A. (2017). *The rhetoric of seeing in attic forensic oratory.* University of Texas Press.

Rhys, W. R., & Bywater, I. (1954). *Rhetoric.* Modern Library.

Ryan, E. E. (1984). *Aristotle's theory of rhetorical argumentation.* Les Editions Ballarmin.

Sciullo, N. J. (2014). Using hip-hop music and music videos to teach Aristotle's three proofs. *Communication Teacher, 28,* 165–169.

Sparrow, R., Harrison, R., Oakley, J., & Keogh, B. (2015). Playing for fun, training for war: Can popular claims about recreational video gaming and military simulations be reconciled? *Games and Culture, 13,* 174–192.

Tindale, C. W. (2015). *The philosophy of argument and audience reception.* Cambridge University Press.

Waresh, M. (2012). Morality, trust, and illusion: Ethos as relationship. *Legal Communication and Rhetoric, 9,* 229–272.

Woodfin, R., & Groves, J. (2012). *Introducing Aristotle: A graphic guide.* Icon Books.

CHAPTER 21
Dramatism

*Based on the research of **Kenneth Burke***

> *"Stories are equipment for living."*
>
> —Kenneth Burke

Karl Nelson

Karl Nelson looked forward to this part of his morning routine. He settled down with his first cup of coffee and the morning paper. Although he used his tablet and smartphone a lot, he loved the feel of a newspaper in his hands. He knew that made him a dinosaur; still, Karl happily allowed himself an hour to read the news and savor his caffeine fix. In many ways, this was his favorite part of the day, and he got up extra early to make sure he'd have enough time for it after his workout and before he left for the office. But, today he wasn't happy. He looked at the headlines with disgust. He was so sick of reading about politicians who had no common sense. Today, he was reading about Eric Spellman, the mayor of Grenada, New Mexico, where Karl lived. The article was about the fact that Spellman had campaigned on a "clean" platform, claiming that he'd make government respectable again, and now he was apologizing to his wife and the public for having had an affair with his children's nanny. Undoubtedly, he wouldn't ever have confessed, but the tabloids had just revealed that he'd had a child, who was now 10, with the nanny!

Karl looked up from his paper just as his husband, Max, came into the breakfast room. Karl asked, "Max, have you read about Spellman? That man is such an incredible low-life hypocrite. How could he pretend to be Mr. Clean during his campaign, talking all about family values, when he's cheating on his wife with their nanny?"

Max just shrugged and laughed. He was used to seeing Karl getting worked up over current events. It didn't seem all that important to him, but Karl certainly cared about this stuff. Max grabbed a cup of coffee, kissed Karl goodbye, and left to go to work. Karl went back to reading the paper.

The article described how Spellman admitted the affair and had apologized profusely to his family and to his constituents. Spellman didn't offer much in the way of an explanation, but merely asked to be forgiven for his weaknesses. The nanny hadn't made a public statement. Karl thought the story in the paper made Spellman look pretty bad, and wondered if he should be impeached. He hoped Spellman wouldn't escape punishment. Karl had voted for him, and he felt let down and deceived. Karl now saw Spellman as like so many wealthy political leaders: someone who thought he was above the rules that others had to abide by; someone who felt entitled to whatever he wanted.

Karl noticed that it was getting late, so he put the rest of the newspaper in his briefcase and left for work. When he got to the office, a couple of people were talking about the Spellman scandal. His colleague Diane agreed with him, saying that Spellman was a hypocrite. But another colleague, Randy, disagreed, saying that we're always forgiving people for mistakes, and that the United States was a country of second chances. Randy argued that this whole scandal was between Spellman and his wife anyway.

As Karl drove home from work that night, he listened to the local news on the car radio. A commentator said that people in the United States loved to build up public figures, but then they loved even more to see them fall. The commentator agreed with Karl's co-worker, Randy. She said that the U.S. is a nation that loves a comeback, and maybe Spellman would survive this scandal and be just fine. Karl disagreed. He thought Spellman deserved disapproval because he'd done the wrong thing and because he'd been so pompous about his ability to make government "clean" and honorable. This Spellman case was disgusting, Karl thought, and he hoped the guy resigned and never got into public office again.

Some rhetoricians might analyze Eric Spellman's problems and Karl's responses to them using the theory we discuss in this chapter: Dramatism. Dramatism is a theoretical position that seeks to explain the actions of human life by using the metaphor of a theatrical drama. Kenneth Burke is known as the originator of Dramatism, although he didn't initially use that term himself. Burke, who died in 1993 at the age of 96, was a fascinating person, and he was unlike many of the other theorists whose theories we profile in this book. Burke never earned an undergraduate degree, much less a Ph.D. He was self-taught in the areas of literary criticism, philosophy, communication, sociology, economics, theology, and linguistics. He taught for almost 20 years at several universities, including Harvard, Princeton, and the University of Chicago. His breadth of interests and perhaps his lack of formal training in any one discipline make him one of the most interdisciplinary theorists we'll study. His ideas have been applied widely in various areas including literature, theater, communication, history, and sociology. No doubt one reason Burke is so widely read and applied has to do with his focus on symbol systems. Researchers from different disciplines have used Dramatism to study topics including natural disasters (Darr & Strine, 2017), the sociological impact of a pick-up basketball game (DeLand, 2018), and information and communication technology development in South Africa (Yu, 2022). It's likely, therefore, that regardless of your major, you'll find something useful in Burke's theory.

Dramatism, as its name implies, conceptualizes life as a drama, placing a critical focus on the acts performed by various players. Just as in a play, the acts in life are central to revealing human motives. Dramatism provides us with a method that's well suited to address communication between a text (e.g., the newspaper story about Eric Spellman) and the audience for that text (e.g., Karl), as well as the inner action of the text (e.g., Spellman's motives and choices). According to Dramatism, when Karl reads about Spellman's case, it's as if he sees the politician as an actor in a scene, trying to accomplish purposes because of certain motives. Karl comments on Spellman's motives as he evaluates his act of cheating on his wife with their nanny. Karl thinks that Spellman was motivated by a sense of entitlement. Burke's theory of Dramatism allows us to analyze both Spellman's rhetorical choices in this situation (i.e., how he framed his case) and Karl's responses to his choices.

Drama is a useful metaphor for Burke's ideas for three reasons. First, drama indicates a grand sweep, and Burke doesn't make limited claims; his goal is to theorize about the whole range of human experience. The dramatic metaphor is particularly useful in describing human relationships because it's grounded in interaction or dialogue. Through dialogue, drama provides models of relationships as well as offers a way to understand relationships (Daas, 2011). Secondly, drama tends to follow recognizable types or genres: comedy, musical, melodrama, and so forth. Burke feels the way we structure and use language may be related to the way we put a human drama into a particular type. So, as Karl says he's disgusted and appalled by Spellman's actions, he's beginning to cast the story as a tragedy. Thirdly, drama is always addressed to an audience. In this sense, drama is rhetorical. Burke views literature as "equipment for living," which means that literature or texts speak to people's lived experiences and provide people with responses for dealing with these experiences (Winslow, 2010). Given this focus, Dramatism studies the ways in which language and its usage relate to audiences (Ross, 2017).

Theory at a Glance • Dramatism

Burke's theory compares life to a play and states that, as in a theatrical piece, the events that occur in life require six elements: an **actor** or **agent** who does some type of an **action/act** within a **scene** with a **position** or **attitude** relative to the others in that scene, using some **means** for the action to take place, and with a **purpose** for the action. The theory allows a rhetorical critic to analyze a speaker's motives by identifying and examining these six elements. Furthermore, Burke believes, guilt is the ultimate motivation for people, and Dramatism suggests that social actors are most successful when they provide their audiences with a means for purging their guilt.

Assumptions of Dramatism

Similar to a few other theorists in this book, Kenneth Burke's thinking is so complex that it's difficult to reduce it to one set of assumptions or to a specific ontology, a term we introduced to you in **Chapter 3.** Researchers such as Barry Brummett (1993) have called Burke's assumptions a symbolic ontology because of his emphasis on language. Yet, as Brummett cautions, "The best one can do, in searching for the heart of Burke's thought, is to find a partial ontology, a grounding-for-the-most-part. For Burke, people *mainly* do what they do, and the world is *largely* the way it is, because of the nature of *symbol systems themselves*" (p. xii; emphasis in original). Brummett's comment contextualizes the following three assumptions of Burke's Dramatism Theory:

- Humans are animals who use symbols.
- Language and symbols form a critically important system for humans.
- Humans are choice makers.

The first assumption speaks to Burke's belief that some of what we do is motivated by our animal nature and some of what we do is motivated by symbols. Recall the semiotic tradition we discussed in **Chapter 2** to understand this notion. For example, when Karl drinks his morning coffee, he is satisfying his thirst, an animal need. When he reads the morning paper and thinks about the ideas he encounters there, he is being influenced by symbols. The idea that humans are animals who use symbols represents a tension in Burke's thought. As Brummett (1993) observes, this assumption "teeters between the realizations that some of what we do is motivated by animality and some of it by symbolicity" (p. xii). Of all the symbols that humans use, language is the most important for Burke. To further understand the role of symbols in human interaction, recall our discussion in **Chapter 15** about Symbolic Interaction Theory.

In the second assumption (the critical importance of language), Burke's position is somewhat similar to the concept of linguistic relativity known as the Sapir–Whorf hypothesis (Sapir, 1921; Whorf, 1956). Edward Sapir and Benjamin Whorf both noted that it's difficult to think about concepts or objects without words for them. Thus, people are restricted (to an extent) in what they can imagine by the limits of their language. For Burke, as well as for Sapir and Whorf, when people use language, they're being used by it as well. When Karl tells Max that Spellman is an incredible lowlife hypocrite, he is choosing the symbols he wishes to use, but at the same time his opinions and thoughts are shaped by hearing himself use those symbols. Furthermore, when a culture's language does not have symbols for a given motive, then speakers of that language are unlikely to recognize that motive. Thus, because English doesn't afford many symbols that express much nuance of opinion about Spellman's behavior and motivations, English speakers' discussions are often polarized. When Karl talks with his colleagues Diane and Randy, the discussion is focused on whether Spellman was right or wrong. There isn't much choice in between, and Burke would argue that this is a direct result of our symbol system. Think back to other controversies you have talked about (e.g., the moral implications of artificial intelligence, privacy issues online, contrasts among presidential candidates, the war in Ukraine, etc.). You may remember the discussions as either/or propositions–positions were cast as either right or wrong. Burke argues that symbols shape our either/or approach to these complex issues because the English language has many word pairings for oppositions (hot/cold, good/bad, tall/short, etc.).

Burke asserts that words, thoughts, and actions have extremely close connections with one another. Burke illustrates this by saying that words act as *terministic screens* leading to *trained incapacities*, meaning that people cannot see beyond what their words lead them

> **terministic screens** people can only think about things their language allows them to

to believe (Burke, 1960). **Terministic screens** "illuminate (or select) certain aspects of the rhetorical moment while obscuring (or deflecting) others" (Dunn, 2018, p. 75). For example, despite educational efforts, U.S. public health officials still have difficulty persuading people to think of the misuse of alcohol and prescription drugs when they hear the words *drug abuse*. Most people in the United States think about "drug abuse" as the misuse of illegal drugs, such as heroin and cocaine (Brummett, 1993). The words *drug abuse* are *terministic screens*, screening out some meanings while including others. For instance, in the United States, when people talk about the opioid crisis, they will say "abuse of prescription drugs," not just "drug abuse," to be clear about what they mean. For Burke, language has a life of its own, and "anything we can see or feel is already *in* language, given to us *by* language, and even produced *as us* by language" (Nelson, 1989, p. 169; emphasis in original). This idea is somewhat at odds with the final assumption of Dramatism.

The final assumption of the theory states that human beings are choice makers even though the second assumption says that language exerts a determining influence over people. Burke persistently suggests that behaviorism has to be rejected because it conflicts with human choice.

> **agency** one prong of the pentad: the ability of a social actor to perform an act out of choice

Thus, as Karl reads about Eric Spellman, he forms his opinions about Spellman's behavior through his own free will. Much of what we discuss in the rest of the chapter rests on Burke's conceptualization of **agency,** or the ability of a social actor to act out of choice.

As Charles Conrad and Elizabeth Macom (1995) observe, "The essence of agency is choice" (p. 11). Yet, as Conrad and Macom go on to discuss, Burke grappled with the concept of agency throughout his career, largely because of the difficult task of negotiating a space between complete free will and complete determinism. Despite this, Burke kept agency in the forefront of his theorizing. Researchers continue to struggle with how agency affects people's symbolic actions and allocations of blame (French & Brown, 2011).

To understand Burke's intentions in Dramatism, we need to discuss how he framed his thinking relative to Aristotelian rhetoric.

Dramatism as New Rhetoric

In his book *A Rhetoric of Motives* (1950), Burke is concerned with persuasion, and he provides ample discussion of the traditional principles of rhetoric articulated by Aristotle (see our discussion of Aristotle's *Rhetoric* in **Chapter 20**). Burke maintains that the definition of rhetoric is, in essence, persuasion, and Burke's writings explore the ways in which persuasion takes place. In so doing, Burke proposes a new rhetoric (Nichols, 1952) that focuses on several key issues, chief among them being the notion of *identification*. In 1952, Marie Nichols said the following about the difference between Burke's approach and Aristotle's: "The difference between the 'old' rhetoric and the 'new' rhetoric may be summed up in this manner: whereas the key term for the 'old' rhetoric was persuasion and its stress was upon deliberate design, the key term for the 'new' rhetoric is identification and this may include partially 'unconscious' factors in its appeal" (p. 323). Yet, Burke's purpose wasn't to displace Aristotle's conceptualizations, but rather to supplement the traditional approach. To do this, Burke wrote about two major themes he thought should be added to traditional rhetoric. These include: (1) substance and identification/division and (2) the cycle of guilt and redemption.

Substance and Identification/Division

Burke asserts that all things have **substance,** which he defines as the general nature of something. Substance can be described in people by listing their demographic characteristics as well as background information and facts about the present situation, such as their talents and occupations. Thus, from our opening scenario, we may understand Karl's substance by noting he's a 38-year-old white male, gay, Episcopalian, of Swedish heritage, high school math teacher, collector of rare coins, tennis player, and crossword puzzle enthusiast. In addition, Karl has been married to Max for five years and lives in Grenada, New Mexico. Of course, many other pieces of information make up Karl's substance as well, but these facts give us a starting point.

> **substance** the general nature of something

According to Burke, when there's overlap between two people in terms of their substance, they have **identification.** The more overlap that exists between them, the greater the identification. The opposite is also true, so the less overlap between individuals, the greater the **division** that exists between them. For instance, the fact that Eric Spellman is a wealthy, white, Catholic, cisgender, married man of German heritage and a former attorney who is the mayor of Grenada, New Mexico, and appears in the media frequently provides a sense of division between him and Karl. They are both white, male professionals who live in Grenada, New Mexico, but they overlap on little else.

> **identification** when two people have overlap in their substances
>
> **division** when two people fail to have overlap in their substances

Students Talking: Nina

 How I could have anything in common with singers of the 1930s–1960s is beyond me. I'm a 19-year-old born in the 21st century, and I can't sing at all, but I completely understand the need for protest music. When I read about Burke's idea of identification, I thought about the words of some of those old songs and their messages: safe working conditions, racial justice, economic justice, civil rights, nonviolence, immigration, etc. Singers like Woody Guthrie, Billie Holiday, Pete Seeger, Joan Baez, and Bob Dylan may have sung those songs years ago, but the words resonate for me today.

However, it's also the case that two people can never completely overlap with each other. Burke recognizes this and notes that the *ambiguities of substance* dictate that identification always rests on both unity and division. As Shane Borrowman and Marcia Kmetz (2011) note, identification and division are inevitably paired, and it's difficult to talk about one without the other. Burke observed that individuals will unite on certain matters of substance, but at the same time remain unique, being "both joined and separated" (Burke, 1950, pp. 20-21). Furthermore, Burke indicates that rhetoric is needed to bridge divisions and establish unity. Debora Antunes (2016) emphasized the importance of symbols in the process of identification by stating that symbols are "carriers of identification", underscoring the first assumption of the theory, which we discussed earlier. Sangchul Lee (2022) agrees based on his study of T. J. Park, the founder of a steel company in Korea. Lee shows that the company became supremely successful in large part because of speeches Park delivered to the employees stressing their identification with management and the nation.

Burke claimed that Dramatism was a theory that generalized to all people in all cultures. Some researchers' work provides support for his claim. For example, Meriell Tendean (2022) found Dramatism was a useful framework for analyzing the rhetorical strategies of three male Indonesian influencers who were all rebutting accusations of sexual harassment. Rukhsana Ahmed (2009) demonstrated that the Burkean concept of identification has applications in non-Western discourse by performing a rhetorical analysis of a political speech made by Begum Zia in Bangladesh. In Ahmed's analysis, Zia was able to make rhetorical appeals convincing her audience that the divisions between them could be bridged. Burke refers to this process as **consubstantiation,** or increasing identification between audience and speaker. Ketil Knutsen (2016) applied Burkean analysis to a Norwegian television series that presented viewers' information about life in Norway in the 1700s. Knutsen argues that the series accomplished its goals through identification and showed viewers "how to make the past a resource" in their current lives (p. 457). However, as we discuss later in the critique section of the chapter, not all researchers agree that Dramatism works in all cultures.

> **consubstantiation** when appeals are made to increase identification between people

The Process of Guilt and Redemption

Consubstantiality, or increased identification between audience and speaker, is related to the guilt/redemption cycle because guilt can be assuaged as a result of identification and divisions. For Burke, the process of guilt and redemption undergirds the entire concept of symbolizing. **Guilt** is the central motive for all symbolic activities, and Burke defines guilt broadly to include any type of tension, embarrassment, shame, disgust, or other unpleasant feelings. Central to Burke's theory is the notion that guilt is intrinsic to the human condition. Because we're continuously feeling guilt, we're also continuously engaging in attempts to purge ourselves of guilt's discomfort. This process of feeling guilt and attempting to reduce it finds its expression in Burke's cycle, which follows a predictable pattern: order (or hierarchy), the negative, victimage (mortification or scapegoating), and redemption.

> **guilt** tension, embarrassment, shame, disgust, or other unpleasant feelings

Order or Hierarchy   Burke suggests that society exists in the form of an **order,** or **hierarchy,** which is created through our ability to use language. Language enables us to create categories like richer and poorer, and more and less powerful—the haves and the have-nots. These categories form social hierarchies. Often we feel guilt as a result of our place in the hierarchy. If we are privileged, for example, we may feel we have power at the expense of those with less wealth and power. If we are poor, we may believe it's somehow our fault that we weren't able to achieve material success. These feelings prompt guilt.

> **order** or **hierarchy** a ranking in society created in our ability to use language

The Negative   **The negative** comes into play when people see their place in the social order and seek to reject it. Saying no to the existing order is both a function of our language abilities and evidence of humans as choice makers. When Burke penned his often-quoted definition of Man, he emphasized the negative:

> **the negative** rejecting one's place in the social order; exhibiting resistance

> Man is
> the symbol-using inventor of the negative
> separated from his natural condition by instruments
> of his own making goaded by the spirit of hierarchy
> and rotten with perfection. (1966, p. 16)

When Burke coined the phrase "rotten with perfection," he meant that because symbols allow us to imagine perfection, we always feel guilty about the difference between the real state of affairs and the perfection that we can imagine. Further, "rotten with perfection" also means that our ability with symbols allows us to get stuck in symbolic "ruts" and then fail to see the fact that we're merely constructing our perspectives. This leads us to believe in the "rightness" of our perspectives so strongly that we become close-minded, which is detrimental to ourselves and others (Steiner, 2009).

Victimage   **Victimage** is the way in which we attempt to purge the guilt that we inevitably feel as part of the human condition. There are two basic types of victimage, or two methods to purge our guilt. Burke calls the type of victimage that we turn in on ourselves **mortification.** When we apologize for wrongdoing and blame ourselves, we engage in mortification. When Eric Spellman said he did the wrong thing and apologized, he was engaging in mortification. In the United

victimage the way we attempt to purge the guilt we feel as a result of being human

mortification one method of purging guilt, by blaming ourselves

States in 1998, President Bill Clinton was accused of having had an affair with White House intern Monica Lewinsky. Clinton denied the affair under oath, and later it was revealed that he had lied. This led to his impeachment by the U.S. House of Representatives for perjury. He was later acquitted by the U.S. Senate. Republican leaders said they would have felt more sympathetic about President Clinton's affair if he'd admitted he was wrong and hadn't perjured himself. But Clinton refused to engage in mortification. Instead he turned to the second purging technique, called *scapegoating*.

In **scapegoating,** blame is placed on some sacrificial vessel. By sacrificing the scapegoat, the actor is purged of sin. Clinton attempted to scapegoat the Republicans and others (Oles-Acevedo, 2012) as deserving the

scapegoating one method of purging guilt, by blaming others

real blame for the country's problems after he confessed to an inappropriate relationship with Monica Lewinsky. When the news first broke in 1998, before Clinton admitted his relationship with Lewinsky, Hillary Rodham Clinton appeared on television suggesting that the rumors about her husband were the result of a complex "right-wing conspiracy" that was out to get her and her family. This type of rhetoric illustrates Burke's concept of scapegoating. Interestingly, in 2014—over 15 years later—Monica Lewinsky said that she was the scapegoat in this situation (Fox News, 2014, **http://www.foxnews.com/politics/2014/05/06/monica-lewinsky-speaks-out-says-was-made-scapegoat.html**), suggesting two very different ways to apply the term.

In 2020, President Trump's response to the same situation that Clinton faced (being acquitted by the Senate after being impeached by the House) also leaned on scapegoating. Trump's frequent use of the term "witch-hunt" and his negative characterizations of Nancy Pelosi, Adam Schiff, and others, placed the blame for all the turmoil surrounding the impeachment and trial outside of himself. Both Presidents Trump and Clinton found that blaming others was an effective purging technique, at least as judged by their supporters.

Redemption   The final step in the cycle is **redemption,** which involves a rejection of the unclean and a return to a new order after guilt has been temporarily purged. Inherent in the term *redemption* is the notion of a redeemer. The redeemer in the Judeo Christian tradition is God for the Jews and Christ for Christians. When politicians

redemption a rejection of the unclean and a return to a new order after guilt has been temporarily purged

blame problems on the media or on the opposing party, they offer themselves as potential redeemers—those who can lead the people out of their troubles. A key in the redemption phase is the fact that guilt is only temporarily relieved, through the redeemer or any other method. After any order or hierarchy becomes reestablished, guilt returns to plague the human condition.

The Pentad

In addition to devising the theory of Dramatism, Burke (1945) created a method for applying his theory to a text in order to understand the symbolic activities involved. Researchers have argued about whether

> **pentad** Burke's method for applying Dramatism to analyze a text

Dramatism is a theory or a method, usually concluding that it's both (Anderson & Prelli, 2018). Floyd Anderson and Lawrence Prelli argue that Dramatism has to be both because Burke saw conceptualizing ideas and testing those ideas as inseparable activities. Burke called his method the **pentad** because it originally consisted of five points for analyzing a symbolic text like a speech or a series of articles about a particular topic, for instance. The pentad allows rhetorical analysts to understand texts, and helps them determine why a speaker selects a specific rhetorical strategy for identifying with an audience. The five points making up the pentad include the act, the scene, the agent, agency, and purpose. Over 20 years after creating this research tool, Burke (1968) added a sixth point, attitude, to the pentad, making it a hexad, although most people still refer to it as the pentad (**Figure 21.1**). We will examine each of the points in turn.

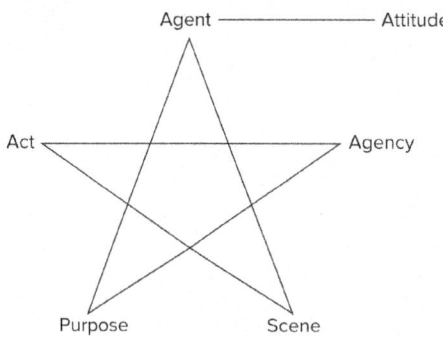

Figure 21.1   Burke's Pentad

The Act   Burke considered the **act** to be what is done by a person. In the case of Eric Spellman, the act would be having an affair that resulted in a child while he was married to someone else.

> **act** one prong of the pentad; that which is done by a person

The Scene   The **scene** provides the context surrounding the act. In Spellman's case, the scene would include a time period in which American politicians are under fire for corruption and hypocrisy. We

> **scene** one prong of the pentad; the context surrounding the act

don't have much information about the way Eric Spellman would contextualize the scene because he has provided very limited information publicly about any contributing factors to the act.

The Agent   The **agent** is the person or persons performing the act. In the case of Eric Spellman, he's the agent. However, if a researcher wished to analyze Karl's act of deciding he didn't support Spellman any longer, then Karl would be the agent.

> **agent** one prong of the pentad; the person performing the act

Agency   **Agency** refers to the means used by the agent to accomplish the act. Possible forms of agency include message strategies, storytelling, apologies, speech making, and so forth. In Spellman's case, the agency includes the justifications and apologies he has made publicly.

Purpose   The **purpose** refers to the goal that the agent had in mind for the act—that is, why the act was done. In Spellman's case, the purpose was unclear. He has provided no reasons. Karl believes he did it because he thought he was entitled.

> **purpose** one prong of the pentad; the goal the agent had for the act

Attitude   **Attitude** refers to the manner in which actors position themselves relative to others. Again, in Spellman's case, this is a contested point. Karl might say that he acted from an attitude of entitlement and superiority. Many of the articles Karl has read have commented that Spellman believed himself to be above the law.

> **attitude** a later addition to the pentad; the manner in which the agent positions himself or herself relative to others

When using the pentad to analyze a text, the analyst first determines all the elements of the pentad and identifies what occurred in a particular act. After labeling the points of the pentad and fully explicating each, the analyst then examines the **pentadic** or **dramatistic ratios,**

> **pentadic** (or **dramatisic**) **ratios** the proportions of one element of the pentad relative to another element

or the proportions of one element relative to another. By isolating any two parts of the pentad and examining their relationship to each other, we derive a ratio. The ratio helps us to determine motives. For instance, "a scene: act ratio may describe the act as a reaction to circumstances, while an agent: act ratio suggests that the motive originates in the agent" (Encheva, 2018, p. 42). In analyzing the ratios in this manner, the researcher is able to discover a dominant element. Is the agent emphasized more than the situation or vice versa? An examination of the dramatistic ratios suggests something about point of view and rhetorical strategies.

Theory-Into-Action

 When Elizabeth Warren ran for the U.S. Senate in Massachusetts in 2012, she was challenging incumbent U.S. Senator Scott Brown. She used identification and division in her campaign rhetoric. Warren was the underdog but was able to stage a big comeback in the polls because of her ability to convince women voters that Brown was not consubstantial with them, while she was. Warren was persuasive to Massachusetts' independent women voters by shifting the focus from Brown's message of working together across differences to her message of identifying with women's issues. This proved a winning strategy for Warren as she defeated Brown in the election. Women running for the U.S. presidency, however, have had difficulty in utilizing this strategy.

Integration and Critique

There is no question that Kenneth Burke has made an immeasurable contribution to the field of communication with his theory of Dramatism. Various researchers have praised Burke in the following terms: "He has become the most profound student of rhetoric now writing in America" (Nichols, 1952, p. 331); "Kenneth Burke is more than a single intellectual worker; he is the ore for a scholarly industry" (Brummett, 1993, p. xi); and in 1981, the *New York Times* recognized Burke as a leading American critic, saying he was "the strongest living representative of the American critical tradition, and perhaps the largest single source of that tradition since its founder, Ralph Waldo Emerson" (cited in Chesebro, 1993, p. xi). Burke's work is widely praised and frequently cited. In fact, the National Communication Association, one of the main organizations for communication teachers, researchers, and professionals, has an entire division devoted to Burkean criticism and there is a journal featuring work based on Burke's theories and ideas (**kbjournal.org**). The research surrounding Dramatism has almost exclusively been qualitative, honoring the notion of "voice" that Burke felt was important.

As we evaluate Dramatism, four criteria are relevant: scope, parsimony, utility, and heurism.

Integration

Communication Tradition	**Rhetorical** | **Semiotic** | Phenomenological | Cybernetic | Socio-Psychological | Socio-Cultural | **Critical**
Communication Context	Intrapersonal | Interpersonal | Small Group | Organizational | **Public/Rhetorical** | Mass/Media | Cultural
Approach to Knowing	Positivistic/Empirical | **Interpretive/Hermeneutic** | Critical

Critique

Evaluation Criteria	**Scope** | Logical Consistency | **Parsimony** | **Utility** | Testability | **Heurism** | Test of Time

Scope

Dramatism has been criticized for being too wide in scope. Burke's goal is no less than to explain the whole of human experience with symbols. This is an extremely broad and ambitious goal, and some critics believe it renders the theory too broad to be meaningful. When you contrast Dramatism with a theory like Uncertainty Reduction Theory, which we discuss in **Chapter 5,** you can see the two extremes of theoretical scope. URT was created to explain the first few minutes of an initial encounter between strangers. Dramatism encompasses all of human symbolic encounters. Some critics argue that when a theory attempts such a lofty goal it's doomed to be overly complex and obtuse. Whether or not you see the range of Dramatism's goal as a weakness is somewhat subjective. For Burke and many who followed him, the wide scope of Dramatism is part of its appeal.

Parsimony

Related to the criticism of scope, Dramatism is seen as having too many concepts and specialized terms, rendering it overly complex and confusing (Foss, Foss, & Trapp, 1991). Herbert Simons (2004) states that Burke was more "convolutionist than revolutionist" (p. 152). Even proponents of Burke acknowledge that he's difficult to read. Marie Hochmuth Nichols concluded her pioneering 1952 essay on Burke by saying:

> Burke is difficult and often confusing. He cannot be understood by casual reading of his various volumes. In part the difficulty arises from the numerous vocabularies he employs. His words in isolation are usually simple enough, but he often uses them in new contexts. To read one of his volumes independently, without regard to the chronology of publication, makes the problem of comprehension even more difficult because of the specialized meaning attached to various words and phrases (p. 330).

However, Nichols also provides a rebuttal to these criticisms by concluding that some of the difficulty arises from "the compactness of his writing, the uniqueness of his organizational patterns, the penetration of his thought, and the breadth of his endeavor" (p. 330). In other words, Burke is a genius and worth the effort it requires to understand his original thinking. When a student is diligent, Burke's theory repays the hard work with many rewards.

Utility

Some researchers (Condit, 1992; Murray, 2003) observe that Dramatism falls short on the criterion of utility. This critique is lodged mainly because of what Burke leaves out of the theory. For example, Celeste Condit (1992) argues that the theory would be more useful if it addressed gender and culture more specifically and expansively. Condit observes that although Burke was supportive of feminism, his support came mainly in the form of including women under the sign of "man." She notes that given the historical context in which Burke wrote, his support for women was not inconsequential. Many writers in Burke's generation completely ignored women, so Burke was making a contribution by including women. Condit maintains, however, that now the scene has altered, and it's inappropriate to subsume women under the word "man." Here Condit is talking both about the use of the generic "man" to represent all people and about our ability as a society to begin to think in new ways about sex and gender. This critique continues to be salient in the 21st century, when issues such as those discussed by the #MeToo movement as well as critiques of gender binaries pervade cultural discourse.

As Condit (1992) notes, "We must extend our language beyond duality to a broad 'humanity' and to 'human beings,' discovering ways to speak that emphasize human plurality" (p. 351). Condit says the definition of man that Burke provides, which we discussed earlier in the chapter, isn't adequate to include women. She recasts Burke's definition to describe how men have relegated women to subordinate roles:

> Woman is
> the symbol-receiving (hearing, passive) animal
> inventor of nothing (moralized by priests and saints)
> submerged in her natural conditions by instruments of man's making
> goaded at the bottom of hierarchy (moved to a sense of orderliness)
> and rotted by perfection (p. 351).

Then she further alters the definition to move beyond the essentialist binary defining men and women as opposites of each other:

> People are
> players with symbols
> inventors of the negative and the possibility of morality
> grown from their natural condition by tools of their collective making
> trapped between hierarchy and equality (moved constantly to reorder)
> neither rotten nor perfect, but now and again lunging down both paths
> (p. 352).

Condit's argument is that Burke's approach needs to be broadened both to include women and to move past a focus on one sex or the other to be truly inclusive of both. But, she asserts that merely broadening the language of "man/his" to include "people/their" will not in itself be sufficient to challenge the hold that language exerts as a terministic screen against women in the United States. We need to change both our language and our thinking about women, men, gender, and inclusivity for significant progress to occur.

Condit also critiques Burke's emphasis on universality among cultures at the expense of particularity. For Condit, this is especially the case in the matter of Burke's contention that victimage is a transcultural experience—a method for purging guilt in all cultures. She argues that cultures other than Western Christian ones (from which Burke draws almost exclusively) might not see victimage as the dominant motive for human conduct. For example, Buddhism might provide different motives than Christianity does. Furthermore, if we examine trickster tales from Native American or African American cultures, we might see victimage characterized in a strikingly different fashion from what Burke describes. The trickster is in a low-power position relative to the rest of the society, but is able to triumph through wit and cleverness. The trickster is not a victim in the Christian sense that we see featured in Dramatism; rather, the trickster emerges victorious by turning the rules of the system against those in power.

In sum, Condit's (1992) critique does not deny the enormous contribution made by Burke's theory. Instead, she suggests some extensions and modifications for improving the theory. Jeffery Murray (2003) agrees with Condit, asserting that although Burke's theory continues to be widely used, it's necessary to expand it to include the voices of those who have been marginalized. Murray uses what he calls an "Other-Burkean" frame in two studies. In one he analyzes Nazi propaganda, and in the other he examines a speech the late U.S. Senator Ted Kennedy made in 1969, after the death of Mary Jo Kopechne. Kopechne was a passenger in Kennedy's car who died when he drove off a bridge in Massachusetts in 1969. In both cases, Murray argues that paying attention to what was omitted in the rhetoric and to "Others" whose concerns are not highlighted by the pentad (such as the Jews in relationship to the Nazis, and Mary Jo Kopechne in relation to Kennedy) provides a richer analysis than Dramatism alone does. If you think back to our opening story, you might notice a lack of emphasis on the nanny's perspective or Eric Spellman's wife's perspective. Neither is named in the story, and neither is presumed to be the focus for an analysis of the situation.

Heurism

With regard to heurism, most critics agree that Dramatism is very successful. For instance, Dramatism was originally used in rhetorical analyses of speeches, but now the focus has widened to other discourse in the public sphere including "editorials, pamphlets and monographs, books, docudramas, radio and television news, movies, music, and even the Internet [*sic*]" (Hunt, 2003, p. 378). Dramatistic principles and concepts have been applied widely. For example, Burke's thinking has been applied to the Syrian Civil War (Bakke & Kuypers, 2016). Marketing on Facebook has been analyzed using the pentad (Casteleyn, Mottart, & Rutten, 2009). Amanda Nell Edgar (2014) employed the Dramatistic lens to understand R&B culture and Chris Brown's assault on Rihanna. Medical researchers (e.g., Bareiss, 2015) utilized Dramatism to understand nonsuicidal self-injuries, such as cutting. Luke Rowland's (2016) assessment of Japanese multilingual signage used Burkean principles. Clearly, Burke has influenced scholars across a wide swath of disciplines examining a great variety of topics.

Students Talking Tech: Emma

I see that some researchers used Dramatism to investigate marketing on Facebook. That's interesting because all the way through the chapter I was wondering how the theory would work with social media or online communication. I think it could, but it's funny that the Facebook study was the only example where social media was mentioned at all. A friend of mine is finishing up her thesis for an M.A., and she's writing about the "Mommy Wars" online. She went to a website for mothers, and they have a chat room for stay-at-home moms and working moms—and they're separate. She's been telling me that members of each group post a lot, justifying why they either work or stay home. They also hate on the other group by saying stuff like, "You can't be a good mom if you work" or "Good mothers must feel fulfilled themselves through a meaningful career." I'm interested in this because I want to work and have children too, so I've been asking my friend a lot about her findings. She's not using Dramatism as her theory, but as I read this chapter, I was thinking she could have. Burke said stories were equipment for living, and there are so many stories being told online now, so I'm surprised rhetoricians aren't using Burke more to study online communication.

Closing

There's general consensus that Burke's theory provides us with imaginative and innovative insights into human motives and interaction. Dramatism is a theory that looks at the big picture and attempts to explain a wide scope of human behaviors. The theory allows us to analyze human motivations and behaviors, and its focus on language as the critical symbol system makes it especially attractive to communication researchers.

Discussion Starters

Case-In-Point: How could Karl talk about Eric Spellman's situation in a less polarized fashion? What are the linguistic barriers to such a discussion? Does a feminist critique of Burke's theory enable you to think of less polarized language? Explain your answer.

Try-It-Your-Selfie: Suggest a research question involving social media that you think could be studied using the framework of Dramatism. Explain how you would proceed with this study and why Dramatism would be useful.

1. Use the pentad to analyze a public figure and the discourse surrounding some current controversy involving this figure.

2. Do you agree with Burke that guilt is the primary human motive? If not, what do you think is the primary human motive?

3. Do you agree with Condit that Burke's theory is culture specific rather than universal? Explain your answer.

4. Apply any element of Dramatism to your life. For instance, have you experienced identification with a public figure? If so, what strategies did they employ that allowed you to feel this way?

References

Ahmed, R. (2009). Interface of political opportunism and Islamic extremism in Bangladesh: Rhetorical identification in government response. *Communication Studies*, 60(1), 82–96.

Anderson, F. D., & Prelli, L. (2018). Kenneth Burke's agonistic theory of knowledge. *Western Journal of Communication, 82*(2), 181–193.

Antunes, D. (2016). Branding cyber-activism: Burke's identification and the visual identity of anonymous. *KB Journal: The Journal of the Kenneth Burke Society, 11*(2). http://kbjournal.org/antunes.

Bakke, P. C., & Kuypers, J. A. (2016). The Syrian Civil War, international outreach and a clash of worldviews. *KB Journal: The Journal of the Kenneth Burke Society, 11*(2).

Bareiss, W. (2015). Adolescent daughters and ritual abjection: Narrative analysis of self-injury in four U.S. films. *Journal of Medical Humanities, 36,* 1–19.

Borrowman, S., & Kmetz, M. (2011). Divided we stand: Beyond Burkean identification. *Rhetoric Review, 30*(3), 275–292.

Brummett, B. (1993). Introduction. In B. Brummett (Ed.), *Landmark essays on Kenneth Burke* (pp. xi–xix). Hermagoras Press.

Burke, K. (1945). *A grammar of motives*. Prentice Hall.

Burke, K. (1950). *A rhetoric of motives*. Prentice Hall.

Burke, K. (1960). *Permanence and change*. Bobbs-Merrill.

Burke, K. (1966). *Language as symbolic action: Essays on life, literature, and method.* University of California Press.

Burke, K. (1968). Dramatism. In D. L. Sills (Ed.), *The international encyclopedia of the social sciences* (vol. 7, pp. 445–452). Macmillan/Free Press.

Casteleyn, J., Mottart, A., & Rutten, K. (2009). How to use Facebook in your market research. *International Journal of Market Research, 51*(4), 439–447.

Chesebro, J. W. (1993). Preface. In J. W. Chesebro (Ed.), *Extensions of the Burkeian system* (pp. vii–xxi). University of Alabama Press.

Condit, C. M. (1992). Post-Burke: Transcending the substance of dramatism. *Quarterly Journal of Speech, 78*(3), 349–355.

Conrad, C., & Macom, E. A. (1995). Revisiting Kenneth Burke: Dramatism/logology and the problem of agency. *Southern Communication Journal, 61*(1), 11–28.

Daas, K. L. (2011). The pieties of death: A Burkean analysis of the Tri-State crematory case. *Texas Speech Communication Journal, 36*(1), 82–93.

Darr, C. R., & Strine IV, H. C. (2017). Natural disasters and the rhetorical construction of American values: Community exceptionalism as representative anecdote. *Atlantic Journal of Communication, 25*(5), 293–304.

DeLand, M. (2018). The ocean run: Stage, cast, and performance in a public park basketball scene. *Journal of Contemporary Ethnography, 47*(1), 28–59.

Dunn, R. C. (2018). "The future is in good hands": A pentadic analysis of president Barack Obama's farewell address. *A Graduate Journal of Qualitative Communication Research, 17*(1), 73–89.

Edgar, A. N. (2014). R&B rhetoric and victim-blaming discourses: Exploring the popular press's revision of Rihanna's contextual agency. *Women's Studies in Communication, 37*(2), 138–158.

Encheva, L. (2018). The grammar and rhetoric of gamification. *Rhetor: Journal of the Canadian Society for the Study of Rhetoric, 7,* 39–48.

Foss, S., Foss, K., & Trapp, R. (1991). *Contemporary perspectives on rhetoric.* Waveland Press.

French, S. L., & Brown, S. C. (2011). It's all your fault: Kenneth Burke, symbolic action, and the assigning of guilt and blame to women. *Southern Communication Journal, 76*(1), 1–16.

Hunt, S. B. (2003). An essay on publishing standards for rhetorical criticism. *Communication Studies, 54*(3), 378–384.

Knutsen, K. (2016). A history didactic experiment: The TV series *Anno* in a dramatist perspective. *Rethinking History, 20*(3), 454–468.

Lee, S. (2022). A Burkean dramatistic analysis of Tae-Joon Park's rhetoric: The Chairman and CEO of POSCO. *Business Communication Research Practice, 5*(2), 68–73.

Murray, J. (2003). An other-Burkean frame: Rhetorical criticism and the call of the other. *Communication Studies, 54*(2), 169–187.

Nelson, C. (1989). Writing as the accomplice of language: Kenneth Burke and poststructuralism. In H. W. Simons & T. Melia (Eds.), *The legacy of Kenneth Burke* (pp. 156–173). University of Wisconsin Press.

Nichols, M. H. (1952). Kenneth Burke and the new rhetoric. *The Quarterly Journal of Speech, 38*(2), 133–134.

Oles-Acevedo, D. (2012). Fixing the Hillary factor: Examining the trajectory of Hillary Clinton's image repair from political bumbler to political powerhouse. *American Communication Journal, 14*(1), 33–46.

Ross, J. M. (2017). *Archbishop Oscar Romero: The last homily and the plight of the poor*. Proceedings from 6th Annual International Conference on Journalism & Mass Communication. Singapore.

Rowland, L. (2016). English in the Japanese linguistic landscape: A motive analysis. *Journal of Multilingual and Multicultural Development, 37*(1), 40-55.

Sapir, E. (1921). *Language: An introduction to the study of speech*. Harcourt, Brace & World.

Simons, H. (2004). The rhetorical legacy of Kenneth Burke. In W. Jost & W. Olmstead (Eds.), *A companion to rhetoric and rhetorical criticism* (pp. 152-168). Blackwell.

Steiner, M. A. (2009). Reconceptualizing Christian public engagement: "Faithful Witness" and the American evangelical tradition. *Journal of Communication & Religion, 32*(2), 289-318.

Tendean, M. (2022). A dramatistic analysis of Indonesian influencers' statements in responding to sexual harassment allegations. *International Journal of Communication and Society, 4*(2), 235-249.

Whorf, B. L. (1956). *Language, thought, and reality*. MIT Press.

Winslow, L. (2010). Rhetorical homology and the caveman mythos: An(other) way to ridicule the aggrieved. *Communication Studies, 61*(3), 257-271.

Yu, K. (2022). Waiting for a hero: Dramatism analysis of South Africa's ICT development. *South African Review of Sociology, 52*(2), 92-111.

CHAPTER 22
The Narrative Paradigm

*Based on the research of **Walter Fisher***

"The narrative paradigm does not deny reason and rationality; it reconstitutes them, making them amenable to all forms of human communication."

—Walter Fisher

Miles Campbell

Miles Campbell rolled over in bed and turned off his screaming alarm. He burrowed under the covers for a minute before he realized he'd better get up or he would miss chem lab. He was tempted to sleep in, but a vision of his mother's face flashed before him, and he thought about how hard she had worked to help him get to college. He didn't want to disappoint her by not doing his very best now that he was here. So, Miles sighed and shrugged off the covers. He slipped out of bed and splashed cold water on his face. By the time he was dressed and headed for the kitchen, he felt better about his day and about life in general.

In the kitchen he heard his housemates, Robert and Carlos, arguing about something. Seems like a normal morning, Miles thought. Those two can never get along. "What has you guys up and yelling so early in the morning?" Miles asked as he began making his breakfast. Both Robert and Carlos looked up and grinned at Miles. "You won't think it's a big deal, Miles," Carlos said, "but we're discussing the candidates who are running for president of the Student Multicultural Association." "Yeah, you're right, Carlos," Miles laughed. "That doesn't seem like something worth arguing about to me!"

Robert handed Miles a copy of two campaign flyers. "Well, you might not think it's all that big a deal, but look at the difference between these two and tell me that Laura Huyge doesn't make more sense than Jorge Vega." Miles glanced at the two flyers that Robert had given him. Huyge was an Asian American graduate student, and she'd presented a list of ten points in her flyer that represented her platform.

She stated her interest in promoting cultural sensitivity and appreciation for diversity within the student body. Her flyer also listed a few ways that she planned to accomplish her goals. Her first big initiative, if she were elected, would be to sponsor a workshop with speakers from outside the university and several hands-on activities to get students of different ethnicities talking to one another about difference and respect. She also mentioned that she wanted to appeal to the administration to hire a VP for diversity issues.

Miles looked up at Robert and Carlos and said, "Well, Laura sounds reasonable enough." Robert clapped Miles on the back, smiling broadly. Carlos jumped in, saying, "Hey, man, you haven't even looked at what Jorge has to say. Keep reading."

Miles put Laura's flyer aside and took up Jorge's. Jorge had chosen a completely different presentation style for his campaign flyer. Instead of laying out a specific platform point by point as Laura had, Jorge's flyer told a series of short stories. In the first one, Jorge related an incident in class when an African American woman couldn't get the professor's attention for anything she had to say. Every time she tried to contribute in class, the professor ignored her and asked a European American student's opinion instead. Because she was the only African American person in the class, this woman believed the professor was prejudiced, but she wasn't sure what she could do about it. Another story in the flyer described a classroom situation in which the only two Latinx students believed that they were called on to give the "Latin perspective" on every issue the professor raised. They were both really tired of being tokens. A third story talked about how certain bars on campus were considered "Black" and others were "Latinx" and others were "white." This story told of two African American students who went to a white bar and felt really isolated. Finally, Jorge had asked in large letters, "Have you ever been told to go back where you came from?" The question lingered in the air as the three roommates thought about it.

As Miles read this flyer, he thought that Jorge had it down cold. His description of life at the university was totally accurate. He himself had been ignored in classes and wondered if it had been because of his race. He'd also experienced professors asking him for the "Black" opinion, and he really resented that. His social life rarely included people outside his own race, except for Carlos and Robert and a couple of Carlos's friends who were also Latinx. He never socialized with the whites on campus. As it happened, when he first came to campus, a group of white students had told Miles to go back to Africa. That had really stung, and he hadn't known what to say to them. He'd yelled back that he'd never even been to Africa. It took him a long time to get over being angry about that incident and he had played it over in his mind a lot wondering what else he might have said back to those students. Jorge had given Miles a lot to think about, and he decided he would vote for him.

Throughout this book we begin each chapter with a story about a person or several people who experience something through which we can illustrate the chapter's theory. The reason we've made this choice may be found within the theory we profile in this chapter: Walter Fisher's Narrative Paradigm. The Narrative Paradigm promotes the belief that humans are storytellers and that values, emotions, and aesthetic considerations ground our beliefs and behaviors more than facts and figures do. In other words, we are more persuaded by a good story than by a good argument. Thus, Fisher would explain Miles's decision to vote for Jorge on the basis of the powerful stories Jorge presented in his campaign flyer. Fisher asserts that the essence of human nature is storytelling.[1]

Fisher is not alone in this belief. Many researchers acknowledge the power and persistence of narrative. Robin Clair and her coauthors (2014) state the fundamental importance of storytelling when they say: "The history of narrative might be traced to the origin of language, to the first symbolic sound or gesture" (p. 2). Other researchers (e.g., Ramsey, Venette, & Rabalais, 2011) have examined what they call the "narrative malleability" of constructs, observing that people's minds can be changed about something based on good stories told by a credible storyteller. Jody Koenig Kellas and Haley Kranstuber Horstman (2015) contend that stories have the potential to make sense of complex ideas, help socialize people into groups (e.g., the family), and can create, reinforce, or challenge an individual's identity. For many years, the discipline of Communication Studies has been influenced by an interest in narration. John Lucaites and Celeste Condit (1985) talk about "the growing belief that narrative represents a universal medium of human consciousness" (p. 90). In addition, scholars (e.g., Koenig Kellas & Trees, 2013) assert that stories do more than explain the world; they shape our world. Further, even as communication modalities change, researchers find support for the power of stories. Mario L. Cassar, Albert Caruana, and Jirka Konietzny (2022) examined people's responses to advertising websites based on facts compared to those based on stories. The researchers found that the story-based websites were more persuasive on people's intention to buy the product.

1. Although some scholars differentiate between stories and narrative, for our purposes here, we will use the terms interchangeably following Koenig Kellas (2016).

Theory-Into-Action

Many political commentators discuss how politicians and political parties try to control the narrative presented to the public. After the presidential debate in 2012 between Barack Obama and Mitt Romney, most observers felt that Romney had soundly defeated Obama. Obama's performance was considered poor and unenthusiastic. Republicans constructed a narrative that featured a weak incumbent who was run over by a forceful challenger. Democrats seemed to be caught off guard and failed to put up a convincing counternarrative, although they could have done so. They might have talked about Obama as calm and reflective and Romney as arrogant and misinformed, because during the debate Romney put forward many flagrant misrepresentations of fact. But the Democrats let the Republicans' narrative stand. Although Obama went on to do better in the subsequent debates and ultimately won reelection, the Republican narrative captured many people's imaginations, which lends support to the Narrative Paradigm's assertions of narrative logic holding sway over rational logic.

It's notable that Fisher calls his approach a paradigm rather than a theory. Fisher uses that term to signal the breadth of his vision because a paradigm is considered bigger than a single theory. Fisher states that "there is no genre, including technical communication, that is not an episode in the story of life" (1985, p. 347). Thus, Fisher has constructed an approach to theoretical thinking that's more encompassing than any one specific theory. In this way, Fisher's goals are similar to Kenneth Burke's goals in Dramatism (see **Chapter 21**).

Furthermore, the use of the term *paradigm* indicates that Fisher's thinking represents a major shift from the thinking that supported most previous theories of communication. Fisher believes he's capturing the fundamental nature of human beings with the insight that we

> **paradigm shift** a significant change in the way most people see the world and its meanings

are storytellers and that we experience our lives in narrative form. He contrasts his approach with what he calls the rational paradigm, which characterizes previous Western thinking. Thus, Fisher presents what can be called a **paradigm shift,** or a significant change in the way people think about the world and its meanings.

Fisher (1987) explains the paradigm shift by recounting a brief history of paradigms that have guided Western thinking. He notes that originally *logos* meant a combination of concepts including story, rational discourse, and thought and was rooted in the discipline of philosophy. Fisher explains that this meaning held until the time of Plato and Aristotle, who distinguished between logos as reason and mythos as story and emotion. In this division, mythos, representing poetical discourse, was assigned a negative status relative to logos, or reason. The concept of rhetoric fell somewhere between the elevated logic of logos and the inferior status of poetics or mythos. Ranking mythos, logos, and rhetoric in this way reinforced the concept that not all discourse is equal. In fact, according to Aristotle (see **Chapter 20**), some discourse is superior to others by virtue of its relationship to true knowledge. Only logos, Aristotle asserted, leads to true knowledge because it provides a system of logic that can be proven valid. Logos was found in the discourse of philosophy. Other forms of discourse lead to knowledge, but the knowledge they produce is probabilistic, not true in an absolute sense, according to Aristotle.

This Aristotelian distinction did not prevent Aristotle himself from valuing all the different forms of communication equally, but it did provide a rationale for later theorists' preference for logic and reason over mythos, or story, and rhetoric. Much subsequent scholarship has focused on a struggle over these forms of discourse. Beginning at the end of the Renaissance period in Europe, the scientific revolution changed people's way of thinking about the world. This period dethroned philosophy as the source of logic, placing logic instead within science and technology. Fisher contends that this change didn't make that big of a change in people's opinion of rhetoric because both philosophy and science privilege a formal system of logic that leaves poetics or rhetoric in a devalued position. The mind-set, employed by many scholars, that regards logical thinking as primary is what Fisher calls the **Rational World Paradigm.**

Struggles among these different branches of knowledge continue today, but Fisher asserts that the Narrative Paradigm finds a way to transcend these struggles. Fisher (1987) argues that "acceptance of the narrative paradigm shifts the controversy from a focus on

> **Rational World Paradigm** a system of logic employed by many researchers and professionals

who 'owns' logos to a focus on what specific instances of discourse, regardless of form, provide the most trustworthy, reliable, and desirable guides to belief and to behavior, and under what conditions" (p. 6). Thus, the Narrative Paradigm represents a different way of thinking about the world than that posited by the Rational World Paradigm. With narrative, Fisher suggests, we move away from an either/or dualism toward a more unified sense of knowledge that embodies science, philosophy, story, myth, and logic. Further, the Narrative Paradigm presents an alternative to the Rational World Paradigm without negating traditional rationality. Fisher's (1984) intent was to provide a synthesis between "two traditional strands in the history of rhetoric: the argumentative, persuasive theme and the literary aesthetic theme" (p. 2).

Fisher (1987) argues the Narrative Paradigm accomplishes all this by recognizing that "some discourse is more veracious, reliable, and trustworthy in respect to knowledge, truth, and reality than some other discourse, but no *form* or *genre* has final claim to these virtues" (p. 19; emphasis in original). In asserting this, Fisher lays the groundwork for reclaiming the importance of the narrative, or story, without denigrating logic and reason, and he establishes a new way of conceptualizing rhetoric. Furthermore, Fisher asserts that story, or mythos, is imbued in all human communication endeavors (even those involving logic) because all arguments include "ideas that cannot be verified or proved in any absolute way. Such ideas arise in metaphor, values, gestures, and so on" (1987, p. 19). Fisher thus attempts to bridge the divide between logos (rational argument) and mythos (story or narrative). Additionally, Fisher applied the concepts of narration to public communication. Most narrative scholars before Fisher used storytelling in intra- and interpersonal communication contexts, rather than in public speaking situations.

Theory at a Glance • The Narrative Paradigm

This approach is founded on the principle that humans are storytelling animals. Furthermore, narrative logic is often more persuasive than the rational logic traditionally used in argument. Narrative logic, or narrative rationality, suggests that people judge the credibility of speakers by whether their stories hang together (have coherence) and ring true (have fidelity). The Narrative Paradigm allows for a democratic evaluation of speakers and a democratic opportunity for audiences because no one has to be specially trained in persuasion to be able to draw conclusions based on the concepts of coherence and fidelity.

Assumptions of the Narrative Paradigm

Although Fisher argues that the Narrative Paradigm fuses logic and aesthetic, he does point out that narrative logic is different from traditional logic and reasoning. We will discuss how these two differ throughout the chapter because this is an important distinction for Fisher and one that he continually refined as his thinking about the Narrative Paradigm evolved. An important aspect of the assumptions of the Narrative Paradigm is that they contrast with those of the Rational World Paradigm. Fisher (1987) stipulated five assumptions of the Narrative Paradigm:

- Humans are naturally storytellers.
- Decisions about a story's worth are based on "good reasons."
- Good reasons are determined by history, biography, culture, and character.
- Rationality is based on people's judgments of a story's consistency and truthfulness.
- We experience the world as filled with stories, and we must choose among them.

We can see how these clearly contrast to the parallel assumptions Fisher highlights in the Rational World Paradigm. This contrast is listed in **Table 22.1.** We briefly discuss each of the assumptions of the Narrative Paradigm, comparing them with their opposites in the Rational World Paradigm.

Table 22.1   Contrast Between Narrative and Rational World Paradigms

NARRATIVE PARADIGM	RATIONAL WORLD PARADIGM
1. Humans are storytellers.	1. Humans are rational beings.
2. Decision making and communication are based on "good reasons."	2. Decision making is based on arguments.
3. Good reasons are determined by matters of history, biography, culture, and character.	3. Arguments adhere to specific criteria for soundness and logic.
4. Rationality is based in people's awareness of how internally consistent and truthful to lived experience stories appear.	4. Rationality is based in quality of knowledge and formal reasoning processes.
5. The world is experienced by people as a set of stories from which to choose among. As we choose, we live life in a process of continual re-creation.	5. The world can be reduced to a series of logical relationships that are uncovered through reasoning.

First, the Narrative Paradigm assumes that the essential nature of humans is rooted in story and storytelling. As our opening example of Miles illustrates, stories persuade us, move us, and form the basis for our beliefs and actions. Miles hadn't heard much about the election for the president of the Multicultural Student Association on campus. In fact, Miles was rather apathetic about the election and had no real interest in, or opinions about, either candidate. Yet, after reading the stories that Jorge included in his campaign literature, Miles decided to vote for Jorge. Miles found Laura's campaign material interesting but not nearly as involving as Jorge's. If the assumption of the Rational World Paradigm held true, we would expect the more rational argument to hold sway over Miles, and he should have decided to vote for Laura. The Narrative Paradigm explains his preference for Jorge. Some commentators gave Pete Buttigieg high marks after the Democratic primary debate in September 2019. Much of the praise came from how Buttigieg wove his personal story of military service and coming out as gay after he returned home into his response to a question about resilience. This offers another example of the power of story.

Fisher also believes in this first assumption because he observes that narrative is universal—found in all cultures and time periods. Fisher asserts, "Any ethic, whether social, political, legal, or otherwise, involves narrative" (1984, p. 3). This universality of narrative prompts Fisher to suggest the term *Homo narrans* as the overarching metaphor for defining humanity. Fisher was influenced in his approach by reading the moral theory espoused by Alasdair MacIntyre (1981). MacIntyre observes that "man [sic] is in his actions and practice, as well as in his fictions, essentially a story-telling animal" (p. 201). Fisher used MacIntyre's ideas as the foundation for the Narrative Paradigm.

James Elkins (2001) agrees with Fisher's assumption about the centrality of stories for humans. Elkins observes that people "turn to stories to both survive and to imagine, as well as for a host of instrumental purposes, for pleasure, and because we must. Stories are part of our human inheritance" (p. 1). Other researchers (e.g., Bute & Jensen, 2011) concur, noting stories provide humans the means to account for their own experiences and behaviors. Kirsten Theye (2008) argues that "narratives are crucial in human communication as a way of explaining the world" (p. 163).

The second assumption of the Narrative Paradigm asserts that people make decisions about which stories to accept and which to reject on the basis of what makes sense to them, or *good reasons*. We will discuss what Fisher means by good reasons later in the chapter, but he doesn't mean strict logic or argument. This assumption recognizes that not all stories are equally effective; instead, the deciding factor in choosing among competing stories is personal rather than an abstract code of argument, or what we traditionally call reason. From Fisher's point of view, in our chapter-opening vignette, Laura has told a story in her campaign flyer, too. Miles simply chooses to reject her story and accept Jorge's because it's more personally involving to him. Debates over issues such as whether we're experiencing a climate crisis or just a cyclical pattern in the climate; whether the events of January 6, 2021, represented insurrection or patriotic protest; whether the use of weight loss drugs is helpful or a sign of weakness; whether university athletes should be paid a salary; and whether people should retire from public service at a certain age or if there is no need to put a maximum age limit on those who serve in government show us how often we are confronted with competing stories.

As people listen to conflicting stories, they choose among them. Their choices do not stem from traditional logic but from narrative logic. When people shift from traditional logic to narrative logic, Fisher believes their lives will be improved because narrative logic is more democratic than formal logic. As Fisher (1984) asserts, "All persons have the capacity to be rational in the narrative paradigm" (p. 10). Whereas formal logic calls for an elite trained in the complexities of the logic system, the Narrative Paradigm calls on the practical wisdom that everyone possesses.

The theory's third assumption deals with what specifically influences people's choices and provides good reasons for them. The Rational World Paradigm assumes that argument is ruled by the dictates of soundness (Toulmin, 1958). For Stephen Toulmin, the anatomy of an argument is the movement from data to a conclusion. This movement needs to be judged by soundness, or an examination of the formal logic that guides the conclusion. In contrast, the Narrative Paradigm suggests that soundness is not the only way to evaluate good reasons. In fact, soundness may not even be an accurate way of describing how people make their choices. The Narrative Paradigm assumes that narrative rationality is affected by history, biography, culture, and character. Thus, Fisher also introduces the notion of *context* into the Narrative Paradigm. People are influenced by the context in which they are embedded. So, what constitutes the appropriate story to choose for Miles today is influenced by the context of his university as well as the emphasis that the culture puts on diversity and giving voice to marginalized positions. If Miles were attending college in the 1950s, the context would be different and his choice might be different as well.

The fourth assumption forms a core issue of the narrative approach. It asserts that people believe stories insofar as the stories seem internally consistent and truthful. We'll discuss this further in the next section when we describe narrative rationality, a key concept of the Narrative Paradigm.

Finally, Fisher's perspective is based on the assumption that the world is a set of stories, and as we choose among them, we experience life differently, allowing us to construct and reconstruct our lives. Miles's choice to support Jorge may cause him to cast his own life story differently. He may no longer see himself as politically apathetic. He may change his sense of political action based on his choice of Jorge's story. You can see how the Narrative Paradigm contrasts with the Rational World Paradigm, which tends to see the world as less transient and shifting and where absolute truth is discovered through rational analysis, not through emotional responses to compelling stories.

Key Concepts in the Narrative Approach

Next, we consider eight key concepts that form the core of the Narrative Paradigm: narration or narratives, narrative rationality, coherence, structural coherence, material coherence, characterological coherence, fidelity, and the logic of good reasons. These central concepts relate to one another. Narratives are judged by narrative rationality. And, narrative rationality is judged by coherence (which has three types) and fidelity (which is assessed by the logic of good reasons). See **Figure 22.1.**

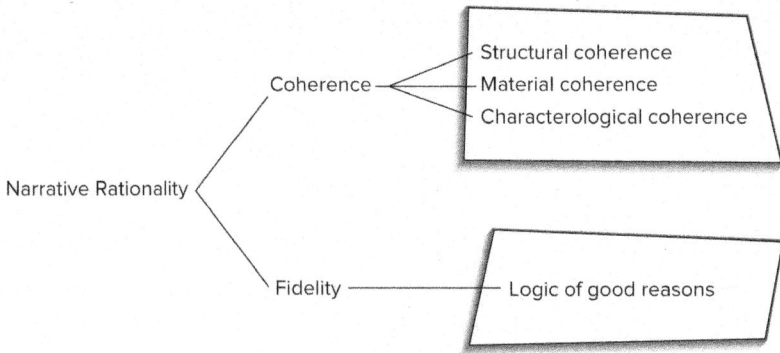

Figure 22.1   **Assessing Narrative Rationality**

Narration (or Narratives)

Narration is often thought of simply as a story, but for Fisher, narration is much more inclusive than a plotted story with a beginning, middle, and end. In Fisher's perspective, narration includes any verbal or

> **narration** an account to which listeners assign meaning

nonverbal account with a sequence of events to which listeners assign a meaning. Specifically, Fisher states, "When I use the term 'narration,' I do not mean a fictive composition whose propositions may be true or false and have no necessary relationship to the message of that composition. By 'narration,' I mean symbolic actions–words and/or deeds–that have sequence and meaning for those who live, create, or interpret them" (1987, p. 58). This definition implies the need for a storyteller and a listener. Jody Koenig Kellas (2016) further defines narratives by saying: "Most stories also include a description of characters, a dramatic tension that moves toward explanation or resolution, temporal sequencing, and a moral or story point" (p. 2).

Fisher's definition is extremely broad and parallels what many people think of as communication itself. This, of course, is Fisher's point: all communication is narrative. He argues that narrative is not a specific genre (stories as opposed to poems, for example), but rather, it's a mode of social influence. When you listen to a class lecture, when you give an excuse to a professor for not turning in a paper on time, when you listen to a podcast, post on X, or talk to your friends, you are hearing and shaping narratives. Given that our lives are experienced in narration, we need a method for judging which ones to believe and which to disregard. This method is found in narrative rationality, which we'll discuss next.

Narrative Rationality

Narrative rationality provides us with a means for judging narratives that's quite different from the traditional methods found in the Rational World Paradigm. As we mentioned previously, traditional tests of rationality include whether claims correspond to actual facts, whether all relevant facts have been considered, whether arguments are internally consistent, and whether the reasoning used conforms to standards of formal and informal logic (Fisher, 1987). Narrative rationality, in contrast to traditional logic, operates on the basis of two different principles: coherence and fidelity.

> **narrative rationality** a standard for judging which stories to believe and which to disregard

Coherence The principle of coherence is an important component of narrative rationality, which will ultimately determine whether a person accepts a particular narrative or rejects it. **Coherence** refers to the internal consistency of a narrative. When judging a story's coherence, the listener asks whether the narrative seems to "hang together" in a consistent manner. Narratives possess coherence when all the pieces of the story are present. Coherence means we believe that the storyteller has included all the important details and hasn't contradicted elements of the story in any way. Coherence is the standard of sensemaking applied to a given narrative and "sensemaking is at the heart of all narrative" (Koenig Kellas, 2016, p. 4). Sensemaking is usually obtained when the characters in a story behave in relatively consistent ways.

> **coherence** a principle of narrative rationality related to the internal consistency of a story

When Miles read the narratives contained in Jorge's campaign literature, he saw a consistent thread running through them: his university has racial problems. If Jorge had presented some problems based on race and then shaped the narrative to conclude that all was well with race relations at the university, Miles would have rejected the story for being inconsistent.

Coherence is often measured by the organizational and structural elements of a narrative. When a storyteller skips around and leaves out important information, interrupts the flow of the story to add elements forgotten earlier, and generally is not smooth in structuring the narrative, the listener may reject the narrative for not possessing coherence. Coherence is based on three specific types of consistency: structural coherence, material coherence, and characterological coherence.

STRUCTURAL COHERENCE   The type of consistency Fisher calls **structural coherence** rests on the degree to which the elements of the story flow smoothly together. When stories are confusing, when one part doesn't seem to lead to the next, or when the plot is unclear, then the narrative lacks structural coherence. If a friend told you a story about breaking up with her boyfriend by texting him, but she failed to explain any of the events leading up to the breakup, or why she chose to break up via text message, and then skipped back to telling you how she and her boyfriend first met, you'd think her story lacked structural coherence.

> **structural coherence** a type of coherence referring to the flow of the story

MATERIAL COHERENCE   **Material coherence** refers to the degree of congruence between one story and other stories that seem related to it. For example, you may have heard several stories about why two friends of yours have stopped speaking to each other. If all the stories but one attribute the problem to one friend having misled the other, causing an embarrassing situation, you're unlikely to believe the one unique story. You'd believe that the different story lacked material coherence.

> **material coherence** a type of coherence referring to the congruence between one story and other related stories

CHARACTEROLOGICAL COHERENCE **Characterological coherence** refers to the believability of the characters in the story. For instance, let's imagine that you have a professor whom you dislike a great deal. This professor ridicules you and other students in the class whenever

characterological coherence a type of coherence referring to the believability of the characters in the story

anyone contributes to class discussions. In addition, the professor makes racist, homophobic, and sexist jokes in class. Your impression is that this professor is a thoroughly objectionable person. Given this background, you'd be unlikely to accept a story in which this professor was shown in an admirable or even heroic light. You'd reject the story for not possessing characterological coherence.

Fidelity The other critical concept constituting narrative rationality is **fidelity,** or the truthfulness or reliability of the story. Stories with fidelity ring true to a listener. When Miles reads the stories that Jorge has in his campaign literature, he thinks to himself that those events have hap-

fidelity a principle of narrative rationality judging the credibility of a story

pened to him at the university. Miles wonders if Jorge has been following him around campus, watching what goes on in his life. This makes the stories powerful to Miles. They possess a great deal of fidelity for him. Fisher (1987) notes that when the elements of a story "represent accurate assertions about social reality" (p. 105) to an audience, they have fidelity.

ASSESSING FIDELITY: THE LOGIC OF GOOD REASONS   Related to Fisher's notion of fidelity is the primary method that he proposes for assessing narrative fidelity: **the logic of good reasons.** Fisher asserts that when narratives possess fidelity, they constitute good reasons for a person to hold a particular belief or take an action. For example, Miles sees Jorge's stories as possessing fidelity, which makes the stories persuasive; the stories form good reasons for Miles to vote for Jorge.

the logic of good reasons a set of values for accepting a story as true and worthy of acceptance; it provides a method for assessing fidelity

Fisher (1987) explains his concept of logic by saying that it's "a systematic set of procedures that will aid in the analysis and assessment of elements of reasoning in rhetorical interactions" (p. 106). Thus, logic for the Narrative Paradigm enables a person to judge the worth of stories. The logic of good reasons presents listeners with a set of values that appeal to them and form warrants for accepting or rejecting the advice advanced by any form of narrative. This doesn't mean that any good reason is equal to any other; it simply means that whatever prompts a person to believe a narrative is bound to a value or a conception of what is good.

As Fisher describes it, this logic is a process consisting of two sets of five questions each that the listener asks about the narrative. The first five questions are the following:

1. Are the statements that claim to be factual in the narrative really factual?
2. Have any relevant facts been omitted from the narrative or distorted in its telling?
3. What are the patterns of reasoning that exist in the narrative?
4. How relevant are the arguments in the story to any decision I may make?
5. How well does the narrative address the important and significant issues of this case?

These questions constitute a logic of reasons. To transform this into a logic of good reasons, there are five more questions introducing the concept of values into the process of assessing practical knowledge. These questions are as follows:

1. What are the implicit and explicit values contained in the narrative?
2. Are the values appropriate to the decision that is relevant to the narrative?
3. What would be the effects of adhering to the values embedded in the narrative?
4. Are the values confirmed or validated in lived experience?
5. Are the values of the narrative the basis for ideal human conduct?

Fisher illustrates the logic of good reasons by referring to a book written by Jonathan Schell (1982) called *The Fate of the Earth*. This book was very popular in the 1980s. In the book, Schell argues that the nuclear weapons race must cease. Fisher asserts that even though experts found the book inaccurate on technical grounds, the narrative it espoused was extremely appealing to the general public. Fisher argues that this was because the book tells a story that meets the criteria of coherence and fidelity. It focuses on a set of values that many people found relevant at that point in history. As the Narrative Paradigm predicts, the well-told story—consisting of narrative rationality—was more compelling to readers than the expert testimony that refuted the factual accuracy of the narrative. The Narrative Paradigm might shed some light on debates between those who demonstrate scientific proof for the efficacy of vaccines against COVID-19 and those who believe these vaccines are useless and even dangerous.

Michael Humphrey (2018) was interested to test whether the concept of narrative rationality, a key part of Fisher's theory, was useful when the narratives involved were what Humphrey calls "quick stories." Quick stories are "a specific type of everyday autobiography: narratives of well-known YouTube vloggers confessing intimate details or turning-point moments about their lives. Examples of videos include coming out as LGBTQ, serious illness, relationship dissolution and depression" (p. 225). In this effort, Humphrey was asking if Fisher's ideas, coined in the 20th century, still held true for 21st-century communication. He concluded that they do, noting that "the narrative rationality we bring into the offline world is useful in the online world as well" (p. 235). Humphrey argues that for the audiences of these quick stories, narrative rationality, especially coherence, is critical to their acceptance of the information presented by the vloggers as well as to which vloggers they choose to follow and create, in Humphrey's words, "digital kinship" with (p. 235).

Students Talking Tech: Courtney

 It was so funny to read that part in the chapter about YouTube vloggers and the idea of narrative rationality influencing digital kinship. I follow a couple of vloggers myself, and the chapter got me thinking about how or whether I accept what they say. I did watch one guy's video about coming out as trans, and I think I was influenced by coherence and fidelity. First of all, he acted like I think someone would act in that situation. He said he had something to tell us, and it was a little hard to say, but it was important to him that we knew about it. So, there was coherence there. But he was also consistent with his personality in the other videos of his that I've watched. It was dramatic and all, but he joked around like he always does. He just seemed authentic—he teared up a couple times, and his voice quavered a little. It all seemed to ring true. He has always made me feel that we have a true, if digital, connection, and that followed through in this video. I just feel like I know him a lot better now, and that makes me happy.

Integration and Critique

Fisher's Narrative Paradigm offers new insights into communication behavior and directs our attention to democratic processes in the area of rhetorical criticism, which is Fisher's intended application. However, the theory is also applicable beyond rhetoric to multiple contexts. Fisher contributes the idea that all people's lived experiences make them capable of analyzing rhetoric. Furthermore, the Narrative Paradigm helps us to see the nature of multiple logics at work in our communication encounters. Thus, the Narrative Paradigm has made a substantial contribution to our understanding of human communication and human nature in general.

As we evaluate the Narrative Paradigm, we'll consider four evaluative criteria: scope, logical consistency, utility, and heurism.

Integration

Communication Tradition	**Rhetorical**	Semiotic	Phenomenological	Cybernetic	Socio-Psychological	Socio-Cultural	Critical
Communication Context	Intrapersonal	Interpersonal	Small Group	Organizational	**Public/Rhetorical**	Mass/Media	Cultural
Approach to Knowing	Positivistic/Empirical	**Interpretive/Hermeneutic**	Critical

Critique

Evaluation Criteria	Scope	**Logical Consistency**	Parsimony	**Utility**	Testability	**Heurism**	Test of Time

Scope

The critique that the Narrative Paradigm is too broad mainly focuses on Fisher's claim that all communication is narrative. Researchers object to that claim for two reasons: first, some have questioned the utility of a definition that includes everything. How meaningful is the definition of narrative if it means all communication behavior? Critics have directed us to consider the following question: is there value in treating a ritual greeting (*"Hi, how was your day?"*) and an involved narrative explaining one's desire for a divorce in the same way? Second, some researchers, notably Robert Rowland (1987, 1989), suggest that some forms of communication are not narrative in the way that Fisher maintains. According to Rowland, science fiction and fantasy don't conform to most people's values. Rather, these genres often challenge existing values. Rowland also questions the utility of considering a novel (such as Arthur Koestler's *Darkness at Noon*) and a political pamphlet (such as one produced by the Committee on the Present Danger) both as narratives as Fisher does. Although both tell stories about the repressive character of the former Soviet system, they do so in such different ways that Rowland believes it does a disservice to both writings to place them in the same category. Furthermore, it complicates our understanding of the definition of narrative when two such disparate examples are both labeled as narrative.

Logical Consistency

The Narrative Paradigm has been faulted for failing to be consistent with some of the claims that Fisher makes about it. For instance, Rowland (1987) finds that the narrative approach doesn't actually provide a more democratic structure compared with the hierarchical system espoused by the Rational World Paradigm, nor does it completely offer an alternative to that paradigm. Rowland says that Fisher overstates the problem of domination of the public by the elite, or by the expert, in the Rational World Paradigm. In addition, Rowland argues that "there is nothing inherent in storytelling that guarantees that the elites will not control a society" (p. 272).

Utility

The Narrative Paradigm has elicited some criticism and praise with respect to its usefulness. Some research suggests that we are not always going to be persuaded be a good story. For example, Patrizia Di Tullio, Matteo La Torre, John Dumay, and Michele Antonio Rea (2022) conducted an empirical study contrasting the Narrative Paradigm with the Rational World Paradigm by looking at the way corporations presented their annual reports. They found an overwhelming preference for the Rational World Paradigm in these reports, suggesting that the context may influence which paradigm is more persuasive. In addition, K. McClure (2009) argues that the Narrative Paradigm is an overly conservative theory because its focus on fidelity actually weds it to normative conceptions of rationality rather than freeing us from them as Fisher claims. William Kirkwood (1992) observes that Fisher's logic of good reasons focuses on prevailing values and fails to account for the ways in which stories can promote social change. In some ways, both Kirkwood and Fisher agree that this observation is more of an extension to the theory than a punishing critique. The idea of extending or modifying the theory to make it more useful is espoused by other researchers as well.

For example, Kirsten Theye (2008) argues that her analysis of Vice President Dick Cheney's apologies after shooting his friend in a hunting accident in 2006 shows that Fisher's distinction between narrative coherence and narrative fidelity is not useful. She states that it's impossible to separate the two. Her suggestion is to forget the two components as separate concepts and instead focus on the basic question underlying narrative rationality: "whether a story rings true to the audience based on their experiences" (p. 174). Again, Theye seems to be modifying the theory rather than arguing to discard it altogether.

Brittany Peterson and Johny Garner (2019) agree with Theye. They studied Mars Hill Church, a mega-church founded in 1996 that at its peak had more than 13,000 members. Mars Hill was led by a polarizing pastor, Mark Driscoll, and was eventually forced to close after allegations of plagiarism, bullying, and misuse of church funds came to a crisis point in 2014. Peterson and Garner use the Narrative Paradigm to analyze the narratives and the resistance to those narratives exhibited by parishioners and church staff. They argue that Fisher's narrative rationality has traditionally been applied to judge the absolute worth of a story, or as a means to distinguish "true" stories from fictional ones. While an audience is implied in this process, Fisher's narrative rationality exists "apart from the real experiences of people living" (p. 15) the stories. Peterson and Garner offer their work as an extension to Fisher's theory. Their approach provides an emphasis on how narrative rationality explains not the truthfulness of stories but how stories resonate with a particular community. They challenge scholars to apply the Narrative Paradigm with attention to "the resonance of narratives and counter narratives for the persons attempting to live with them" (p. 17). In addition, some researchers see the Narrative Paradigm as more useful when combined with another theory. For instance, Danielle Quichocho and Burton St. John (2022) studied pro-fracking messages in a public relations campaign in Colorado launched by those in the industry depending on fracking. They found support for the Narrative Paradigm's assertions, but added what they called the Circuit of Culture to the Narrative Paradigm to make the explanation more complete.

Students Talking: Justin

 I like what Peterson and Garner said about extending Fisher's theory. When I was reading what Fisher said about coherence, I was wondering how the theory explains a story that makes you change your mind about someone or something. If I didn't like a professor but then found out that he was doing some great things for the planet, like starting some big recycling center or something, I might change my mind and start liking him a little bit more. I know in the past, learning more information about someone has made me like them more (if it was good information). When I first met my boyfriend, Joel, I didn't like him right off the bat, but the more I learned about him, the better I liked him. At first, I didn't see how that's explained in the theory, but adding what Peterson and Garner said helps explain it a little. I had one narrative (I didn't like Joel—he seemed geeky), but then as I got to know him other narratives resonated with me more (like how kind he is and how smart, funny, and thoughtful he is all the time).

Heurism

Despite criticisms, which primarily urge refinements of the theory, not its abandonment, Fisher's Narrative Paradigm has contributed a great deal to the study of human communication. For one thing, the idea of people as storytellers has proved captivating and heuristic. Storytelling seems an apt metaphor for understanding how humans use communication to make sense of the world (Johnson, 2016; Primayanti & Puspita, 2022; Suter, Koenig Kellas, Webb, & Allen, 2016; Thompson & Schrodt, 2015). Fisher has provided a new paradigm for understanding human nature, squarely located in the symbolic realm of communication. Some research shows that stories can improve the healing process, particularly with marginalized and vulnerable populations (Lee, Fawcett, & DeMarco, 2016). Further demonstrating the value of storytelling, Jennifer Gray (2015) found that there were differences in stories between women who continued their pregnancies and those who chose not to continue them. Both populations, however, tried to make sense of their experiences through narration.

Interestingly, researchers outside of the field of rhetoric have embraced the Narrative Paradigm. Clair et al. (2014), for instance, concluded that "narrative [theory] has found its permanent spot in health communication literature" (p. 7); health communication literature includes important topics such as medical decision-making, aging, and health care organizations (e.g., Gray, 2015; Wamucii, 2011). Scholars in the field of journalism have used the Narrative Paradigm in analyses of how news stories relate to cultural ideology (e.g., Machill, Kibler, & Waldhauser, 2007; White, 2005) as well as to study tabloid news (e.g., McCartha & Strauman, 2009) and how digital video creates its own narrative (e.g., Bock & Schneider, 2017). Legal scholars use the Narrative Paradigm to illustrate how competing narratives in the courtroom are judged by juries. These scholars have found that a story possessing coherence is more persuasive to a jury than one that has narrative gaps, regardless of the factual grounds for the stories (e.g., Burns, 2001; Rideout, 2008). In a specific illustration of that assertion, Christine Kelly and Michele Zak (1999) argue that former football player O. J. Simpson was acquitted of murdering his wife and her friend in 1995 because of the triumph of narrative argument over rational argument. According to Kelly and Zak's analysis, the defense was victorious because it framed Simpson's story in a manner that resonated with the jury, whereas the prosecution relied on the Rational World Paradigm, directed more toward the judge and the opposing lawyers. The researchers note that the prosecutors "drew on the language of technical expertise and took responsibility for presenting a careful case in a court of law without reference to the lives of the jury" (p. 301).

Other research applies the Narrative Paradigm in a wide variety of contexts, including social media (e.g., Humphrey, 2018; Li, Tang, Liu, & Ma, 2018), political campaigns (e.g., Falk, 2009; Kluver, Cooley, & Hinck, 2019), and advertising campaigns (e.g., Dias & Dias, 2018).

Closing

Future scholarship will undoubtedly extend the framework of the Narrative Paradigm to remediate its shortcomings and capitalize on its strengths. In constructing the Narrative Paradigm, Fisher provided a rich framework for such scholarship to take place. Since both Eastern and Western societies rely upon stories for meaning, we should see this theory utilized by scholars across the globe.

Discussion Starters

 Case-In-Point: Can you think of any other explanations besides those offered by the Narrative Paradigm for Miles's preference for Jorge's candidacy after he read the campaign flyers?

Try-It-Your-Selfie: Choose some type of communication that depends on technology (texts, YouTube, Instagram, etc.) and analyze it based on the Narrative Paradigm. Can you apply narrative rationality to explain its persuasive appeal?

1. When you listen to others' stories, do you evaluate them based on coherence and fidelity? Can you think of any other criteria that you use to evaluate the stories that you hear? Or do you agree with Kirsten Theye that it's not possible to separate coherence and fidelity and they really just form one criterion?

2. The Narrative Paradigm suggests that when an expert argument is compared to a good story, the expert argument will fail because it will lack the coherence and the fidelity that a narrative possesses. Do you agree or disagree with this claim? Give an example where expert testimony failed to persuade you. Give an example of a good story that failed to persuade you.

3. If persuasion is determined, in part, by how compelling and truthful a story seems to a listener, how will it be possible to achieve large-scale agreement on controversial topics? If I am persuaded by a story that seems true to me, but you reject the story as not ringing true to you, how do we bridge the divide between us?

4. How is narrative rationality different from the Rational World Paradigm? Do you think Fisher succeeds in synthesizing the two strands that traditionally characterize rhetoric: the persuasive, argumentative strand and the literary, aesthetic strand? Explain your answer.

References

Bock, M. A., & Schneider, D. A. (2017). The voice of lived experience: Mobile video narratives in the courtroom. *Information, Communication & Society, 20*(3), 335–350.

Burns, R. P. (2001). *A theory of the trial.* Princeton University Press.

Bute, J. J., & Jensen, R. E. (2011). Narrative sensemaking and time lapse: Interviews with low-income women about sex education. *Communication Monographs, 78*(2), 212–232.

Cassar, M. L., Caruana A., & Konietzny, J. (2022) Facts or story? The impact of website content on narrative believability and purchase intention. *Journal of Marketing Communications, 28*(6), 637–656.

Clair, R. P., Carlo, S., Lam, C., Nussman, J., Phillips, C., Sánchez, V., Schnabel, E., & Yakova, L. (2014). Narrative theory and criticism: An overview toward clusters and empathy. *Review of Communication, 14*(1), 1–18.

Di Tullio, P., La Torre, M., Dumay, J., & Rea, M. A. (2022). Accountingisation and the narrative (re)turn of business model information in corporate reporting. *Journal of Accounting & Organizational Change, 18*(4), 592-615.

Dias, L., & Dias, P. (2018). Beyond advertising narratives: Josefinas and their storytelling products. *Anàlisi: Quaderns de Comunicació i Cultura, 58,* 47-62.

Elkins, J. R. (2001). *Narrative theory and literary criticism.* www.wvu.edu/lawfac/jelkins/lawyerslit/theories.html.

Falk, E. (2009). Press, passion, and Portsmouth: Narratives about "crying" on the campaign trail. *Argumentation & Advocacy, 46*(1), 51-63.

Fisher, W. R. (1984). Narration as a human communication paradigm: The case of public moral argument. *Communication Monographs, 51*(1), 1-23.

Fisher, W. R. (1985). The narrative paradigm: An elaboration. *Communication Monographs, 52*(4), 347-367.

Fisher, W. R. (1987). *Human communication as narration: Toward a philosophy of reason, value, and action.* University of South Carolina Press.

Gray, J. B. (2015). "It has been a long journey from first knowing": Narratives of unplanned pregnancy. *Journal of Health Communication, 20*(6), 736-742.

Humphrey, M. (2018). Confession narratives and mass kinship of YouTube celebrities: A narrative rationality analysis. *Studies in Communication & Culture, 9*(2), 225-237.

Johnson, C. (2016). *The way of the writer: Reflections on the art and craft of storytelling.* Scribner.

Kelly, C., & Zak, M. (1999). Narrativity and professional communication: Folktales and community meaning. *Journal of Business and Technical Communication, 13*(3), 301.

Kirkwood, W. (1992). Narrative and the rhetoric of possibility. *Communication Monographs, 59*(1), 30-47.

Kluver, R., Cooley, S., & Hinck, R. (2019). Contesting strategic narratives in a global context: The world watches the 2016 U.S. election. *The International Journal of Press/Politics, 24*(1), 92-114.

Koenig Kellas, J. (2016). Narratives and social interaction. In C. R. Berger & M. E. Roloff (Eds.), *The International Encyclopedia of Interpersonal Communication* (pp. 1-5). John Wiley & Sons.

Koenig Kellas, J., & Kranstuber Horstman, H. (2015). Communicated narrative sense-making: Understanding family narratives, storytelling, and the construction of meaning through a communicative lens. In L. H. Turner & R. West (Eds.), *Sage handbook of family communication* (pp. 76-90). Sage.

Koenig Kellas, J., & Trees, A. R. (2013). Family stories and storytelling: Windows into the family soul. In A. Vangelisti (Ed.), *Handbook of family communication* (pp. 391-406). Routledge.

Lee, H., Fawcett, J., & DeMarco, R. (2016). Storytelling/narrative theory to address health communication with minority population. *Applied Nursing Research, 30,* 58-60.

Li, J., Tang, J., Liu, X., & Ma, L. (2018). How do users adopt health information from social media? The narrative paradigm perspective. *Health Information Management Journal, 48*(3), 116-126.

Lucaites, J. L., & Condit, C. M. (1985). Reconstructing narrative theory: A functional perspective. *Journal of Communication, 35*(4), 9-108.

Machill, M., Kibler, S., & Waldhauser, M. (2007). The use of narrative structures in television news. *European Journal of Communication, 22*(2), 185-205.

MacIntyre, A. (1981). *After virtue: A study in moral theory.* University of Notre Dame Press.

Mario, L. C. M. L., Albert, C. A., & Konietzny, J. (2022). Facts or story? The impact of website content on narrative believability and purchase intention. *Journal of Marketing Communications, 28*(6), 637-656.

McCartha, M., & Strauman, E. C. (2009). Fallen stars and strategic redemption: A narrative analysis of the *National Enquirer. Florida Communication Journal, 37*(2), 71-82.

McClure, K. (2009). Resurrecting the Narrative Paradigm: Identification and the case of Young Earth Creationism. *Rhetoric Society Quarterly, 39*(2), 189–211.

Peterson, B. L., & Garner, J. T. (2019). Tensions of narrative ownership: Exploring the rise of (counter) narratives during the fall of Mars Hill Church. *Western Journal of Communication, 83*(1), 1–19.

Primayanti, N. W., & Puspita, V. (2022). Local wisdom narrative in environmental campaign. *Cogent Arts & Humanities, 9*(1), 1–17.

Quichocho, D., & St. John, B. (2022). Locating a narrative paradigm nexus in the circuit of culture: Articulating the anti-proposition 112 public relations campaign in Colorado. *Journal of Public Relations Research, 34*(3–4), 135–151.

Ramsey, M. C., Venette, S. J., & Rabalais, N. (2011). The perceived paranormal and source credibility: The effects of narrative suggestions on paranormal belief. *Atlantic Journal of Communication, 19*(2), 79–96.

Rideout, J. C. (2008). Storytelling, narrative rationality, and legal persuasion. *The Journal of the Legal Writing Institute, 14,* 53–86.

Rowland, R. C. (1987). Narrative: Mode of discourse or paradigm? *Communication Monographs, 54*(3), 264–275.

Rowland, R. C. (1989). On limiting the narrative paradigm: Three case studies. *Communication Monographs, 56*(1), 39–53.

Schell, J. (1982). *The fate of the earth.* Avon Books.

Suter, E. A., Koenig Kellas, J., Webb, S. K., & Allen, J. A. (2016). A tale of two mommies: (Re)Storying family of origin narratives. *Journal of Family Communication, 16*(4), 303–317.

Theye, K. (2008). Shoot, I'm sorry: An examination of narrative functions and effectiveness within Dick Cheney's hunting accident apologia. *Southern Communication Journal, 73*(2), 160–177.

Thompson, P. A., & Schrodt, P. (2015). Perceptions of joint family storytelling as mediators of family communication patterns and family strengths. *Communication Quarterly, 63*(4), 405–426.

Toulmin, S. (1958). *The uses of argument.* Cambridge University Press.

Wamucii, P. (2011). Walking the extra mile: Navigating slum identities through social activism in Mathare, Kenya. *The Howard Journal of Communications, 22*(2), 183–199.

White, P. (2005). Narrative impulse in mass-media "hard news" reporting. In F. Christie & J. R. Martin (Eds.), *Genre and institutions: Social processes in the workplace and school* (pp. 101–123). Bloomsbury Publishing.

CHAPTER 23
Media Ecology Theory
Based on the research of *Marshall McLuhan*

"Our conventional response to all media, namely that it is how they are used that counts, is the numb stance of the technological idiot."

—Marshall McLuhan

Professor Margaret Randall

As an expert on social media, Professor Margaret Randall is used to traveling around the country. She is an accomplished author, with her most recent book examining the isolation caused by people's reliance upon technology. Professor Randall's research has been celebrated by both scholars and practitioners, and she has received numerous awards and recognitions for her work on the influence of media upon our social lives. Margaret's most recent invitation was to present the keynote speech to an international group of social media professors in Costa Rica. She was both anxious and excited to go.

A week before the trip, Professor Randall was under a lot of pressure to get her speech completed. She had written a few paragraphs, but she had so much more homework to do. She couldn't rely on her "stump" speech with this group, given their expertise. In fact, as Randall sat in her home office with her laptop finding new research in the area, she felt that it was important to include cultural examples of her points to make her speech more compelling. Some materials that she wanted required her to look at the library databases, something she wasn't all that excited to do. But, the professor figured to really be motivated, she would visit the library. So, she went across campus to (literally) check out a few manuscripts on Costa Rica and some of the "unknowns" related to this Central American country. When Margaret returned to her office, she continued her research on the library's databases, and also found stories and pictures on Central and South American youth on Instagram and YouTube.

As Professor Randall sat reviewing the information, she visited chat rooms with younger people from around the globe. She thought she'd ask them about their social media use when the chance arose. And, it occurred to her that perhaps her former colleague, Bella, who was teaching in Central America, could provide some "local perspective" on the topic. It wasn't long before Margaret and her friend were FaceTiming. After some back-and-forth the conversation eventually moved into discussions about their families, their research, and even their reactions to the U.S. presidential debates, which Bella had watched from her Costa Rica home. After a few minutes, the two finally got to the reason for their connection and soon Professor Randall had some firsthand examples to splice into her speech.

Later that evening, as she watched television, Margaret Randall was thinking a lot about her upcoming speech. She knew that the speech was going to be taped and that it would be streamed live. She knew that she couldn't afford any flubs. Her anxiety was palpable. She still had several days to refine the speech and yet, she always became nervous when she stopped to consider that her presentation was not only mediated, but also that it was her first speech given outside of the United States. And despite being a college professor, who has seen a lot of PowerPoint with her students, she was quite nervous about using this technology.

Margaret knew that there was only one person who could calm her down: her 21-year-old daughter, Emma. She texted Emma to ask if she had a few minutes to talk on the phone. Emma has always been credited with getting her mom to FaceTime and text. Before Emma's insistence her mom had simply called her daughter or sent her emails. Now, however, Margaret was relying on technology to make sure the two remained connected. And, it was this connection that the two of them were forging today as Emma answered her cell phone.

Communication theorists often implied technology in their theories, but until recently, rarely embraced it as a fundamental feature in their theories. Throughout this book, we have made a sustained effort to include voices from students who have articulated a technological application of the theory. Yet, many theories were conceptualized well before the advent of the computer. In **Chapter 8,** we underscored one theory—Social Information Processing Theory—that addressed the complex intersection between technology and human relationships. We identified theorists such as Joseph Walther who focus on the coming together of people and how individuals get to know one another online. All of this is done without the benefit of nonverbal cues.

Unlike Social Information Processing Theory, theorists interested in media environments look at the entire "mediated picture"—from its history to how it affects our perceptions or feelings, among other areas. When an interplay between media and their surroundings occurs, the foundation of Media Ecology Theory (MET) has been established (Note: appropriately, unlike in any other chapter, we invoke numerous websites as sources throughout the chapter. We're confident that McLuhan would embrace this technological referencing).

Professor Randall's experiences in our chapter opening would be of interest to Media Ecology theorists. Her inevitable embrace of technology and its effect upon both her professional and personal lives are of interest to scholars in Media Ecology. In addition, the fact that Margaret used to rely only on email, but later used other technologies to communicate with her daughter says a great deal about how technology has influenced their relationship values.

One theorist who could understand and interpret Professor Randall's relational and technological circumstances is Marshall McLuhan. In his groundbreaking book, *Understanding Media* (1964), McLuhan wrote about the influence of technologies, including clocks, televisions, radios, movies, telephones, and even roads and games. Although today we would not classify some of these as technologies, at the time, McLuhan was interested in the social impact of these primal mediated forms of communication. His lifelong work was dedicated to answering the following question: what is the relationship between technology and members of a culture?

It's fair to say that McLuhan himself was part of the culture's media. He appeared regularly on television talk shows, spoke to policymakers, had a cameo role in the Woody Allen film *Annie Hall*, and even was interviewed by *Playboy* magazine. He was so enigmatic and ubiquitous that one writer called him the "high priest of pop cult and metaphysician of media" (McDowell, 2023).

McLuhan was a Canadian scholar of literary criticism who used poetry, fiction, politics, musical theater, and history to suggest that mediated technology shapes people's feelings, thoughts, and actions. He suggests that we have a symbiotic relationship with mediated technology. We create technology, and technology in turn recreates who we are. Interestingly, McLuhan never felt that he was a theorist. As his son Eric posited, his dad would claim that "I don't have a theory of communication. I don't use theories. I just watch what people do. What you do" (E. McLuhan, 2008, p. 25).

Theory at a Glance • Media Ecology Theory

Society has evolved as its technology has evolved. From the alphabet to the internet, we have been affected by, and affect, electronic media. In other words, the medium is the message insofar as technology affects communication. Media Ecology Theory centers on the principles that society cannot escape the influence of technology, technology brings global lands together, and technology will remain central to virtually all walks of life. The dominant media in a particular era shape the culture and the individuals within a culture.

Electronic media have revolutionized society, according to McLuhan. In essence, McLuhan feels that societies are highly dependent on mediated technology and that a society's social order is based on its ability to deal with that technology. Recall that this theory was conceptualized nearly 60 years ago and, especially today, McLuhan's assertion about technology rings true. Indeed, "since the public's growing consciousness of the internet ... starting in the mid-1990s, McLuhan's reputation has experienced an astounding upsurge" (Morrison, 2006, p. 170). Paula McDowell (2023) noted that he was considered the "initiator of discourse" (p. 1391). And, Megan Garber (2011) writes that McLuhan is still a "Media Guru of the first order."

Media, in general, act directly to mold and organize a culture. This is McLuhan's theory. Because the theory centralizes the many types of media and views media as an environment unto itself, scholars aptly term McLuhan's work Media Ecology. The word **ecology,** in this sense, is simply the study of how environments influence individuals. For our purposes, we define **media ecology** as the study of how media and communication processes influence human perception, feeling, understanding, and values. Carlos Scolari (2012–2013) writes that the

> **ecology** the study of environments and their influence upon people
>
> **media ecology** the study of how media and communication processes affect human perception, feeling, emotion, and value

notion of media ecology was borne out of conversations McLuhan had with his colleagues. Given that McLuhan's writing spans a number of different academic disciplines, given that it focuses on a variety of technologies (e.g., radio, television, etc.), and given that it pertains to the intersection of technology and human relationships and how media affect understanding (Postman, 1971), the ecological view of media is appropriate and sensible. (Note: it should be pointed out that it was Postman, not McLuhan, who formally introduced the phrase "media ecology.") Paul Levinson (2000) describes the relationship of Media Ecology to communication this way: "McLuhan's work was startlingly distinct from the others in that he put communications at center stage. Indeed, in McLuhan's schema, there was nothing else on the stage" (p. 18). MET, then, is the communication field's attempt to understand the pervasive influence of media upon cultures across the globe.

McLuhan (1964) based much of his thinking on his mentor, Canadian political economist Harold Adams Innis (1951). Innis believed that major empires in history (Rome, Greece, and Egypt) were built by those in control of the written word. Innis argued

> **bias of communication** Harold Innis's contention that technology has the power to shape society

that Canadian elites used a number of communication technologies to build their "empires." Those in power were given more power because of the development of technology. Innis referred to the shaping power of technology on a society as the **bias of communication.** For Innis, people use media to gain political and economic power and, therefore, change the social order of a society. Innis claimed that communication media have a built-in bias to control the flow of ideas in a society.

McLuhan extended the work of Innis. Philip Marchand (1989) observes that "not long after Innis's death, McLuhan found an opportunity to explore the new intellectual landscape opened up by his [Innis's] work"

(p. 115). McLuhan, like Innis, felt that it's nearly impossible to find a society that is unaffected by electronic media (today, we would expand that thinking to social media). In fact, Susan Greener (2016) posits that the range of media (e.g., screenshots, video, email conversations, etc.) allows for a society of learned individuals, although some may disagree with this premise as we think about the "fake news" infusing all types of media. Our perceptions of the media and how we interpret those perceptions are the core issues associated with MET. We now discuss these themes in the three main assumptions of the theory.

Assumptions of Media Ecology Theory

We have noted that the influence of media technology on society is the main idea behind MET. Let's examine this notion a bit further in three assumptions framing the theory:

- Media infuse nearly every act and action in society.
- Media fix our perceptions and organize our experiences.
- Media tie the world together.

Our first assumption underscores the notion that we cannot escape media in our lives. As we alluded above, media permeate our very existence. We cannot avoid nor evade media, particularly if we subscribe to McLuhan's broad interpretation of what constitutes media. Think about Professor Randall from our chapter-opening story. She clearly has found herself employing mediated technologies in multiple ways. In fact, we could easily argue that her livelihood is contingent upon her ability to navigate multiple media environments and platforms.

Many Media Ecology theorists interpret media in expansive ways. Today, for instance, although McLuhan did not foresee the various (digital) media available for consumption (e.g., Roku, etc.), scholars acknowledge that McLuhan's thinking would have relevance to these digital forms. Still, in addition to looking at traditional forms of media (e.g., radios, movies, phones, and television), McLuhan also looks at the influence that numbers, games, and even money can have on society. He felt that these were basic systems of recordkeeping, dating back about 5,000 years. Let's explore these three in more detail in order for you to understand the breadth and historical thinking of McLuhan's interpretation of media.

McLuhan (1964) views numbers as mediated. He explains: "In the theater, at a ball game, in church, every individual enjoys all those others present. The pleasure of being among the masses is the sense of the joy in the multiplication of numbers, which has long been suspect among the literate members of Western society" (p. 107). McLuhan felt that in numbers a "mass mind" (p. 107) was constructed by the elites in society to establish a "profile of the crowd" (p. 106). Therefore, it may be possible to create a homogenized population, capable of being influenced. Consider, for instance, what happened at the U.S. Capitol on January 6, 2021. Falsely and naively believing that former President Trump's election was "stolen," thousands of like-minded insurrectionists descended on the Capitol and invaded it with violence. Concurrently, Trump watched the events unfold, seemingly appreciating the adoration and justifying the Capitol invasion. The crowd, in turn, clearly appreciated being around like-minded people. Employing theoretical principles, McLuhan would observe that because there was camaraderie among the insurrectionists, "mass-minded" behavior ensued.

In addition to numbers, McLuhan looks at games in society as mediated. He observes that "games are popular art, collective, social reactions to the main drive or action of any culture" (p. 235). Games are ways to cope with everyday stresses and, McLuhan notes, they are models of our psychological lives. He further argues that "all games are media of interpersonal communication" (p. 237), which means that they are extensions of our social selves. Games become mass media because they allow for people to simultaneously participate in an activity that is generally fun and games that represent an individual's identity. Today, as you consider the various online and video games available to you (e.g., Xbox, Minecraft, etc.), it's not surprising that McLuhan's insights still retain relevancy since these games elicit a variety of emotions, including anger, ridicule, joy, among others.

A third mediated form is money. McLuhan concludes that "like any other medium, [money] is a staple, a natural resource" (p. 133). The theorist also calls money a corporate image that relies on society for its status and sustenance. Money has some sort of magical power that allows people access. Money allows people to travel the globe, serving as transmitters of knowledge, information, and culture. McLuhan notes that money is really a language that communicates to a diverse group, including farmers, engineers, plumbers, and physicians.

McLuhan, ultimately, contends that media—interpreted in the broadest sense—are ever present in our lives. These media transform our society, whether through the games we play, the radios we listen to, or the televisions we watch. At the same time, media depend on society for "interplay and evolution" (p. 49).

A second assumption of Media Ecology Theory relates to our previous discussion: we are directly influenced by media. Although we alluded to this influence previously, let's be more specific about how McLuhan views the influence of media in our lives.

Media Ecology theorists believe that media help to alter perceptions and organize our lives. McLuhan suggests here that media are quite powerful in our views of the world. Consider, for instance, what occurs when we watch television or read a news feed on Facebook. If the news reports that the United States is experiencing a "moral meltdown," we may be watching for stories on child abductions, illegal drug use, or teen pregnancies. In our private conversations, we may begin to talk about the lack of morals in society. In fact, we may begin to live our lives according to the types of stories we watch (we address this notion in detail in **Chapter 22**). We may be more suspicious of even friendly strangers, fearing they may try to kidnap our child. We may be unwilling to support laws legalizing medicinal marijuana, regardless of their merits, because we are concerned about possible increases in drug activity. We may also aggressively advocate an "abstinence-only" sex education program in schools, fearing that any other model would cause more unwanted pregnancies.

What occurs with each of these examples is what McLuhan asserts happens all the time: we become (sometimes unwittingly) manipulated by media, namely television. For many during McLuhan's era and even today, television has been viewed as a "vast wasteland." Yet, for McLuhan, television viewing should be complex, given decisions on programming, the various corporate-owned networks, and the (un)predictability of audience viewers, among other items. Our attitudes and experiences are often directly influenced by what we watch on television, and our belief systems apparently can be negatively affected by television.

A third assumption of MET has elicited quite a bit of popular conversation: media connect the world. McLuhan used the phrase **global village** to describe how media tie the world into one great political, economic, social, and cultural system. Recall that although the phrase is almost a cliché these days, (there are hundreds of millions of results on a Google search, for instance), it was McLuhan who argued that the media can organize societies socially. Electronic media, in particular, have the ability to bridge cultures that would not have communicated prior to this connection.

global village the notion that humans can no longer live in isolation, but rather will always be connected by continuous and instantaneous electronic media

The effect of this global village, according to McLuhan (1964), is the ability to receive information instantaneously (an issue we return to later in the chapter). People now have become emotionally engaged in global issues. Consequently, international events have become prominent, sometimes even more so than events in a local community. He observes that "the globe is no more than a village" (p. 5) and that we should feel responsible for others. Others "are now involved in our lives, as we in theirs, thanks to the electric media" (p. 5).

Let's revisit our chapter-opening example of Professor Randall to illustrate this assumption further. It's clear that a presentation to a group of international scholars requires an understanding of various global issues. Further, given that her presentation was to be in Costa Rica, Randall felt that it was important to get some understanding of how local citizens utilized social media. She consulted both the internet and her colleague in Central America. Further, consider the fact that both Bella and Margaret watched the U.S. presidential debates, shrinking the distance between the two countries and the events that take place in those countries. All of these efforts would not have been possible if technology (telephone and computers) was not available.

The global village of Marshall McLuhan follows the General Systems perspective we outlined in **Chapter 3.** You will recall that Systems theorists believe that one part of a system will affect the entire system. Media Ecology theorists believe that the action of one society will necessarily affect the entire global village. Therefore, such things as wars or refugees in the Middle East and economic strife in Europe affect the United States, Australia, and China. According to McLuhan, we can no longer live in isolation because of "electronic interdependence" (McLuhan & Fiore, 1996). Today, some scholars such as Yuezhi Zhao (2022) believe that the global village is becoming anachronistic—that is, out of date. Zhao contends that the rising tide of nationalism and isolationism is rendering this perspective irrelevant. Still, McLuhan's prophetic thinking should not be ignored nor abandoned.

You have now been introduced to the primary assumptions of MET. McLuhan's theory relies heavily on a historical understanding of media. He asserts that the media of a particular time period were instrumental in organizing societies. He identifies four distinct time periods, or **epochs,** in history (**Table 23.1**). We address them next.

epoch era or historical age

Table 23.1 McLuhan's Media History

HISTORICAL EPOCH	PROMINENT TECHNOLOGY/ DOMINANT SENSE	McLUHAN'S COMMENTS
Tribal Era	Face-to-Face Contact/ Hearing	"An oral or tribal society has the means of stability far beyond anything possible to a visual or civilized and fragmented world" (McLuhan & Fiore, 1968, p. 23).
Literate Era	Phonetic Alphabet/ Seeing	"Western man [woman] has done little to study or to understand the effects of the phonetic alphabet in creating many of his [her] basic patterns of culture" (McLuhan, 1964, p. 82).
Print Era	Printing Press/ Seeing	"Perhaps the most significant of the gifts of typography to man [woman] is that of detachment and noninvolvement— the power to act without reacting" (McLuhan, 1964, p. 173).
Electronic Era	Computer/ Seeing, Hearing, Touching	"The computer is by all odds the most extraordinary of all the technological clothing ever devised ... since it is the extension of our central nervous system" (McLuhan & Fiore, 1968, p. 35).

Making Media History and Making "Sense"

McLuhan (1962, 1964) and Quentin Fiore (McLuhan & Fiore, 1967, 1996) claim that the media of an era define the essence of a society. They present four eras, or epochs, in media history, each of which corresponds to the dominant mode of communication of the time. In one of the more provocative claims, McLuhan contends that media act as extensions of the human senses in each era.

The Tribal Era

According to McLuhan, during the **tribal era,** hearing, smell, and taste were the dominant senses. During this time, McLuhan argues, cultures were "ear-centered" in that people heard with no real ability to censor messages. This era was characterized by the oral tradition of storytelling whereby people revealed their traditions, rituals, and values through the spoken word. In this era, the ear became the sensory "tribal chief" and for people, hearing was believing.

> **tribal era** age when oral tradition was embraced and hearing was the paramount sense

The Literate Era

This epoch, emphasized by the visual sense, was marked by the introduction of the alphabet. The eye became the dominant sensory organ. McLuhan and Fiore (1996) state that the alphabet caused people to look at their environment in visual and spatial terms. McLuhan (1964) also maintains that the alphabet made knowledge more accessible and "shattered the bonds of tribal man" (p. 173). While the tribal era was characterized by people speaking, the **literate era** was a time when written communication flourished. People's messages became centered on linear and rational thinking. Out was storytelling; in were mathematics and other forms of analytic logic. This "scribal world" had the unintended consequence of forcing communities to become more individualistic rather than collectivistic (McLuhan & Fiore, 1967). People were able to get their information without help from their communities. This was the beginning of people communicating without the need to be face-to-face.

> **literate era** age when written communication flourished and the eye became the dominant sense organ

The Print Era

The invention of the printing press heralded the **print era** in civilization and the beginning of the Industrial Revolution. Although it was possible to do a great deal of printing by woodcut prior to this era, the printing press made it possible to make copies of essays, books, and announcements. This provided for even more permanency of record than in the literate age. The printing press also allowed people other than the elite to gain access to information. Today, think about self-publishing and its implications as you think about this era. Then, as today, with printing, people didn't have to rely on their memories for information as they had to do in the past. Publication made it possible for permanency of record to be achieved.

> **print era** the age when gaining information through the printed word was customary, and seeing continued as the dominant sense

McLuhan (1964) observes that the book was "the first teaching machine" (p. 174). Consider his words today. Very few courses in college exist without a textbook/e-book. Even with technological teaching approaches such as distance learning or Zoom, the large majority of courses still require books. Books remain indispensable in the teaching–learning process and even in the corporate world because companies such as Starbucks and Zappos require their employees to read various motivational books while employed.

Exemplifying the print era more specifically, McLuhan writes:

> Margaret Mead has reported that when she brought several copies of the same book to a Pacific island there was great excitement. The natives had seen books, but only one copy of each, which they had assumed to be unique. Their astonishment at the identical character of several books was a natural response to what is after all the most magical and potent aspect of print and mass production. It involves a principle of extension by homogenization that is the key to understanding Western power (p. 174).

What McLuhan notes here is that mass production produces citizens who are similar to each other. The same content is delivered over and over again by the same means. This visual-dependent era, however, produced a fragmented population because people could remain in isolation reading their mass-produced media.

The Electronic Era

The age we live in now is electronic. Interestingly, McLuhan (1964) and his colleague (McLuhan & Fiore, 1967) note that this epoch, characterized by the telegraph, telephone, typewriter, radio, and television, has brought us back to tribalization and the art of oral communication. Instead of books being the central repository of infor-

> **electronic era** age in which electronic media pervade our senses, allowing for people across the world to be connected

mation, electronic media decentralized information to the extent that individuals are now one of several primary sources of information. This era has returned us to a primitive-like reliance on talking to one another although as we know, some would argue that the art of conversation has really been lost (Wisner, 2022). Today, though, we define *talking* differently than the way it occurred in the tribal era. We talk through smartphones, television, radio, books on tape, texting, voicemail, blogs, and email, among others. The **electronic era** allows different communities in different parts of the world to remain connected, a concept we discussed previously as the global village.

McLuhan (1964) relates a description of various technologies during this electronic age:

The telephone: speech without walls.
The phonograph: music hall without walls.
The photograph: museum without walls.
The electric light: space without walls.
The movie, radio, and TV: classroom without walls (p. 283).

The electronic era presents unique opportunities to re-evaluate how media influence the people they serve. This age allows for ear and eye and voice to work together.

This historical presentation of media by McLuhan suggests that the primary medium of an age prompts a certain sensory reaction in people. McLuhan and Fiore (1968) theorize that a **ratio of the senses** is required by people, which is a conversation of sorts between and among

> **ratio of the senses** phrase referring to the way people adapt to their environment

the senses. That is, a balance of the senses is required, regardless of the time in history. For instance, with the internet, we reconcile a variety of senses, including visual stimulation of website pictures and the auditory arousal of downloaded music. When we develop online relationships, we already know that our nonverbal communication is severely limited (for more on this information, see **Chapter 8** on Social Information Processing Theory).

Theory Into Practice • Media Ecology Theory

Theoretical Claim: "The Medium is the Message": The media's format—perhaps even more than the message itself—affects and changes people and society.

Practical Claim: Because of her German-Polish heritage, Olga was interested in learning more about the surrender of the Germans in 1945. She started in the library and reviewed some books and periodicals on the topic. Then she sat down and went online to Google to see how various websites introduced the subject. Finally, she went to YouTube and watched a Combat Camera Unit Video of the surrender *as it happened*. Each mediated option was valuable and Olga found one to be more compelling than the next.

The Medium Is the Message

Media Ecology Theory is perhaps best known for the catchphrase *"the medium is the message"* (McLuhan, 1964), a "humble and fascinating" phrase (Hodge, 2003, p. 342). Although followers of McLuhan continue to debate the precise meaning of this equation, it appears to represent McLuhan's scholarly values: the content of a mediated message is secondary to the medium (or communication channel). The medium has the ability to change how we think about others, ourselves, and the world around us. So, for instance, in our opening example of Professor Randall, what she and Bella communicated is less important than that they communicated via a computer, the internet, and email.

> **the medium is the message** phrase referring to the power and influence of the medium—not the content—on a society

McLuhan does not dismiss the importance of content altogether. Rather, McLuhan argues that content gets our attention more than the medium does. McLuhan thinks that although a message affects our conscious state, it is the medium that largely affects our unconscious state. So, for example, we often unconsciously embrace television as a medium while receiving a message broadcast around the world. Consider the fact that the 2001 terrorist attacks in New York City, Hurricane Katrina's devastating effects in 2005, the 2013 Boston Marathon bombings, and the 2021 U.S. Capitol attack were reported not only immediately after the events but, in some cases, *during* the events in real time. Many of us went to TV immediately or watched the online feed and were captivated by the horror and the images as they occurred. We were pretty much unconscious of the medium, but rather consumed with the message. Nonetheless, we turned to television and our laptops again and again for updates as the days and months progressed, rather unaware of their importance in our lives. This behavior underscores McLuhan's hypothesis that the medium shapes the message and it is, ironically enough, our unawareness of the medium that makes a message all the more important. One note: during the height of the COVID-19 pandemic, the medium was confusing. Writers (e.g., Garrett, 2020) noted that with a stock market, disease trolls and liars, and governments unwilling to accept the death knell instigated by the virus, no medium was viewed as fully credible. Further, the message "was chaos" during this time (p. 943).

McLuhan and Fiore (1967) claim that in addition to the medium being the message, the medium is the "massage." By changing one letter, they creatively present readers with another view of media. It's not clear whether the authors were making a pun on the "mass-age," on the need for media to "relax," or whether the team was reinforcing McLuhan's earlier writings on the power of the media. McLuhan and Fiore argue that not only are we influenced by the media, but we can become seduced by it. As a population, we are fascinated with new technologies. For instance, it's customary for media such as *The New York Times* and *USA Today* to feature special sections on technology and culture. New gadgets, applications, and technological inventions (and their prices) are featured for those desiring the latest. Indeed, the medium massages the masses, is part of the "mess-age" (McLuhan & Parker, 1969), and can be understood in a "mass-age" (McLuhan & Nevitt, 1972). James Morrison (2006) sums it up by stating that "'the medium is the message' because the contents of a medium vary and may even be contradictory, but the medium's effects remain the same, no matter what the content" (p. 178).

We have presented several key assumptions and issues associated with MET. We have also discussed media in very broad terms. McLuhan says that some unifying and systematic way of differentiating media is necessary. The result is an interesting analysis of hot and cool media.

Gauging the Temperature: Hot and Cool Media

To understand the "large structural changes in human outlook" (McLuhan, 1964, p. vi) of the 1960s, McLuhan set out to classify media. He explains that media can be classified as either hot or cool, language he borrowed from jazz slang. Today, you might be inclined either to interchange or collapse the two to mean the same thing ("*they're hot and cool*"). This classification system remains a bit confounding to many scholars, and yet it is pivotal to the theory. We distinguish between the two types of media next and provide examples of each in **Figure 23.1.**

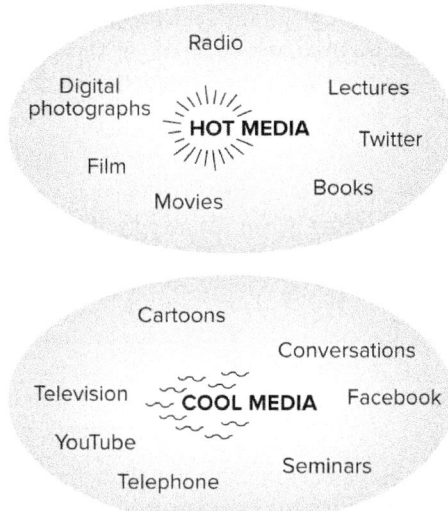

Figure 23.1 **Examples of McLuhan's Hot and Cool Media**

Hot media are described as media that demand little from a listener, reader, or viewer; hot media spoon-feed content to an audience. Hot media are high-definition communications that have relatively complete sensory data; very little is left to the audience's imagination. Hot media, therefore, are low in audience participation. Meaning is essentially provided. An example of a hot medium is a movie, because it requires very little of us. We sit down, watch the film, react, maybe eat some popcorn, and then (only if we have time and are production aficionados!) watch the credits. Hot media provide the audience what they need—in this case, entertainment. Today, one could state that posting multiple characters on X does not require high sensory involvement.

hot media high-definition communication that demands little involvement from a viewer, listener, or reader

McLuhan believes that radio is a hot medium. He acknowledges that radio can serve as background sound, as noise-level control, or for listening pleasure. Little involvement is needed with radio. However, McLuhan wrote before the proliferation of radio talk shows. Given the audience engagement today in much of talk radio, we need to reflect on McLuhan's era rather than our contemporary day.

Unlike hot media, **cool media** require a high degree of participation; they are low definition. Little is provided by the medium, so much has to be filled in by the listener, reader, or viewer. Cool media require audiences to create meaning through high sensory and imaginative involvement. Consider, for instance, cartoons and comics. Generally, we get a few frames of illustrations and perhaps some brief phrases. Overall, cartoons and comics are low definition and provide very little visual information. We usually need to determine the meaning of the words and the pictures, and even supply missing words or ideas that are not provided in the cartoon (Fink, 2022).

cool media low-definition communication that demands active involvement from a viewer, listener, or reader

Let's make one more point about cool media. Interestingly, McLuhan (1964) contends that television can be considered a cool medium. He argues that TV is a medium that requires viewers to be actively involved. In fact, he notes that television "engages you. You have to be with it" (p. 312). Yet with the digital age upon us, television has taken on new meaning that not even McLuhan could have imagined. Would he still consider television a cool medium today? As we think about the current forms of hot media, when we "friend" another person on Facebook, "like" an image or photo on Instagram, or screenshot a link from another person, we are using sensory involvement. In other words, we are exemplifying modern-day cool media.

Let's provide you two different examples of hot and cool media. First, McLuhan believed that a marked difference exists between a lecture and a seminar. A lecture is determined to be hot because generally, it encompasses a person standing before a large group talking. It requires little except to listen to the information. A seminar, however, is cool, smaller in nature with the discussion leader facilitating questions and participants sharing ideas and life experiences.

McLuhan also differentiated between hot and cool media using the 1960 presidential debates between John F. Kennedy and Richard Nixon. The theorist discovered that for those who watched the debate on television, Kennedy had won because he exuded an objective, cool persona, perfect for the cool medium. For those who listened to the debates on the radio (a hot medium), Nixon was the winner. He was considered hot (which, in fact, his sweating showed he was!). So, the medium influences others' perceptions.

Arthur Asa Berger (2007) notes that McLuhan's thinking pertaining to hot and cool media is prompting more research interest as communication theorists try to understand the passion that people have for their cell phones. As the 21st century continues, we need to be cautious in interpreting hot and cool media in straight-forward ways. For instance, multitasking (responding to a text and watching television simultaneously) makes a distinction between hot and cool media more difficult. Moreover, as we try to categorize various forms of (social) media as either hot or cool, we need to remind ourselves of David Bobbit (2015): "McLuhan was not concerned with providing consistent, linear meanings of the terms 'hot' versus 'cool' media. For him, it was the *effect* the medium had that he was trying to get at" (**http://enculturation.net/teaching-mcluhan**).

The Circle Is Complete: The Tetrad

We continue our discussion of Media Ecology Theory by examining a recent expansion of McLuhan's thinking about the media. With his son Eric (McLuhan & McLuhan, 1988), and to respond to those who believed that there was no scientific grounding in his work, McLuhan developed a way to look further into the effects of technology on society. His expansion of the theory included a thorough discussion of the **laws of media** (later, Marshall's grandson [and Eric's son], Andrew, would discuss these laws).

laws of media further expansion of Media Ecology Theory with focus on the impact of technology on society

Although McLuhan's earlier work identified in this chapter does not fully take into account the advent of the computer, his posthumous work with his son takes into consideration the influence of the

tetrad organizing concept to understand the laws of media

internet. Their work was an effort to bring the theory full circle: technology affects communication through new technology, the impact of the new technology affects society, and the changes in society cause further changes in technology. McLuhan and McLuhan offer the **tetrad** as an organizing concept that allows scholars to understand the past, present, and current effects of media. According to Andrew McLuhan (2016), his father and grandfather were interested in two primary questions: "What do the media have in common?" and "What do they do?" (**https://medium.com/@andrewmcluhan/what-is-a-tetrad-ad92cb44d4af**).

To give us a new way of looking at the role of technology in our culture, McLuhan and McLuhan offer four laws, phrased as questions, to understand technology: (1) What do media enhance? (2) What do media make obsolete? (3) What do media retrieve? (4) What do media reverse? Each of these is aligned with the Media Effects research that we articulated in **Chapter 13** (Uses and Gratifications). Let's examine and identify examples of their roles in culture (**Table 23.2**). We pay particular attention to the role of the internet in our discussion.

Table 23.2 Laws of Media

LAWS OF MEDIA	DESCRIPTION
Enhancement	What does the medium enhance or amplify?
Obsolescence	What does the medium push aside or make obsolete?
Retrieval	What does the medium retrieve from the past?
Reversal	When pushed to its limits, what does the medium reverse or flip into?

Enhancement

The first law of media is **enhancement;** that is, media enhance or amplify society. In some ways, McLuhan believes that every new medium enhances a human ability, which, in turn, builds upon

enhancement law that states media amplify or strengthen society

a new medium. The telephone enhanced the spoken word found in face-to-face conversations. Radio, of course, amplified the voice across distance. Television amplified the word and the visual across continents.

The internet has affected society in different ways. First, it has the potential to enhance a number of senses, including sight and sound. People are now creating visual (e.g., Instagram photos) and audible products (e.g., YouTube videos) that have boosted online engagement. Second, the existence of the internet has enhanced the accessibility of information. For instance, we can now obtain birth records, credit card balances, and missing person information online. Third, the internet can exacerbate class division. The "haves and the have-nots" (i.e., those who have access and those who do not) exist along this information superhighway. Finally, and related to the third claim, decentralization of authority is enhanced by the internet. Many communities have access to information and events that years ago were completely out of reach. Further, no longer do our political, scientific, or corporate leaders solely possess information; that information becomes available online.

Obsolescence

McLuhan and McLuhan (1988) note that the second law of media is that media eventually render something obsolete or out of date. Andrew McLuhan (Marshall's grandson) puts it this way: "Obsolete doesn't mean dead—just no longer in charge. Out of date. Out of fashion. Version 1.0 when v. 2.0 is out" (**https://medium.com/@andrewmcluhan/what-is-a-tetrad-ad92cb44d4af**). Television made radio obsolete, although many of us continue to turn to radio each day while we drive. Radio rendered motion pictures obsolete, which, in turn, resulted in fewer movies to watch. DVRs may have also attempted to make movies obsolete, but we know that their efforts have been only somewhat successful. Today, it's all about streaming services.

The internet, too, has brought about **obsolescence**. For example, as we learned earlier, the global village now exists, thanks to the internet. The geographical splits are pretty hard to find; even the remote villages of Africa are becoming accessible by the internet, albeit many, as we noted

obsolescence law that states media eventually render something obsolete or out of date

earlier, have no access. Second, the internet is slowly targeting micromedia (specific audiences) rather than macromedia (large masses), thereby making traditional media outlets such as CBS, NBC, and ABC change their news reporting. Finally, former "tribal" conversations are now electronically derived.

Retrieval

The third law is **retrieval,** meaning that media recover or restore something that was once lost. What older, previously obsolesced media are brought back? Television, for instance, restored the impor-

retrieval law that states media restore something that was once lost

tance of the visual that radio did not achieve, but that was once in face-to-face conversations. Radio retrieved the town crier, the prominent voice of newsworthy events during colonial times in the United States. Print retrieved the tribe's universality of knowledge. And the internet recovers a community that was once lost to other media. For instance, chat rooms visited by people like Professor Randall from our chapter-opening vignette have electronically rekindled conversations that flourished before radio and television.

Reversal

When "pushed to the limit of its potential" (McLuhan & McLuhan, 1988, p. 99), what do media produce or become? What do media reverse into? When too many constraints exist on a medium, it will "overheat" and become ineffective. **Reversal** contains characteristics

reversal law that states media will— when pushed to their limit—produce or become something else

of the system from which it arose. For instance, the public's desire to have access to entertainment in a relatively cheap medium led to the creation of radio dramas and comedy programs. The need to "see" what was heard led to the creation of these programs on television. Yet, we can stream programs and what was once seen by millions of people at the same time is reversed into private "performances." Television, then, reversed into the early days of the print era when people could consume media privately.

Students Talking: Ramone

 I loved the concept of the "global village" because it shows that we're all connected some way. My grandmother came from Mexico, and she didn't know how to speak English when she got here. But there were so many people in Texas who were like her—they wanted to come to America and wanted to have a better life—and they helped her out a lot! Some were from Guatemala; others were from Ecuador, but they all seemed like they understood her. So, the connection among all these countries because of a shared experience made her part of her personal "global village."

The internet—as a medium pushed to its potential—reverses society into a new and unique place. The internet has the potential to bring tribal people together when they discuss websites or chat room conversations with one another. Looking at the number of people who "surf the net" each day alone, the internet can isolate people just as television can. Also, with the ability to download music, television shows, and films, the internet—specifically Netflix or Hulu—has reversed itself into a medium with significant visual and auditory appeal. Finally, the internet is a medium that "flips" on its user. That is, although it can serve to erode power, it can also perpetuate power differences among people. For instance, consider the citizens in countries such as China and Russia, who in 2022 used Facebook and X to protest the decision-making power of political authority. In turn, these powerful politicians exerted power upon the protesters by arresting them and forbidding protest gatherings. The internet provides opportunities for both those in power and those who seek power or who are powerless.

Carrying the McLuhan Banner: Postman and Meyrowitz

The former president of the Media Ecology Association believes that thinking about McLuhan is like taking part in a "moveable feast" in that "if you are lucky enough to have encountered McLuhan's legacy, wherever you go for the rest of your life, his legacy stays with you" (Sternberg, 2011, p. 111). Although scores of scholars identify and accept the tenets of MET, two influential researchers are of particular note. Neil Postman and Joshua Meyrowitz were dedicated scholarly followers of McLuhan. Although each retained many of the values related to McLuhan's writings, each scholar adapted an individual lens as they talked about technology. We close our discussion with an overview of these two "McLuhanites."

Neil Postman's biography is an interesting one and is available elsewhere (**http://neilpostman.org**). Still, let's give you a glimpse into his credentials.

Thomas Gencarelli (2006) observes, "Postman is first and foremost an educator" (p. 239). Indeed, most of his books pertain to public education in the United States and in many cases, his writing is quite satirical. In some cases, he encourages changes to the educational system by stating that it needs to be revitalized. Nowhere can that revitalization occur other than with an infusion of technology, according to Postman. The media environment, he believes, helps shape children's lives. He believed that "infotainment" preoccupied consumers. Television, in particular, is ripe for allowing young people to be exposed to all sorts of information that was originally intended for adults. This conflation of the child and adult worlds is an ongoing concern for Postman.

Postman is clearly a man who practices what he preaches. Jim Benning (2003) relates a relevant anecdote. In shopping for a new car, Postman has been quoted as saying, "Why do I need electric windows? My arm and hand work. If I were paralyzed I could use an electric window." His colleague and friend note that "Neil would always take what he would call an ecological perspective, a balanced view."

Postman has written more than 200 articles for the public and over 20 popular and scholarly books. Postman's research is underscored by a central theme: "All technologies are human impositions into the natural order of things and, as a result, change that order" (Gencarelli, 2006, p. 244). Among the most influential of his published works is *Technopoly: The Surrender of Culture to Technology*. In the book, Postman (1992) hypothesizes that technology negatively changes the fabric of society. Specifically, he believes that culture is subservient to the invisible (e.g., I.Q. scores) and visible (e.g., computers).

Postman coined the term *technopoly*, which means that we live in a culture in which technology dominates our thinking and behaviors. In a technopoly, Postman argues, technological tools serve to take over the culture in which they thrive. We live in a society where being

> **technopoly** a term coined by Postman that means we live in a society dominated by technology

technologically driven may result in being driven mad! We trust that our technology will bring us safety and salvation, and we seem to lose any sense of humility, discipline, and rationality regarding our reliance on and trust in current media. As a result, Postman laments that "tradition, social mores, myth, politics, ritual, and religion have to fight for their lives" (p. 28). Postman, like McLuhan, asks whether we want to live in a culture with such unwavering dependence on technology and such a conclusion has led scholars to call him a "pillar of media ecology" (Strate, 2004, p. 3).

In addition to Postman, Joshua Meyrowitz's (1985) research interconnects with McLuhan's work. Meyrowitz's *No Sense of Place* ushered in a unique way of thinking related to space. First, he argues that space is more than physical. That is, social situations such as contesting a parking ticket at city hall include more than the physical surroundings of the building and courtroom. Meyrowitz contends that the influence that communication has on the situation also needs to be considered.

As a communication scholar, Meyrowitz is interested in uncovering the effects of communication technology, primarily television, on a social situation. Consider, for instance, a private discussion between spouses on marital infidelity. The discussion is likely to be free-wheeling, underscoring the intimacy that the couple shares. Now, put that discussion in front of a live television talk show television audience. A new pattern of communication will likely begin, with different information flow and new rules of conduct. It is this new communication medium that Meyrowitz is interested in and one that has cultural consequences.

Meyrowitz agrees with McLuhan that electronic media have social consequences. Meyrowitz expands the notion that power relations and social class can be traced to electronic media. He draws on sociology research to conclude that media have brought about a blurring of formerly distinct roles or places.

He states that "many Americans may no longer seem to 'know their place' because the traditionally interlocking components of 'place' have been split apart by electronic media. Wherever one is now—at home, at work, or in the car—one may be in touch and tuned in" (p. 308).

As noted earlier and in tune with Media Ecology theorists, Meyrowitz points to television for evidence of the "blurring." For instance, examine talk shows and you can get a sense of how the blurring of place occurs. What was once private (e.g., discussing your mother's alcoholism) is now public on various talk shows. Even television programming has blurred gender lines; shows like RuPaul's *Drag Race* presents gendered images that do not fit neatly into preconceived notions of masculinity and femininity.

The writings of both Postman and Meyrowitz remain effective examples of Marshall McLuhan's legacy. The two carry the banner that boldly proclaims that electronic media have shaken Western society's foundation and many of its core values. Clearly, the two scholars prompt us to consider McLuhan's work in more contemporary ways. Even more than a century after his birth, McLuhan remains relevant.

Students Talking Tech: Grady

I was Snapping the other day and thought about what McLuhan would say about sending pictures that fade quickly! He loved talking about the "medium" but even with his brilliance, I wonder whether he could ever imagine a message disappearing. In Snap's case, the medium AND the message would have probably been confusing to McLuhan.

Integration and Critique

You probably have already figured out that Marshall McLuhan has caused quite a reaction in both academic and public circles. His ideas are provocative, and at times, have been both unilaterally dismissed and embraced. Perhaps Joshua Meyerwitz said it best: "[H]e has both passionate adherents and savage critics" (p. 360). In fact, if you reviewed McLuhan's original work, you may be challenged by his writing style. Some have labeled his thinking "McLuhanacy" (Gordon, 1982), while others feel his writing is equivalent to "genre bending" (Carey, 1998). Without doubt, however, the research undertaken by Media Ecology theorists has been qualitative and it's difficult to find any study that has been undertaken experimentally using the theory as a foundation.

McLuhan's work and reputation have been invoked with considerable regard. Let's provide you a snapshot of the kudos celebrating McLuhan. *Wired* magazine named him their "patron saint," and *Life* magazine called him the "Oracle of the Electronic Age." A concentration in McLuhan studies exists at the University of Toronto. There is a McLuhan Institute as well. In addition, there are a McLuhan newsletter, the *International Journal of McLuhan Studies;* symposia on McLuhan's research; a McLuhan festival; a McLuhan reading club; and even a secondary high school in Canada named the Marshall McLuhan Catholic Secondary School. Finally, the *San Francisco Chronicle* once called him "the hottest academic property around." It's hard to escape his influence both in research and in societies around the world. And, indeed, as we noted earlier, the McLuhan academic legend is extensive, beginning with Marshall, who studied with his son, Eric, who then studied with his son, Andrew. Today, only Andrew is alive. The theory, despite its popularity, has been evaluated by scholars and writers. We examine these critiques using the criteria of testability and heurism.

Integration

Communication Tradition	Rhetorical | Semiotic | Phenomenological | Cybernetic | Socio-Psychological | **Socio-Cultural** | **Critical**
Communication Context	Intrapersonal | Interpersonal | Small Group/Team | Organizational | Public/Rhetorical | **Mass/Media** | Cultural
Approach to Knowing	Positivistic/Empirical | Interpretive/Hermeneutic | **Critical**

Critique

Evaluation Criteria	Scope | Logical Consistency | Parsimony | Utility | **Testability** | **Heurism** | Test of Time

Testability

MET has been criticized because many of its concepts are difficult to understand, thereby making testability of the theory challenging and, indeed, nearly impossible (Gordon & Willmarth, 2012). The question becomes apparent: how does one test something one has trouble understanding?

Criticism pertaining to the testability of the theory is represented in comments that have been offered by media scholars over the years. First, the former president of the Media Ecology Association noted that McLuhan's work is less academic in nature and more "poetic" in nature (Morrison, 2020). This prompted many to perceive MET as impossible to test. George Gordon (1982) is more direct: "Not one bit of sustained and replicated scientific evidence, inductive or deductive, has to date justified any one of McLuhan's most famous slogans, metaphors, or dicta" (p. 42). Dwight Macdonald (1967) also attacked McLuhan's writing, noting that "he has looted all culture from cave painting to *Mad* magazine for fragments to shore up his system against ruin" (p. 203).

A great deal of criticism has been directed at McLuhan's use of words and his clarity. To some, his ideas make little sense. Some writers believe that McLuhan failed to define his words carefully and used too much exaggeration. Yet, some believe that his exaggerations were simply "ploys" to get the attention of his readers and students (Logan, 2011, p. 28). It's true that McLuhan tended to write in a zigzag fashion, weaving in one point after another with no apparent topic sentence or sustained idea. One website author believes that his writing was "overgrown with arcane literary and historic allusions" (**nextnature.net/2009/12/the-playboy-interview-marshall-mcluhan**). Yet another believes that McLuhan simply dismissed any criticism since he felt that "clear prose indicates the absence of thought" (**http://www.thenewatlantis.com/publications/why-bother-with-marshall-mcluhan**). And other writers (e.g., Echelson, Fialka, & Logan, 2022) contend that his decision to "pose" questions rather than to "answer" questions was a hallmark of a legitimate scholar.

Although it's clear that some writers indict his thinking and theorizing, McLuhan (1967) offers no apology: "I don't explain—I explore" (p. i). Or, perhaps McLuhan's one line in *Annie Hall* may offer his ultimate response to critics: "You know nothing of my work!"

Heurism

Media Ecology Theory and McLuhan's thinking have been met with considerable enthusiasm. Because McLuhan was a key figure in popular culture, it's important to keep in mind that his writing prompted all sorts of responses. One indicator of the heuristic value of MET is the fact that there is a Media Ecology Association (**https://www.media-ecology.org/**). This organization is dedicated to promoting the theory in both practical and theoretical ways, thereby ensuring the theory's visibility. The association publishes a journal (*EME: Explorations in Media Ecology*) dedicated to the theory, a testimony to the fact that media communication scholars continue to integrate the theory into their research. There is even an "official estate" of McLuhan "to ensure the integrity of his name and legacy" (**http://www.marshallmcluhan.com**). The site is quite interesting with an entire section devoted to "McLuhanisms" (e.g., "All advertising advertises advertising," etc.).

Researchers have discussed McLuhan and his contributions in a variety of ways. Scholars have provided a comprehensive understanding of the theory and have discussed the theory's influential pioneers (Lum, 2006) and some have "taken up McLuhan's cause" (Anton, Logan, & Strate, 2016). Additional writers over the years have applied several of McLuhan's premises in their feminist research (Sharma & Singh, 2022). Furthermore, many of the theory's concepts have been incorporated into research on such diverse topics as school shootings (O'Dea, 2015), online racism (August & Liu, 2015), and YouTube as "cool" media (Trier, 2007).

Finally, in what could be considered to be a 21st-century response to McLuhan's 20th-century theory, the thinking and research of Sherri Turkle (2015) sustains much of what media ecologists advance. In her book *Reclaiming Conversation: The Power of Talk in the Digital Age,* Turkle contends that technology has allowed us to make connections, but we are rarely communicating. She espouses one of the fundamental tenets of media ecologists: "The forms and biases of our media technologies impact our everyday lives" (Zimmer, 2005, p. 5). Other writers have termed this "digital depersonalization" (Simeon & Abugel, 2023, p. 240). Turkle, for example, contends that social media such as Facebook have resulted in so much "friending" and yet, surveys show that most people report having fewer friends than ever. She is clear in stating that while "mobile technology ... is here to stay," it's time that we also understand that "it is time for us to consider how it may get in the way of other things that we hold dear" (e.g., intimacy, community, etc.) (p. 7). In sum, one cannot ignore Turkle's belief that new technologies are preventing authentic human connections. Further, she believes that "technological promiscuity," or the belief that technology can be introduced into any setting or environment, is already part of corporate life in the United States.

Closing

Marshall McLuhan and Media Ecology Theory will continue to resonate for years to come. Perhaps one day we will revisit McLuhan's original thinking on historical epochs in media history. New media will continue to evolve in our society and so will the application of McLuhan's thinking. Was McLuhan an absurd reactionary? Or was McLuhan a cultural prophet? On his gravestone are the words: "The Truth Will Set You Free." Did McLuhan think he had discovered Truth? Or, even in his death, does he continue to play with our imaginations? Maybe his writing reflects a "cryptic, maddeningly epigrammatic style" (Edwards, 2014, p. 2332). Perhaps one of McLuhan's biographers, Philip Marchand (1989), best illustrates McLuhan's contribution to the study of media: "McLuhan's comments had at least one virtue: They seemed to suggest that the world was more interesting than any of us had previously thought it to be" (p. xiii).

Discussion Starters

 Case-In-Point: Discuss whether you believe that Professor Randall should rely less on technology in preparing her speech and more on her life experiences. Is there too much reliance, for instance, on her consultation of various databases, chat rooms, Zoom, and email?

Try-It-Your-Selfie: If you had a chance to blog about any of the current applications pertaining to MET, what three themes resonate with you today? How would you go about discussing them in relationship to today's global events?

1. Describe how Media Ecology theorists might react to the current news today. What would be their major criticisms and their major objections? What would they be particularly interested in?

2. Given all of the variation related to television today (e.g., Fire TV Stick, streaming services, etc.), do you agree or disagree with McLuhan regarding television being a cool medium? Use examples to defend your view.

3. Interpret and comment on the following statement: "Technology is the end of our beginning." Use terms from MET and various examples to defend your view.

4. Discuss your response to theorists who choose to be part of the popular culture, including participating on talk shows and appearing in films.

References

Anton, C., Logan, R. K., & Strate, L. (2016). *Taking up McLuhan's cause: Perspectives on formal causality.* Intellect LTD.

August, C., & Liu, J. H. (2015). The medium shapes the message: McLuhan and Grice revisited in race talk online. *Journal of Community & Applied Social Psychology, 25,* 232–248.

Berger, A. A. (2007). *Media and society.* Rowman & LIttlefield.

Benning, J. (2003, October 10). *Remembering Neil Postman.* Blog post. Alter-Net Media, Inc.

Bobbit, D. (2015). *Teaching McLuhan: Understanding media.* http://enculturation.net/ teaching-mcluhan.

Carey, J. W. (1998). Marshall McLuhan: Genealogy and legacy. *Canadian Journal of Communication, 23,* 13–21.

Echelson, D., Fialka, G., & Logan, R. K. (2022). Expanding and enriching the McLuhan tetrad. *New Explorations: Studies in Culture and Communication, 2,* 121–127.

Edwards, B. L. (2014). Marshall McLuhan. In F. N. Magil (Ed.), *Dictionary of World Biography* (vol. 8, pp. 2332–2335). Routledge.

Fink, M. (2022). "Cool" media studies: McCloud, McLuhan, and the popification of the Humanities. *Inks: The Journal of the Comics Studies Society, 6*(3), 250–261.

Garber, M. (2011). *Webs and whirligigs: Marshall McLuhan in his time and ours.* http://ffffound.com/image/cfc6c98c37b262a79c7b436e5c9c08d1dad35835.

Garrett, L. (2020). COVID-19: The medium is the message. *The Lancet, 395*(10228), 942–943.

Gencarelli, T. F. (2006). Neil Postman and the rise of media ecology. In C. M. K. Lum (Ed.), *Perspectives on culture, technology, and communication: The media ecology tradition* (pp. 201–25). Hampton.

Gordon, G. (January 1982). An end to McLuhanacy. *Educational Technology, 22*(1), 39–45.

Gordon, T., & Willmarth, S. (2012). *McLuhan for beginners.* For Beginners Press.

Greener, S. (2016). The medium, the message and the memory. *Interactive Learning Environments, 24,* 357–359.

Hodge, B. (2003). How the medium is the message in the unconscious of 'America online.' *Visual Communication, 2,* 341–353.

Innis, H. A. (1951). Industrialism and cultural values. *The American Economic Review, 41*(2), 201–209.

Levinson, P. (2000, Spring). McLuhan and media ecology. *Proceedings of the Media Ecology Association, 1,* 1–7.

Logan, R. K. (2011). McLuhan misunderstood: Setting the record straight. *International Journal of McLuhan Studies, 1,* 27–47.

Lum, C. M. K. (2006). Notes toward an intellectual history of media ecology. In C. M. K. Lum (Ed.), *Perspectives on culture, technology, and communication: The media ecology tradition* (pp. 1–60). Hampton.

Macdonald, D. (1967). "He has looted all culture...." In G. Stearn (Ed.), *McLuhan: Hot and cool* (pp. 56–77). Penguin Random House, LLC.

Marchand, P. (1989). *Marshall McLuhan: The medium and the messenger.* Houghton Mifflin Harcourt.

McDowell, P. (2021). Reading McLuhan reading (and not reading). *Textual Practice, 35*(9), 1391–1417.

McDowell, P. (Ed.). (2023). *Reading McLuhan reading.* Taylor & Francis.

McLuhan, E. (2008). Marshall McLuhan's theory of communication: The Yegg. *Global Media Journal: 1*(1), 25–43.

McLuhan, M. (1962). *The Gutenberg galaxy.* Mentor.

McLuhan, M. (1964). *Understanding media.* McGraw Hill.

McLuhan, M. (1967). Rebuttal. In G. E. Stearn (Ed.), *McLuhan: Hot and cool, a critical symposium.* The Dial Press.

McLuhan, A. (2016, January 8). *What is a tetrad?.* Medium. https://medium.com/@andrewmcluhan/what-is-a-tetrad-ad92cb44d4af.

McLuhan, M., & Fiore, Q. (1967). *The medium is the massage: An inventory of effects.* Bantam Books.

McLuhan, M., & Fiore, Q. (1968). *War and peace in the global village.* McGraw Hill.

McLuhan, M., & Fiore, Q. (1996). *The medium is the massage: An inventory of effects.* HardWired.

McLuhan, M., & McLuhan, E. (1988). *Laws of media: The new science.* University of Toronto Press.

McLuhan, M., & Nevitt, B. (1972). *Take today: The executive as dropout.* Harper & Row.

McLuhan, M., & Parker, H. (1969). *Counterblast.* Harcourt, Brace & World.

Meyrowitz, J. (1985). *No sense of place.* Oxford University Press.

Morrison, J. C. (2006). Marshall McLuhan: The modern Janus. In C. M. K. Lum (Ed.), *Perspectives on culture, technology, and communication: The media ecology tradition* (pp. 163–200). Hampton Press.

Morrison, J. C. (2020). Leonardo's brain: Understanding da Vinci's creative genius. *Explorations in Media Ecology, 19*(3), 381–389.

O'Dea, J. (2015). Media violence: Does McLuhan provide a connection? *Educational Theory, 65,* 405–421.

Postman, N. (1971). *Teaching as a subversive activity.* Delta.

Postman, N. (1992). *Technopoly: The surrender of culture to technology.* Knopf.

Scolari, C. A. (2012–2013). The echoes of McLuhan: Media ecology, semiotics, and interfaces. *International Journal of McLuhan Studies, 2,* 117–123.

Sharma, S., & Singh, R. (Eds.). (2022). *Re-understanding media: Feminist extensions of Marshall McLuhan.* Duke University Press.

Simeon, D., & Abugel, J. (2023). *Feeling unreal: Depersonalization and the loss of the self.* Oxford University Press.

Sternberg, J. (2011, June 23). Proceedings of the media ecology association. *Media Ecology Association, 12,* 110.

Strate, L. (2004). A media ecology review. *Communication Research Trends, 23,* 1–48.

Trier, J. (2007). "Cool" engagements with YouTube. *Journal of Adolescent & Adult Literacy, 50,* 598–604.

Turkle, S. (2015). *Reclaiming conversation: The power of talk in the digital age.* Penguin.

Wisner, C. (2022). *The art of conscious conversations: Transforming how we talk, listen, and interact.* Berrett-Koehler.

Zhao, Y. (2022). Imagining the New global village. *Journal of Transcultural Communication, 2*(1), 107–109.

Zimmer, M. (Spring 2005). Media ecology and value sensitive design: A combined approach to understanding the biases of media technology. *Proceedings of the Media Ecology Association, 6,* 2–15.

CHAPTER 24
Communication Accommodation Theory

Based on the research of **Howard Giles**

> **"As an evolving and adaptive species, we naturally have to adjust to our surroundings; this means accommodating as well as non-accommodating each other."**
>
> —Howard Giles

Luke Merrill and Roberto Hernandez

As an upcoming spring graduate, 22-year-old Luke Merrill was already preparing for the onslaught of interviews he would face graduating with a 3.86 grade point average. Luke knew that his grades were excellent, and as a Communication major/ Spanish minor, he felt that his chances of landing a good job were pretty good.

Luke's chance to show what he's got finally came. He got an interview with a large accounting firm looking to hire someone in their client relations department. The position preferred a candidate with modest conversational ability in two languages, so he was even more excited. Luke flew to Denver for his first face-to-face interview with Roberto Hernandez, the department's director. The interview would be one that both men would remember for quite some time.

"Good morning, Luke. It's great to meet you," Roberto said.

"Good morn ... I mean, *buenos días* to you, Señor Hernandez," Luke replied, realizing that he was already interrupting his possible boss. And yet, the initial tension seemed to ease.

Roberto continued, "Please, have a seat. I hope your flight last night was fine. I know that this time of the year can get a little bumpy, especially over the Rockies."

"Oh, it was great. I like to fly, Mr. Hernandez. This time, I got a chance to watch a movie that I hadn't seen yet. Personally, I love movies while I fly. It distracts me from the other noise and loud engine noises," Luke responded.

Their conversation continued. Luke was certainly a bit on edge, but noticed that his accelerated speech rate finally slowed to match that of his interviewer. Yet, Luke still felt awkward because he did not know exactly when Mr. Hernandez would change the subject from flying to the job.

Roberto began to talk further about his two sons, both of whom love to fly. "Like you, I love to fly," Roberto related. "My wife, though, is a different story. I wish she had the same attitude as you."

Luke replied, "Hey, maybe it's simply *machismo*, but I'm never afraid to fly."

Roberto was struck by Luke's use of the term *machismo*. Was he using Spanish simply because Roberto was Mexican? Did he know that many people find the word rather insulting and biased? As an interviewer, he knew that job candidates get nervous, and maybe this was simply a nervous habit of his. It was definitely odd, Roberto thought. Very odd and rather offensive. Of course, it didn't escape Roberto's notice that Luke was speaking Spanish to a Mexican American man who happened to sound like a native Spanish native from Mexico. Maybe because the job description asked for competency in "conversational Spanish"? Yet, he didn't want to make anything out of it. "I trust you've come here with a lot of questions, Luke. Let me begin by answering one that we haven't really addressed over the phone—the job's travel schedule."

"I'm cool with pretty much anything, Mr. Hernandez. You might say that I'm comfortable with nearly any assignment you'd give to me. We are definitely *simpatico* on that one."

Roberto was now getting visibly agitated by Luke's frequent insertion of Spanish words into the conversation. He could handle the initial greeting; after all, he does have a Latin background. But Luke's continuing in this manner only made Roberto more uneasy. He didn't know whether to ask Luke to leave or simply to say how confused he was over Luke's persistent behavior. Of course, he could choose to remain silent and simply send Luke a rejection email. Roberto decided to confront the young college student instead.

"Luke, I have to admit that I think it's a bit provocative that you're including Spanish here. Ok, I realize that you're interviewing with someone who is from Mexico. But, frankly, it's a bit much, don't you think? Maybe I'm reaching here, but it's...." Roberto stared at the young job candidate.

"Mr. Hernandez," Luke explained, wishing he could start the entire interview over, "I'm sorry if I've offended you. I realize that I broke into Spanish, and to be honest with you, yeah, I guess it was strange. I should have known better. I was using Spanish to show you some respect, to ah ... to show that I can weave Spanish into our conversation.... I'm really sorry if I was out of place. I blew this, right?"

Luke was extremely nervous. He wasn't sure what Mr. Hernandez would say next. He was mad at himself for sounding like an idiot and assumed that he had lost the job. Maybe he had tried too hard to adapt to his interviewer. Maybe an interview was the worst place for him to demonstrate his fluency in Spanish. Maybe he had misread the situation.

"Listen, Luke," Roberto advised. "I've been around this firm for almost 15 years. I've seen people come and go. I...."

"... and I think," Luke interrupted, "that I should keep my mouth shut and not pretend I'm some expert in a language! Really, I'm sorry about this."

Roberto smiled awkwardly. "Now, let's get on with the rest of this interview."

When two people speak, they sometimes mimic each other's speech and behavior. We may talk to another who uses the same language we do, gestures similarly, and even speak at a similar rate. We, in turn, may respond in kind to the other communicator. We refer to this as a kind of conversational echo, where someone repeats—both verbally and nonverbally—what has been presented earlier in the conversation sequence.

Imagine, for instance, situations where you speak to someone who has gone to college. You both are probably going to use phrases and jargon that are unique to college life, such as "prereqs," "electives," "independent study," "gen ed," etc. If we are speaking to someone who has not gone to college, we may make efforts to clarify our speech or use examples that the other person will understand.

Although we all have these types of experiences at the interpersonal level, sometimes there are group- or culture-based differences, such as perceived differences in age group, in accent or ethnicity, or in the pace and rate of speech. Whether in an interpersonal relationship, in a small group, or across cultures, people tend to adjust their communication to others. This adaptation is at the core of Communication Accommodation Theory (CAT), developed by Howard Giles. Formerly known as Speech Accommodation Theory (but later conceptualized more broadly to include nonverbal behaviors and speaking patterns), CAT rests on the premise that when speakers interact, they modify their speech, their vocal patterns, and their gestures to accommodate others. Giles believe that social differences between and among people are usually problematic and can create noise in the communication channel (Popan, 2023). Giles and his colleagues believe that speakers have various reasons for being accommodative to others. Some people wish to (1) evoke the approval of a listener with higher status, (2) achieve communication efficiency, (3) assert a dominant position, and (4) maintain a positive social identity (Giles, Mulac, Bradac, & Johnson, 1987; Hogg & Giles, 2012). As you can see in our chapter-opening story, however, we do not always achieve what we aim for.

Communication Accommodation Theory had its beginnings in the 1970s, almost in tandem with the "accent mobility" model (Giles, 1973), which is based on various accents heard in interview situations (similar to the situation with Luke and Roberto). As the mobility model was being researched, Giles clarified and conceptualized CAT. Much of the subsequent theory and research since then has remained sensitive to the various communication accommodations undertaken in conversations among diverse cultural groups, including senior citizens, people of color, immigrants, and the visually impaired. For Giles and other accommodation scholars, *culture* is an expansive term and includes much more than, say, race and gender. In fact, accommodation theorists believe that culture is essentially "meaning—what a society of people considers meaningful in conducting their social behavior" (Su, 2019, p. 34). Further, the researchers argue that CAT has not only attracted international scholars, but has also been the subject of a body of research published in multiple languages, making this theory truly global in nature. Although the theory resonates in the interpersonal and group and team contexts, we frame much of our discussion within a cultural framework, honoring Giles's scholarly legacy (Harwood, Gasiorek, Pierson, Nussbaum, & Gallois, 2019).

To get a sense of the central characteristic of CAT, we first delineate what is meant by the word *accommodation*. For our purposes, **accommodation** is defined as the ability to adjust, modify, or regulate one's behavior in response to another. Further, accommodation can

> **accommodation** adjusting, modifying, or regulating behavior in response to others

also take place as people try to manage the emotional tenor of a conversation (Bernhold & Giles, 2022). Accommodation, to a large extent, requires multiple levels of communication (Gallois & Giles, 2015) in that it also includes the motivations behind accommodation and the consequences related to the adjustment (Soliz & Giles, 2016). Further, since accommodation effectively requires adjustment, it's important to understand that inappropriate or insufficient adjustment can result in conversational misunderstandings (Giles, 2016a).

Accommodation is usually done unconsciously (Jackson, 2023). We tend to have internal cognitive scripts that we draw on when we find ourselves in conversations with others. In a conversation with a 15-year-old girl, you might find yourself using teen vocabulary (e.g., "whatever") and while having a discussion with an 85-year-old, you might slow your speech. This is all done without much thought and can occur overtly or covertly.

Like several other theories in this book, Communication Accommodation Theory had its infancy in another field and in this case, in social psychology. To this end, it's important to address the theoretical vehicle that launched Giles's thinking: Social Identity Theory.

Social Psychology and Social Identity

Much of the research and theory in the field of social psychology pertain directly to how people search for meaning in the behaviors of others and how this meaning influences future interactions with others. One of the key concepts discussed in the research of social psychology is identity. According to Jessica Abrams, Joan O'Connor, and Howard Giles (2003), "accommodation is fundamental to identity construction" (p. 221). Recognizing the importance of the self and its relationship to group identity, Henri Tajfel and John Turner (1986) developed **Social Identity Theory.** This theory suggests that a person's self-concept comprises personal identity (e.g., body char-

> **Social Identity Theory** a theory that proposes a person's identity is shaped by both personal and social characteristics
>
> **in-groups** groups in which a person feels they belong
>
> **out-groups** groups in which a person feels they don't belong

acteristics, psychological behaviors) as well as a social identity (e.g., affiliation with a group). Researchers and theorists in Social Identity suggest that people are "motivated to join the most attractive groups and/ or give an advantage to the groups to which one belongs (in-group)" (Worchel, Rothgerber, Day, Hart, & Butemeyer, 1998, p. 390). When people are given an opportunity, Worchel and colleagues contend, they will provide more resources to their own groups, rather than to out-groups. And when **in-groups** (groups to which a person feels they belong) are identified, an individual decides the extent to which the group is central to their identity. Social identity, then, is primarily based on the comparisons that people make between in-groups and **out-groups** (groups to which a person feels they don't belong).

Theory at a Glance • Communication Accommodation Theory

This theory considers the underlying motivations and consequences of what happens when two speakers, usually with different cultural backgrounds, languages, or skill sets different cultural backgrounds shift their communication styles—both vocally and nonlinguistically. With Communication Accommodation Theory, during communication encounters, people will try to accommodate or adjust their style of speaking to others. This is primarily done in two ways: divergence and convergence. Groups with strong cultural pride often use divergence to highlight group identity. Convergence occurs when there is a strong need for social approval, frequently from powerless individuals. The words we say and how we say them all have social consequences.

People strive to acquire or maintain positive social identity (Hughes, Kiecolt, Keith, & Demo, 2015; Tajfel & Turner, 1986), and when social identity is perceived as unsatisfactory, they will either join a group they feel more at home in or make the existing group a more positive experience. Summing up the Social Identity perspective, Tajfel and Turner observe that in-group/out-group comparisons lead social groups to differentiate themselves from one another. They note that communication exists on a continuum from "interindividual" to "intergroup." So, according to the theory, the more that Luke perceives Roberto to be either a "Mexican" or a "boss" rather than simply "Mr. Hernandez," the more Luke will rely on stereotypes or group-level impressions to understand Roberto's behavior.

Giles drew from some of the thinking related to Social Identity Theory. He felt that individuals accommodate not only to specific others but also to those they perceive as members of other groups. Thus, intergroup variables and goals influence the communication process. Giles, like many social psychologists, believes that people are influenced by a number of behaviors. Specifically, he argues that an individual's speech style (accent, pitch, rate, interruption patterns) can affect the impressions that others have of the individual. Giles and Smith (1979) also comment that the nature of the setting, the conversation topic, and the type of person with whom one communicates will all intersect to determine the speech manner one adopts in a given situation. In its simplest form, if an individual is viewed favorably, communicator A will shift their speech style to become more like that of communicator B; that is, communicator A has accommodated communicator B. Giles and Smith maintain that people adjust their speech style to accommodate how they believe others in the conversation will best receive it. In our opening, you can see how Luke shifts his speech to accommodate what he believes (incorrectly) Mr. Hernandez will appreciate and which Luke believes will increase his job prospects with the company.

Giles was influenced by the belief that when members of different groups come together, they compare themselves. If their comparisons are favorable, a positive social identity will result. Jake Harwood (2006) sums up the importance of social identity and group membership: "Even our closest interpersonal relationships are imbued with group identifications that both join us to those within our groups and separate us from those not in our groups" (p. 89).

With this theoretical footing in place, we now turn our attention to the assumptions guiding the development of Communication Accommodation Theory. As we discuss these, you will be able to sense the influence of social psychology, and Social Identity Theory in particular.

Assumptions of Communication Accommodation Theory

Accommodation Theory proponents would be interested in the accommodation taking place between Luke Merrill and Roberto Hernandez. Their conversation exemplifies a number of issues that underlie the basic assumptions of the theory. Recalling that accommodation is influenced by a number of personal, situational, and cultural circumstances, we identify several assumptions below:

- Speech and behavioral similarities and dissimilarities exist in all conversations.
- The manner in which we perceive the speech and behaviors of another will determine how we evaluate a conversation.
- Language and behaviors impart information about social status and group belonging.
- Accommodation varies in its degree of appropriateness, and norms guide the accommodation process.

Many concepts of CAT rest on the first assumption, that there are similarities and dissimilarities between communicators in a conversation. Past experiences, you may recall, form a person's field of experience, a concept we discussed in **Chapter 1.** Whether in speech or behaviors, people bring their various fields of experiences into a conversation (West & Turner, 2023). These varied experiences and backgrounds will determine the extent to which one person will accommodate another. The more similar our attitudes and beliefs are to those of others, the more we will be attracted to and accommodate those others.

Let's look at a few examples to illustrate this assumption. Consider our opening scenario with Luke and Roberto. They are clearly from different professional backgrounds with different levels of work experience. Presumably, they are products of different family backgrounds with different beliefs and values. The two are clearly dissimilar in some ways, yet, they are similar in others—for example, they both like to fly and they both have an interest in working at the firm.

To illustrate this assumption further, consider the following dialogue between a grandparent and a teenage granddaughter as the two talk about the girl's prom dress:

GRANDMOTHER: I don't know why you're wearing all black. You'll have people talking. You look like you're at your own funeral!

GRANDDAUGHTER: Whatever. People talked about your generation wearing weird stuff, too. It's my style now.

GRANDMOTHER: But, we only wore black during very solemn times.

GRANDDAUGHTER: Oh yeah, Nana. Sure you did. Your generation also never had any fun at all! When you look at all the old photographs, no one seems to be smiling!

GRANDMOTHER: You would be surprised how much fun we had. It was different from what you think now. We did a lot when I was younger, but you just wouldn't understand.

In this conversation, the granddaughter draws conclusions about the grandparent using group-based expectations: older people can't understand teenagers. This expectation influences the teenager's communication with her grandmother.

The second assumption rests on both perception and evaluation. Communication Accommodation is a theory concerned with how people both perceive and evaluate what takes place in a conversation (Popan, 2023). **Perception** is the process of attending to and interpreting a message, whereas **evaluation** is the process of judging a conversation. People first perceive what takes place in a conversation

perception process of attending to and interpreting a message

evaluation process of judging a conversation

(e.g., the other person's speaking abilities) before they decide how to behave in a conversation. Consider, for example, Luke's response to Roberto. The interviewer's informality at the beginning of the interview is perceived by Luke as a good way to break the ice and eliminate some of his tension. Luke's subsequent behavior reflects a relaxed style (too relaxed, we might conclude). Luke is like most people in an interview: he gets a sense of the interview atmosphere (perception) and then reacts accordingly (evaluation).

Motivation is a key part of the perception and evaluation process in Communication Accommodation Theory. That is, we may perceive another person's speech and behaviors, but we may not always choose to evaluate them. This often happens, for instance, when we greet another person, engage in small talk, and simply walk on. We normally don't take the time to evaluate such a conversational encounter.

Yet, there are times when perceiving the words and behaviors of another leads to our evaluation of the other person. We may greet someone, for instance, and engage in small talk, but then be surprised when we hear that the other person recently got divorced. According to Giles and colleagues (1987), it is then that we decide our evaluative and communicative responses. We may express our happiness, our sorrow, or our support. We do this by engaging in an accommodating communication style.

The third assumption of CAT pertains to the effects that language has on others. Specifically, language has the ability to communicate status and group belonging between communicators in a conversation. Consider what occurs when two people who speak different languages try to communicate with each other. Giles and John Wiemann (1987) discuss this situation:

> In bilingual, or even bidialectal, situations, where ethnic majority and minority peoples coexist, second language learning is dramatically unidirectional: that is, it is very uncommon for the dominant group to acquire the linguistic habits of the subordinate collectivity.... Indeed, it is no accident that cross-culturally what is "standard," "correct," and "cultivated" language behavior is that of the aristocracy, the upper or ruling classes and their institutions (p. 361).

The language used in a conversation, then, will likely reflect the individual with the higher social status. In addition, inferred in this quotation is a desire to be part of the "dominant" group. And, dominant groups, especially when dealing with language use and policies, determine how things are going to be said and whether or not any "minority" views will be entertained (Bourhis, Sioufi, & Sachdev, 2012).

To understand this assumption better, let's return to our opening story. In the interview situation with Mr. Hernandez, Luke's language and behaviors are guided by the interviewer. This is what normally occurs in interviews: the individual with the higher social status sets the tone through their language and behaviors. Although Roberto is a member of a co-culture that has been historically oppressed, he, nonetheless, has the power to establish the interview's direction. Those wishing to identify with or to become part of another's group—for example, Luke, wishing to be offered a job by Mr. Hernandez—will typically accommodate.

The fourth and final assumption focuses on norms and issues of social appropriateness. We note that accommodation can vary in social appropriateness and that accommodation is rooted in norm

norms expectations of behavior in conversations

usage. Norms have been shown to play a pivotal role in Giles's theory. **Norms** are expectations of behaviors that individuals feel should or should not occur in a conversation. The varied backgrounds of communicators like Luke and Roberto, for instance, or those of a grandparent and grandchild, will influence what they expect in their conversations. The relationship between norms and accommodation is made clear by Cynthia Gallois and Victor Callan (1991): "Norms put constraints of varying degree ... on the accommodative moves that are perceived as desirable in an interaction" (p. 253). Therefore, the general norm that a younger person is deferential to an older person or that an interviewee is deferential to a hiring supervisor suggests that Luke will be more accommodative in his communication to Mr. Hernandez.

It's important to understand that accommodation may not always be worthwhile and beneficial. For instance, Melanie Booth-Butterfield and Felicia Jordan (1989) found that people from marginalized cultures are usually expected to adapt (accommodate) to others and at times, these cultures lose their identity (Bourhis et al., 2012).

These four assumptions form the foundation for the remainder of our discussion of the theory. We now examine the ways that people adapt in conversations.

Ways to Adapt

CAT suggests that in conversations people have options. It's important to acknowledge that accommodation is an optional process in which two communicators decide to accommodate, one does, or neither does. That is, we may either accommodate strategically (conscious) or we may be doing it instinctively (unconscious). Further, conversants may be accommodating, but it does not mean that they necessarily agree with things. We may disagree with someone in a conversation but for the sake of argument (e.g., to make a point), we may accommodate the other.

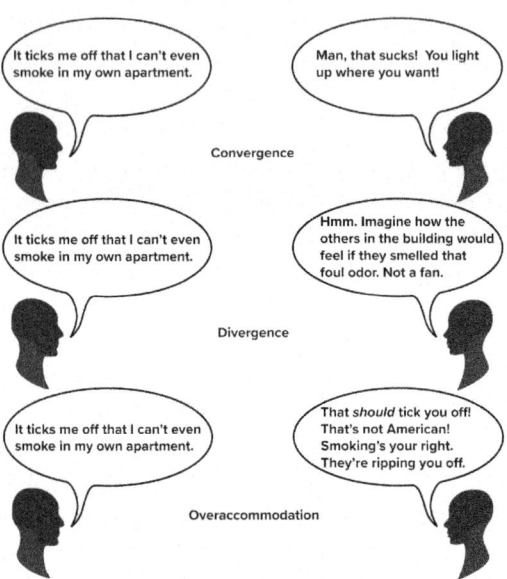

Figure 24.1 Ways to Adapt or Accommodate in Conversation
Oleksii Arseniuk/Shutterstock

Individuals may foster a conversational community that includes using the same or similar language or nonverbal system, they may distinguish themselves from others, or they may try too hard to adapt. These choices are labeled (a) convergence, (b) divergence, and (c) overaccommodation (Gallois & Giles, 2015). We examine each next and identify them in **Figure 24.1.**

Convergence: Merging Thoughts Ahead

The first process associated with CAT is termed **convergence.** Giles, Nikolas Coupland, and Justine Coupland (1991) define convergence as "a strategy whereby individuals adapt to each other's communicative behaviors" (p. 7). People may adapt to speech rate, pause, smiling, eye gaze, and other verbal and nonverbal behaviors. Convergence is a selective process; as we noted earlier, we may perceive but choose not to act with people. When people do converge, they rely on their perceptions of the other person's speech, behaviors, field of experience, or cultural background.

> **convergence** strategy used to adapt to another's behavior

In addition to the perception of the other's communication, convergence is also based on attraction (Giles, 2016a). Usually, when communicators are attracted to others, they will converge in their conversations. *Attraction* is a broad term that encompasses a number of other characteristics, such as liking, charisma, and credibility. Giles and Smith (1979) believe that a number of factors affect our attraction for others—for example, the possibility of future interactions with the listener, the speaker's ability to communicate, and the status differential between communicators. Having similar beliefs, having a similar personality, or behaving in similar ways causes people to be attracted to each other and is likely to prompt convergence. Remember, however, that uncovering similarities occurs over time. People may not instantly know whether they are attracted to each other and whether this will lead to identifying possible similarities. And the relational history between communicators may be a critical issue in convergence. For instance, Richard Street's (1991) research indicates that physicians differ in their convergence patterns with first-time patients and with repeat patients. Some doctors, he found, will vary in how much they tell their patients during medical consultations, and these decisions are influenced by patient anxiety, educational level, and question-asking skills, among others. Street cautions, however, that differences in convergence may be explained by looking at the traditional roles of doctor and patient as well as the time lag between visits.

At first view, convergence may be seen as a favorable way to accommodate another person, and it usually is. Our discussion so far implies that the other person is viewed as similar (in reputation, status, etc.) to the individual or that at least one is attracted to the other in a conversation. We already learned that attraction is a necessary precursor to convergence. Yet, as we explained previously, convergence may be based on stereotypical perceptions. As Giles and colleagues (1987) conclude, "Convergence is often cognitively mediated by our stereotypes of how socially categorized others will speak" (p. 18). What this means is that people may converge toward stereotypes rather than toward real speech and behaviors.

There are obvious implications of stereotypical convergence. For example, many gay fathers and lesbian mothers report that too many people—including educators—rely on outdated stereotypes of gay men and lesbians when they communicate with them (Morrison, Gallardo, & Fuster, 2023). Examining the cultural experiences of African Americans, Mark Orbe (1998) notes that African Americans are often identified in stereotypical ways. He points out that **indirect stereotyping** exists; that is, stereotyping when European Americans talk to their African American friends about what they believe to be African American "subjects" (sports, music, etc.). Some African Americans report that if they speak nonstandard dialect, they are especially prone to stereotypical reactions. Marsha Houston (2016/1992) agrees. She notes that when describing themselves, white women in particular identify their speech as appropriate and standard and describe African American female speech as nonstandard, incorrect, or deviant.

> **indirect stereotyping** imposing outdated and rigid assumptions of a cultural group upon that group

Other cultural groups have also been the target of stereotyping. Nigel Morris (2022), for example, contends that blind people are often viewed as the freak, the outsider, or the marginalized by others via mediated representations. Chin-Hui Chen (2019) observes that some teachers who oversee senior educational opportunities embrace communication patterns that suggest seniors need "baby talk." Brenda Allen (2016) believes that many biracial lesbian relationships rely upon outdated and narrow views of relational intimacy. Shobha Pais (1997) maintains that Asian Indian women in the United States are often perceived as strange because of their sari (fabric draped over the shoulder and head) or *salwar kameez* (pantsuit). And Charmaine Shutiva (2016) bemoans the fact that the Native American culture is erroneously viewed as stoic and unemotional, when in reality, it involves a great deal of humor and joy. These examples demonstrate that a number of cultural groups continue to be stereotyped, and these stereotyped perceptions often influence the extent to which an individual will converge.

Before we leave our discussion of convergence, let's briefly discuss how to assess it. In other words, what occurs when people attempt to converge in their conversations? How do people respond? We have already illustrated a number of examples that demonstrate both positive and negative consequences of convergence. But how do we know how our convergence will be met?

Students Talking Tech: Delsi

I may be wrong here but i think when I look at my Twitter posts and think about CAT, there's a lot of convergence going on. I posted pictures of a #BLM rally with a "thumbs up" emoji and had a lot of likes. I also had posted something about all the gun violence and wrote "#stoptheNRA." I had a lot more people agree than disagree. I know it's not a perfect example, but Twitter posts can show convergence with others.

First, we need to consider that an evaluation of convergence usually depends on whether the convergence is done thoughtfully. When convergence is perceived as good, it can enhance the dialogue; when it is perceived as bad, it can cause noise in the communication channel (see **Chapter 1** for our earlier discussion of this). If a communicator speaks or behaves in a style similar to a listener's, the convergence will probably be favorably perceived. But converging to ridicule, tease, or patronize will usually be perceived negatively. There is a fine line between whether or not convergence will be viewed in the intended way. Consider, for instance, what happens when a nurse speaks to a patient in a nursing home about eating lunch (Ryan, Maclean, & Orange, 1994):

NURSE: It's time to eat Mrs. Tonelli.

RESIDENT: Not this morning. I'm still not a fan of the oatmeal here.

NURSE: Jeannie, the CNAs have a lot of work to do, so we need you to come down to the dining area and eat something. I'm sure you know that we can't just have a special breakfast for you each morning.

RESIDENT: I'd really rather just rest in my room.

NURSE: Now, honey, if you don't eat now, you won't be able to eat until lunch. Let's stop being so fussy, all right?

RESIDENT: I'll only eat some fruit, and that's it.

In this conversation, you can see how convergence by the nurse might be construed as condescending by the patient. In fact, it's clear that this type of communication is less respectful, less nurturing, and more frustrating for the resident. There are other standards of evaluating convergence, including the norms of the situation (Did the speaker converge in an offensive way?), the ability to pull it off effectively (How does a 50-year-old use the "cool" talk of a high school student?), and the value of a language to a community (Should European Americans employ "Black English" vernacular in their interactions with African Americans?).

Divergence: Vive la différence

Giles (1980) believes speakers sometimes accentuate the verbal and nonverbal differences between themselves and others. He terms this *divergence* or *nonaccommodation*. **Divergence** differs greatly from convergence in that it is a dissociation process (Popan, 2023). Instead of showing how two speakers are alike in speech rate, gestures, or

divergence strategy used to accentuate the verbal and nonverbal differences between communicators

posture, divergence is when there are no attempts to demonstrate similarities between speakers. That is, two people speak to each other with no concern about accommodating each other. No effort is made to reduce interpersonal distance. Divergence has not received as much research attention as convergence, and so our knowledge about the process is limited to a few claims about its function in Communication Accommodation Theory.

First, divergence should not be misconstrued as an effort to disagree or not respond to another communicator. Divergence is not the same as inattentiveness. When people diverge, they have simply chosen to dissociate themselves from the communicator and the conversation. The reasons for divergence vary, including asserting "one's own identity, making a statement or fulfill[ing] personal preferences" (Yoneoka, 2011, p. 93). And yet, according to Giles (2016a), not all divergences are negatively perceived.

Divergence is a way for members of various cultural communities to maintain social identity. Giles and his colleagues (1987) observe that there are occasions when people—namely, racial and ethnic groups—"deliberately use their language or speech style as a symbolic tactic for maintaining their identity, cultural pride, and distinctiveness" (p. 28). Individuals may not wish to converge in order to preserve their cultural heritage. Let's give a classic example that many of you may understand: imagine that you are traveling in France; everywhere you go the French people you encounter encourage you to speak French. You are surprised at that until you realize that you, as a visitor, should not expect the French to converge to your language.

We've already learned that some cultural groups are immediately stereotyped and that people communicate with this categorization in mind. It's no wonder, then, that some cultural groups remain committed to divergence in their conversations with others. To illustrate this point, consider the conclusions of Richard Bourhis and Giles (1977). In this classic study, the research team studied Welsh people who were very proud of their ethnic identity, but who did not know the Welsh language (Giles was a Welsh social psychologist). As they learned the language, the researchers asked several questions in a standard English format. During the question-and-answer period, the researchers asked the group why they wanted to learn Welsh since it is "a dying language with a dismal future." The Welsh sample rebutted with not only a strong Welsh dialect, but also Welsh words and phrases. Remarkably, the group could link together difficult Welsh words! The group, then, began to diverge from the English that was spoken to them, ostensibly out of ethnic pride.

A second reason people diverge pertains to power and role differences in conversations. In this case, think of the cultural issue of socio-economic class. Divergence frequently occurs in conversations when there is a power difference between the communicators and when there are distinct role differences in the conversation (e.g., physician–patient; surgeons–athletes) (Street, 1991; Street & Giles, 1982; Womble, 2022). Street (1991), for instance, remarked that "interactants having greater status may speak for longer periods, initiate most of the conversational topics, speak more slowly, and maintain a more relaxed body posture than does the less powerful" (p. 135).

Finally, although not as often as for the reasons cited previously, divergence is likely to occur because the other in the conversation is viewed to be a "member of undesirable groups, considered to hold noxious attitudes, or display a deplorable appearance" (Street & Giles, 1982, p. 195). To this end, Giles and his colleagues (1987) contend that divergence is used to contrast self-images in a conversation. To understand this point, consider the number of so-called undesirable groups in society today. David Pilla (2023), for instance, notes that homeless people are historically part of what society labels as unappealing or hapless. Employing the accommodation principle of divergence, then, a homeless man asking for money outside a movie theater may find himself in a conversation with a communicator (we'll call this person Pat) who wishes to diverge to demonstrate differences between the two. Pat's divergence may take the form of an increased rate of speech or a more abrupt manner. Pat may also use vocabulary and pronunciation that clearly mark the communicator as a member of an employed class and may even shout out that the homeless man should get a job. In this case, the divergence is carried out by the individual who wishes to imply a status difference between the two. A new and contrasting identity exists that illustrates the (in)tangible space between the two (Nguyen & Hamid, 2019).

 During election season, regardless of whether it's at the local or national level, many politicians do what they can to "connect" with the voters. Some, it may seem, will use convergent language that is aligned with the geographical region in which they find themselves—for example, a presidential candidate speaking with a "Southern dialect" while in Mississippi. Others find themselves using words or phrases that are "red flag" words that align with local values, such as "no new taxes" in a particularly high-tax area. Yet, other politicians will diverge, such as a candidate who is gay embracing their husband in public appearances in a politically conservative location. Throughout elections, nearly all politically savvy people will find ways to look and talk like their voters and in other ways provide divergent paths to demonstrate uniqueness.

Overaccommodation: Miscommunicating with a Purpose

Jane Zuengler (1991) observes that **overaccommodation** is a term attributed to the behavior of people who, although acting from good intentions, are perceived, instead, as patronizing or demeaning. Over-accommodation has the effect of making the target feel worse and

overaccommodation attempt to overdo efforts in regulating, modifying, or responding to others

is also viewed as a nonaccommodation technique (Speer, Giles, & Denes, 2013). Consider, for example, the overaccommodation taking place in Luke Merrill's interview. Luke's efforts in speaking Spanish are undermined by Roberto's perception that Luke is patronizing him. In this case, and as some researchers (Coupland, Coupland, Giles, & Henwood, 1988) believe, too much overaccommodation yields poor communication. Luke is not trying to patronize Roberto Hernandez. Although the speaker apparently has the intention of showing respect, the listener perceives it as distracting and disrespectful.

Overaccommodation can be done instinctively (as we noted earlier), but much of it is strategic. Further, it can exist in three forms: sensory overaccommodation, dependency overaccommodation, and intergroup overaccommodation (Zuengler, 1991). Let's define these and present an example of each.

Sensory overaccommodation occurs when a speaker overly adapts to another who is perceived as limited in some way. There is an assumption that a person has a physical or sensory limitation/handicap, such as a hearing impairment. Limitation, in this sense, refers to either a linguistic or a physical limitation. That is, a speaker may

sensory overaccommodation overly adapting to others who are perceived as limited in their abilities (physical, linguistic, or other)

believe that they are sensitive to another's language disability or physical disability, but overdoes the accommodation. For example, in a study on seniors conducted in Singapore, researchers found that the communication practices of those who communicate with the elderly include an exaggerated high pitch, childlike prose, and infantilized rhetoric (Cavallaro, Seilhamer, Chee, & Ng, 2016). Framing questions or comments by being other-centered should help in co-creating meaning in conversations with those who are older citizens. Otherwise, patients may be seen as more incompetent than they really are.

The second type of overaccommodation is **dependency overaccommodation,** which occurs when a speaker places the listener in a lower status role, and the listener is made to appear dependent on the speaker. In dependency overaccommodation, the listener also

dependency overaccommodation a behavior that occurs when speakers place listeners in a lower status role

believes that the speaker controls the conversation to demonstrate higher status. This can be seen by examining the treatment of a number of immigrant populations in the United States.

Many cultural groups are marginalized in the United States, and dependency overaccommodation, it appears, may be one reason for this ostracizing. For instance, during assimilation into their new communities, many refugees are made to feel subordinate when conversing with others. Although government workers may believe that during their conversations with refugees they are doing what is right (helping refugees understand various procedures and rules associated with documentation), refugees may feel quite dependent on the speaker (immigration official). Given that many newly arrived strangers do not know the English language, do not have a basic understanding of cultural values or norms, and do not have a clear sense of their job skills, their perceptions of dependency are warranted.

In addition to sensory and dependency overaccommodation, there is a third type of overaccommodation called **intergroup overaccommodation.** This involves speakers lumping listeners into a particular group, failing to treat each person as an individual. At the heart of this overaccommodation is stereotyping, and as we learned earlier, there can be far-reaching consequences to this perceptual bias. Although maintaining racial and ethnic identity is critical, individual identity is equally important.

intergroup overaccommodation a behavior that occurs when speakers place listeners in cultural groups without acknowledging individual uniqueness

Consider when a speaker uses language that narrowly assigns a listener to a particular cultural group. The speaker may feel comfortable suggesting, for instance, that Vietnamese citizens have never been given a chance to succeed in the United States because they have been busy raising their families. To a Vietnamese individual, this generalization may be perceived negatively. Communicating with this perception in mind may cause some Vietnamese living in the U.S. to accommodate negatively.

Students Talking: Haseeb

The idea of overaccommodating senior citizens is everywhere. One of my unpaid internships was with a senior center. It didn't matter whether it was a doctor, a nurse, a custodian, or even a guy who delivered flowers—they all overaccommodated the residents. I heard people literally scream when they talked to a resident, use baby phrases as if they didn't have a brain, and even ask "Do you understand?" nearly every time. Yes, there are some who didn't get it and needed those communication styles. But, so many—especially in the rehab wing—didn't need to be talked down to—overaccommodated—simply because they had reached a certain age.

Overaccommodation usually results in listeners perceiving that they are less than equal. There are serious implications to overaccommodation, including losing motivation for further language acquisition, avoiding conversations, and forming negative attitudes toward speakers and society (Zuengler, 1991). Since one primary goal of communication is achieving intended meaning, overaccommodation is a significant roadblock to that goal.

Integration and Critique

Communication Accommodation Theory focuses on the role of conversations in our lives and how people's communication influences those dialogues. The cultural backgrounds and expectations of the communicators remain important sources as we try to unpack the theory. Further, over several decades of investigation, the research has consistently followed a quantitative approach. We now turn our attention to three criteria as we evaluate this communication theory: scope, logical consistency, and heurism.

Integration

Communication Tradition	Rhetorical | Semiotic | Phenomenological | Cybernetic | **Socio-Psychological** | Socio-Cultural | Critical
Communication Context	Intrapersonal | Interpersonal | Small Group/Team | Organizational | Public/Rhetorical | Mass/Media | **Cultural**
Approach to Knowing	**Positivistic/Empirical** | Interpretive/Hermeneutic | Critical

Critique

Evaluation Criteria	Scope | **Logical Consistency** | Parsimony | Utility | Testability | **Heurism** | Test of Time

Scope

The boundaries of the theory are rather expansive. You may recall that the theory originally examined speech; it was later expanded to include the nonverbal arena. Giles (2008) underscores the broad scope of Communication Accommodation Theory by stating that over the years, it "began to take propositional forms that became increasingly more complex and, arguably, more demanding on readers" (p. 166). Further, the theory was first concerned with shifting accents and dialects and moved to a broader examination in shifts in communication style (Giles, Edwards, & Walther, 2023). Its scope, however, does not necessarily undermine the theory's integrity. The cross-disciplinary interest in the theory suggests that efforts to delimit it may not be the best course of action (Giles, 2016b). The changing nature of culture in Western society suggests that a theory of this nature may need to be extensive in order to understand multiple populations, particularly those who have recently migrated to the United States.

Logical Consistency

The strengths of the theory may be quite significant because CAT has elicited little scholarly criticism. Still, a few concerns pertaining to the logical consistency of the concepts have been identified. In short, some scholars contend that a few of the central features of the theory warrant further examination. Judee Burgoon, Leesa Dillman, and Lesa Stern (1993), for example, question the convergence–divergence frame advanced by Giles. They believe that conversations are too complex to be reduced simply to these processes. They also challenge the notion that people's accommodation can be explained by just these two practices. For instance, what occurs if people both converge and diverge in conversations? In what circumstances will this happen?

Are there consequences for the speaker? The listener? Giles and his colleagues (Dragojevich, Gasiorek, & Giles, 2016) have responded to some of these issues. They assert that there are "complex accommodative dilemmas" (p. 5) which are often unpredictable, thus accepting the complexity of conversations that often take place between people.

One might also question whether the theory relies too heavily on a rational way of communicating. That is, although the theory acknowledges conflict between communicators, it also rests on a reasonable standard of conflict. Perhaps you have been in conflicts that are downright nasty and with people who have no sense of reason. It appears that the theory ignores this possible negative side of communication (Gilchrist-Petty & Long, 2016). Further, we know from conflict research, too, that conflicts are often handled and managed differently across cultures, prompting some questions into the notion of what to do when convergence and divergence "collide." Still, Giles (Gallois, Ogay, & Giles, 2005) does not ignore this perception and believes a great deal of work is still to be done before the process of accommodation can be fully understood.

Heurism

Without doubt, Giles and his colleagues have conceptualized a theory rich in heuristic value. The theory has been incorporated in a number of different studies. For instance, accommodation has been studied with populations that, until this theory, were rarely researched, including blind massage therapists (Sumalinog, Sambrana, Diaz, & Bebero, 2023), patient–providers (Farzadnia & Giles, 2014), native and nonnative speakers (Rogerson-Revell, 2010), immigrant women (Marlow & Giles, 2013), gay men (Hajek, 2016), and cancer survivors (Wanzer, Simon, & Cliff, 2022). In addition, it has been studied with aphasia (Simmons-Mackie, 2018), in conjunction with texting (Adams, Miles, Dunbar, & Giles, 2018), in business meetings (Rogerson-Revell, 2010), in families (Guntzviller, 2015), and even with messages left on telephone answering machines (Buzzanell, Burrell, Stafford, & Berkowitz, 1996). There is no doubt that the theory is heuristic and has lasting scholarly value in the communication field.

Closing

In his earlier writings on the theory, Giles challenged researchers to apply Communication Accommodation Theory across the life span and in different cultural settings. His suggestions have been heeded and even expanded to a comprehensive examination of both the theory and his life. His research has broadened our understanding of why conversations are so complex, especially with individuals from different cultural backgrounds. Through convergence, Giles sheds light on why people adapt to others in their interactions. Through divergence, we can understand why people choose to ignore adapting strategies. He has championed a theory that has helped us better understand the culture around us and the influences of accommodation on the diverse relationships within society.

Discussion Starters

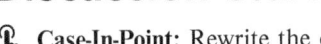 **Case-In-Point:** Rewrite the opening example of Luke Merrill and Roberto Hernandez. This time, place the two individuals in the context of teacher and student.

Try-It-Your-Selfie: If you were asked to write an email to the CEOs of X, LinkedIn, and Facebook, how would you go about asking them to allow divergent points of view on such issues as marriage equality, immigration, and climate change?

1. Explain how convergence might function in the following relationships: police officer–driver, therapist–client, and supervisor–subordinate. In your response, choose a conversational topic that might be unique to the relationship and integrate convergence accordingly.

2. Giles contends that self-perception can influence the accommodation process. How does a person's self-perception affect how one chooses whether or not to accommodate? Provide examples to illustrate your thoughts.

3. CAT has been framed in the intercultural communication context. Based on your understanding of the theory, under what additional context(s) can the theory be understood?

4. Discuss a situation in which accommodation might backfire on an individual.

References

Abrams, J., O'Connor, J., & Giles, H. (2003). Identity and intergroup communication. In W. B. Gudykunst (Ed.), *Cross-cultural and intercultural communication* (pp. 209–224). Sage.

Adams, A., Miles, J., Dunbar, N. E., & Giles, H. (2018). Communication accommodation in text messages: Exploring liking, power, and sex as predictors of textisms. *The Journal of Social Psychology, 158*(4), 474–490.

Allen, B. J. (2016). Sapphire and Sappho: Allies in authenticity. In A. Gonzalez & Y.-W. Chen (Eds.), *Our voices: Essays in culture, ethnicity, and communication* (6th ed.). Oxford University Press.

Bernhold, Q. S., & Giles, H. (2022). Emotional mimicry: A communication accommodation approach. *Cognition and Emotion, 36*(5), 799–804.

Booth-Butterfield, M., & Jordan, F. (1989). Communication adaptation among racially homogeneous and heterogeneous groups. *Southern Communication Journal, 54,* 253–272.

Bourhis, R. Y., & Giles, H. (1977). The language of intergroup distinctiveness. In H. Giles (Ed.), *Language, ethnicity and intergroup relations* (pp. 119–135). Academic Press.

Bourhis, R. Y., Sioufi, R., & Sachdev, I. (2012). Ethnolinguistic interaction and multilingual communication. In H. Giles (Ed.), *The handbook of intergroup communication* (pp. 100–115). Routledge.

Burgoon, J. K., Dillman, L., & Stern, L. A. (1993). Adaption in dyadic interaction: Defining and operationalizing patterns of reciprocity and compensation. *Communication Theory, 3,* 295–316.

Buzzanell, P. M., Burrell, N. A., Stafford, R. S., & Berkowitz, S. (1996). When I call you up and you're not there: Application of Communication Accommodation Theory to telephone answering machine messages. *Western Journal of Communication, 60,* 310–336.

Cavallaro, F., Seilhamer, M. F., Chee, Y. T. F., & Ng, B. C. (2016). Overaccommodation in a Singapore eldercare facility. *Journal of Multilingual and Multicultural Development, 37*(8), 817–831.

Chen, C. H. (2019). Exploring teacher–student communication in senior-education contexts in Taiwan: A communication accommodation approach. *International Journal of Ageing and Later Life, 12,* 1–47.

Coupland, N., Coupland, J., Giles, H., & Henwood, K. (1988). Accommodating the elderly: Invoking and extending a theory. *Language in Society, 17,* 1–41.

Dragojevich, M., Gasiorek, J., & Giles, H. (2016). Accommodative strategies as core of the theory. In H. Giles (Ed.), *Communication accommodation theory: Negotiating personal relationships and personal identities across contexts* (pp. 36–59). Cambridge University Press.

Farzadnia, S., & Giles, H. (2014). Patient-provider interaction: A communication accommodation theory perspective. *International Journal of Society, Culture & Language, 3,* 17–34.

Gallois, C., & Callan, V. J. (1991). Interethnic accommodation: The role of norms. In H. Giles, J. Coupland, & N. Coupland (Eds.), *Contexts of accommodation: Developments in applied sociolinguistics* (pp. 245-269). Cambridge University Press.

Gallois, C., & Giles, H. (2015). Communication Accommodation Theory. *The International Encyclopedia of Language and Social Interaction*, 1-18.

Gallois, C., Ogay, T., & Giles, H. (2005). Communication Accommodation Theory. In W. Gudykunst (Ed.), *Theorizing about intercultural communication* (pp. 121-148). Sage.

Gilchrist-Petty, E., & Long, S. (Eds.) (2016). *Contexts of the dark side of communication*. Peter Lang.

Giles, H. (1973). Accent mobility: A model and some data. *Anthropological Linguistics, 15,* 87-105.

Giles, H. (1980). Accommodation Theory: Some new directions. *York Papers in Linguistics, 9,* 30.

Giles, H. (April 2008). Accommodating translational research. *Journal of Applied Communication Research 36*(2), 121-127.

Giles, H. (2016a). *Communication Accommodation Theory: Negotiating personal relationships and social identities across contexts*. Cambridge University Press.

Giles, H. (2016b). The social origins of CAT. In H. Giles (Ed.), *Communication Accommodation Theory: Negotiating personal relationships and personal identities across contexts* (pp. 36-59). Cambridge University Press.

Giles, H., Coupland, N., & Coupland, J. (1991). Accommodation Theory: Communication, context, and consequence. In H. Giles, J. Coupland, & N. Coupland (Eds.), *Contexts of accommodation: Developments in applied sociolinguistics* (pp. 1-68). Cambridge University Press.

Giles, H., Edwards, A. L., & Walther, J. B. (2023). Communication Accommodation Theory: Past accomplishments, current trends, and future prospects. *Language Sciences, 99.*

Giles, H., Mulac, A., Bradac, J. J., & Johnson, P. (1987). Speech Accommodation Theory: The first decade and beyond. In M. L. McLaughlin (Ed.), *Communication yearbook 10* (pp. 13-48). Sage.

Giles, H., & Smith, P. M. (1979). Accommodation Theory: Optimal levels of convergence. In H. Giles & R. N. St. Clair (Eds.), *Language and social psychology* (pp. 231-244). Blackwell.

Giles, H., & Wiemann, J. (1987). Language, social comparison, and power. In S. Chaffee & C. R. Berger (Eds.), *Handbook of communication science* (pp. 350-384). Sage.

Guntzviller, L. M. (2015). Testing multiple goals theory with low-income, mother-child Spanish-speakers: Language brokering interaction goals and relational satisfaction. *Communication Research, 42,* 4-18.

Hajek, C. (2016). Social and psychological creativity in gay male midlife identity management. *Social Psychology, 55,* 227-243.

Harwood, J. (2006). Communication as social identity. In G. Shepherd, J. St. John, & T. Striphas (Eds.), *Communication as ... : Perspectives on theory* (pp. 84-90). Sage.

Harwood, J., Gasiorek, J., Pierson, H., Nussbaum, J., & Gallois, C. (Eds.). (2019). *Language, communication, and intergroup relations: A celebration of the scholarship of Howard Giles*. Routledge.

Hogg, M., & Giles, H. (2012). Norm talk and identity in intergroup communication. In H. Giles (Ed.), *The handbook of intergroup communication* (pp. 373-388). Routledge.

Houston, M. (2016/1992). The politics of difference: Race, class, and women's communication. In L. F. Rakow (Ed.), *Women making meaning: New feminist direction in communication* (pp. 45-59). Routledge.

Hughes, M., Kiecolt, K. J., Keith, V. M., & Demo, D. H. (2015). Racial identity and well-being among African Americans. *Social Psychology Quarterly, 78,* 25-48.

Jackson, J. (2023). *Introducing language and intercultural communication*. Routledge.

Marlow, M. L., & Giles, H. (2013). "I don't know how to speak so I just stay silent": Uncertainty management among Chinese immigrant women seeking health care in the United States. In F. Sharifian & M. Jamarani (Eds.), *Language and intercultural communication in the new era* (pp. 245–262). Routledge.

Morris, N. (2022). Stereotypes, focalization, and modality: Mainstream cinematic images of disability. In L. Friedman & T. Jones (Eds.), *Routledge handbook of health and media* (pp. 143–179). Routledge.

Morrison, R., Gallardo, S., & Parra Fuster, F. (2023). Heteronormative representations of the family and parenting in public policies: Implications for LGBTIQ+ families. *Social Sciences, 12*(2), 66.

Nguyen, T. T. T., & Hamid, M. O. (2019). Language choice, identity and social distance: Ethnic minority students in Vietnam. *Applied Linguistics Review, 10*(2), 137–161.

Orbe, M. P. (1998). From the standpoint(s) to traditionally muted groups: Explicating a co-cultural communication theoretical model. *Communication Theory, 8*(1), 1–26.

Pais, S. (1997). Asian Indian families in America. In M. K. DeGenova (Ed.), *Families in cultural context* (pp. 173–190). Mayfield.

Pilla, D. (2023). *Stereotypes and public stigma against people experiencing homelessness* [Doctoral dissertation, Fordham University Press].

Popan, E. (2023). Communication Accommodation Theory. *Salem Press*, 1–2.

Rogerson-Revell, P. (2010). "Can you spell that for us nonnative speakers?" Accommodation strategies in international business meetings. *Journal of Business Communication, 47*, 432–454.

Ryan, E. B., MacLean, M., & Orange, J. B. (December 1994). Inappropriate accommodation in communication to elders: Inferences about nonverbal correlates. *International Journal of Aging and Human Development, 39*(4), 273–291.

Shutiva, C. (2016). Native American culture and communication through humor. In A. Gonzalez, & Y.-W. Chen (Eds.), *Our voices: Essays in culture, ethnicity, and communication* (6th ed.). Oxford University Press.

Simmons-Mackie, N. (2018). Communication partner training in aphasia: Reflections on Communication Accommodation Theory. *Aphasiology, 32*(10), 1215–1224.

Soliz, J., & Giles, H. (2016). Relational and identity processes in communication: A contextual and meta-analytical review of Communication Accommodation Theory. *Annals of the International Communication Association, 38,* 107–144.

Speer, R. B., Giles, H., & Denes, A. (2013). Investigating stepparent–stepchild interactions: The role of communication accommodation. *Journal of Family Communication, 13,* 218–231.

Street, R. L. (Spring 1991). Information-giving in medical consultations: The influence of patients' communicative styles and personal characteristics. *Social Science and Medicine, 32*(5), 541–548.

Street, R. L., & Giles, H. (1982). Speech Accommodation Theory: A social cognitive approach to language and speech behavior. In M. E. Roloff & C. R. Berger (Eds.), *Social cognition and communication* (pp. 193–226). Sage.

Su, D. (2019). Culture is essentially meaning. In J. Harwood, J. Gasiorek, H. Pierson, J. NussBaum, & C. Gallois (Eds.), *Language, communication, and intergroup relations: A celebration of the scholarship of Howard Giles* (pp. 34–51). Routledge.

Sumalinog, E., Sambrana, G., Diaz, W. D., & Bebero, L. K. (2023). Emerging communication gap to the lives of differently abled individuals working in a blind massage service. *American Journal of Multidisciplinary Research and Innovation, 2*(1), 1–8.

Tajfel, H., & Turner, J. C. (1986). The social identity theory of intergroup behavior. In S. Worchel & W. Austin (Eds.), *The psychology of intergroup relations* (pp. 7–24). Nelson Hall.

Wanzer, M. B., Simon, K. G., & Cliff, N. J. (2022). Interpreting cancer survivors' perceptions of the survivor label through social identity and communication accommodation theories. *Health Communication, 37*(13), 1600-1608.

West, R., & Turner, L. H. (2023). *Interpersonal communication.* Sage.

Womble, E. (2022). *Communication accommodation of surgeons with student-athletes.* [Master's Thesis, Abilene Christian University Press].

Worchel, S., Rothgerber, H., Day, E. A., Hart, D., & Butemeyer, J. (1998). Social identity and individual productivity within groups. *British Journal of Social Psychology, 37,* 389-413.

Yoneoka, J. (2011). The importance of language negotiation in initial intercultural encounters: The case of the service industry employee. *Intercultural Communication Studies, 20,* 90-103.

Zuengler, J. (1991). Accommodation in native-nonnative interactions: Going beyond the "what" to the "why" in second-language research. In H. Giles, J. Coupland, & N. Coupland (Eds.), *Contexts of accommodation: Developments in applied sociolinguistics* (pp. 223-244). Cambridge University Press.

CHAPTER 25
Relational Dialectics Theory

Based on the research of **Leslie Baxter** *and*
Barbara Montgomery

"RDT focuses on the struggles in meaning—the discursive tensions—that frequent interpersonal communication. However, the position of the theory is not that competing discourses are negative. Instead, they are the heart of the meaning-making enterprise."

—Leslie Baxter and Dawn Braithwaite

Eleanor Robertson and Jeff Meadows

Eleanor Robertson and Jeff Meadows worked together to clean up the mess left from the dinner party they'd just given for Jeff's friend Mary Beth's 35th birthday. They both agreed that the party had been a great success and that everyone had a wonderful time. As Eleanor and Jeff cleaned up, they were having fun talking about their friends —who was breaking up and who was getting together.

Eleanor smiled as she thought about how much she and Jeff had learned about each other and their relationship in the 2 years they'd lived together. Eleanor used to get upset when Jeff wanted to be with friends and not spend all his time with her alone. Now she thought she understood Jeff's desire to have others in their lives. She also found that the more she was able to let go of her posses- sive feelings and behaviors, the more Jeff wanted to be close.

Jeff came over and hugged Eleanor. He said, "Ellie-Bear, that was a great party. The food was perfect—I'm glad we decided to do Italian. Thanks for all your help to make everything go so well. Mary Beth really appreciated it, I know. And since she and I have been friends forever, it really means a lot to me."

Eleanor smiled at Jeff—she loved when he called her by his pet name for her. "I didn't do all that

much, sweetheart," she said. "You were the one who did all the cooking. But I am glad Mary Beth had fun. I really like her a lot, too." And, it's always nice to have good friends come over for dinner.

Jeff and Eleanor finished cleaning up and started talking about what they wanted to do tomorrow. They decided they would take a picnic to Mayer Markham Park and then maybe go to a movie— they agreed they'd decide on the movie after the picnic depending on how they felt. It sounded like a fun Sunday—they had a plan, but they could change their minds and skip the movie if the park turned out to be too nice to leave.

Eleanor was very happy, and she thought about telling Jeff how much she loved and needed him, but she decided to keep quiet about the depth of her feelings for now. Jeff probably knew how she felt, and she was a little afraid of revealing all her feelings to him at the moment. It had been a perfect day, and she didn't want to spoil it by saying something that might make Jeff withdraw a little. And, actually, she wasn't sure she wanted to be so vulnerable to Jeff at this point in their relationship. It was kind of funny—when they first started going out together, she had told Jeff she loved him all the time. But, after 3 years together, she found herself being a bit more strategic and self-protective even though she loved him more now than she had at first. She wondered what that meant and worried that maybe she wasn't being spontaneous enough in their relationship. Before she could think too much about it, though, Jeff called her over and told her a funny story about one of their dinner guests. Eleanor forgot about her worries as she and Jeff laughed about the story together.

When researchers examine our opening vignette about Eleanor and Jeff, they might speculate that their relationship is moving through stages. Researchers who work within the framework of Social Penetration Theory (see **Chapter 7**), for example, would point to the fact that Jeff and Eleanor have worked through some of their earlier issues and now interact with each other at a deeper level of intimacy than they once did. These researchers might point to the fact that the relationship between Eleanor and Jeff is more coordinated and less conflictual than it once was as an indication that they've moved to a more intimate stage of relational development. Yet, other researchers would look at Jeff and Eleanor and see their story as best explained by a different theoretical position, called Relational Dialectics Theory (RDT), which we discuss in this chapter.

RDT maintains that people in close relationships experience ongoing tensions between contradictory impulses (Baxter & Norwood, 2016). Although that may sound confusing and somewhat messy, researchers who take the dialectical position believe it accurately depicts the way that life is experienced by people. People aren't always able to resolve the contradictory elements of their beliefs, and they can hold inconsistent beliefs about relationships. For example, the adage "absence makes the heart grow fonder" seems to coexist easily with its opposite, "out of sight, out of mind." And, as Pravav Malhotra, Kristina Scharp, and Lindsay Thomas (2022) note, when communication researchers are interested in how meaning fluctuates in relationships, they often turn to RDT.

Leslie Baxter and Barbara Montgomery (1996) formulated the initial statement of RDT in their book *Relating: Dialogues and Dialectics,* although both of them had been writing about dialectical thinking for several years prior to that book's publication. Baxter and Montgomery's work was directly influenced by Mikhail Bakhtin, a Russian philosopher who developed a theory of personal dialogue called *dialogics*. Social life for Bakhtin was an open dialogue among many voices, and its essence was the "simultaneous differentiation from yet fusion with another" (Baxter, 2004, p. 1). According to Bakhtin, the individual self is only possible when it's in context with another. This is similar to Sandra Petronio's contention (see our discussion of Communication Privacy Management Theory in **Chapter 17**) that we can only understand the terms *privacy* and *openness* in relationship to one another. Neither term is meaningful in isolation. Bakhtin also believes that human experience is constituted through communication with others. Of course, this is an attractive notion for communication researchers.

From Bakhtin's thinking, Baxter and Montgomery (1996) shaped the dialectic approach. We can best explain this vision of human behavior by contrasting it to two other common approaches: the monologic and **monologic approach** an approach framing contradiction as either/or

dualistic. The **monologic approach** frames contradictions through either/or thinking. For instance, monologic thinking would advance that Jeff and Eleanor's relationship was *either* close *or* distant. In other words, the two parts of a contradiction are mutually exclusive in monologic thinking, and as you move toward one extreme, you retreat from the other. See **Figure 25.1** for a visual representation of this approach.

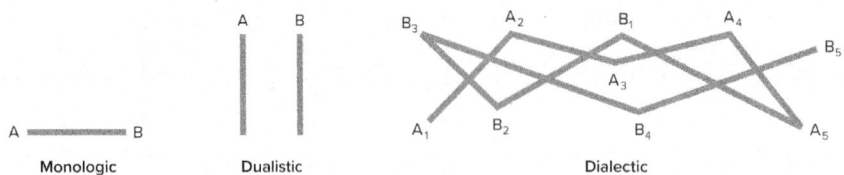

Figure 25.1   Monologic, Dualistic, and Dialectic Approaches to Relational Contradictions

In contrast, the **dualistic approach** pictures the two parts of a contradiction as separate entities, somewhat unrelated to each other. In our example of Eleanor and Jeff, dualistic thinkers might choose to evaluate them separately, rating how close each one feels compared

> **dualistic approach** an approach framing contradiction as two separate entities

to the other. Furthermore, dualism allows for the idea that relationships may be evaluated differently on these scales at different times, so when Eleanor and Jeff first met or when they first decided to live together, their degree of closeness would be different than it is today (see **Figure 25.1**).

Alternatively, thinkers with a **dialectic approach** maintain that multiple points of view play off one another in every contradiction. Although a contradiction involves two opposing poles, the resulting

> **dialectic approach** an approach framing contradiction as both/and

situation expands beyond these two poles. As Baxter and Montgomery (1996) observe, "dialectical thinking is not directed toward a search for the 'happy mediums' of compromise and balance, but instead focuses on the messier, less logical, and more inconsistent unfolding practices of the moment" (p. 46). A simple way of putting it is that dialectic thinking substitutes *both/and* for *either/or,* recognizing that people want multiple (and often contradictory) goals in their relational lives. Leslie Baxter and Dawn Braithwaite (2008) give the example of a college student saying, "Well, I'm kinda like, seeing him, but we're not, ya know, serious" (p. 349) to illustrate a struggle where voices of both closeness and distance compete with each other. Further, in RDT, there are often more than two different elements involved in a relational situation (Baxter & Scharp, 2016). There may be the struggle for both closeness and distance as well as the recognition that romantic couples are supposed to be close and additional voices cautioning that it's important to be independent, for instance (see **Figure 25.1**).

As Eleanor and Jeff interact, many voices contribute to their sense of their relationship: their memories of their past conflicts about Jeff's friends, their current sense of confidence in their relationship, their hopes and concerns about their future, their friendships with others, and so forth. RDT holds that it isn't accurate to say that only two positions exist in relational contradictions.

We now turn our attention to the assumptions underlying RDT.

Theory at a Glance • Relational Dialectics Theory

Relational Dialectics Theory pictures relational life as constant process and motion. Throughout the life of a relationship, people continually feel the push and pull of conflicting discourses. This push and pull causes a tension called a dialectic. Basically, the theory asserts that both/and, not either/or, is a more accurate way to explain how people deal with opposing goals. For instance, people in relationships want to be both connected and autonomous, open and self-protective, and to have both predictability and spontaneity in their interactions. As people communicate in relationships, they attempt to reconcile these conflicting desires, but they never eliminate their need for both parts of the opposition. Relational life is composed of a constant struggle among competing discourses, and RDT 2.0 is especially concerned with examining these multiple, competing discourses with an eye to illuminating which discourses are central (the centripetal) and which are marginalized (the centrifugal). RDT 2.0 broadens its lens to examine more than just the competing desires of the individuals in the relationship to include voices that come from the culture as well. For instance, a given culture might value a specific discourse (e.g., in the United States, openness is generally valued above privacy), and this gives that discourse a certain amount of power. This emphasis on power makes RDT 2.0 a critical theory.

Assumptions of Relational Dialectics Theory

RDT is grounded in five main assumptions that reflect its contentions about relational life:

- Relationships are not linear.
- Relational life is characterized by change.
- Contradiction is the fundamental fact of relational life.
- Communication is central to organizing and negotiating relational contradictions.
- Human beings are choice makers.

The first assumption underlying RDT is the notion that relationships don't progress in a linear fashion. Rather, relationships consist of oscillation between contradictory desires. In fact, Baxter and Montgomery (1996) suggest that we should rethink our language and our metaphors about relationships. They note that the phrase *relational development* connotes some linear movement or forward progress, when, in fact, relationships do not steadily move in one direction.

The second assumption of RDT promotes the notion of process or change, without framing it as linear change. Thus, Jeff and Eleanor are different now than they were a year ago. But that difference is not a linear move toward intimacy as much as it's a difference in the way they express their togetherness and their independence. Change relates to an important concept in RDT: motion. **Motion** refers to the processual nature of relationships and their change over time. Think back to one of your own relationships. Compare how you relate with your relational partner now to how you did when you first met. No doubt you will see motion at work.

motion the processual nature of relationships and their change over time

The third assumption states that contradiction is the fundamental fact of relational life. **Contradiction** refers to "the dynamic interplay between unified oppositions" (Baxter & Montgomery, 1996, p. 8). Furthermore, this assumption posits that contradictions or tensions between opposites never go away and never cease to provide tension.

contradiction the ongoing interplay between opposing desires such as the desire for closeness and the desire for autonomy

People manage these tensions and oppositions in different ways, as we'll discuss later in the chapter, but they're continuously present in relational life. The push and pull represented by dialectic tensions construct relational life, and one of our main communication tasks in relationships is managing these tensions. This approach differs from

dialectics the tensions that result when two relational forces are both interdependent and mutually opposing

other types of relational theories in that it considers homeostasis to be unnatural: change and transformation are the natural characteristics of relational life (Carr & Wang, 2012; Semlak & Pearson, 2011). Contradiction is the foundation of **dialectics,** which are the tensions that result when two relational forces are both interdependent and mutually opposing (Baxter, 1988). When Eleanor desires to tell Jeff that she loves him but also wishes to withhold that information to protect herself, she's experiencing a dialectic.

The fourth assumption of RDT pertains to communication. Specifically, RDT gives a central position to communication. It's through communication practices that people achieve **dialectical unity,** or the way they're able to make contradictions feel complete and meaningful (Harrigan & Braithwaite, 2010).

dialectical unity the way people use communication to make sense of contradictions in their relationships

When Jeff thanks Eleanor for what she did for the party and uses his pet name for her, for example, he's expressing affection and negotiating closeness. When Eleanor keeps silent instead of saying "I love you" to Jeff, she's protecting herself and keeping some distance between them. As Eleanor and Jeff plan their picnic for the next day, they're discussing how much predictability and how much novelty they wish to have. Thus, Jeff and Eleanor's communication practices revolve around the three basic dialectics we'll discuss later in the chapter: autonomy and connection, openness and protection, and novelty and predictability.

Finally, the fifth assumption of RDT asserts that human beings have the ability to choose among alternatives. Although we don't have complete free choice in all instances—we're restricted by our previous choices, by the choices of others, and by cultural and social conditions—yet, we're still conscious and active choice makers in many ways. For example, Eleanor chooses to be with Jeff, illustrating her ability to choose among alternatives. However, this choice restricts other choices she might make. She must get along with Jeff's parents and siblings when they get together for the holidays, for instance. She might not ever have chosen to be with these people on her own, but because she's in love with Jeff, she must spend time with them. Jeff and Eleanor didn't choose the time in which they live, but the time period shapes some of their choices. If they had lived in the 1950s, for example, they'd be less likely to be living together as an unmarried couple.

Central Propositions of Relational Dialectics Theory

In 2008, Leslie Baxter and Dawn Braithwaite said that RDT rested on three interrelated central propositions:

1. "Meanings emerge from the struggle of different, often opposing, discourses" (p. 350).
2. "The interpenetration of discourses is both synchronic and diachronic" (p. 352).
3. "The interpenetration of competing discourses constitutes social reality" (p. 354).

We'll discuss each of these propositions briefly. Baxter and Braithwaite argued that the first proposition directly flowed from Bakhtin's notion that meaning is produced through dialogue, both the literal conversation that relational partners engage in as well as cultural and contextual discourses that the partners are more or less aware of through their socialization. Although this means that conversations are messy processes of dialectical flux, it doesn't mean that people are unable to make meaning from them. We use the interplay of different discourses to make meaning. In our story of Eleanor and Jeff, you can see the cultural discourse of romantic love in Eleanor's interior conversation about her feelings for Jeff. There's also the competing discourse of individuality evident in Jeff's desires to keep his friends from before he knew

Eleanor. Additionally, there's a discourse of community in the act of inviting old friends over to share dinner with them. When Jeff invites his friends over, he "goes public" and alerts his community of friends about his relationship with Eleanor. Going public is apparent when people make their relationships "social media official" by posting pictures of themselves holding hands or kissing. All these discourses are sanctioned in U.S. culture, which values romantic love, individuality, and community. Yet, they compete with each other, and Eleanor and Jeff have to struggle to make meaning from them.

The second proposition refers to the role of time in the meaning-making process. *Synchronic* means occurring in the moment, while *diachronic* means occurring over a longer period of time. In stating this proposition, Baxter and Braithwaite acknowledge that meanings might become fixed in the moment only to change and be renegotiated over time. So, while Eleanor might feel that closeness is achieved through finding separate activities today, later she might be less content with that meaning and seek to change it.

Finally, the third proposition frames meaning as something constructed between people through communication, not something that exists in people's heads before they begin to interact with another. When Jeff invokes a pet name for Eleanor and she accepts it, they're creating a sense of their relationship. When Jeff says of Mary Beth, "She and I have been friends forever," his communication provides a social identity for his relationship with Mary Beth. As some researchers put it, communication constitutes relationships.

Interpretive Research in Relational Dialectics Theory

After Baxter and Montgomery published their book in 1996, researchers began investigating specific tensions that animated relationships. Most of the research found three basic dialectics that seemed characteristic of interpersonal relationships: autonomy and connection, openness and protection, and novelty and predictability (e.g., Baxter, 1990; Erbert, 2000). Additionally, interpretive researchers also found other dialectics, beyond the basic three, that characterized a variety of relationships. Further, scholars identified other tensions arising from the place of the relationship in the culture. They labeled these contextual dialectics. Finally, researchers in the interpretive tradition dealt with the question of how people responded to dialectics or how they tried to manage them. We discuss each of these four topics below, beginning with an explanation of the three basic dialectics.

Basic Dialectic 1: Autonomy and Connection

The dialectic between **autonomy and connection** refers to our simultaneous desires to be independent of our significant others and to find intimacy with them. As our story of Eleanor and Jeff illustrates, relational life involves the conflicting desires to be both close to as well as separate from relational partners. Jeff appreciates Eleanor's efforts on the dinner party, which makes him feel close to her. Yet the reason for the party in some sense celebrates his separateness from Eleanor. The fact that Jeff has a long-standing friendship with Mary Beth, separate from his relationship with Eleanor, illustrates his autonomy. Seeing both autonomy and closeness as constants in relational life is a hallmark of RDT, which makes it unique from many other theories of communication in relationships.

autonomy and connection an important relational tension that shows our conflicting desires to be close and to be separate

For example, other theories like Social Penetration Theory (**Chapter 7**) and Uncertainty Reduction Theory (**Chapter 5**) advance a rather static view of closeness; partners move toward closeness or they move away from it. From this vantage point, Eleanor and Jeff may be seen as growing closer and closer as they share more experiences and affection. But, RDT holds that contradictions are inherent in all relationships and that the dynamic interplay between autonomy and closeness is important to understanding a relationship. So Jeff and Eleanor are seen by RDT as moving *between* closeness and distance throughout their relationship. They aren't understood as moving toward or moving away from either of these two competing needs.

Basic Dialectic 2: Openness and Protection

A second important tension in relational life has to do with openness and protection. The **openness and protection** dialectic focuses on our conflicting desires to be *both* open and vulnerable, revealing personal information to our relational partners, *and* to be strategic and protective in our communication. As Eleanor mulls over how much to tell Jeff about her feelings, she is wrestling with the tension between disclosure and silence, or the openness and protection dialectic.

> **openness and protection** an important relational tension that shows our conflicting desires to tell our secrets and to keep them hidden

RDT is interested in communication behaviors that seem to simultaneously honor the both/and approach. For instance, Leslie Baxter and Erin Sahlstein (2000) mention how gossiping can honor both openness and privacy as gossipers simultaneously disclose about others while keeping silent about themselves.

Basic Dialectic 3: Novelty and Predictability

The dialectic between **novelty and predictability** refers to the conflict between the comfort of stability and the excitement of change. The dialectic position focuses on the interplay of certainty and uncertainty in relationships. Jeff and Eleanor's planning behavior illustrates this interplay. When they make a plan together, they are

> **novelty and predictability** an important relational tension that shows our conflicting desires to have both stability and change

accomplishing at least two things with reference to predictability. First, their plan defines them as in a relationship, because planning is a relational activity. Second, the plan also establishes a routine so they know what they will be doing in the short-term future. Yet, they leave the plan open-ended to allow for creativity and novelty. With their plan for Sunday, Eleanor and Jeff have dealt simultaneously with their contradictory needs for routine and spontaneity.

Beyond Basic Dialectics

The three basic dialectics we've just reviewed characterize many interpersonal relationships, but as researchers continued applying RDT, they began to uncover several additional tensions in different contexts. For instance, Joshua Pederson (2014) studied online narratives of forgiveness and found three new dialectics: "(a) outward practice versus inward process, (b) liberation for self versus liberation for others, and (c) long process versus epiphany" (p. 353). In a study about Black women's friendship, Marnel Goins (2011) found support for the logic of RDT, but saw completely different tensions, such as "good" English versus "bad" English, and spending versus saving money. William (1992) found in his study of friendship that the novelty–predictability dialectic didn't seem to exist. Instead, he observed a dialectic focusing on the tensions between judgment and acceptance. In studying friendships in the workplace, Ted Zorn (1995) did find the three main dialectics, but he also found some additional tensions that were specific to the workplace context. Julie Apker, Kathleen Propp, and Wendy Zabava Ford (2005) studied nurses working on health-care teams, and discovered three new dialectics related to how nurses negotiated their roles relative to physicians and other medical professionals. These included: the equal–subordinate role dialectic, the superior–equal role dialectic, and the detached–attached role dialectic.

In examining people's participation in a community theater group, Michael Kramer (2004) advanced 11 dialectic tensions ranging from commitment to group versus commitment to other life activities to tolerance versus judgment (of the other members). Michaela Meyer (2003) examined the relationship between two characters in the television show *Dawson's Creek* and found a new dialectic: informing versus constituting identity formation.

Other researchers examined issues of illness and death in the context of the family and also uncovered additional dialectics. They found that the dialectic of presence versus absence was an important consideration for families dealing with members with Alzheimer's (Baxter, Braithwaite, Golish, & Olson, 2002), and for those experiencing the death of a parent, as well as for parents who experienced the death of a child (Toller, 2005). Tamara Golish and Kimberly Powell (2003) advanced a further dialectical tension for parents experiencing a premature birth: joy versus grief. A study on stepfamilies' communication (Baxter, Braithwaite, Bryant, & Wagner, 2004) found the dialectic of one parent versus two parents in authority to be important to children. Also, in the context of stepfamilies, other researchers (Braithwaite, Toller, Daas, Durham, & Jones, 2008) found children voicing the dialectic of control versus restraint. And for those becoming part of a stepfamily, the dialectic of presence versus absence (also found in the context of death and illness) was enacted in one study (Bryant, 2003).

Contextual Dialectics

The various dialectics we have just discussed are called **interactional dialectics** because they're located within the relationship itself—they're part of the partners' interactions with each other (Rawlins, 1992). Researchers have also found another type of tension that affects relational life. William Rawlins calls this type **contextual dialectics,** which means that they arise from the place of the relationship in the culture.

Contextual dialectics are formed from the tension between the public definitions of a given relationship—friendship, for example—and the private interactions within a specific friendship. Rawlins discusses two contextual dialectics—between the public and the private and between the real and the ideal. Although perhaps a little less important to communication students than interactional dialectics, these two do affect interpersonal communication in relationships. The **public and private dialectic** refers to the tension between having both a private relationship and public life. Rawlins points out that in the public realm friendship occupies a rather marginal position. He notes that friendship suffers in comparison to other relationships because there's no institution to sanction it. Eleanor and Jeff's cohabiting relationship may have the same marginal status because, unlike marriage, it's not legally sanctioned.

interactional dialectics tensions resulting from and constructed by communication between relational partners

contextual dialectics tensions resulting from the place of the relationship within the culture

public and private dialectic a contextual dialectic resulting from navigating both a private relationship and a public life

"I can't wait 'til we're so close we don't have to talk."

Victoria Roberts/CartoonStock Ltd.

Rawlins (1992) argues that tension arises in a close friendship between this marginal public status and the friendship's deep private character. Rawlins states that this dialectic results in friendships (and by implication, other unsanctioned relationships) acting with what he calls *double agency*. By this he means that these relationships fulfill both public and private functions. Rawlins observes that sometimes the public functions constrain the private ones. For instance, people who form friendships in the workplace may encounter negative feedback from their significant others who see the friendships as a threat to their relationship. When Jeff and Eleanor first became serious, for instance, Eleanor felt threatened by Jeff's female friends.

The second contextual dialectic is between the real and the ideal. The tension of the **real and ideal dialectic** is featured when we think of media that present an idealized vision of relationships, such as the picture of family that comes from old family sitcoms (1950s/ 1960s) including *Leave It to Beaver* or the image of romantic love

> **real and ideal dialectic** a contextual dialectic resulting from the difference between idealized relationships and lived relationships

that we get in romcoms like *You People* or *Always Be My Maybe*. We receive an idealized message of what relationships are like in these shows and then when we look at our own relationships we have to contend with troublesome realities. The tension between these two images forms this dialectic. If Eleanor reads a lot of romance novels emphasizing complete openness between romantic partners, she may feel a tension between that vision and her lived experience with Jeff, which involves some sharing but not complete openness.

Cultural and contextual factors influence these two dialectics. In cultures where friends are elevated to the status of family (e.g., some Middle Eastern cultures), the tension between public and private will be experienced quite differently than in Rawlins's description, if it's felt at all. In addition, social mores and expectations vary over time, as we discussed previously, and dialectics are influenced by these changes. For example, Ernest Burgess and Harvey Locke (1953) distinguished between an institutional and a companionate marriage. Prior to the 1950s, marriage was expected to be an economic institution, critical to the survival of the human race. More recently in the United States, marriage has been viewed as a love relationship where one's partner functions as one's best friend. The tensions between the ideal and the real may shift depending on the contours of the socially prescribed ideal. The institutional marriage didn't raise the expectations for the relationship nearly as much as the current ideal. See **Table 25.1** for a summary of the basic interactional dialectics and the contextual dialectics.

Table 25.1 Interactional and Contextual Dialectics

BASIC INTERACTIONAL DIALECTICS	CONTEXTUAL DIALECTICS
Autonomy–Connection: desiring a separate identity and wanting a bond between partners	**Public–Private:** contrasting the public and private aspects of a relationship
Openness–Protection: wanting to tell personal information while also wanting privacy	**The Real–The Ideal:** comparing a lived relationship to a fantasy of that relationship presented in media
Novelty–Predictability: wanting spontaneity while desiring stability	

Responses to Dialectics

Dialectic tensions are ongoing in relational life, as we mentioned previously, and so people make efforts to manage them. Baxter (1988) identifies four specific strategies for this purpose: cyclic alternation, segmentation, selection, and integration (see **Table 25.2**).

> **cyclic alternation** a coping response to dialectical tensions; refers to changes over time

Cyclic alternation occurs when people choose one of the opposites to feature at particular times, alternating with the other. For instance, when sisters are very young, they may be inseparable, highlighting the closeness pole of the dialectic. As adolescents, they may favor autonomy in their relationship, seeking separate identities. As adults, they may favor closeness again.

Table 25.2   Responses to Dialectic Tensions

RESPONSE	DESCRIPTION
Cyclic alternation	Choosing different poles for different times. Being close when young and more distant with age, for example.
Segmentation	Choosing different poles for different contexts. Being close at home and more distant at work, for example.
Selection	Choosing one pole and acting as though the other does not exist. Being an extremely close family, for example.
Integration	Synthesizing the oppositions in dialectic tensions; composed of three substrategies.
Neutralizing	A substrategy of integration; involves choosing a compromise between the oppositions. Being moderately close, for example.
Disqualifying	A substrategy of integration; involves exempting certain issues from the general pattern. Deciding to be open on all topics except sex, for example.
Reframing	A substrategy of integration; involves transforming the oppositions so they no longer appear to oppose one another. Deciding that closeness can only be achieved if there's a little distance too, for example.

Segmentation involves isolating separate arenas for emphasizing each of the opposites. For example, a married couple who work together in a family business might stress predictability in their working relationship and novelty for times when they're at home. The third strategy, **selection,** refers to making a choice between the opposites. The couple who choose to be close at all times, ignoring the partners' needs for autonomy, use selection.

Finally, **integration** involves some kind of synthesis of the opposites. Integration can take three forms: neutralizing, disqualifying, or reframing. **Neutralizing** means compromising between the polarities. People who choose this strategy try to find a happy medium between the two extremes of the tension. Jeff and Eleanor may decide they cannot really be as close as Eleanor would like, yet they cannot be as autonomous as Jeff might want either. Thus, they forge a moderately close relationship. **Disqualifying** manages the tensions by exempting certain issues from the general pattern. A family might be very open in their communication in general, yet keep a few topics off limits for discussion, such as sex and finances. **Reframing** refers to transforming the dialectic in some way so that it no longer seems to contain an opposition. When Eleanor notices that she and Jeff are closer when they have some separate activities, she's redrawing the dialectic between autonomy and connection as a unity rather than as an opposition and redefining what it means to be close.

segmentation a coping response to dialectical tensions; refers to changes due to context

selection a coping response to dialectical tensions; refers to prioritizing one of the oppositions over the other

integration a coping response to dialectical tensions; refers to synthesizing the opposition; composed of three substrategies

neutralizing a substrategy of integration; refers to compromising between the oppositions

disqualifying a substrategy of integration; refers to exempting certain issues from the general pattern

reframing a substrategy of integration; refers to transforming the oppositions

Baxter and Montgomery (1996) argue that any techniques that people use for responding to dialectic tensions share three characteristics: they are improvisational, they are affected by time, and they are possibly complicated by unintended consequences. Let's look more closely at these three characteristics.

Improvisational, according to Baxter and Montgomery, means that whatever people do to deal with a particular tension of relational life won't alter the ongoing nature of the tension. For example, Jeff and Eleanor have come to a moderately close relationship to deal with the dialectical tension, but they haven't changed the fact that autonomy and connection continue to be an issue in their relationship, and they'll have to continue dealing with it, perhaps altering their strategy at some point.

The aspect of time refers to the notion that, when dealing with dialectics, communication choices made by relational partners are affected by the past, enacted in the present, and filled with anticipation for the future. When Jeff praises Eleanor for the party they gave for Mary Beth, he does so knowing that they've argued about his friends in the past and hoping that these arguments won't continue in the future. William (1992) comments that "dialectical inquiries are intrinsically historical investigations" (p. 8), concerned with developmental processes over time.

Finally, Baxter and Montgomery point out that relational partners may enact a strategy for coping with a tension, yet it may not work out as they intended. For example, the husband and wife who work together and employ segmentation as we described earlier may feel they are coping with the novelty and predictability tension. But they could become dissatisfied with this coping strategy because they spend a lot of time at work and so the strategy has the unintended consequence of restricting the novelty in their relationship.

Relational Dialectics Theory 2.0: A Critical Turn

After a substantial body of research had been generated using RDT from an interpretive perspective, Leslie Baxter (2011) argued that it wasn't useful to continue discovering distinctive dialectic tensions to advance RDT. Instead, she called for a critical turn in the theory. This "new" theory was named RDT 2.0. In some ways, although RDT 2.0 marks an evolution in the theory, it also represents a return to where Baxter and Montgomery had started in 1996. Baxter and Montgomery began with Bakhtin's idea that language is "ideologically saturated" (Bakhtin, 1981, p. 271) with competing voices that people struggle with while making meaning. This competition among differing ideological positions characterizes RDT 2.0. Also, RDT 2.0 focuses on power and makes space for voices that previously had been marginalized (Suter & Norwood, 2017).

In reworking the theory, Baxter (2011) addressed five basic issues. These include: (1) the false distinction between public and private, (2) the bias in favor of certainty over uncertainty, (3) the illusion of the individual actor, (4) an acknowledgment of power differences, and (5) the false image of relationships as containers. We discuss each of these briefly below.

With regard to the false distinction between the public and private realms, Baxter observes that we have long organized social life into either the public or the private arena. She defines the two as follows: the **public sphere** is where people come together to discuss issues of mutual interest and people take on roles such as worker, citizen, and so forth. In contrast, the **private sphere** houses interpersonal and family concerns. Emotions reside in the private sphere, as opposed to rationality, which is presumed to characterize the public sphere.

public sphere where people come together to discuss issues of mutual interest and people take on roles such as worker, citizen, and so forth

private sphere where interpersonal and family concerns are managed

Baxter points out that the tradition of bifurcating these two spheres ignores the ways in which they are deeply interrelated. RDT 2.0 acknowledges this "interpenetration" between the spheres.

In a study examining how young Asian women navigate discussions with family members about their desire for a career in engineering, Debalina Dutta (2017) illustrates how the public and private spheres interrelate. The results of Dutta's study show that the families' discourse both reinforces and challenges the cultural discourse that frames engineering as a profession unsuited for young women. Dutta concludes that RDT 2.0 makes an important contribution to the field of career studies. Similarly, Lynsey Romo and Jenna Abetz (2016) found that cultural discourse around the value of money in the United States (i.e., money is everything) made private financial discussions between romantic partners a struggle. The partners often created a competing discourse (i.e., money isn't as important as relational life) to resist the dominant cultural message. The researchers show that private and public spheres aren't mutually exclusive when they state, "Cultural values and romantic relationships exist as a web, constituted by interwoven threads of meaning that emerge in everyday interaction" (p. 106). In a study using RDT 2.0 to investigate online letters people wrote seeking birth mothers willing to allow them to adopt their babies (*"Dear Birth Mother"* letters), Kristen Norwood and Leslie Baxter (2011) illustrate the ways in which the public (online letters) and the private (adoption) interrelate.

Students Talking Tech: Kellye

I actually went online and downloaded a copy of the study the chapter talked about with the "Dear Birth Mother" letters. I'm adopted, and I found it pretty interesting. The things the authors talked about made a lot of sense to me, and I do think it's true, like they said, that adoption is a stigmatized thing. There's something about your birth parents giving you up that is a little hard to think about, much less talk about. It also sounds right, like the chapter said, that putting a letter online asking for a kid does show how the public and private are mixed together. But otherwise, I didn't think that the online context made any difference to the ideas in the letters. My parents didn't adopt me online, but I think the things that were in that article were all things they have talked about with me—especially the part about adoption as a gain or a blessing to contrast with the idea of adoption as loss. My mom and dad tell me all the time that I was a blessing and the family gained so much when they adopted me. I like hearing that.

The second basic issue that RDT 2.0 reworks is the bias against uncertainty that permeates most interpersonal communication research. For instance, Uncertainty Reduction Theory, which we discuss in **Chapter 5,** assumes that people don't like to be uncertain, and when they feel uncertainty, they're motivated to do something to reduce it as soon as possible. Baxter (2011) argues that the creative properties of uncertainty have been marginalized as a result of approaches such as URT. RDT 2.0 takes the position that meaning making is never finalized, and creativity can only flourish when there's a lack of certainty. RDT 2.0 sees identity as a dialogic struggle that's fluid and ever changing. So, as Jeff and Eleanor talk while they're cleaning up from the party, they're creating their identities together. Yet, in future conversations, these identities will change as they communicate differently with each other.

Thirdly, RDT 2.0 addresses the illusion of the individual actor. Baxter notes that a great deal of interpersonal research begins with the assumption that the individual is central. RDT 2.0 critiques that assumption and instead places social relationships as central. Baxter comments that this illusion is a difficult one to deconstruct because of the strong hold the narrative of the individual has in U.S. culture. But she argues that there's no objective self outside of relationships and the communication that takes place within social encounters. In this regard, RDT 2.0 is much like Symbolic Interaction Theory, which we discuss in **Chapter 15.**

The fourth basic issue, the inattention to power, focuses on how past research fails to acknowledge that some of the competing discourses in an interaction are dominant, or **centripetal discourses,** while others are marginalized, or **centrifugal discourses.** Dominant discourses are those sanctioned by the culture, such as discourses of individualism in U.S. culture and those of collectivism in Asian culture (see **Chapter 14**). Further, when discourses are dominant, they have the

centripetal discourses those that are dominant in, and valued by, the culture

centrifugal discourses those that are marginalized by the culture

power to become the standard, making centrifugal discourses seem abnormal. As Baxter (2011) states: "The center is easily legitimated as normative, typical, and natural, and thus it functions as a baseline against which all else is somehow positioned as a deviation. By contrast, the centrifugal margins are positioned as non-normative, off-center, unnatural, and somehow deviant" (p. 123). In a study framed by RDT examining heterosexual dating from women's perspectives (Wolfe & Scharp, 2022), researchers found two competing discourses: dating is romantic and dating is restrictive. The first fits with the dominant U.S. cultural discourse and thus has marginalized the second. The researchers conclude that despite major cultural shifts around dating, heterosexual women's resistance to the meaning of dating hasn't changed much over time. If anything, the resistance has merely delayed the idea of marriage and family as an end-goal for women, not altered it.

When researchers examine how dominant discourses are either accepted or resisted in people's talk, they illuminate the process of meaning making. For example, some research has investigated how both adult children and parents navigate family estrangement (Scharp & Thomas, 2016; Scharp & Thomas, 2018). The researchers found that when family members needed to resist the powerful cultural narrative of "families are forever," they created other narratives around discourses like "families take effort." In a similar vein, Benjamin Baker (2019) investigated how gay male parents create and sustain their family identity against the centripetal discourses of the heteronormative family. A study of how lesbian and bisexual mothers disrupt dominant narratives of motherhood using RDT 2.0 also attends to power in a similar way (Suter, Seurer, Webb, Grewe, & Koenig Kellas, 2015). Valerie Cronin-Fisher and Erin Sahlstein Parcell (2019) take this same approach to an examination of how women experiencing dissatisfaction in the transition to motherhood resist the centripetal discourses of the "Good Mother" who's never dissatisfied with her role.

Finally, RDT 2.0 reworks the metaphor of relationships as containers. This metaphor suggests that people are communicating within their relationship (container) and different relationships/containers (i.e., friendships, marriages, lesbian co-parenting, and so forth) can be compared to see how communication differs within them. RDT 2.0 argues that relationships are meanings, not containers. As Baxter (2004; 2011) observes, relationships are constructed in communication; they aren't just contexts where communication takes place. Bobbi Van Gilder and Michael K. Ault (2018) adopted this approach in an examination of a religious group in Arizona practicing plural marriage. Van Gilder and Ault conclude the communication the families engaged in created the families. They observe that there was a "profound influence of intersecting discourses (e.g., religious, cultural, political and legal) in the shaping of family. As such, this study provided important insights into the many intersecting and conflicting discourses that work to shape notions of family" (p. 93). This element of RDT 2.0 is consistent with the third proposition of the theory, stating that communication constitutes relationships.

Integration and Critique

RDT adds a great deal to our explanatory framework for relational life. First, it allows us to think specifically about how relational partners construct meaning. Second, it removes a static frame and puts emphasis on the interplay between change and stability. With RDT, we don't have to choose between observing patterns or observing unpredictability because we recognize the presence of both within relationships. Likewise, dialectical thinking directs people to observe the interactions within a relationship, among its individual members, as well as those outside a relationship, as its members interact with the larger social and cultural systems in which they are embedded. This approach helps us focus on power and multicultural diversity. Finally, communication researchers are especially impressed with RDT because communication practices are central to the theory.

With respect to its methodological approach, Baxter and Norwood (2016) state the following: "Although the 1996 version of RDT was ecumenically receptive to both quantitative and qualitative methods, the 2011 version of the theory favors the latter" (n.p.). Baxter and Scharp (2016) also posit that the theory has now taken a critical turn and needs to be applied in that paradigm alone in future research. As you evaluate RDT, parsimony, utility, and heurism are three important criteria to consider.

Integration

Communication Tradition	Rhetorical | Semiotic | Phenomenological | Cybernetic | Socio-Psychological | **Socio-Cultural** | **Critical**
Communication Context	Intrapersonal | **Interpersonal** | Small Group/Team | Organizational | Public/Rhetorical | Mass/Media | Cultural
Approach to Knowing	Positivistic/Empirical | **Interpretive/Hermeneutic** | **Critical**

Critique

Evaluation Criteria	Scope | Logical Consistency | **Parsimony** | **Utility** | Testability | **Heurism** | Test of Time

Parsimony

With respect to parsimony, some researchers believe the original conception of RDT might have been too parsimonious in listing only the three basic dialectics of autonomy and connection, openness and protection, and novelty and predictability. As we've noted throughout the chapter, many studies have generated additional dialectics (e.g., Baxter et al., 2002; Bryant, 2003; Goins, 2011; Meyer, 2003; Pederson, 2014; Toller, 2005). Yet, this proliferating list of dialectic tensions raises the opposite concern about how well the theory does on the criterion of parsimony. Endless lists of new dialectic tensions make the theory problematic. It's possible that some of these newer tensions may enable RDT to explain relational life in a more complete manner yet the criterion of parsimony suggests that a theory generating a lengthy list of possible options has some problems. Baxter's response to this critique is twofold. First, she argues that the theory has now made a critical turn and should, consequently, focus less on generating lists of dialectic tensions and more on examining the power differentials that support different voices in relationships (Baxter, 2011). Secondly, she notes that parsimony suggests that a theory should be only as complex as needed to fully explicate the phenomenon of interest. Because meaning-making is an extremely complicated process, RDT must, of necessity, be complex as well.

Utility

RDT is especially appealing because it so clearly describes the push and pull people experience in relationships. Further, it does this much better than some other, more linear, theories of relational life. Most people intuitively experience their relationships in ebb-and-flow patterns, whether the issue is intimacy, self-disclosure, or something else. That is, relationships don't simply become more or less of something in a linear, straight-line pattern. Instead, they often seem to be both/and as we live through them. Dialectics offers a compelling explanation for this both/and feeling.

Heurism

Leslie Baxter (2006) observed the most important criterion for a theory like RDT is heurism because the theory's goal is to shed light on the "complex and indeterminate process of meaning-making" (p. 130). She argues that the theory shouldn't be evaluated based on how well it supports hypotheses (or predictions) but rather on how well it "functions as a heuristic device to render the communicative social world intelligible" (Baxter, 2011, p. 7). RDT seems to have been successful in this regard. Researchers have examined how RDT explains communication processes in a number of different contexts, including parent–child relationships (Scharp & Thomas, 2016), traumatic birth stories (Cronin-Fisher & Timmerman, 2023), romantic relationships (Faulkner & Ruby, 2015), Alcoholics Anonymous members (Thatcher, 2011), cohabiting couples (Moore, Kienzle, & Flood Grady, 2015), adult students (O'Boyle, 2014), nonvoluntary family relationships (Carr & Wang, 2012), postsecondary classrooms (Thompson, Rudick, Kerssen-Griep, & Golsan, 2018), Chinese corporations (Ngai & Singh, 2018), and mentoring relationships among professional psychologists (Johnson, Jensen, Sera, & Cimbora, 2018), among others. A few studies have examined online communication using RDT as a framework (e.g., Herrmann, 2018; Ngai & Singh, 2018), and one study specifically examined discussions of the meaning of suicide on online chat rooms (Conrad & Coohey, 2023). The sustained use of RDT as a theoretical framework across so many different interpersonal contexts speaks well to its performance as a heuristic theory.

Closing

Baxter and Montgomery (1996) observe that dialectics is not a theory in the empirical tradition because it doesn't offer axioms or predictive statements. This lack of predictive ability results from its specific goals. An empirical or positivistic theory seeks prediction and final statements about communication phenomena; dialectics is an interpretative/critical theory and it operates from an open-ended emphasis on the process of meaning-making. RDT seeks to describe what coping strategies people might use to deal with the major dialectic tensions in their relationships; it doesn't seek to make predictions. Baxter and Montgomery end their 1996 book with a personal dialogue between themselves about the experience of writing about a theory that encourages conversation rather than provides axiomatic conclusions. They agree that in some ways it's difficult to shake the cultural need for consistency and closure.

Many researchers believe that the dialectic approach is an exciting way to conceive of communication in relational life. Although quite a bit of research has been undertaken using RDT and RDT 2.0, expect to see more refinements of this theory and more studies testing its premises. Leslie Baxter (2011) notes that "growing a theory is a process akin to raising a child" (p. 1). In that vein, RDT is a theory that continues on a journey of exploration and investigation.

Discussion Starters

 Case-In-Point: Can you think of other dialectic tensions that might pervade the relational life of Eleanor and Jeff besides those discussed in the chapter?

Try-It-Your-Selfie: Only a few studies were mentioned in the chapter that utilized RDT in an online context. Why do you suppose that's the case? Propose a study you could do that focuses on online communication and uses RDT as the theoretical frame.

1. Do you think relationships are better explained through stage theories or dialectics? Why? Provide support for your position.

2. Only a few of the articles cited in the chapter have to do with RDT applied in ethnicities other than European American. How do you think culture and ethnicity impact the theory? Is RDT universal across cultures? Or do you see its claims as culture specific?

3. Explain the differences between the original applications of RDT and RDT 2.0. How do you see the transformation of the theory? Do you think it's more productive to generate lists of the tensions that characterize relational life or to examine the power struggles underlying the many voices in relational life? Explain your answer.

4. Provide an example of the multiple discourses that animate a specific relationship you are in now or have been in the past. (How) has power played a role in these discourses?

References

Apker, J., Propp, K. M., & Ford, W. S. Z. (2005). Negotiating status and identity tensions in healthcare team interactions: An exploration of nurse role dialectics. *Journal of Applied Communication Research, 33*(2), 93–115.

Baker, B. M. A. (2019). "We're just family, you know?" Exploring the discourses of family in gay parents' relational talk. *Journal of Family Communication, 19*(3), 213–227.

Bakhtin, M. M. (1981). Discourse in the novel. In M. Holquist (Ed.), *The dialogic imagination: Four essays by M. M. Bakhtin* (pp. 259–422). University of Texas Press.

Baxter, L. A. (1988). A dialectical perspective on communication strategies in relationship development. In S. Duck (Ed.), *Handbook of personal relationships* (pp. 257–273). Wiley.

Baxter, L. A. (1990). Dialectical contradictions in relationship development. *Journal of Social and Personal Relationships, 7*(1), 69–88.

Baxter, L. A. (2004). Distinguished scholar article: Relationships as dialogues. *Personal Relationships, 11,* 1–22.

Baxter, L. A. (2006). Relational dialectics theory: Multivocal dialogues of family communication. In D. O. Braithwaite & L. A. Baxter (Eds.), *Engaging theories in family communication: Multiple perspectives* (pp. 130–145). Sage.

Baxter, L. A. (2011). *Voicing relationships: A dialogic approach.* Sage.

Baxter, L. A., & Braithwaite, D. O. (2008). Relational dialectics theory. In L. A. Baxter & D. O. Braithwaite (Eds.), *Engaging theories in interpersonal communication: Multiple perspectives* (pp. 349–361). Sage.

Baxter, L. A., Braithwaite, D. O., Bryant, L., & Wagner, A. (2004). Stepchildren's perceptions of the contradictions in communication with stepparents. *Journal of Social and Personal Relationships, 21*(4), 447–467.

Baxter, L. A., Braithwaite, D. O., Golish, T. D., & Olson, L. N. (2002). Contradictions of interaction for wives of elderly husbands with adult dementia. *Journal of Applied Communication Research, 30*(1), 1-26.

Baxter, L. A., & Montgomery, B. M. (1996). *Relating: Dialogues and dialectics*. Guilford Press.

Baxter, L. A., & Norwood, K. M. (2016). Relational dialectics theory. In C. R. Berger & M. E. Roloff (Eds.), *The international encyclopedia of interpersonal communication* (pp. 1-9). John Wiley & Sons.

Baxter, L. A., & Sahlstein, E. M. (2000). Some possible directions for future research. In S. Petronio (Ed.), *Balancing the secrets of private disclosures* (pp. 289-300). Erlbaum.

Baxter, L. A., & Scharp, K. M. (2016). Dialectical tensions in relationships. In C. R. Berger & M. E. Roloff (Eds.), *The international encyclopedia of interpersonal communication* (pp. 1-6). John Wiley & Sons.

Braithwaite, D. O., Toller, P. W., Daas, K. L., Durham, W. T., & Jones, A. C. (2008). Centered but not caught in the middle: Stepchildren's perceptions of dialectical contradictions in the communication of co-parents. *Journal of Applied Communication Research, 36*(1), 33-55.

Bryant, L. E. (2003). *Stepchildren's perceptions of the contradictions in communication with stepfamilies formed post bereavement*. [Unpublished doctoral dissertation, University of Nebraska Press, Lincoln, NE].

Burgess, E. W., & Locke, H. J. (1953). *The family: From institution to companionship* (2nd ed.). American Book Company.

Carr, K., & Wang, T. R. (2012). "Forgiveness isn't a simple process: It's a vast undertaking": Negotiating and communicating forgiveness in nonvoluntary family relationships. *Journal of Family Communication, 12*(1), 40-56.

Conrad, J. B., & Coohey, C. (2023). The constructed meaning of suicide: A relational dialectics theory analysis of online suicide chats. *Journal of Communication Inquiry, 47*(2), 168-186.

Cronin-Fisher, V., & Sahlstein Parcell, E. (2019). Making sense of dissatisfaction during the transition to motherhood through relational dialectics theory. *Journal of Family Communication, 19*(2), 157-170.

Cronin-Fisher, V., & Timmerman, L. (2023). Redefining "healthy mom, healthy baby": Making sense of traumatic birth stories through relational dialectics theory. *Western Journal of Communication, 87*(1), 1-21.

Dutta, D. (2017). Cultural barriers and familial resources for negotiation of engineering careers among young women: Relational dialectics theory in an Asian perspective. *Journal of Family Communication, 17*(4), 338-355.

Erbert, L. A. (2000). Conflict and dialectics: Perceptions of dialectical contradiction in marital conflict. *Journal of Social and Personal Relationships, 17*(4), 638-659.

Faulkner, S. L., & Ruby, P. D. (2015). Feminist identity in romantic relationships: A relational dialectics analysis of e-mail discourse as collaborative found poetry. *Women's Studies in Communication, 38*(2), 206-226.

Goins, M. N. (2011). Playing with dialectics: Black female friendship groups as a homeplace. *Communication Studies, 62*(5), 531-546.

Golish, T. D., & Powell, K. A. (2003). "Ambiguous loss": Managing the dialectics of grief associated with premature birth. *Journal of Social and Personal Relationships, 20*(3), 309-334.

Harrigan, M. M., & Braithwaite, D. O. (2010). Discursive struggles in families formed through visible adoption: An exploration of dialectical unity. *Journal of Applied Communication Research, 38*(2), 127-144.

Herrmann, A. F. (2018). Working more and communicating less in information technology: Reframing the EVLN via relational dialectics. In L. Turner, N. P. Short, A. Grant, & T. E. Adams (Eds.), *International perspective on autoethnographic research and practice* (pp. 84-95). Routledge.

Johnson, W. B., Jensen, K. C., Sera, H., & Cimbora, D. M. (2018). Ethics and relational dialectics in mentoring relationships. *Training and Education in Professional Psychology, 12*(1), 14–21.

Kramer, M. W. (2004). Toward a communication theory of group dialectics: An ethnographic study of a community theater group. *Communication Monographs, 71*(3), 311–332.

Malhotra, P., Scharp, K., & Thomas, L. (2022). The meaning of misinformation and those who correct it: An extension of relational dialectics theory. *Journal of Social and Personal Relationships 39*(5), 1256–1276.

Meyer, M. D. E. (2003). "It's me. I'm it": Defining adolescent sexual identity through relational dialectics in *Dawson's Creek. Communication Quarterly, 51*(3), 262–276.

Moore, J., Kienzle, J., & Flood, J. G. (2015). Discursive struggles of tradition and non-tradition in the retrospective accounts of married couples who cohabited before engagement. *Journal of Family Communication, 15*(2), 95–112.

Ngai, C. S., & Singh, R. G. (2018). Using dialectics to build leader-stakeholder relationships: An exploratory study on relational dialectics in Chinese corporate leaders' web-based messages. *International Journal of Business Communication, 55*(1), 3–29.

Norwood, K. M., & Baxter, L. A. (2011). "Dear birth mother": Addressivity and meaning-making in online adoption-seeking letters. *Journal of Family Communication, 11*(3), 198–217.

O'Boyle, N. (2014). Front row friendships: Relational dialectics and identity negotiations by mature students at university. *Communication Education, 63*(3), 169–191.

Pederson, J. R. (2014). Competing discourses of forgiveness: A dialogic perspective. *Communication Studies, 65*(4), 353–369.

Rawlins, W. K. (1992). *Friendship matters: Communication, dialectics, and the life course.* Aldine De Gruyter.

Romo, L. K., & Abetz, J. S. (2016). Money as relational struggle: Communicatively negotiating cultural discourses in romantic relationships. *Communication Studies, 67*(1), 94–110.

Scharp, K. M., & Thomas, L. J. (2016). Family "bonds": Making meaning of parent–child relationships in estrangement narratives. *Journal of Family Communication, 16*(1), 32–50.

Scharp, K. M., & Thomas, L. J. (2018). Making meaning of the parent–child relationship: A dialogic analysis of parent-initiated estrangement narratives. *Journal of Family Communication, 18*(4), 302–316.

Semlak, J. L., & Pearson, J. C. (2011). Big Macs/Peanut butter and jelly: An exploration of dialectical contradictions experienced by the sandwich generation. *Communication Research Reports, 28*(4), 296–307.

Suter, E. A., & Norwood, K. M. (2017). Critical theorizing in family communication studies: (Re)reading relational dialectics theory 2.0. *Communication Theory, 27*(3), 290–308.

Suter, E. A., Seurer, L. M., Webb, S., Grewe, B., & Koenig Kellas, J. (2015). Motherhood as contested ideological terrain: Essentialist and queer discourses of motherhood at play in female-female co-mothers' talk. *Communication Monographs, 82*(4), 458–483.

Thatcher, M. S. (2011). Negotiating the tension between the discourses of Christianity and spiritual pluralism in alcoholics anonymous. *Journal of Applied Communication Research, 39*(4), 389–405.

Thompson, B., Rudick, C. K., Kerssen-Griep, J., & Golsan, K. (2018). Navigating instructional dialectics: Empirical exploration of paradox in teaching. *Communication Education, 67*(1), 7–30.

Toller, P. W. (2005). Negotiation of dialectical contradictions by parents who have experienced the death of a child. *Journal of Applied Communication Research, 33*(1), 46–66.

Van Gilder, B., & Ault, M. K. (2018). Disrupting dominant discourses of the idealized nuclear family: A study of plural families in Centennial Park, Arizona. *Journal of Communication & Religion 41*(4), 77–96.

Wolfe, B. H., & Scharp, K. M. (2022). Resisting and reifying meaning in everyday life: Using relational dialectics theory to understand the meaning of heterosexual dating. *Journal of Social and Personal Relationships 39*(9), 2680–2700.

Zorn, T. E. (1995). Bosses and buddies: Constructing and performing simultaneously hierarchical and close friendship relationships. In J. T. Wood & D. Duck (Eds.), *Understudied relationships: Off the beaten track* (pp. 122–147). Sage.

CHAPTER 26
Cultivation Theory

Based on the research of **George Gerbner**

> *"The longer we live with television, the more invisible it becomes. As the number of people who have never lived without television continues to grow, the medium is increasingly taken for granted as an appliance, a piece of furniture, a storyteller, a member of the family. Ever fewer parents and even grandparents can explain to children what it was like to grow up before television."*

—George Gerbner, Larry Gross, Michael Morgan, and Nancy Signorielli

Joyce Jensen

Joyce Jensen was preparing to vote for the very first time. She had been looking forward to this privilege since she was 12 years old. She considered herself a news junkie, and she was always reading news from a variety of sources such as the *Huffington Post*, the *New York Times*, the *Wall Street Journal,* the *American Spectator, Apple News*, and the *Weekly Standard*. She liked to read liberal and conservative takes on the same issue so she could make up her mind about which position she thought made the most sense. She continued to read her local morning newspaper and watch both local TV news and MSNBC. She made it a point to watch C-SPAN, a cable station dedicated to the world of politics. Joyce got news alerts on her phone from a variety of sources and first thing every morning, she looked through all of them. She knew that she was one of only a handful of her friends who could identify all of the U.S. Supreme Court justices and knew all her state and local representatives. She was ready to take some flak for being a news nerd because the world fascinated her. She wanted to be prepared for the right and responsibility of voting. And, when she got stressed out by reading and watching all the news, she relaxed by watching Netflix, Hulu, and even network TV. When she thought about it, she had to admit she was a media junkie!

Now she was about to vote in her first local election. She was choosing between two candidates for her state's governor. She had read a lot about the candidates and watched all the debates. Although she was still undecided, she was leaning toward Roberta Johndrew, the tough-on-crime candidate. Johndrew favored greater use of the death penalty, limits on appeals by people convicted of crimes, and putting more police on the street. Yet, Joyce thought that Frank Milnes, the education candidate, had some good ideas as well. Crime—in Joyce's state and in the country as a whole—was down for the eighth consecutive year. And Milnes argued that high-profile crimes like the tragic killings in schools, movie theaters, and at music events and parades took people's attention away from the facts; according to FBI statistics, all types of violent crime had declined for the past several years. Milnes argued that money being spent for more police, more prisons, and more executions would be better spent on improving schools, providing training for students, and offering mental health support. After all, Milnes asserted, more dollars in their state were being spent on incarceration than on educating young people. Better schools and mental health support, he argued, would mean even less crime in the future. "What kind of state do we live in," he demanded in his campaign speeches, "when we pay prison guards more than teachers and school counselors?" Milnes also pointed out the racial disparity in incarceration rates. Their state wasn't alone in this, but they definitely had a problem. Black men in Joyce's community were a much higher percentage of the prison population than their percent of the overall population would suggest they should be.

Joyce thought Milnes made some powerful points and raised important issues. She regretted that teachers were not getting paid commensurate with their expertise and responsibilities. She knew that she wanted to have children eventually, and she wanted them to get the best education possible. She could see how paying teachers more might help achieve that. She also knew that racial justice was an issue that needed to be addressed in her community. She volunteered at a homeless shelter, and the population there included many people of color. A few of the men had spoken to her about their experiences in prison and how hard it was to get back on their feet after their release.

But as a young, single woman, safety considerations were also at the top of her mind. There just seemed to be so much crime in the city. Every night when she watched the news on television, there were more crimes reported. She was often uncomfortable when she was out at night. At times, she even felt uneasy being at home alone. Maybe it's an irrational fear, she thought to herself, but it was there, and it felt real. When she watched her favorite movies, she saw a lot of violence too, and this intensified her feelings that her safety was at risk and that it should be a priority in her vote.

As Joyce pondered how she would cast her vote, so much was going through her mind. She considered her present situation as a single woman as well as her desire for future children. She reflected on both Roberta Johndrew and Frank Milnes and their comments from the past several months. As she contemplated her options for another moment or two, she felt she would be able to make a good decision. So much depended on citizens making informed choices. Joyce was thrilled to exercise her right as a U.S. citizen.

As Walter Lippmann (1922) noted over a century ago, people's opinions transcend their lived experiences. George Gerbner (1999) agreed with this and observed, "Most of what we know, or think we know, we have never personally experienced" (p. ix). This is possible, in large part, because of the impact of media and the stories told through the media that bring events and ideas to media consumers that are beyond their own realities (Northrup, 2010). We "know" many things from the stories we see and hear in the media.

And today, regardless of the influence and pervasiveness of social media, television still holds a central place in our lives. Although the upswing in internet usage cannot be denied, it's also interesting to note that there has actually been an uptick in television usage, as well. For instance, in 2016, there were approximately 116 million TV households in the United States. In the 2022-2023 season, that number rose to 123.8 million (Stoll, 2023). Total media usage for adolescents (ages 8-18 years old) averages 7 hours 38 minutes daily, and watching TV constitutes almost half of that screen time, at 3 hours 16 minutes daily (Guttmann, 2023). Most kids in this age group have TVs and computers in their bedrooms and smart phones that they take with them everywhere. Additionally, as adolescents increase their media use, the increase is especially apparent in the time they spend watching TV programming, both live and streaming. Adults (19 years and older) spend approximately 2.8 hours per day watching TV, accounting for nearly half of their daily leisure activities (U.S. Bureau of Labor Statistics, 2023). Clearly, television remains pervasive and potentially influential in the average U.S. home. Television programming regularly ushers in "mass-mediated storytelling," making it the "dominant entertainment medium" (Romer, Jamieson, Bleakley, & Jamieson, 2014, p. 115). Or, as Erica Scharrer and Greg Blackburn (2018a) term it, television is "the primary storyteller of the modern environment" (p. 235).

The theory profiled in this chapter, Cultivation Theory (CT), began as a way to test the impact that television had on viewers, particularly with regard to violence. It's true that television has changed a great deal since it became widely available to U.S. viewers in 1948. Yet, despite dramatic transformations in technology and social systems, many researchers believe that the model of cultivation developed by George Gerbner in the

1960s remains healthy and thriving, even if it may need a few tweaks to keep it completely relevant in today's media climate (Arth & Billings, 2019; Morgan & Shanahan, 2017; Ruddock, 2020).

In initiating what would become known as Cultivation Theory, Gerbner and his colleagues were making a **causal argument** (television cultivates, or causes, conceptions of social reality in people's minds). Cultivation Theory is a theory that predicts and explains the long-term formation and shaping of perceptions, understandings, and beliefs about the world as a result of consumption of media messages. Gerbner's line of thinking in Cultivation Theory suggests that mass communication, particularly television, is a socializing agent for consumers. One of the ways media socialize audiences is by offering consistent pictures of life that over a period of time of repeated exposure shift the audience's perceptions toward mediated *versions* of reality rather than real life itself (Scharrer & Blackburn, 2018b).

> **causal argument** an assertion of cause and effect, including the direction of the causality

Cultivation researchers can easily explain Joyce Jensen's quandary. Official statistics that indicate that some violent crimes are in steady decline are certainly real enough. But so, too, is Joyce's feeling of unease and insecurity when she's alone. CT refers to these feelings of insecurity as a reflection of her mediated reality. Moreover, mediated reality is as real as anything Joyce actually experiences.

Iver Peterson (2002) makes a similar observation relative to the anthrax scares in the United States post–September 11, 2001. He notes that although the media-fueled fears about anthrax were very pervasive and real, the actual cases of anthrax contamination are rare. Peterson quotes Clifton R. Lacy, commissioner of the New Jersey Department of Health and Senior Services, as saying that the risks to the citizens of New Jersey by anthrax spores are "vanishingly small" (p. B5) despite people's persistent fears. In the 1970s, Gerbner's view that media messages alter traditional notions of time, space, and social groupings was a direct challenge to the prevailing thought that media had little, if any, effect on individuals and on the culture. Like Uses and Gratifications Theory, which we discuss in **Chapter 13,** Cultivation Theory was developed in response to beliefs about the media's limited effects that were dominant at the time. CT also reflects media theory's slow transformation from reliance on the transmissional perspective to greater acceptance of the ritual perspective of mass communication.

The **transmissional perspective** sees media as senders of messages—discrete bits of information—across space (Baran & Davis, 2016). This perspective makes a comfortable partner for limited effects theories. If all media do is to transmit bits of information, people can choose to use or not use that information as they wish. In the **ritual perspective,** however, media are conceptualized not as a means of transmitting "messages in space," but as central to "the maintenance of society in time" (Carey, 1975, p. 6). Mass communication is "not the act

> **transmissional perspective** a position depicting the media as senders of messages across space
>
> **ritual perspective** a position depicting the media as representers of shared beliefs

of imparting information, but the representation of shared beliefs" (Carey, p. 6). The ritual perspective is represented in Gerbner's idea that people don't watch television selectively but simply watch at certain times, like a ritual. This occurs when people mindlessly turn on the set first thing in the morning or when they walk in the house after class or work. Gerbner likened the ritual function of TV to a religion, with a pulpit in every home. The consequence of this ritual viewing, according to Gerbner, is that television provides a wide variety of diverse viewers holding different opinions with a rather homogeneous "reality." We'll discuss this idea further a bit later in the chapter when we talk about a process in CT called mainstreaming.

Theory at a Glance • Cultivation Theory

Television and other media play an extremely important role in how people view their world because media tell us stories in a compelling fashion. In today's society, most people get their information from mediated sources rather than through direct experience. Therefore, mediated sources can shape a person's sense of reality. This is especially the case with regard to violence. Television presents at least ten times more violent acts in its programming than exist in reality. Heavy television viewing cultivates a sense of the world as a violent place, and heavy television viewers perceive that there is more violence in the world than there actually is or than lighter viewers perceive. Further, watching television has the effect of bringing together diverse people who receive the same messages and cultural imagery from the shows they watch in common (a process known as mainstreaming). The effects of television are increased when a viewer's actual experience resonates with the images presented (a process known as resonance).

Developing Cultivation Theory

Gerbner first used the term *cultivation* in 1969; however, Cultivation Theory, as a developed theory, didn't emerge until a number of years later. CT evolved over time through a series of methodological and theoretical steps by Gerbner and his colleagues and, as such, reflects that evolution. The method that Gerbner and others use to investigate questions of mass media cultivation is called Cultivation Analysis, and sometimes people use the terms *Cultivation Analysis* and *Cultivation Theory* interchangeably.

During the 1960s in the United States, interest in media effects, particularly the effects of television, ran very high. The federal government was concerned about media's influence on society, especially media's possible contribution to rising levels of violence among young people. In 1967, U.S. President Lyndon Johnson ordered the creation of the National Commission on the Causes and Prevention of Violence. It was followed in 1972, by the surgeon general's Scientific Advisory Committee on Television and Social Behavior. Both groups examined media (especially television) and their impact (especially the effects of aggression and violence as shown in the media). Gerbner, a respected social scientist, was involved in these efforts.

Gerbner's task was to produce an annual **Violence Index,** a yearly content analysis of a sample week of network prime-time television content that would show, from season to season, how much violence was actually present on television. The value of the Violence Index to those interested in the media violence issue was obvious: if the link

Violence Index a yearly content analysis of prime-time network programming to assess the amount of violence on TV

between television programming and subsequent viewer aggression was to be made, the presence of violence on television needed to be demonstrated first. Moreover, observers would be able to correlate annual increases in the amount of violent television content with annual increases in the amount of real-world violent crime. But, the index was immediately challenged by both those in the media industry and media researchers who believed in limited effects. They asked: how was violence defined? Was verbal aggression violence? Was obviously fake violence on a comedy counted the same as more realistically portrayed violence on a drama? Why examine only prime-time network television, because children's heaviest viewing occurs at other times of the day? Why focus on violence? Why not examine other social ills, such as racism and sexism?

Gerbner and his associates have continued to refine the Index in response to the complaints of its critics, and what their annual counting demonstrated was that violence appeared on prime-time television at levels unmatched in the real world. The 1982 Index, for example, showed that "crime in prime time is at least ten times as rampant as in the real world [and] an average of five to six acts of overt physical violence per hour involves over half of all major characters" (Gerbner, Gross, Morgan, & Signorielli, 1982, p. 106).

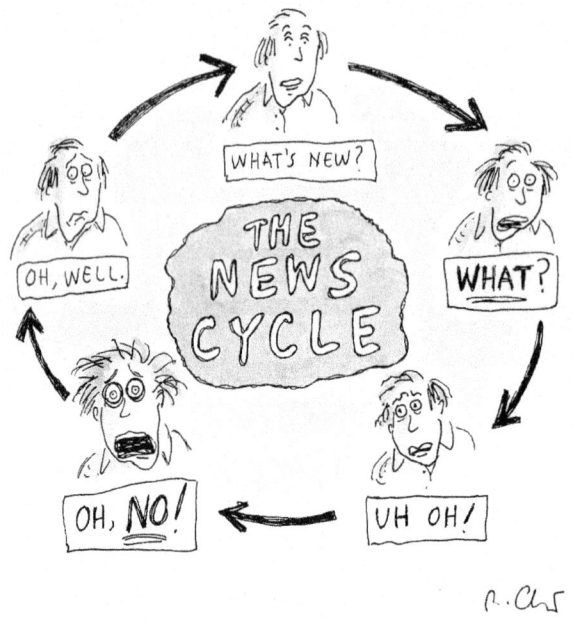

Roz Chast/CartoonStock Ltd.

In response to the critiques of the theory that asked whether comedies and dramas would cultivate the same ideas about violence in viewers, some cultivation researchers began focusing on specific TV genres (i.e., news, drama, and so forth) rather than simply overall TV viewing (Potter & Chang, 1990). According to Lennert Coenen and Jan Van den Bulck (2018), "approaches of this sort have come to dominate cultivation research, with numerous studies testing relationships between exposure to specific genres and attitude or belief outcomes across various topics" (p. 614).

Assumptions of Cultivation Theory

In advancing the position that "the more time people spend 'living' in the television world, the more likely they are to believe social reality is congruent with television's reality" (Riddle, 2010, p. 156), Cultivation Theory makes a number of assumptions. Because it was and still remains primarily a television-based theory, these three assumptions speak to the relationship between that medium and the culture:

- Television is essentially and fundamentally different from other forms of mass media.
- Television shapes our society's way of thinking and relating.
- The influence of television is relatively limited.

Students Talking: Milly

I'm one of those few 19-year-olds who watches TV almost constantly. I have Hulu so I can watch a lot of TV shows online. And, streaming means it doesn't matter where I am or when I want to watch (except in class and at my job, of course). So, according to Gerbner (and I know I'm simplifying the theory here), because I watch TV a lot, I'm getting exposed to a world that's more violent than reality. And that's supposed to make me afraid. I don't agree with Gerbner on that. I love watching old *Law & Order* reruns and even some of the horror shows on FX. But, I'm not afraid to go out at night. I know TV is just fictional and for entertainment, so it doesn't make me think that it really represents the world I live in. Maybe I'm an exception to the rule, but I don't think the theory describes my experience.

The first assumption of Cultivation Theory underscores the uniqueness of television. First, it requires no literacy, as do print media. Unlike the movies, it can be free (beyond the initial cost of the set, the addition of streaming services if a person wishes to buy those, and the cost of advertising added to the products we buy). Unlike radio, it combines pictures and sound. It requires no mobility, as do going to the movies or the theater. Television is ageless—that is, people can use it in the earliest and latest years of life, as well as all those years in between.

Because it's accessible and available to everyone, television is the "central cultural arm" of our society (Gerbner, Gross, Jackson-Beeck, Jeffries-Fox, & Signorielli, 1978, p. 178). Television draws together dissimilar groups and can make them forget their differences for a time by providing them with a common experience. For example, in 2014, 21.3 million people on average watched each of the 18 nights of the Winter Olympics in Sochi. Although in 2018, the viewership for the Winter Games in Pyeongchang, Korea, dipped to an average of 19.8 million per night, it's still the case that an enormous number of diverse people shared the experience of the Winter Olympics. In 2020, 2.05 billion people watched the Summer Olympics in Tokyo. Over 5 billion people worldwide watched Queen Elizabeth II of England's funeral in September of 2022. In May of 2023, 20.4 million people tuned in for King Charles's coronation as the new monarch of England. Regardless of their nationality, ethnicity, gender, politics, or other potentially divisive identities, these people had a common experience through their television viewing. In addition, no one can doubt the role that media have played as the United States worked through 9/11, the wildfires in Northern California, and the shootings at schools, concerts, grocery stores, and other public places, among other national tragedies.

The second assumption pertains to the influence of television. Gerbner and Gross (1976) comment that "the substance of the consciousness cultivated by TV is not so much specific attitudes and opinions as more basic assumptions about the 'facts' of life and standards of judgment on which conclusions are based" (p. 175). That is, television doesn't so much persuade us (it didn't try to convince Joyce Jensen that the streets are unsafe) as paint a more or less convincing picture of what the world is like (Riddle, Potter, Metzger, Nabi, & Linz, 2011). Gerbner agrees with Walter Fisher, whose Narrative Paradigm we discuss in **Chapter 22,** that people live in stories. Gerbner, however, asserts that most of the stories in current society come from television.

Television is a medium of socialization and enculturation. Gerbner and his colleagues state the following:

> The repetitive pattern of television's mass-produced messages and images forms the main-stream of the common symbolic environment that cultivates the most widely shared conceptions of reality. We live in terms of the stories we tell—stories about what things exist, stories about how things work, and stories about what to do—and television tells them all through news, drama, and advertising to almost everybody most of the time. (Gerbner et al., 1978, p. 178)

Where did Joyce Jensen's—and other voters'—shared conceptions of reality about crime and personal safety come from? Cultivation researchers would immediately point to television, where, despite a nationwide 20 percent drop in the homicide rate between 1993 and 1996, for example, the number of murder stories on the network evening news soared 721 percent (Kurtz, 1998). This distortion has resulted in effects that the theory would predict. Barbara Wilson and her colleagues (Wilson, Martins, & Marske, 2005) found that parents who paid a great deal of attention to television news thought their children were more at risk for kidnapping than those parents who watched less TV. Yet, the Bureau of Justice Statistics rate of violent crimes among 12- to 17-year-olds since 1994 does not support the fear engendered by TV. The Bureau of Justice Statistics indicates that from 1994 to 2010 the rate of violent crimes against children ages 12–17 decreased for youth in married households by 86 percent and for those in unmarried households by 65 percent (White & Lauritsen, 2012). Further, kidnapping makes up less than 2 percent of all violent crimes against youth (Finkelhor & Ormrod, 2000) and that number seems to be going down (Finkelhor, 2013).

Based on the second assumption, Cultivation Theory also supplies an alternative way of thinking about TV violence. Some theories, like Social Learning Theory (Bandura, 1977), assume that we become more violent after being exposed to violence. Other approaches, like the notion of catharsis, would suggest that watching violence purges us of our own violent impulses and we actually become less violent. Cultivation Theory does not speak to what we will *do* based on watching violent television; instead, it assumes that watching violent TV makes us *feel* afraid because it cultivates within us the image of a mean and dangerous world.

The third assumption of Cultivation Theory states that television's effects are relatively limited. This may sound counterintuitive, given the fact that the theory states television is capable of creating so much fear within its audience. Yet, the observable, measurable, and independent contributions of television to the culture are relatively small. This isn't a restatement of minimal effects thinking, because Gerbner uses an ice age analogy to distance Cultivation Theory from a limited effects approach. The **ice age analogy** states that "just as an average temperature shift of a few degrees can lead to an ice age or the outcomes of elections can be determined by slight margins, so too can a relatively small but pervasive influence make a crucial difference. The 'size' of an 'effect' is far less critical than the direction of its steady contribution" (Gerbner, Gross, Morgan, & Signorielli, 1980, p. 14). The researchers' point isn't that television's impact is inconsequential. Rather, they argue that although television's measurable, observable, and independent effects on the culture at any point in time might be small, that impact is nonetheless present and significant. Further, Gerbner and his associates argue that it isn't the case that watching a specific television program causes a viewer to engage in a specific behavior (e.g., that watching *NCIS* will cause someone to kill a naval officer) but rather that watching television in general has a cumulative and pervasive impact on our vision of the world. Gerbner illustrates how television can be so powerful and yet have effects that move as slowly as the ice age through the three Bs of television. Television, he wrote, *blurs* traditional distinctions of people's views of their world, *blends* people's realities into television's cultural mainstream, and *bends* that mainstream to the institutional interests of television and its sponsors (Gerbner, 1999) (see **Table 26.1**).

> **ice age analogy** a position stating that television doesn't have to have a single major impact, but influences viewers through steady limited effects

Table 26.1 The Three Bs of Television

TERM	DEFINITION	EXAMPLE
Blurring	Traditional distinctions are blurred.	Educated people see the world similarly to those who have less education.
Blending	"Reality" is blended into a cultural mainstream.	We all agree on what's real.
Bending	The mainstream reality benefits the elite.	We all want to buy more products.

Processes and Products of Cultivation Theory

Cultivation Theory has been applied to a wide variety of media effects issues, as well as to different situations in which television viewers find themselves. In doing so, researchers refined the theory and developed specific processes and products related to it, including the four-step process, mainstreaming, resonance, and the Mean World Index.

The Four-Step Process

To empirically demonstrate their belief that television has an important causal effect on the culture, cultivation researchers developed a four-step process for examining questions of violence on TV. The first step, message system analysis, consists of detailed content analyses of television programming in order to demonstrate its most recurring and consistent presentations of images, themes, values, and portrayals. For example, it's possible to conduct a message system analysis of the number of episodes of bodily harm on such shows as *Game of Thrones*.

The second step, formulation of questions about viewers' social realities, involves developing questions about people's understandings of their everyday lives. For example, a typical Cultivation Theory question is, "In any given week, what are the chances that you'll be involved in some kind of violence? About one in ten or about one in one hundred?" Another is, "Of all the crime that occurs in the United States in any year, what proportion is violent crime like rape, murder, assault, and robbery?" The third step, surveying the audience, requires that the questions from step two be posed to audience members and that researchers ask these viewers about their levels of television consumption.

Finally, step four entails comparing the social realities of light and heavy viewers. For Gerbner, a *cultivation differential* exists between light and heavy viewers and perceptions of violence. The **cultivation differential** can be defined as the percentage of difference in responses between light and heavy television viewers. Gerbner (1998) explains

cultivation differential the percentage of difference in responses between light and heavy television viewers

that "amount of viewing" is used in relative terms. Thus, heavy viewers are those who watch the most in any sample of people that are measured, whereas light viewers are those who watch the least.

Mainstreaming and Resonance

How does television contribute to viewers' conceptions of reality? The process of cultivation occurs through two subprocesses: mainstreaming and resonance. **Mainstreaming** occurs when, especially for heavier viewers, television's symbols dominate other sources of information and ideas about the world. As a result of heavy viewing, people's constructed social realities move toward the mainstream— not a mainstream in any political sense, but a culturally dominant

> **mainstreaming** the tendency for heavy viewers across differing co-cultural groups to perceive a similar dominant reality to that pictured on the media although this differs from actual reality

reality that is more similar to television's reality than to any measurable, objective external reality. Heavy viewers tend to believe the mainstreamed realities that the world is a more dangerous place than it really is, that immigrants are ruining their country, that teen crime is at record high levels, that all poor families are on welfare, that illegitimate births are skyrocketing, and so forth. Jennifer Good (2009) found that a mainstreaming effect occurred when people who were concerned about the natural environment were also heavy TV consumers. The mainstreaming effect had the result of decreasing these viewers' concerns about the environment.

Mainstreaming means that heavy television viewers of different co-cultures are more similar in their beliefs about the world than their varying group memberships might suggest. Thus, Latinx and European Americans who are heavy television viewers would perceive the world more similarly than might be expected. As Gerbner (1998) states, "Differences that usually are associated with the varied cultural, social, and political characteristics of these groups are diminished in the responses of heavy viewers in these same groups" (p. 183). Jerel Calzo and Monique Ward (2009) found support for mainstreaming effects in their study of television viewing and attitudes toward homosexuality. They found that more exposure to media representation of gay characters tended to promote more acceptance of homosexuality across a variety of groups. Michael Morgan and James Shanahan (2017) found an indirect mainstreaming effect when they studied authoritarianism attitudes and support for Donald Trump in the 2016 U.S. presidential election. In general, the amount of television watched didn't relate to an intention to vote for Trump, but when the researchers factored in authoritarianism, they found a significant effect. People who endorsed authoritarianism attitudes and were heavy TV viewers intended to vote for Trump. This indirect mainstreaming effect held true across a number of diverse demographics of the viewers.

The second way cultivation operates is through **resonance.** Resonance occurs when things on television are, in fact, congruent with viewers' actual everyday realities. Some urban dwellers, for example, may see the violent world of television resonated in their deteriorating neighborhoods. As Gerbner (1998) notes, this provides "a 'double dose' of messages that 'resonate' and amplify cultivation" (p. 182).

> **resonance** a process that occurs when a viewer's lived reality coincides with the reality pictured in the media amplifying the effects of cultivation

The reality that's cultivated for these viewers may in fact match their objective reality, but its possible effect is to preclude the formation of a more optimistic social reality; it denies them hope that they can build a better life. See **Figure 26.1** for a representation of the effects of mainstreaming and resonance.

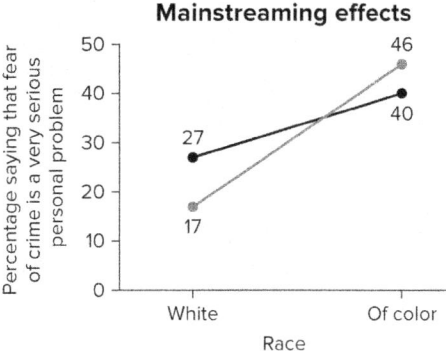

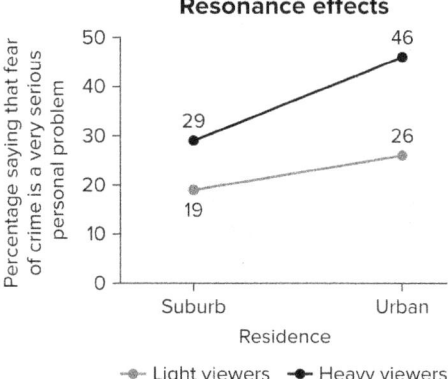

Figure 26.1   Effects of Mainstreaming and Resonance

Source: Gerbner, Gross, Morgan, & Signorielli, 1980.

Theory-Into-Action

 In cities like Chicago, Illinois, where crime is unevenly distributed, resonance is evident. For instance, in 2012 in Chicago, although overall crime dropped by 9 percent, the homicide rate actually increased by 16 percent. But over 80 percent of those homicides occurred on the South and West Sides of the city. The area of the business district downtown did not report a single homicide in 2012. Therefore, the resonance between people's experiences on the South and West Sides of Chicago and televised violence would be much greater than it would in other parts of the city of Chicago. As CT would predict, on the South and West Sides of the city, people felt more afraid and took precautions to keep their children and businesses safe. When people experience both the mediated and external reality of violence, they are convinced the world is violent, and they have little hope for a better future.

Cultivation, either as mainstreaming or as resonance, produces effects on two levels. **First-order effects** refer to learning facts from the media such as how many employed males are involved in law enforcement or what proportion of marriages end in divorce. For example,

first-order effects a method for cultivation to occur; refers to learning facts from the media

Joyce Jensen knew from candidate Milnes's television spots that the amount of crime in her state was in decline, providing a first-order effect. **Second-order effects** involve "hypotheses about more general issues and assumptions" that people make about their environments (Gerbner, Gross, Morgan, & Signorielli, 1986, p. 28). Questions like, "Do you think people are basically honest?" and "Do you think the police should be allowed to use greater force to subdue criminals?" are aimed at these second-order effects. Second-order effects result in Joyce's feelings of fear and uneasiness about her personal safety.

> **second-order effects** a method for cultivation to occur; refers to inferring values and making assumptions from the what's presented in the media

The Mean World Index

A product of Cultivation Theory is the Mean World Index (Gerbner et al., 1980; Morgan, Shanahan, & Signorielli, 2016), which consists of a series of three statements:

1. Most people are just looking out for themselves.
2. You can't be too careful in dealing with people.
3. Most people would take advantage of you if they got the chance.

Cultivation Theory predicts that heavy and light viewers will differ in their agreement with these statements, with heavy viewers seeing the world as a meaner place than light viewers. It also predicts that amount of television viewing is the best predictor of people's answers, overwhelming other kinds of distinctions among different people, such as income and education, for example (Gerbner et al., 1980). Cultivation researchers see this as evidence that television content is a factor in the construction of reality for heavy viewers, regardless of individual or social differences.

Gerbner and his associates showed that heavy viewers were much more likely to see the world as a mean place than were light viewers. In their 1980 study, they found that people with light viewing habits believed that about one in a hundred people will be a victim of violence; heavy viewers of television predicted that about one in ten will be involved in violence. They also found that heavy viewers overestimated the amount of violent crime in terms of what crime statistics indicated actually existed. Finally, heavy viewers thought that 5 percent of the culture is involved in law enforcement, whereas light viewers felt that 1 percent is involved in law enforcement. Important to the logic of Cultivation Theory is that the responses of the heavy viewers mirror quite accurately the results of content analyses of television, where violence is usually recorded in heavy doses: because violence is so common on television, heavy viewers are more likely to be fearful or mistrustful of the real world. Given what we've presented here, Joyce Jensen's viewing habits may be influencing her thinking about her choice between Milnes and Johndrew.

The relevancy of the Mean World Index is important to consider because it might be that fear of crime can lead to self-protective behaviors possibly increasing crime (Custers, Hall, Smith, & McNallie, 2017). Further, its impact on politics is especially important; media may have cultivated, in heavy viewers, an overall perception of "meanness," leading to polarized, uncivil political discussions. Morgan et al. (2016, p. 257) summed up this notion by describing an "in your face" political environment where "talk radio, cable news channels, internet blogs, discussion boards, partisan websites, online videos, and pervasive social media" that has cultivated a sense of "political incivility." This lack of civility ultimately negatively affects our political perceptions and voting patterns.

Learning from television produces not only perceptions of a mean world (which researchers in Cultivation Theory argue becomes a self-fulfilling prophecy as people's distrust of others breeds an atmosphere of further distrust) but also a warping of political, social, and cultural discourse. How many political candidates, they ask, have the courage to argue against the building of more prisons or against the death penalty? The issue is not the validity of these positions but the absence of meaningful, objective debate on them. The argument here is similar to that offered in **Chapter 12,** when we explained the Spiral of Silence Theory: politicians, and people in general, may be less willing to speak out about alternative approaches to crime and crime prevention because the media, especially television, cultivate a dominant social reality that renders these conversations out of step with voters.

Cultivation Theory as Critical Theory

Television's power rests in its utilization by powerful industries and elites to meet their own, rather than the culture's, interests. This perspective has led some researchers away from examining violence on TV and toward investigations of how heavy TV viewing affects materialism in viewers (e.g., Harmon, 2001; Harmon, Fontenot, Geidner, & Mazumdar, 2019). These studies generally support the idea that heavy TV viewing cultivates materialism in its audience. Recall George Gerbner's three Bs; the last one states that television *bends* the mainstream to its institutional interests and those of its sponsors. As we noted, this means that television is engaged in selling products and cultivating resistance to change. Thus, viewers should be compliant and buy the material goods that television is selling. Of course, this generally benefits the television executives and the advertisers rather than the audience.

Cultivation Theory examines an important social institution (television) in terms of how it uses its storytelling function to serve ends other than the benefit of the larger society. Cultivation Theory, like other critical theories, is political and oriented toward action; that is, in accepting its assumptions, its proponents must commit to doing something about the situation. George Gerbner has taken to heart the critical researcher's call to action. In 1996, Gerbner helped found the worldwide Cultural Environment Movement (CEM) to assist people in their struggle against powerful media industries. Its Viewers' Declaration of Independence reads, in part:

> Let the world hear the reasons that compel us to assert our rights and take an active role in the shaping of our common cultural environment.... Humans live and learn by stories. Today they are no longer hand-crafted, home-made, community-inspired. They are no longer told by families, schools, or churches but are the products of a complex mass-production and marketing process. (Viewers' Declaration of Independence, **http://web.asc.upenn.edu/ gerbner/Asset.aspx?assetID=542**)

Theory Into Practice • Cultivation Theory

Theoretical Claim: People who watch a lot of television are influenced by the way the world is shown in television programs.

Practical Implication: As an elderly man who uses a wheelchair to get around, Clarence watches a great deal of television because it's difficult for him to navigate the wheelchair to get out of his apartment. He spends a lot of time watching newscasts that focus on violent crimes, refugee deaths, world poverty, gun violence, financial crises, abuse, and other unfavorable stories. Clarence doesn't double-check his television sources and, thus, has cultivated a "world view" that is very violent and he is afraid to go out of the apartment even when a friend volunteers to help him with his wheelchair.

Gerbner was deeply concerned with the effects created by stories told by agencies that do not aim to teach but rather aim to sell.

Around the same time as he participated in the CEM, Gerbner developed the PROD (Proportional Representation of Diversity) index as another way of doing something to change the problems he saw in television. The goal of the index was to examine the distortion in representation of various co-cultures "across the demography of the media landscape" (Shanahan & Morgan, 1999, p. 223). The index determined how well or poorly groups were represented on television relative to their numbers in the population. The first index Gerbner produced surveyed broadcast network programming and major Hollywood films for 1995–1996. Almost every group (women, African Americans, Latinx, Asian Americans, Native Americans, people under age 18, people over age 65, gay men and lesbians, people with disabilities, and people living in poverty or those of lower socioeconomic class) listed in the diversity index was grossly underrepresented in the media. The only group that was not was Native Americans, and this is probably explained by their relatively low population proportionally.

Gerbner took his critical role seriously and stated in a press release associated with the presentation of the index:

> Far from being "quotas" to be imposed on creative people, the Index reflects the limitations on creative freedom in the television and motion picture industries. This is a "report card" of industry performance. We look forward to steady improvement in the diversity and equity of the cultural environment into which our children are born and in which they come to define themselves and others. (Gerbner, 1997, cited in Shanahan & Morgan, 1999, p. 223)

Gerbner believed it's important to highlight how the media reflect the needs and perspectives of dominant groups, and in the process, marginalize and silence all other groups.

Integration and Critique

Cultivation Theory has made an important contribution to contemporary thinking about mass communication. Horace Newcomb (1978), an early commentator about CT, wrote of Gerbner and his colleagues: "Their foresight to collect data on a systematic, long-term basis, to move out of the laboratory and away from the closed experimental model, will enable other researchers to avoid costly mistakes. Their material holds a wealth of information" (p. 281). Gerbner and his colleagues have been influential in identifying television as a shaping force in society. Cultivation Theory helps explain the implications of viewing habits, and it has been a very popular theory in mass communication research. In a study conducted by Jennings Bryant and Dorina Miron (2004) surveying almost 2,000 articles published in the three top mass communication journals since 1956, Cultivation was the third most frequently utilized theory. In addition, since 2013, more than 600 studies have been published using CT, all authored by researchers unrelated to Gerbner and his team (Morgan et al., 2016). And, from its inception in the 1960s–1970s to 2016, CT was cited in over 1,200 publications (Coenen & Van den Bulck, 2018). The methods employed in CT studies have been primarily experimental in nature, although, as we've noted, the theory utilizes a critical lens. As we evaluate Cultivation Theory, the following criteria are addressed: logical consistency, utility, heurism, and test of time.

Integration

Communication Tradition	Rhetorical | Semiotic | Phenomenological | Cybernetic | Socio-Psychological | **Socio-Cultural** | **Critical**
Communication Context	Intrapersonal | Interpersonal | Small Group | Organizational | Public/Rhetorical | **Mass/Media** | Cultural
Approach to Knowing	**Positivistic/Empirical** | Interpretive/Hermeneutic | **Critical**

Critique

Evaluation Criteria	Scope | **Logical Consistency** | Parsimony | **Utility** | Testability | **Heurism** | **Test of Time**

Logical Consistency

Critics who fault the logical consistency of CT note that the methods employed by Cultivation Theory researchers don't match the conceptual reach of the theory. They note that the research supporting Cultivation Theory employs social scientific methods typically identified with the limited effects findings. Yet, Cultivation Theory examines larger cultural questions most often raised by humanists. Horace Newcomb (1978) writes, "More than any other research effort in the area of television studies the work of Gerbner and Gross and their associates sits squarely at the juncture of the social sciences and the humanities" (p. 265). In some ways, Cultivation Theory offends many humanists, who feel that their turf has been improperly appropriated and misinterpreted. In a related critique, some scholars complain that many CT studies don't actually test the propositions of the theory itself, causing confusion about the meaning of the concept of cultivation (Coenen & Van den Bulck, 2018). However, Lennert Coenen and Jan Van den Bulck argue that this criticism doesn't mean the theory should be abandoned but rather reconceptualized somewhat.

Utility

Cultivation Theory is also criticized because its claims are not always useful in explaining the phenomenon of interest: how people see the world (Mutz & Nir, 2010). Newcomb (1978) argues that violence isn't presented as uniformly on television as the theory assumes, so television cannot be reliably responsible for cultivating the same sense of reality for all viewers. In addition, Cultivation Theory is criticized for ignoring other issues such as the perceived realism of the televised content, which might be critical in explaining people's understanding of reality (Minnebo & Van Acker, 2004). Furthermore, other researchers (Wilson et al., 2005) found that attention to television might be more important to cultivating perceptions than simply the amount of TV viewing. The fact that the theory seems to ignore cognitive processes such as attention or rational thinking style renders it less useful than is desired, according to some researchers (Berger, 2005). Further, because in earlier versions of the theory, Gerbner failed to establish a clear interpretation of what constitutes "violence," some writers (e.g., Hanson, 2016) contend that the theory is problematic (and subsequently, not useful for researchers). For example, as Ralph Hanson points out: there is a difference between "a fantasy violence of a *Road Runner* cartoon and the more graphic gore of a *Saw* or *Hostel* movie" (p. 42). Gerbner agreed that not all violence was the same. Further, genre approaches respond to this critique by restricting their studies to violence in a particular TV genre, such as newscasts or dramas, as we've mentioned. Elena Pelzer and Patric Raemy (2022) discuss how a great deal of television is a combination of news and entertainment or, as they call it, "infotaining," such as shows like *The Daily Show*. They argue that Cultivation Theory can be very useful in showing how these programs influence viewers' conceptualizations of the world. In the end, however, the utility of Cultivation Theory may be problematic given that researchers could have a hard time effectively determining some basic foundations of the theory, such as the definition of violence.

Some researchers (e.g., Katz & Fialkoff, 2017) have gone so far as to argue that the theory should be retired altogether because media and audience changes since the 1960s have made CT obsolete. Yet, others conclude that Cultivation Theory still is a very useful framework. In the words of Michael Morgan and James Shanahan (2017):

> Even in an era when television is increasingly less "mass" than it used to be, the cultivation perspective suggests that message commonalities still exist that are more or less resistant to change and that contribute to the cultivation of insecurity, fear, and mistrust. Where the minorities who are tagged as outgroups may change (women, gays, blacks), new ones (Hispanics, Muslims) emerge that provide the grist for the fear factory that television can still embody. (pp. 440–441)

Further, in contradiction to the claim that CT isn't useful in a changed media environment, Mina Tsay-Vogel, James Shanahan, and Nancy Signorielli (2018) found that it was a useful framework for examining Facebook users' attitudes and behaviors relative to self-disclosure and privacy.

Students Talking Tech: Bree

It was funny when I read that reference about using CT to study Facebook users. I was actually thinking about Facebook when I was reading the chapter. I was thinking about FOMO and the studies I've seen that show that people who are on Facebook and social media a lot are actually depressed because the pictures people post are always so beautiful and show people having so much fun. It can be depressing to see all that. I've sometimes felt that way when I've spent a bunch of time looking at Facebook and Instagram. I think CT does work to explain why that happens.

Additionally, Rong Ma and Zexin Ma (2023) note that Gerbner never confined CT only to television, and it's appropriate to examine other media or to combine other media with television when using Cultivation Theory. In their study of how U.S. people perceived Chinese people during the COVID-19 pandemic, they examined a combination of both television and social media consumption. Their results were consistent with Cultivation Theory in that heavy media consumers perceived Chinese people more negatively and as more of a threat than did light media consumers.

Heurism

When we examine Cultivation Theory against our criteria from **Chapter 3,** we find that it measures up quite well with regard to heurism. Indeed, research that has employed the theory has been prolific by any measure (Morgan et al., 2016) as well as diverse. For example, the theory has been applied to prime time dramas (Jamieson & Romer, 2014), cultural relations in the United States (Ortiz & Behm-Morawitz, 2016), adolescents' cooperation tactics with law enforcement (Dirikx & Van den Bulck, 2014), how crime reporting in the media impacts an individual's desire to become a police officer (Pollock, Tapia, & Sibila, 2022), video gaming (Breuer, Kowert, Festl, & Thorsten, 2015), women's desires to start smoking (Johnson, Len-Rios, Shoenberger, & Han, 2019), representations of gender (Sink & Mastro, 2017), and perceptions of immigrants (Seate & Mastro, 2016). Additionally, one study (Mastro & Figueroa-Caballero, 2018) examined the presentation of body type on TV. In this study, about half the women on TV were characterized as underweight, while three-fourths of the male TV characters were in the average, healthy weight range. Dana Mastro and Andrea Figueroa-Caballero suggest that this disparity "could be indicative of a growing gender chasm in body type portrayals" (p. 331) on television, and thus, cause for concern from a Cultivation Theory perspective. Gerbner et al. (1980) utilized CT to investigate how viewers thought about a variety of careers. They found, for example, that heavy TV viewers thought 5 percent of the population had a career in law enforcement, whereas light viewers felt that only 1 percent did. In fact, in 2023, fewer than 1 percent of the population of the United States is employed in law enforcement.

Test of Time

As we've noted, Cultivation Theory is heuristic and long lasting, but two issues may be working against it more than 50 years since its inception. First, some studies based on its tenets fail to find results consistent with the theory's predictions. Leo Jeffres, David Atkin, and Kimberly Neuendorf (2001), for instance, found that heavy television viewing seemed to be cultivating more diversity of opinion about public issues rather than mainstreaming people's perceptions as Cultivation Theory predicts. In other words, the three Bs that Gerbner and his colleagues discussed were not found in Jeffres, Atkin, and Neuendorf's study. Jeffres and his colleagues called the effect they found "scatter-streaming" and noted that it provided weak support for Cultivation Theory. Consistent with the Mean World Hypothesis, however, they did find that heavier users of TV expressed a greater need for gun control than did lighter users.

Second, as James Shanahan and Michael Morgan (1999) observe, times and media use are changing: "As more and more people grow up with TV, it is possible that it will become increasingly difficult to discern differences between light and heavy viewers" (p. 161). In addition, as streaming video services, digital cable, and other technologies alter our manner of TV viewing, it's likely that some of the theory's contentions will no longer hold true. For instance, if viewers can organize programming for themselves, it's unlikely that heavy viewing will mean the same thing for all viewers. Heavy viewing of cooking shows such as *Chopped*, for example, would be expected to cultivate a different reality from heavy viewing of crime shows such as *Law and Order SUV* or *Criminal Minds*. Kathleen Beullens, Keith Roe, and Jan Van den Bulck (2011) concur, finding that *what* people watch on television is more important to the cultivation effect than simply the *amount* they watch. Their research showed that teens who watched more TV news programs appeared to take fewer risks as drivers, but teens who watched more action shows exhibited more risk-taking behaviors while driving.

Cultivation offers responses to these criticisms. First, although there may be many more channels and people may have greater control over selectivity than they once had, television's dramatic and aesthetic conventions produce remarkably uniform content within as well as across genres. Second, because most television watching is ritual—that is, selected more by time of day than by specific program or the availability of multiple channels—heavy viewers will be exposed overall to more of television's dominant images. Third, most viewers, even with dozens of channels available to them, primarily select from only five or six, evidencing a very limited range of selection. Fourth, the genre approach that CT researchers have adopted can account for the problems some critics point out.

Closing

Cultivation Theory has been and remains one of the most influential theories of mediated communication of the last several decades. It's the foundation of much contemporary research and, as we've seen, has even become an international social movement. Another source of its influence is that it can be applied by anyone. It asks people to assess their own media use alongside the media-constructed reality of the world they inhabit. Imagine yourself as Joyce Jensen preparing to cast an important vote. You may well undergo the same mental debate as she does. Now, think of how even a passing understanding of Cultivation Theory might help you arrive at your decision and understand your motivations.

Discussion Starters

 Case-In-Point: Are you like Joyce Jensen in that you do not feel safe walking in your neighborhood at night? How much television do you watch? Do you fit the profile offered by Cultivation Theory? Why or why not?

Try-It-Your-Selfie: How do you respond to the criticism that more television channels and more divisions among viewers mean that the assumptions of Cultivation Theory are no longer valid?

1. How do you define violence on television? Do you think it's possible to calculate violent acts as Gerbner and his colleagues have done? Explain your answer.

2. People post a lot of pictures on social media of themselves and their friends drinking. Do you think that people who spend a lot of time on social media would cultivate a view of reality that overemphasizes drinking behavior, as CT would predict?

3. How do you respond to the criticism that more television channels and more divisions among viewers mean that the assumptions of Cultivation Theory are no longer valid?

4. How do you think AI might change some of the assertions of CT, if at all? If the media landscape changes drastically as a result of technology like AI will that mean that CT will no longer be valid?

References

Arth, Z. W., & Billings, A. C. (2019). Touching racialized bases: Ethnicity in Major League Baseball broadcasts at the local and national levels. *Howard Journal of Communications, 30*(3), 230–248.

Bandura, A. (1977). *Social learning theory*. Prentice Hall.

Baran, S. J., & Davis, D. K. (2016). *Mass communication theory: Foundations, ferment, and future*. Cengage.

Berger, C. R. (2005). Slippery slopes to apprehension: Rationality and graphical depictions of increasingly threatening trends. *Communication Research, 32*(1), 3–28.

Beullens, K., Roe, K., & Van den Bulck, J. (2011). The impact of adolescents' news and action movie viewing on risky driving behavior: A longitudinal study. *Human Communication Research, 37*(4), 488–508.

Breuer, J., Kowert, R., Festl, R., & Thorsten, Q. (2015). Sexist games=sexist gamers? A longitudinal study on the relationship between video game use and sexist attitudes. *Cyberpsychology, Behavior, and Social Networking, 18*(4), 197–202.

Bryant, J., & Miron, D. (2004). Theory and research in mass communication. *Journal of Communication, 54*(4), 662–704.

Calzo, J. P., & Ward, L. M. (2009). Media exposure and viewers' attitudes toward homosexuality: Evidence for mainstreaming or resonance? *Journal of Broadcasting & Electronic Media, 53*(2), 280-299.

Carey, J. W. (1975). A cultural approach to communication. *Communication, 2*(2), 1-22.

Coenen, L., & Van den Bulck, J. (2018). Reconceptualizing cultivation: Implications for testing relationships between fiction exposure and self-reported alcohol use evaluations. *Media Psychology, 21*(4), 613-639.

Custers, K., Hall, E. D., Smith, S. B., & McNallie, J. (2017). The indirect association between television exposure and self-protective behavior as a result of worry about crime: The moderating role of gender. *Mass Communication & Society, 20*(5), 637-662.

Dirikx, A., & Van den Bulck, J. (2014). Media use and the process-based model for police cooperation. *British Journal of Criminology, 54*(2), 344-365.

Finkelhor, D. (2013, May 10). Five myths about missing children. *The Washington Post.* https://www.washingtonpost.com/opinions/five-myths-about-missing-children/2013/05/10/efee398c-b8b4-11e2-aa9e-a02b765ff0ea_story.html.

Finkelhor, D., & Ormrod, R. (2000, June). Kidnapping of juveniles: Patterns from NIBRS. In *Juvenile Justice Bulletin.* U.S. Department of Justice, Office of Justice Programs: Office of Juvenile Justice and Delinquency Prevention. https://www.ojp.gov/pdffiles1/ojjdp/181161.pdf.

Gerbner, G. (1998). Cultivation analysis: An overview. *Mass Communication and Society, 1*(3/4), 175-194.

Gerbner, G. (1999). What do we know? In J. Shanahan & M. Morgan (Eds.), *Television and its viewers: Cultivation theory and research* (pp. ix-xiii). Cambridge University Press.

Gerbner, G., & Gross, L. (1976). Living with television: The violence profile. *Journal of Communication, 26*(2), 173-199.

Gerbner, G., Gross, L., Jackson-Beeck, M., Jeffries-Fox, S., & Signorielli, N. (1978). Cultural indicators: Violence profile No. 9. *Journal of Communication, 28*(3), 176-207.

Gerbner, G., Gross, L., Morgan, M., & Signorielli, N. (September 1980). The "mainstreaming" of America: Violence profile no. 11. *Journal of Communication, 30*(3), 10-29.

Gerbner, G., Gross, L., Morgan, M., & Signorielli, N. (1982). Charting the mainstream: Television's contributions to political orientations. *Journal of Communication, 32*(2), 100-127.

Gerbner, G., Gross, L., Morgan, M., & Signorielli, N. (1986). Living with television: The dynamics of the cultivation process in J. Bryant & D. Zillman (Eds.), *Perspectives on media effects* (pp. 17-40). Lawrence Erlbaum.

Good, J. E. (2009). The cultivation, mainstreaming, and cognitive processing of environmentalists watching television. *Environmental Communication, 3*(3), 279-297.

Guttmann, A. (2023). Daily entertainment screen time among teens in the U.S. 2021, by activity. Statista. https://www.statista.com/statistics/1312624/average-daily-entertainment-screen-time-teens-us-activity/.

Hanson, R. E. (2016). *Mass communication: Living in a media world.* Sage.

Harmon, M. D. (2001). Affluenza: Television use and cultivation of materialism. *Mass Communication & Society, 4*(4), 405-418.

Harmon, M. D., Fontenot, M., Geidner, N., & Mazumdar, A. (2019). Affluenza revisited: Casting doubt on cultivation effects. *Journal of Broadcasting & Electronic Media, 63*(2), 268-284.

Jamieson, P. E., & Romer, D. (2014). Violence in popular US prime time TV dramas and the cultivation of fear: A time series analysis. *Media and Communication, 2*(2), 31-41.

Jeffres, L. W., Atkin, D. J., & Neuendorf, K. A. (2001). Expanding the range of dependent measures in mainstreaming and cultivation analysis. *Communication Research Reports, 18*(4), 408-417.

Johnson, E. K., Len-Ríos, M., Shoenberger, H., & Han, K. J. (2019). A fatal attraction: The effect of TV viewing on smoking initiation among young women. *Communication Research, 46*(5), 688-707.

Katz, E., & Fialkoff, Y. (2017). Six concepts in search of retirement. *Annals of the International Communication Association, 41*(1), 86-91.

Kurtz, H. (1998, August 13). *Homicide rate down, except on the evening news* (p. A8). San Francisco Chronicle.

Lippmann, W. (1922). *Public opinion.* Harcourt Brace.

Ma, R., & Ma, Z. (2023). How are we going to treat Chinese people during the pandemic? Media cultivation of intergroup threat and blame. *Group Processes & Intergroup Relations, 26*(3), 515-533.

Mastro, D., & Figueroa-Caballero, A. (2018). Measuring extremes: A quantitative content analysis of prime-time TV depictions of body type. *Journal of Broadcasting & Electronic Media, 62*(2), 320-336.

Minnebo, J., & Van Acker, A. (2004). Does television influence adolescents' perceptions of and attitudes toward people with mental illness? *Journal of Community Psychology, 32*(3), 257-275.

Morgan, M., & Shanahan, J. (2017). Television and the cultivation of authoritarianism: A return visit from an unexpected friend. *Journal of Communication, 67*(3), 424-444.

Morgan, M., Shanahan, J., & Signorielli, N. (2016). Cultivation theory. In G. Mazzoleni, K. G. Barnhurst, K. Ikeda, R.C.M., Rousiley, & H. Wessler (Eds.), *The international encyclopedia of political communication* (pp. 1-5). John Wiley & Sons.

Mutz, D. C., & Nir, L. (2010). Not necessarily the news: Does fictional television influence real-world policy preferences? *Mass Communication and Society, 13*(2), 196-217.

Newcomb, H. (1978). Assessing the violence profile studies of Gerbner and Gross: A humanistic critique and suggestion. *Communication Research, 5*(3), 264-283.

Northrup, T. (2010). Is everyone a little bit racist? Exploring cultivation using implicit and explicit measures. *Southwestern Mass Communication Journal, 26*(1). 29-41.

Ortiz, M., & Behm-Morawitz, E. (2016). Latinos' perceptions of intergroup relations in the United States: The cultivation of group-based attitudes and beliefs from English- and Spanish-Language television. *Journal of Social Issues, 71*(1), 90-105.

Pelzer, E., & Raemy, P. (2022). What shapes the cultivation effects from infotaining content? Toward a theoretical foundation for journalism studies. *Journalism, 23*(2), 552-568.

Peterson, I. (2002, Aug. 14). Anthrax finding prompts questions in Princeton about scientist. *The New York Times,* B5.

Pollock, W., Tapia, N. D., & Sibila, D. (2022). Cultivation theory: The impact of crime media's portrayal of race on the desire to become a U.S. police officer. *International Journal of Police Science & Management, 24*(1), 42-52.

Potter, W. J., & Chang, I. C. (1990). Television exposure measures and the cultivation hypothesis. *Journal of Broadcasting & Electronic Media, 34*(3), 313-333.

Riddle K. (2010). Always on my mind: Exploring how frequent, recent, and vivid television portrayals are used in the formation of social reality judgments. *Media Psychology, 13*(2), 155-179.

Riddle, K., Potter, W. J., Metzger, M. J., Nabi, R. L., & Linz, D. G. (2011). Beyond cultivation: Exploring the effects of frequency, recency, and vivid autobiographical memories for violent media. *Media Psychology, 14*(2), 168-191.

Romer, D., Jamieson, P. E., Bleakley, A., & Jamieson, K. H. (2014). Cultivation theory. In R. S. Fortner & P. M. Fackler (Eds.), *The handbook of media and mass communication theory* (pp. 113-136). Wiley & Sons.

Ruddock, A. (2020). *Digital media influence: A cultivation approach.* Sage.

Scharrer, E., & Blackburn, G. (2018a). Cultivating conceptions of masculinity: Television and perceptions of masculine gender role norms. *Mass Communication and Society, 21*(2), 149–177.

Scharrer, E., & Blackburn, G. (2018b). Is reality TV a bad girls club? Television use, docusoap reality television viewing, and the cultivation of the approval of aggression. *Journalism & Mass Communication Quarterly, 95*(1), 235–257.

Seate, A. A., & Mastro, D. (2016). Media's influence on immigration attitudes: An intergroup threat theory approach. *Communication Monographs, 83*(2), 194–213.

Shanahan, J., & Morgan, M. (1999). *Television and its viewers: Cultivation theory and research.* Cambridge University Press.

Sink, A., & Mastro, D. (2017). Depictions of gender on primetime television: A quantitative content analysis. *Mass Communication and Society, 20*(1), 3–22.

Stoll, J. (2023). *Number of TV households in the U.S. 2000–2023.* https://www.statista.com/statistics/243789/number-of-tv-households-in-the-us/#:~:text=According%20to%20estimates%2C%20there%20are,the%202022%2D2023%20TV%20season.

Tsay-Vogel, M., Shanahan, J., & Signorielli, N. (2018). Social media cultivating perceptions of privacy: A 5-year analysis of privacy attitudes and self-disclosure behaviors among Facebook users. *New Media & Society, 20*(1), 141–161.

U.S. Bureau of Labor Statistics. (2023). *American time use survey summary.* http://www.bls.gov/news.release/atus.nr0.htm.

White, N., & Lauritsen, J. L. (2012). *Violent crime against youth, 1994–2010.* Bureau of Justice Statistics. http://www.bjs.gov/.

Wilson, B. J., Martins, N., & Marske, A. L. (2005). Children's and parents' fright reactions to kidnapping stories in the news. *Communication Monographs, 72*(1), 46–70.

> *"There is no understanding Englishness without understanding its imperial and colonial dimensions."*
>
> —Stuart Hall

Luisa and John Petrillo

Luisa and John Petrillo have lived in the same trailer court for four years with their two young children.

They realize that the wages they earn as migrant workers will make it difficult for the two to own their own home. They appreciate that Mr. DeMoss, the owner of the egg farm where they work, has provided housing for them, but they wish that they could have more privacy so that their neighbors would not be able to hear every word they say in the evening. Some evenings are very quiet, and the two often find themselves whispering to each other to avoid having their neighbors overhear their conversations. At times, they've even resorted to texting each other, even though they are less than 10 feet away from each other. The Petrillos do not have any desire to leave their tiny town because they realize that jobs are not that plentiful in northern New England. So they get by in the trailer community and still dream about a big backyard where their two kids and their dog, Scooter, can play.

The Petrillo's dream often unfolds on the television shows they watch at night. Even though cable TV is expensive and not a necessity, they both love to watch shows that help them escape their daily routines. There is no possible way they could purchase a bunch of streaming services, so watching cable is their default. While watching TV, Luisa and John are bombarded with messages about interest rates being at an all-time low accompanied by relatively low home selling prices. Whether watching network or cable TV, they keep seeing the same commercials over and over again promoting

home ownership as the "American Dream." They watch infomercials that promote "Five Ways to Get Your American Dream." They look at each other, wondering why they continue to live the way they do—two children, two bedrooms, and a common bathing area for all the migrants who are part of this living collective camp. They both know that they don't have the money to purchase a home, but they also know that they aren't happy with the way things are.

Recently, DeMoss's farm was investigated by the government for unsanitary living and working conditions. DeMoss was told to clean up the place and was threatened with daily stiff fines unless he improved the situation. Immediately, he authorized spending a lot of money for individual lavatory facilities and began talking to the local grocery chain to work on establishing discounts for his employees at the egg farm. As a gesture that would likely increase employee productivity and deter media attacks, DeMoss was also prepared to increase each worker's paycheck by 15 percent by the end of the month. He also promised to help relocate families with children to more suitable accommodations, once the rooms became available.

The Petrillos were cautiously optimistic, but very happy. They hated the "communal" bathing area and welcomed more privacy for their children and themselves. They were very excited about the opportunity to save money on food, and, of course, they were thrilled that their paychecks would increase almost immediately. This, they thought, was the beginning of saving for their Dream. They knew that they made just enough money to pay all of their bills, and now with the raise, the extra money could go into a "rainy day fund" that could eventually be used to purchase a home. For now, though, Luisa was excited about the chance to move to what DeMoss called "suitable" housing. "It has to be better than this," she thought. Indeed, it had to be.

We cannot overstate how much the U.S. culture relies on media. Each day, for instance, millions of homes tune in to dozens of different "news" programs on television. In fact, a Pew Research poll provides a number of important pieces of information about where people get their news (see **Figure 27.1**):

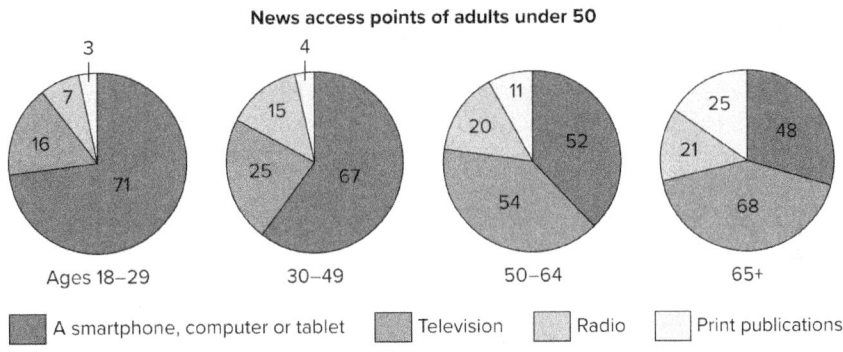

Figure 27.1   Television Dominates as a News Source for Older Americans

Pew Research Center, 2020.

Although the media have become the primary sources for how individuals learn about events around the globe (Baran, Davis, & Striby, 2020), it's the manner in which the media report events that varies significantly. Some journalists depend on fact-finding. Others rely on personal testimony or stories. Still others seek out experts to comment on events and topics as they unfold, whether they pertain to celebrity trials, natural disasters, refugees' plight, war, terrorist attacks, or school shootings. In fact, a common template for following a school shooting, for instance, resulted from the coverage of the Marjory Stoneman Douglas High School shooting in Parkland, Florida, in which 17 were killed and nearly 20 injured. We can now predict the evolution of coverage: first, the tragedy is usually reported in "real time," that is, as it happens, via social media such as X. Posts by students and those covering the event as it happens occupy the social media ether. Next, reporters interview witnesses to get firsthand accounts of the shootings. Finally, journalists gather experts on both sides of the gun control issue to assess whether gun control laws are sufficiently tough. Regrettably, this formulaic approach to media coverage has been undertaken over the years as over 350,000 students experience gun violence each year (Cox & Rich, 2023). This last effort involving experts seems to beg other questions about the media's role in such events: are they trying to convey a larger message about society in general? Is the reporting of images and stories done thoughtfully and conscientiously or is sensationalism a primary goal?

Reporting events with hidden or disingenuous ambitions has several implications. When the media fail to report all aspects of a story, someone or some group is inevitably affected. Nowhere is this perhaps more profound than in the early coverage of AIDS, which was first diagnosed in the gay community. Edward Alwood (1997) notes that because most news editors did not consider gay deaths to be newsworthy, major news outlets (including the esteemed *New York Times*) failed to provide coverage of the disease. In fact, the disease was killing far more people than the 34 who died from Legionnaires' disease in 1976 and the 84 women who died of toxic shock syndrome in 1980. Yet, it wasn't until the death of A-list actor Rock Hudson in 1985 that major news stories were devoted to the subject of AIDS. By that time, however, more than 6,000 people had died from the disease. The media's message related to the reporting of AIDS was implied, but significant: gay men are aberrant and their deaths are not newsworthy.

As you think about Parkland, school shootings, and *The New York Times*, you should also be thinking about the contributions of a theorist named Stuart Hall. Hall questioned the role of the media and their frequently sensational, false, misleading, and gut-wrenching images. Unlike other communication theorists, however, Hall focused on the role of the media and their ability to shape public opinions of marginalized populations, including people of color, the poor, and others who do not reflect a white, male, heterosexual (and wealthy) point of view.

For Hall, the personal is the political. At times throughout this book, we have explained a little about the personal background of theorists because that field of experience influenced their thinking. Understanding Hall's past allows us, again, to understand some of what motivated his theoretical contributions. He was a former high school teacher who taught English, math, and geography. Ultimately, he won a Rhodes scholarship and received his master's degree at Oxford University. Through his thinking, writing, and speaking, Hall, as a Jamaican, continued to try to understand his Jamaican culture and how it influenced his thinking and behavior. He believed there was a "pigmentocracy" in Jamaican society (Morley, 2019b, p. 1), meaning that skin color was paramount in his mind and family throughout his childhood. Hall was very much concerned with how "cultural forces" influenced the culture at large. And, many scholars who embraced Hall's thinking simultaneously embraced the scholar himself. Ann Curthoys and John Docker (2016) offered that Hall was a "public intellectual, academic leader, writer, editor, teacher, political activist, family man and friend" (p. 302). David Morley (2019a) even more succinctly states that Hall's "voice has been central in shaping many of the cultural and political debates of our time" (p. 1). This sort of posturing and orientation resonates throughout Cultural Studies.

Cultural Studies is a theoretical perspective that focuses on how culture is influenced by powerful, dominant groups. The theory is rooted in politics, but not the sort of electoral approach that characterize much of your understanding of politics. Rather, the theory is aligned with the politics of identity, namely the interplay between and among culture and race, ethnicity, gender, sexual and gender identity, and other markers of one's identity. Unlike several other theoretical traditions in this book, Cultural Studies does not refer to a single doctrine of human behavior. In fact, Stuart Hall (1992) persuasively argues that "Cultural Studies has multiple discourses; it has a number of different histories. It is a whole set of formations; it has its own different conjunctures and moments in the past.... I want to insist on that!" (p. 278).

Cultural Studies has its background and its beginnings in Britain, although scholars in the United States have also advanced our understanding of Stuart Hall's theory. As a cultural theorist and the former director of the Center for Contemporary Cultural Studies (CCCS) at the University of Birmingham in England, Hall (1981, 1989) contends that the media are powerful tools of the elite. Media serve to communicate dominant ways of thinking, regardless of the efficacy of such thinking. The theory emphasizes that the media keep the powerful people in control while the less powerful absorb what is presented to them. Luisa and John Petrillo, for instance, exemplify a marginalized group (the poor) who have been enamored with the "American Dream" of owning a home. Of course, Cultural Studies theorists would argue that the media—in this case, the infomercial sponsors—are (indirectly) taking advantage of a couple who will probably never have enough money to own a home. Yet, the message from the popular media is that it is possible. All that is needed, according to the message, is "good sense and good money."

Theory at a Glance • Cultural Studies

Cultural Studies is a theoretical perspective rooted in the notion of equality. The media represent ideologies of the dominant class in a society. Because media are controlled by corporations (the elite), the information presented to the public is consequently influenced and targeted with profit in mind. The media's influence and the role of power must be taken into consideration when interpreting a culture because those without equal influence are usually subordinate to those with disproportional influence.

Cultural Studies is a tradition based upon the writings of German philosopher Karl Marx. Because Marxist principles form the foundation of the theory, let's look further into this backdrop. We'll then examine two assumptions of Cultural Studies.

The Marxist Legacy: Power to the People

Philosopher Karl Marx (1818–1883) is generally credited with identifying how the powerful (the elite) exploit the powerless (the working class). He believed that being powerless could lead to **alienation,** or the psychological condition whereby people begin to feel that they have little control over their future. For Marx, though, alienation is most destructive under capitalism. Specifically, when people lose control over their own means of production (as happens in capitalism) and must sell their time to some employer, they become alienated. Capitalism results in a profit-driven society, and workers in a capitalistic society are measured by their labor potential.

alienation perception that one has little control over *their* future

Students Talking: Alexandra

Of the different ideas related to Cultural Studies, the one that sticks out the most for me is that there is a perpetuation of power by those in power. I saw it in my town newspaper when I was home for break. On the front page, there was a huge story on the high school football team losing in the playoffs. The article talked about their wins and losses through the year, and they even had a picture of the coach and his family. That same weekend, though, the speech and debate team won state finals, and there was a small boxed article on page 8. There were no pictures, and the coach of the team certainly didn't have her family pictured. I think this goes to show that football and sports perpetuate a dominant patriarchal script (complete with boys and a male coach) and that things like the speech and debate team (with a female coach and mostly female debaters) get shortchanged. Oh: the editor of the paper is a father of a boy who's on the football team.

Marx believed that the class system—a monolithic system that pervades all society—must be unearthed by the collective working class, or proletariat. He felt that laborers were often subjected to poor working and living conditions because the elite were unwilling to yield their control. As with the Petrillos, laborers across society are constantly relegated to secondary status. The elite, or ruling, class's interests become socially ingrained, and therefore people become enslaved in society. One of Marx's principal concerns was ensuring that some revolutionary action of the proletariat working class undertaken to break the chains of slavery and ultimately to subvert alienation under a capitalistic society. The capitalistic society, according to Marxist tradition, shapes society and the individuals within it.

Marxist thinkers who believed the working class was oppressed because of corporate-owned media have been called the **Frankfurt School theorists.** These thinkers and writers believed that the media's messages were constructed and delivered with one goal in mind: capitalism. That is, although the media might claim that they are delivering information for the "common good," the bottom line (money)

> **Frankfurt School theorists** a group of scholars who believed that the media were more concerned with making money than with presenting news

frames each message. Those affiliated with the Frankfurt School felt that the media could be considered an "authoritarian personality," which meant that they were opposed to the male-centered/male-owned media. In fact, Herbert Marcuse, a Frankfurt thinker, was the leader of a group of social revolutionaries whose goal was to break down this patriarchal system.

The application of Marxist principles to Cultural Studies is more subtle than direct. This has prompted some scholars to consider the theory to be more **neo-Marxist,** which means the theory diverges

> **neo-Marxist** limited embracement of Marxism

from classical Marxism to some extent. First, unlike Marx, those in Cultural Studies have integrated a variety of perspectives into their thinking, including those from the arts, the humanities, and the social sciences. Second, theorists in Cultural Studies expand the subordinate group to include additional powerless and marginalized people, not just laborers. These groups include, among others, the GLBT communities, various racial/ethnic minority groups, women, and even children. Third, everyday life for Marx was centered on work and the family. Writers in Cultural Studies have also studied recreational activities, hobbies, and sporting events in seeking to understand how individuals function in society. Marx's original thinking now requires clarification, elaboration, and application to a diverse society. The theory moves beyond a strict, limited interpretation of society toward a broader conception of culture.

Now that you have a brief understanding of how Hall and other theorists in Cultural Studies were influenced by the writings of Karl Marx, we examine two primary assumptions of the theoretical perspective.

Assumptions of Cultural Studies

Cultural Studies is essentially concerned with how elite groups such as the media exercise their power over subordinate groups. And, as you may know, power can be communicated both in person and online. The theory is rooted in a few fundamental claims about culture and power:

- Culture pervades and invades all facets of human behavior.
- People are part of a hierarchical structure of power.

The first assumption pertains to the notion of culture, a concept we addressed in **Chapter 2,** and which is embedded in nearly all of the theories covered in this text. To review, we identified culture as a

> **ideology** framework used to make sense of our existence

community of meaning. In Cultural Studies, "culture is such an alluring, and such a confusing concept" (Szulc, 2023, p. 4). Therefore, we need a slightly modified interpretation of the word, one that underscores the nature of the theory. The various norms, ideas, values, and forms of understanding in a society that help people interpret their reality are part of a culture's **ideology.** According to Hall (1981), ideology refers to "those images, concepts and premises which provide the frameworks through which we represent, interpret, understand, and 'make sense' of some aspect of social existence" (p. 31). Hall believes that ideologies include the languages, the concepts, and the categories that different social groups collect in order to make sense of their environments.

To a great extent, cultural practices and institutions permeate our ideologies. We cannot escape the cultural reality that, as a global community, actions are not vacuous. Graham Murdock (1989)

culture wars cultural struggles over meaning, identity, and influence

emphasizes the pervasiveness of culture by noting that "all groups are constantly engaged in creating and remaking meaning systems and embodying these meanings in expressive forms, social practices, and institutions" (p. 436). Still, being part of a diverse cultural community often results in struggles over meaning, interpretation, identity, and control. These struggles, or **culture wars,** suggest that there are frequently deep divisions in the perception of the significance of a cultural issue or event. Individuals often compete to help shape a nation's identity. For example, during the peak of the arguments related to marriage equality several years ago, both nonsupporters and supporters of same-sex marriage wanted to interpret what "marriage" is. One wanted to define marriage in traditional (man–woman) ways and the other defined it in less tangible (love) ways. Both groups strove to make their meanings dominant. Ultimately, in 2015, the nine members of the U.S. Supreme Court ruled that the fundamental right to marriage is guaranteed in the Constitution, culminating a decades-long struggle that occurred both in the legal courts and in the "court" of public opinion. At the time, one could certainly conclude that there was a culture war taking place, as in those countries around the globe that are also wrestling with this fundamental right. Currently, it's fair to say that diverse sexual identity continues to be the target of dissenters as trans individuals struggle to achieve the equality that Hall envisioned.

Dreama Moon (2008) notes that culture includes a number of diverse activities of a population. In the United States, there are many behaviors, some done daily and others less frequently. For instance, men asking women out on a date, families visiting one another during holidays, and people attending religious services are all Western-centered practices. (As we noted in **Chapter 2,** keep in mind that many of these behaviors are being modified to a large extent because of the influx of new cultural communities.) There are also more mundane behaviors, such as getting your driver's license renewed, running on the treadmill, pulling weeds from a garden, posting online your reaction to a news story, or listening to the radio while driving home from school. For those interested in Cultural Studies, it's crucial to examine these activities to understand how the ideology of a population is maintained. Paul du Gay, Stuart Hall, Linda Janes, Hugh Mackay, and Keith Negus (1997) explain that these practices intersect to help us understand the production and dissemination of meaning in a culture. At the same time, the meaning of a culture is reflected by such practices. Culture, then, cannot be separated from meaning in society, underscoring our transactional and holistic models of communication, which we articulated in **Chapter 1.**

Meaning in our culture is profoundly sculpted by the media. The media could simply be considered the technological carrier of culture, but as this chapter points out, the media are so much more. Consider the words of Michael Real (1996) regarding the media's role in U.S. culture: "Media invade our living space, shape the taste of those around us, inform and persuade us on products and policies, intrude into our private dreams and public fears, and in turn, invite us to inhabit them" (pp. xiii–xiv). A more current interpretation of media is offered by Humphreys (2023). The author embraced "mobile social media" as paramount in communication. Specifically, these media include software, applications, and services that allow users to connect with others. Further, these media provide opportunities to share information with others, too. No doubt, for example, the media contain both intentional and unintentional messages that entice the Petrillos to accept mediated interpretations of what constitutes the dream of home ownership. If the Petrillos go online, they will see that mobile social media will be even more replete with messages about home buying.

A second assumption of cultural theory pertains to people as an important part of a powerful social hierarchy. Power operates at all levels in society. However, power in this sense is not role-based. Rather, Hall is interested in the power held by social groups or the power between groups. Meaning and power are intricately related, for as Hall (1989) contends, "meaning cannot be conceptualized outside the field of play of power relations" (p. 48). In keeping with the Marxist tradition, power is something that subordinate groups desire but cannot achieve. Often there is a struggle for power, and the victor is usually the person at the top of the social hierarchy. An example of what we are discussing here can be observed in the U.S. culture's preoccupation with beauty. Theorists in Cultural Studies would contend that because beauty, with limited exceptions, is often narrowly defined as being thin and physically attractive, anyone not matching these qualities might be considered unattractive. Hall would advance that the attractive people—those usually at the top of the social hierarchy—are able to wield more power than those at the bottom (the unattractive).

The foregoing suggests that the ultimate source of power in our society is the media. Hall (1989) maintains that the media are simply too powerful. He is not shy in his indictment of the media's character by calling the media dishonest and "fundamentally dirty" (p. 48). In a diverse culture, Hall argues, no institution should have the power to decide what the public hears. In some ways, journalists and television anchors serve as guardians of a country's cultural activities: if the media deem something to have importance, then something has importance; consequently, an otherwise unimportant event suddenly carries importance.

Let's revisit our story of Luisa and John Petrillo. Cultural Studies supporters would argue that as members of a minority population, the Petrillos have been inherently relegated to a subordinate position in society. Their work environment—as migrant workers on a large egg farm—is the product of a capitalistic society, one in which laborers work under difficult conditions. Although they will inevitably have difficulty owning their own home because of their low wages, writers in Cultural Studies would point to the media's barrage of images and stories touting the "American Dream" (consider, for instance, all of the dire mediated images of migrant "camps" in Texas, Arizona, and other U.S. locations along the border). Although the message may convey hope for the Petrillos, their dream of owning their own home might better be called a fantasy because the elite power structure (the media) fails to convey the reality of their circumstances. Likely unknown to the Petrillos is the fact that the media are a tool of the dominant class. The future of Luisa and John Petrillo, then, will overtly and covertly be influenced by the ruling class.

Theory-Into-Action

 The use of children around the world as labor is a long-term problem and a topic in which Cultural Studies theorists are keenly interested. Because children as young as six are employed in industries considered to be inhumane or illegal, these researchers wish to uncover and reveal these episodes fraught with problems. Whether it's big tobacco companies causing children to get chest pains or nicotine poisoning because they work in a "tobacco district" in Malawi, or children ages 5–14 serving as "little maids" in Morocco for domestic industries, exploitation and abuse of young children continues to happen across the globe. Despite laws in the United States, child farmworkers as young as 12 are asked to lift heavy machinery, climb tall ladders, and clean up toxic spills. Some may argue, therefore, that the United States cannot begin to "point a finger" at child labor around the world unless/until it aggressively enforces protection for its own children.

Hegemony: The Influence on the Masses

The concept of hegemony is an important feature of Cultural Studies, and much of the theory rests on an understanding of this term. In fact, for Hall, hegemony always remained on the "theoretical and political horizon" (Colpani, 2022, p. 239). Scott Lash (2007) contends that "from the beginnings of cultural studies in the 1970s, 'hegemony' has been perhaps the pivotal concept" (p. 55). **Hegemony** can be generally defined as the influence, power, or dominance of one social group over another. Embedded in hegemony is a *master-narrative,* which is a "grand story told by the dominant groups to legitimate and justify their actions and policies" (Campbell & Kean, 2016, p. 18). The idea is a complex one that can be traced back to the work of Antonio Gramsci, one of the founders of the Italian Communist Party who was later imprisoned by the Italian fascists. Writers in Cultural Studies have called Gramsci a "second progenitor Marxist" (Inglis, 1993, p. 74) because he openly questioned why the masses never revolted against the privileged class:

> **hegemony** the domination of one group over another, usually weaker, group

> The study of hegemony was for him [Gramsci], and is for us, the study of the question why so many people assent to and vote for political arrangements which palpably work against their own happiness and sense of justice. What on earth is it, in schools or on the telly [television], which makes rational people accept unemployment, killing queues [wards] in hospitals, ludicrous waste on needless weaponry, and all the other awful details of life under modern capitalism (Inglis, 1993, p. 76)?

Gramsci's notion of hegemony was based on Marx's idea of **false consciousness,** a state in which individuals are unaware of the domination in their lives. Gramsci contended that audiences can be exploited by the same social system they support (financially). From popular culture to religion, Gramsci felt the dominant groups in society manage to direct people into complacency. Consent is a principal component of hegemony. Consent is given by populations if they are given enough "stuff" (e.g., freedoms, material goods, etc.). Ultimately, people will prefer to live in a society with these "rights" and consent to the dominant culture's ideologies.

> **false consciousness** Gramsci's belief that people are unaware of the domination in their lives

Gramsci's thinking on hegemony is quite applicable to today's society. Under a hegemonic culture, some profit (literally) while others lose out. What happens in hegemonic societies is that people become susceptible to a subtle imbalance in power. That is, people are likely to support tacitly the dominant ideology of a culture. The complexity of the concept is further discussed by Hall (1989), although he never fully embraced all of the elements of hegemony articulated by Gramsci. He notes that hegemony can be multifaceted in that the dominant, or ruling, class is frequently divided in its ideologies. That means that during the subtle course of being influenced, the public may find itself pushed and pulled in several directions. Unraveling such complexity is one goal of researchers in Cultural Studies. Think about, for instance, the water crisis that happened in Flint, Michigan, in 2016 (with consequences still being felt several years later). During that time, state officials were blaming the EPA for the contamination found in the city's drinking water. The EPA, however, pointed fingers at the state officials. In the meantime, nearly 12,000 children who were overwhelmingly African American were exposed to very high levels of lead in their water and may have long-term problems as a result. This sort of crisis, sadly, continued in 2022 in Jackson, Mississippi—a city with a population majority of people of color. At the time, as a result of flooding, there was a breakdown in the city's water and wastewater system—all because there had been little to no maintenance undertaken in decades. This "water democracy" would be of particular interest to Hall and his associates. Although the "audience" was given vastly different interpretations to the problem, in the end, the dominant (political) culture was trying to influence the discourse and the public's opinion of the crisis. As Stacia Ryder (2016) observed: the contamination debacle was "covered up and informing the public was drastically delayed" (p. 23).

Students Talking Tech: Olivia

 I can easily see hegemony when I'm looking at Twitter. I see so many hashtags that are nothing but hegemony-related, like those that say one race is better than the other or that straight is better than gay or ... the list goes on and on. It's obvious from their hashtag what a person's values are, and I can definitely say that it's easy for me to see how they want to oppress others with what they write. If Hall had the internet when he thought about this theory, he would have been able to write about a hundred more articles on hegemony.

Hegemony can be further understood by looking at today's corporate culture, where—using Marx's thinking—ruling ideas are ideas of the ruling class. In most corporate cultures, decision-making is predominantly made by white, heterosexual males. Hall challenges this dominant way of thinking and relating and argues that a homogenous leadership may simply lead to a subordination of the people (workers). How is consciousness raised and how is new consciousness presented? Perhaps it is the language used in an organization, for as Hall (1997) states, "Language in its widest sense is the vehicle of practical reasoning, calculation, and consciousness, because of the ways by which certain meanings and references have been historically secured" (p. 40). People must share the same way of interpreting language; however, Hall notes that meanings change from one culture or era to another. So what exists in one organizational setting may not exist in another.

What all this means is that there are multiple ideologies in a society as complex as the United States. This translates into what Hall calls a **theater of struggle,** which means that various ideologies in society

> **theater of struggle** competition of various cultural ideologies

compete and are in temporary states of conflict. Thus, as attitudes and values on different topics shift in society, so do the various ideologies associated with these topics. For example, think about what it meant to be a woman before 1920 and what it means to be a woman today. Before 1920, women were unable to vote and were generally regarded as subordinate and subservient to men. Then in August 1920, the amendment giving women the right to vote was ratified. Today, of course, women not only vote but also hold high political office (one even received the most number of votes in a U.S. Presidential election!). Although U.S. society still does not provide complete equal opportunity for women, and women continue to be targets of discrimination, the culture and ideology pertaining to women's rights have changed with the times (there used to be a time in which women's voices were not even heard in a college classroom!).

Hegemony is but one component of the intellectual currents associated with Cultural Studies. Although people (audiences) are frequently influenced by dominant societal forces, at times hegemonic tendencies will emerge in the population. We explore this notion further.

Counter-Hegemony: The Masses Start to Influence the Dominant Forces

We have noted that hegemony is one of the core concepts associated with Cultural Studies. Yet, audiences are not always tricked into accepting and believing everything presented by the dominant forces. At times, audiences will use the same resources and strategies of dominant social groups. To some extent, individuals will use the same practices of hegemonic domination to challenge that domination. This is what Gramsci called **counter-hegemony.**

counter-hegemony when, at times, people use hegemonic behaviors to challenge the domination in their lives

Counter-hegemony becomes a critical part of Cultural Studies thinking because it suggests that audiences are not necessarily willing and compliant. In other words, we—as audience members—are not always the dumb and submissive people some have us made out to be! Danny Lesh of the Counter Hegemony Project, writes in *Counter Heg* (a blog dedicated to the counter-hegemonic movement) that part of the goal of counter-hegemony "is to understand history from other lenses, particularly from women's, workers', and racial minorities' perspectives." That is, in counter-hegemony, researchers try to raise the volume on voices that have been voiceless. Think of counter-hegemony as a point where individuals recognize their consent and try to do something about it. In **Chapter 28** on Muted Group Theory, you are introduced to how silenced voices deal with dominant groups.

Counter-hegemonic messages, interestingly enough, occur quite a bit in television, affecting viewers in subtle ways. Over the years, television has been a major vehicle for counter-hegemonic messages. For instance, counter-hegemony can be found in daytime television talk shows. Since many of the shows (e.g., *The View*) are comprised of solely women, viewers find topics that have been normally relegated to men (e.g., sex, politics, power, etc.). This genre embraces audience-centeredness as well as topics that relate to everyday people. This approach counters many dominant views; such shows (starting with Oprah many years ago) "challenge authority and expertise" in ways that were unseen prior to this new female-only genre. While these shows may appear to be nothing but entertainment exemplars, they, in their own ways, demonstrate how television content challenges the priorities established by the dominant forces, and subsequently, influences viewer attitudes and behaviors.

Before leaving this discussion, let's address one more point related to counter-hegemony and use one of the most popular television shows in history. *The Simpsons*, the longest-running comedy on television, also contains satiric counter-hegemonic messages aimed at showing that individuals who are dominated use the same symbolic resources to challenge that domination. Let's detail this multiple Emmy award winning show using elements of Cultural Studies. The relevance of *The Simpsons* to the lives of people has been persuasively argued. Barbara Hernandez (2021) observes that the show reveals so much about us because so much of the show intersects with the social institutions (family, school, jobs, houses of worship, etc.) to which we belong. The show has included references to such diverse topics as talk shows, Kafka, the Beatles, Tennessee Williams, Donald Trump, the invention of television, and hormone therapy! The core cast of characters—Marge (mother), Homer (father), Bart (son), Lisa (daughter), and Maggie (infant daughter)—all present different counter-hegemonic messages. For Marge, although cultural representations of a homemaker/housewife suggest a doting and supportive wife and mother, she is arguably the most independent of all the characters. She has tried a number of other professions, from police officer to protester against handgun violence. Homer, an employee at the local nuclear power plant, shows the viewer that despite governmental regulations on competency in these facilities, his bumbling approach to safety is still not a reason to fire him! And then there's Lisa, who shows that she is intellectually curious, artistically savvy, and environmentally aware.

One of the more central characters of the show is Bart Simpson. Interestingly, although society tends to shut down boys of Bart's age (and girls to some extent), Bart manages to shut down the same society that tries to subdue him. His pranks range from harassing a local bar with sophomoric phone calls to disrespecting his grade school principal to calling his father by his first name. In the end, despite the 20 minutes of chaos, the family members usually show that they have high regard for one another in personal ways. As Carl Matheson (2001) notes, the show advocates "a moral position of caring at the level of the individual, one which favors the family over the institution" (p. 4). Counter-hegemony clearly pervades this television series.

Scholars contend that the show has managed to transcend traditional images of the family that demonstrated and embraced patriarchal control. Indeed, a close examination of the series shows that children are frequently viewed as subverting or overthrowing parental control and asserting dominance in their family. *The Simpsons* continues to challenge prevailing religious, political, and cultural notions that the family is weakening.

Audience Decoding

No hegemonic or counter-hegemonic message can exist without an audience's ability to receive the message and compare it with meanings already stored in their minds (O'Donnell, 2017). This is called **decoding,** the final topic of Cultural Studies we wish to address. When we receive messages from others, we decode them according to our perceptions, thoughts, and past experiences. So, for instance, when Luisa Petrillo, from our chapter-opening story, interprets information on purchasing a home, she is relying on several mental behaviors. These include her desire to have a home, her conversations with people who have already purchased a home, her library visits, and the fact that she and her family have never owned a home. Luisa will store the information she receives pertaining to a new home and retrieve it when someone engages her in a conversation on the topic. All of this is done instantaneously; that is, she will make immediate decisions about how to interpret a message once she receives it.

> **decoding** receiving and comparing messages

The issues surrounding the decoding process are central to Cultural Studies. But, before we delve further into this and explain its importance let's review the gist of the theory up to this point. You will recall that the public receives a great deal of information from the elite and that people unconsciously consent to what dominant ideologies suggest. Theorists reason that the public should be envisioned as part of a larger cultural context, one in which those struggling for a voice are oppressed. As we discussed previously, hierarchical social relations (between the elite bosses and the subordinate workers, for example) exist in an uneven society. This results in subordinate cultures decoding the messages of the ruling class. Usually, according to Hall, the media connote the ruling class in Western society. Hall (1980) elaborates on how decoding works in the media. He recognizes that an audience decodes a message from three vantage points, or positions: dominant, negotiated, and oppositional. We explore each of these next.

Hall claims that individuals operate within a code that dominates and exercises more power than other codes. He terms this the **dominant position.** The professional code of television broadcasters, for instance, will always operate within the hegemony of the dominant code. Hall relates that professional codes reproduce hegemonic interpretations of reality. This is done with subtle persuasion. Consider John and Luisa Petrillo from our chapter opener. The television images of owning a home prompt the Petrillos to believe that owning a home is within their reach. The selection of words, the presentation of pictures, and the choice of spokespeople in infomercials are all part of the staging in the professional code. Audiences, like the Petrillos, are prone to either misunderstanding a message or selectively perceiving only certain parts of a message. Why? Hall writes, "The viewer does not know the

> **dominant position** operating within a code that allows one person to have control over another

terms employed, cannot follow the complex logic of argument or exposition, is unfamiliar with the language, finds the concepts too alien or difficult or is foxed by the expository narrative" (1980, p. 135). Television producers are worried that people like the Petrillos will not accept the intended and preferred media message of owning a home. They (the media) therefore contextualize their professional code in the larger, dominant cultural code of meaning. This ensures that John and Luisa Petrillo will seriously consider the value of working hard to buy a home.

The second position is a **negotiated position;** audience members are able to accept dominant ideologies, but will operate with some exceptions to the cultural rule. Hall holds that audience members always reserve the right to apply local conditions to large-scale events. This

negotiated position accepting dominant ideologies, but allowing for cultural exceptions

happens frequently when the media report on laws that are enacted at the national level and interpreted at the state or community level. For example, Hall might argue that although audiences may accept the elite's interpretation of a welfare reform bill in Washington, DC ("*All people should work if they are able to*"), they may have to negotiate when it does not coincide with a local or personal principle ("*Children need parents at home*"). Hall notes that due to the difficulty of negotiations, people are prone to communication failures.

The final way in which audiences decode messages is by engaging in an oppositional position. An **oppositional position** occurs when audience members substitute an alternative code for the code supplied by the media. At times, critical consumers reject the media's intended

oppositional position substituting alternative messages presented by the media

and preferred meaning of the message and instead replace it with their own way of thinking about a subject. Consider, for instance, the manner in which the media communicate feminine images of beauty. To many, the media present feminine beauty as a way to serve the sexual desire of men, reinforcing the hegemonic nature of feminine beauty standards (Dang, 2022). Some consumers, however, reject this capitalistic message and substitute more realistic portrayals.

Hall accepts the fact that the media frame messages with a covert intent to persuade. Audience members have the capacity to avoid being swallowed up in the dominant ideology, yet, as with the Petrillo family, the messages the audience receives are often part of a more subtle campaign. Theorists in Cultural Studies do not suggest that people are gullible, but rather that they often unknowingly become a part of the agenda of others.

Integration and Critique

Although Cultural Studies began at the CCCS in England, its influence in the United States has been vast. And, despite its infancy in the 20th century, issues pertaining to power and classism resonate today— in the United States and around the globe. The theory has attracted the attention of critical theorists in particular because it is founded on the principles of criticism. Its Marxist influence has also drawn scholars from philosophy, economics, and social psychology, and its emphasis on underrepresented groups in society has enticed writers in sociology and women's studies to take notice (Steiner, 2017). In addition, Cultural Studies theorists would/could never consider using experimental methods, making this theory reliant upon qualitative investigation. For additional criticism, we discuss three criteria for evaluating a theory: logical consistency, utility, and heurism.

Integration

Communication Tradition	Rhetorical | Semiotic | Phenomenological | Cybernetic | Socio-Psychological | Socio-Cultural | **Critical**
Communication Context	Intrapersonal | Interpersonal | Small Group/Team | Organizational | Public/Rhetorical | **Mass/Media** | Cultural
Approach to Knowing	Positivistic/Empirical | **Interpretive/Hermeneutic** | **Critical**

Critique

Evaluation Criteria	Scope | **Logical Consistency** | Parsimony | **Utility** | Testability | **Heurism** | Test of Time

Logical Consistency

Despite some glowing endorsements, the logical consistency of the theory has been challenged. This criticism relates to the audience. Even though some audiences resist the role of dupe, are they able to become interpretive and active resisters? In other words, to what extent can audiences be counter-hegemonic? Mike Budd, Robert Entman, and Clay Steinman (1990) suggest that some cultural and critical theorists overestimate the ability of oppressed and marginalized populations to escape their culture. Particularly for those communities that lack the skills, insights, and networks, escaping is very difficult. In fact, as we consider the television-viewing habits of people like the Petrillos from our chapter opening, counter-hegemony may not even be considered. That is, the family may simply be repackaging and reframing the information they receive about home ownership and critically applying relevant information to their lives. This dialogue is not likely to go away because "debates over the audience were once, and continue to be, a major field of contestation in cultural studies" (Kellner & Hammer, 2004, p. 79).

Utility

Cultural Studies "makes up a vehicle that can alter our self-image" (Carey, 1989, p. 94). Therefore, it's possible to translate some of the theory into daily life, making it useful to some extent. Its utility can also be found in its dedication to studying the cultural struggles of the underprivileged. According to Hall, these populations have remained subordinate for too long. By concentrating on these marginalized social groups, a number of subfields have emerged, namely, ethnic studies and gay, lesbian, bisexual, transgender studies (Surber, 1998). Hall's theory has been called "empirically elegant" (Carey, 1989, p. 31), and its usefulness beyond the written page has been widely articulated (e.g., Grossberg, 2010). The theory continues to be of importance in the academic world, with many of its elements spilling over into popular culture, including politics and culture.

Heurism

Many of the principles and features of Cultural Studies have been investigated. Ideology has been examined (Lewis & Morgan, 2001; Soar, 2000), and the concept of hegemony has been applied to episodes of television shows, including megahits such as *The Mary Tyler Moore Show* (Dow, 1990), *Saturday Night Live* (Davis, 2012), and *Sex in the City* (Brasfield, 2007). Hegemony has even been applied to the singing of the national anthem at sporting events (Molnar & Kelly, 2012), homophobia in football (Willson & Magrath, 2023), and Nike products (Moore, 2018), as well as employed in a study to understand educational reform in Brazil (Fischman & Gandin, 2016). Research by Janice Radway (1984, 1986) focused on romance novels and the women who read them. She discovered that many women read these books silently to protest male domination in society. Some feminists have embraced Cultural Studies and the efforts by Stuart Hall as he helped shape an understanding of voices that usually go unnoticed (e.g., Driscoll, 2016).

One final piece of heuristic evidence of Cultural Studies can be found in the expansive undertakings by scholars who have honored the tradition and the concepts of Cultural Studies. For instance, it's common to see entire issues of journals dedicated to Stuart Hall (e.g., *International Journal of Cultural Studies*) and volumes of scholarly essays devoted to his words and ideas (e.g., *Essential Essays, Vol. 1: Foundations of Cultural Studies*) as well as written pieces showering praise and admiration on his work (Jordan, 2016). It's remarkably clear that thinkers, writers, researchers, and essayists continue to laud a theorist who had a profound effect on research and society, a "spellbinding orator and a teacher of enormous influence" ("Stuart Hall obituary," 2014).

Closing

Cultural Studies remains one of the few theoretical traditions that has attracted the attention of scholars from a variety of disciplines outside communication. For that reason alone, it is uniquely both interdisciplinary and multidisciplinary. Researchers interested in understanding the thinking, experiences, and activities of historically oppressed populations usually endorse Cultural Studies as a model. Although some critics have faulted the theory for a number of reasons, Stuart Hall, identified as "the godfather of multiculturalism," is credited with criticizing the elite and drawing attention to oppressed voices in society.

Discussion Starters

 **Case-In-Point:** Do you think that the Petrillo family is responsible for not trying to leave their current situation? If they are not able to achieve the "American Dream" of owning a home, should the media be blamed? Include examples when expressing your opinion.

Try-It-Your-Selfie: Suppose you were asked to differentiate between Cultural Studies and social media. What would similarities and differences would you note between the two?

1. Discuss how hegemony functions in world events, such as the refugee crisis and political oppression. Now apply the concept to your campus. Identify any similarities and differences between the two applications. Use examples in your response.

2. What other cultural artifacts exist in our society that could be studied within a Cultural Studies framework?

3. British Cultural Studies is strongly focused on class differences. What do you think about applying the thinking of British Cultural Studies to Cultural Studies in North America? Do you believe that the concepts and principles are relevant to all countries? Why or why not?

4. Do you agree or disagree with the belief that oppressed populations have little voice in the United States? How does this view relate to how you feel about the theory?

References

Alwood, E. (1997, October 14). *The power of persuasion: The media's role in addressing homosexual issues in the future* (p. 54). The Advocate.

Baran, S. J., Davis, D. K., & Striby, K. (2020). *Mass communication theory: Foundations, ferment, and future.* Oxford University Press.

Brasfield, R. (2007). Rereading sex and the city: Exposing the hegemonic feminist narrative. *Journal of Popular Film and Television, 34,* 130–139.

Budd, M., Entman, R. M., & Steinman, C. (1990). The affirmative character of U.S. cultural studies. *Critical Studies in Mass Communication, 7,* 169–184.

Campbell, N., & Kean, A. (2016). *American cultural studies: An introduction.* Routledge.

Carey, J. W. (1989). *Communication as culture: Essays on media and society.* Unwin Hyman.

Colpani, G. (2022). Two theories of hegemony: Stuart Hall and Ernesto Laclau in conversation. *Political Theory, 50*(2), 221–246.

Cox, J. W., & Rich, S. (2023, May 3). *Are there warning signs? What we've learned from covering school shootings.* Washington Post. https://www.washingtonpost.com/dc-md-va/2023/05/03/questions-answers-gun-violence-us/.

Curthoys, A., & Docker, J. (2016). Stuart Hall: Reflections, memories, appreciations. *Cultural Studies Review, 22*(1), 302–306.

Dang, D. (2022). *Artificial intelligence: AI in fashion and beauty e-commerce.* [Master's Thesis, LAB University Press of Applied Sciences].

Davis, A. J. (2012). *Defining mixed race on television: Defining Barack Obama and Saturday Night Live.* [Master's Thesis. California State University Press].

Dow, B. J. (1990). Hegemony, feminist criticism and The Mary Tyler Moore Show. *Critical Studies in Mass Communication, 7,* 261–274.

Driscoll, C. (2016). Teaching cultural studies; Teaching Stuart Hall. *Cultural Studies Review, 22,* 269–276.

du Gay, P., Hall, S., Janes, L., Mackay, H., & Negus, K. (1997). *Doing cultural studies: The story of the Sony Walkman.* Sage/The Open University Press.

Fischman, G. E., & Gandin, L. A. (2016). The pedagogical and ethical legacy of a "successful" educational reform: The citizen school project. *International Review or Education, 62,* 63–89.

Grossberg, L. (2010). On the political responsibilities of Cultural Studies. *Inter-Asia Cultural Studies, 11*(2), 241–247.

Hall, S. (1980). Encoding/decoding. In S. Hall, D. Hobson, A. Lowe, & P. Willis (Eds.), *Culture, media, language* (p. 135). Taylor & Francis.

Hall, S. (1981). The whites of their eyes: Racist ideologies and the media. In G. Bridges & R. Brunt (Eds.), *Silver linings: Some strategies for the eighties* (p. 31). Lawrence and Wishart.

Hall, S. (1989). Ideology and communication theory. In B. Dervin, L. Grossberg, B. J. O'Keefe, & E. Wartella (Eds.), *Rethinking communication: Paradigm issues* (pp. 40–51). Sage.

Hall, S. (1992). Cultural Studies and its theoretical legacies. In L. Grossberg, C. Nelson, & P. Treichler (Eds.), *Cultural studies* (pp. 277–294). Taylor & Francis.

Hall, S. (1997). The problem of ideology: Marxism without guarantees. In S. Hall (Ed.), *Representation: Cultural representations and signifying practices* (p. 40). Sage.

Hernandez, B. (2021). *The Simpsons fun facts: An incredibly fascinating book for not only relaxing but also getting to know more about The Simpsons.* Self-published.

Humphreys, L. (2023). Mobile social media: The challenges and opportunities continue. *Mobile Media & Communication, 11*, 74-79.

Inglis, F. (1993). *Cultural studies.* John Wiley & Sons, Inc.

Jordan, G. (2016). Beyond essentialism: On Stuart Hall and Black British arts. *International Journal of Cultural Studies, 19*, 11-27.

Kellner, D., & Hammer, R. (2004). Critical reflections on Mel Gibson's "The Passion of the Christ." *Logos, 3*, 3-4.

Lash, S. (2007). Power after hegemony: Cultural Studies in mutation. *Theory, Culture, & Society, 24*, 55-78.

Lewis, J., & Morgan, M. (2001). He may not be a liberal but he plays one on TV: Imagining the ideology of President Clinton. *The Communication Review, 4*, 327-346.

Matheson, C. (2001). *The Simpsons and philosophy: The d'oh! of Homer.* Open Court.

Molnar, G., & Kelly, J. (2012). *Sport, exercise, and social theory: An introduction.* Routledge.

Moon, D. (2008). Concepts of "culture": Implications for intercultural communication research. In M. K. Asante, Y. Miike, & J. Yin (Eds.), *The global intercultural communication reader* (pp. 11-26). Routledge.

Moore, R. C. (2018). Islamophobia, patriarchy, or corporate hegemony?: News coverage of Nike's pro sport hijab. *Journal of Media and Religion, 17*(3-4), 106-116.

Morley, D. G. (2019a). General introduction. In S. Hall & D. Morley (Eds.), *Stuart Hall: Essential essays Volume 2–Identity and diaspora* (pp. 1-17). Duke University Press.

Morley, D. G. (2019b). General introduction: A life in essays. In S. Hall & D. Morley (Eds.), *Stuart Hall: Essential essays Volume 1–Foundations of Cultural Studies* (pp. 1-27). Duke University Press.

Murdock, G. (1989, May). Cultural Studies: Missing links. *Critical Studies in Mass Communication, 6*(4), 436-440.

O'Donnell, V. (2017). *Television criticism.* Sage.

Radway, J. (1984). *Reading the romance: Women, patriarchy, and popular literature.* University of North Carolina Press.

Radway, J. (1986). Identifying ideological seams: Mass culture, analytical methods, and political practice. *Communication, 9*, 93-123.

Real, M. R. (1996). *Exploring media culture.* Sage.

Ryder, S. (2016). The Flint water crisis and beyond. *Natural Hazards, 60*, 23.

Soar, M. (2000). Encoding advertisements: Ideology and meaning in advertising production. *Mass Communication & Society, 3*, 415-437.

Steiner, L. (2017). "Wrestling with the angels": Stuart Hall's theory and method. *Howard Journal of Communications, 27*, 102-111.

Stuart Hall obituary. (2014, February 10). *The Guardian.* Retrieved from https://www.theguardian.com/politics/2014/feb/10/stuart-hall.

Surber, J. P. (1998). *Culture and critique: An introduction to the critical discourses of Cultural Studies.* Westview Press.

Szulc, L. (2023). Culture is transnational. *International Journal of Cultural Studies, 26*, 3-15.

Willson, J., & Magrath, R. (2023). English (men's) football, masculinity and homophobia. In W. Roberts, S. Whigham, A. Culvin, & D. Parnell (Eds.), *Critical issues in football: A sociological analysis of the beautiful game* (pp. 149-162). Routledge.

CHAPTER 28
Muted Group Theory

*Based on the research of **Cheris Kramarae***

"Women (and members of other subordinate groups) are not as free or as able as men are to say what they wish, when and where they wish, because the words and the norms for their use have been formulated by the dominant group."

—Cheris Kramarae

Patricia Fitzpatrick

Patricia Fitzpatrick is sitting in the back row in her political science classroom; she closes her eyes, letting the professor's words wash over her. She is wondering what in the world possessed her to go back to college at the age of 40. It seemed like a good idea when she first thought about finishing the BA she had begun over 20 years ago, but now she thinks she must have been out of her mind to try this. She feels like she doesn't understand a word her professors say, and she is having an extremely difficult time keeping up with her assignments. She is beginning to think she isn't cut out for school. Most of the other students are much younger and seem to have much more disposable income, too. She really hasn't spoken to many of them beyond a quick hello. All in all, this college idea is turning into a big hassle.

It wasn't supposed to be this way. Initially, Patricia thought finishing her degree would improve her self-esteem. She also hoped it would get her out of the dead-end job she currently has at Hudson's Department Store. She works nights as a shipping clerk for the store, sending out orders for their mail-order business. She's been working there for the past seven years, and it doesn't seem as if she will ever get anywhere higher than where she is now. It's a pretty boring job with low pay, and Patricia was hopeful that a degree might give her more opportunities. As a single mom with three kids, she really needs a better job.

But Patricia is starting to despair about her prospects. One of the most disturbing things so far has been how tongue-tied she feels in class. She has so much trouble articulating what she wants to say about the material that she ends up not saying anything at all in most of her classes. Patricia worries that this will count against her, especially in her smaller classes, where the professors really encourage class participation. She had expected that she'd have a lot to say as she learned new things, but she's finding it difficult to express her opinions. So, instead of increasing her self-esteem, college is making her feel worse about herself. She spends a lot of time trying to figure out what she's doing wrong and why she can't make herself successful at school.

This isn't the first time that Patricia's felt inarticulate. She admits to herself that she's had a hard time speaking up since long before she returned to college. At work, she's the only woman on her shift, and the men don't pay attention to her when she speaks. Once she tried to compare an experience at work to teaching her son to read, and her coworkers looked at her like she was nuts. Then they all started laughing and calling her "Mom." Another time she was the chair of a committee at her daughter's school. The committee was supposed to research ways the school could raise money for more computer equipment. Patricia had to get up in front of representatives from the parents' group, teachers, and administrators and report on the committee's findings. Although her presentation had been okay, Patricia thought she had paused too much while she spoke. Even though she'd spent a great deal of time preparing for the speech, she still found herself occasionally groping for words.

She had hoped that college would help her become more articulate. It would be wonderful to be able to speak up and say what's on her mind. Patricia isn't ready to give up, but things certainly aren't going as planned.

Some researchers analyzing Patricia's situation might begin by examining her individual traits, noting that she might have problems with shyness or a fear of communicating, called *communication apprehension*. However, the theory that we discuss in this chapter offers a different approach to explaining Patricia's situation. Muted Group Theory (MGT) asserts that, as a female, as a single mom, and as a person with a low income, Patricia is a member of several groups whose power base doesn't allow her to express her voice, or even to acknowledge her own voice in her head. The groups to which Patricia belongs have experiences that aren't easily expressed in her language system—a language system that was devised primarily by well-to-do men to represent their own experiences. According to MGT, Patricia and others are muted because their native language often does not provide a good fit with their life experiences. As Dennis Nangabo (2015) notes, it's quite challenging to communicate in a language that another group has created to express their experiences when you're not a member of that group.

MGT focuses on who has the power to name experiences and explains that women trying to use man-made language to describe their own experiences is somewhat like native English speakers learning to converse in Spanish or another nonnative language. To do so, they have to go through an internal translation process, scanning the foreign vocabulary for the best word to express themselves. This process makes them hesitant and often inarticulate because they're unable to use the language fluently for their purposes. In the process, muted groups metaphorically lose their voice.

Yet, it isn't the case that all women are silenced, and all men have voice. MGT allows us to understand that any non-dominant group member may be silenced by the inadequacies of their language (e.g., Ayinla & Amenaghawon, 2021; Hechter, 2004; Herakova, 2009; Orbe, 1998b, 2005). Kami Kosenko (2010) notes that MGT offers a heuristic explanation for how the transgender individuals in her study were "rendered mute by a biomedical discourse that fails to represent the transgender body or sexual experience" (p. 140).

Similarly, Lance Young (2017) argues that military veterans trying to reintegrate into civilian society are also a muted group. In his study of why the "thank a veteran for their service" campaign is problematic, Young suggests that military vets actually have to learn or relearn civilian language and struggle to find their voice in the process. Young asserts that unlike the groups that MGT usually refers to, vets are objectified and muted while being elevated to hero status. As one of Young's sources notes, thanking a veteran for their service accomplishes muting by being "dangerously dismissive" (p. 7).

Radhika Chopra (2001) examines the issue of muting nurturing fathers. Chopra argues that the discourse of mothering reduces the father to a "depersonalized cipher" whose presence is "posited as an absence, in contrast with the hands-on vital involvement of the mother" (p. 447). Chopra notes that this discourse not only silences the father but also fails to account for fathering in other societies or periods of history where the distant, silent father did not exist. Chopra points to fathers in the early American colonial period who were very involved in teaching their children both reading and writing skills as well as moral and religious lessons. Although some Muted Group theorists might disagree that fathers (or any men) are ever in a low-power position, Chopra concludes that giving voice to "father-care" is a critical process that will remove the category of caregiving from female ownership and in so doing transform the concept of gender identity completely.

Some people have argued that current sensitivities to non-dominant groups are muting dominant group members even when they try to be allies. In 2018, a Black woman and a white woman witnessed two Black businessmen being arrested at a Starbucks after one asked to use the restroom. Both women, who didn't know each other previously, protested the men's arrest (which led to Starbucks instituting diversity training throughout the company). After this event, the women, Michelle Saahane and Melissa DePino, became friends and partners who established a website and spoke to groups about racism. Ultimately Melissa considered Michelle her best friend, but over time their friendship began to fray and Melissa, who is white, felt she was unable to find the words to correctly express her positions. She felt Michelle unfairly cast her as racist no matter how she tried to express herself. Melissa thought words failed her. They dissolved their partnership and are no longer friends.

Furthermore, muting may take place as a result of the unpopularity of the views that a person is trying to express. Muted Group theorists argue that within a group, even a marginalized group, a faction may become dominant and further marginalize a subgroup. Anita Taylor and M. J. Hardman (2000) advance the example of Matilde Joslyn Gage, a 19th-century feminist, who argued views that made her unpopular with the other feminists of the day, such as Elizabeth Cady Stanton. As a result of advancing ideas that were considered too radical, "she became a persona non grata among the women of the movement" (pp. 2–3) who had more power; consequently, Gage's voice was muted and virtually lost to future generations. In addition, Marsha Houston and Cheris Kramarae (1991) observe that silencing occurs not only through preventing talk but also by shaping and controlling the talk of others. When white women criticize Black women's talk as "confrontational," "the message to Black women is, 'Talk like me, or I won't listen'" (p. 389).

Theory at a Glance • Muted Group Theory

Language serves the dominant group who created it (and those who belong to the same groups as its creators) better than those in non-dominant groups who have to learn to use the language to describe their different experiences as best as they can. This is the case because the experiences of the creators are named clearly in language, whereas the experiences of other groups are not. Due to their problems adapting to a language they didn't create, people from other groups appear less articulate than those from groups like the creators. As a result, they are often belittled or ridiculed. Sometimes these muted groups create their own language to compensate for the problems they have using the dominant group's language.

As you can probably guess from our discussion thus far, MGT points out problems with the status quo and suggests ways to remediate these problems, making it a critical theory. Later in this chapter, we describe some of the action steps advocated by Muted Group's proponents. Further, MGT is a theory that examines power issues (Nangabo, 2015). As Cheris Kramarae (2005) observes, "people attached or assigned to subordinate groups may have a lot to say, but they tend to have relatively little power to say it" (p. 55). Or if they do venture to speak, those in a greater power position may ignore, ridicule, or disrespect their contribution in a variety of ways. Kramarae (2009) notes that muted groups "get in trouble" when they speak out in their own voices.

History of Muted Group Theory

MGT originated with the work of Edwin and Shirley Ardener, social anthropologists who were concerned with social structure and hierarchy. In 1975, Edwin Ardener noted that groups making up the top end of the social hierarchy determine the communication system for the culture. Lower-power groups in the society, such as women, children, the poor, and people of color, have to learn to work within the communication system that the dominant group has established. Turning this generalization to the specific case of women in a culture, Shirley Ardener observes that social anthropologists have studied women's experiences by talking almost exclusively to men. Thus, not only do women have to contend with the difficulties of a language that doesn't completely give voice to their thoughts, but their experiences are represented in the social science literature from a male perspective (Ardener, 2005).

Edwin Ardener (1975) commented on why researchers (who at the time were mainly men) tended to speak and listen to men in the cultures they studied. Ardener believes that these scholars "have a bias towards the kinds of model that men are ready to provide (or to concur in) rather than towards any that women might provide. If the men appear 'articulate' compared with the women, it is a case of like speaking like" (p. 2). Further, Shirley Ardener (1978) observes that women's mutedness is the counterpart to men's deafness. She explains that women (or members of any subordinate group) do speak, but their words fall on deaf ears, and when this happens over time, they tend to stop trying to articulate their thoughts, and they may even stop thinking them. In Ardener's view, "Words which continually fall upon deaf ears may, of course, in the end become unspoken, or even unthought" (p. 20).

The Ardeners might argue that Patricia Fitzpatrick is twice muted: once by the failure of her language to act as a reliable tool for her use and once by anyone who judges her as unintelligent or inarticulate rather than acknowledging she's poorly served by her language. Muted groups are rendered inarticulate by the dominant group's language system, which grows directly out of the dominant group's worldview and experience. When a member of a muted group wishes to speak, what they say first has to shift out of their own worldview and be compared to the experiences of the dominant group. Thus, articulations for the muted group are indirect and broken (see **Figure 28.1**). Others have applied the Ardeners' ideas to a variety of contexts and situations, including classrooms (e.g., Gatere, Kiumi, & Ngugi, 2018; Jule, 2018), students with disabilities (Cubbage, 2019), and people experiencing domestic violence (e.g., Owusu, 2016).

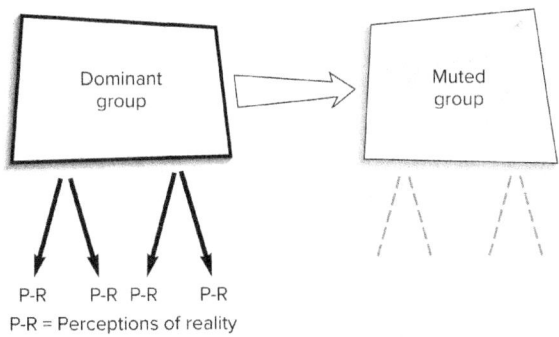

Figure 28.1 Muted Group Theory

Adapted from Ardener, Edwin, 1975.

Cheris Kramarae took the Ardeners' ideas and applied them to the field of communication and made the theory more explicitly feminist. In 1981, Kramarae published an influential book called *Women and Men Speaking: Frameworks for Analysis.* In this book, Kramarae profiles several theories that come from disciplines outside communication studies and suggests how these theoretical frameworks can help explain questions of communication. In her discussion of MGT, Kramarae states that her aims are somewhat more limited than those of the Ardeners, who were concerned with applying MGT across many cultures. Kramarae's interest is in the English language and so she restricts the theory to Great Britain and the United States. Kramarae asserts that MGT's assertions are especially true for the English language because English was developed and formalized by male clerics and academics. Because women were not instrumental in formalizing the English language and were in a lower power position in the culture that created it, Kramarae asserts that women are a muted group.

Speaking about MGT, Anita Taylor and M. J. Hardman (2000) observe that "English does not name concepts important to women, but to men (e.g., one can have a 'seminal' idea; but was one ever described as 'ovular'?)" (p. 8).

Dominant and Non-Dominant Groups

Mark Orbe (1998a & b, 2005) suggests that in the United States and several other cultures, society privileges specific characteristics and perspectives, which include: male, European American, heterosexual, able-bodied, youthful, middle and upper class, and Christian. People with these characteristics form the **dominant group,** or the group that holds the power in the culture. Other groups that coexist with the dominant group are generally **non-dominant groups** and are sub-

dominant group the group that holds the power in a given culture

non-dominant groups groups that have less power and are subordinate to the dominant group

ordinate to the dominant group. They don't have access to as much power as do members of the dominant group. Thus, non-European groups such as African Americans or Asian Americans, gay men and lesbians, the elderly, those in lower socioeconomic classes, people with disabilities, and non-Christians all can be members of muted groups in the same way that women are. Orbe (1998a & b, 2012) developed Co-Cultural Theory, based on Muted Group Theory, to capture these ideas. We discuss Co-Cultural Theory in **Chapter 30.** However, because most of the theorizing and application of Kramarae's version of MGT has focused on women as a muted group, that's how we'll direct our attention in this chapter.

In training our focus on women as a non-dominant group, we need to discuss two terms: sex and gender. Generally, researchers have used the term **sex** to mean biological categories, male and female, determined by the presence of XX chromosomes for females and XY chromosomes for males. As a biological concept, sex is presumed to be fixed and unchanging. In contrast, **gender** is defined as the learned behaviors that constitute femininity and masculinity in a given culture. Thus, the definition of gender is changeable and reflects what-

sex biological category divided into male and female

gender social category consisting of the learned behaviors that constitute masculinity and femininity for a given culture

ever the culture accepts at a given time for these roles. Currently, in the United States, it's within the definition of masculinity to have pierced ears, and it's within the definition of femininity to have tattoos. At an earlier time, these attributes would not have been deemed appropriate for those gender roles. It's also the case that many laypeople and researchers now view sex as malleable. This leads to a move away from the fixed, binary perspective that allows only two categories for sex and presumes those categories are unchanging. Researchers argue that it's more productive to view sex on a continuum and not bifurcate it (e.g., Foss, Domenico, & Foss, 2013). While transgender people challenge the idea that biological sex is a constant, they tend to abide by the binary that creates two sexes only. The terms *sex* and *gender* are often blurred because in most cultures, women are socialized to be feminine and men are encouraged to be masculine. Therefore, we will use the terms interchangeably in this chapter. Yet, it's useful to keep clear when a researcher is talking about a social construction like the learned behaviors constituting gender roles or a biological construction such as sex.

Assumptions of Muted Group Theory

Cheris Kramarae (1981) isolated three assumptions she believes are central to MGT. We review each of them below:

- Women perceive the world differently than men because of women's and men's different experiences and activities rooted in the division of labor.
- Because of their political dominance, men's system of perception is dominant, impeding the free expression of women's alternative models of the world.
- In order to participate in society, women must transform their own models in terms of the received male system of expression (p. 3).

The first assumption begins with the premise that the world is a different place for women and men and that their experiences differ. Furthermore, the assumption posits an explanation for these differences. The explanation lies in the division of labor that allocates work on the basis of sex, such that women are responsible for tasks in the home and men are responsible for work outside the home. The division of labor began in Western countries in the 18th and 19th centuries as a result of social transformations, in large part related to the Industrial Revolution (Coontz, 1988). The Industrial Revolution took work out of the home and made it a paid activity. Prior to that, work had been intertwined with home life because all members of the family contributed to the family's survival, usually on subsistence farms. No one was literally paid for their specific labor; the money the family realized came from selling their produce or livestock, the result of collective work by the entire family.

The separation of the workplace from the home led to a recognition of the two as separate spheres; the conceptualization of public and private came about, where the workplace was seen as public life, and the family was classified as private life (Butler & Modaff, 2015). The result of this division was to cast women's role in the home, or private life, and men's role in the workplace, or public life. This had the effect of sharply dividing what women's responsibilities were in contrast to men's. Stephanie Coontz (1988) notes that this trend occurred in all classes and ethnic groups in the United States at the time except African Americans.

You can see the logic of attributing the division of labor as the cause of men's and women's differing worldviews. When people's occupations differ greatly, they tend to see the world in different ways. If **gender polarization lens** viewing men and women as polar opposites your day is spent caring for children and the home, your experiences will vary a great deal from those of someone whose day is spent selling merchandise to others. Sandra Bem (1993) argues that this division also created what she labels a **gender polarization lens** that causes people to see women and men as very different from each other. In 1909, Clara E. Haase, a student at Milwaukee Downer Women's College, wrote a senior essay that illustrates the lens of gender polarization. She wrote about the topic of single-sex schooling, saying, "It were far better not to class girls with boys while receiving their education, the most important reason being that they are so entirely different and therefore should be taught diversely." Further, when people believe that women and men are radically different from each other, we might expect that they get treated differently. Different treatment certainly results in different experiences.

As we have written elsewhere with Judy Pearson (Pearson, Turner, & West, 1995), "From birth it is clear that male and female babies are treated differently.... Male [infants] are more likely than female infants to be described as 'strong,' 'solid,' or 'independent.' Female infants, on the other hand, are often described as 'loving,' 'cute,' and 'sweet'" (p. 49). Males and females are treated differently and expected to do different activities. Even when women work outside the home, they are often still expected to take primary responsibility for the home and care for children or elderly parents as the need arises (Wood & Fixmer-Oraiz, 2017).

Arlie Hochschild's research (Blair-Loy, Hochschild, Pugh, Williams, & Hartmann, 2015; Hochschild, 1989) talks about the phenomenon of the **second shift,** where working mothers put in 8 hours at their paid job and then come home to do a second shift there.

Sometimes the differing experiences of men and women are subtle. One study (Turner, 1992) asked respondents to describe situations that they believe are uniquely experienced by their own sex and for which, currently, no word exists and then to give those experiences a name. This was patterned after an activity Judy Pearson (1985) calls

> **second shift** the phenomenon of working women putting in 8 hours on the job and another day's work at home

creating "genlets" or "sexlets." The results were not completely supportive of MGT because men generated essentially the same number of "genlets" as women, indicating that language fails men as well as women.

However, the types of words created did differ between men and women. Men coined words concerned with drinking and competition, whereas women created words focusing on relationships and personal issues such as appearance. Women also noted experiences where they felt fearful or uncertain, and this concern was not included in the men's words. For instance, women coined the word *herdastudaphobia*, which meant feeling fear when passing a group of strange men on the street, and they created the word *piglabelphobia*, to designate a woman's tendency to limit what she eats in front of a date for fear of being labeled a pig. Men created words such as *scarfaholic*, an eating contest among men, and *gearheaditis*, an obsession with fixing up one's car to make it the best on the road. Interestingly, men's focus was frequently on competition, whereas women were most often focused on their own insecurities and their desire to please others. Thus, the first assumption of MGT, that men and women have different experiences, seemed to be supported by these results.

The second assumption of MGT goes beyond simply noting that women and men have different experiences. This assumption states that men are the dominant group, and their experiences are given preference over women's. Specifically, men are in charge of naming and labeling social life, and women's experiences are often unnamed as a result. Women then have difficulties talking about their experiences.

Patricia Fitzpatrick from our chapter-opening vignette is experiencing problems speaking out in class, talking to the other students, speaking to her coworkers, and generally articulating her ideas. She's uncomfortable about her difficulties, and she blames herself because things aren't working out as she planned. MGT takes a different perspective, however, noting that these problems are not the result of women's inadequacies, but rather are caused by the unresponsiveness of the language women have to use when expressing themselves. MGT argues that any speaker would be inarticulate if there were no words in their language to describe their experiences and thoughts.

MGT asserts that men's political dominance allows their perceptions to be dominant. This dominance forces alternative perceptions—those that women hold as a result of their different experiences—into a subordinate position. Women's communication is constrained because of this subordinate position. Ann Burnett and her colleagues (2009) used MGT and Co-Cultural Theory to investigate men's dominance on college campuses. They found that the way rape was talked about on campus tended to mute female students, potentially contributing to the creation and perpetuation of a campus rape culture.

Cindy Reuther and Gail Fairhurst (2000) discuss the "glass ceiling" for women in organizational hierarchies and comment on how (white) men's experiences dominate the world of work. They observe that patriarchal values tend to reproduce themselves in U.S. organizations to men's advantage. Even though Kramarae put boundaries around MGT that focused on the U.S. and the U.K., this finding seems to hold across cultures because Sayed Mahrukh, Ayaz Ahmad, and Liaqat Iqbal (2017) used MGT in their study and also found women to be muted in workplaces in the Pashtun society. The researchers found that although women and men tended to share an equal workload and worked well together, this only provided a facade of equality. From their interviews, the authors concluded that male workers are the dominant group and use authoritative language to reinforce their dominance. Women reported being fearful of ridicule at work and often chose to remain silent. Their silence results in a negative spiral wherein men believe that women cannot speak logically and then marginalize women workers even more.

Theory Into Practice • Muted Group Theory

Theoretical Claim: Men have created and controlled language that has rendered them as dominant and women as subordinate.

Practical Implication: It didn't take long for Alicia to see the "man's world" in tech. During breaks with her coworkers, she heard the guys telling her to "man up." Alicia found some of the "everyday" terms like "hard drive", "joy stick", and "RAM" very offensive, and yet, she was rather disempowered because she was a new employee in a male-dominated company. It wasn't long, though, before a female VP was hired, and one of her first directives was related to throwing out any perceived or real sexist language in policies and procedures as well as establishing a statement discouraging and prohibiting its usage. That was really helpful and made a huge cultural change at the company. It wasn't perfect—there was still sexism—but Alicia thought it was a big improvement.

Overall, these results support MGT's second assumption, that men's, especially white men's, experiences are dominant, and women and people of color need to subordinate their own experiences to the extent that they can in order to partake in social and organizational success.

The final assumption of MGT speaks to the process of translation that women must go through in order to participate in social and work life. Women's task is to conceptualize a thought and then scan the vocabulary, which is really better suited to men's thinking, for the best words for encoding that thought. You can see how this would render women less fluent speakers than men. Tillie Olsen (1978), the author of *Silences*, says that although men are supposed to tell it straight, women have to "tell it slant" (p. 23). When subordinate group members undertake this translation process, their talk is slowed and less powerful.

The pauses that worry Patricia Fitzpatrick in our chapter-opening scenario are attributable, according to MGT, to this cumbersome process of translation that women have to engage in when speaking. Some researchers (e.g., Hayden, 1994) suggest that women's groups engage in a great deal of overlaps and simultaneous speaking because they are helping one another cope with a language system that isn't well suited to their tasks. Thus, when women speak with one another, they collaborate on storytelling—not so much because women are collaborative by nature, but rather because they need to help one another find the right words to encode their thoughts.

Some of the problems inherent in the translation process are highlighted by examining instances when women's experiences are named and do become part of the general vocabulary. For example, before the 1970s, the term *sexual harassment* didn't exist, but the behaviors, nonetheless, existed. Monica Hesse (2016) writes in the *Washington Post* that sexual harassment was common as early as the 19th century. She writes: "The issue had been addressed in the 19th century, when an 1887 report on 'women wage-workers' declared that household service 'has become synonymous with the worst degradation that comes to a woman,' because male employers so commonly assaulted their maids." Ultimately, in the 1970s, activists at Cornell University coined the term *sexual harassment* (Reed, 2013). Prior to that time, women who experienced what we now call sexual harassment had no clear way to label their experiences. As Gloria Steinem has said, before the term was accepted into the vocabulary, women simply accepted harassment as part of life. For instance, Angie had a job in a library years ago, and her boss was a man who liked to comment on her figure and talk about how it was so much better than his wife's. He frequently noted that her breasts, for instance, were a lot larger than his wife's. At the time, Angie could only say that she had a bad job with a horrible boss. Without a label, the experience seems individual and the inappropriateness of that behavior is minimized. Labeling it sexual harassment places it in a category, suggests some coping strategies, and points to its seriousness. Furthermore, giving the experience a term, *sexual harassment,* allows us to see that it exists broadly and is supported at many levels of society. In the same way, the #MeToo movement is an effort to give voice to those who've experienced sexual harassment, abuse, rape, and assault. Social change is possible when we recognize the phenomenon with a label. Similarly, the terms *domestic violence, date rape, marital rape,* and *stalking* all name crimes that without the labels might simply be seen as individual problems and not recognized as serious offenses. Before these words existed, women had difficulty when they wanted to talk about their experiences, and they were at a loss for instituting social change.

Students Talking: Ethan

When I started reading about Muted Group Theory, it really seemed wrong to me. I mean there are a lot of times I'm not sure exactly how to put my thoughts into words and I'm a white, cisgender guy, so I don't qualify for any of the non-dominant muted groups that the book talks about. And, to be honest, that part about some MGT researchers never thinking a man could be muted or have a problem, got me bummed. But, when I read that sexual harassment didn't have a name for a long time, it did make me think that maybe the theory had a point. I guess it would be pretty hard to talk about something if there were literally no words to describe it. And I can see how you'd think it was something that just happened to you and not some big thing that was going on with a lot of people if there were no words for it.

The Process of Silencing

The central premise of MGT is that members of marginalized groups are silenced and rendered inarticulate as speakers. Silencing doesn't rely on explicit enforcement or coercion. As Robin Sheriff (2000) observes, silencing of muted groups is a socially shared phenomenon. According to Sheriff, "unlike the activity of speech, which does not require more than a single actor, silence demands collaboration and the tacit communal understandings that such collaboration presupposes. Although it is contractual in nature, a critical feature of this type of silence is that it is both a consequence and an index of an unequal distribution of power" (p. 114). Thus, silence is accomplished through a social understanding of who holds the power and who doesn't (Gendrin, 2000). When other workers are silent (or join in) while a woman is ridiculed at work or school, for instance, the social group shares in the muting process (e.g., Robertson, Zuniga, Christenson, & Young, 2022). This is why bystanders are encouraged to speak up and not allow bullying to go uncontested. The process of silencing flourishes when the culture allows it to.

We'll talk briefly about some specific ways the dominant group silences the non-dominant group below, but first we provide some general ways that muting occurs (c.f. Poapst & Harper, 2017). First, the dominant group makes the rules that govern how talk should occur, what vocabulary is appropriate, and what topics are out of bounds. When you're not a member of the rule-making group, it's difficult to know how to speak, and you must spend time figuring out what's allowable and what's not. Secondly, the dominant group gets to define experiences, their own and those of members of non-dominant groups. For instance, when a woman reports a rape to a male-dominated police force, the police, not the woman, define the parameters of the crime. The police's definition is unlikely to be the same as the woman's definition, but she's not in charge, so it's difficult for her to challenge the police. Finally, the dominant group members believe that they are representative of the population at large, and so they don't stop to think that their perspectives might be skewed. When a group doesn't acknowledge the limits of its perspective, it's difficult for change to take place. If you think about Patricia from our chapter-opening story, you can see how all these factors work in concert to silence her. At work and school and even on the committee she chaired, she didn't make the rules for contributing. Teachers, administrators, and bosses set the expectations for her participation based on their own experiences and their assessment of everyone else's experiences. Patricia has to figure this out and adjust herself to these rules and definitions. Because her situation as a low-income, single mom returning to school and working in a male-dominated work environment is so different from the dominant group's perspective, she's struggling. Given this general backdrop, it should be easy to see why Patricia can't communicate as effectively as members of the dominant group. Even when the police, the schools, and so forth have women employees and bosses, usually the women have achieved their positions through socialization. Although they bring women's bodies to the organizations, they have shifted their thinking to a male perspective in order to succeed. Think about women and men bosses you have worked for or men and women professors you have taken classes from. How did they differ? If they were significantly different, were they also different in their rank? That is, did one sex tend to hold higher power positions?

Now, we briefly review four specific methods that actualize the general atmosphere of silencing we've just discussed. These include: ridicule, ritual, control, and harassment. This isn't an exhaustive list of methods for silencing. Perhaps you can think of other ways you have experienced silencing yourself or observed when someone else has been silenced.

Ridicule

Houston and Kramarae (1991) point out that women's speech is often trivialized and disrespected: "Men label women's talk chattering, gossiping, nagging, whining, bitching (Cut the cackle!)" (p. 390). Men often say that women talk about meaningless things and claim that they cannot understand how women spend so much time talking to their girlfriends. Women themselves often refer to their own talk as gabbing or gossiping. Women are also told that they have no sense of humor, and this is another opportunity to ridicule women. Also, women's concerns are often trivialized by men as not being important enough to listen to, yet women are expected to be supportive listeners for men. Even in the medical context, women report that their doctors may ridicule or trivialize their concerns (Brann & Mattson, 2004; Ehrenreich & English, 2005).

Ritual

Some people have pointed out that many social rituals have the effect of silencing women or advocating that women are subordinate to men. One such ritual is the wedding ceremony. There are a number of aspects in a traditional wedding ceremony that silence the bride. First, the groom stands at the front while the bride is "delivered" to him on the arm of her father. The father then "gives her away" to the groom. Furthermore, the groom stands at the right hand of the minister (or whoever officiates), and this is traditionally a place of higher status than the bride's position, to the left. The groom says the vows first. The bride wears a veil and a white gown to indicate that she has been "preserved" for the groom. The groom is told at the end that he may kiss the bride. The couple is pronounced "man" and "wife." Although many couples work to individualize the ceremony and have changed some of these aspects, the traditional ritual implies subordination of the bride. In addition to the wedding ritual itself, the ritual of a woman taking her husband's name as her own and losing her birth name after marriage casts women in a non-dominant position. In her 1981 book, Kramarae talks about the process she went through concerning her name changes. When she married, her husband's name was Kramer and her name upon marriage became Cheris Rae Kramer. At the time of her marriage, it wasn't legal for a wife to keep her own name. Later the laws changed, and Kramarae combined her husband's last name with her middle name and created her own name. This change was accomplished with a great deal of discussion and controversy, and many people questioned her judgment. Kramarae points out that no one questioned her husband throughout the entire process, and his name remained unchanged. According to the Pew Research Center (Lin, 2023) although many things may have changed about the traditions surrounding marriage, one thing that hasn't is the ritual of a woman taking her husband's last name. Their survey found almost 80 percent of women in heterosexual marriages said they took their husband's name when they got married. Fourteen percent of the respondents said they kept their birth names while 5 percent said they hyphenated their names with their spouse's last name.

Control

Researchers have noted that men control many decisions, such as what goes into history books, leaving women's history unrecorded (Fixmer-Oraiz & Wood, 2019). In some patriarchal cultures, control is established through the practice of subordinating women to men regardless of what material value women may bring to the society (Sari, 2016). In addition, most media are controlled by men, and women's talk and contributions get less coverage in mainstream media. Furthermore, many communication practices place men as central and women as on the margins. For instance, men talk more than women in mixed-sex interactions despite conventional wisdom to the contrary. In an analysis of 66 previous studies examining U.S. men's and women's talk in a variety of contexts, men's talk time exceeded women's in 61 of the cases (James & Drackich, 1993).

One communication behavior that keeps men in control is interruptions (Cannon, Robinson, & Smith-Lovin, 2019). When men interrupt women, the women often switch to talk about whatever topic the men raise. When women interrupt men, this is usually not the case. Men frequently go back to what they originally were talking about (DeFrancisco, 1991). In addition, Victoria DeFrancisco has concluded that men often fail to attend to their partners' talk; they refuse to consider what the women are speaking about and shift the conversation to a topic of their preference.

Harassment

Holly Kearl (2015) writes about street harassment, noting that women don't have free access to public streets. Men dominate public spaces in that women walking there may receive verbal threats (sometimes couched as compliments). Sexual harassment in the workplace is another method of telling women that they do not belong outside of the domestic sphere. Mary Strine (1992) shows how some talk in universities naturalizes harassment, making it seem like an acceptable practice. When women who have experienced sexual harassment are labeled as hysterical, overly sensitive, or troublemakers, their concerns are generally dismissed and defined as unimportant.

Strategies of Resistance

As we mentioned at the beginning of this chapter, MGT is a critical theory; as such, it goes beyond explaining a phenomenon—such as women's muting—to advocating change in the status quo. Houston and Kramarae (1991) offer several strategies for this purpose. One strategy of resistance consists of *naming the strategies of silencing* we've just reviewed. Through this process, silencing is made accessible and a topic for discussion. Women might do this by pointing out when someone is ridiculing what they've said or interrupting them, for example. When couples discuss the traditional marriage ceremony and reject it in favor of a ceremony that better reflects their values, they're using this resistance strategy. A second approach advocated by Houston and Kramarae is to *reclaim, elevate, and celebrate women's discourse.* Houston and Kramarae mention that women are beginning to celebrate and study oral histories, diaries and journals, and the so-called alternative means of expression like sewing, weaving, and other handwork that's often done by women. Through examining these forums for expression, women are recognizing the "effectiveness, impact, and eloquence of women's communicative experiences as well as men's" (Foss & Foss, 1991, p. 21). Other researchers (Hoover, Hastings, & Musambira, 2009) have examined communication venues where women's voices are celebrated and their particular outlook is valorized, such as on websites for the bereaved. Still others have investigated how women's music might provide a way to give women voice literally and figuratively (Huber, 2010; Moody, 2011).

A third approach to resistance involves creating a *new and more representative language* to capture women's experiences. Although changing the language is an ambitious task, language is malleable, and as new concepts enter our culture, new words are created to describe them. Although much of our computer jargon has rather sexist roots (e.g., joystick, floppy, hard drive, etc.), there are newer tech words that are gender neutral (e.g., firewall, screenshot, etc.). Think of all the words we now have to talk about computers and computer-mediated communication (itself a new term!). Suzette Elgin (1988) invented a whole language, which she calls Laadan, focusing on women's experiences. Kramarae and Paula Treichler (1992) compiled a feminist dictionary to give woman-centered definitions to words of importance in women's experiences.

In addition to Houston and Kramarae's (1991) suggestions, Ezster Hargittai and Aaron Shaw (2015) note that new media offer opportunities to give voice to previously muted groups. Yet, they caution that even communication conducted via the internet does not take place in a vacuum, and when using new media people have to be mindful of structural inequities that can keep some groups muted even online. They note that there are "profound gender inequalities" (p. 424) that exist which must be reduced and ultimately eliminated. Bridget Alichie (2023) concurs, arguing that despite the supposed neutrality of the internet, there is actually a great deal of online misogyny. Alichie's study examined feminists' social media presence in Nigeria and found that online misogyny muted women's voices and visibility. Women developed strategies to resist this muting, but they required a great deal of energy and emotional labor and ultimately were largely unsustainable.

Through a variety of resistance strategies including naming strategies of silencing; reclaiming, elevating, and celebrating women's discourse; creating new words for uniquely gendered experiences; and using new media, muting can be resisted. In short, there are many approaches to changing the situation that MGT delineates and explains, although it's not an easy matter to enact them. Changing the status quo is a difficult process.

Students Talking Tech: Skyler

I like the idea that women could resist being muted by using social media. But sometimes I feel like quitting Instagram because I see so many sex-stereotypic images on it. I guess there's some beefcake too, but the number of women you see on Insta in tiny bathing suits, and flashing skin is unbelievable. Makes me wonder how much women have bought into the "be silent and sexy" stereotype. But, on the flip side, it's great when women like Camila Cabello clap back at body shamers and post how bad that can make a person feel. Camila said that a generation of young girls shouldn't grow up trying to embody inhuman standards of beauty. I liked that, and it showed me that she's using her voice for good. And social media gives her voice a wide audience, I think.

Integration and Critique

MGT has many adherents, but as you might guess, because it's a critical feminist theory, it has also prompted critiques. As we evaluate MGT, we'll consider the criteria of utility and test of time.

Integration

Communication Tradition	Rhetorical	**Semiotic**	Phenomenological	Cybernetic	Socio-Psychological	Socio-Cultural	**Critical**
Communication Context	Intrapersonal	Interpersonal	Small Group	Organizational	Public/Rhetorical	Mass/Media	**Cultural**
Approach(es) to Knowing	Positivistic/Empirical	Interpretive/Hermeneutic	**Critical**

Critique

Evaluation Criteria	Scope	Logical Consistency	Parsimony	**Utility**	Testability	Heurism	**Test of Time**

Utility

First, like Feminist Standpoint Theory (**Chapter 29**), MGT has been accused of **essentialism,** or the belief that all women are essentially the same and all men are essentially the same and the two groups are opposites of one another. These critics note that this makes the theory less useful because there's a great deal of difference within

> **essentialism** the belief that all women are essentially the same, all men are essentially the same, and the two differ from each other

groups; sometimes the difference within a group (such as women) can be greater than the difference between groups (such as women and men). Further, essentialist thinking obscures differences among women based on socioeconomic status, age, ethnicity, or upbringing, for instance, which also have enormous impact on how people are treated and how they communicate. Thinking about all women as essentially the same fails to allow for how certain identities interact with one another to make the experiences of Black women, for example, quite different from those of white women. Other critics disavow the notion of influences altogether, claiming that both individuals and groups are constantly changing through communication. Therefore, any attempt to state unequivocally what women or men are like falsely "freezes" those groups in time, as if they have a natural, unchangeable essence (Ivy, 2017).

Supporters of MGT agree there are many differences among women and acknowledge these may cause some women to be closer to the dominant group than others (white, wealthy women, for instance). However, they argue that to be female is to be part of a central group in our culture, and thus, even though MGT acknowledges that women aren't all alike and there's no essential womanness that all women possess, women in the United States are often treated alike. This treatment forms a common set of experiences that allows MGT to make generalizations about men and women.

Some critics fault the usefulness of MGT because they note that women do speak out in public forums, and they point to women such as Kamala Harris, Nancy Pelosi, and Taylor Swift as examples of women who don't seem muted. Muted Group theorists would agree that some women have gained a public forum, but they would also argue that these women have done so by becoming extremely adept at translating and speaking the language of the dominant group. Women who bring a uniquely female perspective to the table have not fared as well, according to these theorists. Linda Lee Smith Barkman (2018) uses her own experience as an example of this rebuttal when she notes how, as a beginning Ph.D. student, she used language that sounded like how many women speak and got graded down. It was only when she learned the language of the dominant group that she began to succeed in academia. Proponents of MGT argue that until we are able to hear from a wide diversity of voices rather than forcing all who wish to speak out to conform to a narrow range of options defined by the dominant group, we will still need the critical commentary offered by MGT.

Test of Time

Some critics of MGT assert that the theory has not been utilized much in communication research and occasionally when it has been employed, its tenets have not been supported. Elspeth Tilley (2010), for instance, found that men were not more likely than women to speak out in public when confronted with an employee who had breached ethical standards. Related to this notion is the fact that the theory was derived many decades ago, and the theory's assumptions haven't been updated. However, Heather Kissack (2010) tested some of MGT's assumptions and found that they still held true in the context of organizational emails. Further, because researchers are using MGT to examine contemporary topics such as athlete's concussions (Sanderson, Weathers, Snedaker, & Gramlich, 2016), women prisoners (Smith Barkman, 2018), online reporting of Hillary Clinton's 2008 campaign (Luecht, 2016), human trafficking (Matzke-Fawcett, 2021), and cervical cancer (Kutto & Mulwo, 2015), it seems as though the theory may be standing the test of time.

Closing

Certainly MGT is provocative and causes us to think about biases in language. It also shines a light on what we accept and what we reject from speakers. MGT also explains some problems women and those in other non-dominant groups experience in speaking out in many settings. It's up to us to decide if these issues form a systematic bias against subordinate groups and in favor of the dominant group, as MGT asserts.

Discussion Starters

 Case-In-Point: Do you think MGT provides a good explanation for the difficulties faced by Patricia Fitzpatrick from our chapter-opening story? Why or why not? What other explanations might you offer to understand her situation?

Try-It-Your-Selfie: Think about some strategies that you might try online to resist muting. How might you enact these if you believe yourself to be a member of a non-dominant group or if you are a member of a dominant group and you wish to be an ally?

1. Can you coin terms for experiences that women have and men do not? What about the reverse? Do you agree with MGT that the English language fits the male experience of the world better than the female experience? Explain your answer.

2. We mentioned that one criticism of MGT is that some women do manage to speak out on the public stage and so the theory overstates the problem that women have with language. Do you agree or disagree with this criticism?

3. Would it make MGT a stronger theory if it dropped the feminist perspective and just focused on any non-dominant groups that are muted in the English language? Does MGT provide a good explanation for groups, other than women, who are silenced? Why or why not?

4. If MGT is a critical theory, what actions has it stimulated? How has it advocated for change?

References

Alichie, B. O. (2023). "You don't talk like a woman": The influence of gender identity in the constructions of online misogyny. *Feminist Media Studies, 23*(4), 1409–1428.

Ardener, E. (1975). The problem revisited. In S. Ardener (Ed.), *Perceiving women* (pp. 2–20). Malaby Press.

Ardener, S. (1978). Introduction: The nature of women in society. In S. Ardener (Ed.), *Defining females* (pp. 9–48). Wiley.

Ardener, S. (2005). Muted group theory excerpts. *Women and Language, 28*(2), 50–72.

Ayinla, F. I., & Amenaghawon, F. (2021). Music advocacy and gender-based violence (GBV) during COVID-19 pandemic. *Gender & Behaviour, 19*(3), 18520–18529.

Bem, S. (1993). *The lenses of gender: Transforming the debate on sexual inequality.* Yale University Press.

Blair-Loy, M., Hochschild, A., Pugh, A. J., Williams, J. C., & Hartmann, H. (2015). Stability and transformation in gender, work, and family: Insights from the second shift for the next quarter century. *Community, Work & Family, 18*(4), 435–454.

Brann, M., & Mattson, M. (2004). Reframing communication during gynecological exams: A feminist virtue ethic of care perspective. In P. M. Buzzanell, H. Sterk, & L. H. Turner (Eds.), *Gender in applied communication contexts* (pp. 147–168). Sage.

Burnett, A., Mattern, J. L., Herakova, L. L., Kahl, D. H., Tobola, C., & Bornsen, S. E. (2009). Communicating/muting date rape: A co-cultural theoretical analysis of communication factors related to rape culture on a college campus. *Journal of Applied Communication Research, 37*(4), 465–485.

Butler, J. A., & Modaff, D. P. (2015). The personal SWOT: An opportunity for student reflection, growth, and planning. *Journal of the Speech and Theatre Association of Missouri, 45*, 85–93.

Cannon, B. C., Robinson, D. T., & Smith-Lovin, L. (2019). How do we "do gender"? Permeation as over-talking and talking over. *Socius: Sociological Research for a Dynamic World, 5*, 1–20.

Chopra, R. (2001). Retrieving the father: Gender studies, "father love" and the discourse of mothering. *Women's Studies International Forum, 24*, 445–455.

Coontz, S. (1988). *The social origins of private life.* Verso.

Cubbage, J. (2019). How will they learn without access? Ending the exclusion of disabled students in media production courses and programs in higher education. *Journal of Media Literacy Education, 9*(2), 114–121.

DeFrancisco, V. (1991). The sounds of silence: How men silence women in marital relationships. *Discourse and Society, 2*(4), 355–370.

Ehrenreich, B., & English, D. (2005). *For her own good: Two centuries of the experts' advice to women.* Anchor Books.

Elgin, S. (1988). *A first dictionary and grammar of Laadan* (2nd ed.). Society for the Furtherance and Study of Fantasy and Science Fiction.

Fixmer-Oraiz, N., & Wood, J. T. (2019). *Gendered lives: Communication, gender, & culture* (13th ed.). Cengage.

Foss, S. K., Domenico, M. E., & Foss, K. A. (2013). *Gender stories: Negotiating identity in a binary world.* Waveland Press.

Foss, K., & Foss, S. (1991). *Women speak: The eloquence of women's lives*. Waveland Press.

Gatere, J. W., Kiumi, P. J. K., & Ngugi, D. M. (2018). To assess the influence of selected community based factors on girls' participation rate in public day secondary schools in Naivasha subcounty, Kenya. *International Journal of Scientific Research and Management*, 6(7), 517–527.

Gendrin, D. M. (2000). Homeless women's inner voices: Friends or foes? In M. J. Hardman & A. Taylor (Eds.), *Hearing many voices* (pp. 203–219). Hampton Press.

Hargittai, E., & Shaw, A. (2015). Mind the skills gap: The role of internet know-how and gender in differentiated contributions to Wikipedia. *Information, Communication, & Society*, 18(4), 424–442.

Hayden, S. (1994). Interruptions and the construction of reality. In L. H. Turner & H. M. Sterk (Eds.), *Differences that make a difference: Examining the assumptions in gender research* (pp. 99–106). Bergin & Garvey.

Hechter, M. (2004). From class to culture. *The American Journal of Sociology*, 110(2), 400–445.

Herakova, L. L. (2009). Identity, communication, inclusion: The Roma and (new) Europe. *Journal of International and Intercultural Communication*, 2(4), 279–297.

Hesse, M. (2016). *Anita Hill's testimony compelled America to look closely at sexual harassment*. https://www.washingtonpost.com/lifestyle/style/anita-hills-testimony-compelled-america-to-look-closely-at-sexual-harassment/2016/04/13/36999612-ea2e-11e5-bc08-3e03a5b41910_story.html.

Hochschild, A. R. (1989). *The second shift: Working parents and the revolution at home*. Viking Penguin.

Hoover, J. D., Hastings, S. O., & Musambira, G. W. (2009). Opening a gap in culture: Women's uses of the compassionate friends website. *Women & Language*, 32(1), 82–90.

Houston, M., & Kramarae, C. (1991). Speaking from silence: Methods of silencing and of resistance. *Discourse and Society*, 2(4), 387–399.

Huber, J. L. (2010). Singing it out: Riot grrrls, Lilith Fair, and feminism. *Kaleidoscope: A Graduate Journal of Qualitative Communication Research*, 9, 65–85.

Ivy, D. K. (2017). *Genderspeak: Communicating in a gendered world* (6th ed.). Kendall-Hunt.

James, D., & Drackich, J. (1993). Understanding gender differences in amount of talk. In D. Tannen (Ed.), *Gender and conversational interaction* (pp. 281–312). Oxford University Press.

Jule, A. (2018). All together now: Using conversational analysis and Muted Group Theory to understand gendered classroom discourse in a Cameroonian primary classroom. *African Journal of Teacher Education*, 7(1), 1–18.

Kearl, H. (2015). *Stop global street harassment: Growing Activism around the World*. ABC-CLIO.

Kissack, H. (2010). Muted voices: A critical look at email in organizations. *Journal of European Industrial Training*, 34(6), 539–551.

Kosenko, K. A. (2010). Meanings and dilemmas of sexual safety and communication for transgender individuals. *Health Communication*, 25(2), 131–141.

Kramarae, C. (1981). *Women and men speaking: Frameworks for analysis*. Newbury House.

Kramarae, C. (2005). Muted Group Theory and communication: Asking dangerous questions. *Women and Language*, 28(2), 55–61.

Kramarae, C. (2009). Muted Group Theory. In S. W. Littlejohn & K. A. Foss (Eds.), *Encyclopedia of communication theory* (vol. 2, pp. 667–669). Sage.

Kramarae, C., & Treichler, P. A. with assistance from Russo, A. (1992). *Amazons, bluestockings and crones: A feminist dictionary* (2nd ed.). Pandora.

Kutto, V. C., & Mulwo, A. K. (2015). Communication challenges of culture and stigma in the control of cervical cancer among rural women in Kenya. *Advances in Social Sciences Research Journal*, 2(2), 178–184.

Lin, L. (2023, September 7). About 8 in 10 women in opposite sex marriages say they took their husband's last name. *Pew Research Center*. https://www.pewresearch.org/short-reads/2023/09/07/about-eight-in-ten-women-in-opposite-sex-marriages-say-they-took-their-husbands-last-name/#:~:text=Most%20women%20in%20opposite%2Dsex,they%20kept%20their%20last%20name.

Luecht, J. A. (2016). *Gendered discourse on the trail to the White House: A quantitative analysis of media coverage during Hillary Clinton 2015/16 campaign to become Democratic nominee.* [Master's Thesis, University of Oregon Press].

Mahrukh, S., Ahmad, A., & Iqbal, L. (2017). Silencing the silence: A study of women at workplace. *Global Social Sciences Review*, 2(2), 162–176.

Matzke-Fawcett, A. (2021). *"We call them survivors": Human trafficking and the Muted Group Theory* [Unpublished]. https://digitalcommons.odu.edu/cgi/viewcontent.cgi?article=1005&context=arts_and_letters.

Moody, M. (2011). A rhetorical analysis of the meaning of the "independent woman" in the lyrics and videos of male and female rappers. *American Communication Journal*, 13(1), 43–58.

Nangabo, D. (2015). *The Muted Group Theory: An overview.* GRIN Verlag.

Olsen, T. (1978). *Silences.* Delacorte Press.

Orbe, M. P. (1998a). From the standpoint(s) of traditionally muted groups: Explicating a co-cultural communication theoretical model. *Communication Theory*, 8(1), 1–26.

Orbe, M. P. (1998b). An outsider within perspective to organizational communication: Explicating the communicative practices of co-cultural group members. *Management Communication Quarterly, 47*(3), 157–176.

Orbe, M. P. (2005). Continuing the legacy of theorizing from the margins: Conceptualizations of co-cultural theory. *Women and Language, 28*(2), 65–66.

Orbe, M. (2012). Researching biracial/multiracial identity negotiation: Lessons from diverse contemporary U.S. public perceptions. In N. Bardhan & M. Orbe (Eds.), *Identity research and communication: Intercultural reflections and future directions* (pp. 165–178). Lexington Books.

Owusu, D. A. (2016). Mute in pain: The power of silence in triggering domestic violence in Ghana. *Social Alternatives*, 35(1), 26–32.

Pearson, J. C. (October 1985). *Innovation in teaching gender and communication: Excluding and including women and men.* Paper presented at the annual meeting of the Organization for the Study of Communication, Language, and Gender, Lincoln, NE.

Pearson, J. C., Turner, L. H., & West, R. L. (1995). *Gender and communication.* Brown & Benchmark.

Poapst, J., & Harper, A. (2017). Reflections on the 2014 celebration of women in debate tournament at George Mason University Press. *Argumentation and Advocacy*, 53(2), 127–137.

Reed, A. (2013). *A brief history of sexual harassment in the United States.* NOW blog http://now.org/blog/a-brief-history-of-sexual-harassment-in-the-united-states/.

Reuther, C., & Fairhurst, G. T. (2000). Chaos theory and the glass ceiling. In P. Buzzanell (Ed.), *Rethinking organization and managerial communication from feminist perspectives* (pp. 236–253). Sage.

Robertson, S., Zuniga, P., Christenson, H., & Young, J. (2022). Gender dynamics in high school policy debate: Propagating gender hierarchies in advocating 'better' futures. *Gender and Education, 34*(8), 1025–1040.

Sanderson, J., Weathers, M., Snedaker, K., & Gramlich, K. (2016). "I was able to still do my job on the field and keep playing": An investigation of female and male athletes' experiences with (not) reporting concussions. *Communication and Sport*, *5*(3), 267–287.

Sari, W. P. (2016). *Women as muted group: Case study of women in Sasak ethnic group*. Proceeding from 2nd International Conference on Transformation in Communication. Bandung, West Java, Indonesia.

Sheriff, R. E. (2000). Exposing silence as cultural censorship: A Brazilian case. *American Anthropologist*, *102*(1), 114–132.

Smith Barkman, L. L. (2018). Muted Group Theory: A tool for hearing marginalized voices. *Priscilla Papers*, *32*(4), 3–7.

Strine, M. S. (1992). Understanding how things work: Sexual harassment and academic culture. *Journal of Applied Communication Research*, *20*(4), 391–400.

Taylor, A., & Hardman, M. J. (2000). Introduction: The meaning of voice. In *Hearing many voices* (p. 8). Hampton Press.

Tilley, E. (2010). Ethics and gender at the point of decision making: An exploration of intervention and kinship. *Prism*, *7*(4), 1–19.

Turner, L. H. (1992). An analysis of words coined by women and men: Reflections on the Muted Group Theory and Gilligan's model. *Women and Language*, *15*(1), 21–26.

Wood, J. T., & Fixmer-Oraiz, N. (2017). *Gendered lives: Communication, gender, & culture* (12th ed.). Cengage.

Young, L. B. (2017). "Thank a veteran": The elevation and instrumentation of U.S. military veterans. *Journal of Veterans Studies*, *2*(2), 58–75.

CHAPTER 29
Feminist Standpoint Theory

Based on the research of **Nancy C. M. Hartsock** and **Julia T. Wood**

"Because standpoint resists, or opposes, the dominant worldview, it is necessarily political. It follows that the dominant worldview, or perspective, cannot be a standpoint. It is a position; it is a social location; it provides a perspective on social life; but it is not a standpoint because it does not grow out of critical reflection on its genesis and character and is not oppositional to itself."

—Julia Wood

Angela Coburn and Latria Harris

Angela Coburn banged her books together and left class with a frown. Usually she enjoys her linguistics class and she likes Professor Townsend. But today she was upset and offended. Professor Townsend had explained the concepts of denotation and connotation by discussing a case that was in the news a long time ago. The case concerned a European American aide to the mayor of Washington, DC, who had lost his job for using the word *niggardly* to mean "sparingly" in a meeting with two others, one European American and one African American. Angela gasped out loud listening to the lecture. As an African American, the sound of that word is extremely offensive to her. It doesn't matter what the aide had said it meant; she knew a racial slur when she heard one. She is older than most of the students at Mead University, and she remembers her mother being involved in the civil rights struggles of the 1960s. The way her mother talked about it, those struggles didn't seem that long ago, and Angela knew some of her family members had been beaten in Selma, Alabama. Further, she knows that racism didn't disappear from the U.S. after the 1960s. She didn't have to look beyond George Floyd, Breonna Taylor, Daunte Wright, Christian Cooper, and countless others to know that. So Angela couldn't believe it when Professor Townsend told that story to illustrate the lesson. There are racist

students at Mead—even in class with her—and Angela feels that using that example sent the wrong message.

She was still fuming later that evening when she met Latria Harris at the coffee shop for their study group. Latria is also in Professor Townsend's linguistics class, and she and Angela study together every Wednesday. Angela enjoys their meetings; they help her keep on top of the assignments for a challenging class, and it's nice to socialize with another African American woman at the university. Angela and Latria get along well despite the difference in their ages and circumstances. Angela is a 48-year-old mother of three grown children; her husband left her many years ago, and she raised the kids alone. She has always struggled financially and has been on welfare, but now things are looking up. She began her studies toward a BA 6 years ago, and she's hoping to finish this year. Latria is a 21-year-old single woman who came to college straight from a college-prep high school, and her parents are quite wealthy.

As soon as Angela sat down, she burst out with her displeasure at Professor Townsend and the linguistics class. She spoke for some time about how upset she was until she noticed a blank look on Latria's face. "Don't you agree with me, Latria?" Angela asked. "Well," Latria responded,

"I saw how upset it made you, but I can't say that I feel the same. My family doesn't use that word, but it doesn't really bother me to hear it. I know a lot of people who use it, and rappers seem to have made it mainstream. Also, I understood Townsend's point. The words sound alike but are defined completely differently. I guess Townsend was trying to say that people's connotations about the meaning of the one word may poison their responses to the other word."

Angela looked at Latria in silence for a while. Finally, she spoke: "Latria, I understand the point, too. I just think a different example should have been used. Those words sound too much alike for me. And that's all some of the racists around here need to hear—a professor speaking a word like that. They will start to think it's okay to say such a thing. And it is not okay. No way. People have died because of that word."

Latria looked at Angela in some disbelief. "I know you're upset, but I think you're carrying this too far. I want to stand by you, but I just can't see it as a big deal."

Angela and Latria fell silent, each pursuing her own thoughts. Finally, Angela said, "Girl, I guess we aren't going to agree about this. Maybe we should just get to studying this week's assignments." Angela smiled as she spoke and looked warmly at her friend, so Latria opened her book with relief. The last thing she wanted to do was to offend or disrespect Angela. Latria gives Angela a lot of credit for everything she has gone through and for how hard she works at all her classes. Really, Angela is an inspiration to Latria. She grinned back at Angela as they began their work.

Feminist Standpoint Theory (FST) provides an entry point for understanding some of the dynamics illustrated in this story because it offers a framework for understanding women's positions relative to systems of power. This framework is built on knowledge generated from the everyday lives of people—acknowledging that individuals are active consumers of their own reality and that individuals' own perspectives are the most important sources of information about their experiences (Chakraborty, 2023; Fixmer-Oraiz & Wood, 2019; Johnston, Friedman, & Peach, 2011). FST gives authority to women's own voices.

The theory claims that women's experiences, knowledge, and communication behaviors are shaped in large part by the social groups to which they belong. Thus, insofar as Angela and Latria share membership in social groups, they may share similar experiences. The fact that they are both African American women students gives them common ground, but standpoints are derived from multiple social locations (Jaggar, 2013), and Angela and Latria's economic differences and ages make a difference in their experiences and positions of power (Cox & Ebbers, 2010). Furthermore, standpoints are achieved through reflection, introspection, and struggle. While Angela has engaged in this process, Latria hasn't, so although they are both Black women only Angela has achieved a standpoint, according to FST.

Everyday people, not the elite, provide the framework for FST because everyday people possess knowledge different from that of those in power. The knowledge they possess forges standpoints in opposition to the perspectives of powerful people. Standpoints arise when everyday people resist those in power and refuse to accept the way society defines their group (Wood, 2004). FST advocates criticizing the status quo because the status quo represents a power structure of dominance and oppression. Thus, FST points to problems in the social order and suggests new ways of organizing social life so that it will be more equitable and just. Some research (e.g., Blackwell, 2010; Davis, 2018; Kia, MacKinnon, & Göncü, 2023) focuses specifically on how women enact strategies of resistance to the inequities in the social system. In this regard, the theory belongs to the category of theories we call critical theories (see **Chapters 2** and **3**).

Evolution of Feminist Standpoint Theory

Although communication researchers have only relatively recently begun to apply FST to studies of communication behavior (e.g., Wood, 2008), it's a theoretical framework with a long history. We'll now briefly review that history to examine how FST has evolved over time.

Standpoint Theory

FST is derived from Standpoint Theory, which originated in the 19th century, when the German philosopher Georg Wilhelm Friedrich Hegel discussed how the master–slave relationship created different standpoints in its participants. Hegel wrote that although slaves and masters live in a common society, their knowledge of that society is vastly different because of the very different positions they occupy within it. Hegel argued that there isn't one single vision concerning social life. Each social group perceives a partial view of society that comes from their specific social position. Karl Marx adapted Hegel's approach to focus specifically on capitalism. As Ismail Demirezen (2018) explains, Marx, like Hegel, claimed that the position of the worker shapes their access to knowledge. Marx argued that "people are what they do, not what they think" (Demirezen, p. 34). The person who works as a server in a restaurant sees things very differently than the person who owns the restaurant. Thus, Marx advanced that a capitalist society creates a sort of double vision, one vision held by workers and the other held by the ruling class. For example, someone in the ruling class might think of capitalism as a system advocating freedom and material abundance while those in the working class view the same system as violent, unfair, and offering only deprivation. Furthermore, Marx emphasized that workers are marginalized in the capitalist system because the vision of the ruling class is taken to be reality (Cockburn, 2015). Marx argued, however, that the standpoint of the proletariat should be the privileged standpoint, given that it includes a fuller picture of what actually happens to the majority of people in capitalist societies.

Feminist Standpoint Theory

Nancy Hartsock and other feminist writers (e.g., Harding, 1993) reacted to the fact that although Marx understood the marginalization of workers, in Standpoint Theory, he completely ignored the marginal position of women, as well as their oppression. Hartsock's interest was to "make women present" in Marx's theory and, in so doing, to forge a feminist–Marxist theory. Thus, Hartsock (1983) drew on Hegel's ideas and Marxist theory to adapt Standpoint Theory for use in examining relations between women and men. In doing this, she developed FST.

Additionally, Hartsock was interested in expanding Marx's account to include all human activities rather than simply focusing on paid labor, or what was primarily a male activity within capitalism. She used Marx's claim that "a correct vision of class society is available from *only one* of the two major class positions in capitalist society" (Hartsock, 1983, p. 106, emphasis added). On the basis of this idea, Hartsock observed that Marx developed a powerful critique of class structure. Hartsock suggested that it's Marx's critique of class relationships rather than his critique of capitalism that is most helpful to feminists.

Hartsock applied Hegel's concepts about masters and slaves and Marx's notions of class and capitalism to issues of sex (the biological categories of male and female) and gender (the behavioral categories of masculinity and femininity) to create FST. However, it's worth noting that there is no consensus on the exact meaning of the term *feminist*. Many authors have observed that there are different kinds of feminism (Barker & Jane, 2016). For our purposes, we need to acknowledge this diversity, yet stipulate that the defining characteristic that unites all types of feminism is a focus on women's particular social position and a desire to end any oppression based on sex or gender.

Hartsock approached FST with the belief that women were oppressed as a result of their sex/gender. Other researchers using FST agree that women are an oppressed group. For instance, Sarah Amer and Guowei Jian (2018) wrote about women's difficult experiences in assimilating into military culture in the United States. Amer and Jian note that women do work in male-dominated fields; in 2015, there were over 200,000 active women troops in the U.S. military, for instance. Yet, despite the presence of women, a male perspective continues to dominate most organizations, including the military, and this perspective is "used to form the rules, cultural norms, and expectations" (p. 48) governing organizations. Perhaps as a consequence, women are subjected to marginalization, sexual harassment, sexual assault, and discrimination while they serve in the military. Amer and Jian call the military "a man's world" (p. 48).

In a different context, Laura Parson (2019) focuses on women's oppression online. She writes about her experience as a woman academic who received a great deal of negative online commentary in response to a scholarly article she published about the problems of gendered language in syllabi for STEM classes. In the original article, Parson suggested that women students would be better served in STEM classes by changing the syllabi to include more cooperative language, and reducing the amount of competitive and individualistic language that currently exists. She reports that the onslaught of online negative commentary about her article and about her personally seemed geared to silence her voice. She uses FST to suggest ways to resist this type of oppression.

Students Talking Tech: Jenny

That part in the chapter where they talked about how a woman got a lot of online hate from an article she wrote reminded me of my situation as a gamer. It's funny, I read somewhere that almost half of all gamers are women, but you wouldn't know it from the backlash I get all the time online. Some of it is laughable, but sometimes it's a little frightening when people write about what they think should happen to me if I don't stop playing. It doesn't help when some gamers like Ninja say they won't stream with women because it might cause problems with his wife. Come on. I thought the online world was supposed to be gender-neutral. Right, that's a joke! It should be gender-neutral, though, and if you're a good player, it shouldn't matter if you're male or female. A woman's standpoint as a gamer is definitely different than a man's.

Finally, there's an interesting difference between FST and many other theories in this book. As we've mentioned, FST is a critical theory, and in many ways, it goes beyond a general social critique to express a critique of other mainstream theories and approaches to research. Feminist Standpoint theorists begin their critique with the observation that most academic research comes from one common perspective: that of the white, middle-class male. This has the effect of blocking women's unique perspectives (Chafetz, 1997).

In this critique, Feminist Standpoint Theory highlights the relationship between power and knowledge. As Lynn O'Brien Hallstein (2000) comments, FST

> tries to hold together two tensions: the search for better knowledge and the commitment to the idea that knowledge is always intertwined with issues of power and politics. As a consequence, the foundational tenet of feminist standpoint theory is that knowledge always arises in social locations and is structured by power relations. (p. 3)

In other words, FST begins with imperfect methods, devised by men in power, and seeks to develop better approaches while recognizing that knowledge is never completely separable from politics.

Black Feminist Standpoint Theory

While white scholars such as Hartsock developed FST because ST overlooked the validity of women's standpoint, other researchers (e.g., Hill Collins, 2009; hooks, 2001) argued that FST ignored how other parts of women's identities, such as race and social class, intersected with each other to shape women's experiences. This realization led to the establishment of Black Feminist Standpoint Theory (BFST). In an example of why BFST was needed, Deanna Blackwell (2010) notes that even when teachers try to foster antiracism in the classroom, this is done with white students in mind (i.e., with the goal of raising the white students' awareness about race) and in so doing, they render students of color less visible or visible only as mentors to their white peers. Blackwell turns to Black Feminist Standpoint Theory to explain and protest this phenomenon.

Patricia Hill Collins (2009) describes a specific type of Black standpoint when she describes herself as an African American woman academic. This social position places her as an **outsider within,** or a person who normally would be marginalized but who has somehow gained access to a site of power. In the case of outsiders moving into

> **outsider within** a person in a normally marginalized social position who has gained access to a more privileged location

a social location that historically excluded them, Hill Collins suggests that a particular clarity of vision occurs. When someone is an outsider within, they can detect racism and sexism that people who are only insiders or only outsiders would miss. As a non-traditional college student, Angela, from our opening story, is an outsider within and so is sensitized to racism that Latria misses.

One of the contributions of BFST is to make clear the divisions that exist within the common experience of being a woman or being a Black woman. For instance, Katrina Bell, Mark Orbe, Darlene Drummond, and Sakile Camara (2000) used Black Feminist Standpoint Theory to examine African American women's communication practices. Bell and her colleagues purposely collected information from very diverse women in an effort to capture different lived experiences. This resulted in a sample of women who shared a racial identity but differed in terms of many of their other social locations, including age, religion, sexual identity, and so forth.

The researchers found that despite their diversity, the participants in their study did share a multiple consciousness of oppression as a result of racism, sexism, classism, and heterosexism. Yet, the participants also talked about ways in which the issue of diversity existed within their group of African American women. Throughout their responses, the participants described other aspects of their identity that became salient in their interactions with others and in many instances impeded their sense of connection with some African American women. As such, the researchers saw multiple consciousness as a "dynamic process of constant negotiation" (p. 50). The researchers conclude that Black Feminist Standpoint Theory allows for the assumption of commonality of oppression while acknowledging various diverse expressions of this common experience.

A second contribution of BFST (or, more generally, Black Feminist Thought [Davis, 2018]) lies in its ability to guide studies of Black women's resistance to oppressive circumstances. Shardé Davis examines how Black women think about and enact resistance communicatively. Davis argues that an understanding of how Black women "use language to navigate their particular standpoint in U.S. society allows communication scholars to bridge macrolevel ideas (e.g., identity, domination, resistance) with microlevel behaviors (e.g., self-presentation, code-switching, affiliation)" (p. 313). Davis concludes that "Black women have the agency to [*speak*] against harmful forces" (p. 313). Davis's work provides an introduction to a discussion of how FST and BFST have been used in the field of communication studies.

Feminist Standpoint Theory and Black Feminist Standpoint Theory in Communication Studies

Julia Wood (1992) brought the tenets of FST to the field of communication studies. Wood argues that it isn't women's essential nature that affects their communication behaviors as much as it is their shared standpoints, and she was interested in how social location influences women's interactions. Wood and her colleague (Dennis & Wood, 2012) explored how Black mothers' reluctance to discuss sex with their daughters could be a result of their knowledge of how Black women were exploited sexually.

Julia Wood noted that FST advances a reciprocal relationship between communication behavior and standpoints. Communication is responsible for shaping our standpoints to the extent that we learn our place in society through interaction with others. When we think about Angela and Latria from our opening story, we can imagine that when Angela's mother told her stories about her African American heritage and about her ancestors who were slaves, Angela learned much about her standpoint. Every time a teacher told Angela that she couldn't go to college and receive a BA degree, communication shaped her standpoint. Latria's situation was different, and she probably didn't hear stories about slavery in her family's past. She might have been encouraged to go to college by her teachers. Thus, the communication Latria engaged in didn't encourage the struggle that Angela experienced. This resulted in two African American women who view the world from different social locations.

Theory at a Glance • Feminist Standpoint Theory

Women are situated in specific social locations; they occupy different places in the social hierarchy based on their membership in social groups (poor, wealthy, European American, African American, Latinx, uneducated, well educated, etc.). Despite these differences, however, all women share in the experience of "femaleness," which provides them a specific social location, subordinate to men's position. When an individual woman reflects on her subordinate power position and struggles in opposition to those in power, resisting the social definition given to her by those in power, her perspective becomes a standpoint. No standpoint allows a person to view the entire social situation completely—all standpoints are partial—but people on the lower rungs of the social hierarchy do see more than their own position, whereas those in power rarely see beyond their own position. This expanded vision allows women to be involved in changing society for the better and creating a more just and equitable world. FST comes from Hegel and Marx's Standpoint Theory and has motivated others to created allied theories such as Black Feminist Standpoint Theory and Co-Cultural Theory.

Similarly, FST and Black Feminist Standpoint Theory assert that those who share a standpoint will also share certain communication styles and practices. For example, we expect that women who take care of children will communicate in a maternal fashion, whereas men who aren't responsible for such caregiving won't develop the same communication behaviors (Ruddick, 1989). Marsha Houston (1992) cites the example of communication behaviors among Black women—like loud talking and interrupting—that are shaped by their standpoints and misunderstood by those outside the group. Houston points out that some white researchers interpret these communication behaviors differently (and usually more negatively) than do members of the group themselves.

FST and BFST both illustrate the centrality of communication in shaping and transmitting standpoints. These theories also point to the use of communication as a tool for challenging the status quo and producing change (Zaytseva, 2012). FST and BFST focus on communication practices as a change agent. The concepts of voice, speaking out, and speaking for others are important to Feminist Standpoint Theory and Standpoint epistemology in general, and they are all concepts rooted in communication.

In this chapter, we explain FST as Nancy Hartsock originally detailed it, and communication scholars such as Julia Wood approached it, focusing on women's particular standpoint. Yet, it's wise to remember that a variety of standpoints, such as those provided by status, race, and so on, can be explained by the theory, as BFST illustrates. Theorists such as Mark Orbe (e.g., Orbe & Harris, 2015) build on Standpoint Theory to examine race and culture, and we discuss Orbe's approach, called Co-Cultural Theory, in **Chapter 30.**

The social hierarchy is not fixed. There are continual struggles in society to determine which group is dominant and who has the right to speak for its members (see **Figure 29.1**). For instance, the ongoing global debate over undocumented immigrants reflects this struggle. Who has the right to "speak" on behalf of this population? And, what does their standpoint allow them to see and voice that others cannot?

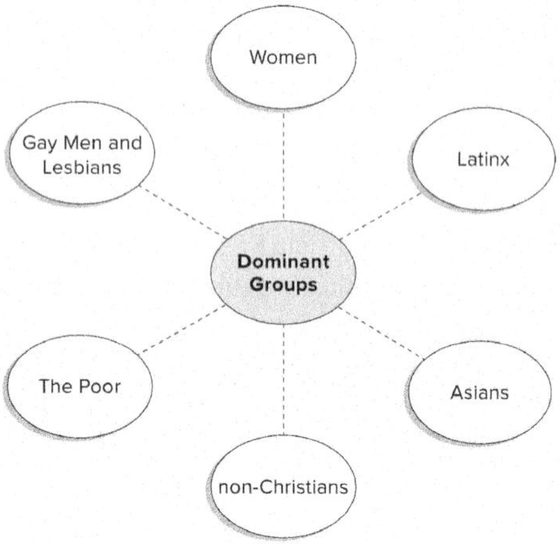

Figure 29.1   Relationship of Multiple Groups in U.S. Society

Adapted from Wood, J. C. (July 2004). Lecture delivered at the Hope Summer Institute, Decorah, Iowa.

Assumptions of Feminist Standpoint Theory

FST rests on four general axioms that Janet Saltzman Chafetz (1997) says characterize any feminist theory: (1) Sex or gender is a central focus for the theory; (2) sex or gender relations are viewed as problematic, and the theory seeks to understand how sex or gender is related to inequities and contradictions; (3) sex or gender relations are viewed as changeable; and (4) feminist theory can be used to challenge the status quo when the status quo debases or devalues women.

In addition, FST is based on five specific assumptions about the nature of social life:

- Material life (or class position) structures and limits understandings of social relations.
- When material life is structured in two opposing ways for two different groups, the understanding of each will be an inversion of the other, providing standpoints that are both partial. When there is a dominant and a subordinate group, the understanding of the dominant group will be both partial and harmful, while the standpoint of the subordinate group is more complete.
- The vision of the dominant group structures the material relations in which all groups are forced to participate.
- The vision available to an oppressed group represents struggle and an achievement.
- The potential understanding of the oppressed (the standpoint) makes visible the inhumanity of the existing relations among groups and moves us toward a better and more just world.

We'll discuss each of these assumptions briefly. They are framed in the modified Marxist perspective that Hartsock favors.

Theory-Into-Action

 Much has been written in the popular press about differences between women and men in the United States and worldwide. One such difference has to do with crying and depression. Worldwide, women cry more than men, and there are more documented cases of depression in women than in men. There might be several explanations for why this is the case (i.e., biology or the greater social acceptability for women admitting to depression and seeking help than for men, for example). FST would suggest, however, that it might have something to do with the differing social locations (and resulting access to power) occupied by women and men. Women, who despite decades of social advances still suffer from economic inequality, discrimination, abuse, and violence probably have more reason to cry and be depressed than men do.

The first assumption about the nature of social life that FST rests on, sets forth the notion that individuals' location in the class structure shapes and limits their understandings of social relations. Our chapter-opening story of Angela and Latria illustrates the power of one's location in society for shaping understanding. Because of Angela's standpoint, based on her difficult circumstances raising her children alone, her struggles to finish her degree, and her age, her response to the class discussion differs sharply from Latria's perspective, whose class background and age have somewhat shielded her from Angela's experience. Second, FST assumes that dominant and subordinate groups have opposing standpoints, both of which are incomplete, or **partial.** Yet, the standpoints of people in the subordinate groups result in a clearer, more accurate vision than those possessed by the dominant group. Ruth Frankenberg (1993) says that "the oppressed can see, with the greatest clarity, not only their own position but ... indeed the shape of social systems as a whole" (p. 8). This clarity of vision suggests that those in the lower positions on the power hierarchy possess the greatest accuracy in their standpoints, where **accuracy** refers to the ability to see beyond one's own specific social location. This is the case because the dominant group (or the ones in power) can make rules that actually harm those of the subordinate group, making it in the best interest of those in the subordinate groups to figure out the dominant group enough so they can "get by" successfully. A woman at work needs to know what her male boss is likely to do because of the power he exerts over her; he has less need to know about her. This point leads naturally to the third assumption, which asserts that the dominant group structures life in such a way as to remove some choices from the subordinate group. Hartsock comments that in the United States people have very little choice about participating in a market economy, which is the preferred mode for the dominant group. As Hartsock (1998) observes, the vision of the dominant group structures social life and forces all parties to participate in this structure. She states: "Truth is, to a large extent, what the dominant groups can make true; history is always written by the winners" (p. 96). Furthermore, the dominant group promotes propaganda that describes the market as beneficial and virtuous. Some of the controversies in the 2016 U.S. presidential campaign illustrate many of these assumptions. When Donald Trump talked about his business experience, for example, some people questioned whether his experience of dismantling companies that weren't succeeding economically or buying up foreclosed homes helped him to understand the experience of those that he put out of their homes and work in the process.

partial recognition that no one has a complete view of the social hierarchy

accuracy the ability to see more than what's available to one's own specific social location

Students Talking: Ama

 I thought Feminist Standpoint Theory was interesting, but I noticed that none of the examples referred to my specific social location. I am a Cherokee Indian, and I am very proud of my heritage. But it's certainly true that we Native Peoples are on the very bottom rungs of the social hierarchy. If anyone should have an accurate picture of the whole social structure by being on the bottom looking up, it'd be us. I'm sure I know a million times more about the white students at this university than they know about me. Most of them have zero idea of what life is like for Indians in the United States. I also was thinking about the tensions between commonality and difference mentioned in the chapter. I know I get annoyed when white students assume that all Indians are the same and think that I know everything about the Crow or the Blackfeet, even though I am Cherokee and we have very different cultures.

The fourth assumption asserts that the subordinate group has to struggle for their vision of social life. This leads to the final assumption, which claims that this struggle results in a relatively clear vision of the social order (though still a partial one). With this clear vision, the subordinate group can see the inherent inequalities in the social order and can thus attempt to change the world for the better.

Added to these assumptions that characterize Hartsock's Marxist view of FST, most conceptions of the theory also embody a set of four conclusions about knowledge and knowledge gathering (ontology and epistemology).

1. All knowledge is a product of social activity, and thus no knowledge can be truly objective.
2. Cultural conditions "typically surrounding women's lives produce experiences and understandings that routinely differ from those produced by the conditions framing men's lives" (Wood, 1992, p. 14). These different types of knowledge often produce distinct communication patterns.
3. It's a worthwhile endeavor to understand the distinctive features of women's experience.
4. We can only know women's experience by attending to women's interpretations of this experience.

The first conclusion states that knowledge is not an objective concept, but rather is shaped subjectively by knowers. Knowledge evolves from subjective experience. This suggests an approach to knowing that's much different from that suggested by a belief in objective truth.

The second conclusion points to the different social locations that men and women inhabit in the United States even when they work and live in what seem to be similar situations. In a study examining sexual harassment in the workplace, Debbie Dougherty (2001) begins with an idea grounded in this conclusion. She postulates that while sexual harassment is dysfunctional for women, it may serve some functions for men. And in her study, she did find that men interpreted sexual harassment as a form of coping behavior for work-related stressors, a mode of therapy, and a means for demonstrating camaraderie. Women saw none of these functions in sexual harassment. Dougherty concludes that her findings indicate how different social locations have shaped men's and women's reactions. Other researchers (e.g., Richardson & Taylor, 2009) point out that this is true among women as well as between women and men. They note that sexual harassment is a different experience for Black and Latinx women than for white women.

The third conclusion deals with ontology, or what is worth knowing. This conclusion places marginalized people (women) at the starting place for theorizing and research. Sandra Harding (1991) comments on this by saying, "What 'grounds' feminist standpoint theory is not women's experiences, but the view from women's lives ... we start our thought from the perspective of lives at the margins" (p. 269). Anne Johnston and her colleagues (2011) agreed and found that women bloggers, contrary to the dominant opinion, blog a great deal about politics, but they do so in quite different ways from their male counterparts.

Finally, FST operates from the belief that the only way to honor women's standpoints is by having women talk about their experiences and interpret them for themselves. Thus, researchers using FST should approach research participants as active partners in the research endeavor. As Beth Driscoll and DeNel Rehberg Sedo (2019) state in their study of women's book reviews posted on Goodreads, "As researchers, we are critics, but we are not the only critics" (p. 249). Driscoll and Sedo note that FST allows them to grant agency to the readers whose reviews they're studying and legitimize their reading practices. It's a continuing goal for Feminist Standpoint theorists to develop new methodologies that give voice to those who have been silenced.

Through these axioms, assumptions, and conclusions, we get a picture of FST as an evolving framework, grounded in Marxism, but rejecting some of the central tenets of that perspective in favor of a feminist approach. FST seeks to understand the influence that a particular social location exerts on people's views of the world and on their communication. In this quest, researchers in FST wish to begin with those who are marginalized in a society and focus on their stories and interpretations. As Feminist Standpoint theorists work with research participants, they recognize the limited view of their own vision and they acknowledge the subjective nature of truth(s).

"All of those in favor of looking at pictures of grandchildren before we discuss *Feminist Theory* say aye."

Key Concepts of Feminist Standpoint Theory

FST rests on several key concepts, including: *voice, standpoint, situated knowledges,* and *sexual division of labor.* We will discuss each of these four concepts below.

Voice

Voice is a concept that shows FST's similarity to Muted Group Theory, which we discuss in **Chapter 28.** Many scholars have observed that voice relates to identity. When we find our voice or use our voice,

> **voice** the ability to project to others who we are and what we're all about

that means that we project to others who we are and what we are all about. **Voice** may be best defined by its opposition to silence; when we have voice we're self-actualizing, and when we're silent, especially against our inclination, we're denying our self. Jason Stanley (2011) wrote that in service of silencing others, politicians misappropriate words and twist meanings. He calls this "linguistic strategies for stealing the voices of others."

Stanley notes that silencing is robbing others of the ability to engage in speech acts, such as assertion. He goes on to observe that "there is another kind of silencing familiar in the political domain" (n.p.) as well. "It is possible to silence people by denying them access to the vocabulary to express their claims" (n.p.), which is similar to the tenets of Muted Group Theory. Further, Stanley's comment points to a key aspect for understanding voice: it's a relational concept. We have it to the extent that others listen to us and don't actively work to rob us of our ability to express ourselves. Many authors who utilize FST in their work talk about "giving voice to the voiceless" or sharing power with those who have previously been marginalized (e.g., Joseph, 2022).

Standpoint

We have discussed the central concept of the theory, standpoint, previously in the chapter, but here we'll define it formally. A **standpoint** is a location, shared by a group experiencing outsider status within the social structure, that lends a particular kind of sense making to a person's lived experience. Furthermore, from Hartsock's (1998)

> **standpoint** an achieved position based on a social location that lends an interpretative aspect to a person's life

perspective, a standpoint is an engaged position. The concept of engagement is amplified by researchers who distinguish between a standpoint and a perspective (Hirschmann, 1997; O'Brien Hallstein, 2000). It's easy to confuse the two, but there's a critical difference. A perspective is shaped by experiences that are structured by a person's place in the social hierarchy. A perspective may lead to the achievement of a standpoint, but only through effort. As O'Brien Hallstein argues, standpoints are only achieved after thought, interaction, and struggle. Standpoints must be actively sought; they're not possessed by all people who have experienced oppression. Standpoints are achieved when experiences of oppression are added to active engagement, reflection, and recognition of the political implications of these experiences. Standpoints are achieved only when a person critically and consciously thinks about how their social location is affected by the prevailing power structures in their culture (Bohrman, Tennille, Levin, Rodgers, & Rhodes, 2017). Thus, as we've noted in our chapter-opening story, Latria has a perspective while Angela has a standpoint.

Furthermore, standpoints are not free of their social and political contexts. As Welton (1998) states, developing a standpoint requires "active, political resistance to work against the material embodiment of the perspective and experience of the dominant group. It is the act of having to push against the experience-made-reality of the hegemonic group, that makes it a political standpoint and potentially liberating" (p. 11).

Moreover, as O'Brien Hallstein (1999, 2000) argues, standpoints are political because they are achieved in collaboration and dialogue with others rather than in isolation.

Finally, standpoints aren't rigid and fixed. Rather they refer to "a fluid and dynamic negotiation of experience and point of view that can be temporarily stabilized in order to interrogate dominant ideologies" (Lenz, 2004, p. 98). Standpoints can change over time and with new experiences. When a person becomes a parent, for instance, their standpoint changes to incorporate this new identity. Further, as some of what we've discussed relative to Black Feminist Standpoint Theory suggests, different groups of parents create differing standpoints related to their particular social location. Patrice Buzzanell and her colleagues (2017) operate within this perspective when they study how pink-collar (i.e., service jobs typically held by women such as administrative assistants and customer service representatives) employees adhere to and resist how their employers talk about maternity leave.

Situated Knowledges

Donna Haraway (1988) contributes the term **situated knowledges** to FST, meaning that any person's knowledge is grounded in context and circumstances. Haraway's concept suggests that knowledges are multiple and are situated in experience. For example, what one person learns from her position as a caregiver for her ailing parents is different from the knowledge that another person develops from her position as an electrical engineer. The notion of situated knowledges reminds us that what we know and do isn't innate, but rather is the result of learning from our experiences. So, if the engineer cared for elderly family members, she also would learn caregiving knowledge. Some research (e.g., Hine, 2011) suggests that aging women share a standpoint, but also diverge from one another based on other variables such as social, health, and lifestyle factors that affect them individually at any given point in time. Situated knowledges also explain how a person's social location both constrains and enables what they can know (Steiner, 2018). In a specific application of FST, Hendar Putranto (2022) uses the theory, and especially the concept of situated knowledges, to understand, critique, and ultimately propose action against the practice of Female Genital Mutilation (FGM) in Indonesia. Putranto argues that "Women's collective agency based on situated knowledge will empower their communicative skills as enablers of transformation to eradicate FGM" (n.p.).

> **situated knowledges** what anyone knows is grounded in context and circumstance

Sexual Division of Labor

Hartsock's Marxist-inspired FST rests on the notion that men and women engage in different occupations based on their sex, which results in a **sexual division of labor.** Not only does this division simply assign people to different tasks based on sex, but it also exploits women by demanding work without providing wages (e.g., the work that women do in the home and as mothers). In addition, Julia Wood (2005) states that "patriarchy naturalizes male and female divisions, making it seem natural, right, unremarkable that women are subordinate to men" (p. 61). Furthermore, the inequities that women suffer in the workplace when involved in labor for wages are linked to their responsibility for unpaid domestic work. In addition, as Nancy Hirschmann (1997) points out, a feminist standpoint allows women to label the activities performed at home *work* rather than just the essential tasks their sex requires they do. Thus, FST highlights the exploitation and distortion that result when labor is divided by sex.

> **sexual division of labor** allocation of work on the basis of sex

Integration and Critique

FST has generated a great deal of research, interest, and spirited controversy. This theory has a rich history employing primarily qualitative methods. As you examine the value of this theory, keep in mind the criterion of utility.

Integration

Communication Tradition	Rhetorical | Semiotic | Phenomenological | Cybernetic | Socio-Psychological | Socio-Cultural | **Critical**
Communication Context	Intrapersonal | Interpersonal | Small Group/Team | Organizational | Public/Rhetorical | Mass/Media | **Cultural**
Approach to Knowing	Positivistic/Empirical | Interpretive/Hermeneutic | **Critical**

Critique

Evaluation Criteria	Scope | Logical Consistency | Parsimony | **Utility** | Testability | Heurism | Test of Time

Utility

The complaint most commonly leveled against FST revolves around essentialism, which we discussed in **Chapter 28** with reference to Muted Group Theory. As we stated in **Chapter 28, essentialism** refers to the practice of generalizing about all women (or any group) as though they were essentially the same. Essentialism obscures

> **essentialism** the belief that all women are essentially the same, all men are essentially the same, and the two differ from each other

the diversity that exists among women. Our chapter-opening story of Angela and Latria illustrates the mistake we make if we engage in essentialism. Although they are both African American women attending the same university, the differences between them cause them to interpret the classroom discussion differently. Because FST focuses on the location of social groups, many researchers have argued that it is essentialist. For example, Catherine O'Leary (1997) argues that although FST has been helpful in reclaiming women's experiences as suitable research topics, it contains a problematic emphasis on the universality of this experience at the expense of differences among women's experiences.

Implicit (and often explicit) in this critique are the ways that many white women researchers have excluded the standpoints and voices of women of color, women who are disabled, lesbians, poor women, and women from developing nations (Blackwell, 2010). Yet, this criticism may be unfair to Hartsock's conceptualization of the theory (Intemann, 2016). Hartsock (1983) has stated that although there are many differences among women, she's pointing to specific aspects of women's experience that are shared by many: unpaid household labor and provision of caregiving and nurturance. Hirschmann (1997) argues that FST really does accommodate difference by allowing for a multiplicity of feminist standpoints. Hirschmann suggests that standpoints can be developed from Hartsock's framework that will bring a useful approach to the tension between shared identity and difference. Wood (1992) concurs with this rebuttal, and she suggests that FST is different from essentialist views of women in one important way. The theory doesn't suggest that men and women are fundamentally different (or have different essences); rather, FST begins with the assumption that the social and cultural conditions that typically surround women's lives produce different experiences and understandings from the social and cultural conditions typically surrounding men's lives. Wood concludes that there's nothing in FST that says intersections among race, class, and gender can't be studied.

For example, in a study of academic women, Debbie Dougherty and Kathleen Krone (2000) sought to apply FST to a group of diverse women to capitalize on the creative tensions between similarity and difference. They interviewed four women in the same department in an academic institution and from the interviews they created a narrative that, although fictional, attempted to capture the standpoints voiced in the interviews. Then they had the participants comment on the narrative and identify areas for change and develop action plans. In this process, the researchers found both commonality and difference among the standpoints of the participants.

The women agreed that their standpoints were shaped by a sense of isolation, a strong desire for community, and the feeling of invisibility. Yet, they differed in their consciousness of their own oppression. The women used these differences to strengthen their relationship. The researchers note that the critique focused on essentialism may miss the point. Dougherty and Krone observe that "differences and similarities create and recreate each other, becoming so intertwined that they are difficult to separate" (p. 26).

A second area of criticism related to utility concerns the notion of **dualisms,** or dualistic thinking. Feminists (Cirksena & Cuklanz, 1992) note that much of Western thought is organized around a set

> **dualisms** organizing things around pairs of opposites

of oppositions, or dualisms. Reason and emotion, public and private, and nature and culture are just a few of the pairs of opposites that are common organizing principles in Western thinking. Feminists have been concerned with dualisms for two related reasons: first, dualisms usually imply a hierarchical relationship between the terms, elevating one and devaluing the other. When we suggest that decisions should be made rationally, not emotionally, for example, we are showing that reason holds a higher value in our culture than does emotion. Related to this issue is the concern that dualisms often become gendered in our culture. In this process, men are associated with one extreme and women with the other. In the case of reason and emotion, women are identified with emotion. Because our culture values emotion less than reason, women suffer from this association. Feminist critics are usually concerned with the fact that dualisms force false dichotomies onto women and men, failing to see that life is less either/or than both/and, as Relational Dialectics Theory (**Chapter 25**) tells us.

O'Leary (1997) argues that FST does not present us with a sufficiently complex understanding of experience, and as a result, it still rests on a dualism between subjective experience and objective truth. O'Leary suggests that Hartsock's framework cannot accommodate the complexities of multiple knowledges. Yet, Hartsock specifically allows for this within her original theory, arguing for clusters of standpoints. As Hartsock (1998) notes, the controversies engendered by FST indicate a "fertile terrain" for debate, opening possibilities for expanding and refining the theory so that it's more responsive to diversity and clearer about the distinctions between subjectivity and objectivity.

Closing

FST presents us with another way of viewing the relative positions, experiences, and communication of people within various social groups. It has a clear political, critical bent, and it locates the place of power in social life. It's generated controversy as people find it either offensive or compatible with their own views of social life. The theory may be compatible with other theories, enabling us to combine them to get more useful explanations for human communication behaviors. Specifically, Amanda Kennedy (2016) combined FST with Carol Gilligan's (1982) ethic of care to examine how global public relations campaigns can maintain global consistency while also being tailored effectively to local situations. Overall, FST holds much promise for illuminating differences in the communication behaviors of different social groups, and suggesting tools for remediating the inequities caused by the power structure in which these groups function.

Discussion Starters

Case-In-Point: Do you accept the argument of Feminist Standpoint Theory that all standpoints are partial? Explain your answer. How does the case of Angela and Latria, from the beginning of the chapter, illustrate this claim?

Try-It-Your-Selfie: This chapter doesn't talk too much about how the theory can be applied to questions of online communication. Why do you think that is? Construct a question concerning online communication that you could frame with FST. Discuss how you would research it.

1. Do you agree that women form a subordinate group in the United States or in any culture worldwide given the chapter's definition of a subordinate group? Explain your answer.

2. If all truths are understood as coming from some subjective standpoint, how is it possible for people to communicate? If there is no objective truth, how do we reach agreement among people with different standpoints?

3. What are the limits of difference that ground a standpoint? Might we eventually end up with standpoints that are so particular that only relatively few people share them? How would this help us to understand communication between and among people?

4. Take a position on the essentialist argument relative to Feminist Standpoint Theory, and defend your position.

References

Amer, S., & Jian, G. (2018). It is a man's world: A qualitative study on females in a masculine military. *The Pennsylvania Communication Annual, 74*(1), 48–74.

Barker, C., & Jane, E. (2016). *Cultural studies: Theory and practice.* Sage.

Bell, K. E., Orbe, M. P., Drummond, D. K., & Camara, S. K. (2000). Accepting the challenge of centralizing without essentializing: Black feminist thought and African American women's communicative experiences. *Women's Studies in Communication, 23*(1), 41–62.

Blackwell, D. M. (2010). Sidelines and separate spaces: Making education anti-racist for students of color. *Race, Ethnicity & Education, 13*(4), 473–494.

Bohrman, C., Tennille, J., Levin, K., Rodgers, M., & Rhodes, K. (2017). Being superwoman: Low income mothers surviving problem drinking and intimate partner violence. *Journal of Family Violence, 32*(7), 699–709.

Buzzanell, P. M., Remke, R. V., Meisenbach, R., Liu, M., Bowers, V., & Conn, C. (2017). Standpoints of maternity leave: Discourses of temporality and ability. *Women's Studies in Communication, 40*(1), 67–90.

Chafetz, J. S. (1997). Feminist Theory and sociology: Underutilized contributions for Mainstream Theory. *Annual Review of Sociology, 23*, 97–120.

Chakraborty, A. (2023). "Symbolic violence" and Dalit feminism: Possibilities emerging from a Dalit feminist standpoint reading of Bourdieu. *International Feminist Journal of Politics, 25*(2), 160–178.

Cirksena, K., & Cuklanz, L. (1992). Male is to female as is to: A guided tour of five feminist frameworks for communication studies. In L. Rakow (Ed.), *Women making meaning: New feminist directions in communication* (pp. 18-44). Routledge.

Cockburn, C. (2015). Standpoint Theory. In S. Mojab (Ed.), *Marxism and feminism* (pp. 331-346). Zed Books.

Cox, E. M., & Ebbers, L. H. (2010). Exploring the persistence of adult women at a Midwest community college. *Community College Journal of Research & Practice, 34*(4), 337-359.

Davis, S. M. (2018). Taking back the power: An analysis of Black women's communicative resistance. *Review of Communication, 18*(4), 301-318.

Demirezen, I. (2018). Gadamer's hermeneutics as a model for the Feminist Standpoint Theory. *Journal of Divinity Faculty of Hitit University Press, 17*(34), 31-44.

Dennis, A. C., & Wood, J. T. (2012). "We're not going to have this conversation, but you get it": Black mother-daughter communication about sexual relations. *Women's Studies in Communication, 35*(2), 204-223.

Dougherty, D. S. (2001). Sexual harassment as [dys]functional process: A feminist standpoint analysis. *Journal of Applied Communication Research, 29*(4), 372-402.

Dougherty, D. S., & Krone, K. J. (2000). Overcoming the dichotomy: Cultivating standpoints in organizations through research. *Women's Studies in Communication, 23*(1), 16-40.

Driscoll, B., & Rehberg Sedo, D. (2019). Faraway, so close: Seeing the intimacy in Goodreads reviews. *Qualitative Inquiry, 25*(3), 248-259.

Fixmer-Oraiz, N., & Wood, J. T. (2019). *Gendered lives: Communication, gender, & culture* (13th ed.). Cengage.

Frankenberg, R. (1993). *White women, race matters: The social construction of whiteness.* University of Minnesota Press.

Gilligan, C. (1982). *In a different voice: Psychological Theory and women's development.* Harvard University Press.

Haraway, D. (1988). Situated knowledges: The science question in feminism and the privilege of partial perspective. *Feminist Studies, 14*(3), 575-599.

Harding, S. (1991). *Whose science, whose knowledge? Thinking from women's lives.* Cornell University Press.

Harding, S. (1993). Rethinking standpoint epistemology: What is strong objectivity? In L. M. Alcoff & E. Potter (Eds.), *Feminist epistemologies* (pp. 49-82). Routledge.

Hartsock, N. C. M. (1983). The feminist standpoint: Developing the ground for a specifically feminist historical materialism. In S. Harding & M. B. Hintikka (Eds.), *Discovering reality* (pp. 283-310). Ridel.

Hartsock, N. C. M. (1998). Standpoint theories for the next century. *Women & Politics, 18*(3), 93-101.

Hill Collins. P. (2009). *Black feminist thought: Knowledge, consciousness, and the politics of empowerment.* Routledge.

Hine, R. (2011). In the margins: The impact of sexualised images on the mental health of ageing women. *Sex Roles, 65*(7), 632-646.

Hirschmann, N. J. (1997). Feminist standpoint as postmodern strategy. *Women and Politics, 18*(3), 73-92.

hooks, b. (2001). *Salvation: Black people and love.* William Morrow.

Houston, M. (1992). The politics of difference: Race, class, and women's communication. In L. F. Rakow (Ed.), *Women making meaning: New feminist directions in communication* (pp. 45-59). Routledge.

Intemann, K. (2016). Feminist standpoint. In L. Disch & M. Hawkesworth (Eds.), *The Oxford handbook of feminist theory* (pp. 261-282). Oxford University Press.

Jaggar, A. M. (2013). Feminist ethics. In H. LaFollette & I. Persson (Eds.), *The Blackwell guide to ethical theory* (pp. 433-460). Wiley-Blackwell.

Johnston, A., Friedman, B., & Peach, S. (2011). Standpoint in political blogs: Voice, authority, and issues. *Women's Studies, 40*(3), 269-298.

Joseph, S. (2022). Standpoint Theory and trauma: Giving voice to the voiceless. In R. Alexander & W. McDonald (Eds.), *Literary journalism and social justice* (pp. 99-115). Palgrave Macmillan.

Kennedy, A. (2016). Landscapes of care: Feminist approaches in global public relations. *Journal of Media Ethics, 31*(4), 215-230.

Kia, H., MacKinnon, K. R., & Göncü, K. (2023). Harnessing the lived experience of transgender and gender diverse people as practice knowledge in social work: A standpoint analysis. *Affilia, 38*(2), 190-205.

Lenz, B. (2004). Postcolonial fiction and the outsider within: Toward a literary practice of Feminist Standpoint Theory. *NWSA Journal, 16*(2), 98-120.

O'Brien Hallstein, D. L. (1999). A postmodern caring: Feminist standpoint theories, revisioned caring and communication ethics. *Western Journal of Communication, 63*(1), 32-56.

O'Brien Hallstein, D. L. (2000). Where standpoint stands now: An introduction and commentary. *Women's Studies in Communication, 23*(1), 1-15.

O'Leary, C. M. (1997). Counteridentification or counterhegemony? Transforming Feminist Standpoint Theory. *Women and Politics, 18*(3), 45-72.

Orbe, M., & Harris, T. (2015). *Interracial communication: Theory into practice.* Sage.

Parson, L. (2019). Digital media responses to a feminist scholarly article: A critical discourse analysis. *Feminist Media Studies, 19*(4), 576-592.

Putranto, H. (2022). Criticizing female genital mutilation practice from Feminist Standpoint Theory: A view from communication science perspective. *Humaniora, 34*(2), 95-107.

Richardson, B. K., & Taylor, J. (2009). Sexual harassment at the intersection of race and gender: A theoretical model of the sexual harassment experiences of women of color. *Western Journal of Communication, 73*(3), 248-272.

Ruddick, S. (1989). *Maternal thinking: Toward a politics of peace.* Beacon Press.

Stanley, J. (2011, June 25). The ways of silencing. Opinionator blog. *The New York Times.* http://opinionator.blogs.nytimes.com/2011/06/25/the-ways-of-silencing/?_r=0s

Steiner, L. (2018). Solving journalism's post-truth crisis with feminist standpoint epistemology. *Journalism Studies, 19*(13), 1854-1865.

Wallace, R. A., & Wolf, A. (1995). *Contemporary Sociological Theory: Continuing the classical tradition.* Prentice Hall.

Welton, K. (1998). Nancy Hartsock's Standpoint Theory: From content to "concrete multiplicity." *Women and Politics, 18*(3), 7-24.

Wood, J. T. (1992). *Spinning the symbolic web: Human communication as symbolic interaction.* Ablex Publishing.

Wood, J. T. (2004). Monsters and victims: Male felons' accounts of intimate partner violence. *Journal of Social and Personal Relationships, 21*(5), 555-576.

Wood, J. T. (2005). Feminist Standpoint Theory and Muted Group Theory: Commonalities and divergences. *Women and Language, 28*(2), 61-64.

Wood, J. T. (2008). Critical feminist theories. In L. A. Baxter & D. O. Braithwaite (Eds.), *Engaging theories in interpersonal communication* (pp. 323-334). Sage.

Zaytseva, O. (2012). *Evolving issues and theoretical tensions: A revised Standpoint Theory for 21st century.* http://www.academia.edu/916643/Evolving_Issues_and_Theoretical_Tensions_A_Revised_Standpoint_Theory_for_21st_Century.

CHAPTER 30
Co-Cultural Theory

*Based on the research of **Mark Orbe***

> *"Culture and communication are inextricably linked. The ability to comprehend one concept is contingent on understanding its relationship to the other."*
>
> —Mark Orbe

Donna Blackhawk and Mark Warne

Donna Blackhawk had arrived at North Dakota State University a week ago, and now she was feeling lonely and frustrated. She hadn't met a single American Indian yet. She'd imagined that moving from Detroit, Michigan, to college in North Dakota would give her more opportunities to be around Indigenous people. She had hoped that going to college in a state with a large Indigenous population would give her a chance to learn more about her heritage. At home her parents hadn't really talked a lot about American Indians—her mother was white, and her father was a member of the Te-Moak Tribe of the Western Shoshone. By the time she was born, her grandparents had passed away and her father had fully assimilated into the European culture around him. He hardly ever talked about his culture to Donna or her four brothers. In her senior year of high school Donna had done some research on the Shoshone and had started asking her father questions. She'd learned a lot and decided she was going to study medicine and work on a reservation after becoming a doctor. She wanted to help solve some of the health problems that Native people had. It was weird; all through school, Donna hadn't felt at all different from her white friends, but as she began learning about Native people, she noticed huge gaps between her and her lifelong friends. She began withdrawing from them, and was more silent than usual when she was with them. She had been hoping to find more people like her in North Dakota, but so far she hadn't, and that was disappointing. Tonight, she was going to a talk about diversity on campus and she thought she'd finally meet some Native students.

Her hopes were realized as she gathered her stuff to leave after the speech; she turned around and practically bumped into Mark Warne. After some laughing and apologies, the two walked out together and began talking. It turned out that Mark, a sophomore, was a member of the Oglala Lakota Nation. He was also a member of the Native Students Action (NSA) group and he promised Donna that there were many Native students on campus that he'd introduce her to. Donna was excited, and they made plans to go out for dinner the next week and then attend the kick-off meeting for NSA's fall semester.

But when Donna and Mark approached the meeting the following week, she was having a few second thoughts. At dinner Mark had sounded so militant and he said that NSA had marches planned for fall, and they might even want to stage a sit-in at the administration building. Mark explained that the university hadn't delivered on several promises it made to NSA last year, including hiring more Native professors, establishing a Native Studies program, and increasing the number of Native students admitted to the school. Mark got pretty passionate when he told Donna that even when American Indian and Native Alaskan students were combined, they still comprised less than 1 percent of the student body. That made them the smallest ethnic group at the university. Donna was a little surprised at the number, but she was used to being a minority, even among minorities, and she was taken aback by Mark's vehemence. She wanted to know more

American Indians than she had in high school, but she wasn't sure she wanted to protest, and she was positive she didn't want to participate in a sit-in. Donna felt uncomfortable and she questioned what she was getting herself into. But as she listened at the meeting, she began to feel better. Some people agreed with Mark, but others urged caution. They thought they'd be more successful if they increased their visibility on campus in less threatening ways, like bringing Native speakers to campus and establishing a clear Native identity among themselves. Donna agreed with this, and she thought she'd join NSA. This kind of activity was what she'd hoped for in coming to North Dakota State.

To understand the different communication goals that Mark and Donna have in relation to their status as members of a minority, or co-cultural group, one theory is especially useful: Co-Cultural Theory (CCT) (Orbe, 1996, 1998a, 1998b, 2004). CCT illuminates how members of marginalized groups "negotiate their positioning as outsiders within mainstream organizations on a highly conscious level..." (Orbe, 1998a, p. 273). From the perspective of CCT, we see that Donna has practiced assimilation during most of her life; she simply tried to blend in with the white dominant group in her high school, and didn't draw attention to her Native heritage. Then, she started to separate from the dominant group as she learned more about the Shoshone Nation. When she formulates her plan to become a physician on an Indian reservation, she's thinking about creating solidarity with other Indigenous people and separating from the dominant group that defined so much of her upbringing. When she gets to college and meets Mark, she observes that he has forged another path; Mark seems to want the dominant group to make accommodations for Native students and he's talking about a series of communication events (sit-ins and demonstrations) to agitate for the accommodations he wants. CCT elaborates on these three communication goals (assimilation, separation, and accommodation) and explains how each of them supports a variety of communication strategies that members of co-cultural groups use when communicating with those in the dominant group.

Noura Ibrahim, Kasey Windels, and Lincoln Lu (2023) found those same three communication goals mentioned when talking with members of co-cultural groups who worked as advertising professionals. When these advertising workers were asked what they tried to accomplish when interacting with their dominant group member colleagues, these goals were mentioned. Orbe defined **co-cultures** as those groups consisting of people whose identities are "traditionally marginalized in dominant societal structures" (Groscurth & Orbe, 2006, p. 126), such as persons of color, women, persons with disabilities, LGBTQ+ people, those with a lower socioeconomic status than average, and so forth.[1] We now turn our attention to the foundations of Co-Cultural Theory: Muted Group Theory (MGT), Feminist Standpoint Theory (FST), and phenomenology.

> **co-cultures** those groups consisting of people whose identities are traditionally marginalized in dominant social structures

Foundations of Co-Cultural Theory

Orbe noted that his theory was grounded in existing feminist research framed by Muted Group Theory (**Chapter 28**) and Feminist Standpoint Theory (**Chapter 29**) and informed by a phenomenological approach to research. MGT is a useful underpinning for CCT because it explains that in any society where power isn't distributed equally, lower-power groups have to contend with tools that weren't created with their needs in mind. Feminist Standpoint Theory (FST) is also useful for CCT because it begins with the premise that people understand themselves and their world based on their specific social position. Finally, Orbe used phenomenology in his

---

1. Orbe (1996; 1998b) used the term "co-cultural" to accomplish two goals. First, he wished to avoid the negative connotations that come with a term like "subcultural." Orbe noted that while some groups may be dominant in U.S. culture, they aren't superior to the non-dominant groups. Second, he wanted to draw attention to the diversity of cultures in the United States even if the experiences of some groups are rendered invisible by those in the dominant culture.

research because it's a method that focuses on the lived experiences of people. In phenomenology, researchers think of themselves as conduits for the voices of their interviewees; they want to alter or manipulate those voices as little as possible in their reports. We briefly review each of these foundations of CCT below.

Muted Group Theory

Muted Group Theory (MGT) originated with the work of Edwin and Shirley Ardener. The Ardeners were social anthropologists studying social structure and hierarchy. Edwin Ardener (1975) argued that groups at the top end of the social hierarchy determine the communication system for the whole culture; they are the dominant group. Any lower-power groups in a culture, such as women, people living in poverty, and people of color, have to learn to work within the communication system established by the dominant group. Further, Shirley Ardener (2005) explained that when social anthropologists studied women's experiences, they rarely spoke directly to women. Rather, they talked almost exclusively to men. This results in a distorted perspective on women's experiences. In a specific example, when researchers study the experience of giving birth, instead of asking women about it, they examine it primarily in terms of the doctor's perspective. When a woman is told she's 3 cm dilated, that helps a doctor (the person in power) understand the process, but it doesn't represent what a woman giving birth feels or experiences. Even if the doctor is female, the perspective being accessed is outside of the person who's giving birth.

Cheris Kramarae (1981) adapted the Ardeners' ideas specifically to the English language and the communication discipline. Kramarae asserts that English-speaking women, and other marginalized groups, have difficulty articulating their thoughts and making themselves heard because their experiences aren't well described by their language system; the English language was primarily devised by well-to-do white men, and thus, English best represents their experiences. Further, as Gina Castle Bell, Mark Hopson, Melinda Weathers, and Katy Ross (2015) observe, over time this communication system is "(re)produced by both DGMs' [dominant group members] and NDGMs' [non-dominant group members] discourse and, thus, the dominant communication systems remain in place" (p. 3). So, according to MGT, when Donna and her father learn how to speak like the dominant group, they have a hand in reproducing a system that isn't satisfactory for them as Native people. This would explain how women who teach English pass on the system that really doesn't serve them well, without even thinking about it most of the time.

MGT focuses on how well a language serves its speakers is directly related to the power those speakers possess in the culture. Cheris Kramarae (2005) observes, "people attached or assigned to subordinate groups may have a lot to say, but they tend to have relatively little power to say it" (p. 55). Or if they do venture to speak, those in greater power positions may ignore, ridicule, or disrespect their contributions in a variety of ways. Muted groups "get in trouble" when they speak out in their own voices (Kramarae, 2009). MGT is a critical theory in that it points out this inequity in English-speaking cultures and provides some avenues for muted groups to change that situation, such as by creating their own languages.

Standpoint Theory

The second source that Orbe utilized to develop CCT was Standpoint Theory (or Feminist Standpoint Theory) (FST). As we discussed in **Chapter 29,** a standpoint is a person's perspective, or way of thinking, resulting from their experiences and the view they have from their specific social location. However, FST explains that a standpoint is more than just a perspective and not everyone has a standpoint. First of all, people in higher-power positions don't often achieve standpoints because standpoints require critical reflection on, and opposition to, the power structure. Usually those in positions of power don't critique their own position or question why they have access to power. And, they don't usually wish to oppose or resist the social structure that's providing them with benefits. Further, even people in positions without much power don't always have standpoints. This is true because a standpoint is achieved only when someone engages in critical thought about power structures and social locations. Standpoints are the result of struggle and reflection. Until Donna Blackhawk, from our opening story, started to think critically about her position as a Native American in a predominantly white culture, she had a perspective, but not a standpoint. She began developing her standpoint when she started questioning her position and her social location.

FST has three core aspects that make it a good foundation for CCT. First, FST acknowledges that individuals have multilayered identities because a person belongs to more than one social group and thus occupies more than one social location (Hecht, Jackson, & Ribbeau, 2005). This insight comes from critical race theory and the work of Kimberlé Crenshaw (1989) on *intersectionality*. Crenshaw published an article in the *University of Chicago Legal Forum* in 1989, arguing that the court took too narrow a view of discrimination when they reduced it to a single issue such as racism or sexism. **Intersectionality** means that you must consider how race *and* sex, for instance, combine to create a different type of discrimination for Black women than for either Black men or white women. One of the contributions of intersectionality is that it clarifies the divisions existing within the common experience of being a woman, or being a Black woman. It allows for the assumption of common oppression while acknowledging various ways of experiencing it.

> **intersectionality** the recognition that identities like race and sex, for instance, combine to create a different type of discrimination for Black women than for either Black men or white women

Second, FST clearly points out how power plays a role in the social world (e.g., Allen, 1998; Camara & Orbe, 2010; Orbe, 1998b; Wood, 1992, 2008). From the perspective of FST, power privileges members of the dominant group (DGMs) and oppresses members of non-dominant, or co-cultural, groups (NDGMs). NDGMs aren't allowed authority and the issues raised by their standpoints are ignored. As a result, every interaction occurring between those in power and those who aren't is influenced by this power differential.

Third, FST advances a reciprocal relationship between communication behavior and standpoints. Communication shapes our standpoints because we learn our place in society through interaction with others. Similarly, FST asserts that those who share a standpoint will also share certain communication styles and practices. Further, FST notes that if you're a member of a marginalized group, you'll be a strategic communicator when talking with members of the dominant group, if for no other reason than self-preservation. FST also points to the use of communication as a tool for producing change. By giving voice to those whose standpoints are infrequently heard, the theory focuses on communication practices as a change agent (Jaggar, 2013), making FST a critical theory like MGT.

Phenomenology

Phenomenology (Husserl, 1973; Lanigan, 1988) is a qualitative research method focusing on people's lived experiences. Phenomenology takes the position that researchers are best when they interfere least with what their *co-researchers* tell them. The term **co-researcher** means that the people researchers interview are active in shaping the finished product. Research findings should ideally represent a true interaction among all involved. Orbe (1996, 1998a, 1998b) consciously chooses the term *co-researcher* because of these connotations rather than terms that interpretative researchers use like *respondents* or *participants,* or the term *subjects* used by post-positivist

> **phenomenology** a qualitative research method focusing on people's lived experiences
>
> **co-researcher** a term acknowledging that the people researchers interview are partners with the scholar in shaping the finished research product

researchers. The latter term indicates that people are manipulated by an objective researcher while the former two terms don't accord people an active role in shaping research outcomes.

Phenomenology involves a three-step process: first, researchers collect descriptions of experiences from co-researchers, usually through in-depth interviewing or focus group discussions or both, and record them; second, they spend a lot of time examining the data, looking for repetitions, patterns, and themes; finally, they interpret the themes and suggest what overall meanings they derived from hearing the co-researchers' descriptions. Sometimes, they may consult the co-researchers in stages two and three to check if their perceptions agree. But phenomenology is more than a method for collecting and interpreting data. It's also a philosophy of research that acknowledges that co-researchers are multidimensional and complex, and they have particular social, cultural, and historical positions. Further, phenomenology takes the position that the researcher isn't the expert—co-researchers are the experts on their own lives.

Orbe and his colleagues (e.g., Orbe, 1996, 1998a, 1998b; Roberts & Orbe, 1996) used phenomenological methods to create CCT. They found 89 co-researchers representing a variety of non-dominant marginalized groups (based on sexual/affectional orientation, ethnicity, sex/gender, ability, and social class). Then the researchers listened to their stories about interactions with people in the dominant culture, and followed the three-step process outlined above. These narratives, and the themes derived from them, form the essence of CCT's theoretical framework. Orbe and his colleagues' work was inductive as we defined it in **Chapter 3;** they began with specific narratives and induced the overarching framework that became CCT from them.

Assumptions of Co-Cultural Theory

CCT rests upon several assumptions that intersect the three areas (culture, power, and communication) that frame the theory. Specifically, the following assumptions guide CCT:

- When a culture is organized in a hierarchy, certain groups are favored over others and power is awarded based on this preference.
- The "lived experiences" of underrepresented groups are valuable and must be identified and embraced.
- Members of underrepresented groups, although differing from each other in many ways and representing widely different lived experiences, share an "outsider" view of the dominant culture.

The first assumption reflects Orbe's belief that Westernized cultures are rooted in hierarchical thinking and structures. This type of thinking divides groups and gives certain cultural groups dominance over others. Dominant group members have the power to control decision making, wealth allocation, naming practices, and a variety of other important areas for a culture. The power to control all this both puts them and maintains them at the top of a culture's hierarchy.

Corporate, academic, medical, religious/spiritual, and political contexts are replete with examples where those in power make decisions that benefit them while, perhaps, negatively affecting those with less power. For example, what would be the result of a large donation to a U.S. politician's campaign from a pharmaceutical company? Would it give the corporation the power to get legislation that's more beneficial for them than for individual customers? Would the corporation's large donation give them more influence than would an individual's much smaller donation? Questioning democracy in this way is aligned with the Marxist ideology we explained in **Chapter 27.**

Let's apply this first assumption to a specific group in society—the elderly. It's reasonable to conclude that a person who has lived for many years has acquired a great deal of knowledge, life experience, and wisdom. Therefore, the opinions and concerns of seniors who have lived 70+ years, for instance, should be valuable to the society as a whole. Thus, you'd think that seniors exert significant influence upon a culture. However, that conclusion would be somewhat inaccurate.

Particularly in the United States, the elderly form a co-cultural group that may be dismissed by those in decision-making roles. Words such as *feeble, sick, decrepit, over-the-hill* are often attached to people in this age group. And this negative response seems to be increasing over time. In fact, in a study of over 400 million words over 200 years, researchers found that with very little exception, the words applied to the elderly have become increasingly more negative and off-putting (Ng, Allore, Trentalange, Monin, & Levy, 2015). Further, many older individuals don't have the disposable income that would make them capable of making large donations to political campaigns or causes. As a result, they remain marginalized, and it's no surprise that those in power find little reason to solicit their input, expertise, or wisdom.

The second assumption states that the "lived experiences" and voices of a marginalized group must be both considered and embraced. This claim underscores the ethical imperative (Martin & Nakayama, 2018) for welcoming non-dominant groups into the cultural conversation. CCT suggests that non-dominant group members have very little say in creating the overall structure of society (Orbe, 2016), even though they are essential components of that society. Further, although dominant groups silence those on the margins, marginalized groups' voices have great value. Individuals in these groups are able to communicate various issues, themes, and "lived experiences" that are unknown to those in dominant groups.

For instance, immigrants are among the least powerful groups in a culture like the United States. Yet, the show *Taste the Nation* which debuted on Hulu in 2020, presents another perspective on immigrants in the United States. The show focuses on the foods of various immigrant groups in different cities in the United States, such as El Paso, Texas, and Milwaukee, Wisconsin. In each episode, Padma Lakshmi, the host and producer, illustrates how much immigrants have shaped U.S. culture and influenced what it means to be American.

The third assumption underscores the idea that NDGMs share an "outsider's" perception. They do this, despite the fact that they all have differing experiences and their specific social locations are diverse. This assumption contends that it doesn't matter if an individual is, for example, transgender (Ross & Castle Bell, 2017), an African American female pilot (Zirulnik & Orbe, 2019), or physically disabled (Cohen & Avanzino, 2010); they all understand what it means to be on the cultural periphery. Any member of a marginalized community recognizes they aren't in a position of power; they know their vision of the culture differs from the DGMs' vision. In some cases, this may create a situation where marginalized groups connect with each other for solidarity and to take action. So, for example, despite beliefs about antagonism between the Black and Asian communities in the United States, we have seen Asian American activists march in support of Black Lives Matter, especially after the killing of George Floyd by a Minneapolis police officer, and Black activists rallying against anti-Asian violence that intensified during COVID-19.

To further illustrate this assumption, think about how some groups utilize the experiences of other marginalized groups in their own struggles. For example, some have compared the gay rights movement in the 21st century to the struggle of Black Americans for civil rights in the 20th century. Media studies scholar David Craig in an interview on National Public Radio (Bates, 2015) contends that had it not been for the Supreme Court case in 1967 allowing interracial marriage (*Loving v. Virginia*), same-sex marriage wouldn't be legal today. Specifically, Craig states: "We owe the African-American community and the Loving couple incredible gratitude ... And we are only able to be where we are today because of the civil rights struggle" (Bates, 2015, para. 10). However, it's also true that sharing an outsider perspective doesn't always make different groups allies. Members of one group may clearly see that their group is marginalized while not recognizing that another group's plight is equally problematic. Yet, despite the inability at times for members of marginalized groups to become allies with one another, it is the case that being an outsider prompts a specific perspective on the dominant culture.

Theory at a Glance • Co-Cultural Theory

 Co-Cultural Theory (CCT) provides a framework for understanding how individuals from marginalized groups select strategies for interacting with those from the dominant group. CCT is unique because it's grounded in the experiences of the persons it seeks to describe. Specifically, CCT asserts that co-cultural group members make decisions about how they'll communicate with members of their culture's dominant group based on their *preferred outcomes* (assimilation, accommodation, or separation) and *communication approaches* (nonassertive, assertive, or aggressive). Combining each of the preferred outcomes with each of the approaches yields nine communication orientations (nonassertive assimilation, nonassertive accommodation, nonassertive separation, assertive assimilation, assertive accommodation, assertive separation, aggressive assimilation, aggressive accommodation, and aggressive separation). In addition to their communication orientation, co-cultural group members make choices based on the following factors: their *field of experience*, their perceptions of the *costs and rewards* associated with the communication choice, their *capability* to engage in the choice, and the specific *situational context* in which they're interacting. Research has identified multiple communicative strategies that are associated with each orientation. CCT affirms that co-cultural group members engage in trials of the strategies, after which they evaluate the results of using a specific strategy in a given situation. Thus, CCT posits that communication between nondominant group members (NDGMs) and members of the dominant group (DGMs) isn't static; it's a dynamic negotiation and adjustment of the performance of NDGMs' identities in interaction with DGMs.

Communication Orientations and Strategies

Using the information obtained from his co-researchers, Mark Orbe (1996, 1998a, 1998b) asserted that NDGMs may adopt one or more of nine different communication orientations when in communication with DGMs. Orbe derives these nine orientations by pairing three communication goals or preferred outcomes (assimilation, accommodation, and separation) with three communication approaches (nonassertive, assertive, and aggressive). Each goal intersects with each approach to form nine orientations (see **Table 30.1**). We'll now review each of the goals and approaches that the co-researchers told Orbe about and then discuss the communication orientations and strategies that CCT advances.

Table 30.1 Communication Orientations

		Goals		
		Assimilation	Accommodation	Separation
Approaches	Nonassertive	Nonassertive Assimilation	Nonassertive Accommodation	Nonassertive Separation
	Assertive	Assertive Assimilation	Assertive Accommodation	Assertive Separation
	Aggressive	Aggressive Assimilation	Aggressive Accommodation	Aggressive Separation

Goals/Preferred Outcomes

Co-cultural group members define **assimilation** as "attempts to eliminate cultural differences, including the loss of any distinctive characteristics in order to fit in with dominant society" (Orbe, 1998b, p. 10).
Co-cultural members who have the goal of assimilation try to adopt
the dominant group's culture, and in the process try to eliminate, or
greatly reduce, any verbal and nonverbal cues that would mark them as
members of a different culture. Overall, assimilation means fitting into
the dominant group at the expense of your own cultural identity. For

> **assimilation** a possible goal of non-dominant group members involving attempts to eliminate cultural differences between themselves and dominant group members in an effort to fit in with dominant society

example, Donna Blackhawk's father, and to some extent Donna and her brothers, assimilated into the dominant
white culture. They didn't practice Native rituals or celebrate Native holidays. They didn't speak the language of
the Te-Moak Tribe of the Western Shoshone. In their dress, manner, language, and habits they blended in with
the white people in their community.

Orbe's co-researchers defined **accommodation** as working to change
the rules of the dominant culture so that the specific experiences of
co-cultural group members are honored. For example, Mark Warne,
from our opening story, wanted the university administration to make
accommodations for the Native American students attending North
Dakota State. He, along with other members of the NSA, wanted
more Native professors, a Native Studies program, and so forth at the
university. In advocating for these changes, Mark and his colleagues

> **accommodation** a possible goal of non-dominant group members consisting of working to change the rules of the dominant culture so that the specific experiences of co-cultural group members are honored

were following the path that other co-cultural groups have forged, such as disabled people who fought to
have specific accommodations in the workplace and succeeded in getting the Americans with Disabilities
Act passed in 1990. Similarly, women's fight for the right to vote, women's work to get sexual harassment
recognized as a crime, and the #MeToo movement all represent efforts to change the rules of the dominant
culture to achieve accommodation. Movements like Black Lives Matter and the Human Rights Campaign
operated in much the same way for African Americans and LGBTQ+ people.

Separation was defined by the co-researchers as a rejection of efforts
either to assimilate into the dominant culture or to demand that the
dominant culture make accommodations for them. Instead, those
whose goal is separation seek to promote group solidarity among
co-cultural members, and to create a distinct identity that reflects
their differences from DGMs. Then, based on their unique qualities
and their solidarity, NDGMs are encouraged to retreat from the
dominant culture to maintain their own culture. The Amish, especially the Old Order Amish, provide an extreme example of separation because they believe the Bible and their religion instruct them to
keep themselves separate from mainstream culture (Romans 12:2).

> **separation** a possible goal of NDGMs that involves rejecting efforts either to assimilate into the dominant culture or to demand that the dominant culture make accommodations for them. Instead, this goal is to promote group solidarity among co-cultural members, and to create a distinct identity that reflects their differences from DGMs

For instance, the clothing worn by the Amish people (modest dress in black and white with head coverings
for both men and women and hooks and eyes instead of zippers) is meant to reflect both their humility as
well as their separation from the rest of the world.

Approaches

In addition to having an overall goal for a communication encounter with DGMs, Orbe's co-researchers told him that they also employed three general approaches in these encounters: a nonassertive approach, an aggressive approach, and an assertive approach. Each approach is defined as follows: a **nonassertive approach** occurs when the person is nonconfrontational and seems to put their own needs in the background while paying more attention to the needs of DGMs. For instance, Greg is a 50-year-old gay man who works for a conservative accounting firm in Cincinnati, Ohio. When he's at work he feels quite inhibited and consciously censors himself, never mentioning his husband or any of their social activities. He spends a lot of energy being sure his work colleagues feel comfortable around him.

> **nonassertive approach** a general approach that NDGMs may take in interactions with DGMs involving being non-confrontational and putting their own needs in the background while paying more attention to the needs of the DGMs

An **aggressive approach** goes to the opposite extreme and consists of behaviors that are "hurtfully expressive, self-promoting, and controlling" (Orbe, 1998b, p. 14). For instance, in an extreme example of aggression as an approach for dealing with the dominant culture, in 1973, 200 members of the American Indian Movement (AIM), led by Russell Means, and their supporters occupied the town of Wounded Knee, South Dakota on the Pine Ridge Indian Reservation. They were protesting the U.S. government's failure to fulfill treaties from the 19th and early 20th centuries. The protest lasted for 71 days and involved armed conflict; at least three people died. The occupation drew a great deal of national attention to the cause of Native people, and celebrities of the time, such as Marlon Brando, Johnny Cash, and Jane Fonda, supported the Indians' cause. Ultimately, however, the incident wasn't completely successful in improving the lot of Native Americans in general, or specifically that of the Oglala Lakota people who lived on the Pine Ridge Reservation. The aggressive approach shone a light on the situation but in 2012, when Means died, the unemployment rate on the reservation was still estimated to be between 70 and 80 percent (Chertoff, 2012).

> **aggressive approach** a general approach that NDGMs may take in interactions with DGMs consisting of behaviors that are hurtful, controlling, and self-promoting

Between the nonassertive and the aggressive approach is the assertive approach—a kind of "happy medium." Orbe defines the **assertive approach** as expressing "self-enhancing, expressive communication that takes into account the needs of both self and others" (1998b, p. 14). When NDGMs use an assertive approach, the co-researchers said, they're sensitive to their own culture and they wish to explain it to members outside their group. They recognize the power differences that exist between NDGMs and DGMs, but they believe that these can be reduced over time through cooperative efforts. They want to be sure their cultural practices are honored while not dismantling the dominant culture entirely. If Donna Blackhawk continues to work with the NSA she'll probably be using an assertive approach.

> **assertive approach** a general approach that NDGMs may take in interactions with DGMs involving self aware, expressive communication that also takes into account the needs of those in the dominant group

Communication Orientations

The approaches and the preferred outcomes or goals work together, and, as we've explained, nine orientations result when you pair each goal or preferred outcome with each approach: nonassertive assimilation, aggressive assimilation, assertive assimilation, nonassertive accommodation, aggressive accommodation, assertive accommodation, nonassertive separation, aggressive separation, and assertive separation (see **Table 30.1**). We'll discuss each of the nine orientations briefly now. In **nonassertive assimilation,** members of co-cultural groups tend to ignore their own needs for cultural identity and instead try to blend into the dominant culture while drawing little attention to their cultural differences. As we've mentioned, Donna Blackhawk adopted this orientation for most of the time she was in high school. A different orientation is expressed in **assertive assimilation** because although co-cultural members are still trying to blend in, they are much more self-aware during this process. In high school, Donna Blackhawk simply went along with the dominant culture without giving it a lot of thought. But, when NDGMs choose assertive assimilation, they are aware that they have to "play the game" to get along and that means they need to downplay the things that make them culturally distinct from DGMs. In the **Students Talking** box below, Christopher comments on this orientation, noting that this is a tiring orientation to practice, and one that might not be popular with friends in his co-cultural group. **Aggressive assimilation** represents a person's strong demand to be included as part of the dominant group and a rejection of their co-cultural membership. When someone adopts aggressive assimilation, they're more concerned with being included in the dominant group than they are with the rights of other NDGMs around them. When Lloyd came to work at an engineering firm after he graduated from college, he was so intent on assimilating into the dominant culture (90 percent of the engineers there were white men) that he rejected friendship overtures from Bruno, the only other African American man at the firm.

nonassertive assimilation a communication orientation favored by some members of non-dominant groups where they tend to ignore their own needs for cultural identity and instead try to blend into the dominant culture while drawing little attention to their cultural differences

assertive assimilation a communication orientation favored by some members of non-dominant groups where they are still trying to blend in, but are much more aware of the process and strategic than those favoring nonassertive assimilation

aggressive assimilation a communication orientation favored by some members of non-dominant groups involving a strong demand to be included as part of the dominant group and a rejection of their co-cultural membership

Students Talking: Christopher

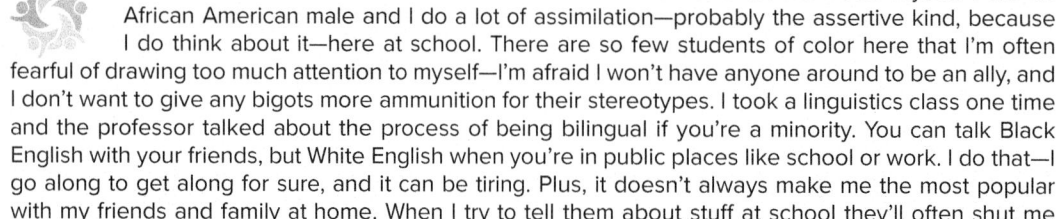

When we talked in class about those communication orientations I saw myself. I am an African American male and I do a lot of assimilation—probably the assertive kind, because I do think about it—here at school. There are so few students of color here that I'm often fearful of drawing too much attention to myself—I'm afraid I won't have anyone around to be an ally, and I don't want to give any bigots more ammunition for their stereotypes. I took a linguistics class one time and the professor talked about the process of being bilingual if you're a minority. You can talk Black English with your friends, but White English when you're in public places like school or work. I do that—I go along to get along for sure, and it can be tiring. Plus, it doesn't always make me the most popular with my friends and family at home. When I try to tell them about stuff at school they'll often shut me down and tell me I'm putting on airs. They laugh at me and call me "college boy" and I'm pretty sure that's no compliment.

When members of co-cultural groups employ **nonassertive accommodation,** they ask for small changes in the dominant culture. They request these in a nonconfrontational manner and hope to show, by their demeanor, that they aren't a threat, but they want some changes made. When Lindsey, a lesbian, moved to her new apartment building, she went out of her way to show up for social events in the common room. She was friendly and hoped to show the others in the building that although she wanted acceptance as a lesbian, they had nothing to fear from her. She usually waited until she thought she'd established a friendly relationship with someone before she told them she was gay. **Assertive accommodation** is an orientation adopted by co-cultural members when they are actively working for the rights of everyone. They'll make connections with those in the dominant group while also stressing networking within their own group. For example, Nolan, who was functionally deaf, attended Baker College in New York City. He was a member of a large number of student groups, including one specifically for disabled students. Nolan's mission was to help all Baker's students succeed and he often was seen leafletting on campus for a variety of student causes. Nolan was known as a tireless advocate for students, no matter what their ability. Co-cultural members may also adopt a communication orientation called **aggressive accommodation,** which means they work within the dominant culture, but their goal is to shine a light on the structures and practices of that culture that oppress NDGMs. Mark Warne, from our opening story, seems to have this orientation. He isn't advocating leaving the dominant culture, but he wants the NSA to engage in active and, possibly, radical demonstrations and public displays that will draw attention to his cause and bring about his desired changes.

Co-cultural group members adopting an orientation of **nonassertive separation** believe that it's best for people to stick with those of their own kind. Nonassertive separation involves avoiding members of the dominant culture whenever possible. If a person with this orientation can't avoid an interaction, they'll leave quickly and may isolate themselves. When Teddy came to college at a prestigious Northeastern U.S. university, he felt like an imposter. He was the first member of his family to attend college and he came from a lower socio-economic class than most of his classmates. His aunt lived in a nearby town and whenever he could he fled to visit her so he could avoid the people at school. **Assertive separation** means making a strategic decision to separate from the dominant culture and take, or create, opportunities to confidently express the strengths that make their co-cultural group admirable. When Marta thought about college, she made the conscious decision to go to a historically Black institution. As an African American student, she didn't want to waste her energy trying to fit in at a predominantly white university. She loved school and never regretted her decision. Finally, the

nonassertive accommodation a communication orientation favored by some members of non-dominant groups involving asking for small changes in the dominant culture. NDGMs request these in a non-confrontational manner and hope to show, by their demeanor, that they aren't a threat, but they want some changes made

assertive accommodation a communication orientation favored by some members of non-dominant groups who are actively working for the rights of everyone. They'll make connections with those in the dominant group while also stressing networking within their own group

aggressive accommodation a communication orientation favored by some members of non-dominant groups when they work within the dominant culture, but their goal is to shine a light on the structures and practices of that culture that oppress NDGMs

nonassertive separation a communication orientation favored by some members of non-dominant groups who believe that it's best for people to stick with those of their own kind. Nonassertive separation involves avoiding members of the dominant culture whenever possible

assertive separation a communication orientation favored by some members of non-dominant groups involving making a strategic decision to separate from the dominant culture and confidently expressing the strengths that make their co-cultural group admirable

orientation of **aggressive separation** is often used when the dominant group seems oppressive and members of the co-cultural group need to attack or intimidate DGMs to strengthen the solidarity of the co-cultural group. Rhonda and Sasha had been dating for 5 months and they found the best way to get along was to hang out mainly with other lesbians. They enjoyed going to the local lesbian bar and spending time making fun of straight people and laughing at the provincial attitudes they heard from members of the dominant culture. Sometimes they got in screaming matches with people who disagreed with their right to love one another. It gave Rhonda and Sasha some satisfaction to see they could intimidate them when they yelled back. But, it didn't always feel safe, they had to admit.

> **aggressive separation** a communication orientation favored by some members of non-dominant groups when they want to hang out mainly with members of their own group and when they may feel the need to attack or intimidate DGMs to strengthen the solidarity of their co-cultural group

Testing the Communication Orientations

In a somewhat unusual collaboration, Orbe, a critical researcher, teamed with Maria Knight Lapinski, a quantitative researcher, to test the reliability and validity of the goals, approaches, and resulting communication orientations proposed in CCT (Lapinski & Orbe, 2007). Lapinski and Orbe developed self-report scales for measuring the goals and approaches of CCT. Their scales contained statements that respondents were supposed to rate on a Likert-type scale (1 = never to 5 = always) indicating how much they endorsed what the statement said. They conducted two studies, one with 243 mainly white women and a second study with a sample of 72 African American or biracial women and men. Their results confirmed the three goals of assimilation, accommodation, and separation as well as the three approaches of nonassertiveness, aggression, and assertiveness. Their respondents saw them as distinct from one another. Further, the researchers saw a pattern where those whose goal was assimilation preferred a nonassertive approach and those who wanted accommodations or separation endorsed assertive or aggressive communication approaches. All these findings are consistent with the claims CCT advances about communication orientations.

Strategies

As you probably noticed, when we explained the communication orientations in the previous section, we illustrated each of them with an example of what a co-cultural member did to enact the orientation. In CCT, Orbe argues that non-dominant group members understand that they're often muted by dominant groups and that their social location affords them a standpoint that's unique. With this knowledge, they actively select specific communication strategies to be effective in dominant society (e.g., Orbe, 1996, 1998a, 1998b, 2004). Each strategy is associated with one of the nine orientations. Initially, Orbe (1996) advanced 12 communication strategies. Later, he expanded the list to 26 (Orbe, 1998b). Other researchers have added to this list until it now includes 38 strategies (Camara, 2002; Camara & Orbe, 2010; Castle Bell et al., 2015; Gates, 2003; Herakova, 2012). **Table 30.2** presents each of the 38 strategies, briefly defines them, provides a communication example, and lists them under the communication orientation associated with each.

Table 30.2   Co-Cultural Strategies Associated with Communication Orientations

NONASSERTIVE ASSIMILATION	ASSERTIVE ASSIMILATION	AGGRESSIVE ASSIMILATION
Emphasizing commonalities–Focusing on similarities that characterize all humans while ignoring co-cultural differences. *Ali, a Muslim man, likes to talk with his Christian classmates about all the ways their religions are alike.*	**Extensive preparation**–Doing your homework to prepare for interactions with DGMs. *Domingo, a Latinx man, reports that before he talks with any white person at work he works out a script for what he'll say.*	**Dissociating**–Making a concerted effort to set yourself apart from your co-cultural group. *Melanie, a Black woman, said she made an effort to speak more quietly and use "White English" when she was with her white roommates at school. "I stifle how I'd normally talk with my Black friends, so I can sound more 'white' for these girls."*
Developing positive face–Being a gracious, polite, considerate communicator with DGMs. *Tina, a Black woman, is always super polite to her co-workers and she often lets them take credit with their boss for ideas she originally brought up.*	**Overcompensation**–Conscious attempts to excel and be a "superstar." *Paige, a woman with cerebral palsy, knows she has to give more than 100 percent to be accepted by DGMs. She works extra hard.*	**Mirroring**–Adopting the dress, speech, and so forth of the dominant group to hide your association with your co-cultural group. *Angel, a Latinx man, learned how to play golf so he could hang out with the DGMs that worked at his office. He made sure to wear the same type of golf attire they favored for their outings.*
Censoring self–Remaining silent when DGMs make inappropriate or insulting comments about NDGMs. *At a local bar after work Leslie just smiled when two men in her group started making jokes about women's bodies. She was privately offended, but thought it was better to stifle the comments that came to her mind.*	**Manipulating stereotypes**–Conforming to stereotypes as a way to achieve something you want. *Jen is a good-looking blond woman and she knows people think she's a "dumb blond." Sometimes she acts like she doesn't know how to do something just to watch a man rush in to do it for her. She thinks that's funny.*	**Strategic distancing**–Avoiding any associations with other members of your co-cultural group. *Rachel, a lesbian, started to wear skirts and dresses when she went to work at a conservative law firm. She said, "I did everything I could to look just like the others at the firm. I didn't want anyone calling me a dyke."*
Averting controversy–Diverting conversations away from controversial subjects. *Mika, a Black man, was in a study group with three white men from his sociology class. When they were talking about what to do their project on, one suggested racism. Mika steered the conversation toward the topic of health care.*	**Bargaining**–Striking a deal with DGMs, where you both agree to ignore co-cultural differences. *Neela, a woman from India, states that she and the other teachers at her school who are DGMs just "don't see color–everyone is treated the same."*	**Ridiculing self**–Demeaning your co-cultural group yourself in a joking manner. *Bennet, a Black man, says he always jumps in to make a joke before his white friends can–when they were at a picnic, he made a big deal of picking up the watermelon and mugging for a selfie with a big slice.*
Remaining silent**–Refusing to speak up when something bothers you because you know there's a greater good (like staying employed) to be attained by being quiet. *Tamara, a Japanese woman, doesn't like it when she catches the white women in her office staring at her,*	**Rationalization******–Prejudice or racist actions are justified, diminished, or rationalized by providing an alternate explanation for them. *Mona was the only Black woman in her communication class. Once the professor turned to her during the lecture and said, "Mona probably understands this concept because African Americans use slang in this way frequently."*	

NONASSERTIVE ASSIMILATION	ASSERTIVE ASSIMILATION	AGGRESSIVE ASSIMILATION
but she doesn't say anything, figuring it's better not to rock the boat. **Journaling****–Keeping a diary that records all the problems encountered because you belong to a co-culture. *Justine, a Black woman, goes home from work and writes in her journal about what the white women at the factory have said to her. She finds it's therapeutic to write it out and she has a record if she ever decides to complain.* **Checking yourself****–Taking some time to think about whether the problem might reside with you or be the result of something you've done. *Antonio, a Latinx man, owns a stucco business. He reports that he never says the first thing that comes to his mind when he thinks a white person has disrespected him. He always takes a day to think about whether his own behavior had anything to do with the negative interaction.* **Internalizing*****–Thinking about discriminatory remarks and feeling bad about oneself as a result. *Tina, a Chinese woman, was shopping in a grocery store and overheard two white women whispering about how the Asians were taking over their neighborhood. Tina left the store feeling depressed.*	*Mona was shocked but labeled her professor's behavior as ignorant, saying, "She's not really racist, but that was an ignorant comment to make."*	**Showing appreciation****–Complimenting members of the dominant group to make it clear they are appreciated. *Viola, a Korean woman who works in a small office, frequently compliments her boss, Karen, telling her, "Thanks so much for establishing such a positive climate in the office" or "I appreciate your help, Karen, it means a lot."*

NONASSERTIVE ACCOMMODATION	ASSERTIVE ACCOMMODATION	AGGRESSIVE ACCOMMODATION
Increasing Visibility—Strategically, but covertly, maintaining a co-cultural presence in interactions with DGMs. *Will, a gay man, often wears a rainbow lapel pin at work, just to remind his co-workers that he's an NDGM.*	**Utilizing liaisons**—Identifying DGMs who can be allies to support, guide, and assist you. *Marco, an immigrant to the United States, seeks out his boss to help him bring his wife to the United States and help him navigate some of the complicated laws about immigration.*	**Confronting**—Using necessary methods to make your voice heard even if it involves trampling on the rights of others. *Lola, a lesbian, knows people think she's pushy, but she thinks she has to yell and dominate a conversation to get her voice heard.*
Dispelling stereotypes—Just being oneself to disabuse DGMs of negative stereotypes surrounding your group. *Molly has found that her housemates at college have a better opinion of lesbians since she moved in. She's an outgoing woman who makes friends easily, so her housemates have accepted her.*	**Educating others**—Explaining to DGMs about your culture's history, norms, and values. *Greta, a woman from a lower socio-economic status who's the first in her family to graduate college, takes the time to tell her co-workers how her family supports her and how much they value hard work. Greta always mentions her family at work, and lets others know that although she's forging a different path, she values her roots.*	**Gaining advantage**—Bringing up co-cultural oppression to provoke reactions from DGMs. *Andrew uses a wheelchair and when he was elected to the State Senate he immediately began to lobby for the legislature to make the meeting rooms more accessible. He never passed up an opportunity to point out how he, and others in his situation, were oppressed by a culture and by buildings built for able-bodied people. Finally, he wore them down and accommodations were made.*
		Speaking out**—Standing your ground and addressing problems when you see them occur. *Nikki, a woman in a male-dominated firm, is very clear about telling her co-workers when she hears them go too far in their jokes about women. She goes over to the group and tells them to knock it off and they usually do.*
		Regulating interactions*****—Trying to create situations where you can communicate in your preferred manner and perhaps change the dominant practices. *Tim, a male nurse, knows that the majority of his co-workers prefer to conduct conflict indirectly, but when he discovers another nurse is angry with him and has been talking behind his back, he meets her for coffee and tells her that he wants to deal with their conflict head-on.*

NONASSERTIVE ACCOMMODATION	ASSERTIVE ACCOMMODATION	AGGRESSIVE ACCOMMODATION
		Reporting the incident*—Telling someone in authority about the situation. *Amelia, a Latinx woman, noticed a salesperson in a clothing store watching her suspiciously while she was shopping. When she asked for some help the salesperson was rude and said, "I don't think you can afford this." Amelia reported the interaction to a store manager.*

NONASSERTIVE SEPARATION	ASSERTIVE SEPARATION	AGGRESSIVE SEPARATION
Avoiding—Maintaining distance from DGMs; refraining from activities where interaction would be likely. *Melanie is a visually impaired person, and she spends most of her time at home or with her family.*	**Communicating self**—Interacting with DGMs in an authentic and open manner. *Layla, a Mexican American woman, found that being her open, friendly self helped her get along with the DGMs in her office.*	**Attacking**—Engaging in personal attacks on DGMs' self-concept to inflict psychological pain. *David, a Native American man, stated that at a demonstration he carried a sign that said, "In 1492 Native Americans discovered Columbus lost at sea"; his friend went to a demonstration protesting the Washington Redskins and held a sign saying "Washington Racists."*
Maintaining interpersonal barriers —Using verbal and nonverbal cues to establish psychological distance from DGMs. *Todd, a transgender college student, finds if he doesn't make eye contact with his classmates, they leave him alone.*	**Intragroup networking**—Identifying and working with others of your co-cultural group. *When Candida, a lesbian, got to college one of the first places she looked for was the Gay Students Center. She wanted to connect with other gay students on campus to work with them on a variety of projects.*	**Sabotaging others**—Undermining DGMs' ability to take full advantage of their privilege. *Tanya uses a walker and she says that she's not above bumping into an able-bodied person at the grocery store and using the walker to kind of push them aside. She says, "Usually they're too shocked to say anything, so they let me go ahead of them."*
Leaving the situation*—Taking yourself out of the situation completely. *Jules, an Orthodox Jewish woman, reports entering a laundromat in her college town. She was wearing a Jewish star around her neck. She says she noticed everyone in the laundromat just got quiet and started staring at her. Jules says, "I felt uncomfortable and just took my laundry bag and went to another place to clean the clothes."*	**Exemplifying strengths**—Promoting the strengths, contributions, and accomplishments of those in your co-cultural group. *Carlos, a student from Puerto Rico, took every chance he could to do a report or a presentation on the contributions of Puerto Ricans to the United States. He liked showing the white students and instructors the positive attributes of his culture.*	**Intimidation****—Using something about yourself to strike fear in a DGM. *Rafe, a Black man, admits using his size to intimidate the white people he works with. He also frowns and scowls a lot*

NONASSERTIVE SEPARATION	ASSERTIVE SEPARATION	AGGRESSIVE SEPARATION
Isolation**–Separating yourself from others. *Jake, a veteran, simply walks away from others as they ridicule the "naive" young people who enlist in the military.*	**Embracing stereotypes**– Explaining how the perceptions of DGMs can be seen in a positive light. *Juanita, a housekeeper at a local hotel, explains to her children that despite the difficult workload, long hours, and low pay, as recent asylum-seekers in town, taking any job allows them to show others that they are "hard workers."*	*so they think he's angry. He's laughing on the inside because he wouldn't hurt a fly, but they don't know that.*

\# The strategies are examples of ways to enact each orientation. However, some strategies can be used to promote more than one communication orientation.

Strategies without asterisks from Orbe, 1998.

*from Camara & Orbe, 2010;

**from Gates, 2003;

***from Camara, 2002;

****from Castle Bell et al., 2015;

*****from Herakova, 2012

Table adapted from Castle Bell, Hopson, Weathers, & Ross (1988).

Theory Into Practice • Co-Cultural Theory

Theoretical Claim: Members of co-cultural groups give some thought (conscious or subconscious) to the preferred outcome they want relative to the dominant group (i.e. assimilation, accommodation, or separation) and ask themselves: "What communication behavior will lead to the effect that I desire?" When co-cultural members want the dominant group to make accommodations for their needs, they'll often try to educate someone so that person will be able to act as a liaison between them and others in the dominant group. This is called assertive accommodation.

Practical Implication: Moira uses a wheelchair and works for the county. Her office is in the county courthouse, a beautiful old building built in 1881. However, there are some spots in the building that are inaccessible to her. Moira has thought a lot about how to broach her concerns about this with her bosses. At first she thought she should just ignore it and try to get by without drawing attention to herself. But, lately she's been thinking it's impeding her ability to advance in her job because she cannot attend some informal meetings that are held in the inaccessible parts of the building. She's decided to talk to Joe, the man who originally hired her. He's been an ally in the past and Moira thinks he may be able to help her bring some attention to this problem and eventually get it solved. But, before she talks to Joe, Moira plans to prepare what she'll say. She wants to spend enough time preplanning her remarks so that Joe will really understand the problem. She knows she's in a position to teach Joe a little about what life is like for a disabled worker. Also, she wants to come to the meeting with him with a few possible ideas for a solution.

Other Factors Influencing NDGMs' Communication Choices

Orbe (1998b) found six factors influenced NDGMs' communication choices. The two factors of *preferred outcomes or goals* and *communication approaches* combine to form the nine communication orientations we've discussed. Now, we turn to the four other factors that CCT posits affect a co-cultural group member's choice of communication strategies: field of experience, perceived costs and rewards, capability, and situation.

Field of Experience

In **Chapter 1** we mentioned that **field of experience,** which refers to the totality of an individual's life experiences, plays a role in the communication process. It's also a key concept in CCT because experiences such as family interactions and influences, regional influences (e.g., where a person grew up or has lived most of their life), educational encounters, both inside and outside of classrooms,

> **field of experience** a factor in how NDGMs choose communication strategies when interacting with DGMs, it is the totality of an individual's life experiences

interpersonal exchanges, and so forth, exert an influence on what strategies NDGMs choose to enact in a communication encounter with a DGM. Obviously, field of experience is dynamic as people keep experiencing new things throughout their lives. Mark Orbe and Tabatha Roberts (2012) comment on this dynamism when they state that "over time, each co-cultural group member engages in a dynamic process of constructing, and subsequently deconstructing, the perceptions of what constitutes appropriate and effective communication with dominant group members" (p. 297). Field of experience exerts an influence on strategy choice, and from their past experiences, NDGMs learn which strategies are most effective and how to modify those that aren't as effective. Field of experience is unique to each individual and in this way CCT allows for diversity within co-cultural groups, based on differing field of experience. This is like the idea of intersectionality that we discussed earlier. If Hallie and Cassandra are both Latinx teens but Hallie comes from a well-to-do household and is attending a private high school while Cassandra lives with a widowed mother who's on a fixed income and goes to the public high school, they share an ethnic background but their socio-economic backgrounds differ and so their fields of experience will differ as well.

Perceived Costs and Rewards

As we discussed in our description of Social Exchange Theory (**Chapter 6**), people engage in a mental calculus trying to decide how costly and how rewarding a particular behavior will be. CCT says NDGMs think about the costs and rewards of using a specific strategy in their communication with DGMs. Co-cultural group members reflect on how likely (or unlikely) it is that a given strategy, used in a particular context, will result in their preferred goals. This forms their **perceived costs and rewards.** And, as we stated above, each co-cultural

> **perceived costs and rewards** a factor in how NDGMs choose communication strategies when interacting with DGMs, it involves reflecting on how likely (or unlikely) it is that this strategy, used in this context, will result in their preferred goals

group member will use their field of experience to determine how rewarding or how costly a strategy will be for them. It's important to note that no one strategy (or orientation) is rewarding all the time. The constant negotiation between an individual and the needs of the situation determines perceived costs and rewards (Orbe, 1998b).

Capability

Capability refers to a person's ability to enact a given strategy. This factor suggests that not all co-cultural group members have the ability to enact every one of the 38 strategies listed on **Table 30.2.** For example, perhaps Donna Blackhawk is able to express herself clearly and use the assertive accommodation strategy of *communicating self,* but another person might not possess Donna's communication skills.

capability a factor in how NDGMs choose communication strategies when interacting with DGMs, it refers to a person's ability to enact a given strategy

It's also the case that when Donna was in high school, she didn't really know anyone who was Native outside of her family. This would restrict her ability to engage in the strategy of *intragroup networking.*

Situation

Situation or situational context refers to details such as where the interaction occurs, who's present, and a variety of important environmental issues surrounding the interaction. Co-cultural group members don't just use one strategy in all their encounters with DGMs (Orbe, 1998b). Situation is the main factor influencing the appropriate choice of strategy. So, it's possible that a member of a co-cultural group may use different strategies in different settings, such as when speaking in class when there are several other members of their co-cultural group present, at work where they're the only member of their group present, or at the home of a friend of a different co-cultural group, and so forth.

situation or situational context a factor in how NDGMs choose communication strategies when interacting with DGMs, it refers to details such as where the interaction occurs, who's present, and a variety of important environmental issues surrounding the interaction

Integration and Critique

Mark Orbe's CCT provides an important framework for understanding the communication between NDGMs and DGMs. It's grounded in the perceptions and experiences of NDGMs, and well suited to examining how communication produces, reproduces, and ultimately can change systems of discrimination. The majority of the research conducted using CCT is qualitative because that's seen by researchers as the most appropriate method for pursuing questions in the critical tradition using phenomenology. There has been one quantitative study that successfully checked the reliability and validity of the building blocks (approaches and goals) for the communication orientations (Lapinski & Orbe, 2007). Using quantitative methods to establish the validity of a critical theory is unusual, but it suggests that researchers from several different perspectives might find CCT useful. To evaluate the overall effectiveness of CCT, we'll discuss the following criteria: scope, parsimony, and heurism.

Integration

Communication Tradition	Rhetorical | Semiotic | Phenomenological | Cybernetic | Socio-Psychological | Socio-Cultural | **Critical**
Communication Context	Intrapersonal | Interpersonal | Small Group/Team | Organizational | Public/Rhetorical/Mass/Media | **Cultural**
Approach to Knowing	Positivistic/Empirical | Interpretive/Hermeneutic | **Critical**

Critique

Evaluation Criteria	**Scope** | Logical Consistency | **Parsimony** | Utility | Testability | **Heurism** | Test of Time

Scope

The scope of CCT has been critiqued in at least two ways. First, critics note that there's a lack of attention in the theory to DGMs. This is important because as Cerise Glenn and Dante Johnson (2012) point out, the way a DGM responds to a strategy may cause a co-cultural member to change their preferred strategy. Glenn and Johnson give the example of a Black man who wants to use aggressive accommodation but hesitates because of past reactions to that strategy on the part of DGMs. So instead he employs a strategy of nonassertive assimilation (censoring self) while he thinks about his next move. Without understanding the dominant group, it's harder to understand this behavior. Further, Robert Razzante (2018) observes that CCT essentializes DGMs, presenting them all as committed to maintaining their privilege. He and Orbe (Razzante & Orbe, 2018) responded to that critique by creating Dominant Group Theory, which explains "how dominant group members come to challenge and/or reinforce structures of oppression" (Razzante, 2018, p. 344). Thus, this new theory expands the scope of CCT and when used together, the two theories provide a more nuanced understanding of the production and reproduction of privilege. One study (Rudnick & Munz, 2022) combined Dominant Group Theory and Co-Cultural Theory in an examination of an LGBTQ discussion panel in an educational setting. The authors concluded that DGMs do challenge structures of oppression in attempts to be allies of NDGMs, but it isn't always a straightforward process. Even when DGMs work diligently to be allies, they may reinforce structures of oppression and the authors note, "We must also be wary of such practices in consideration of the ways they also maintain our complicity in relegating issues of diversity and inclusion to 'diverse' populations" (n.p.). By this they mean the real work of changing the system may still be left to the NDGMs.

Second, the concept of national culture isn't included in CCT according to some critics, and that narrows the scope of the theory considerably. For instance, Bijie Bie and Lu Tang (2016) found that culture was the most important factor mentioned by their co-researchers in a study of how Chinese gay men described their coming out process. A large percentage of their narratives mentioned that being gay is contrary to Chinese cultural expectations. Bie and Tang found that "Chinese gay men demonstrated different communication orientations when coming out to different types of relational partners ... and their choice of orientation was determined by the rules about proper behavior in dealing with different social relations as dictated by the Chinese culture" (p. 352). A similar argument was made by Eun-Jeong Han and Paula Price (2018) in their study of Korean husbands and their immigrant wives. Han and Price found that both husbands' and wives' communication orientations were determined as much by the cultural influence of Confucianism as by their preferred outcomes and approaches. Although CCT's advocates might argue that culture is a part of situational context and the field of experience, Bie and Tang's and Han and Price's work indicates that cultural expectations should be featured more prominently in the theory.

Parsimony

Some researchers have argued that CCT is too parsimonious and that it would benefit from taking a few other concepts into consideration. For instance, Ruben Ramírez-Sánchez (2008) argues that calling co-cultural groups by one monolithic name (people of color, disabled, and so forth) obscures the lived experiences of those who form a co-culture within a larger co-culture. As an example, he studies African American punks who are a minority within the overall minority punk co-culture. He argues that intersectionality can't be ignored when examining people's experiences. His critique isn't damning to CCT; he's actually urging an expansion of the theory. As he notes, "theories should not be static tools: they can only be useful within the constant flux of our observations and the changing cultural contexts that call for their development in the first place" (p. 103).

An opposite critique states that CCT runs the risk of failing on the parsimony criterion because of the proliferation of strategies that's occurred since 1996. As we noted, in 1996 Orbe proposed 12 strategies that members of co-cultural groups use to communicate with DGMs. By 2020, that number had increased to 38. This burgeoning list of strategies might cause the theory to suffer the same problem noted in Relational Dialectics Theory (**Chapter 25**), where the theory becomes less explanatory, more descriptive, and potentially the source of an infinite number of strategies. This situation is, in large part, what motivated Leslie Baxter to change RDT to RDT 2.0 and discard the search for more tensions characterizing relationships. It remains to be seen if Orbe will do the same with CCT.

Heurism

CCT scores well on the criterion of heurism. Studies framed in CCT range from examinations of date rape on college campuses (e.g., Burnett et al., 2009) to investigations of Chinese international students studying in the United States dealing with anti-Asian discrimination during COVID-19 (Ji & Chen, 2022). CCT has been an effective theoretical framework in the context of the workplace. For example, Michael Zirulnik and Mark Orbe (2019) utilized CCT to investigate the ways that Black women airline pilots negotiate workplaces governed by white men's norms. Some studies examine how people with disabilities navigate workplaces that don't offer accommodations for their needs (e.g., Cohen & Avanzino, 2010), while Liliana Herakova (2012) interrogated the strategies used by men when they're the minorities in the workplace (e.g., male nurses). Elena Gabor and Patrice Buzzanell (2012) investigated the Roma culture as a co-culture with respect to career development. CCT has been used to study educational institutions (e.g., Glenn & Johnson, 2012; Orbe & Groscurth, 2004; Razzante, 2018; Urban & Orbe, 2007). A group of studies has looked at online contexts using CCT (e.g., Fox & Warber, 2015; Ju, 2017; Weathers & Hopson, 2015). For instance, Melinda Weathers and Mark Hopson investigated the co-cultural communication strategies of young women (18 to 24 years old) who were in digitally abusive heterosexual romantic relationships. Weathers and Hopson observe that technological communication provides opportunities for intimate partner abuse to occur online or through phone calls, texts, and instant messaging. They call this *digital dating abuse* and note that it also occurs when a partner breaks into social media accounts, engages in cyberbullying, or threatens to share private information or photos to a wider social network. Weathers and Hopson found that co-researchers described their responses in terms of the three preferred outcomes (assimilation, accommodation, and separation) and two of the approaches (nonassertive and assertive). Finally, some co-cultural research has expanded the theory beyond its initial focus on interpersonal communication and moved it into rhetorical analyses (e.g., Bridgewater & Buzzanell, 2010; Groscurth & Orbe, 2006; Opdycke, 2013; Orbe, 2016), showing the heurism of CCT.

Students Talking Tech: Lauren

 Wow, that section in the chapter about digital dating abuse kind of got to me. I was a victim of that, even though I didn't know the term. My boyfriend at the time, Vince, used to text me 10 to 20 times a day and he always wanted to know where I was when I wasn't with him. When it started, I was a little flattered, but then it got crazy. When I said I didn't want to see him anymore, he said he had pictures of me that he'd post on Instagram or Facebook and I'd be sorry. I got scared, but I didn't say much to anyone—I think I censored myself. Then I used the strategy of avoidance too—I got a new phone number, and whenever I saw a message from Vince on Facebook I deleted it without reading it. Luckily for me, Vince got a new girlfriend and he stopped digitally stalking me. Unlucky for her, though. I'm sure he's doing the same thing with her.

Closing

CCT is an important theory in communication. It deals with communication behaviors and helps us think about how power is negotiated through communication exchanges. It builds on two other theories (MGT and FST) and extends them to offer a fuller picture of how members of marginalized co-cultures interface with members of the dominant culture. Although some researchers have found flaws in the theory, most of the critiques point to extensions and refinements of CCT; they don't advocate abandoning the theory. Given the age of many of the theories we've discussed in this text, CCT is relatively young, and its future seems promising.

Discussion Starters

 Case-In-Point: Do you think Co-Cultural Theory provides a good explanation for the ways Donna Blackhawk experienced communicating with her friends in high school and then how she communicated when she made the transition to college? Specifically, how does CCT help you understand Donna's experiences?

Try-It-Your-Selfie: How could you use co-researchers to find out whether NDGMs have different goals and use different strategies when interacting with DGMs online than they do offline?

1. What do you think about phenomenology as a method for discovering what marginalized people think about their communication strategies? What might be the pitfalls in asking people to tell you what they think? Are people always honest? Do people always know what they think? What are the strengths in this method?

2. What do you think Co-Cultural Theory offers that's distinct from what we can learn just from Muted Group Theory or Feminist Standpoint Theory alone? Do you think CCT contributes enough to make it worth a separate theory? Explain your answer.

3. What do you think about Orbe's decision to label the people who participated in his research "co-researchers"? How does that label represent the philosophy of phenomenology? Do you think it actually makes any difference what you call those who provide answers to researchers' questions? Explain your perspective.

4. Provide examples of what a person might say or do utilizing the nine communication orientations beyond those in the chapter. Have you ever observed people communicating using these orientations? Have you ever adopted any of them? What do these orientations tell you about the communication between dominant and non-dominant groups?

References

Allen, B. J. (1998). Black womanhood and feminist standpoints. *Management Communication Quarterly, 11*(4), 575–586.

Ardener, E. (1975). The problem revisited. In S. Ardener (Ed.), *Perceiving women* (pp. 2–20). Malaby Press.

Ardener, S. (2005). Muted group theory excerpts. *Women and Language, 28*, 50–72.

Bates, K. G. (2015, July 2). African-Americans question comparing gay rights movement to civil rights. *All Things Considered,* Transcript for 3-minute listen. *NPR News.* https://www.npr.org/2015/07/02/419554758/african-americans-question-comparing-gay-rights-movement-to-civil-rights.

Bie, B., & Tang, L. (2016). Chinese gay men's coming out narratives: Connecting social relationship to co-cultural theory. *Journal of International and Intercultural Communication, 9*(4), 351–367.

Bridgewater, M. J., & Buzzanell, P. M. (2010). Caribbean immigrants' discourses. *Journal of Business Communication, 47*(3), 235–265.

Burnett, A., Mattern, J. L., Herakova, L. L., Kahl, D. H., Tobola, C., & Bornsen, S. E. (2009). Communicating/muting date rape: A co-cultural theoretical analysis of communication factors related to rape culture on a college campus. *Journal of Applied Communication Research, 37*(4), 465–485.

Camara, S. K. (2002). Ideological uncertainty: Exploring racism as a social issue in communication. *Journal of Intergroup Relations, 29*(3), 16–38.

Camara, S. K., & Orbe, M. P. (2010). Analyzing strategic responses to discriminatory acts: A cocultural communicative investigation. *Journal of International and Intercultural Communication, 3*(2), 83–113.

Castle, B. G., Hopson, M. C., Weathers, M. R., & Ross, K. A. (2015). From "laying the foundations" to building the house: Extending Orbe's (1998) co-cultural theory to include "rationalization" as a formal strategy. *Communication Studies, 66*(1), 1–26.

Chertoff, E. (2012, October. 23). *Occupy Wounded Knee: A 71-day siege and a forgotten civil rights movement.* The Atlantic. https://www.theatlantic.com/national/archive/2012/10/occupy-wounded-knee-a-71-day-siege-and-a-forgotten-civil-rights-movement/263998/.

Cohen, M., & Avanzino, S. (2010). We are people first: Framing organizational assimilation experiences of the physically disabled using co-cultural theory. *Communication Studies, 61*(3), 272–303.

Crenshaw, K. (1989). Demarginalizing the intersection of race and sex: A Black feminist critique of antidiscrimination doctrine, feminist theory, and antiracist politics [1989]. *University of Chicago Legal Forum.* http://chicagounbound.uchicago.edu/uclf/vol1989/iss1/8.

Fox, J., & Warber, K. M. (2015). Queer identity management and political self-expression on social networking sites: A co-cultural approach to the spiral of silence. *Journal of Communication, 65*(1), 79–100.

Gabor, E., & Buzzanell, P. M. (2012). From stigma to resistant career discourses: Toward a co-cultural career communication model for nondominant group members. *Intercultural Communication Studies, 21*(3), 1–17.

Gates, D. (2003). Learning to play the game: An exploratory study of how African American women and men interact with others in organizations. *The Electronic Journal of Communication, 13*(2/3).

Glenn, C. L., & Johnson, D. L. (2012). "What they see as acceptable": A co-cultural theoretical analysis of Black male students at a predominantly White institution. *The Howard Journal of Communications, 23*(4), 351–368.

Groscurth, C. R., & Orbe, M. P. (2006). The oppositional nature of civil rights discourse: Co-cultural communicative practices that speak truth to power. *Atlantic Journal of Communication, 14*(3), 123–140.

Han, E-J., & Price, P. G. (2018). Communicating across difference: Co-Cultural Theory, capital and multicultural families in Korea. *Journal of International and Intercultural Communication, 11*(1), 21–41.

Hecht, M., Jackson, R. L., & Ribbeau, S. (2005). *African American communication: Exploring identity and culture*. Lawrence Erlbaum Associates.

Herakova, L. L. (2012). Nursing masculinity: Male nurses' experiences through a co-cultural lens. *The Howard Journal of Communications, 23*(4), 332–350.

Husserl, E. (1973). *The idea of phenomenology* (W. P. Alston & G. Nakhnikian, Trans.). The Hague: Martinus Nijhoff. (Original work published 1931.).

Ibrahim, N., Windels, K., & Lu, L. (2023). Examining the ad industry's race and ethnicity problem: Application and extension of Co-Cultural Theory. *Journal of Current Issues & Research in Advertising, 44*(4), 411–428.

Jaggar, A. M. (2013). Feminist ethics. In H. LaFollette & I. Persson (Eds.), *The Blackwell guide to ethical theory* (pp. 433–460). Wiley-Blackwell.

Ji, Y., & Chen, Y.-W. (2022). "Spat on and coughed at": Co-cultural understanding of Chinese international students' experiences with stigmatization during the COVID-19 pandemic. *Health Communication, 38*(3), 1964–1972.

Ju, R. (2017). Communicating homosexuality online in China: Exploring the blog of a lesbian organization through the lens of Co-Cultural Theory. *Intercultural Communication Studies, 26*(2), 79–94.

Kramarae, C. (1981). *Women and men speaking: Framework for analysis*. Newbury House.

Kramarae, C. (2005). Muted Group Theory and communication: Asking dangerous questions. *Women and Language, 28*(2), 55–61.

Kramarae, C. (2009). Muted Group Theory. In S. W. Littlejohn & K. A. Foss (Eds.), *Encyclopedia of communication theory* (vol. 2, pp. 667–669). Sage.

Lanigan, R. L. (1988). *Phenomenology of communication: Merleau-Ponty's thematics in communicology and semiology*. Duquesne University Press.

Lapinski, M. K., & Orbe, M. P. (2007). Evidence for the construct validity and reliability of the cocultural theory scales. *Communication Methods and Measures, 1*(2), 137–164.

Martin, J., & Nakayama, T. (2018). *Intercultural communication in contexts* (7th ed.). McGraw Hill.

Ng, R., Allore, H. G., Trentalange, M., Monin, J. K., & Levy, B. R. (2015). Increasing negativity of age stereotypes across 200 years: Evidence from a database of 400 million words. *PloS One, 10*(2), e0117086.

Opdycke, K. (2013). A comic revolution: Comedian Bassem Youssef as a voice for oppressed Egyptians. *Colloquy, 9*, 1–20.

Orbe, M. P. (1996). Laying the foundation for cocultural communication theory: An inductive approach to studying "non-dominant" communication strategies and the factors that influence them. *Communication Studies, 47*(3), 157–176.

Orbe, M. P. (1998a). An "outsider within" perspective to organizational communication: Explicating the communicative practices of co-cultural group members. *Management Communication Quarterly, 12*(2), 230–279.

Orbe, M. P. (1998b). From the standpoint(s) of traditionally muted groups: Explicating a co-cultural communication theoretical model. *Communication Theory, 8*(1), 1–26.

Orbe, M. P. (2004). Co-Cultural Theory and the spirit of dialogue: A case study of the 2000–2002 community-based Civil Rights Health Project. In G. M. Chen & W. J. Starosta (Eds.), *Dialogue among diversities* (pp. 191–213). National Communication Association.

Orbe, M. P. (2016). The rhetoric of race, culture, and identity: Rachel Dolezal as co-cultural group member. *Journal of Contemporary Rhetoric, 6*(1/2), 23–35.

Orbe, M., & Groscurth, C. R. (2004). A co-cultural theoretical analysis of communicating on campus and at home: Exploring the negotiation strategies of first generation college (FGC) students. *Qualitative Research Reports in Communication, 5*(1), 41–47.

Orbe, M. P., & Roberts, T. L. (2012). Co-cultural theorizing: Foundations, applications & extensions. *The Howard Journal of Communications, 23*(4), 293–311.

Ramírez-Sánchez, R. (2008). Marginalization from within: Expanding Co-Cultural Theory through the experience of the *Afro Punk. The Howard Journal of Communications, 19*(2), 89–104.

Razzante, R. J. (2018). Intersectional agencies: Navigating predominantly White institutions as an administrator of color. *Journal of International and Intercultural Communication, 11*(4), 339–357.

Razzante, R. J., & Orbe, M. P. (2018). Two sides of the same coin: Conceptualizing dominant group theory in the context of Co-Cultural Theory. *Communication Theory, 28*(3), 354–375.

Roberts, G., & Orbe, M. (1996). *Creating that safe place among family: Exploring intergenerational gay male communication.* Paper presented at the annual meeting of the International Communication Association, Chicago, Illinois.

Ross, K. A., & Castle Bell, G. (2017). A culture-centered approach to improving healthy trans-patient-practitioner communication: Recommendations for practitioners communicating with trans individuals. *Health Communication, 32*(6), 730–740.

Rudnick, J. J., & Munz, S. M. (2022). Speaking out, speaking up: Co-cultural communication through an LGBTQ discussion panel. *Southern Communication Journal, 88*(4), 379–393.

Urban, E., & Orbe, M. P. (2007). "The syndrome of the boiled frog:" Exploring international students on US campuses as co-cultural group members. *Journal of Intercultural Communication Research, 36*(2), 117–138.

Weathers, M. R., & Hopson, M. C. (2015). "I define what hurts me": A co-cultural theoretical analysis of communication factors related to digital dating abuse. *Howard Journal of Communications, 26*(1), 95–113.

Wood, J. T. (1992). *Spinning the symbolic web: Human communication as symbolic interaction.* Ablex Publishing.

Wood, J. T. (2008). Critical feminist theories. In L. A. Baxter & D. O. Braithwaite (Eds.), *Engaging theories in interpersonal communication* (pp. 323–334). Sage.

Zirulnik, M. L., & Orbe, M. (2019). Black female pilot communicative experiences: Applications and extensions of co-cultural theory. *Howard Journal of Communications, 30*(1), 76–91.

Afterword ← *ConnectingQuests*

A primary goal of this book is introducing you to the richness, usefulness, and comprehensiveness of communication theories. Each of the 27 theories in the book approaches communication and its numerous functions and trajectories in multiple exciting ways. Whether, for example, a theorist emphasizes technology, culture, or gender; is concerned about an interpersonal, organizational, or mediated environment; or is situated in the post-positive, interpretative, or critical tradition, there can be no mistake that the area known as "communication theory" is multi-faceted and wide-reaching.

We hope that our efforts in elucidating a theory's background, assumptions, key concepts, and features have given you an understandable and workable framework for learning about communication theory. All too often, textbook writers forget what it's like to absorb challenging subject matter. We're hopeful and, yes, confident that the template we followed in each chapter presented the material in a way that allowed you to manage complicated information effectively and to apply it to your own life and culture.

We wish to close our book by presenting one additional way to make communication theory more compelling and useful: connecting and integrating various theories. Communication theory isn't created in a vacuum. Indeed, as you've read, theorists rely upon—and are influenced by—other theories to conceptualize and develop their own unique thinking. Sometimes theories are created to build on the ideas of previous theories, and other times theories are created in direct contrast to the thinking of earlier theorists. In some cases, it's an interesting intellectual exercise to pit two theories against each other to examine how they conceptualize similar phenomena in different ways. To illustrate connections among theories, we return to each chapter of the text and pose questions to demonstrate the interrelationship between and among various theories of communication. We hope that not only will these questions show connections that you otherwise might not have considered but also that they prompt further reflection about the theories. Making connections in this way should cause you to recall different theories, and ideally, you'll think about the theories in a different light once you begin making these connections. Finally, such an effort will result in a better appreciation for the contexts, traditions, and approaches that are embedded in and manifested by each theory.

We term this unique feature *ConnectingQuests*. You might respond to each question yourself and then compare your responses with those of others in the class. And, because you likely have not reviewed every theory in the book, you might also develop your own *ConnectingQuests* for those chapters that you did review and now understand.

Below you'll find each theory in the book with accompanying connections and questions. These Connecting-Quests provide you one more effort at applying and understanding the theories in this text:

Chapter 4: Expectancy Violations Theory (Uncertainty Reduction Theory)

Expectancy Violations Theory is primarily concerned with our expectations for other people's behavior, whereas Uncertainty Reduction Theory is concerned with our desire for predictability in attitudes and behavior. Discuss the relationship between the two theories using expectations, attitudes, and behavior as your overarching concepts.

Chapter 5: Uncertainty Reduction Theory (Organizational Information Theory)

How does Uncertainty Reduction Theory's view of uncertainty (and the world) relate to, or differ from, the viewpoint of the same concept taken in Organizational Information Theory? Do the two theories operate with similar or dissimilar assumptions about human behavior?

Chapter 6: Social Exchange Theory (Social Penetration Theory)

Compare how individuals assess their relationships using Social Exchange Theory and then using Social Penetration Theory. Explain how (or if) you see the two theories as different.

Chapter 7: Social Penetration Theory (Social Information Processing Theory)

Discuss how the principles of Social Penetration Theory and Social Information Processing Theory overlap. We know that SIP is concerned with technology. (How) might the principles of SPT be influenced by technology?

Chapter 8: Social Information Processing Theory (Muted Group Theory)

Discuss the differences and similarities in self-presentation in Social Information Processing Theory and Muted Group Theory. Although one theory relies on technology and another on FtF communication, what similarities do you see?

Chapter 9: Structuration Theory (Groupthink)

Decision making is an essential component in both Structuration Theory and Groupthink. Differentiate between the two theories as to how they explain the decision-making process, how the theorists interpret the term, and what parameters should be used when discussing the decision-making process from the perspective of each theory.

Chapter 10: Organizational Information Theory (Structuration Theory)

Organizational Information Theory and Structuration Theory each deal with the role of communication activities in an organization. Differentiate between the two theories by discussing how communication functions in both theories. In particular, interpret what is meant by communication activities in each theory and how the theorists might interpret the information and relationships inherent in an organization.

Chapter 11: Agenda Setting Theory (Social Exchange Theory)

A fundamental issue underscoring Agenda Setting Theory is the influence of the "agenda" of the media. Social Exchange Theory has its roots in economics, sociology, and psychology. Explore and explain how economic, sociological, and psychological principles might influence the media's agenda.

Chapter 12: Spiral of Silence Theory (Cultural Studies)

Compare and contrast the role of the media in Spiral of Silence Theory and Cultural Studies. Drawing on each theorist's interpretation of the media, be sure to address how media content and structure influence the communication process.

Chapter 13: Uses and Gratifications Theory (Cultivation Theory)

Theories such as Uses and Gratifications and Cultivation Theory highlight the use of media such as television in people's lives. Explain the role of television in communicating a "social reality" in both theories. How do the two theories differ in the way they see the interactions between televised content and human interpretation of that content?

Chapter 14: Face-Negotiation Theory (Dramatism)

Ting-Toomey's theory addresses a concern for individual identity. Dramatism, too, focuses on identity, but the concept is situated very differently than the way it's proposed in FNT. Differentiate between the theories based on each theorist's perspective on the use of identity.

Chapter 15: Symbolic Interaction Theory (Communication Accommodation Theory)

How does Mead's concept of the *self* in Symbolic Interaction Theory relate to the understanding of the role of the self in Communication Accommodation Theory?

Chapter 16: Coordinated Management of Meaning (Face-Negotiation Theory)

Culture is an important theme in many theories. Discuss how culture functions in CMM and in Face-Negotiation Theory. Do you think the different treatment of culture in the two theories makes a difference in how the two theoretical explanations function?

Chapter 17: Communication Privacy Management Theory (Relational Dialectics Theory)

Discuss possible interpretations for the role of silence in Communication Privacy Management Theory and Relational Dialectics Theory. Do you see these two as separate theories or would you argue that CPM is just an expanded part of RDT?

Chapter 18: Groupthink (Expectancy Violations Theory)

Speculate about the role of expectations in Groupthink and Expectancy Violations Theory. What sort of conversation might Janis and Burgoon have regarding expectations of behavior?

Chapter 19: Organizational Culture Theory (Coordinated Management of Meaning)

According to Organizational Culture Theory, stories are considered an important element in the culture of an organization. The Coordinated Management of Meaning centers on how meaning is achieved. Compare the two theories in regard to using stories and achieving meaning.

Chapter 20: The Rhetoric (Cultural Studies)

How does "the audience" function in the Rhetoric and in Cultural Studies? Discuss how technology would be, and perhaps change, both theoretical models.

Chapter 21: Dramatism (Narrative Paradigm)

Compare Burke's concept of the dramatic structure of life with Fisher's Narrative Paradigm. Would it be possible for a researcher to work comfortably with both theories without having ideational conflicts?

Chapter 22: The Narrative Paradigm (Agenda Setting Theory)

Compare the process of persuasion as it's explained through the Narrative Paradigm to how it functions in Agenda Setting Theory.

Chapter 23: Media Ecology Theory (Uses and Gratifications Theory)

Compare McLuhan's view of the audience in Media Ecology Theory with Katz, Blumler, and Gurevitch's view of the audience in Uses and Gratifications Theory. What conclusions can you draw regarding how active or passive the audience is in each theory?

Chapter 24: Communication Accommodation Theory (Feminist Standpoint Theory)

Communication Accommodation Theory emphasizes, among other things, the importance of perception in diverse communities. Feminist Standpoint Theory, too, emphasizes perception in a different, but equally important way. Differentiate between the two theories employing the concept of perception.

Chapter 25: Relational Dialectics Theory (Uncertainty Reduction Theory)

Do you think that the concepts of Relational Dialectics Theory intersect with the concepts of Uncertainty Reduction Theory as Berger suggested? Do you think the metatheoretical assumptions of the two theories are compatible? Explain your answer. How does RDT's critical turn affect your response?

Chapter 26: Cultivation Theory (Media Ecology Theory)

Cultivation Theory concentrates on media effects that also involve an investigation of media messages (primarily messages representing violence). How does this compare with the focus in McLuhan's Media Ecology Theory about the media in general? Can these two theories inform each other?

Chapter 27: Cultural Studies (Organizational Culture Theory)

Examine the role of culture in Cultural Studies and compare it to the role of culture in Organizational Culture Theory. What parameters exist on the interpretation of culture? What would Hall and Pacanowsky and O'Donnell-Trujillo agree about? What would they disagree about?

Chapter 28: Muted Group Theory (Social Penetration Theory)

How would the assumptions of a researcher using Muted Group Theory compare to those of a researcher using Social Penetration Theory? Do you think that one researcher would be comfortable using both theories? Why or why not?

Chapter 29: Feminist Standpoint Theory (Communication Privacy Management Theory)

When discussing the Feminist Standpoint Theory, power becomes an essential concept to consider. In Communication Privacy Management Theory, power may be a driving force in the decision to self-disclose. What role do you believe power plays in a discussion of both theories of communication?

Chapter 30: Co-Cultural Theory (Face-Negotiation Theory)

Although they both deal with questions of culture, Co-Cultural Theory is classified as a critical theory while Face-Negotiation Theory is considered post-positive. What differences do these classifications make in the way the two theories deal with the topic of culture? Do the two theories define culture in the same ways, for instance?

APPENDIX
Theory Summaries

Agenda Setting Theory

In choosing and displaying news, editors, newsroom staff, webcasters, and anchors play an important part in shaping social and political reality. When readers and viewers consume news, they not only learn about a given issue, but also learn how much importance to attach to that issue by the amount of coverage and position relative to other stories it's given by the media. When examining what candidates say during a campaign, Agenda Setting Theory suggests that the mass media may well determine the important issues—that is, the media may set the "agenda" of the campaign. How influential the media are in this agenda setting function depends on several factors including media credibility, the extent of conflicting evidence, shared values, and the audience's need for guidance.

Co-Cultural Theory

Co-Cultural Theory (CCT) provides a framework for understanding how individuals from nondominant groups select strategies for interacting with those from the dominant group. CCT is unique because it's grounded in the experiences of the persons it seeks to describe. Specifically, CCT asserts that co-cultural group members make decisions about how they'll communicate with members of their culture's dominant group based on their preferred outcomes (assimilation, accommodation, or separation) and communication approaches (nonassertive, assertive, or aggressive). Combining each of the preferred outcomes with each of the approaches yields nine communication orientations (nonassertive assimilation, nonassertive accommodation, nonassertive separation, assertive assimilation, assertive accommodation, assertive separation, aggressive assimilation, aggressive accommodation, and aggressive separation). In addition to their communication orientation, co-cultural group members make their choices based on the following factors: their fields of experience, their perceptions of the costs and rewards associated with the communication choice, their capability to engage in the choice, and the specific situational context in which they're interacting. Research has identified multiple communicative strategies that are associated with each orientation. CCT affirms that co-cultural group members engage in trials of the strategies, after which they evaluate the results of using a specific strategy in a given situation. Thus, CCT posits that communication between nondominant group members (NDGMs) and members of the dominant group (DGMs) isn't static; it's a dynamic negotiation and adjustment of the performance of NDGMs' identities in interaction with DGMs.

Communication Accommodation Theory

This theory considers the underlying motivations and consequences of what happens when two speakers, usually with different cultural backgrounds, shift their communication styles—whether verbally or nonlinguistically. During communication encounters, people will try to accommodate or adjust their style of speaking to others. This is primarily done in two ways: divergence and convergence. Groups with strong cultural pride often use divergence to highlight group identity. Convergence occurs when there is a strong need for social approval, frequently from powerless individuals. The words we say and how we say them all have social consequences.

Communication Privacy Management Theory

People believe that they own private information about themselves and their families and loved ones. Because they own this information, they can choose whether and with whom they will share it. And, people experience enduring tension over the decision of whether to share private information or to withhold it; this ongoing tension between opposing positions makes CPM a dialectic theory. When someone tells another something private, the other becomes an authorized co-owner of the information. Even after sharing, however, the original owner believes they should still retain control over the information. People create rules for sharing or withholding their private information based on a variety of criteria that can be classified as either core (consistent to the person) or catalyst (variable based on the situation) criteria. When these rules break down for some reason (either intentionally or through a mistake), privacy turbulence occurs. When this happens, people try to address the turbulence in some way in order to restore the system.

Coordinated Management of Meaning (CMM)

In conversations and through the messages we send and receive, people co-create meaning. As we create our social worlds, we understand that rules exist and we employ them to construct and coordinate meaning. That is, rules guide communication between people. CMM focuses on the relationship between an individual and their society. Through a hierarchical structure, people come to organize meaning for the literally hundreds of messages they receive throughout the day.

Cultural Studies

Cultural Studies is a theoretical perspective rooted in the notion of equality. The media represent ideologies of the dominant class in a society. Because media are controlled by corporations (the elite), the information they present to the public is consequently influenced and targeted with profit in mind. The media's influence and the role of power must be taken into consideration when interpreting a culture because those without equal influence are usually subordinate to those with disproportional influence.

Cultivation Theory

Television and other media play an important role in how people view their world because media tell us stories in a compelling fashion. In today's society, most people get their information from mediated sources rather than through direct experience. Therefore, mediated sources can shape a person's sense of reality. This is especially the case with regard to violence. Television presents at least ten times more violent acts in its programming than exist in reality. Heavy television viewing cultivates a sense of the world as a violent place, and heavy television viewers perceive that there's more violence in the world than there actually is or than lighter viewers perceive. Further, watching television has the effect of bringing together diverse people who receive the same messages and cultural imagery from the shows they watch in common (a process known as mainstreaming). The effects of television are increased when a viewer's actual experience resonates with the images presented on TV (a process known as resonance).

Dramatism

Burke's theory compares life to a play and states that, as in a theatrical piece, the events that occur in life require several elements: an actor who has a position or attitude relative to the others in the drama, a scene, an action, some means for the action to take place, and a purpose. The theory allows a rhetorical critic to analyze a speaker's motives by identifying and examining these six elements. Furthermore, Burke believes, guilt is the ultimate motivation for people, and Dramatism suggests that social actors are most successful when they provide their audiences with a means for purging their guilt.

Expectancy Violations Theory

Expectancy Violations Theory is concerned primarily with the structure of nonverbal messages. It asserts that when communicative norms are violated, the violation may be perceived either favorably or unfavorably, depending on the perception the receiver has of the violator. Violating another's expectations may be used strategically rather than conforming to another's expectations.

Face-Negotiation Theory

How do people in individualistic and collectivistic cultures negotiate face in conflicts? Face-Negotiation Theory is based on face management, which describes how people from different cultures manage conflict negotiation to maintain face. Self-face and other-face concerns explain the conflict negotiation between and among people from various cultures.

Feminist Standpoint Theory

Women are situated in specific social locations; they occupy different places in the social hierarchy based on their membership in social groups (poor, wealthy, European American, African American, Latinx, uneducated, well educated, etc.). Despite these differences, however, all women share the experience of "femaleness," which places them as subordinate to men. When an individual woman reflects on her subordinate power position and struggles in opposition to those in power, resisting the social definition given to her by those in power, she achieves a standpoint. No standpoint allows a person to view the entire social situation completely—all standpoints are partial—but people on the lower rungs of the social hierarchy do see more than their own position, whereas those in power rarely see beyond their own position. This expanded vision allows women to be involved in changing society for the better and creating a more just and equitable world. FST comes from Hegel and Marx's Standpoint Theory and has motivated others to created allied theories, such as Black Feminist Standpoint Theory and Co-Cultural Theory.

Groupthink

A psychological drive for getting along and for consensus drives groups to become highly cohesive. Highly cohesive groups frequently fail to consider alternatives to their course of action. When group members think similarly and do not entertain contrary views, they are also unlikely to share unpopular or dissimilar ideas with others. Groupthink suggests that these groups typically make premature decisions, some of which will have lasting and tragic consequences.

Media Ecology Theory

Society has evolved as its technology has evolved. From the alphabet to the internet, we have been affected by, and affect, media. In other words, the medium is the message insofar as technology affects communication. Media Ecology Theory centers on the beliefs that society cannot escape the influence of technology, technology brings people together across great distances, and technology will remain central to virtually all walks of life. The dominant media in a particular era shape the culture and the individuals within the culture.

Muted Group Theory

Language serves the dominant group who created it (and those who belong to the same group as its creators) better than those in nondominant groups who have to learn to use the language to describe their different experiences as best as they can. This is the case because the experiences of the creators are named clearly in language, whereas the experiences of other groups are not. Due to their problems adapting to a language they didn't create, people from other groups appear less articulate than those from groups like the creators. As a result, they are often belittled or ridiculed. Sometimes these muted groups create their own language to compensate for the problems they have using the dominant group's language.

The Narrative Paradigm

This approach is founded on the principle that humans are storytelling animals. Furthermore, narrative logic is often more persuasive than the rational logic traditionally used in argument. Narrative logic, or narrative rationality, suggests that people judge the credibility of speakers by whether their stories hang together (have coherence) and ring true (have fidelity). The Narrative Paradigm allows for a democratic evaluation of speakers and a democratic opportunity for audiences because no one has to be specially trained in persuasion to be able to draw conclusions based on the concepts of coherence and fidelity.

Organizational Culture Theory

According to Organizational Culture Theory, stories are considered an important element in the culture of an organization. Further, Organizational Culture Theory likens people to spiders who are suspended in webs that they create at work. An organization's culture is composed of shared symbols, each of which has a unique meaning. Organizational stories, rituals, values, and rites of passage manifest the culture of an organization.

Organizational Information Theory

The main activity of organizations is the process of making sense of equivocal and ambiguous information. Organizational members accomplish this sensemaking process through various behaviors, including enactment, selection, and retention of information. Organizations are successful to the extent that they are able to reduce equivocality through these means.

Relational Dialectics Theory

Relational Dialectics Theory pictures relational life as constant process and motion. Throughout the life of a relationship, people continually feel the push and pull of conflicting discourses. This push and pull causes tension called dialectics. Basically, the theory asserts that both/and, not either/or, is a more accurate way to explain how people deal with opposing goals. For instance, people in relationships want to be both connected and autonomous, open and self-protective, and to have both predictability and spontaneity in their interactions. As people communicate in relationships, they attempt to reconcile these conflicting desires, but they never eliminate their need for both parts of the opposition. Relational life is composed of a constant struggle among competing discourses, and RDT 2.0 is especially concerned with examining these multiple, competing discourses with an eye to illuminating which discourses are central (the centripetal) and which are marginalized (the centrifugal). RDT 2.0 broadens its lens to examine more than just the competing desires of the individuals in the relationship to include voices that come from the culture as well. For instance, a given culture might value a specific discourse (e.g., in the United States, openness is generally valued above privacy), and this gives that discourse a certain amount of power. This emphasis on power makes RDT 2.0 a critical theory.

The *Rhetoric*

Aristotle's theory centers on the notion of rhetoric, which he calls all the available means of persuasion. That is, a speaker who is interested in persuading their audience should consider three rhetorical proofs: logic (logos), emotion (pathos), and ethics/credibility (ethos). Audiences are key to effective persuasiveness, and rhetorical syllogisms, requiring audiences to supply missing pieces of a speech, are used in persuasion. Audiences are not viewed as passive, but as active agents in the speechmaking process.

Social Exchange Theory

Social Exchange Theory posits that the major force in interpersonal relationships is the satisfaction of both people's self-interest. Self-interest is not considered necessarily bad and can be used to enhance relationships. Interpersonal exchanges are thought to be somewhat analogous to economic exchanges where people are satisfied when they receive what they consider to be a fair return for their effort in a relationship. Thibaut and Kelley's Theory of Interdependence, a specific type of SET, argues that people's feelings of satisfaction or dissatisfaction in a relationship rest on the interdependence between the partners. This means that you are able to be happy only depending on your partner's happiness and the actions they take relative to you.

Social Penetration Theory

Interpersonal relationships evolve in some gradual and predictable fashion. Social Penetration theorists believe that self-disclosure is the primary way that superficial relationships progress to intimate relationships. Although self-disclosure can lead to more intimate relationships, it can also leave one or more persons vulnerable. Effective relationships not only progress toward more intimacy, but this intimacy is directly related to the extent to which a person discloses.

Social Information Processing Theory

Individuals have the ability to establish online relationships and these relationships are equal to or greater in intimacy than what's achieved in face-to-face (FtF) relationships. Even without nonverbal cues, through various technologies (e.g., email, texting, etc.), online relationships have the potential to be significant in people's lives. Individuals use computer-mediated communication (CMC) to get to know each other and use this information to form impressions of each other. Some of these impressions get verified by others, which is called warranting. Because messages travel via one primary channel, it takes longer for relationships to achieve the same level as those that are FtF. In some cases, online relationships may be viewed as more important than FtF relationships.

Spiral of Silence Theory

Because of their enormous power, media have a lasting and profound effect on public opinion. Spiral of Silence theorists believe that media (and social media) work simultaneously with majority opinion to silence minority beliefs on cultural and social issues in particular. A fear of isolation prompts those with minority views to examine the beliefs of others. Individuals who fear being socially isolated are prone to conform to what they perceive to be the majority view. Every so often, however, the silent majority raises its voice in activist ways.

Structuration Theory

Organizations create structures, which can be interpreted as an organization's rules and resources. These structures, in turn, create social systems in an organization. Organizations achieve a life of their own because of the way their members use their structures. Power structures guide the decision making that takes place in these organizations.

Symbolic Interaction Theory

People are motivated to act based on the meanings they assign to people, things, and events. These meanings are created in the language that people use both in communicating with others (interpersonal context) and in self-talk (intrapersonal context), or their own private thoughts. Language allows people to develop a sense of self and to interact with others in the community. As people interact with others, they work together co-constructing a sense of reality.

Uncertainty Reduction Theory

When strangers meet, their primary focus is on reducing their level of uncertainty in the situation because uncertainty is uncomfortable. People use the processes of proactive uncertainty reduction (prediction) and retroactive uncertainty reduction (explanation) to reduce their discomfort. There are two main types of uncertainty that people may experience. They may be unsure of how to behave (or how the other person will behave), and they may also be unsure of what they think of the other person and what the other person thinks of them. High levels of uncertainty are related to a variety of verbal and nonverbal behaviors.

Uses and Gratifications Theory

People are active in choosing and using particular media to satisfy specific needs. Although Uses and Gratifications takes a limited effects position, it does so by positing that the media have that limited effect on audiences because users are able to exercise choice and control. People are self-aware, and they're able to understand and articulate the reasons they use media. They see media use as a good way to gratify the needs they have. Uses and Gratifications Theory is primarily concerned with the following question: what do people do with media?

Name Index

Subject Index

A

abstract symbols, 5
accent mobility model, 428
accommodation
   See also Communication
     Accommodation Theory (CAT);
     overaccommodation
   aggressive, 545
   assertive, 545
   defined, 428, 541
   nonassertive, 544
   norms and, 431-432
   varying in social
     appropriateness, 431
accountability, 334
accounting fraud, 15
accuracy, 525
acquired immunodeficiency syndrome
   (AIDS), 223,
   485-486
act, 382
active audience, 235, 236, 239, 241
activeness, 241
active strategies, 95
activity, 241
actual self, 147
adaptability, 260
Adaptive Structuration
   Theory, 165
Adelphia Communications, 15
adjustment, 189
affective exchange
   stage, 138-139
affective needs, 238
affiliative constraints, 327
African American women, 519, 522,
   523, 530
agency, 169-170, 382
agenda, 204
   media, 202, 206
   policy, 202
   public, 202
agenda setting, 198
   attribute, 204

levels of, 204-206
three-part process
   of, 202-204
Agenda Setting Theory,
   197-212, 564
   in action, 206
   assumptions of, 201-202
   critique of, 206-209
   elaborating, 200-201
   establishing, 199-200
   at a glance, 200
   heurism, 209
   history of research on,
     198-201
   integration of, 206-209
   levels of agenda setting,
     204-206
   pretheoretical conceptualizing,
     198-199
   three-part process of agenda setting,
     202-204
   utility of, 207-208
agent, 169, 382
aggressive accommodation, 545
aggressive assimilation, 545
aggressive separation, 546
AIDS. *See* acquired immunodeficiency
   syndrome (AIDS)
alienation, 487
allocative resources, 171
Altman, Irwin, 128
American Dream, 490
American Society of Law,
   Medicine, and Ethics, 17
Animal Liberation Front, 224
animal rights activism, 224
antecedent conditions, 95
anthrax scares, 466
anti-bullying laws, 202
antiwar activism, 224
Anxiety-Uncertainty Management
   (AUM), 100
applied research, 62
approbation facework, 254
archetypes, 293

Aristotle
   Golden Mean principle, 15-16
   life of, 359-360
   paradigm shift, 392
   Rhetoric, 20, 38, 357-362
arousal
   cognitive, 79
   defined, 79
   physical, 79
arousal value, 79
arrangement, 364,
   365-366
assertive accommodation, 545
assertive approach, 543
assertive assimilation, 544
assertive separation, 545
assimilation, 542
asymmetric dependence, 116
asymmetric information, 100
asynchronous communication,
   150, 155
attitude, 383
attraction, 432-433
attribute agenda setting, 204
attributions, 154-155
audience, 495
   active, 234-235, 239, 241
   Aristotle's views on, 358,
     360-362
   counter-hegemony, 493-494
   decoding, 494-495
   effective persuasiveness and, 365
   needs and gratifications of,
     237-238
   oppositional position, 495
   rhetorical tradition and, 28
   speaker–audience relationship, 360
audience analysis, 360
AUM. *See* Anxiety-Uncertainty
   Management (AUM)
authoritative resources, 171
autonomy and connection,
   449-450, 454
avoiding (AV) style, 261
axiology, 50-51